DIESEL

FUNDAMENTALS, SERVICE, REPAIR

by

WILLIAM K. TOBOLDT

Author of Goodheart-Willcox Automotive Encyclopedia, Auto Body Repairing and Repainting, Fix Your Ford, Fix Your Chevrolet, Member Society of Automotive Engineers, Associate Member Automotive Engine Rebuilders Association, Associate Member Association of Diesel Specialists, Formerly Editor of Motor Service Magazine.

South Holland, Illinois

THE GOODHEART-WILLCOX COMPANY, INC.

Publishers

 Printed in U.S.A. Library of Congress Catalog Card Number 79–24013. International Standard Book Number 0–87006–287–5.

23456789–80–54321

Library of Congress Cataloging in Publication Data

Toboldt, William King,
Diesel: fundamentals, service, repair.

Includes index.
1. Diesel motor. I. Title.
TJ795.T62 1980 621.43'6 79–24013
ISBN 0–87006–287–5

INTRODUCTION

This text is designed to provide students in the rapidly growing diesel engine field with a thorough knowledge of diesel mechanics. It contains up-to-date information and basic instruction on DIESEL FUNDAMENTALS, SERVICE AND REPAIR.

Excellent coverage is given in the areas of basic diesel engine design, classification, construction and operation. Included is extensive information on the various systems which provide diesel engine lubrication, exhaust, filtering, starting, cooling and governing, with special emphasis on fuel injectors and injection systems. The service and repair, tune-up, trouble shooting and engine reconditioning units are presented in an understandable and interesting manner.

Note that the text features diesel engines for commercial vehicles and passenger cars. Also covered are diesels ranging in power from 1.5 to 55,000 hp (1.25 to 41,000 kW), with applications in power generating plants, farming, deep well drilling, pleasure boats, huge ships, locomotives and earthmoving equipment.

The many illustrations, a large number of which have been specially drawn, make DIESEL FUNDAMENTALS, SERVICE, REPAIR an ideal text for beginning students. It also will be invaluable for those now engaged in diesel mechanics who want to increase their skills.

Because of tougher standards for the emission of toxic gases and the soaring price of fuel and the demand for greater fuel mileage, more diesel engines will be produced. This, in turn, means that more diesel engines will need service and repair.

William K. Toboldt

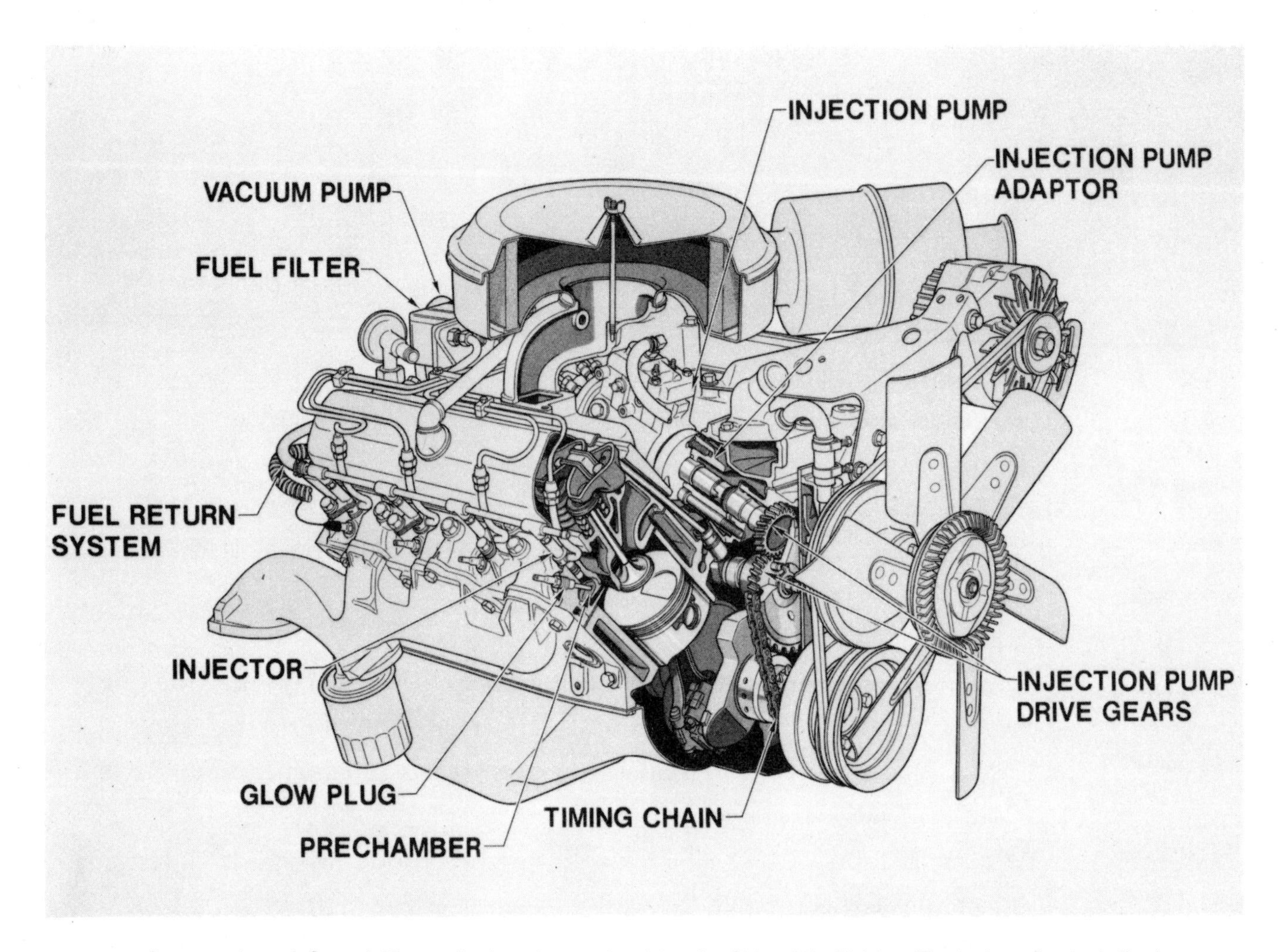

Cutaway view of General Motors diesel engine produced by the Oldsmobile Division. The basic engine is similar in construction to the 5.7 litre (350 cu. in.) gasoline engine. The diesel engine is of the precombustion chamber type with a glow plug to aid in starting. The fuel injection pump is a Roosa Master of the distributor type. The Oldsmobile diesel engine is either standard or optional equipment in all General Motors lines.

CONTENTS

International Harvester's mid-range diesel engine family includes the 9.0 litre V-8 (left front) and the DT-466 in-line Six (right rear). The V-8 is available in 165 and 180 hp ratings and is naturally aspirated with an open combustion chamber. The Six is turbocharged and is available in 160, 180 and 210 hp models.

Chapter 1
DIESEL APPLICATIONS AND ADVANTAGES

A diesel engine is an internal combustion engine (engine which burns fuel inside the engine) which operates on liquid fuel. It depends on heat developed by compressing air to ignite the fuel which is forced into the combustion chamber at the instant of maximum compression and heat.

TWO, FOUR-CYCLE DIESELS

Diesel engines are available in both two and four-cycle types; and in horsepowers up to 55,000 hp (41,000 kW).

Diesel engines are used as a source of power for many

Fig. 1-1. Sectional view of series 92 V-type Detroit Diesel engine. (Detroit Diesel Allison Div., General Motors Corp.)

Fig. 1-2. A six cylinder in-line 504 cu. in. diesel tractor engine. (Minneapolis Moline, Inc.)

Fig. 1-3. Ford's 256 cu. in. diesel engine is designed for industrial, marine and agricultural applications.

purposes: Automotive vehicles, both passenger car and commercial; Farm tractors; Road building equipment; Mobile compressor and power plants; Marine propulsion (small pleasure craft up to ocean liners); Electric generating plants; Railroad locomotives; Submarines; Power hoists; Drag lines; Bulldozers; Power shovels; Graders; Scrapers; and Irrigation pumps.

The diesel engine is frequently referred to as a compression ignition engine, to distinguish it from the familiar automobile engine which uses a spark to ignite the fuel mixture and is known as a spark ignition engine.

Diesel engines vary considerably in appearance and construction. See Figs. 1-1 to 1-8 inclusive.

ADVANTAGES OF DIESELS

The diesel engine offers a number of advantages including:

High reliability in operation.
Low fuel cost.
High power per pound of engine.
Low fuel consumption per hp hour.
Low fire hazard.
High sustained torque.

The diesel engine is highly reliable. When supplied with clean fuel, it can be depended on to operate continuously for

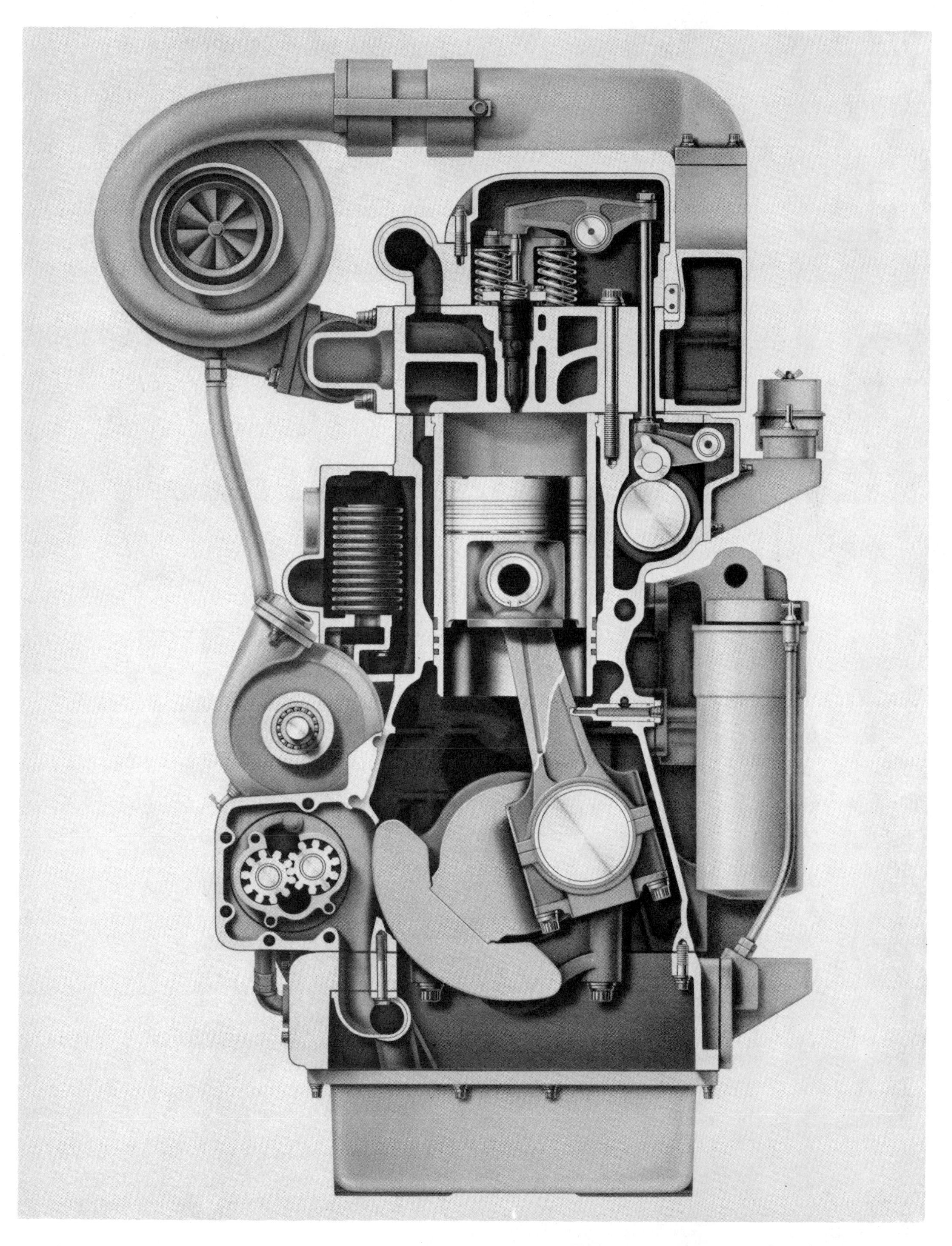

Sectional view of Cummins K-6 turbocharged diesel engine with after cooling. Displacement is 1150 cu. in. (18.9 liters). Individual cylinder heads are a feature. The K series is also available in V-12 and V-16 types.

Fig. 1-6. This two-cycle, 45 deg. V-type diesel has a bore of 9 1/16 in. and a stroke of 10 in. and is of the type used in railway locomotives, marine applications, electric generating plants and industrial power units. (Electro-Motive Div. GM)

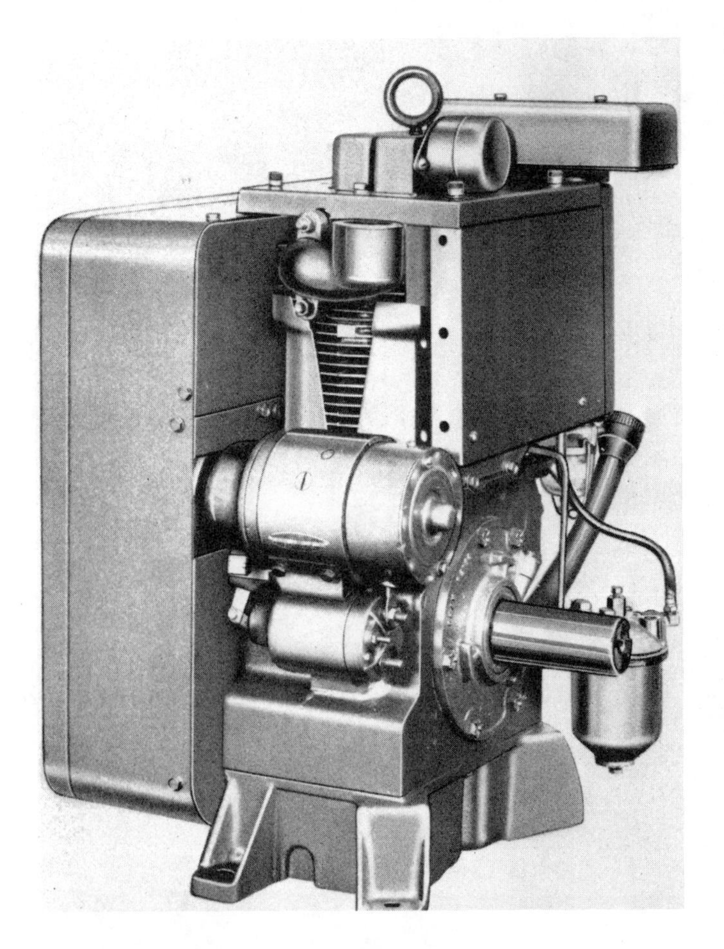

Fig. 1-7. Diesel engines are also available in sizes less than 10 hp. This model weighs 228 lbs. and develops 8.1 hp at 2400 rpm. (Onan)

long periods of time with minimum danger of breakdown. It has no ignition breaker points to replace. The diesel fuel lubricates the injector and has long life.

Diesel fuel costs less per gallon than gasoline. In the operation of passenger cars and trucks the saving in the cost of fuel is particularly apparent. Diesel fuel has a heat value of 139,500 heat units per gallon, compared to 124,500 heat units for gasoline. The maximum air to fuel ratio for the diesel is 40 to 1; for gasoline engines, it is 18 to 1. The diesel burns more air than spark ignition engines and is remarkably free from exhaust emissions of hydrocarbons and carbon monoxide.

SAFETY

Diesel fuel is more difficult to ignite than gasoline and is safer to handle. This feature is particularly important in boat installations.

Also important is the fact that diesel exhaust gases are not as toxic as exhaust fumes from gasoline engines. Neither do they tend to form as much smog. The odor of the diesel exhaust is unpleasant, but both carbon monoxide and hydrocarbons are less in the case of the diesel than in a gasoline engine. These characteristics are of particular interest because of the attention now being focused on exhaust, crankcase and fuel tank emission and their contribution to smog.

DIESEL PERFORMANCE

Because of the higher compression ratio used by the diesel engine, and the accurate injection of fuel into each cylinder, the combustion of fuel is more complete and the efficiency is

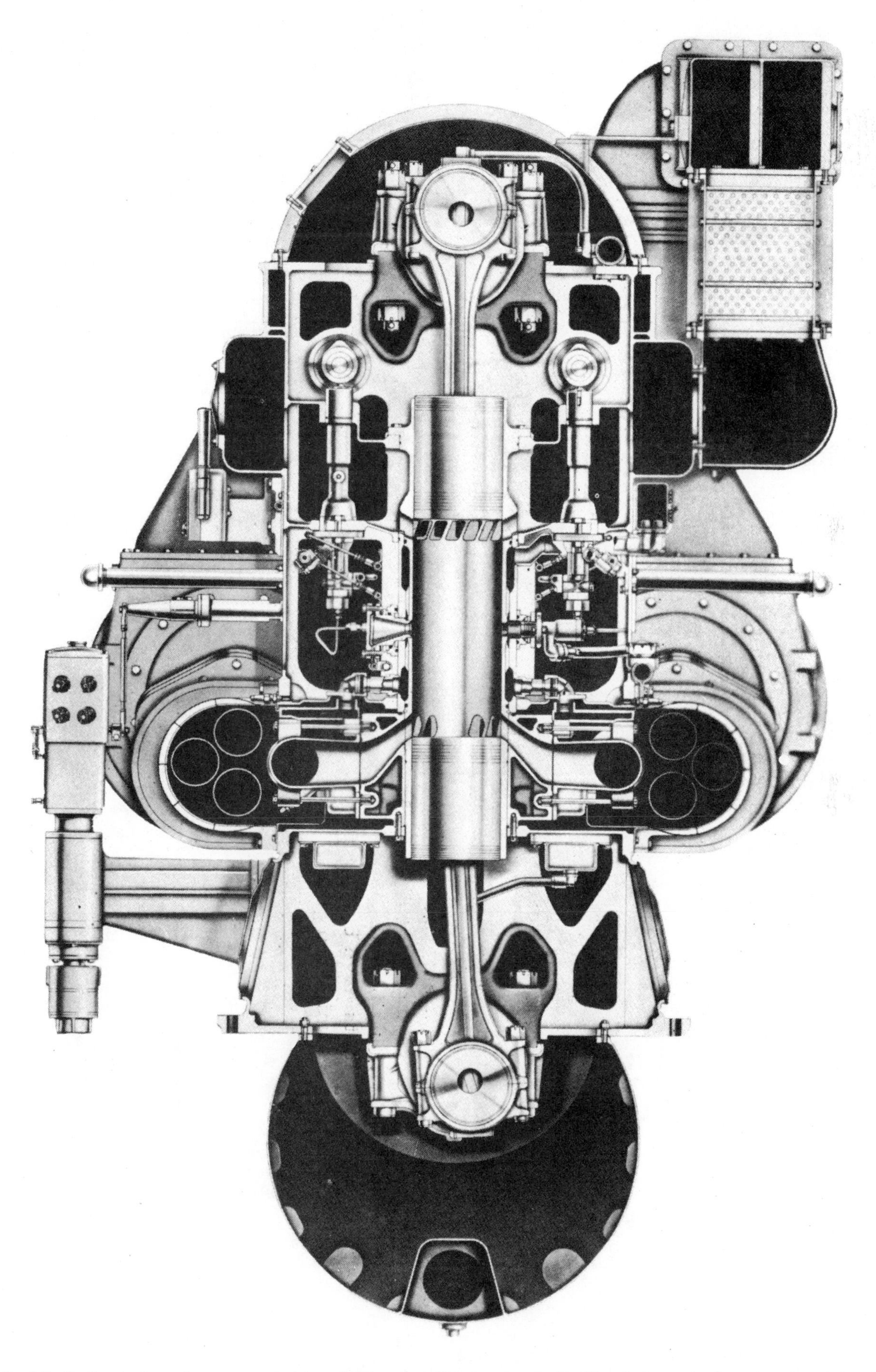

Fig. 1-8. Sectional view of an opposed piston diesel engine. This is a two-cycle, 12 cylinder, turbocharged engine designed for industrial and marine, stationary and mobile installations. (Fairbanks Morse Power Systems Div., Colt Industries Inc.)

higher than in a gasoline engine.

A gasoline engine reaches its peak horsepower at higher speed than a diesel. A diesel operates at lower speed with a higher torque, or turning effort output.

REVIEW QUESTIONS – CHAPTER 1

Write your answers on a separate sheet of paper. Do not write in this book.

1. List five applications of diesel engines as a power source, other than passenger cars, trucks and tractors.
   a. ______________________.
   b. ______________________.
   c. ______________________.
   d. ______________________.
   e. ______________________.
2. Diesel engines are available in both two and four cycle types that develop up to ________ horsepower.
3. Five advantages that the diesel engine offers are:
   a. ______________________.
   b. ______________________.
   c. ______________________.
   d. ______________________.
   e. ______________________.
4. The maximum air to fuel ratio for the diesel engine is ________.
5. Two advantages of diesel engines compared with gasoline engines as far as exhaust fumes are concerned are:
   a. ______________________.
   b. ______________________.
6. Compared to gasoline, diesel fuel is:
   a. Less difficult to ignite.
   b. Just about the same to ignite.
   c. More difficult to ignite.
7. The diesel engine is often referred to as a ________ ignition engine.

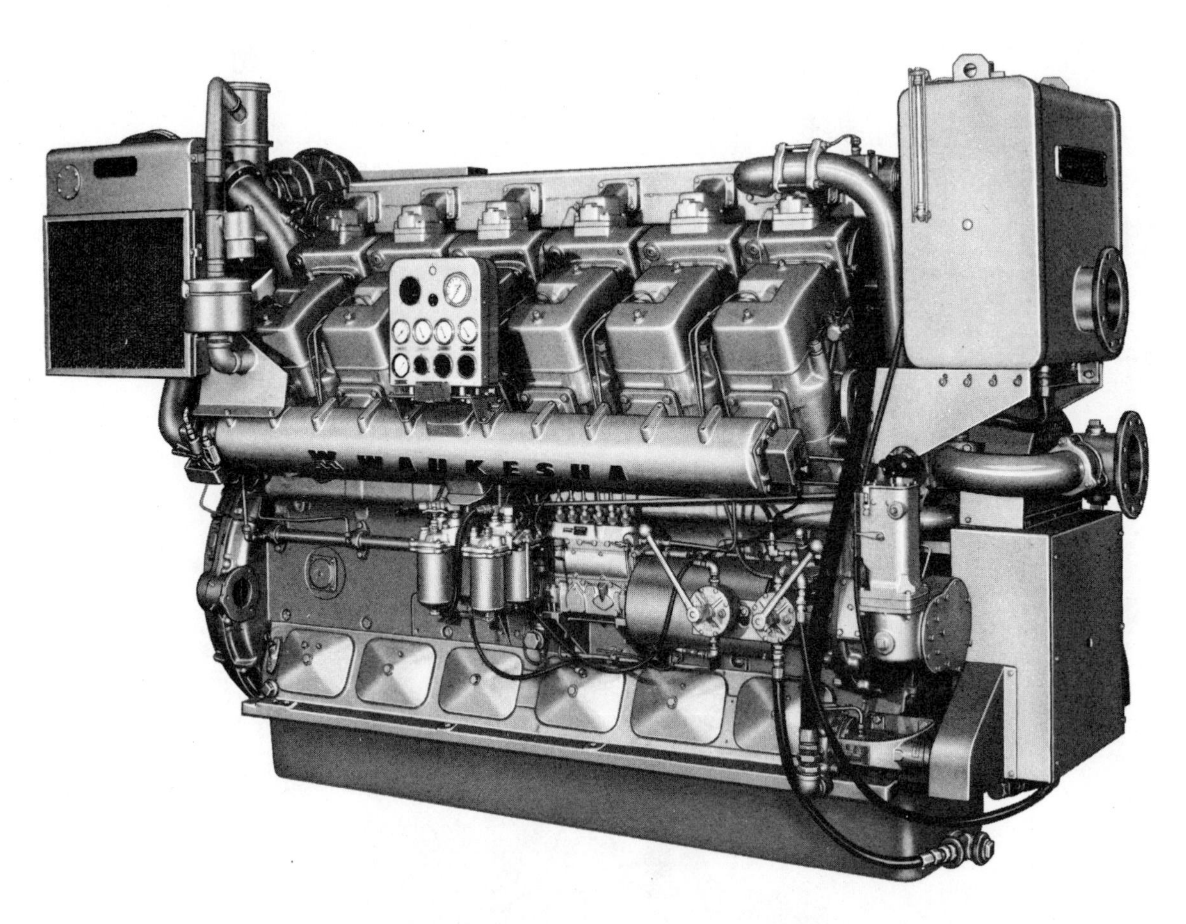

This L5792DSI marine diesel fuel engine has a displacement of 5788 cu. in. (94.9 litres). (Waukesha Engine Div., Dresser Industries, Inc.)

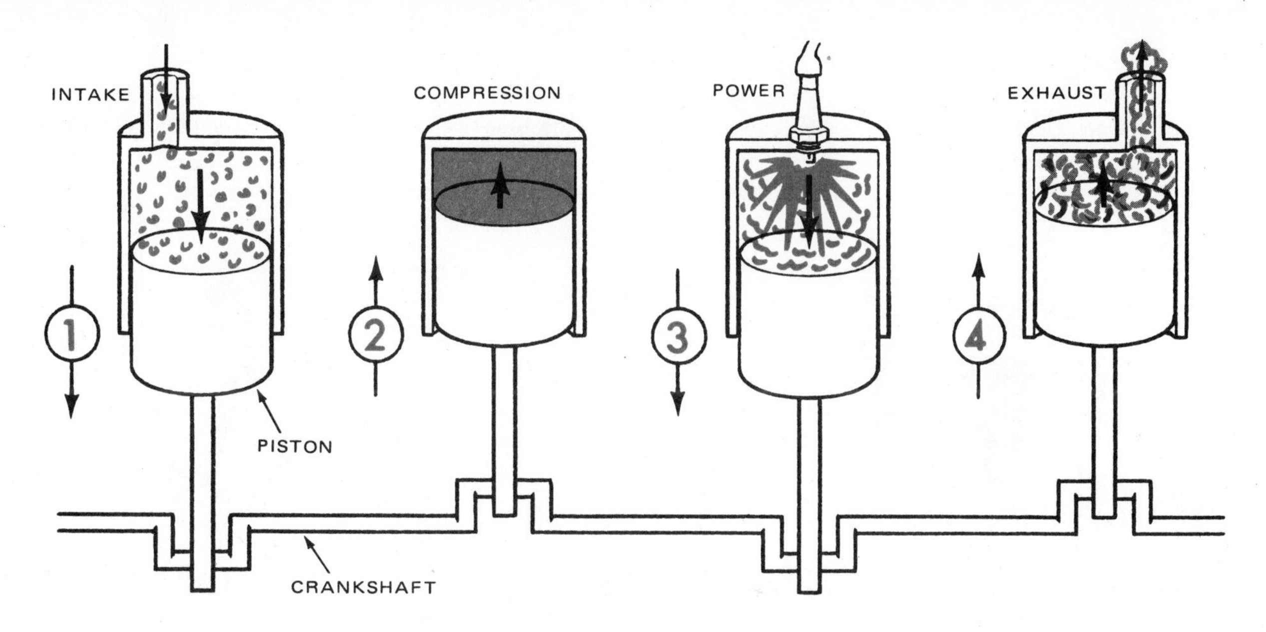

Fig. 2-1. In a four-stroke cycle engine, four strokes of the piston are required to complete one cycle, or sequence of events.

Chapter 2
BASIC TYPES OF ENGINES

In this Chapter, we will review the fundamental principles of gasoline engine operation; then see how the diesel engine differs in both design and operation.

There are two basic types of gasoline engines and diesels, four-stroke cycle engines, and two-stroke cycle engines.

A cycle is a complete sequence of events, which is repeated over and over again. A piston moves up and down in the cylinder. Each movement of the piston is called a stroke.

GASOLINE ENGINES

FOUR-STROKE CYCLE ENGINES

In the four-stroke cycle engine, four strokes of the piston are required to complete one cycle: intake, compression, power, exhaust, Fig. 2-1. Note from the arrows that the piston moves down on strokes 1 and 3, and up on strokes 2 and 4.

Start your observation with the piston in the raised position, Fig. 2-2. As the piston moves down, a mixture of gasoline and air flows into the cylinder. This first downward stroke is called the "Intake Stroke."

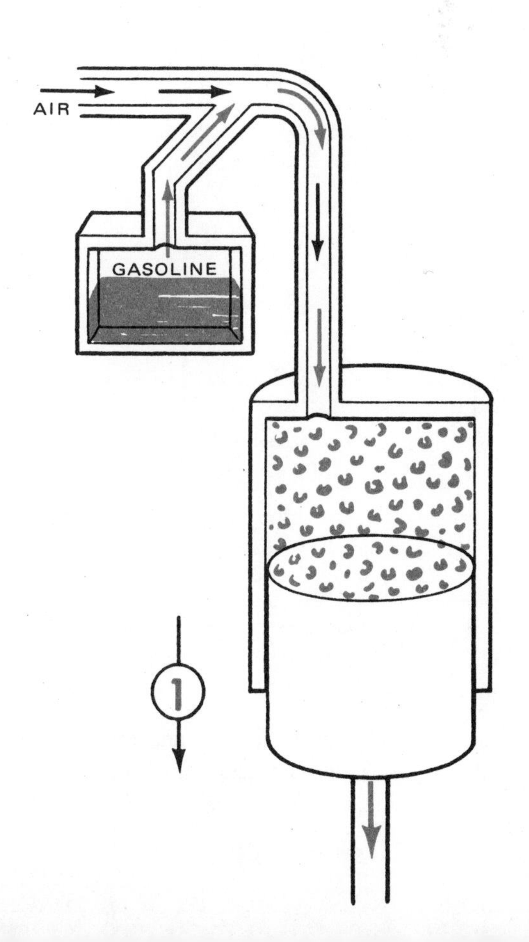

Fig. 2-2. Intake stroke. A fuel-air mixture flows into the cylinder, as the piston moves downward.

On the next (second) stroke the piston moves upward into the closed upper part of the cylinder, to squeeze the gasoline-air mixture into a tiny space. This is called the Compression Stroke, Fig. 2-3.

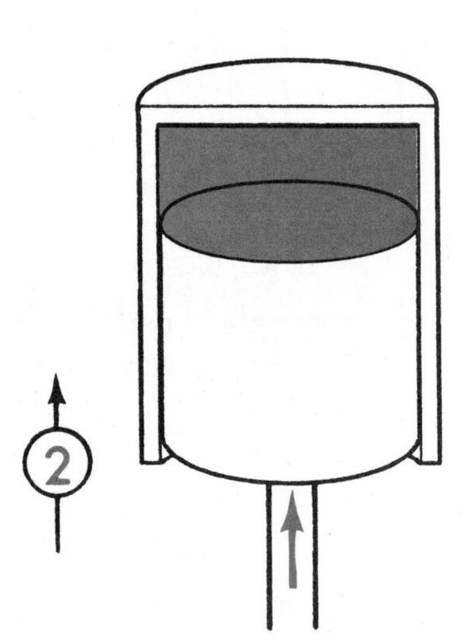

Fig. 2-3. Compression stroke. Piston moves upward to squeeze gasoline-air mixture into tiny space.

The rapid buildup of burning gases in the cylinder forces the piston down, giving us a Power Stroke, Fig. 2-4.

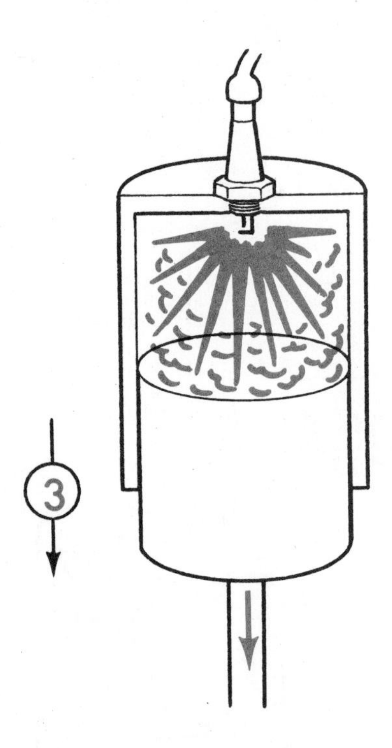

Fig. 2-4. Power stroke. Expansion caused by burning gases forces piston down.

When the piston reaches the bottom of its stroke, the cylinder is filled with burned gases, and the piston moves upward to push them out. This is called the Exhaust Stroke, Fig. 2-5.

The four-stroke cycle has been completed, and the cycle is repeated.

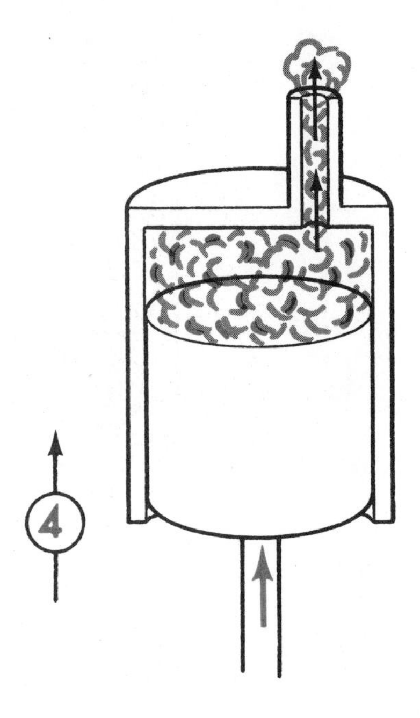

Fig. 2-5. Exhaust stroke. Piston moves upward to remove burned gases.

Only one stroke out of four is a power stroke. Turning effort provided by a heavy, rotating flywheel keeps the crankshaft turning through the other three strokes, Fig. 2-6.

Fig. 2-6. Turning effort provided by rotation of heavy flywheel keeps crankshaft turning between power strokes.

ENGINE VALVES

Earlier in this chapter, we showed how a fuel mixture (gasoline and air) enters the engine cylinders and the burned gases leave during the exhaust stroke, but we did not discuss the valves which open and close the fuel and exhaust passageways, Fig. 2-7.

INTAKE: On the intake stroke, Fig. 2-8, the intake valve is open and the exhaust valve is closed. The fuel-air mixture flows into the cylinder.

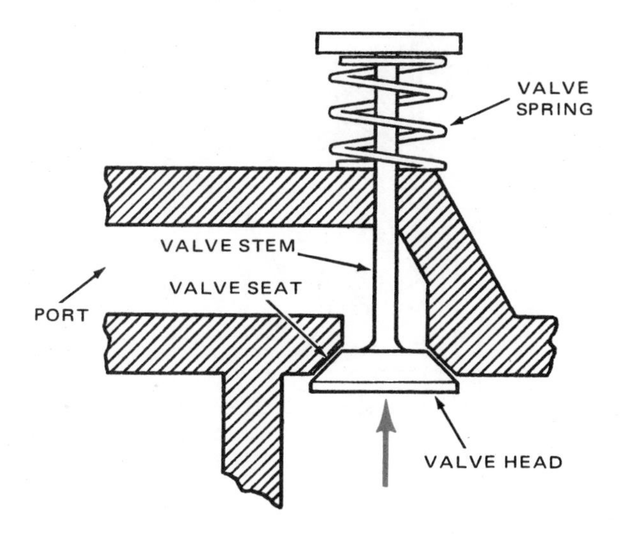

Fig. 2-7. Valves open and close the fuel and exhaust passageways. The valve is closed by the action of the valve spring. Cylinder pressure keeps it sealed.

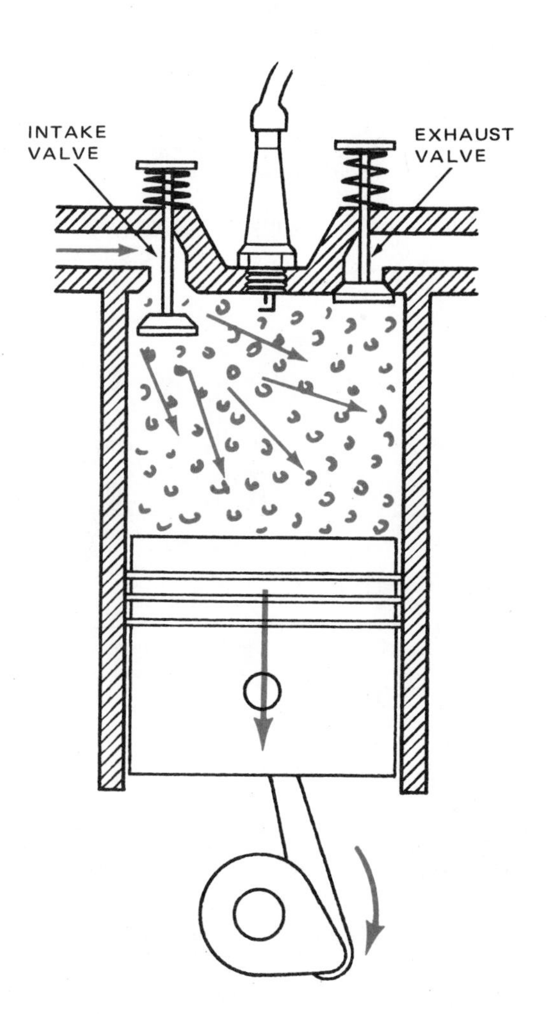

Fig. 2-8. On the intake stroke, the intake valve is open and the exhaust valve is closed.

COMPRESSION: On the second or compression stroke, Fig. 2-9, both valves are closed. Valve spring pressure and pressure of gases within cylinder keep valves tightly closed.

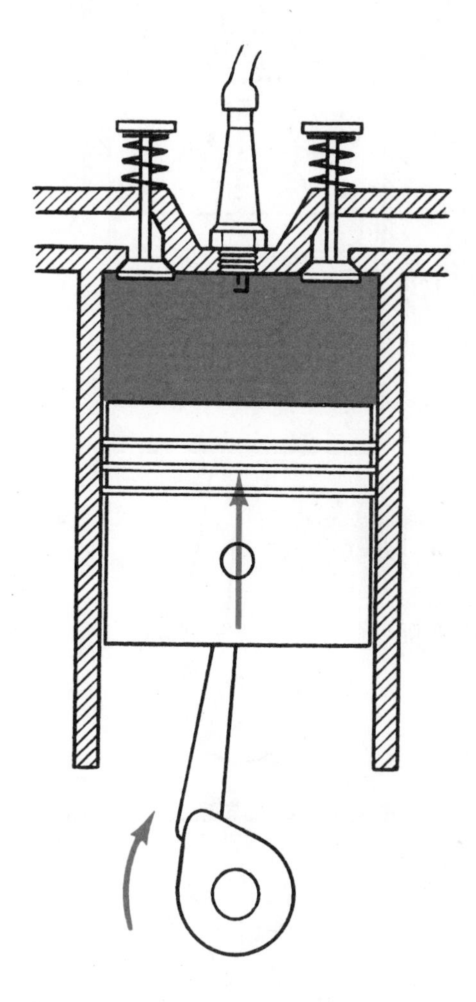

Fig. 2-9. On the compression stroke, both valves are closed.

POWER: On the third or power stroke, both valves are closed. The fuel-air mixture is ignited, and the piston moves downward, Fig. 2-10.

EXHAUST: The exhaust valve opens and the piston moves upward to remove the burned gases. The intake valve is closed. See Fig. 2-11.

CRANKSHAFT

Four-stroke automotive and diesel type engines generally have four or more cylinders. The cylinders are connected to the crankshaft so each delivers one power stroke during a complete cycle, as in Fig. 2-12. For each cycle the crankshaft must make two turns. During the first revolution of the crankshaft the inlet valve is open on the intake stroke. During the second revolution, the exhaust valve is open on the exhaust stroke. Both valves remain closed during the compression and power strokes.

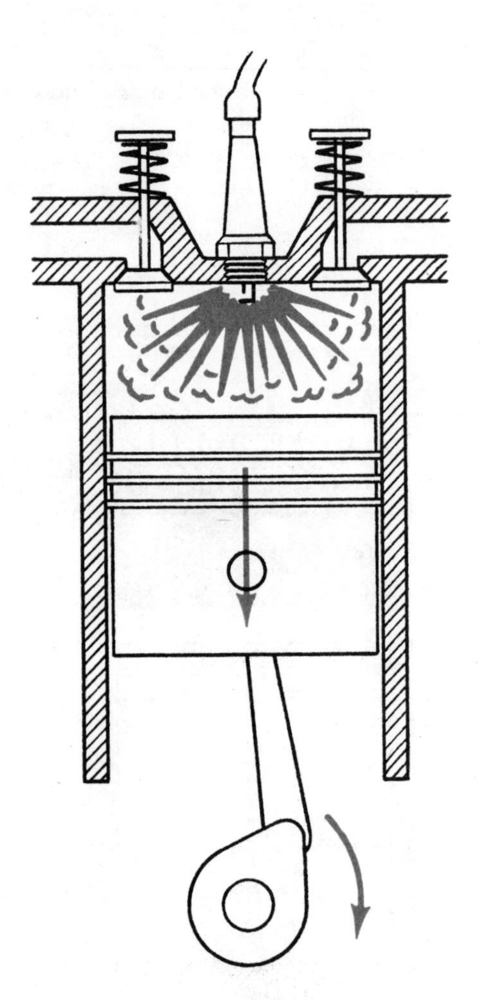

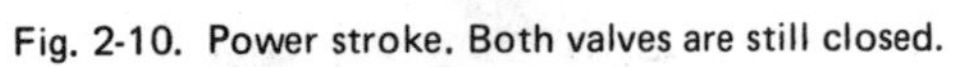

Fig. 2-10. Power stroke. Both valves are still closed.

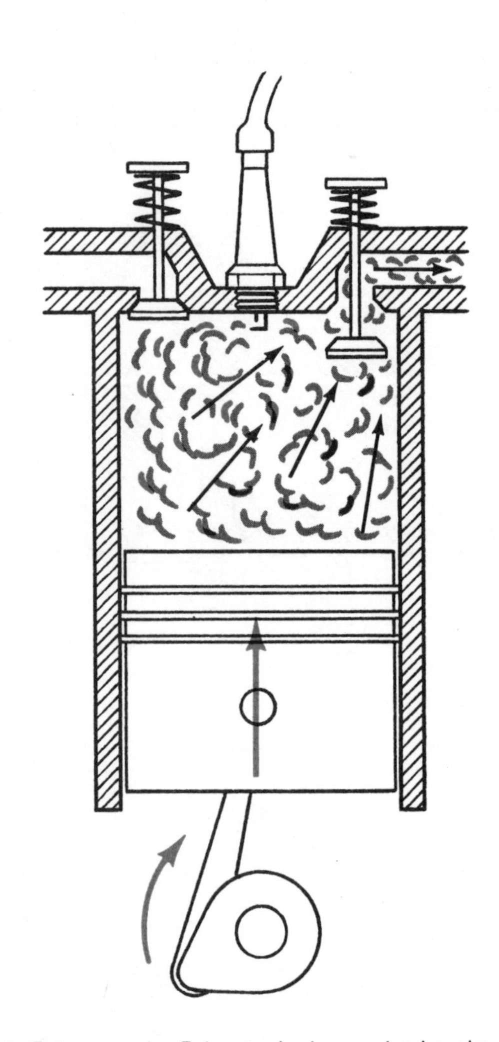

Fig. 2-11. Exhaust stroke. Exhaust valve is open; intake valve is closed.

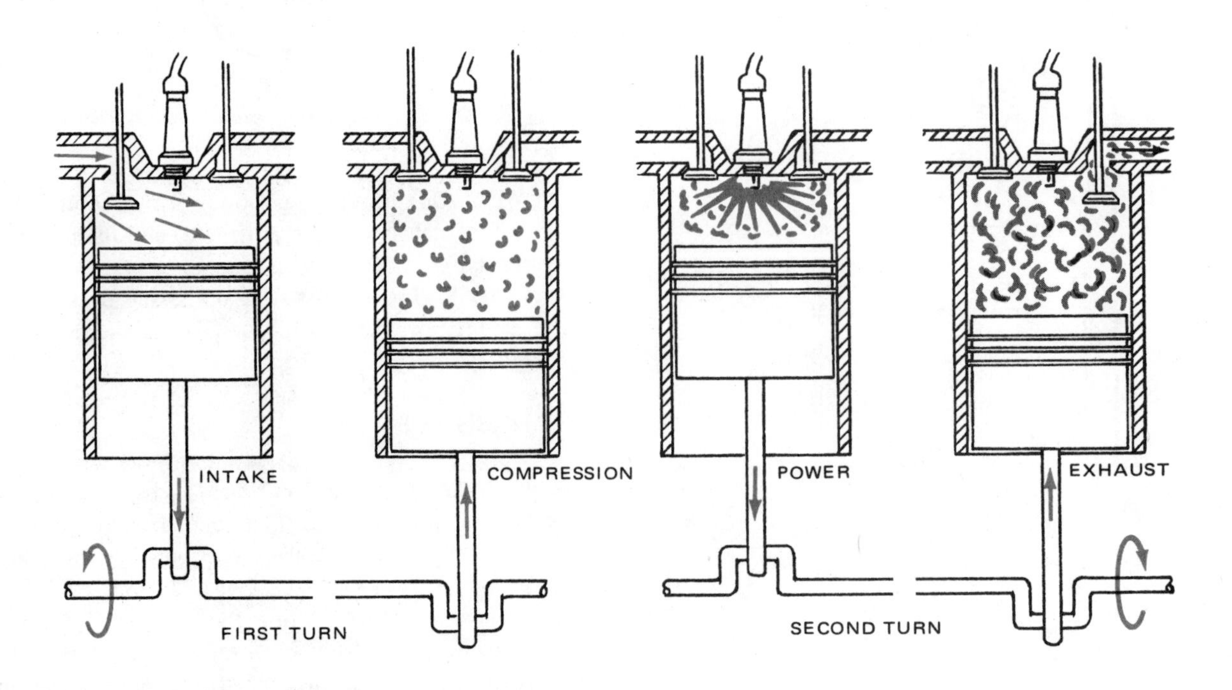

Fig. 2-12. For each cycle (intake, compression, power, exhaust) the crankshaft turns twice.

VALVE OPERATING MECHANISM

Engine valves are closed by springs, and are opened by cam lobes (ear-shaped projections) on a rotating camshaft. The cams push down on the valves to open them as the camshaft turns, Figs. 2-13 and 2-14. On four cycle engines, the small gear on the crankshaft goes around twice every time the larger gear on the camshaft turns once. Lobes on the camshaft operate both the intake and the exhaust valves.

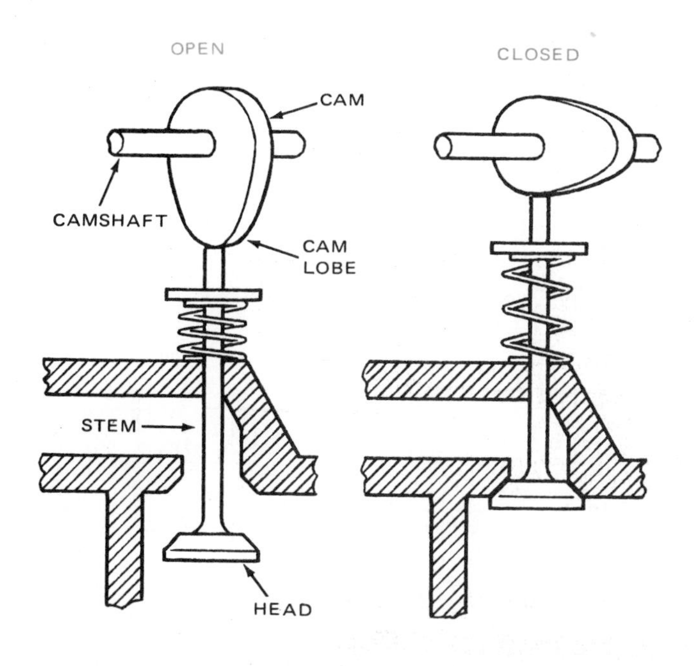

Fig. 2-13. The position of the cam determines when the valve will be open.

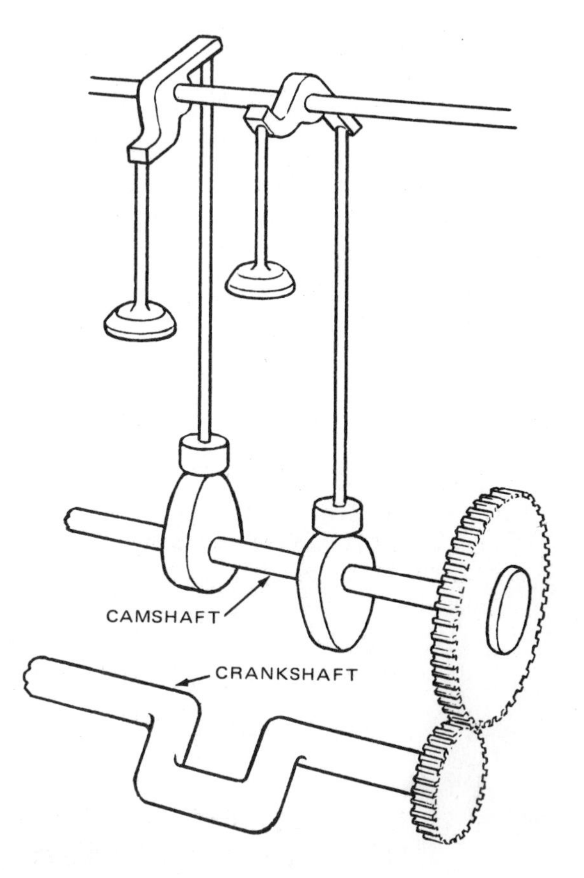

Fig. 2-14. When the small gear on the crankshaft goes around twice, the large gear on the camshaft turns once.

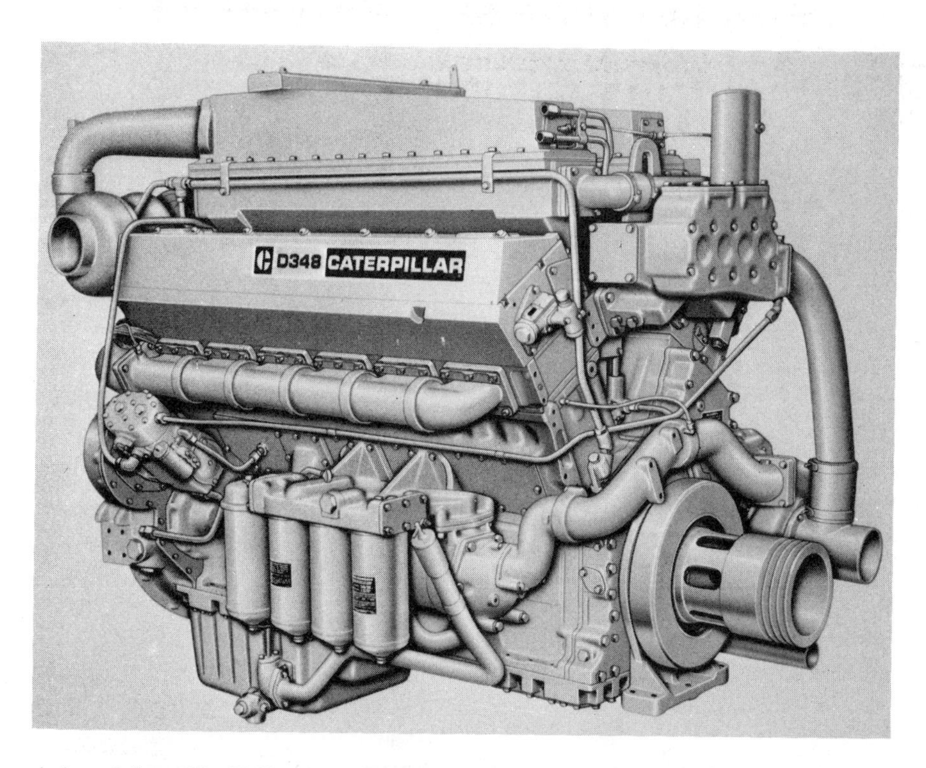

Industry Photo. External view of Caterpillar V-12 diesel engine used in marine applications. Notice the lack of ignition wires and spark plugs found on spark ignition engines used in most automobiles. Approximate fuel consumption for this engine is given as 28.03 gallons per hour.

DIESEL ENGINE FUNDAMENTALS

A diesel engine is similar to a gasoline engine in that it gets its power from the expansion of burning gases. The diesel depends on the heat of compression of the air to ignite the fuel, instead of an electric spark, as used in gasoline engines.

On the compression stroke, air without fuel, is compressed until it becomes hot enough (about 1,000 degrees), so liquid diesel fuel forced into the cylinder burns instantly, as shown in Fig. 2-15. The piston in the fuel pump forces the fuel under high pressure through the fuel line and nozzle. The fuel pump piston is pushed upward by a cam on the camshaft. The nozzle valve closes when fuel injection stops. Each cylinder has its own fuel pump, and each pump is operated by a separate cam on the camshaft, Fig. 2-16.

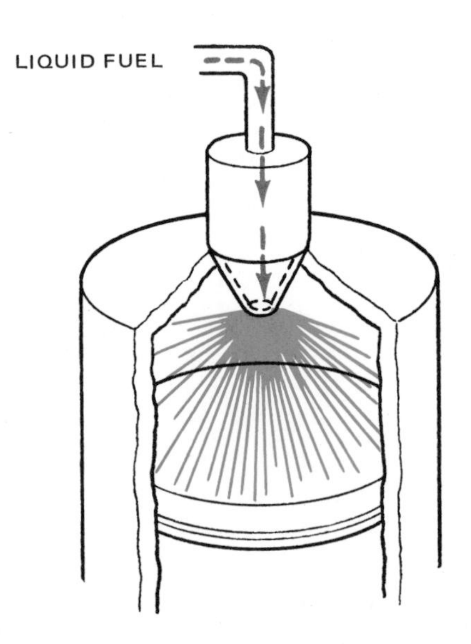

Fig. 2-15. Diesel fuel is forced through a nozzle and sprayed into compressed air in the cylinder.

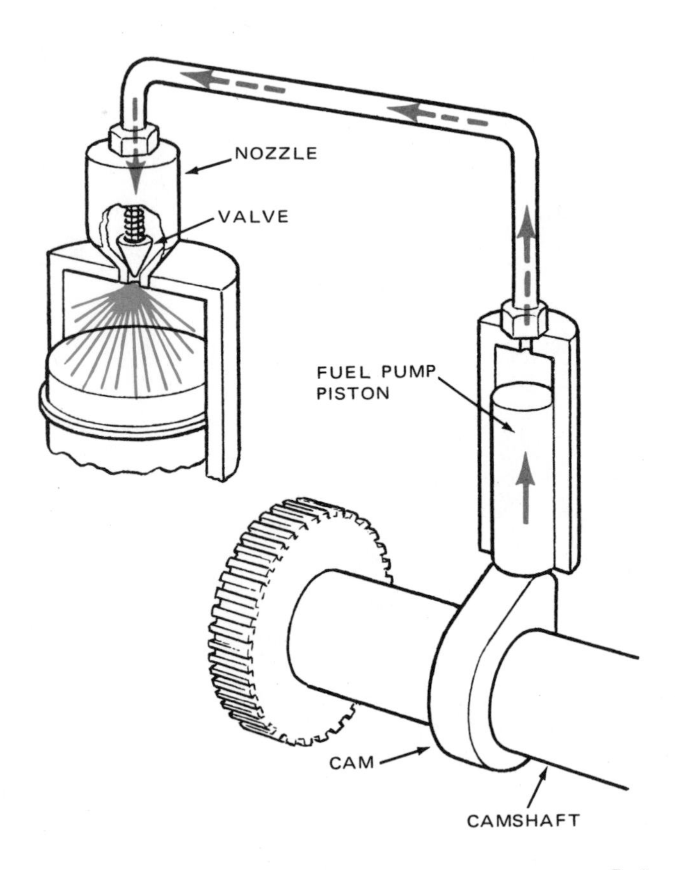

Fig. 2-16. The nozzle valve closes when the fuel injection stops. Each cylinder has its own fuel pump.

TWO-STROKE CYCLE DIESELS

In a two-stroke cycle, as the name indicates, there are two strokes to each cycle. One stroke is up, one down. Each down stroke is a power stroke.

In the two-cycle diesel, Fig. 2-17, both valves are exhaust valves. Ports (holes) in the cylinder wall which are opened and closed by the movement of the piston, permit air to be blown into the cylinder. When the piston is at the bottom of the stroke, the ports are open, and air is forced into the cylinder under high pressure, by a blower (air pump). At the same time exhaust gases are being blown out through the open valves at the top of the cylinder.

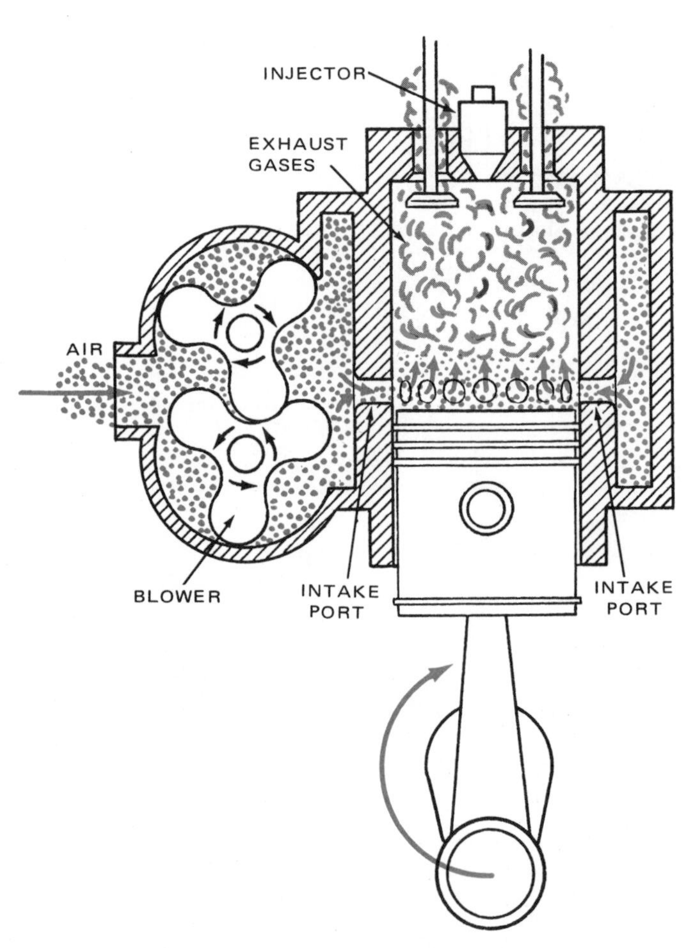

Fig. 2-17. Two-stroke diesel. Piston is at the bottom of the stroke. The intake ports are open, air is coming in, and the burned gases are being expelled through the valves at the top of the cylinder.

As the piston rises, the intake ports are covered, the

exhaust valves close, and air in the cylinder is compressed. See Fig. 2-18.

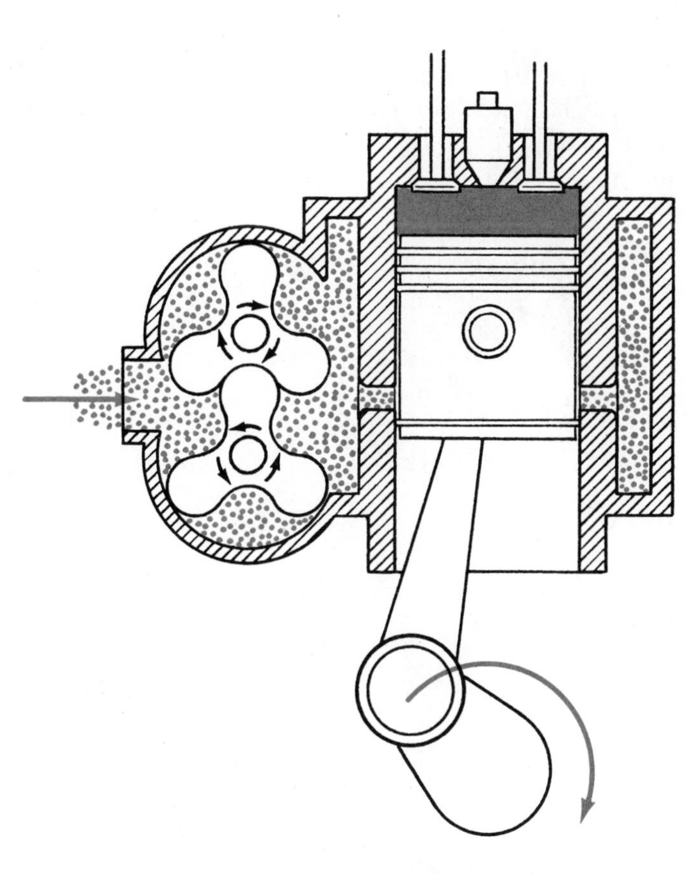

Fig. 2-18. Two-stroke diesel. As the piston rises, the intake ports are covered, the exhaust valves close and air in the cylinder is compressed.

Fuel injected when the piston is near the top of the stroke is ignited by heat developed from compression of the air, and the expanding gases force the piston down to develop power, Fig. 2-19.

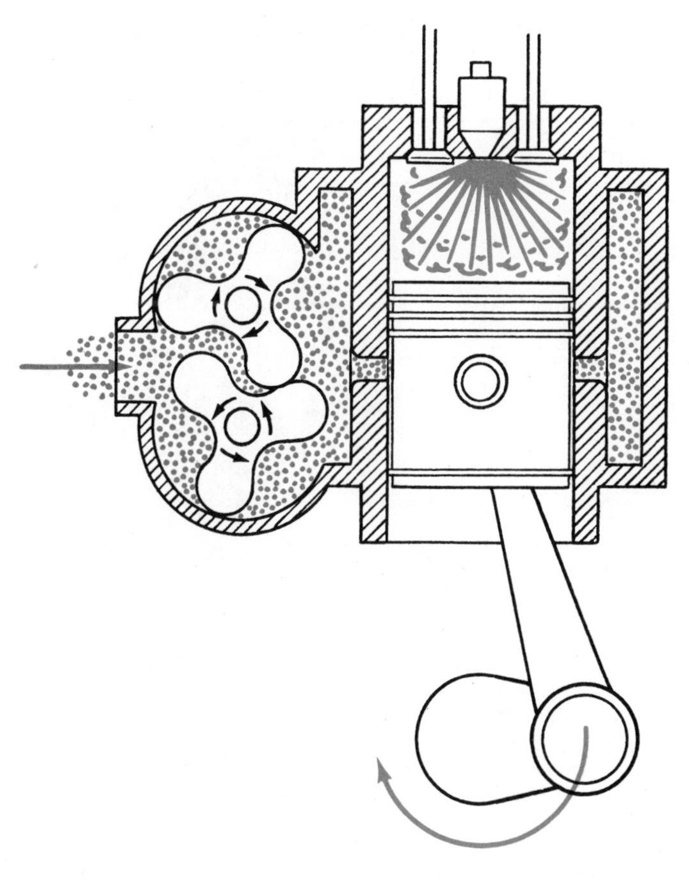

Fig. 2-19. Two-stroke diesel. Heat developed from compression of air in the cylinder ignites fuel and expanding gases force piston down to develop power.

ENGINE VARIATIONS

There are many variations in the design of both automotive and industrial diesel engines. However, the basic fundamentals of both four-cycle and two-cycle engines as covered in this chapter are applicable to all types.

REVIEW QUESTIONS – CHAPTER 2

1. In the four-stroke cycle engine, the strokes of the piston required to complete one cycle are:
   a. ______________________________.
   b. ______________________________.
   c. ______________________________.
   d. ______________________________.
2. Engine valves are opened by cam lobes on the rotating camshaft and closed by ________________.
3. The diesel engine depends on the heat of compression of the air to ____________ the fuel.
4. Air without fuel is compressed in the engine so the liquid diesel fuel forced into the cylinder burns instantly at about:
   a. 500 degrees.
   b. 1,000 degrees.
   c. 1,500 degrees.
   d. None of the above.
5. In a two-stroke cycle engine, each down stroke is a _________ stroke.
6. Air enters the cylinder of a two-cycle diesel engine through _________ in the cylinder wall.

Peugeot four cylinder engine, model XD 90. Note side-mounted fuel injection pump.

This 8.2 litre (500 cu. in.) Detroit Diesel Allison Div. diesel engine is installed in Chevrolet C70 medium-duty trucks. Naturally aspirated (165 hp) and turbocharged (205 hp) versions are available.

Chapter 3
BASIC MEASUREMENTS

In your study of diesels . . . how they are constructed, and how they work; it is important that you understand these terms:

HEAT

Heat as developed in the operation of diesels, may be described as a nonmechanical transfer of energy between the parts of a diesel, as indicated by temperature differences.

Heat may also be described as the random movement of the units of which matter is composed; molecules and atoms.

In studying and working with diesels, we will be concerned with the generation of heat as a result of compressing air, and burning diesel fuel.

TEMPERATURE

Temperature may be defined as the measurement of the hotness or coldness of an object or substance. When two objects are placed in close contact, the one which is hotter will begin to heat the other, and is said to have the higher temperature.

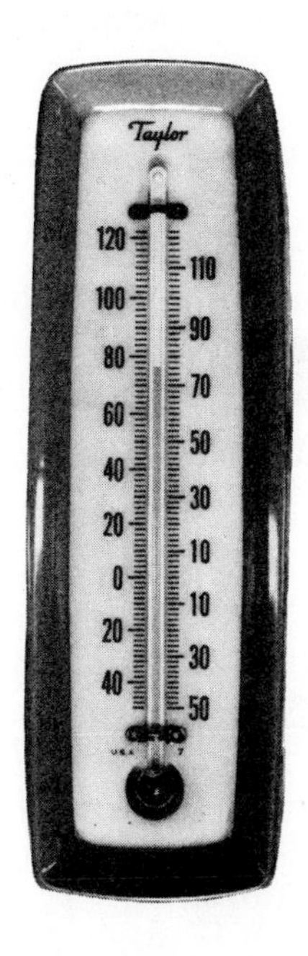

Fig. 3-1. Thermometer used for measuring atmospheric temperature, graduated in Fahrenheit degrees.

Temperature may be measured by using a thermometer. Some thermometers are graduated in Fahrenheit degrees, some in Celsius. In the conventional type thermometer, Fig. 3-1, temperature change is indicated by the expansion and contraction of a column of mercury or alcohol enclosed in a tiny glass tube from which most of the air has been removed.

Fahrenheit and Celsius scales are compared in Fig. 3-2. You will note the Fahrenheit scale shows that water freezes at 32 deg. and boils at 212 deg. The Celsius scale indicates that water freezes at 0 deg., and boils at 100 deg.

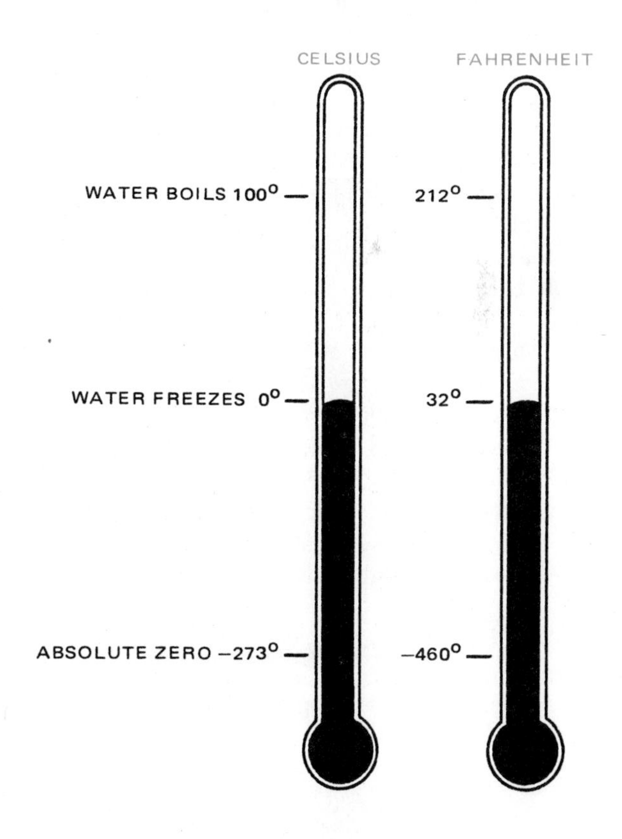

Fig. 3-2. Comparison of Celsius and Fahrenheit scales.

Two methods of converting Fahrenheit degrees to Celsius or Celsius to Fahrenheit are by the use of the following simple formulas or by the use of a table, Fig. 3-3.

$$F = \frac{9}{5}(C) + 32 \qquad C = \frac{5}{9}(F - 32)$$

C	F	C	F	C	F	C	F
−40	−40.0	+40	+104.0	+120	+248.0	+350	+662.0
−30	−22.0	+45	+113.0	+130	+266.0	+400	+752.0
−20	− 4.0	+50	+122.0	+140	+284.0	+450	+842.0
−15	+ 5.0	+55	+131.0	+150	+302.0	+500	+932.0
−10	+14.0	+60	+140.0	+160	+320.0	+550	+1022.0
− 5	+23.0	+65	+149.0	+170	+338.0	+600	+1112.0
0	+32.0	+70	+158.0	+180	+356.0	+650	+1202.0
+ 5	+41.0	+75	+167.0	+190	+374.0	+700	+1292.0
+10	+50.0	+80	+176.0	+200	+392.0	+800	+1472.0
+15	+59.0	+85	+185.0	+225	+437.0	+900	+1652.0
+20	+68.0	+90	+194.0	+250	+482.0	+1000	+1832.0
+25	+77.0	+95	+203.0	+275	+527.0	+1500	+2732.0
+30	+86.0	+100	+212.0	+300	+572.0	+2000	+3632.0
+35	+95.0	+110	+230.0	+325	+617.0	+2400	+4352.0

Fig. 3-3. Converting temperature readings from Celsius to Fahrenheit.

THERMOCOUPLE PYROMETER

Another temperature measuring device with which you should be familiar, is the thermocouple pyrometer. A thermocouple makes use of the principle that an electric current is produced when wires or rods made from two different metals, such as iron and copper, are joined together and the junction is heated, Fig. 3-4. The amount of current produced is in direct proportion to the temperature being measured.

Thermocouple pyrometers are used in checking the temperature of exhaust gases and engine parts, such as valves, bearings, and crankshafts.

THERMAL ENERGY

Heat, which is a form of mechanical energy, is often referred to as thermal energy. Thermal energy is measured in British thermal units (Btu's). A British thermal unit is the amount of heat required to raise the temperature of one pound of water one degree Fahrenheit.

Heat values of various fuels are measured in Btu's. For example, the heat value of gasoline per gallon is 124,500 Btu's; for diesel fuel, 139,500 Btu's.

THERMAL EXPANSION

When metals and gases are heated they expand. For example, pistons expand in the cylinders of a diesel engine. It is necessary to provide enough clearance between the piston and the cylinder so the piston will not seize (grab and fail to move) due to expansion.

In a diesel engine, expansion due to heat, is the basis on which the diesel operates.

HEAT TRANSFER

Heat is a form of energy in motion. When there is a difference in temperature between the areas of a body, or between neighboring bodies, heat will move from the area of higher temperature to that of the lower temperature. As an example, in a water-cooled diesel engine heat from fuel combustion transfers through the cylinder walls to the coolant (water). The coolant is heated, and the heat is transferred to the radiator where it is dissipated (scattered in various directions).

BORE-STROKE-DISPLACEMENT

The diameter of an engine cylinder is called the bore, Fig. 3-5. The distance the piston moves from the bottom dead center (position at which piston is not moving in either direction) to the top dead center is called the stroke, Fig. 3-6.

Displacement of an engine cylinder is a measurement of size, and is equal to the number of cubic inches its piston

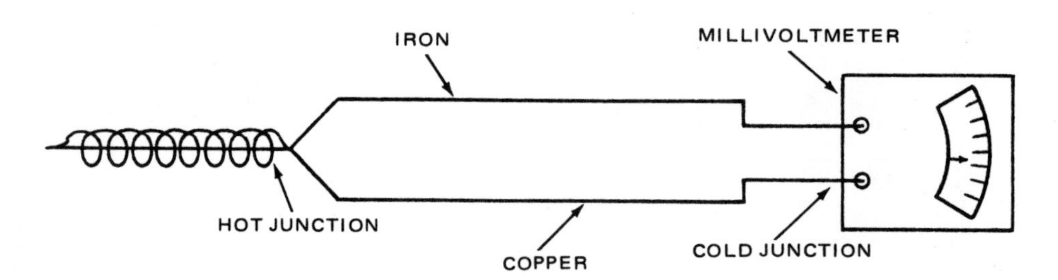

Fig. 3-4. Schematic drawing of thermocouple used to measure high temperatures.

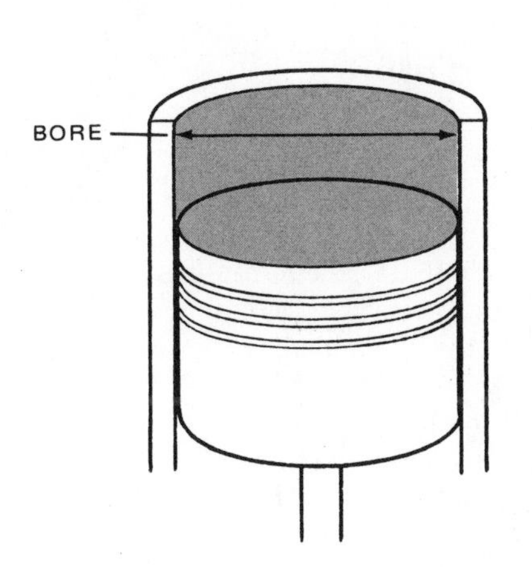

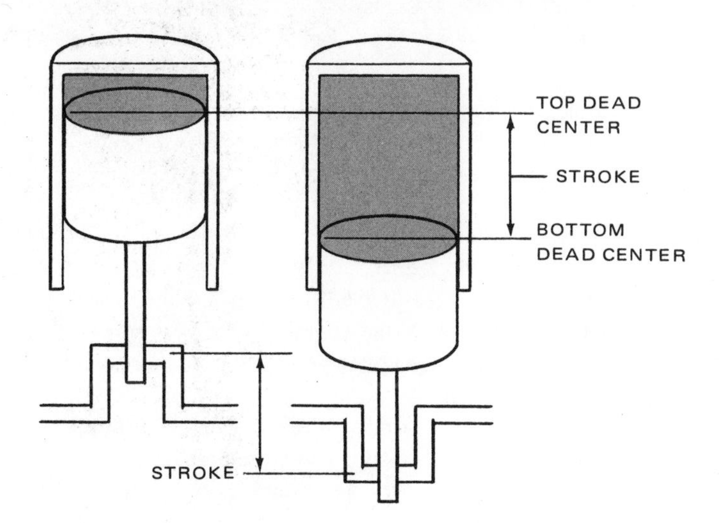

Fig. 3-5. Left. The term used to indicate the diameter of an engine cylinder is the "bore."
Fig. 3-6. Right. Determining engine stroke.

displaces as it moves from bottom dead center to top dead center. In other words, it is equal to the area of the piston times the stroke. If the engine has several cylinders, it is necessary to multiply the displacement of one cylinder by the number of cylinders.

COMPRESSION RATIO

Compression ratio is the relationship between the cylinder volume when the piston is at top dead center, and the cylinder volume when the piston is at bottom dead center, Fig. 3-7.

THERMAL EFFICIENCY

The thermal (heat) efficiency of an engine is based on how much of the energy in the burning fuel is converted into usable power. Heat developed by the burning fuel drives the piston down to produce the power stroke. In an internal combustion engine such as a diesel, a great deal of the heat energy is lost to the exhaust, cooling, and lubricating systems of the engine.

HORSEPOWER

Horsepower is a measurement of the ability of an engine to do work. One horsepower may be described as the ability to lift 33,000 pounds one foot in one minute. The term was originated in the days that strong horses were used to lift coal from mines.

INDICATED HORSEPOWER

Indicated horsepower is a measure of the power developed by burning of the fuel in the cylinders of the engine. To

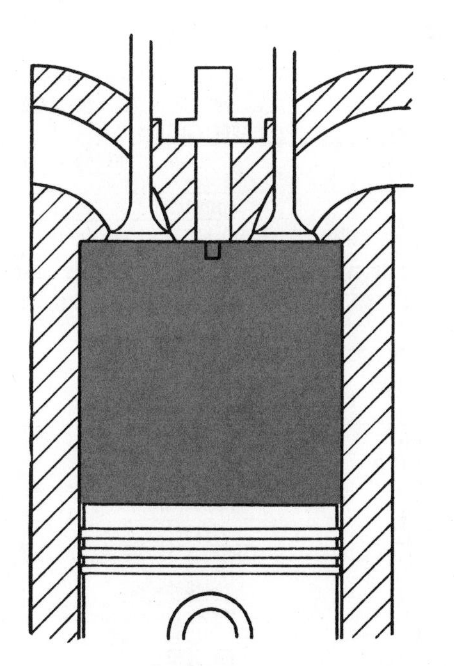

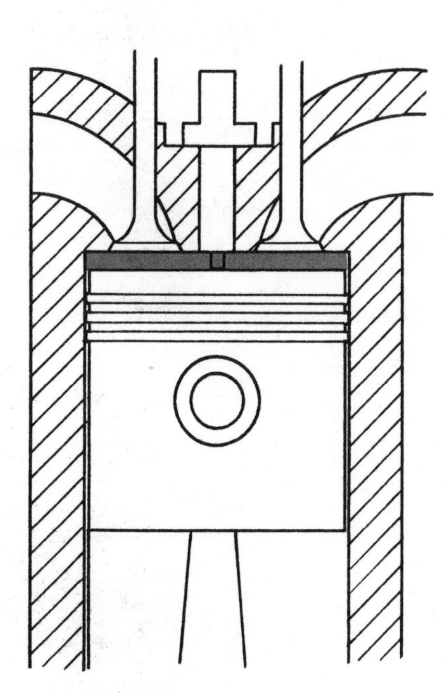

Fig. 3-7. Compression ratio is the relationship between the cylinder volume when the piston is at top dead center, and the cylinder volume when the piston is at bottom dead center.

determine indicated horsepower, it is necessary to determine the pressures within the cylinders for the intake, compression, power and exhaust strokes.

BRAKE HORSEPOWER

Brake horsepower is a measurement of actual usable power delivered to the crankshaft of the engine. Brake horsepower may be measured with a dynamometer.

A dynamometer includes a resistance creating device, such as an electric generator, or a paddle wheel revolving in a fluid, which is arranged so it will absorb and dissipate the power produced by the engine. Gages are provided to show the amount of power absorbed.

FRICTION HORSEPOWER

Friction horsepower is the power needed to overcome friction within the engine. Friction results from pressure of the piston and rings against the cylinder walls, rotation of the crankshaft and camshaft in their bearings, and friction developed by other moving parts.

ENGINE TORQUE

Torque may be defined as a turning or twisting force. An example of torque is the turning force exerted on the crankshaft by connecting rods attached to the pistons.

REVIEW QUESTIONS – CHAPTER 3

1. Thermal energy is measured in ____________ units.
2. In a diesel engine, a great deal of the heat energy is lost to:
   a. ______________________.
   b. ______________________.
   c. ______________________.
3. One horsepower is the ability to lift ________ pounds one foot in one minute.
4. The relationship between the cylinder volume when the piston is at top dead center, and the cylinder volume when the piston is at bottom dead center is called the ____________.
5. The centigrade scale indicates that water freezes at 0 degrees and boils at ______ degrees.
6. To determine indicated horsepower, it is necessary to determine the pressures within the cylinders for the following strokes.
   a. ______________________.
   b. ______________________.
   c. ______________________.
   d. ______________________.
7. Brake horsepower may be defined as: ______________ ______________________________.
8. To convert inches to millimeters, multiply the inches by:
   a. 3.14.
   b. 25.4.
   c. 2.54.

FACTORS FOR CONVERTING FREQUENTLY USED UNITS

Multiply	by	to get equivalent number of:
	LENGTH	
Inch	25.4	millimetres (mm)
Foot	0.304 8	metres (m)
Yard	0.914 4	metres
Mile	1.609	kilometres (km)
	AREA	
Inch²	645.2	millimetres² (mm²)
	6.45	centimetres² (cm²)
Foot²	0.092 9	metres² (m²)
Yard²	0.836 1	metres²
	VOLUME	
Inch³	16 387.	mm³
	16.387	cm³
	0.016 4	litres (l)
Quart	0.946 4	litres
Gallon	3.785 4	litres
Yard³	0.764 6	metres³ (m³)
	MASS	
Pound	0.453 6	kilograms (kg)
Ton	907.18	kilogram
Ton	0.907	tonne (t)
	FORCE	
Kilogram (force)	9.807	newtons (N)
Ounce	0.278 0	newtons
Pound	4.448	newtons

Multiply	by	to get equivalent number of:
	ACCELERATION	
Foot/sec²	0.304 8	metre/sec² (m/s²)
Inch/sec²	0.025 4	metre/sec²
	TORQUE	
Pound-inch	0.112 98	newton-metres (N·m)
Pound-foot	1.355 8	newton-metres
	POWER	
Horsepower	0.746	kilowatts (kW)
	PRESSURE OR STRESS	
Inches of mercury	3.38	kilopascals (kPa)
Pounds/sq. in.	6.895	kilopascals
	ENERGY OR WORK	
BTU	1 055.	joules (J)
Foot-pound	1.355 8	joules
Kilowatt-hour	3 600 000 or 3.6x10⁶	joules (J = W·s)
	LIGHT	
Footcandle	10.764	lumens/metre² (lm/m²)
	FUEL PERFORMANCE	
Miles/gal	0.425 1	kilometres/litre (km/l)
Gal/mile	2.352 7	litres/kilometre (l/km)
	TEMPERATURE	
Degree Fahrenheit (°F-32) ÷ 1.8 = degree Celsius (°C)		

Chapter 4

DIESEL FUELS

Many people think that all diesels operate on diesel fuel oil. This is not entirely true because some are dual-fuel engines, and are designed to operate on natural gas and fuel oil. Others operate only on gas.

DIESEL FUEL PRODUCTION

Diesel fuel is obtained from crude oil, which is a mixture of hydrocarbons (compounds consisting of hydrogen and carbon) such as benzine, pentane, hexane, heptane, toluene, propane and butane.

Compounds which form crude oil vaporize at different temperatures. That is, they have different boiling points. To separate the hydrocarbons, the crude oil is heated and the different hydrocarbons are given off as vapor. The hydrocarbon with the lowest boiling point is given off first. This is natural gas used by homes and industry. After the natural gas has been separated from the crude oil, the temperature is raised to drive off the hydrocarbon having the next highest boiling point. This is high octane aviation gasoline.

When this separation is complete, the temperature is again raised to obtain the hydrocarbon with the next highest boiling point, and so on until all the commercial gasoline, kerosene, diesel fuel, domestic heating oil, industrial fuel oil, lubricating oil, paraffin, etc. are obtained, leaving coke and asphalt.

The process of distilling diesel fuel from crude oil is a highly complicated process involving precision control of temperatures and pressures. The chart, Fig. 4-1, traces the flow of crude oil from the well to the finished product.

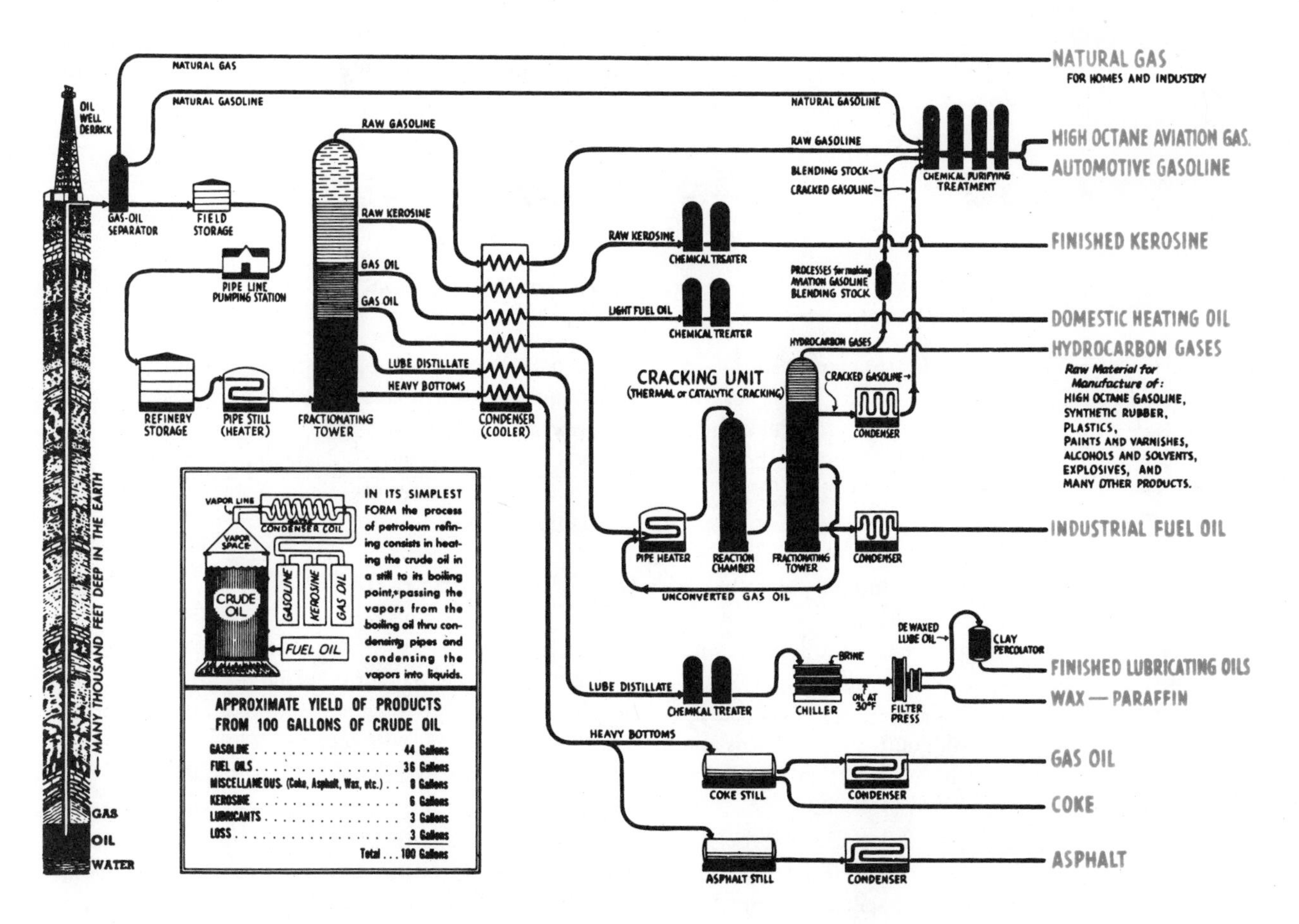

Fig. 4-1. Typical flow chart tracing crude oil from its source in the oil well to the finished products.

DIESEL FUEL PROPERTIES

As crude oil is refined, approximately 44 percent is gasoline, 36 percent is fuel oil, and the balance is kerosene, lubricants, etc.

The characteristics or properties of diesel fuel include:

1. Heat value
2. Specific gravity
3. Flash point
4. Pour point
5. Viscosity
6. Volatility
7. Ignition quality (cetane)
8. Carbon residue
9. Sulphur content
10. Oxidation and water

HEAT VALUE

The heat value of a fuel is of primary importance as it is an indication of how much power the fuel will provide when burned. The heat value of fuel oil can be determined by burning the oil in a special device known as a calorimeter. With such equipment a measured quantity of fuel is burned and the amount of heat is carefully measured in Btu's per lb. of fuel.

SPECIFIC GRAVITY

Specific gravity of a liquid such as diesel fuel, is the ratio of the density of the fuel (oil) to the density of water. Specific gravity may be measured by using a hydrometer, Fig. 4-2.

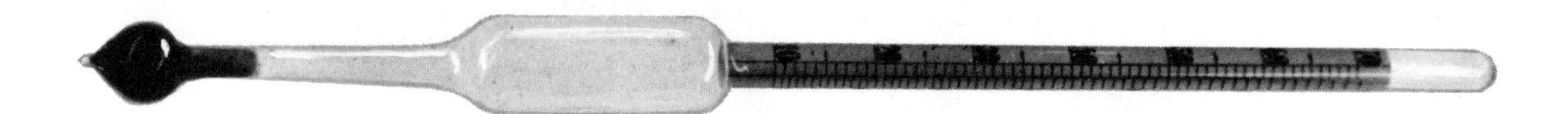

Fig. 4-2. Typical hydrometer float used in testing specific gravity.

Specific gravity of a fuel affects the spray penetration as the fuel is injected into the combustion chamber. It is also, to a degree, a measure of the heat content of the fuel. Fuel which is heavy usually has greater heat value per gallon than light oil.

FLASH POINT

The flash point of an oil is the temperature at which the oil must be heated until sufficient flammable vapor is driven off to flash (ignite) when brought into contact with a flame. Fire point is the higher temperature at which the oil vapors will continue to burn after being ignited. The fire point is usually 50 to 70 deg. F. higher than the flash point. Flash point is also an indication of fire hazard; the lower the flash point, the greater the fire hazard.

The flash point of diesel fuels is 100 deg. F. for type 1-D fuel, 125 deg. F. for type 2-D fuel, and 130 deg. F. for type 4-D fuels. In some states there are laws specifying the flash point of diesel fuels which may be used. In general, the flash point of diesel fuel should be high enough so fuel in transit will not produce flammable vapors.

CLOUD POINT, POUR POINT

The cloud point of diesel fuels is the temperature at which the hydrocarbon components of the fuel becomes insoluble (cannot be dissolved).

Pour point is the temperature at which enough of the fuel becomes insoluble to prevent flow under specified conditions. A high pour point implies that in cold weather the fuel oil will not flow easily through the filters and fuel system of the engine. Also, its spraying characteristics will not be satisfactory.

VISCOSITY

Viscosity is the property of a fluid that resists the force which causes the fluid to flow. Viscosity is measured by observing the time required for a certain volume of the fluid to flow, under stated conditions, through a short tube of small bore. The flow is measured by using a device called a viscometer. Saybolt universal viscosities are normally taken at 100 deg. F, 130 deg. F and 210 deg. F.

Viscosity of the diesel fuel affects the pattern of spray in the combustion chambers. Low viscosity produces a fine mist, while high viscosity tends to result in coarse atomization.

VOLATILITY

Volatility of a liquid is its ability to change into a vapor. The volatility of a liquid fuel is indicated by the air-vapor ratio that can be formed at a specific temperature. In the case of diesel fuels, volatility is indicated by a 90 percent distillation temperature (temperature at which 90 percent of the fuel is distilled off). As volatility is decreased, carbon deposits, and in some engines, wear are increased. Some engines will produce more smoke as volatility is decreased.

CETANE NUMBER RATING

The ignition quality of diesel fuel, ease with which the fuel will ignite, and the manner in which it burns may be expressed in cetane numbers. A cetane number rating is obtained by comparing the fuel with cetane, a colorless, liquid hydrocarbon which has excellent ignition qualities, and is rated at 100. The higher the cetane number, the shorter the lag between the instant the fuel enters the combustion chamber and the instant it begins to burn, Fig. 4-3. By comparing the performance of a diesel fuel of unknown quality with cetane, a cetane rating may be obtained.

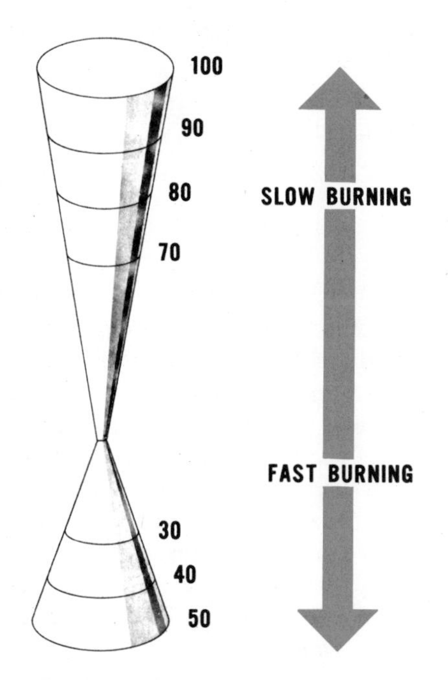

Fig. 4-3. Diagram comparing octane and cetane. Note that the higher the cetane number the faster the fuel burns.

CARBON RESIDUE

The carbon residue of a diesel fuel (matter left after combustion) is an indication of the amount of combustion chamber deposits that will be formed when the fuel is consumed in the engine. This can be measured in the laboratory by heating a sample of the fuel in a closed container in the absence of air. The carbon residue will remain in the container.

The amount of carbon residue which is considered permissible in diesel fuel oil depends somewhat on characteristics of the engine. This is more critical in small high speed engines than in large, low speed industrial engines.

SULPHUR CONTENT

The presence of sulphur in diesel fuel in excessive quantities is objectionable, as it increases piston ring and cylinder wear. In addition, it causes the formation of varnish (hard coating) on piston skirts, and oil sludge (mushy solution) in the engine crankcase.

When fuel containing sulphur is burned in an engine, the sulphur combines with water, which results from the fuel combustion, to form corrosive acids. These acids tend to etch finished surfaces, add to the deterioration of engine oil, and the production of sludge.

NITROGEN COMPOUND

Fuels which have a high sulphur content often contain considerable quantities of various nitrogen compounds. There is evidence that the high wear rate of engine parts is, in part, caused by the nitrogen compounds.

OXIDATION AND WATER

Many fuel filter plugging problems result from particles of rust, dirt and water contamination caused by careless storage or transportation conditions. Some injector sticking and filter stoppage may result from the instability (variation of content) of the fuel.

The American Society of Testing Materials has a test for determining the stability of fuel oil.

Elimination of water from fuel storage tanks is quite difficult. Some fuels reject water easily; others retain it and have a cloudy appearance for an extended period after being mixed with water. Some fuels containing as little as .01 percent of water will appear cloudy. Other fuels can suspend nearly 1 percent water for an extended period of time.

A fuel should provide good rust protection so it will minimize the formation of iron oxide particles in storage tanks and the rusting of fuel lines and injection equipment when the engine is idle.

COMMERCIAL DIESEL FUEL

Commercially, by American Society of Testing Materials Standards, there are three grades or classifications of automotive diesel fuels. There are known as Grade 1-D, Grade 2-D, and Grade 4-D. See Fig. 4-4. At one time there was a Grade 3-D, but that has been discontinued.

	1–D	2–D	4–D
Minimum flash point, deg. F.	100 or legal	125 or legal	130 or legal
Viscosity, 100 F.			
Minimum	1.4	2.0	5.8
Maximum	2.5	5.8	26.4
Carbon residue, wt. percent maximum	0.15	0.35	. . . .
Ash, wt. percent maximum	0.01	0.02	0.10
Sulphur, wt. percent maximum	0.50	1.0	2.0
Ign. quality, cetane No. Minimum	40	40	30
Distillation temp. deg. F. 90 percent evaporated			
Minimum	. . . .	540	. . . .
Maximum	550	576	. . . .

Fig. 4-4. Grades of fuel oil.

Grade 1-D, is a volatile fuel for engines in service requiring frequent changes in load and speed.

Grade 2-D, is a low volatility fuel for engines in industrial and large mobile service.

Grade 4-D, is a fuel oil for low and medium speed engines.

As engine speed increases, cleanliness and viscosity become more critical. Chemical, physical and engine tests may be used to determine the quality of diesel fuel.

GRADES OF DIESEL FUEL

In addition to the grades of diesel fuels listed in Fig. 4-4, there are special fuels for railroad locomotives, buses, the military, etc.

PURCHASING FUEL OIL

Fuel oil for diesel engine trucks is usually purchased by the gallon. When purchased in large quantities for industrial use, the oil is generally purchased by the barrel. Each barrel contains 42 gallons. When purchasing oil in quantities, it is necessary to know the temperature as the oil is always billed as if it is 60 deg. F. As oil temperature rises it expands in volume and it is, therefore, necessary to convert the volume back to 60 deg. F. gallonage. The expansion rate of diesel fuel oil is 0.0004 per degree of temperature rise above 60 deg. F. The chart, Fig. 4-5, can be used in determining the corrected volume or gallonage of fuel oil.

GASEOUS FUELS

Almost all automotive type diesel engines use liquid petroleum fuels. Many large industrial engines use gaseous fuels.

There are eight major types of gaseous fuels used in diesel engines:

1. Natural Gas
2. Producer Gas
3. Blast Furnace Gas
4. Blue Water Gas, from coke
5. Carbureted Water Gas
6. Coal Gas
7. Coke Oven Gas
8. Refinery Oil Gas

Natural gas is found in commercial quantities throughout the United States and Canada, and is distributed by pipe lines to the various metropolitan centers.

Natural gas, like petroleum, is a mixture of hydrocarbons. Its chief constituents are methane, ethane and nitrogen. In addition, it usually contains some hydrogen sulphide. Its constituents and characteristics vary in different areas.

Other gaseous fuels listed previously may be classified as either manufactured or by-product gases.

With the exception of refinery oil gas, which has a heat value of 1426, the other gases mentioned have a relatively low heat value, ranging from 91 for blast furnace gas to 574 for coke oven gas. These gases are generally used in the area in which they are produced.

DUAL-FUEL ENGINES

Dual-Fuel engines are designed to burn both gas and oil and are used in some localities to take advantage of low cost natural gas.

The compression ratio of a dual-fuel engine is in the same range as the conventional diesel. The gas fuel and air are mixed providing a lean mixture before compression. The mixture is so lean (heavy on air, light on gasoline) it is not ignited by the heat of compression. A jet of oil is injected into the compressed air-gas mixture. The injected oil (also known as pilot oil) ignites as soon as it strikes the heated mixture in the combustion chamber. This in turn ignites the lean air-gas mixture.

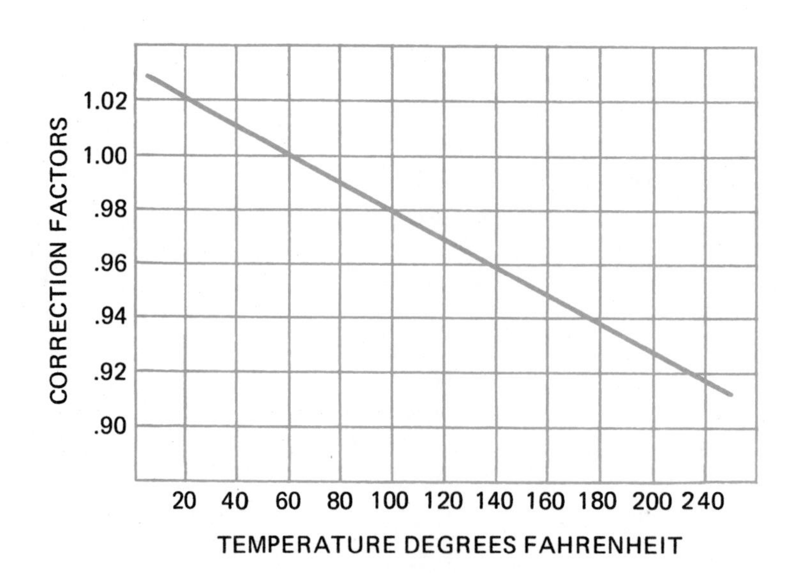

Fig. 4-5. Chart used in obtaining correction factor for volume of fuel oil at different temperatures.

FUEL ADDITIVES, SMOKE

As in the case of gasoline fuels, various additives are used in diesel fuels to reduce smoke, reduce knocking, and improve cold weather operation.

One of the problems in connection with the operation of diesel engines is smoke. When an engine is over-fueled, a heavy black smoke is emitted by the exhaust. Not only is this a factor in polluting the air we breathe, but it produces excessive accumulations of carbon in the combustion chamber.

Special additives are now used in most diesel fuels to counteract the production of such smoke.

POUR DEPRESSANTS

Under certain conditions of engine operation, automotive diesel fuels of high cetane number are not needed. Under such conditions, lower cost fuels of higher density can be used in automotive diesel engines.

In cold weather operations, some fuels have a serious drawback as wax separates from the fuel and obstructs the oil flow through critical filters and lines. This type of problem generally can be overcome by the addition of kerosene. Adding kerosene to the fuel lowers its density, at a sacrifice of the heat value of the fuel.

Another method of combating the wax problem is to add a suitable depressant (substance which lessens undersireable properties). Depressants improve the flow of fuel through the filters by reducing the size of the wax crystals so they will pass through the filters.

The degree to which vehicles will benefit from the improved filter ability when a depressant is added to the fuel, depends largely on the design of the fuel system.

Insoluble residues tend to form while fuel is in storage. Residues in untreated fuels may clog the filters and the fuel lines. Additives have been developed which contain dispersing agents which help to control the size of residue particles and reduce the rate of formation.

IGNITION ACCELERATORS

In diesel fuels it is desirable to expedite (speed up) the ignition of the fuels, whereas in gasoline spark ignition engines, it is desirable to slow the rate of burning. In the case of diesel fuels, the ignition of the fuel is expedited by reducing the temperature required for self-ignition. Adding a small amount of ethyl nitrate is one way to produce that effect.

POLYMERIZATION

Another problem is that diesel fuel is unstable, causing "polymerization." As fuel temperature rises, the molecules polymerize or tend to combine to form larger molecules that clog filters and cause sticking injectors.

STORING FUEL

A major cause of trouble in diesel engines is contaminated fuel. It is apparent that great care must be taken in handling and storing diesel fuel.

Delivery tanks as well as storage tanks must be kept clean and free from water and residue. Hoses used for fuel delivery must be in good condition. Tanks on delivery vehicles must be carefully inspected and flushed at regular intervals. Keeping the storage tank as nearly filled as practical will help to reduce moisture condensation.

Storage tanks (steel) should be pitched so any condensed moisture will collect at the low point making it possible to drain it off at regular intervals. The storage tank filler cap should be provided with a fine (80-mesh) screen to prevent foreign matter from entering. Fuel should not be used from the storage tank until it has had time to settle.

It is not advisable to use copper and galvanized tanks and lines for diesel fuel because of chemical action between those metals and the fuel.

FUEL COMBUSTION

Combustion is the burning of the fuel-air mixture in the cylinders of the engine.

Oxygen used in the combustion of diesel fuel is obtained from the air. Air is a mixture of various gases, including oxygen, nitrogen, argon, carbon dioxide, hydrogen, neon, and other gases. With the exception of oxygen and nitrogen, the remaining gases are in such small quantities that for normal combustion they are ignored.

Atmospheric air has a composition by volume of 20.94 percent oxygen, and 79.06 percent nitrogen. By weight the composition is 23.14 percent oxygen, and 76.86 percent nitrogen.

COMBUSTION OF CARBON

When carbon is burned with excess air, carbon dioxide is formed. If the air (oxygen) is limited, carbon monoxide is formed.

The air-fuel ratio required to burn diesel fuel is usually between 14.5 and 15 lbs. of air per pound of fuel.

The hydrogen component of the hydrocarbons burns first, and if only enough oxygen is present to burn the hydrogen, the carbon will deposit as soot.

In a spark ignition, carburetor type engine, the fuel and air are mixed prior to entering the combustion chamber. Consequently, the ratio of fuel to air can be closely controlled so fuel economy can be obtained. In that way substantially all of the fuel is burned.

In the diesel engine, the mixing of the fuel with air and combustion occur almost simultaneously. As a result, it is hardly possible for all of the fuel molecules to find the oxygen molecules needed for combustion.

So that ample oxygen for complete combustion of the fuel will be available, extra air is provided. Such excess air not only helps to insure complete combustion of the fuel, but also helps to eliminate heat loss due to the formation of carbon monoxide and depositing of carbon in the combustion chamber.

FUEL CONSUMPTION

In the operation of diesel engines it is important to know how much fuel will be consumed. In the case of industrial applications, such as electrical generating plants, where the engines are confined in small areas, it is also necessary to know the amount of air that will be required.

The quantity of air needed is expressed in engine displacement per minute, or, volume in cubic feet per minute (cfm).

Ignition quality (cetane number) of a fuel has a neglibile influence on output and economy of engine operation. Low cetane fuels (as long as they satisfy the cetane requirement of an engine) tend to give slightly more power at maximum load. This results in lower fuel consumption than with high cetane fuels. This is because low cetane fuels are heavier and, therefore, contain more heat units per gallon.

Combustion (burning) of fuel in a diesel engine is much slower than in a gasoline engine. After injection, the first effect on the fuel is partial vaporization. This reduces the temperature of the fuel and the air in the immediate vicinity of each fuel particle. This effect is only momentary since the very hot compressed air quickly raises the temperature of the sprayed fuel to the self-ignition point.

Then, combustion begins. Smaller particles are the first to ignite, and they burn rapidly. Larger particles are slower to ignite because heat must be transferred to bring them to the point of ignition. The slight delay between the time the fuel is injected and the time it reaches the self-ignition point is called ignition delay or "lag." Duration of "lag" is dependent on the characteristics of the fuel and the spray, the temperature and pressure of the compressed air in the combustion chamber, and the amount of turbulence (irregular motion) present. Shape of the combustion chamber is also a factor.

As combustion progresses, the temperature and pressure within the combustion chamber rise; thus, the ignition delay of fuel particles injected into the combustion process later is less than those injected early. The delay period between the start of injection and the start of self-ignition is sometimes referred to as the first phase of combustion. The second phase of combustion includes ignition of fuel injected during the first phase, and the spread of the flame through the combustion phase, as the injection continues. The resulting increases in the temperature and pressure reduce the ignition lag for fuel particles entering the combustion chamber, during the remainder of the injection period.

Only a portion of the fuel is injected during the first and second phases. As the remainder of the fuel is injected, the third or final phase of combustion takes place. The third phase is sufficient to cause most of the remaining fuel particles to ignite, as they come from the injection equipment. The rapid burning during the final phase of combustion causes an additional, rapid increase in pressure, which is accompanied by a distinct knock. This knock is characteristic of normal diesel operation, particularly with light loads.

DETONATION

Detonation (condition where fuel burns too fast) in a diesel engine usually results from too long a delay in ignition. The greater the delay, the greater the accumulation of fuel that accumulates in the combustion chamber before ignition.

When the ignition point of the excess fuel is reached, all of the fuel ignites and the resulting extreme high pressure produces the undesirable knock. Detonation in a diesel generally occurs at what is normally considered the start of the second phase of combustion instead of the final phase.

Detonation may occur in a diesel under any condition that permits excessive fuel to reach the combustion chamber, particularly when the engine has not reached operating temperature.

SMOKE AND AIR POLLUTION CONTROL

There are a number of methods used to reduce diesel smoke. These include using additives in the fuel, proper maintenance and operation, using supplementary fuel, avoidance of overloading, engine derating (reducing maximum flow of fuel) and adherence to proper fuel specifications.

SUPPLEMENTARY FUEL

A supplementary fuel such as propane or gasoline is used when the engine is operating under full power conditions. The setting of the diesel pump is reduced to obtain the desired reduction in smoke, and power is reestablished to full rating by pumping the supplementary fuel into the intake manifold. This system requires additional equipment which includes an electric supplementary fuel pump, fuel tank, valves, lines, switch and wiring.

DERATING

Derating the engine is another method of reducing diesel smoke. This is accomplished by setting the fuel pump on the engine to reduce the maximum flow of fuel. While the power is reduced slightly by this method, smoke is materially reduced.

The practice of using a transmission gear too high for operating conditions, causing the engine speed to drop below the recommended speed, may also increase the smoke.

REVIEW QUESTIONS – CHAPTER 4

1. Other than fuel oil, some diesel engines are designed to operate on:
   a. ______________________.
   b. ______________________.
2. When crude oil is heated, which of the following hydrocarbons is given off first?
   a. Kerosene.
   b. Gasoline.
   c. Natural gas.
   d. Paraffin.

3. The ratio between the density of fuel oil and the density of water is known as ____________.
4. Excess quantities of sulphur in diesel fuel is objectionable because it may cause the following:
   a. ________________________________.
   b. ________________________________.
   c. ________________________________.
   d. ________________________________.
5. The main constituents of natural gas are methane, ethane, and _______.
6. Besides natural gas, blue water gas, carbureted water gas, and blast furnace gas, other major gases used as fuel in diesel engines are:
   a. ________________________________.
   b. ________________________________.
   c. ________________________________.
   d. ________________________________.
7. Reduction of moisture condensation in fuel storage tanks can be accomplished by keeping them ______________.
8. Various additives are used in diesel fuels to:
   a. ________________________________.
   b. ________________________________.
   c. ________________________________.
9. Grade 1-D fuel is provided for engines in service requiring frequent changes in load and _________.
10. Burning of the fuel-air mixture in the cylinder of the engine is called:
    a. Ignition.
    b. Flash point.
    c. Exhaust.
    d. Combustion.
    e. Injection.
11. When carbon is burned with a limited amount of air, _________ is formed.
12. The ratio of fuel to air can be closely controlled so that substantially all fuel is burned resulting in fuel ________.
13. The quantity of air required in fuel consumption is usually expressed in either one of the following ways.
    a. ________________________________.
    b. ________________________________.
14. The delay between the time fuel is injected and the time it reaches the self-ignition point is usually called ignition delay or _______.
15. Polymerization is a chemical process in which molecules of a compound become:
    a. Larger.
    b. Slowed down.
    c. Smaller.
    d. Liquid.
16. Detonation may occur in a diesel under any condition that allows excessive fuel to reach the ____________.
17. Some of the methods used to reduce diesel smoke are as follows:
    a. ________________________________.
    b. ________________________________.
    c. ________________________________.
    d. ________________________________.
    e. ________________________________.

A Deere and Co. farm tractor powered by a four cylinder 239 cu. in. (3917 cm^3) diesel engine. Compression ratio is 23.9 to 1.

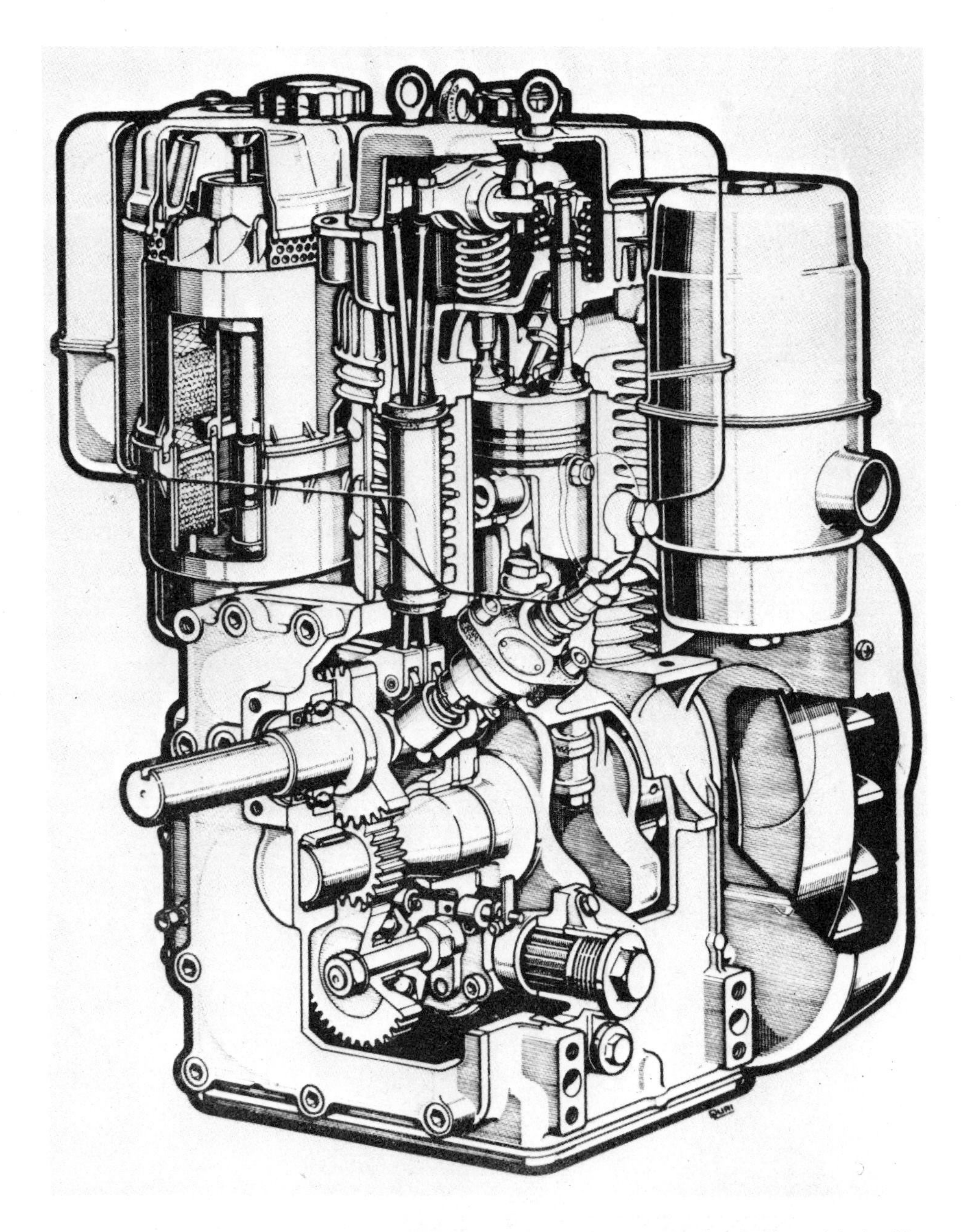

Cutaway view of a single cylinder, air-cooled engine manufactured by Lombardini of Italy. Note cooling fins on cylinder and cylinder head. Also note end of fuel injector between two engine valves.

Chapter 5
COMBUSTION CHAMBER TYPES

In both gasoline and diesel engines, it is essential that air and fuel be properly mixed to obtain maximum efficiency. In the gasoline engine the mixing takes place in the carburetor, which is outside the cylinder. The mixture then passes on to the cylinders where it is compressed. In the diesel engine, the fuel in the form of fine particles is sprayed into the cylinder after the air has been compressed.

To secure complete combustion, each particle of fuel must be surrounded by sufficient air. That condition is obtained by providing turbulence (violent swirling motion) in the combustion chamber.

Fuel injection provides some turbulence. Additional turbulence is provided by the design features of the combustion space.

Combustion chambers of various types are used including:

1. Open combustion chambers.
2. Turbulence chambers.
3. Precombustion chambers.
4. Energy cell or air cell chambers.

Each type has its advantages.

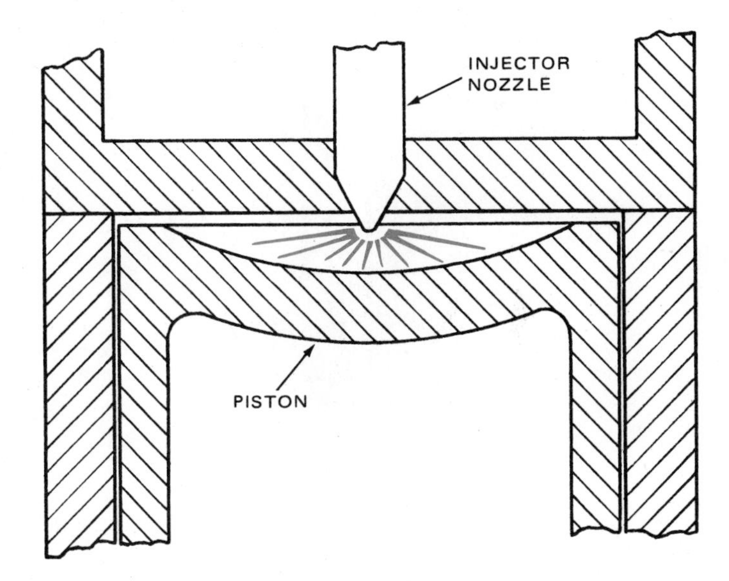

Fig. 5-2. This open combustion chamber is formed by a dished piston and is known as a hemispherical type.

OPEN COMBUSTION CHAMBERS

Various types of modern open combustion chambers are shown in Figs. 5-1, 5-2, and 5-3. Many manufacturers use modifications of these designs. It will be noted that in each of the designs, the motion imparted to the air results primarily from the contour of the top of the piston head.

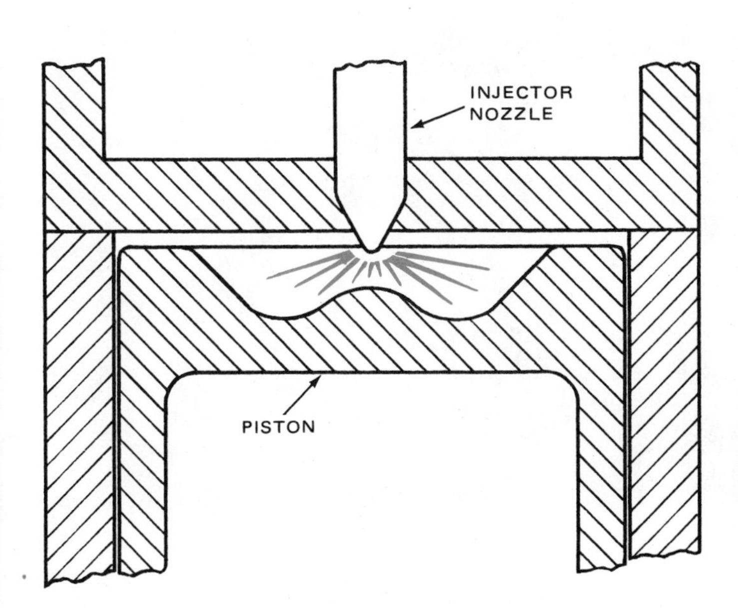

Fig. 5-1. Typical open combustion chamber.

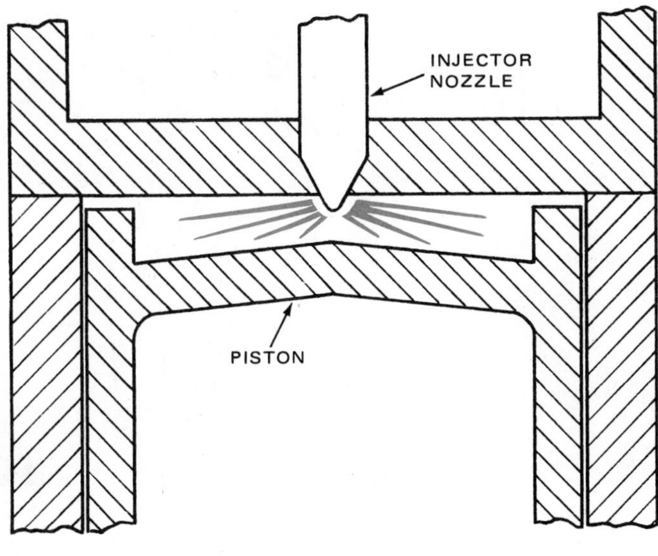

Fig. 5-3. Open combustion chamber formed by a flange on the piston top.

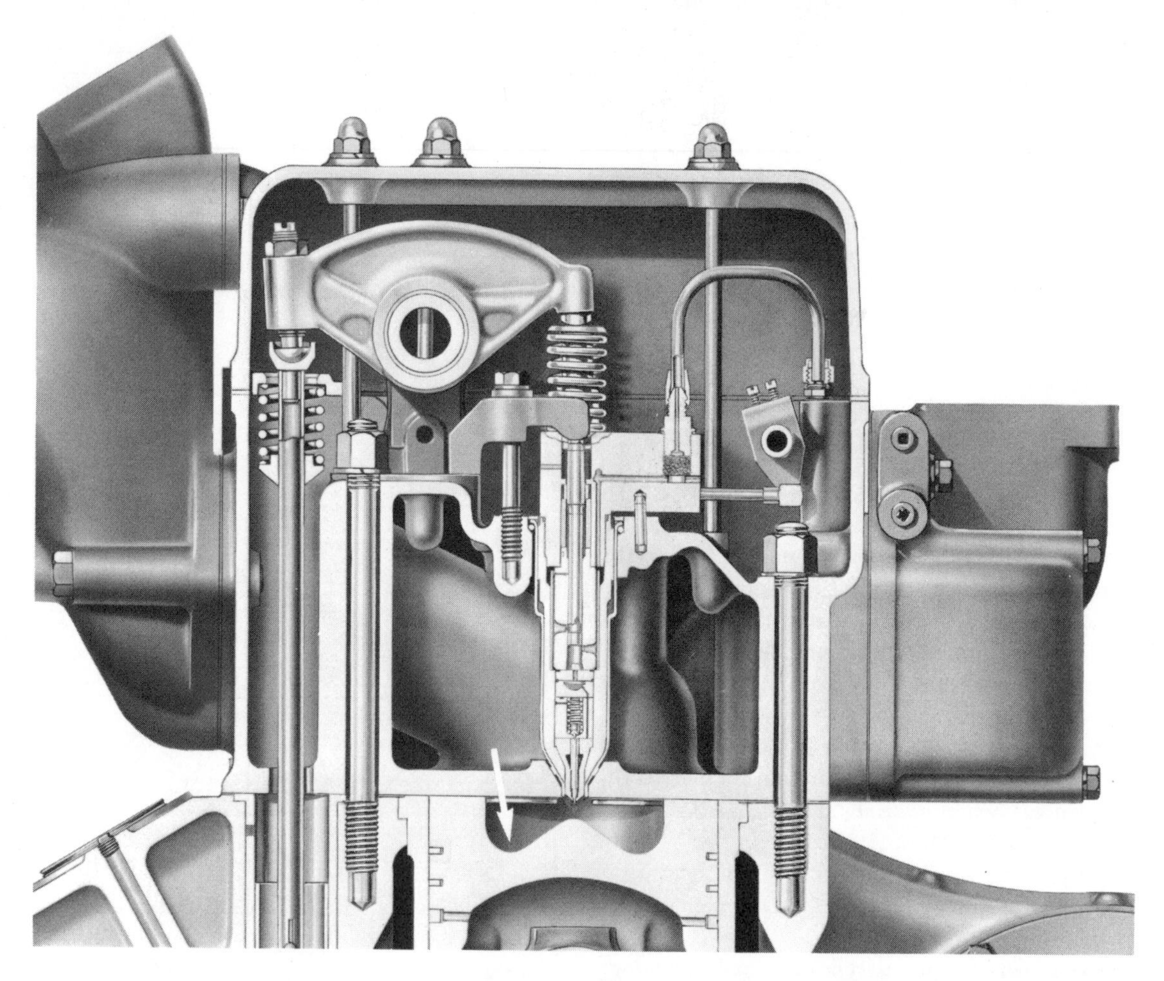

Fig. 5-4. Note contour of piston top forming an open combustion chamber. (Waukesha Motor Co.)

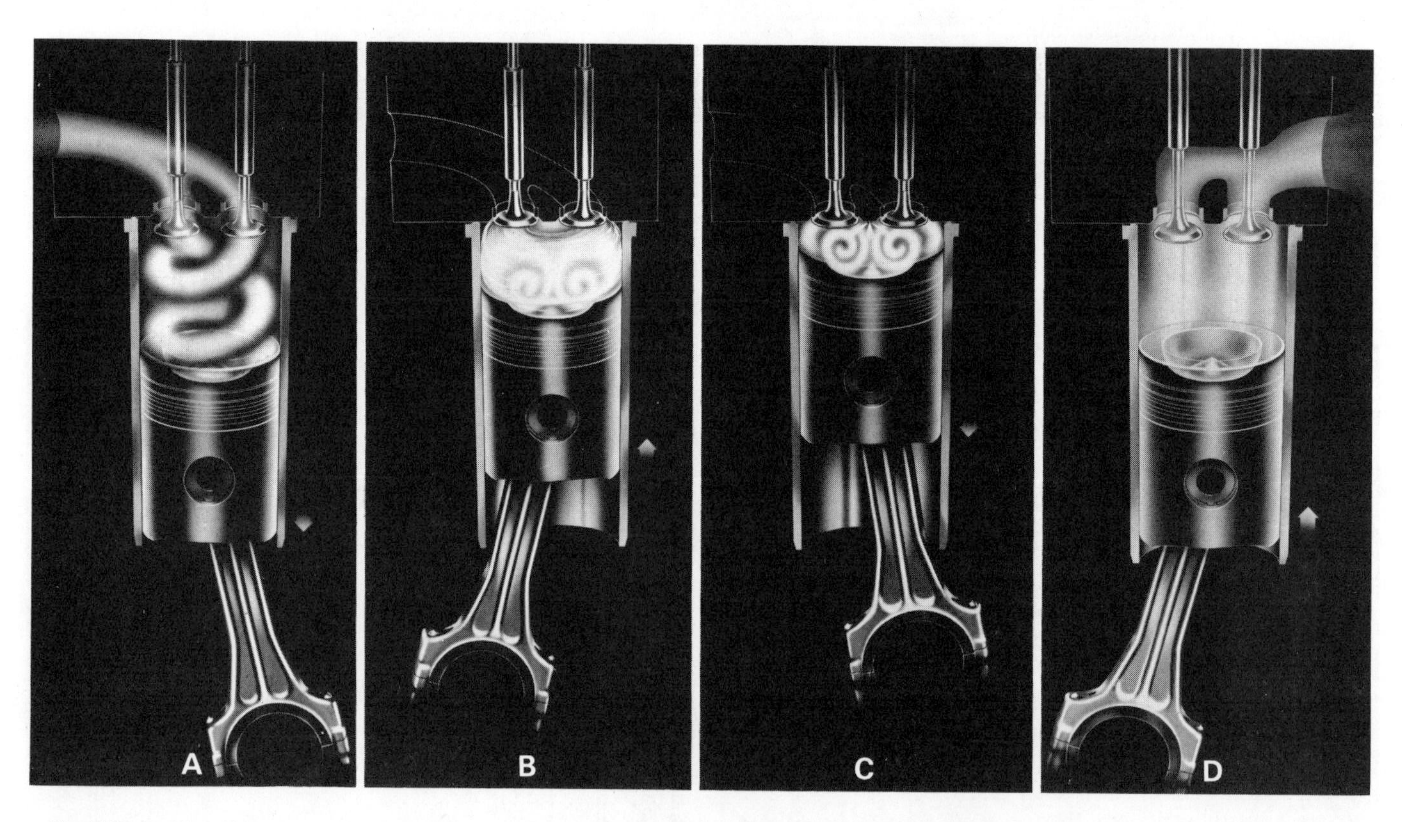

Fig. 5-5. A—Air rushing into cylinder on intake stroke is given a swirling motion by the directional ports in the cylinder head. B—The swirling motion is intensified on the compression stroke. The contoured piston moves the air toward the center giving it an additional rolling motion and greater velocity. C—At point of maximum turbulence, fuel is injected in a four spray pattern under high pressure. D—Atomized fuel mixes with the turbulent air and combustion produces the power stroke. (Allis Chalmers Mfg. Co.)

Fig. 5-4 shows the open combustion chamber design as used in a Waukesha engine. Note that the air on entering the cylinder will be given a swirling motion as it strikes the top of the piston. A cone-shaped spray from the centrally located injector will then be fully dispersed (scattered) through the swirling air.

One method of imparting increased swirl to the air is by means of the port through which the air reaches the valve. When this port is at an angle to the axis of the valve, air will enter the combustion chamber at an angle and attain a swirling action.

An example of such construction is in the Allis Chalmers diesel engines, Fig. 5-5. Air entering the cylinder on the intake stroke is given a swirling motion by the directional ports in the cylinder head. The swirling motion is intensified in the compression stroke. The contoured piston top forces the air toward the center, giving it additional rolling motion and greater velocity. At the point of maximum turbulence, the fuel is injected in a four jet pattern and it mixes thoroughly with the turbulent air.

MASKED VALVE

Use of masked valves is another way in which turbulence is provided in the combustion chamber. In this widely used design, a lip formed on the underside of the valve prevents air from entering the combustion chamber from the area blocked by the lip. See Fig. 5-6.

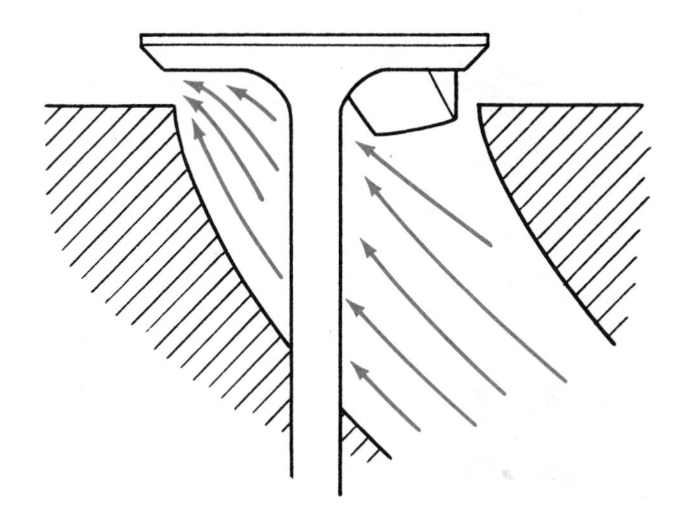

Fig. 5-6. Masked valve design to give turbulence to the incoming air.

While highly successful in imparting a swirling action to the air, a disadvantage of the masked valve is that the valve cannot be rotated.

Proper location of the injection nozzle, together with the pressure at which the fuel is injected, are major features contributing to the thorough mixing of the fuel and air.

In the open combustion chamber, Fig. 5-1, added turbulence is obtained by the wide rim around the top of the piston. This comes close to the cylinder head at top center, and the swirl action of the piston causes an inward flow of the air toward the center of the piston.

Two of the major advantages of the open combustion chamber are its high fuel economy and simplicity of design.

A variation of the open combustion chamber is shown in Fig. 5-7. This European design, known as the M design, is used in the Maybach railroad engine, in addition to engines for the commercial vehicle field. It will be noted that the ball shaped combustion chamber is located entirely within the piston.

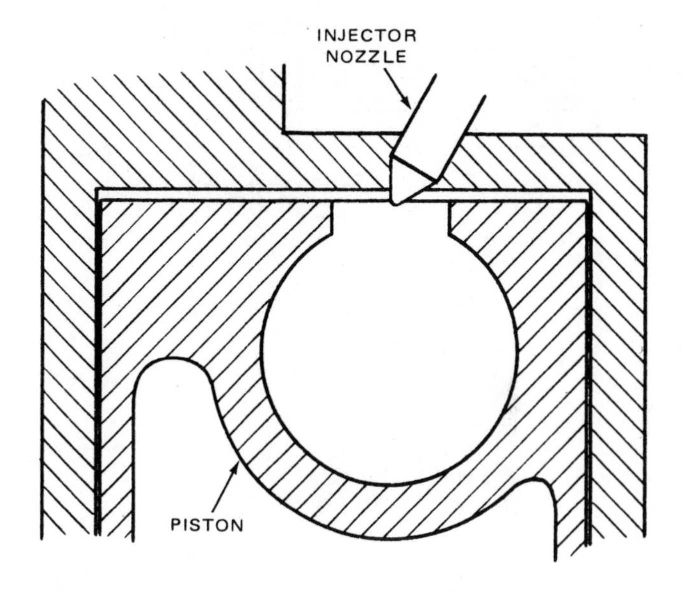

Fig. 5-7. Combustion chamber, known as the M type, is completely within the piston. This is a variation of the open combustion chamber.

A major feature of engines using the M type combustion chamber is the absence of diesel knock, even at low engine speeds. In addition there is less soot formed per pound of fuel consumed than with most other types of diesel engines and fuel consumption is low. This type engine operates on a wide range of fuels including diesel fuel, gasoline, kerosene and jet engine fuel. The ball shaped combustion chamber of the M system has a diameter of approximately one-half diameter of the piston. The system uses masked inlet valves which impart considerable swirl to the air. A squeezing effect is due to the flat area of the piston head which gives still greater rotation to the incoming air.

Fuel is injected at approximately 20 deg. before top center through two holes directly onto the surface of the ball shaped combustion chamber. A film of fuel is formed on the inner surface of the chamber. The temperature of the chamber is closely controlled at about 343 deg. C. by means of an oil jet on the lower side of the piston head.

PRECOMBUSTION CHAMBER

In some designs a portion of the clearance volume is placed in a separate chamber located in the cylinder head or in the cylinder wall. See Fig. 5-8. This chamber is known as a precombustion chamber and is connected to the space over the piston by one or more passages.

As the air is compressed within the cylinder, it is given a

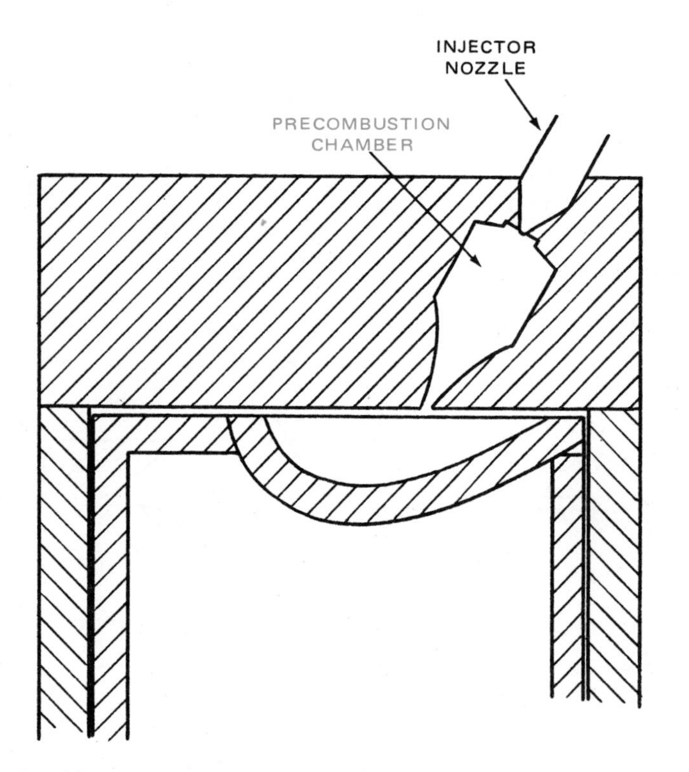

Fig. 5-8. The precombustion chamber contains only part of the full air charge, the remainder is in the cylinder space.

high degree of turbulence as it passes through the narrow passageway into the precombustion chamber.

These chambers hold from 25 to 40 percent of the total clearance volume. As a result of the larger combustion chamber surface, heat losses are increased and thermal efficiency is reduced. However, smooth combustion is attained, and a large variety of fuels may be burned without difficulty.

The precombustion chamber conditions the fuel for final combustion in the cylinder and distributes the fuel through the air so that complete combustion is assured.

At the beginning of fuel injection the precombustion chamber contains a definite volume of air. Combustion is started at that point. Because of the limited amount of air available, combustion is incomplete, but the resulting heat and high pressure forces the fuel at high velocity into the cylinder. There is an ample supply of oxygen to complete the combustion of the fuel. Examples of the precombustion chamber engine designs are the Onan engine, Fig. 5-9, the Mercedes, Fig. 5-10, and the Caterpillar, Fig. 5-11.

CATERPILLAR DESIGN

In the Caterpillar design, Fig. 5-11, the injection nozzle and the combustion chamber are designed to form a single unit. In

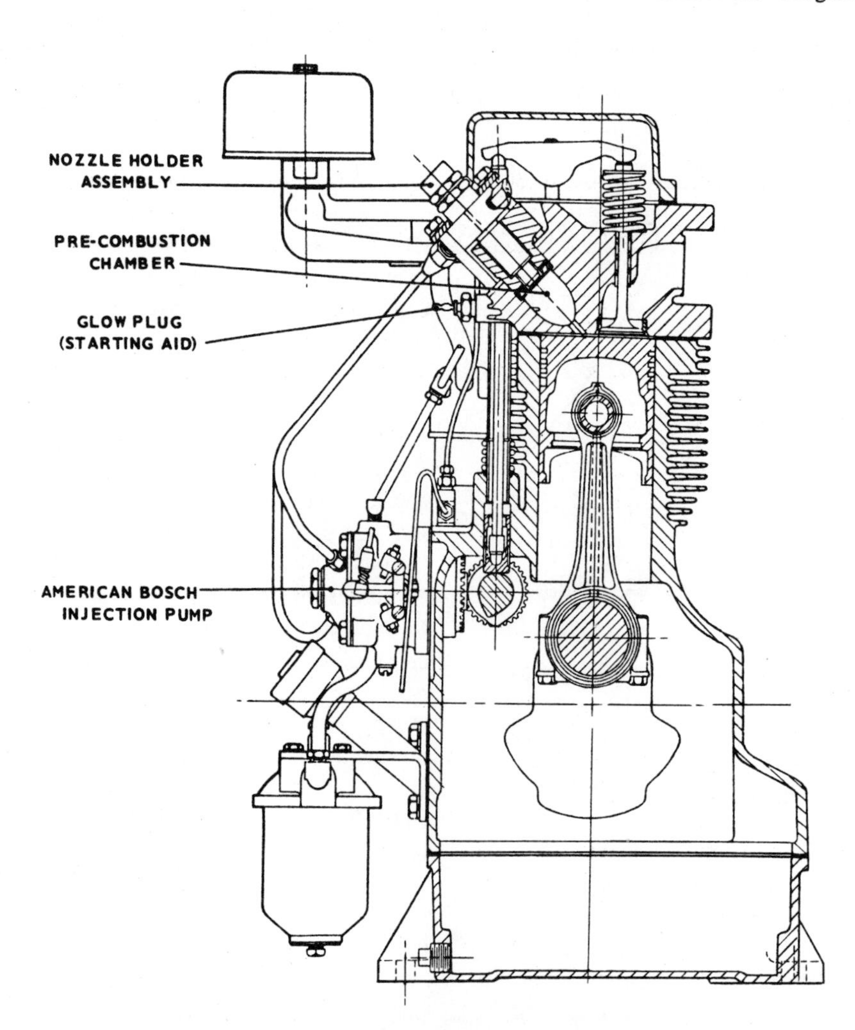

Fig. 5-9. Illustrating the precombustion chamber on this air-cooled Onan engine.

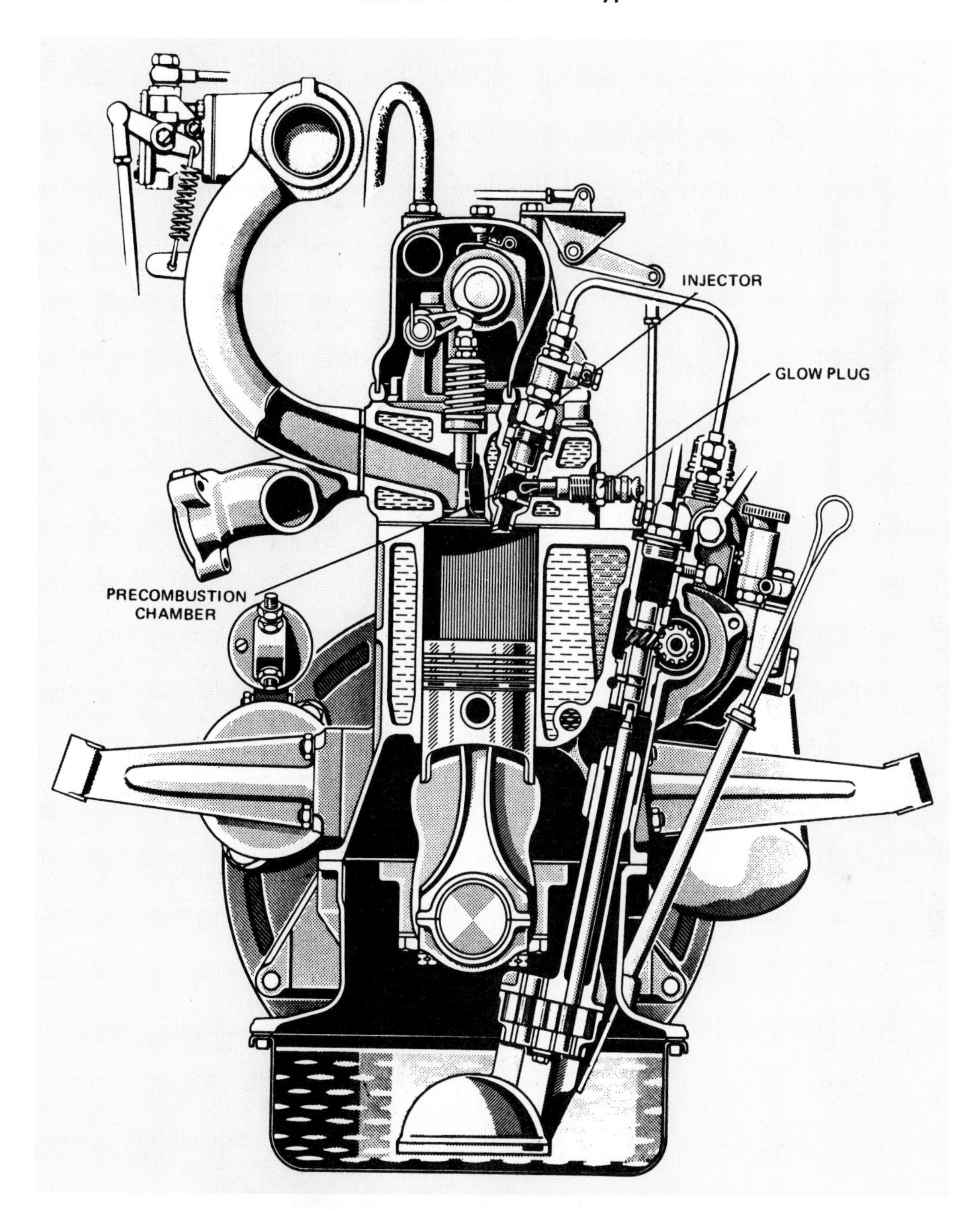

Fig. 5-10. Mercedes diesel with precombustion chamber and glow plug.
(Daimler-Benz Aktiengesellschaft)

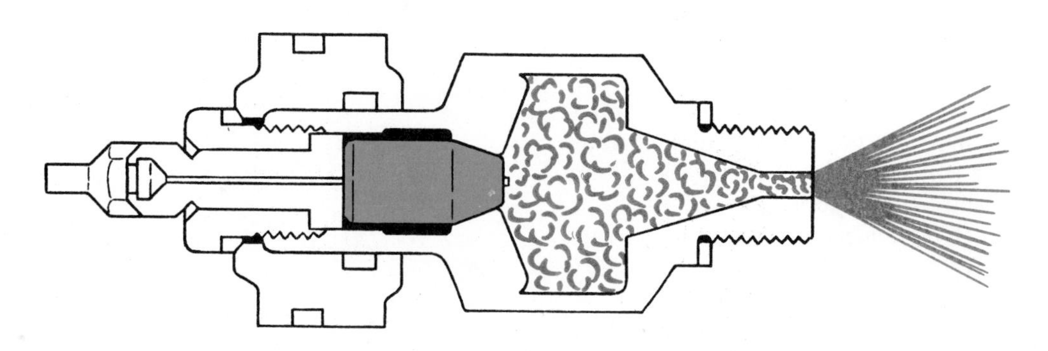

Fig. 5-11. In this construction, the precombustion chamber and the injector form a single unit. (Caterpillar Tractor Co.)

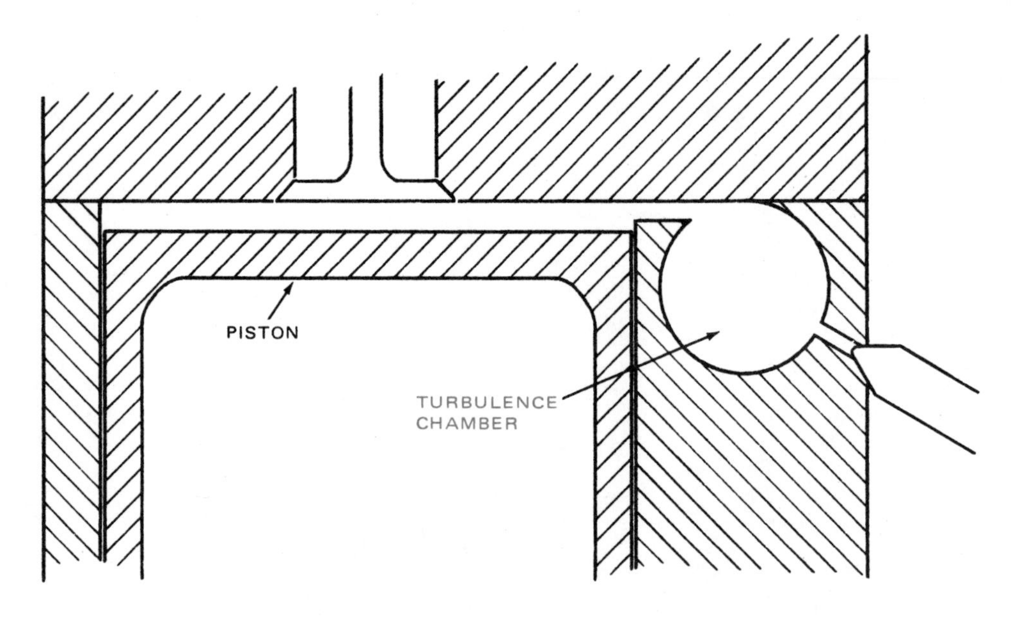

Fig. 5-12. In this turbulence chamber design, the rising piston will partly cover the passage to the chamber.

this design the precombustion chamber occupies approximately 30 percent of the total volume of the clearance space, when the piston is at top of its stroke. In operation, fuel injection occurs before the piston reaches top center. As it reaches top center, approximately one-half of the fuel has been delivered, and ignition of the fuel commences. Fuel is first deposited on the surface of the throat of the chamber. Later the fuel is dispersed by high velocity gases from the precombustion chamber. Maximum pressure occurs soon after the piston passes top center. The air fuel mixture then flows into the cylinder. The combustion chamber is designed so that the highly atomized fuel is forced into the region where it is ignited.

TURBULENCE CHAMBER

In some designs, the turbulence (swirl) chamber is in the cylinder head. In other designs, the chamber is at one side of the cylinder, as shown in Fig. 5-12.

A turbulence chamber is similar to the precombustion chamber shown in Fig. 5-11. Since there is very little clearance above the piston at the top of the compression stroke, the operating principle is different than in the precombustion chamber. When the piston reaches top center virtually all of the air has been compressed within the turbulence chamber.

The chamber is usually spherical in form. The opening through which the air must pass becomes smaller as the piston reaches the top of the stroke, thereby increasing the velocity of the air in the chamber. As in other designs, the fuel is injected at the time of maximum turbulence which insures thorough mixing of fuel and air. This in turn results in a greater portion of combustion in the turbulence chamber.

In the Waukesha engine, Fig. 5-13, the lower half of the chamber, or combustion cup is made of Stellite (metal combining cobalt, tungsten, chromium) for great heat resistance. It is loosely fitted to provide an insulating air space. The cup retains the heat of combustion and operates at a heat of approximately 1700 deg. F. The highly compressed air passing through the heated throat into the combustion chamber picks up considerable additional heat. The heating feature of the combustion cup acts much like an automatic control to adapt combustion to load requirements. The greater the load on the engine, the higher the temperature at which the combustion cup operates. This increases the ability of air in the combustion chamber to provide the greater quantity of fuel

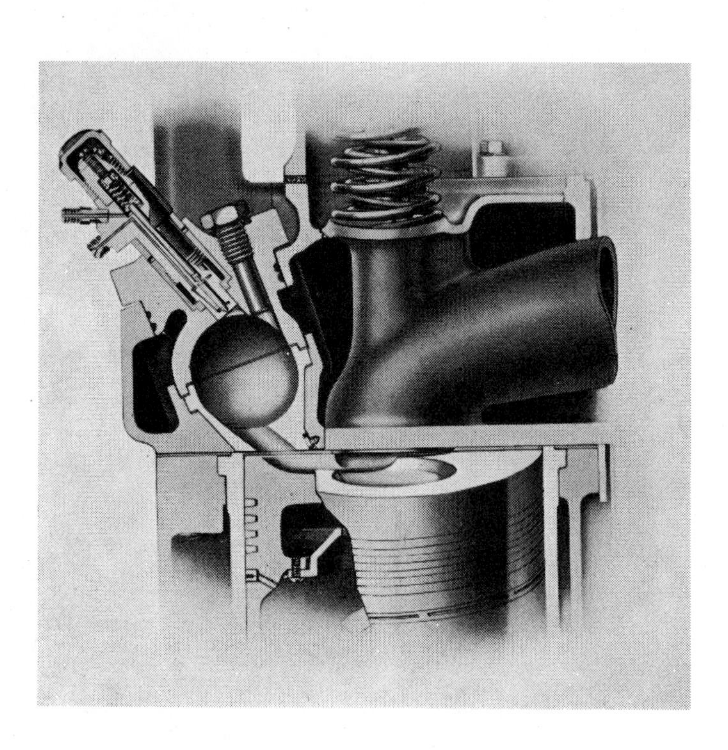

Fig. 5-13. Sectional view of the turbulence chamber as used on some Waukesha engines.

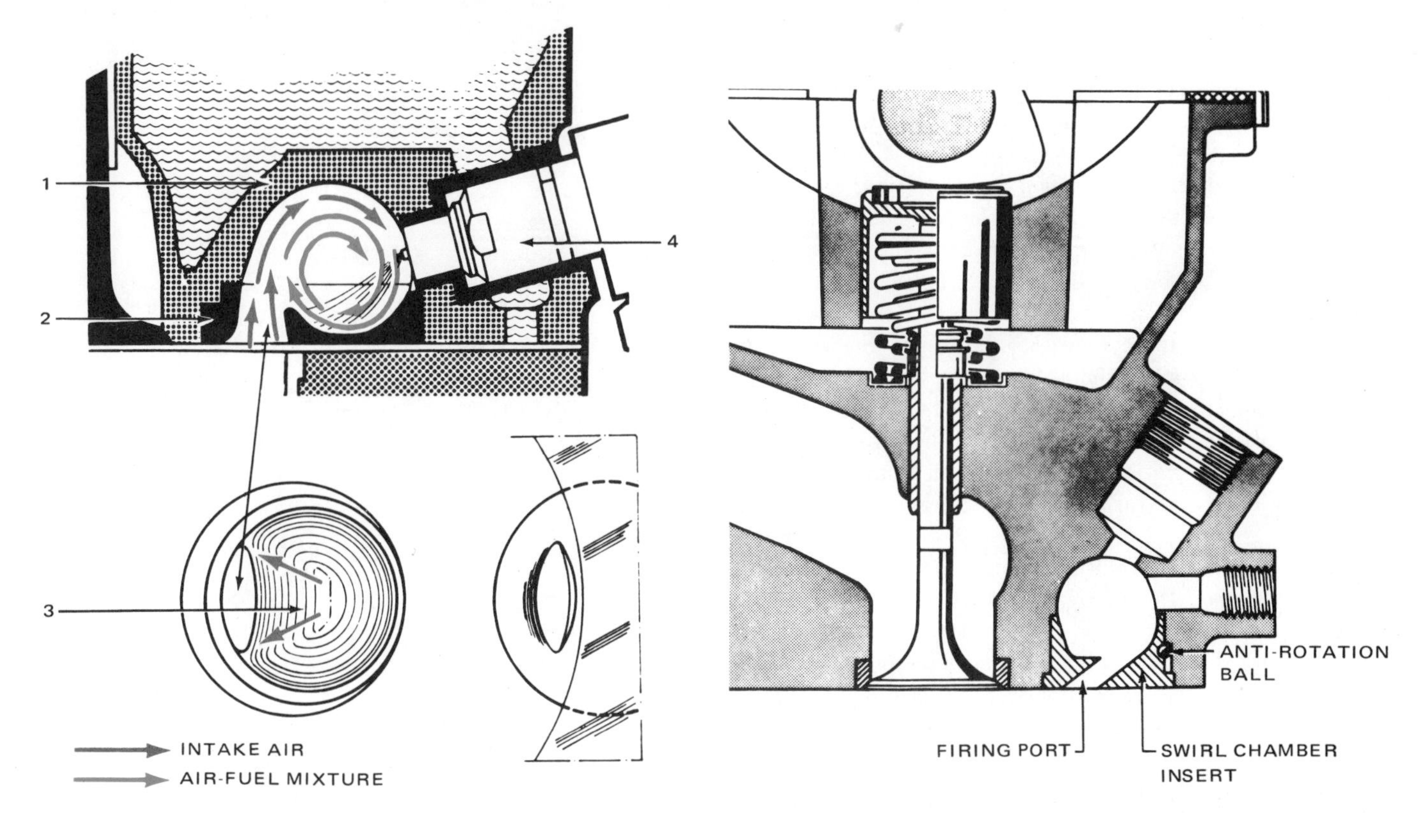

Fig. 5-14. Details of the Hispano-Suiza turbulence chamber. (Hispano-Suiza France)

Fig. 5-15. Volkswagen Rabbit diesel engine utilizes an insert in this unusual swirl chamber design.

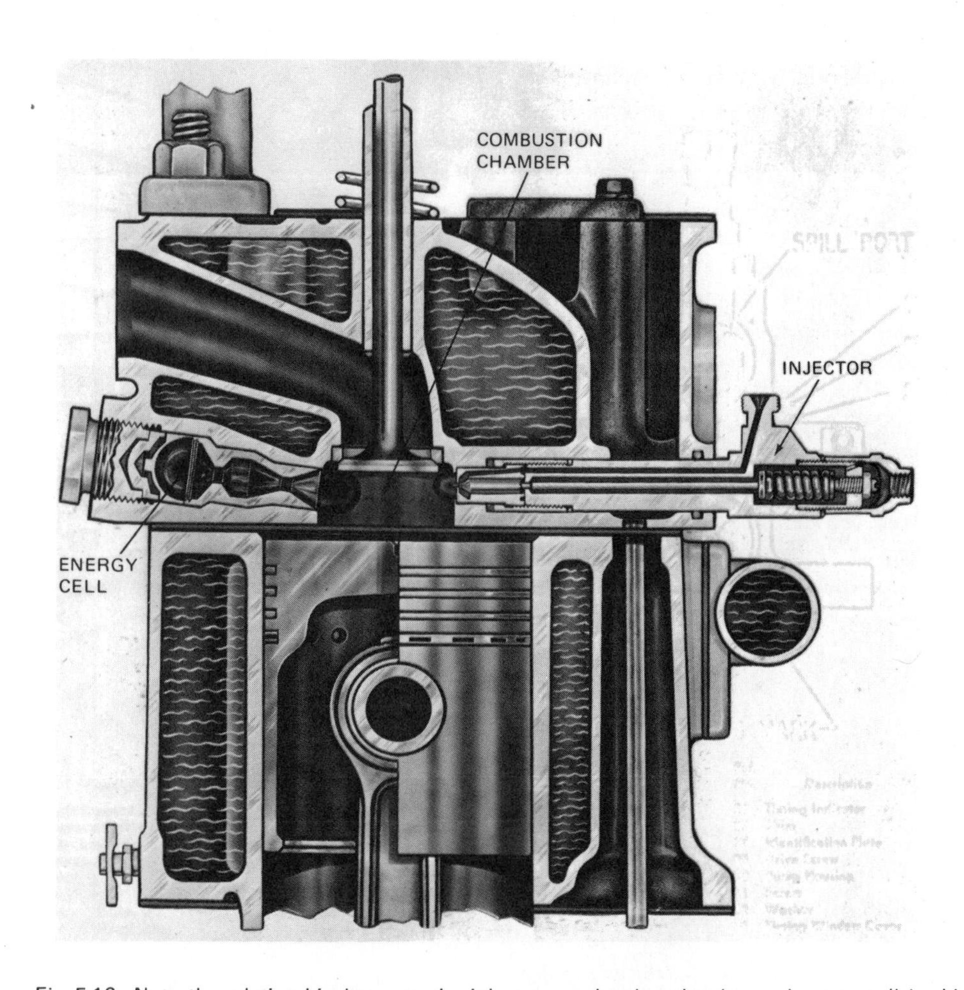

Fig. 5-16. Note the relationship between the injector, combustion chamber and energy cell in this Lanova energy cell. (Minneapolis Moline Inc.)

needed for the heavier loads. The upper half of the combustion chamber is surrounded by a water jacket, which prolongs injector life.

The Hispano-Suiza combustion chamber, Fig. 5-14, is of the turbulence type. It is used on some Hercules and Ford engines. The upper half of the chamber is in the cylinder head and is water jacketed, 1, Fig. 5-14. The lower portion is an insert of special steel, 3, Fig. 5-14, which forms the lower part of the throat leading to the cylinder space. Fuel is directed to the center of the chamber and at 90 deg. to airflow. This causes the fuel spray from the nozzle, 4, Fig. 5-14, to strike the bowl and lip of the hot insert, 2, Fig. 5-14. A high degree of vaporization and ignition are produced. Air from the piston, 3, Fig. 5-14, enters the turbulence chamber as indicated by blue arrows.

The Volkswagen Rabbit diesel engine also uses a turbulence or swirl chamber. Details are shown in Fig. 5-15. The cylinder head is made of aluminum for lightness and improved heat conductivity. This four cylinder engine develops 48 hp at 5000 rpm and has a displacement of 89.7 cu. in. (1471 cc). Shims are provided for adjusting valve lash in the VW diesel engine.

ENERGY CELL

The energy cell, which is of divided chamber construction, basically is a combination of the precombustion chamber and the turbulence chamber designs. This construction is also called the Lanova combustion chamber, after its originator. Fig. 5-16 shows the construction as used by Minneapolis Moline. Fig. 5-17 illustrates the successive steps of combustion in an energy cell system.

The Lanova system has two rounded spaces, shaped like figure 8's, cast in the cylinder head. As in the case of the open combustion chamber system the main volume of air is in that space and the principal combustion takes place there. However, unlike the open combustion chamber system, combustion is controlled. In some designs, both intake and exhaust valves are located directly above the figure 8 shaped combustion chamber. On others, one valve is located in the top of the chamber, while the other is in the flat portion of the head.

The Lanova system depends on a high degree of turbulence to promote thorough mixing and distribution of the air and fuel. In this design, about 90 percent of the combustion chamber is directly in the path of the in and out movement of the valves. The turbulence is dependent upon the thermal action, and not on engine speed, as is the case in the open combustion chamber designs.

The sequence of operation of combustion in the Lanova system involves the combustion in the figure 8 shaped combustion chamber, which is centrally located over the piston, and a small head chamber known as the energy cell, Figs. 5-16 and 5-17. The energy cell has passed through several stages of design development and currently comprises two separate chambers, an inner and an outer.

The inner chamber which is the smaller of the two, opens into the narrow throat between the two lobes of the main combustion chamber through a funnel shaped venturi passage (passage with short, narrow section and wide, tapered ends).

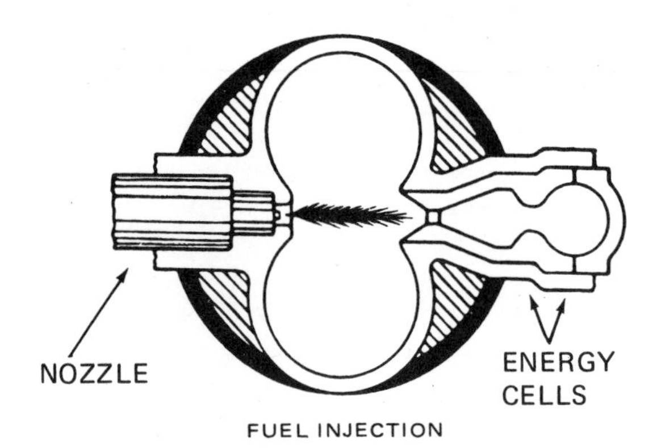

FUEL INJECTION

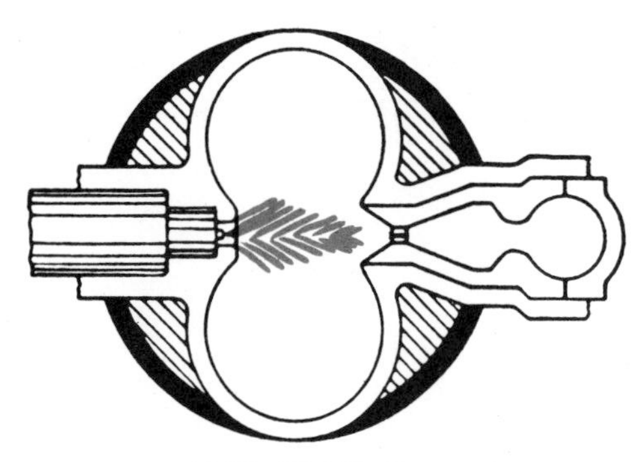
FUEL IGNITION

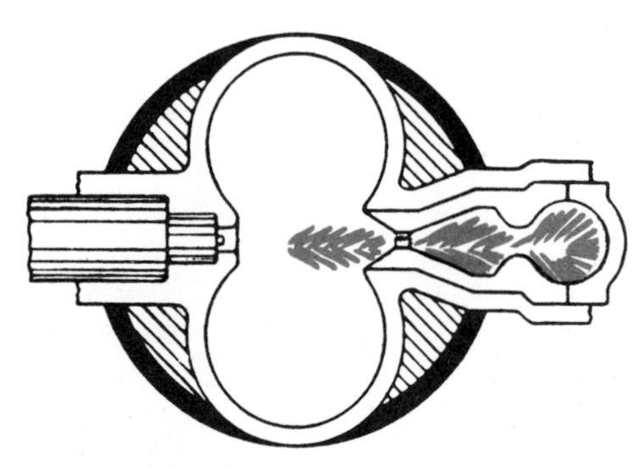
COMBUSTION IN ENERGY CELL

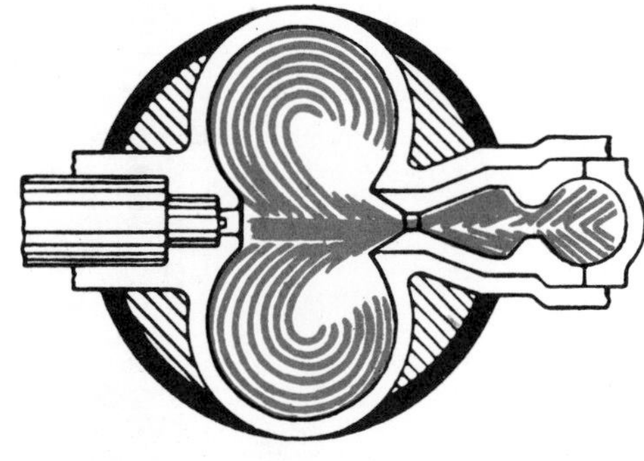
COMBUSTION IN MAIN CHAMBER

Fig. 5-17. Combustion sequence in Lanova energy cell system.

The larger spherically shaped chamber of the energy cell is connected with the smaller chamber through a second venturi. The injection nozzle is directly opposite the energy cell and sprays the fuel directly into the energy cell.

During the compression stroke about 10 percent of the total compressed volume of air passes into the energy cell. The remainder passes into the figure 8 shaped combustion chamber. The fuel is injected in a pencil stream, passing directly across the narrow throat of the combustion chamber so most of the fuel enters the energy cell.

A smaller portion of the boundary layer of fuel follows the curvature of the combustion chamber lobes and swirls within them. Most of the fuel entering the energy cell is trapped in the small outer cell, but a small portion passes into the outer cell where it meets with a sufficient quantity of super-heated air to explode violently.

This in turn causes a rapid rise in pressure within the energy cell. It forces the main quantity of fuel in the inner cell back and into the main combustion chamber so there is a swirling action of fuel and air around the two lobes of the combustion chamber. There is sufficient air, swirling around at a high rate of turbulence, to complete the combustion.

Because of the restrictive action of the two venturi connecting the energy cells, the blow back of fuel into the combustion chamber is on a controlled basis and takes appreciable time. The result is a prolonged and smooth combustion.

In addition to the smooth combustion, high performance, smooth operation and excellent economy of operation are claimed for the design.

REVIEW QUESTIONS – CHAPTER 5

1. After the air has been compressed in the cylinder of a diesel engine, fuel is sprayed into the cylinder in the form of ____________.
2. Four types of combustion chambers in diesel engines are:
   a. ______________________________.
   b. ______________________________.
   c. ______________________________.
   d. ______________________________.
3. The swirling motion of the air as it is compressed in the cylinder is caused by the __________ of the piston head.
4. Two major advantages of the open combustion chamber design are:
   a. ______________________________.
   b. ______________________________.
5. An advantage of the M type combustion chamber engine is that it operates on a variety of fuels including:
   a. ______________________________.
   b. ______________________________.
   c. ______________________________.
   d. ______________________________.
6. The injection nozzle and the combustion chamber are designed to form a single unit in the ____________ type engine.
7. A noted feature of the turbulence type of chamber is its ____________ form.
8. The energy cell differs from other combustion chamber construction in that it is basically a combination of ____________ and ____________ designs.
9. Give four advantages of the energy cell combustion chamber design.
10. The Volkswagen diesel engine uses a turbulence or swirl chamber. True or False?

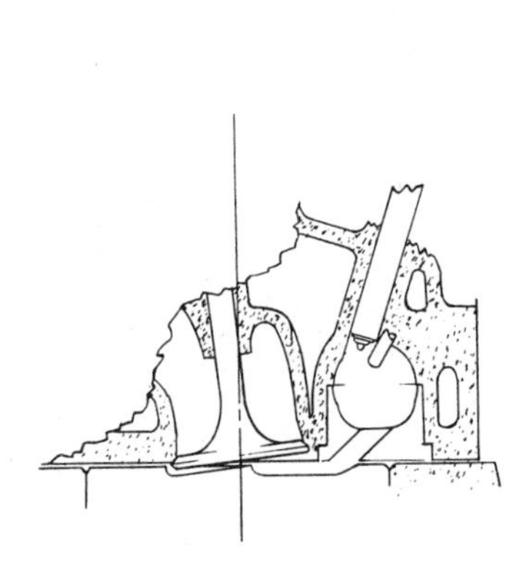

PRECHAMBER INJECTION

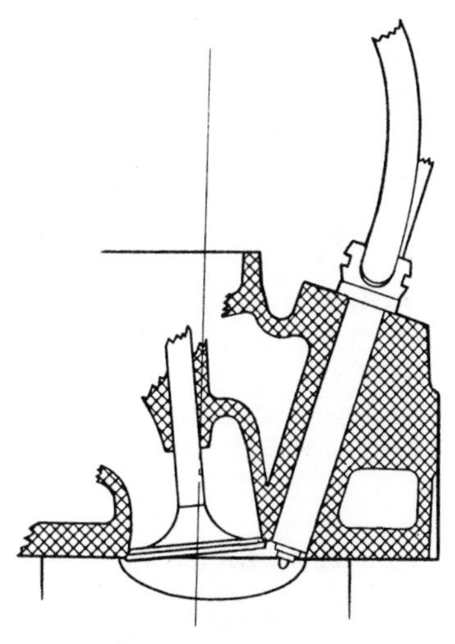

DIRECT INJECTION

Combustion chamber designs: Left. Sectional view of precombustion chamber designed for Oldsmobile's 5.7 litre (350 cu. in.) diesel engine. Right. Example of hemispherical type open combustion chamber for direct injection.

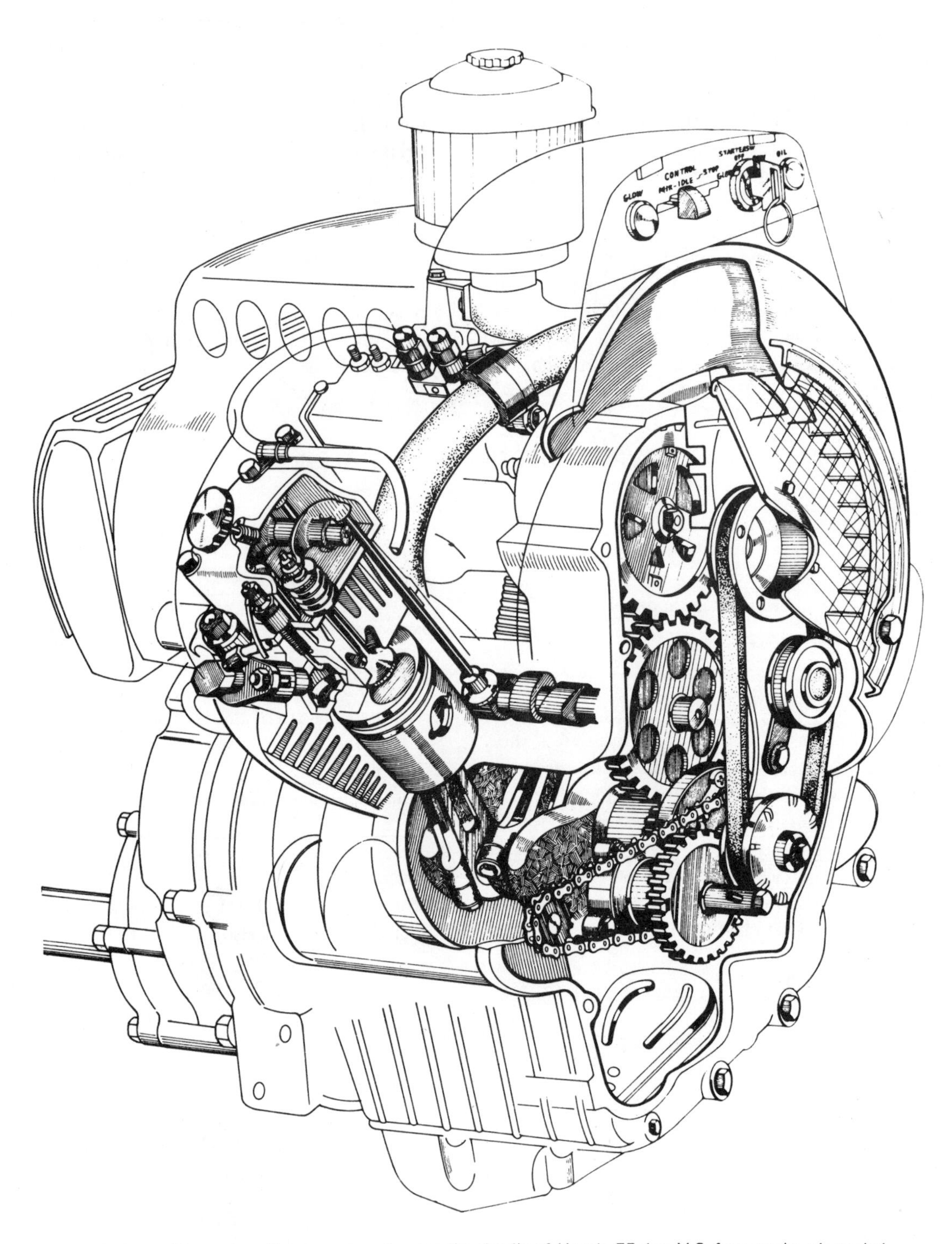

Industry illustration. Cutaway view showing the details of Honda 75 deg. V-2, four-stroke, air-cooled diesel engine. Rated at 7.5 hp at 3000 rpm, the engine has a compression ratio of 23 to 1.

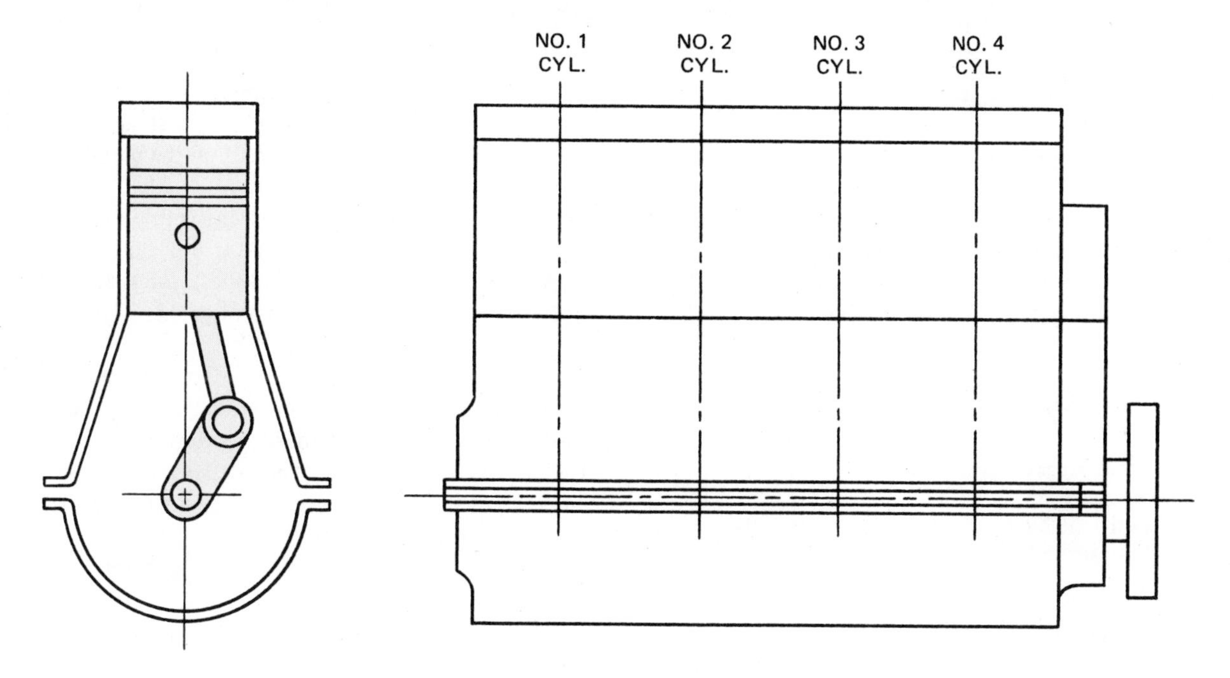

Fig. 6-1. When cylinders are placed in a single line, the engine is called an in-line engine.

Chapter 6
CLASSIFICATION OF DIESEL ENGINES

There are a number of different ways to classify diesel engines. It is important for you to know about the different classes or types and the particular service for which each is best suited.

The broadest classification of diesels is by developed power. Some engines develop as little as 3 hp; others develop as much as 40,000 hp.

Diesel engines are also classified by the number of cylinders, which range from one cylinder up to 24.

Single cylinder engines are often used for portable power and for irrigation. In the commercial vehicle field, 4, 6 and 8 cylinder engines predominate. In the industrial field, such as power generation, and also in ship propulsion, more cylinders are preferred, such as 12, 16, 20 and 24.

Another important way to classify diesel engines is by the cycles of operation; either two or four-cycle. The four-cycle diesel is used extensively for motor vehicle propulsion, while two-cycle engines are used largely for locomotives, ship propulsion and large electric generating plants.

The arrangement of the cylinders is also used to classify diesel engines. The most popular is the familiar vertical arrangement, where the cylinders are arranged in a single line, Fig. 6-1. Next is the V-type, Fig. 6-2. The vertical type is often placed on its side and is then known as a flat or horizontal

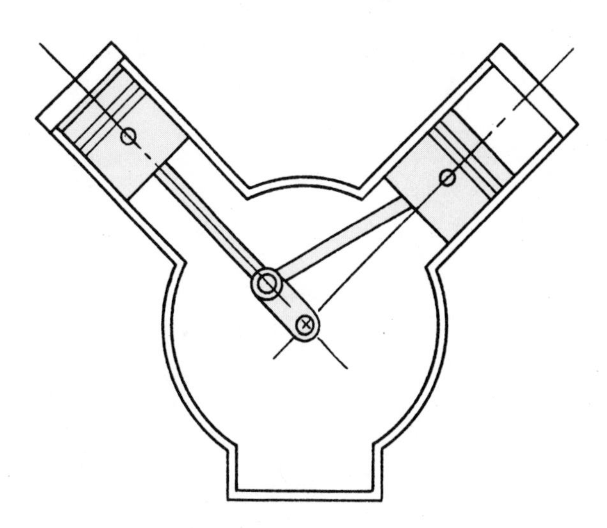

Fig. 6-2. A V-type engine has the cylinder placed in two banks and at an angle to each other. The connecting rods operate on a single crankpin.

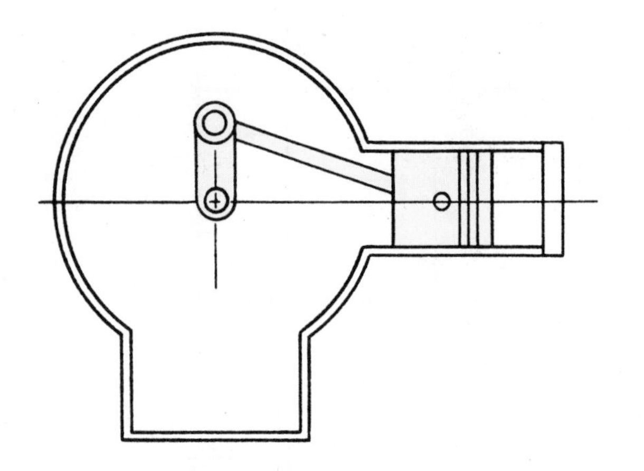

Fig. 6-3. When the motion of the piston is in a horizontal position the engine is known as a horizontal engine.

engine, Fig. 6-3. These engines may have all of the cylinders on one side, or an equal number of cylinders on each side of the crankshaft. With an equal number of cylinders on each side it is known as a flat opposed engine; or V-type, Fig. 6-4.

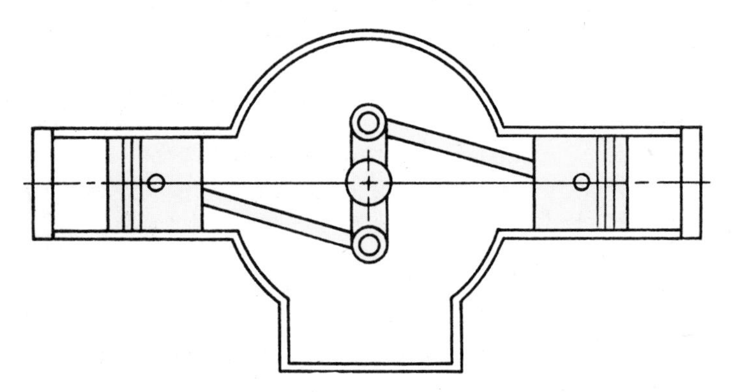

Fig. 6-4. When cylinders are placed on each side of the crankshaft, the engine is of the opposed type. Such engines may have the cylinders placed either vertically or horizontally.

Fig. 6-5. When four banks of cylinders are placed at angles to each other and operate on a single crankshaft, it is a W engine.

In the case of the conventional V-type engine, it is usual practice to specify the angle of the V, such as 45, 50, 55, 60 or 90 degrees. The angle is dependent on the number of cylinders and the design of the crankshaft.

An unusual cylinder arrangement is the W-type engine, Fig. 6-5. This may be thought of as two V-type engines, side by side, operating on a single crankshaft. Such an engine is built in Italy for use in Naval boats and yachts. Cylinders arranged in the form of an X have also been built. This design is basically two V-type engines, one over the other.

Another unusual cylinder arrangement is known as the Delta, in which the cylinders are formed into a triangle, or Delta form, as made by Napier. This is an opposed piston engine and is used in both railroad and marine fields. Details are shown in Fig. 11-20, page 144.

V, W, X and Delta type engines require less space than in-line engines.

The radial type engine, Fig. 6-6, has the cylinders arranged in a circle around a common crankshaft. In this type of engine the connecting rods of all the pistons work on a single crankpin, which rotates around the center of a circle. One type of radial engine places four banks of cylinders one above the other and uses a single crankshaft to form a 16 cylinder engine.

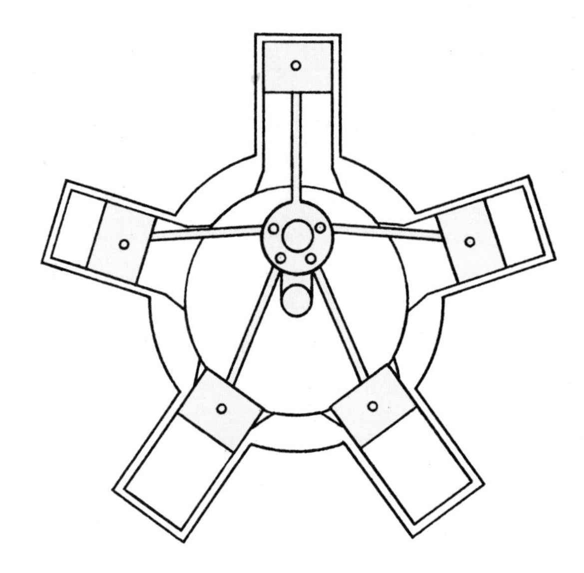

Fig. 6-6. Cylinders placed in a circle, operating on a single crankpin, form a radial engine.

Another method of classifying diesel engines is by action of the pistons. On this basis, engines may be classified as 1–single acting, 2–double acting, and 3–opposed pistons.

A single-acting piston engine may be described as the familiar or conventional engine, with explosive pressures applied to one face (or head) of the piston to produce power.

A double-acting piston engine uses both ends of a cylinder, and applies explosive pressure alternately to the two faces of the piston to produce power. The pressure is applied on both up and down strokes. This requires a complicated design and is used mostly on slow speed engines of large size.

The opposed piston engine, Fig. 6-7, has two pistons, opposed to each other in the same cylinder. The combustion chamber is between the pistons. Each piston, through its connecting rod, is connected to a separate crankshaft which in turn, is connected to gears and shafts. As shown in Fig. 6-7,

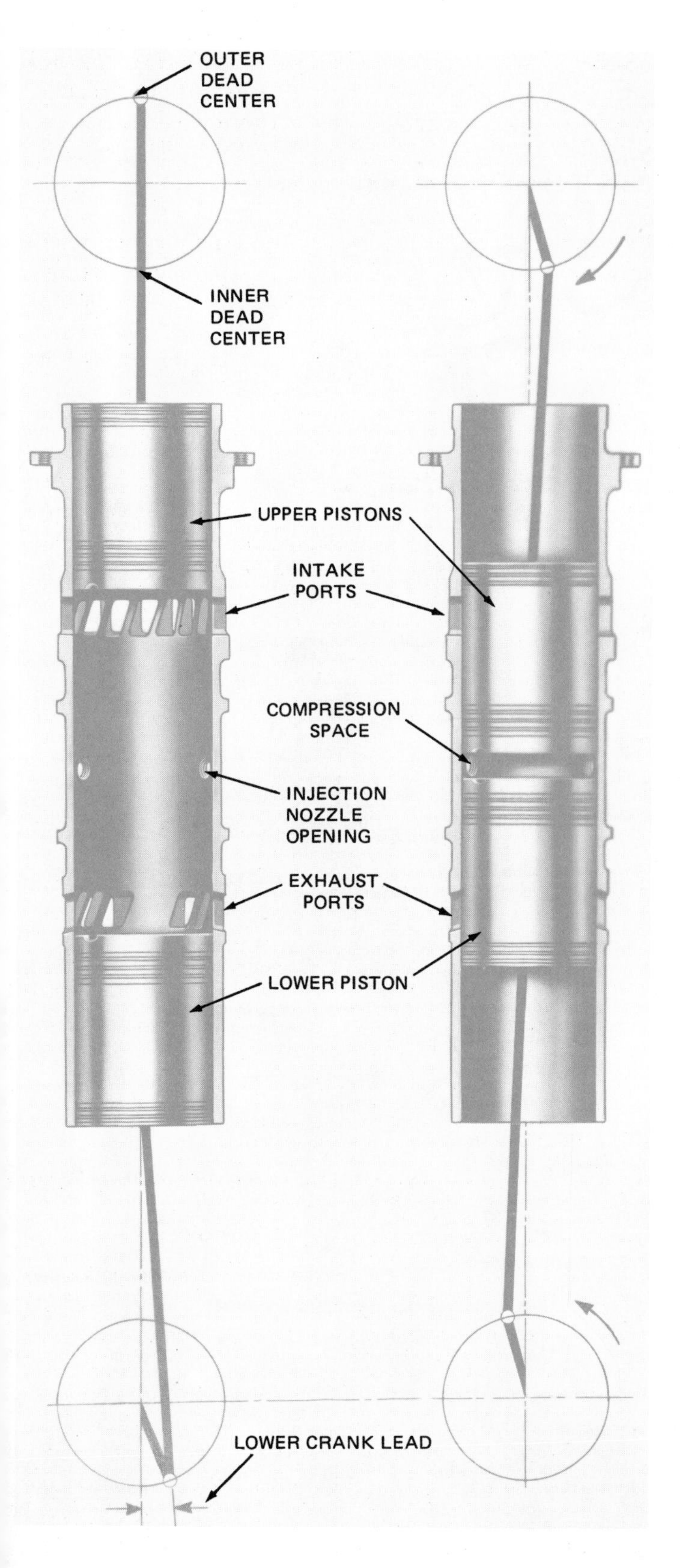

Fig. 6-7. The opposed piston engine has two pistons in the same cylinder, with the combustion chamber between them. (Fairbanks Morse)

this engine operates on the two-cycle principle with the inlet ports at the top, and the exhaust ports at the lower end of the cylinder. The injection nozzle is centrally located. Such engines may be either of the vertical or horizontal type. They are usually turbocharged (a powerful blower is used to force air into the cylinders). In the United States, this engine is produced by the Fairbanks Morse Div. of Colt Industries.

PARALLEL PISTON ENGINES

Another type of engine has pistons operating parallel to each other. Such an engine, Fig. 6-8, has two crankshafts which are parallel. These engines are built by Sulzer Bros. Ltd., in Switzerland, for use in locomotives.

FREE PISTON ENGINE

A free piston engine is illustrated in Fig. 6-9. This engine uses the diesel cycle, along with a turbine which is an essential part of the power plant.

The engine consists of large and small cylinders, each containing a set of two (one large and one small) horizontally opposed pistons. Several pairs of cylinders may be assembled into one power plant and coupled to a single turbine. An air-fuel mixture is fired between the opposed small pistons, with injectors which drives the pistons apart, compressing air in the chambers at the ends of the large pistons.

Air compressed in the large cylinders then forces the pistons back toward the inner center, compressing the mixture in the small cylinders for the next firing stroke. The compression ratio and piston stroke vary with the speed of the engine. As the pistons travel inward, the large pistons also compress air and pump the air into the diesel cylinder through the ports uncovered by the small pistons. The cycle of operations is shown in Fig. 6-10.

The expanding hot gas which is generated goes to the turbine part of the engine to provide usable power. As there are no connecting rods or crankshaft, the pistons are kept in place by means of connecting linkage.

ENGINE SPEED

Another method of classifying engines is according to speed. However there is no distinct separation between low, medium and high-speed engines. As the result of improved design and metallurgy, engine speed has been gradually increasing. What was formerly considered a high-speed engine is now in the intermediate class.

In general, speeds below 1,000 rpm are considered low, while intermediate speeds range from 1,000 to 2500 rpm. High speeds range from 2500 up to about 6,000 rpm.

OTHER ENGINE TYPES

Engines may also be classified according to the type of fuel they burn and the way the fuel is ignited. This classification will be discussed in the following chapter.

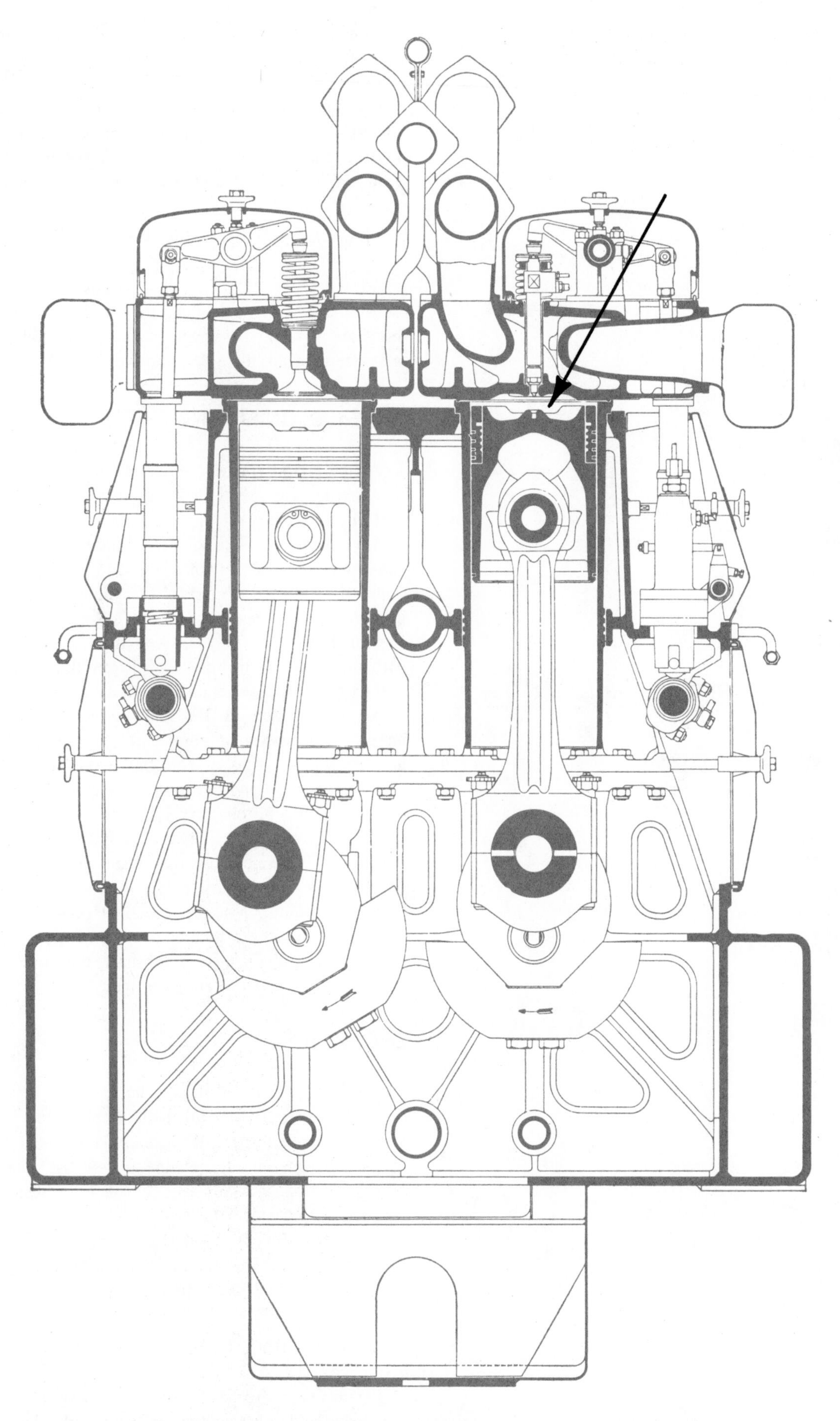

Fig. 6-8. Twin bank or parallel type of engine has two banks of cylinders placed side by side with separate cylinders and crankshafts. Note open type combustion chamber (arrow). (Sulzer Bros. Ltd.)

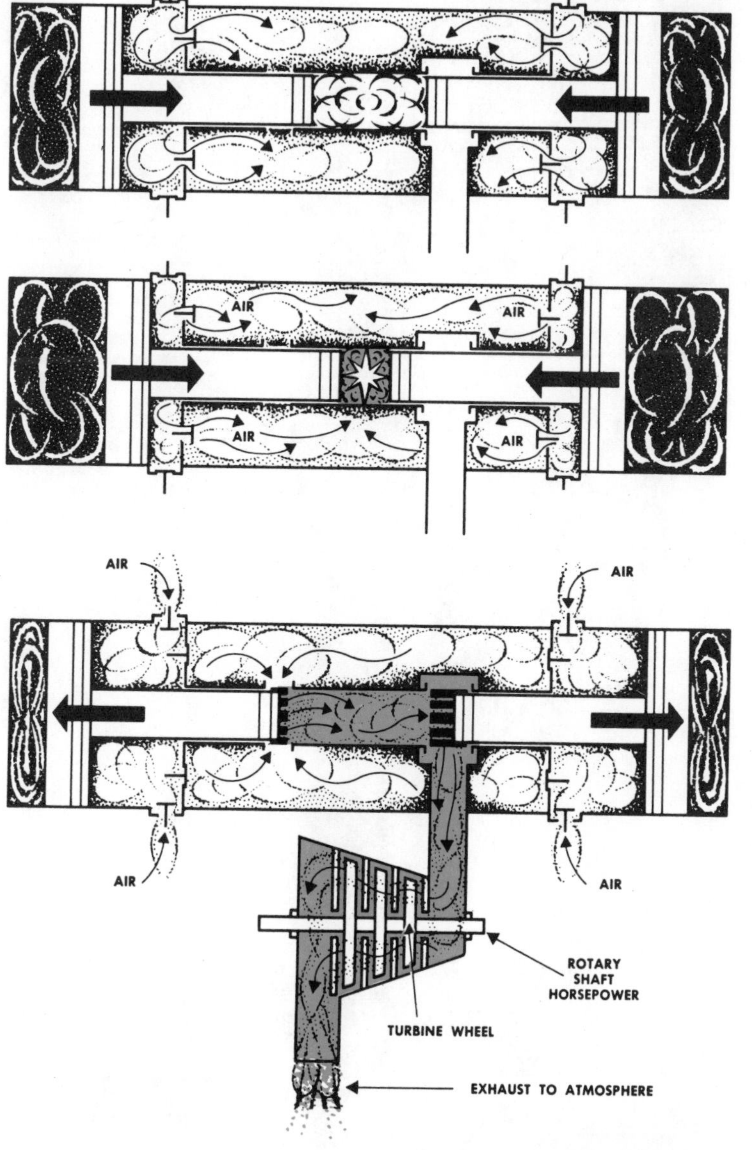

Pistons travel inward pumping air from the compressor cylinder into the air box, trapping air in diesel combustion space. Intake and exhaust ports are closed - - delivery valves are open.

Pistons are completing inward travel. Fuel is injected into cylinder. This is combustion or the beginning of the power stroke. Intake and exhaust ports are still closed - - air delivery valves are open.

End of power stroke compressing air in bounce space to return pistons for next cycle. Exhaust and intake ports are just opening to scavenge diesel cylinder. Exhaust gases escape to turbine, spinning turbine wheels for usable power. Air is being drawn into compressor cylinder.

Fig. 6-9. Diagram of free piston engine. Note absence of crankshaft.

Fig. 6-10. Operating cycle of free piston engine.

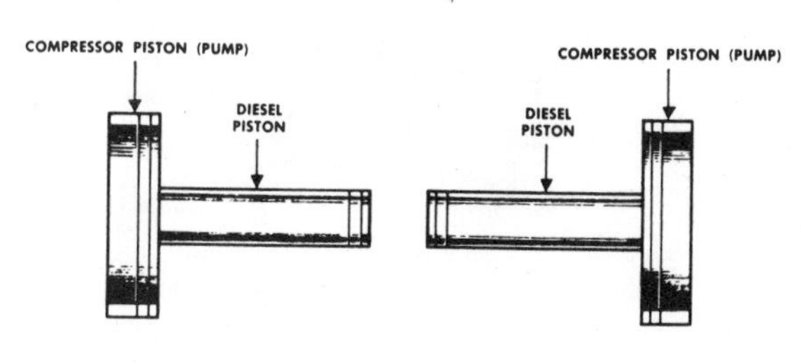

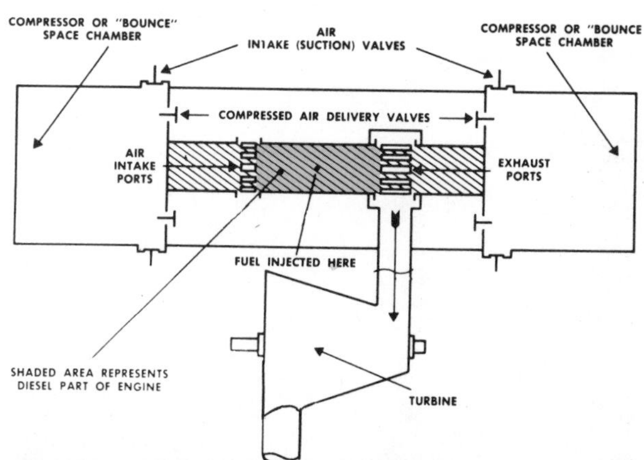

REVIEW QUESTIONS – CHAPTER 6

1. Diesel engine speeds in the range of 1,000 to 2500 rpm are considered to be __________.
2. Two-cycle diesel engines are extensively used as a power source for:
   a. ______________________________ .
   b. ______________________________ .
   c. ______________________________ .
3. The basis of classifying diesel engines by action of the pistons is as follows:
   a. ______________________________ .
   b. ______________________________ .
   c. ______________________________ .
4. A triangular cylinder arrangement for diesel engines is known as the __________.
5. In the opposed piston engine, the combustion chamber is located:
   a. Above the piston.
   b. Below the piston.
   c. Between the pistons.

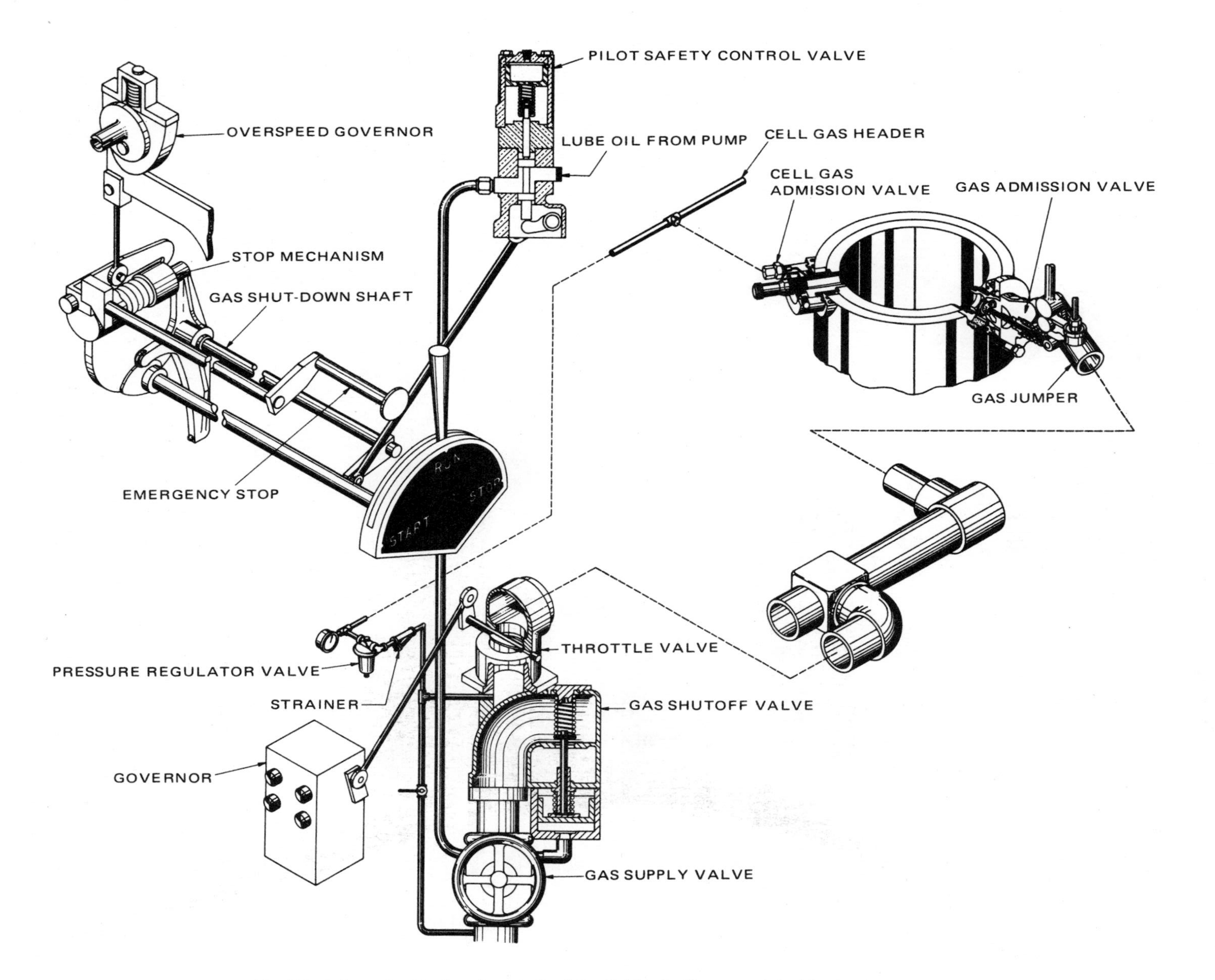

Fig. 7-1. Schematic drawing of controls of the Fairbanks Morse spark ignition engine.

Chapter 7

DIESEL ENGINES CLASSIFIED BY FUELS USED

In this chapter we will discuss diesel engines from the standpoint of the fuels that are used; except the liquid fuel engine which has been covered in preceding chapters.

Types of engines to be discussed:

1. Gas fuel engines.
2. Dual fuel engines.
3. Bi-fuel engines.
4. Multi-fuel engines.

GAS FUEL ENGINES

The gas-diesel engine uses gas such as natural gas or producer gas or sewage gas for fuel. In the true gas diesel, air is compressed in the usual manner, and the gas is injected and ignited by the heat of the compressed air. Such a system requires some means of compressing the gas so it can be injected. Engines of this type which operate satisfactorily are available, but other designs have proven to be more popular.

Another type of gas-diesel engine compresses a mixture of natural or other gas and air at a ratio of approximately 12 or 13 to 1, Fig. 7-1. The operation is much like a gasoline engine. While a spark is used to ignite the fuel, this engine is grouped with the diesels because of the high compression ratio. However, the compression ratio of the gas-diesel is somewhat lower than the conventional diesel.

Gas fuel diesel engines come in two and four-stroke cycle types (up to 22,000 hp). Usually, these are stationary engines since the gas fuel is piped in from a distance.

The control system for gas operation of a typical spark ignition opposed piston engine is shown in Fig. 7-1. With control shaft lever in the "run" position, and main gas valve open as shown, gas will flow through automatic shutoff valve, then through gas emission valves and cell gas valves by way of cell gas header and pressure regulator. See Fig. 7-2. For safety, automatic gas shut down features are included in the system. Note that the spark plug is located in an ignition cell.

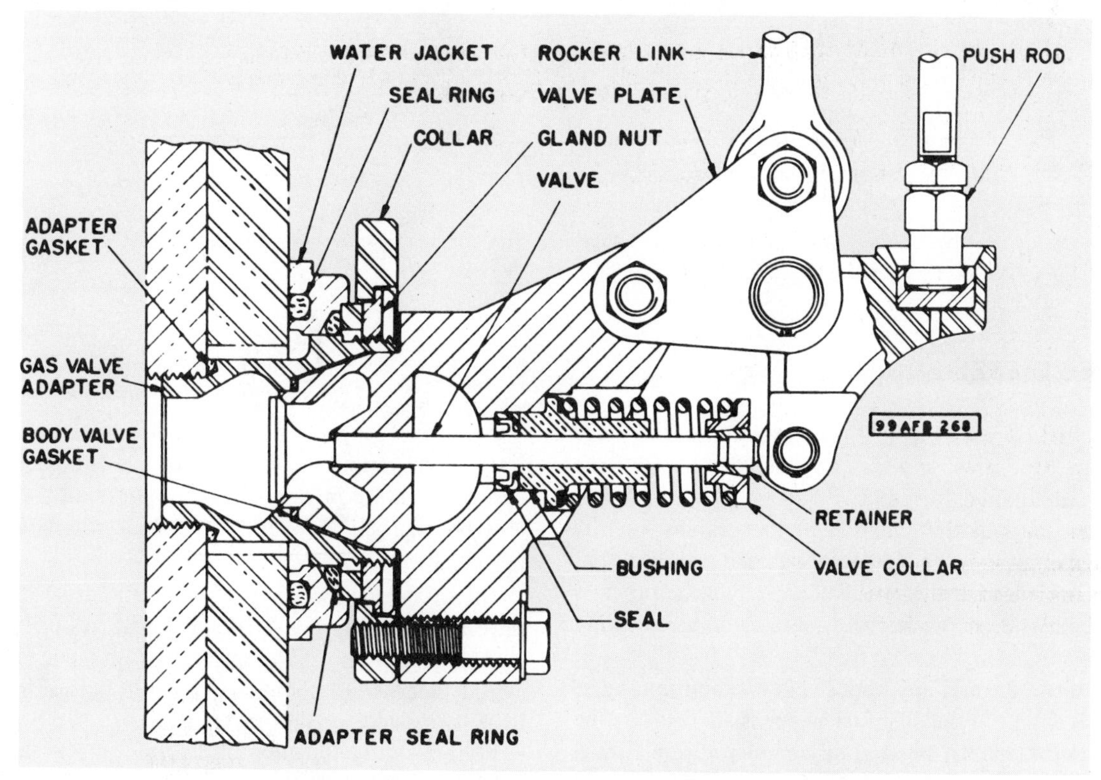

Fig. 7-2. Details of gas admission valves of Fairbanks Morse opposed piston engine.

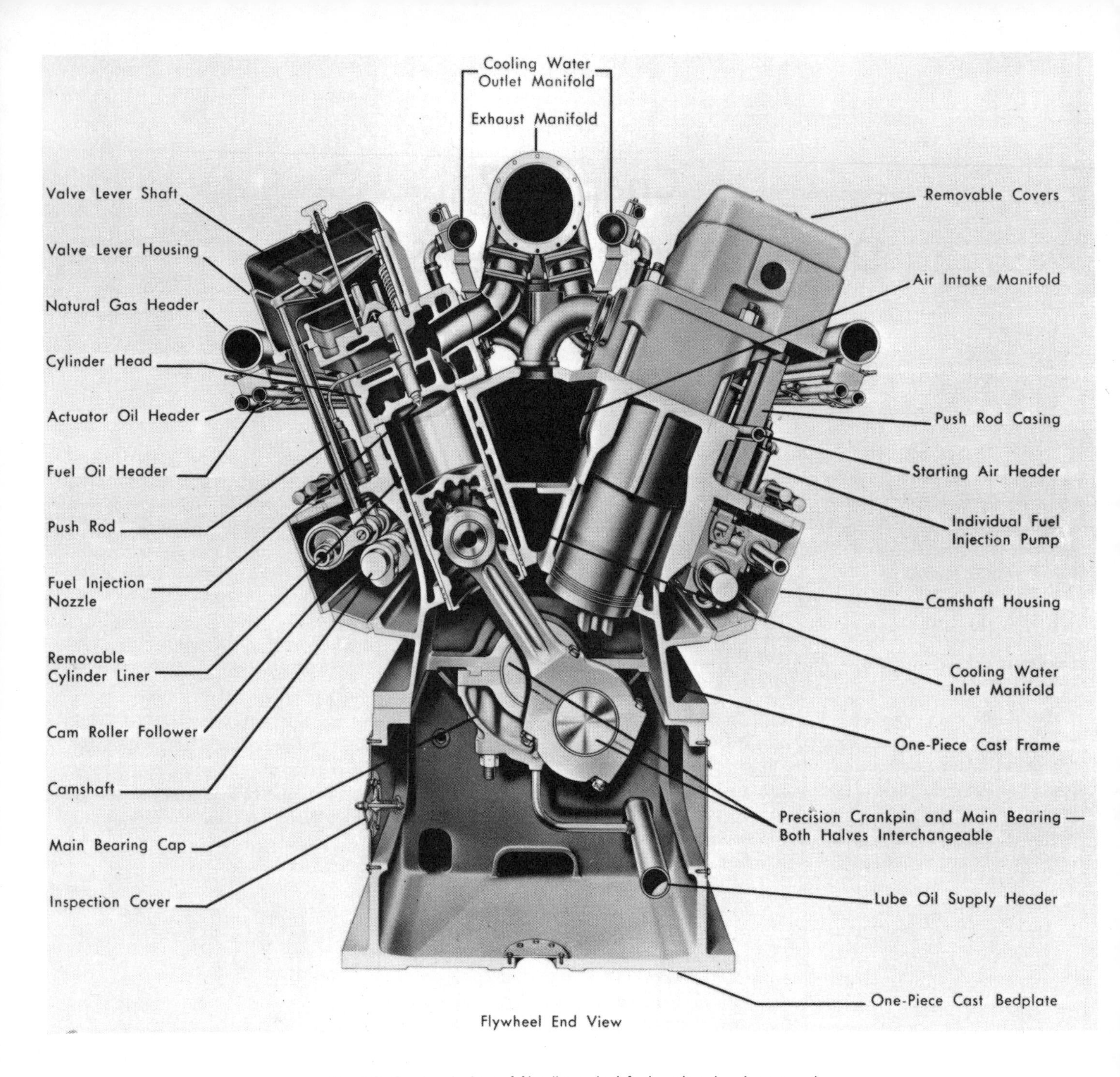

Fig. 7-3. Sectional view of Nordberg dual fuel engine showing natural gas header and fuel injection nozzle.

DUAL FUEL DIESEL ENGINES

The dual fuel engine, Fig. 7-3, compresses natural, producer, sewage or other gas to normal diesel pressures. Excessive air is supplied to reduce the possibility of preignition. Ignition occurs at the injection of the primary fuel. A basic difference between the dual fuel diesel and the conventional oil fuel diesel is that the dual fuel engine compresses a mixture of gas and air, while the oil diesel compresses only air.

Dual fuel diesel engines are produced in two and four-cycle types. The gas charge in the dual fuel engine is admitted to the cylinder at approximately the start of the compression stroke. The pilot charge which supplies 3 to 5 percent of the total heat charge to the cylinder, is injected near the end of the compression stroke, and starts the ignition.

Dual fuel engines can, in most cases, be converted to operate on various percentages of gas and fuel oil. This is a desirable feature in case of gas supply failure. The engine can also be changed to operate as a conventional diesel on 100 percent fuel oil.

In the case of two-cycle engines, the usual design has a gas valve mechanically operated and timed to open only after the cylinder has been scavenged by air and the exhaust port is closed. Because of the time element, the gas must be under sufficient pressure to fill the cylinder.

Four-cycle gas engines often use cam-operated valves to control the entrance of gas and air into the cylinder, Fig. 7-4.

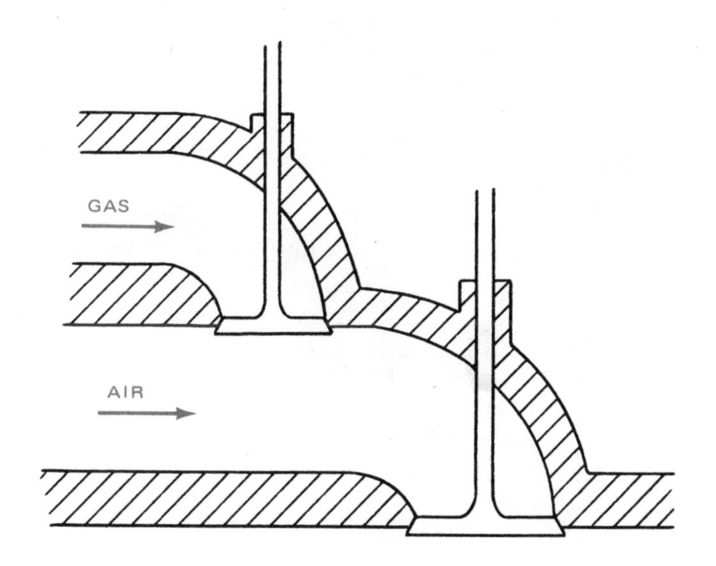

Fig. 7-4. Gas and air have separate valves in this system. Valves can be either cam or linkage operated.

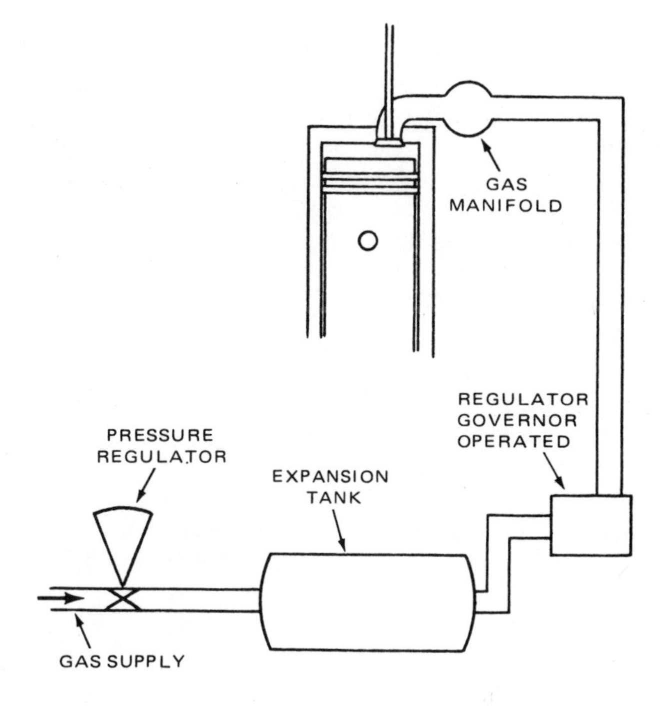

Fig. 7-5. Typical gas supply.

The usual practice is to open the gas valve only after the exhaust valve has closed. Such valves can be cam-operated or hydraulically operated. A typical gas system is shown in Fig. 7-5.

For small four-cycle gas engines, a carburetor and gas regulator are often used to supply the fuel. In such systems, the quantity of fuel for the engine is controlled by the carburetor throttle in much the same manner as in the familiar carburetor used on a gasoline engine.

BI-FUEL ENGINES

In the bi-fuel engine two liquid fuels are used, each having its own injector, Fig. 7-6. In the two-fuel combustion system a small percentage of a suitable auxiliary (second) fuel is injected into the cylinder either during the intake stroke or

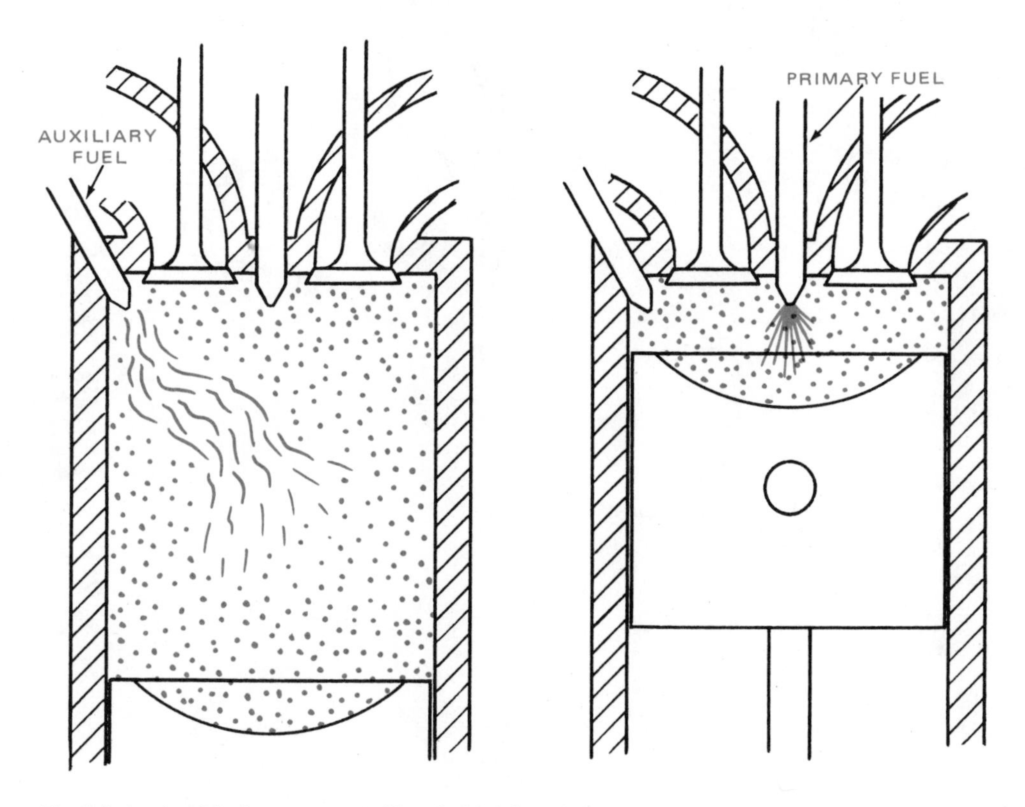

Fig. 7-6. In the bi-fuel system an auxiliary fuel is injected during the intake stroke, which is then ignited by the pilot or primary fuel which is injected at the normal time of ignition.

early in the compression stroke. Slightly later in the stroke the primary diesel fuel is injected.

In two-stroke cycle engines, the auxiliary fuel is injected directly into the cylinder by means of an additional injector. In a four-stroke cycle engine, the same method can be used, or the fuel can be introduced immediately ahead of the intake valves by means of individual low-pressure nozzles.

MULTI-FUEL ENGINES

Experimental engines are being developed to burn a wide variety of fuels and have the ability to start and operate from minus 25 deg. to 150 deg. F. without using power from other sources. Such fuels range from medium octane to middle distillate. While such engines are of primary interest to the military, they are also important to the general power user as undoubtedly many of their findings will eventually find commercial applications.

REVIEW QUESTIONS – CHAPTER 7

1. The compression ratio of a gas operated diesel engine is generally __________ than the conventional liquid fuel diesel.
2. Gas fuel diesel engines are usually stationary because _____.
3. Three gases typically used in the gas fuel diesel engines are:
   a. ______________________________.
   b. ______________________________.
   c. ______________________________.
4. Excessive air is supplied to the dual fuel diesel engine to reduce the possibility of __________.
5. Some small four-cycle gas diesel engines use a __________ and gas regulator to supply the fuel.
6. Bi-fuel diesel engines are operated on:
   a. One gas and one liquid fuel.
   b. Two gas fuels.
   c. Two liquid fuels.

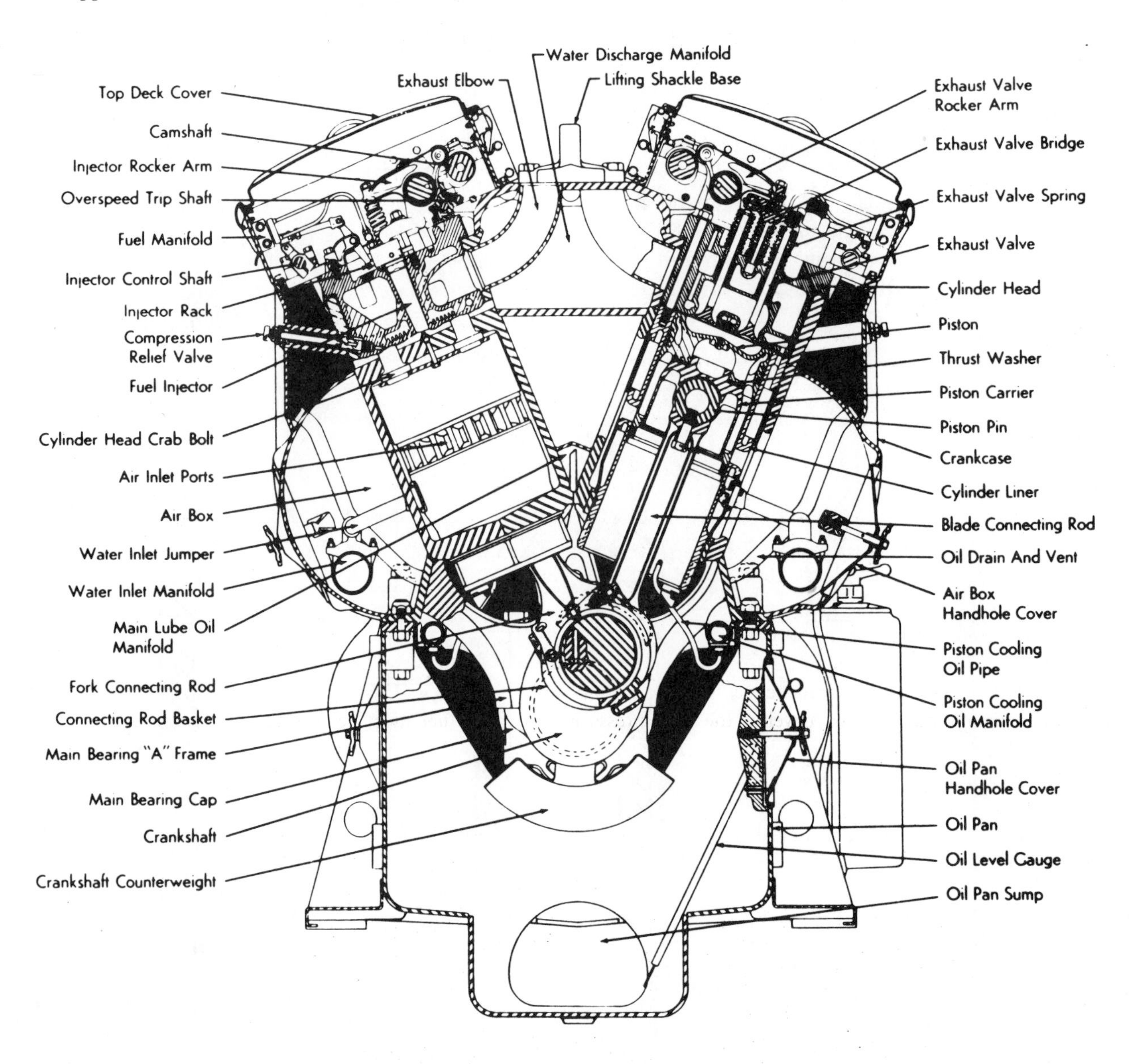

General Motors 645 E series diesel engines produced by the Electro-Motive Division range in size from 800 hp (597 kW) to 3950 hp (2947 kW). They are used in heavy-duty drilling, marine, industrial, off highway and locomotive applications.

Fig. 8-1. Types of spray patterns. (Waukesha Motor Co.)

Chapter 8

FUEL INJECTORS AND INJECTION SYSTEMS

The equipment used to force fuel into the combustion chamber is a very important part of the diesel engine. It is precision made. Considerable research has been done in developing and testing various parts which are included in the complete systems.

This chapter will deal with the different types of systems, fuel injectors and fuel injection pumps. Subsequently, other parts of the complete system will be discussed.

The diesel fuel injection system performs many more functions than the carburetor on the automobile.

Not only must it meter the quantity of fuel required for each cycle of the engine, in accordance with the load and speed of the engine, but it must develop the high pressure required to inject fuel into the cylinder at the correct instant of the operating cycle. It must control the rate at which the fuel is injected and atomize and distribute the fuel throughout the combustion chamber. Fuel injection must start and end abruptly.

PRESSURE

From the foregoing it can be seen that the diesel fuel injection system is not a simple device. To build up the pressure required to inject the fuel into the engine with its compression ratio of approximately 15 to 1, a high degree of precision is required. Some systems develop up to 5,000 psi at the rated load and speed.

METERING

The metering of the fuel must be accurate. Not only must the same amount of fuel be delivered to each cylinder for each power stroke, but the quantity must be varied in accordance with the load on the engine, and its speed. In addition, each cylinder must receive the same amount of fuel. Should the quantity of fuel vary in the different cylinders, the power per cylinder would vary, and rough operation would result.

TIMING

The fuel must be injected at the correct instant. Early or late injection results in loss of power. If the fuel is injected too early in the cycle, compression will not be at the maximum, the temperature will be low and ignition will be delayed. When the fuel injection is late, the piston will be past top center and power will be less because maximum expansion of the burned fuel will not take place. The injection must therefore start instantly, continue for the prescribed time, and then stop abruptly.

RATE OF IGNITION

Fuel is not injected in one single spurt, but extends over a period of time. If the fuel is injected too fast, it has the same effect as too early injection. Similarly, if the injection is too slow, and it extends over too long a period of time, the effect is similar to late injection. The rate of injection varies with different engines, and is affected largely by the type and contour of combustion chambers, together with engine speed and fuel characteristics.

ATOMIZATION OF FUEL

Fuel is spurted into the combustion chamber as a spray. The degree of atomization is dependent on the type of combustion chamber. Proper atomization increases the surface area of the fuel which is exposed to the oxygen of the air, and results in improved combustion, and maximum development of power.

To avoid simultaneous combustion of all droplets of the spray, the spray is usually formed of some fine droplets to start ignition, and larger droplets for further combustion. The extent of atomization is controlled largely by the diameter and form of the nozzle orifice, the injection pressure, and the density of the air into which the fuel is injected. Fig. 8-1 shows the type of spray used by a typical diesel - - the Waukesha, in their open combustion chamber type engine.

TYPES OF INJECTORS

There are two basic methods of injecting fuel into the combustion chamber:

1. Air injection.
2. Mechanical injection.

AIR INJECTION

Basically, air injection equipment consists of a three-stage air compressor with intercooling, which supplies air at a pressure several hundred pounds greater than engine compression pressure. The compressed air is usually contained in two bottles or reservoirs, one for dampening the pressure variations produced by the compressor, and another for storage of air required for starting and standby. Capacity of the reservoir ranges up to 15 times displacement of the engine.

Atomizers for injecting the metered fuel by blast of the compressed air into the combustion chamber are usually operated by a cam mechanism which provides the timing. A fuel pump is also required, which meters and delivers the fuel to each atomizer as required by the engine.

Air injection systems will not be discussed in detail because such systems have been largely replaced by mechanical injection systems.

MECHANICAL INJECTION

A mechanical injection system forces fuel through spray nozzles by hydraulic pressures ranging up to 2,000 psi, Fig. 8-1.

Four general systems of mechanical fuel injection have been developed:

1. Common rail system.
2. Pump controlled (or jerk pump) system.
3. Unit injection system.
4. Distributor system.

COMMON RAIL SYSTEM: The common rail system consists of a high-pressure pump which distributes fuel to a common rail or header to which each injector is connected by tubing, Fig. 8-2.

PUMP CONTROLLED SYSTEM: This is also known as the jerk pump system and provides a single pump for each injector. The pump is separately mounted and is driven by an accessory shaft. Connection to the injectors is made by suitable tubing.

UNIT INJECTOR SYSTEM: This system combines the pump and the injector into a single unit. High-pressure fuel lines are eliminated. Operation of the unit injector is usually by means of push rods and rocker arms.

DISTRIBUTOR SYSTEM: There are several types of distributor systems. One type provides a high-pressure metering pump with a distributor which delivers fuel to the individual cylinders. Another design provides a low-pressure metering and distribution. High pressure needed for injection is provided by the injection nozzles which are cam operated.

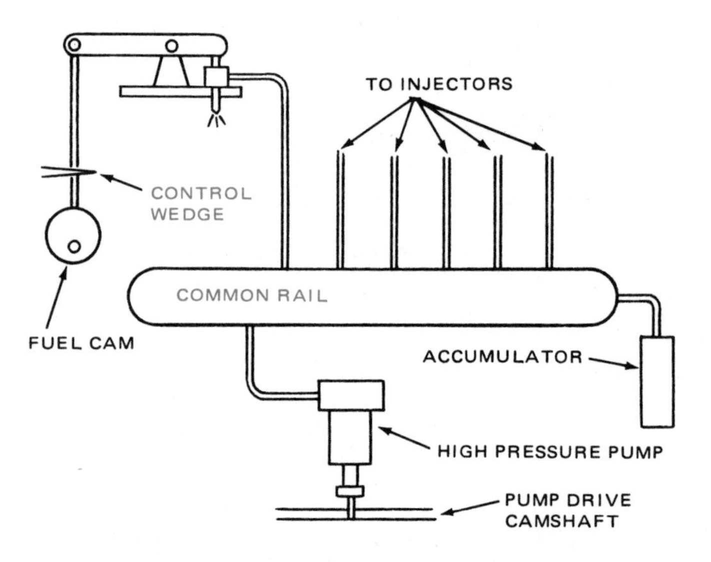

Fig. 8-2. Details of common rail system of fuel injection.

COMMON RAIL SYSTEM

The basic common rail system consists of a high-pressure pump which discharges the fuel into a common rail or header, Fig. 8-2, to which each injector is connected by tubing.

A spring-loaded bypass valve on the header maintains a constant pressure of 5,000 to 8,000 psi on the system and returns all excess fuel to the supply tank. The fuel injectors are operated mechanically. The amount of fuel injected into the cylinder is controlled by the lift of the needle valve in the injector.

The pressure regulator is usually a part of the high-pressure pump. The spray or injection nozzle extends from the top of the cylinder into the combustion area. This consists essentially of a multispray tip, a valve seat and a needle valve extending the full length of the nozzle, held to its seat by a spring. The high pressure is conducted from the common rail to the spray

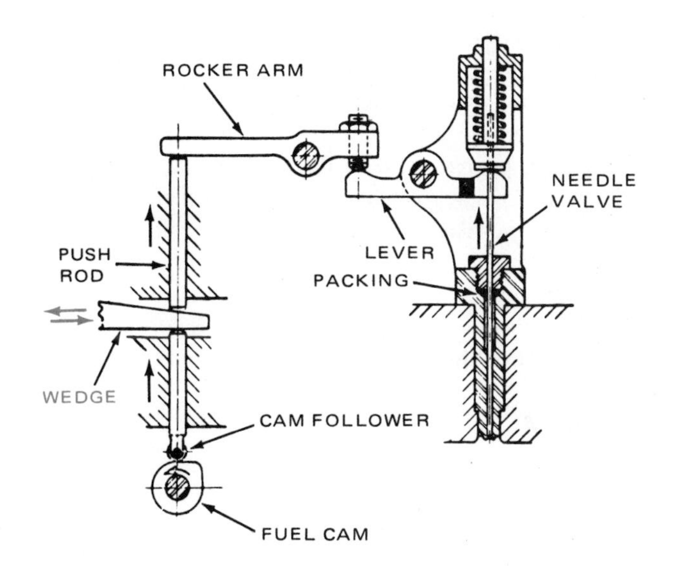

Fig. 8-3. Method of controlling fuel volume in common rail systems.

tip immediately above the valve seat. When the needle valve is lifted vertically from its seat, fuel is sprayed into the combustion chamber.

The duration of the injection period depends on the length of time the valve is off its seat. The quantity of fuel depends on the duration, size and number of holes in the tip, and the operating pressure. The needle valve is lifted mechanically by a system of rods and cross rods actuated by a timed camshaft. The duration period depends on the clearance existing between the cam and push rod mechanism. A minimum clearance provides a greater needle valve lift, and consequently more fuel is injected.

Control of the amount of clearance which affects cam lift is by means of a wedge, Figs. 8-2 and 8-3. The wedge is generally located between the cam and the push rod mechanism. However, on some engines the wedge is located between the cross rod and the injector needle valve. When the wedges are pushed all the way in, a minimum clearance is provided.

As the contour of the closing side of the cam is usually the same as that of the opening side, the needle valve returns to its seat an equal number of crankshaft degrees late as it was early in opening.

The wedges are attached to a single shaft which is coupled to governor or engine throttle. Consequently, wedges are moved in or out as engine load changes to increase or decrease cam-to-push-rod clearance and vary the duration of injection.

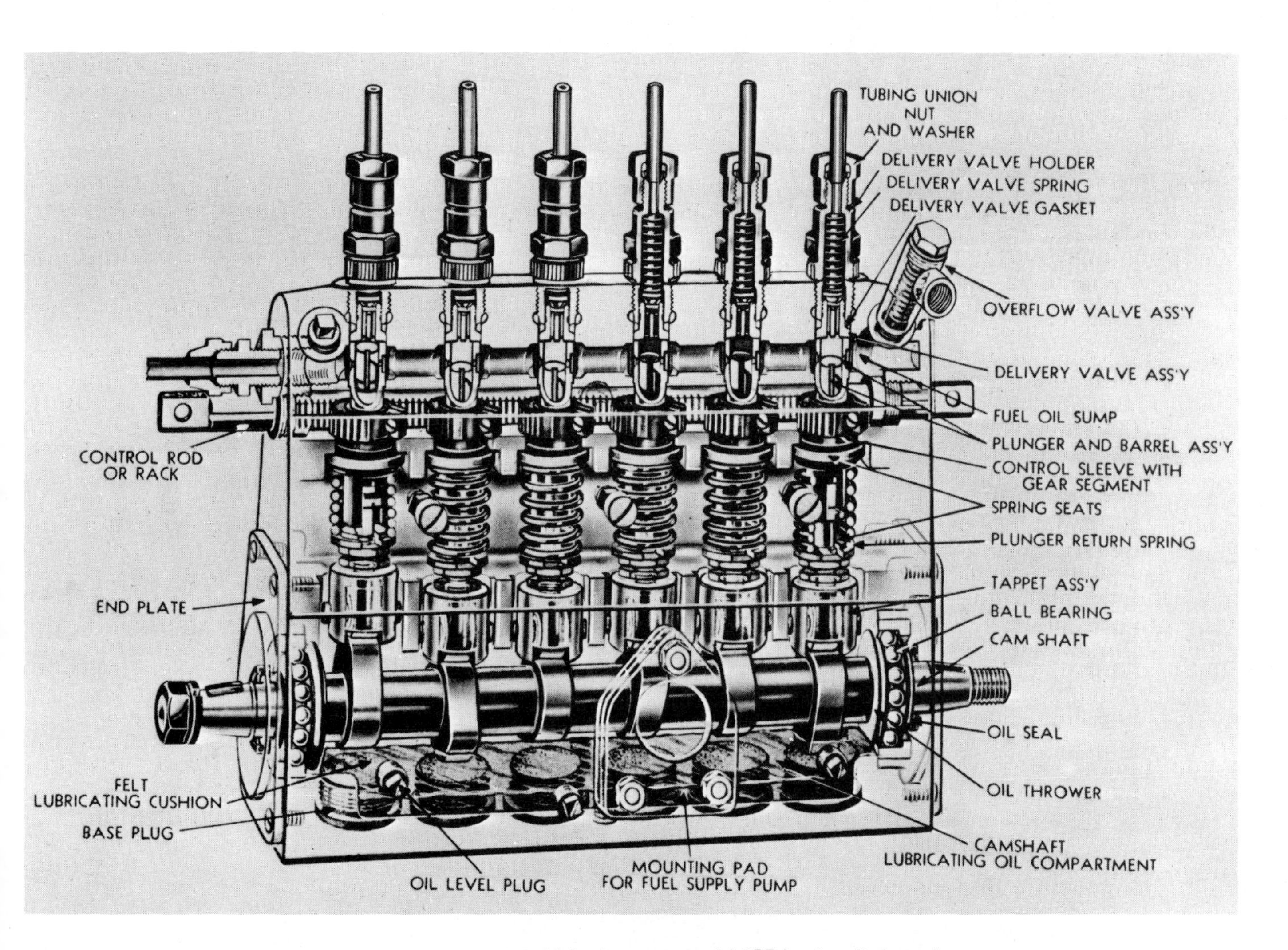

Fig. 8-4. American Bosch fuel injection pump model APE for six cylinder engine. (American Bosch Arma Corp.)

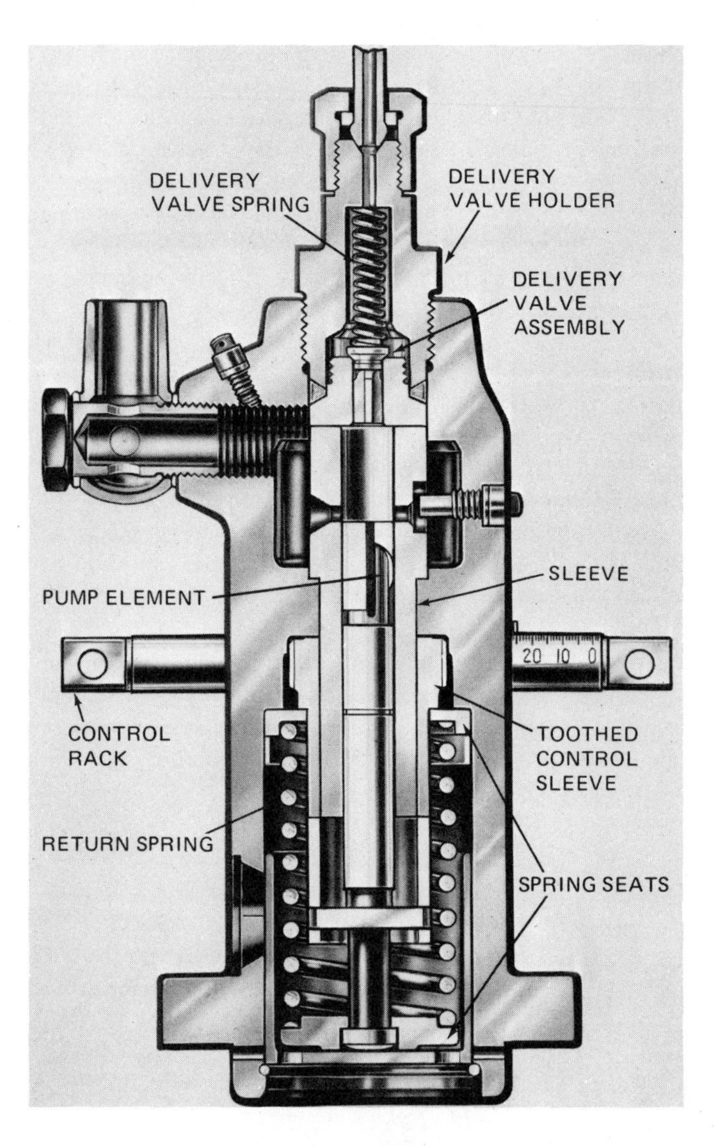

Fig. 8-5. Single cylinder fuel injection pump, type APF. (American Bosch Arma Corp.)

The common rail system is not well adapted for small bore, high-speed engines as it is difficult to control accurately the small quantity of fuel injected at each power stroke.

The modified common rail system sometimes referred to as the controlled pressure system differs from the basic system just described in that mechanically operated fuel injectors are included and nozzles are operated hydraulically instead of mechanically.

The common rail system is used by Atlas Imperial and Cooper Bessemer. In one of the systems produced by Atlas Imperial, injection valves are controlled electromagnetically.

MODERN MECHANICAL INJECTION PUMPS

The mechanical fuel injection pump performs many functions. It times, meters and forces the fuel at high pressure through the spray nozzle.

Most designs are of the plunger type and are cam operated, but there is considerable variation in the method used to control the quantity of fuel delivered. Among the methods used of controlling the quantity of fuel are:

1. Variable stroke.
2. Throttle inlet.
3. Throttle bypass.
4. Timed bypass.
5. Control of port opening.

Current design favors the port opening type of control.

VARIABLE STROKE DESIGN: The stroke in the variable stroke design is changed by sliding a cam plate in or out of its slot in the hollow camshaft. Axial movement of the camshaft is governor controlled which in turn produces radial displacement of the cam plate.

THROTTLE INLET DESIGN: In the throttle inlet design, the flow of fuel into the pumping cylinder is throttled. This is done by rotation of a metering valve, which varies the port opening into the plunger bore.

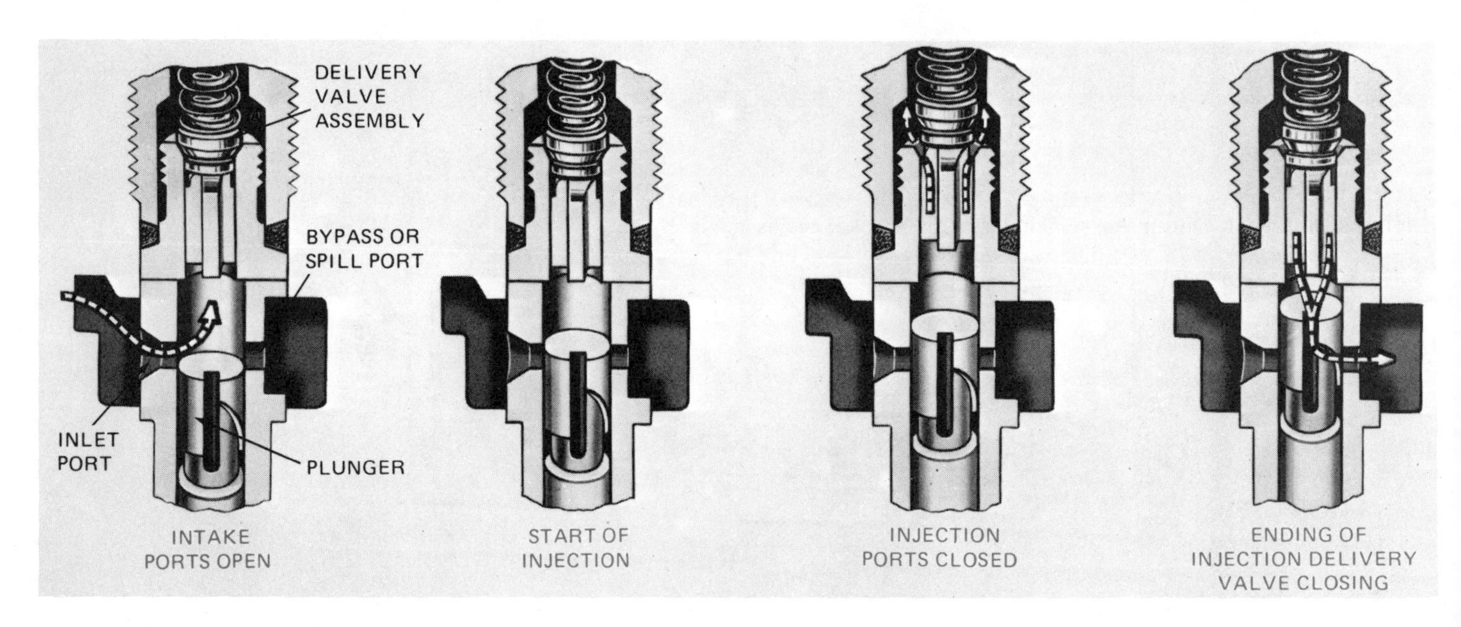

Fig. 8-6. Sequence of operations of American Bosch fuel injection pump. (American Bosch Arma Corp.)

THROTTLE BYPASS DESIGN: In the throttle bypass method of controlling fuel, metered fuel in the plunger chamber is discharged to the nozzle and at the same time is bypassed through a throttle valve back to the inlet. The size of the bypass port opening is varied by governor action controlling a needle valve.

TIMED BYPASS DESIGN: Metering the fuel in the timed bypass method is controlled by spilling excess fuel to a mechanically operated bypass valve. The quantity of fuel discharged is controlled by rotation of an eccentric shaft on which a rocking lever pivots. Fuel delivery starts on the upstroke of the plunger, and ceases when the bypass valve is lifted by contact with the rocking lever.

PORT CONTROL DESIGN: In the port control method of metering fuel, a portion of the plunger functions as a valve to cover and uncover ports in the plunger barrel. A groove on the plunger is designed to rotate so that the plunger stroke can be varied, thus controlling the quantity of fuel delivered on each stroke.

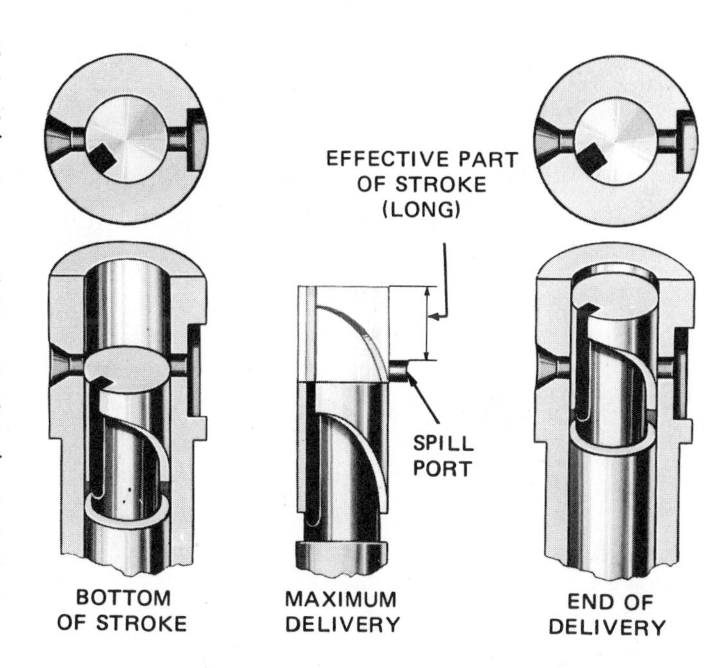

Fig. 8-7. American Bosch pump with plunger in position for maximum fuel delivery. (American Bosch Arma Corp.)

TYPES OF MODERN INJECTION PUMPS

The American Bosch type APE and the APF pumps, Figs. 8-4 and 8-5, are port controlled types and have a constant stroke with lapped plungers. The type shown in Fig. 8-4, is designed for multiple cylinder engines, and has a self-contained drive. The pump illustrated in Fig. 8-5 is of single-cylinder construction and requires a separate drive mechanism.

The pumping principle of the American Bosch type APE and APF fuel injection pumps is illustrated in Fig. 8-6. That illustration shows a cross section of the barrel, with various plungers and delivery valve positions as the plunger travels on upward strokes.

Fuel enters the pump from the supply system through the inlet connection, and fills the fuel pump surrounding the barrel. With the plunger at the bottom of the stroke, Fig. 8-6, fuel flows through the barrel ports, filling the space above the plunger, the vertical slot cut in the plunger, and the cut away area below. As the plunger moves upward, the barrel ports are closed. As it continues to move upward, the fuel is discharged through the delivery valve into the high-pressure line.

Delivery of the fuel ceases when the plunger helix passes and opens the barrel bypass or spill port and the delivery valve returns to its seat. During the remainder of the stroke, fuel is spilled back into the sump. This termination of the fuel delivery controls the quantity of fuel delivered per stroke.

Fig. 8-7 illustrates the plunger rotated in the position for maximum fuel delivery where the effective part of the stroke is relatively long before opening the spill port. Fig. 8-8 illustrates a normal delivery with a much shorter effective stroke. By rotating the plunger still further so the vertical slot is in line with the spill port, the entire stroke is noneffective and no fuel is delivered.

It should be noted that while the effective stroke of the plunger is varied, the plunger always travels the same distance because the cam lift never varies.

The plunger is rotated by means of a control sleeve which in turn is rotated by the control rack, Figs. 8-4 and 8-5. Movement of the control rack, either manually or by governor action, rotates the plunger and varies the quantity of fuel delivered by the pump.

A typical Bosch fuel system as used with the type APE fuel injection pump is shown in Fig. 8-9. Note the fuel pump with its connection to the fuel nozzle. Details of the fuel nozzle, with its holder, are shown in Fig. 8-10. Two types of nozzles are available, one known as the pintle type, and the other as the hole type, Fig. 8-11.

In the case of the pintle type the nozzle valve carries an extension on the lower end in the form on a pin called a pintle, which protrudes through the closely fitting hole in the nozzle bottom. This requires the injected fuel to pass through a round orifice, thus producing a hollow-cone shaped spray.

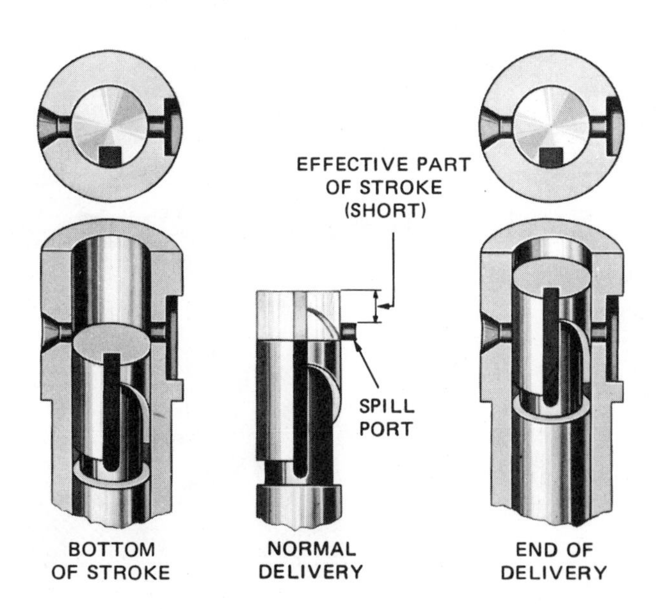

Fig. 8-8. American Bosch pump with plunger in position for normal fuel delivery. (American Bosch Arma Corp.)

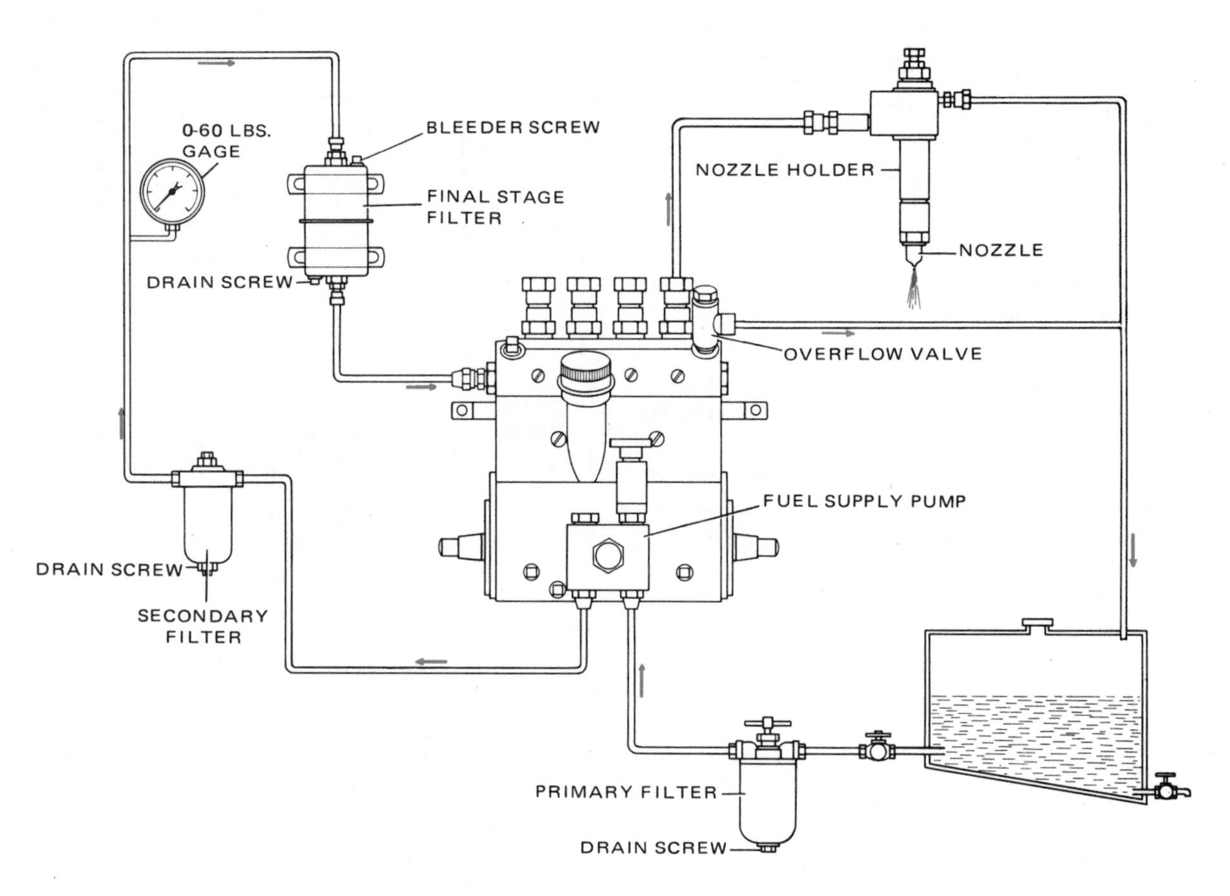

Fig. 8-9. Typical American Bosch fuel supply system. (American Bosch Arma Corp.)

The projection of the pintle through the nozzle induces a self-cleaning effect thereby reducing the accumulation of carbon at that point.

The hole type nozzle has no pintle, but is basically similar in construction to the pintle type. The hole type nozzle has one or more spray orifices which are straight round holes through the tip of the nozzle body beneath the valve seat, Fig. 8-11. Spray from each individual orifice is relatively dense and

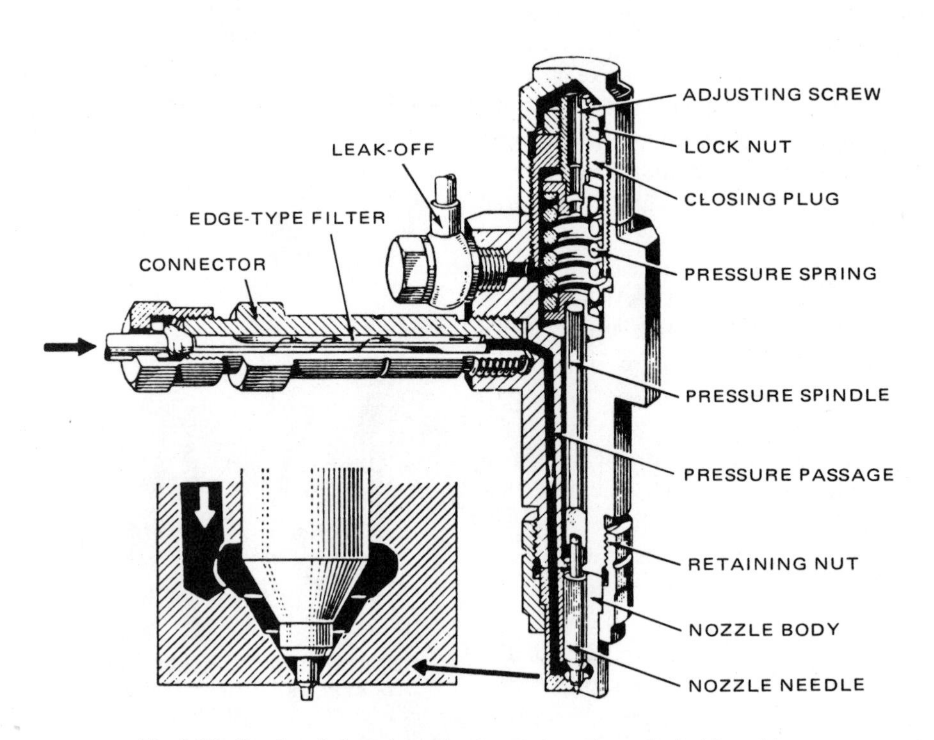

Fig. 8-10. Details of American Bosch nozzle and nozzle holder with pintle type nozzle. (American Bosch Arma Corp.)

compact. The general spray pattern is determined by the number and arrangement of the holes. As many as 18 holes can be provided in the larger nozzles. The diameter of the individual orifices may be as small as .006 in. The spray pattern may or may not be symmetrical (regular in shape), depending on the contours of the combustion chamber, and the fuel distribution requirements.

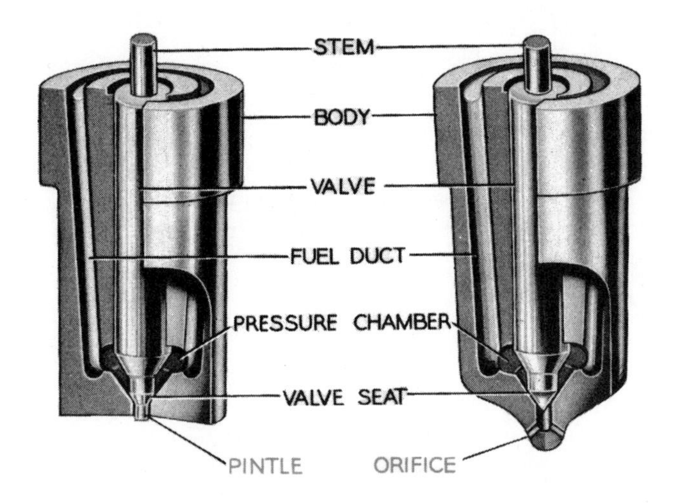

Fig. 8-11. Details of pintle, left, and hole type nozzle, right. (American Bosch Arma Corp.)

AMERICAN BOSCH FUEL SUPPLY PUMP

The fuel supply pump used in connection with the American Bosch APE and APF fuel injection pumps is shown in Fig. 8-12. A complete system showing the location of the individual parts is shown in Fig. 8-9.

The purpose of the fuel supply pump is to draw fuel oil from the fuel tank and pump it through the filtering system into the fuel injection pump. It is mounted on the housing of the injection pump and is driven by the injection pump camshaft. It is a variable, self-regulating plunger type pump which will build up pressure to a predetermined point.

The operation of the fuel supply pump is as follows: As the injection pump cam allows the plunger to be forced by a spring toward the camshaft, the suction effect created opens the inlet valve and permits fuel to enter the plunger spring chamber, Fig. 8-12. As the cam drives the plunger against its spring the fuel is forced by the plunger through the outlet valve, into the chamber created in back of the plunger by its forward movement. As the injection cam continues to rotate, it allows the plunger spring, which is under compression, to press the plunger backward again, thus forcing the fuel oil behind the plunger out into the fuel line leading to the filters, and the injection pump. At the same time, the plunger is again creating a suction effect which allows additional oil to flow through the inlet valve into the spring chamber.

This pumping action continues as long as fuel is being used by the injection pump fast enough to keep the supply pressure from rising to the point where it equals the force exerted by the spring on the plunger. The pressure between the supply pump and the injection pump holds the plunger stationary against the spring and away from the cam. This prevents further pumping action until the pressure drops enough to permit the plunger to resume operation. The entire cycle is automatic and continues as long as the engine is running.

As shown in Fig. 8-9, the system is supplied with a return flow to the fuel tank. In this system, pressure is limited by a spring-loaded overflow valve, usually placed on the injection pump opposite its fuel inlet connection. Such an overflow valve, usually adjusted to approximately 6 to 15 lbs. psi, permits fuel to pass entirely through the injection pump sump back to the supply tank. Any air or gas which may have entered the system will be carried away. An accumulation of air or fumes within the injection pump will result in misfiring of the engine.

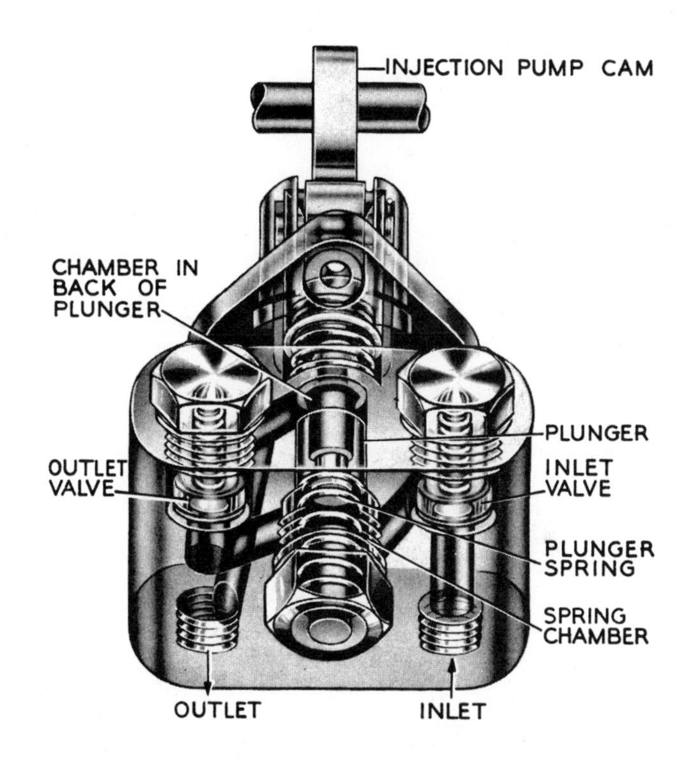

Fig. 8-12. Details of fuel supply pump used in American Bosch APE and APF systems. (American Bosch Arma Corp.)

GOVERNORS

There are two types of governors used in connection with the American Bosch type APE fuel injection pumps. One is a centrifugal mechanical governor and the other is a pneumatic governor operated by air flow into the engine intake manifold. Both types are directly connected to the American Bosch type APE fuel injection pump without external linkage.

The type GVA and GVB mechanical governor is shown in Fig. 8-13. A friction clutch is built into the drive gear to protect the governor parts, dampen vibration and permit the drive gear to slip on its hub whenever sudden change in speed occurs. This feature assures smooth operation of the fly-weights.

As the flyweights of the governor revolve, centrifugal force tends to throw them outward, moving the noses along the shaft. This movement is opposed by the action of the springs. Movements of the flyweights are transmitted to the fuel injection pump rack through appropriate linkage. The fuel quantity injected will be increased or decreased in accordance with the movement of the flyweights and the attached linkage. Fig. 8-14 shows a complete unit with the various adjustments for high speed, idling, and the bumper spring.

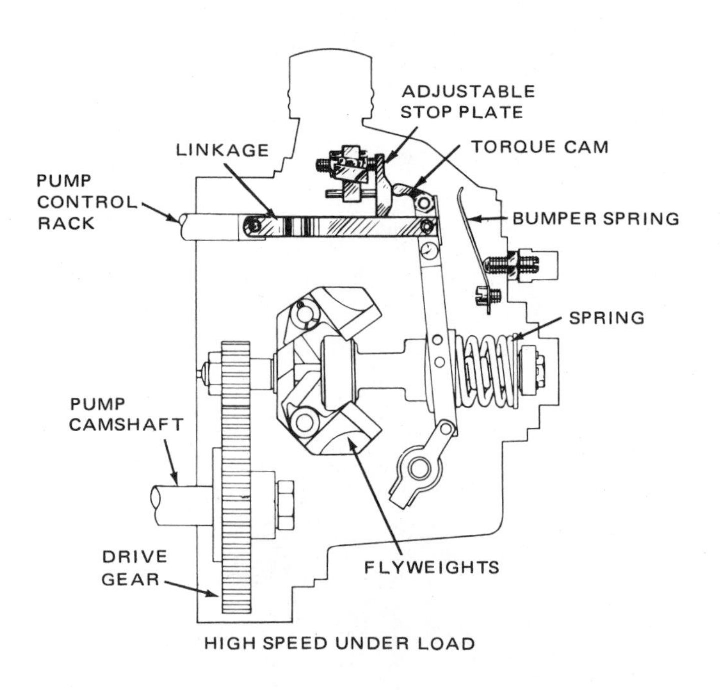

Fig. 8-13. Centrifugal mechanical governor as used with APE type pump. (American Bosch Arma Corp.)

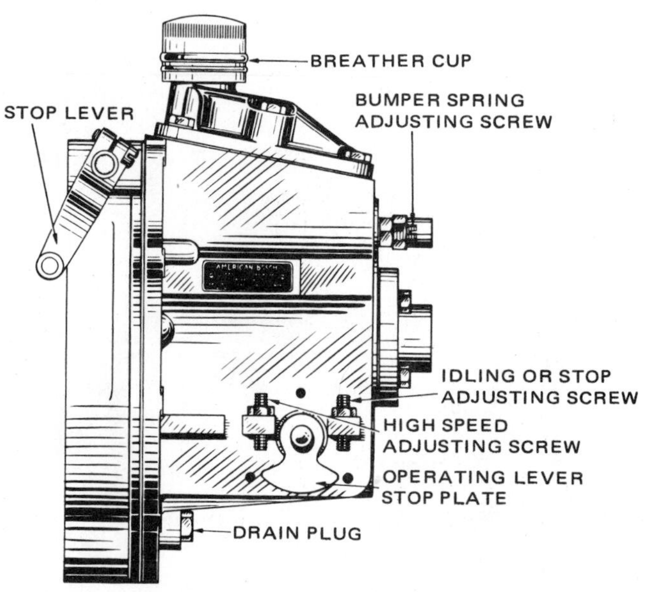

Fig. 8-14. External view of American Bosch type GVA and GVB mechanical governor showing adjustments. (American Bosch Arma Corp.)

PNEUMATIC GOVERNOR

The pneumatic governor (type AEP-MZ) as used with the APE fuel injection pump serves the same purpose as the GV type mechanical governor which has just been described. The pneumatic governor provides means for presetting and maintaining within close regulation any diesel engine speed within the nominal maximum speed range, irrespective of engine load. The governor also controls the engine idling speed to prevent

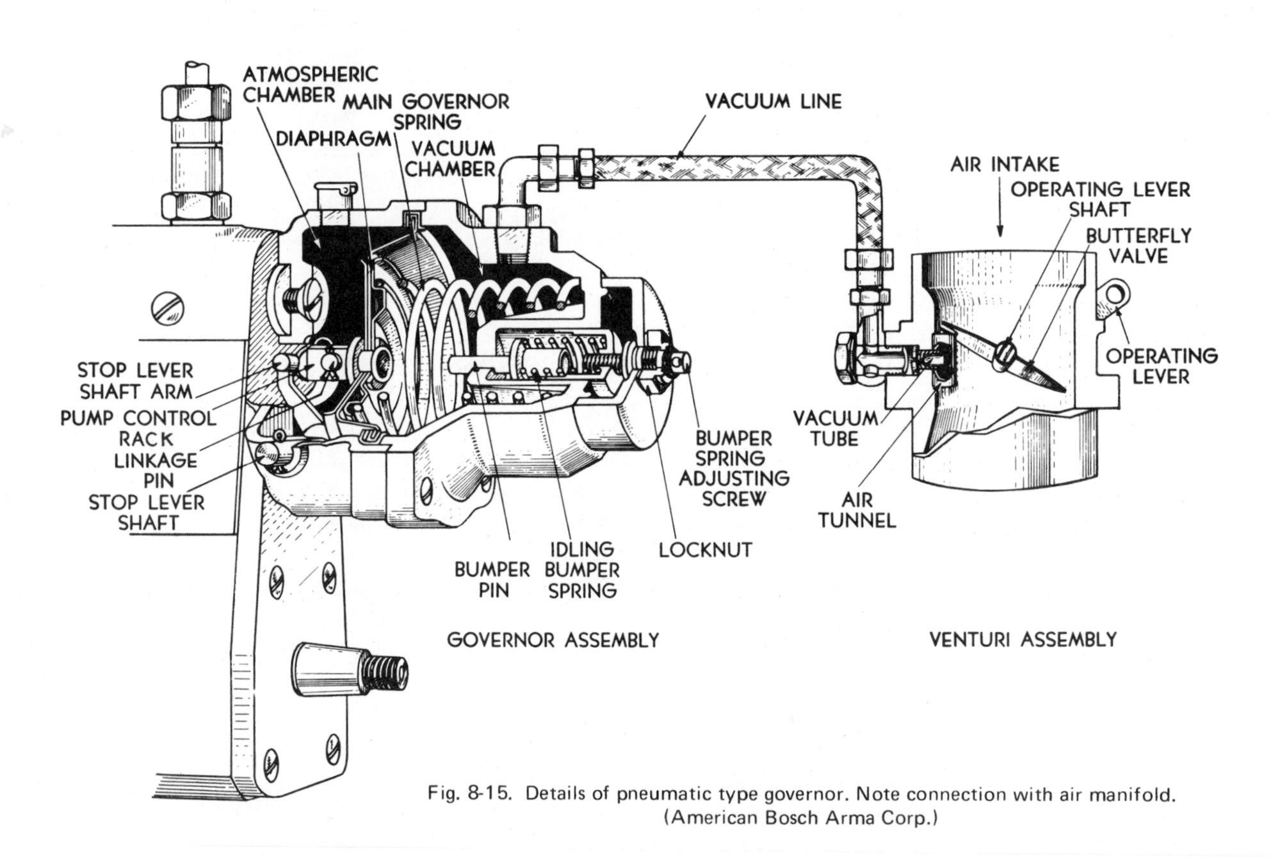

Fig. 8-15. Details of pneumatic type governor. Note connection with air manifold. (American Bosch Arma Corp.)

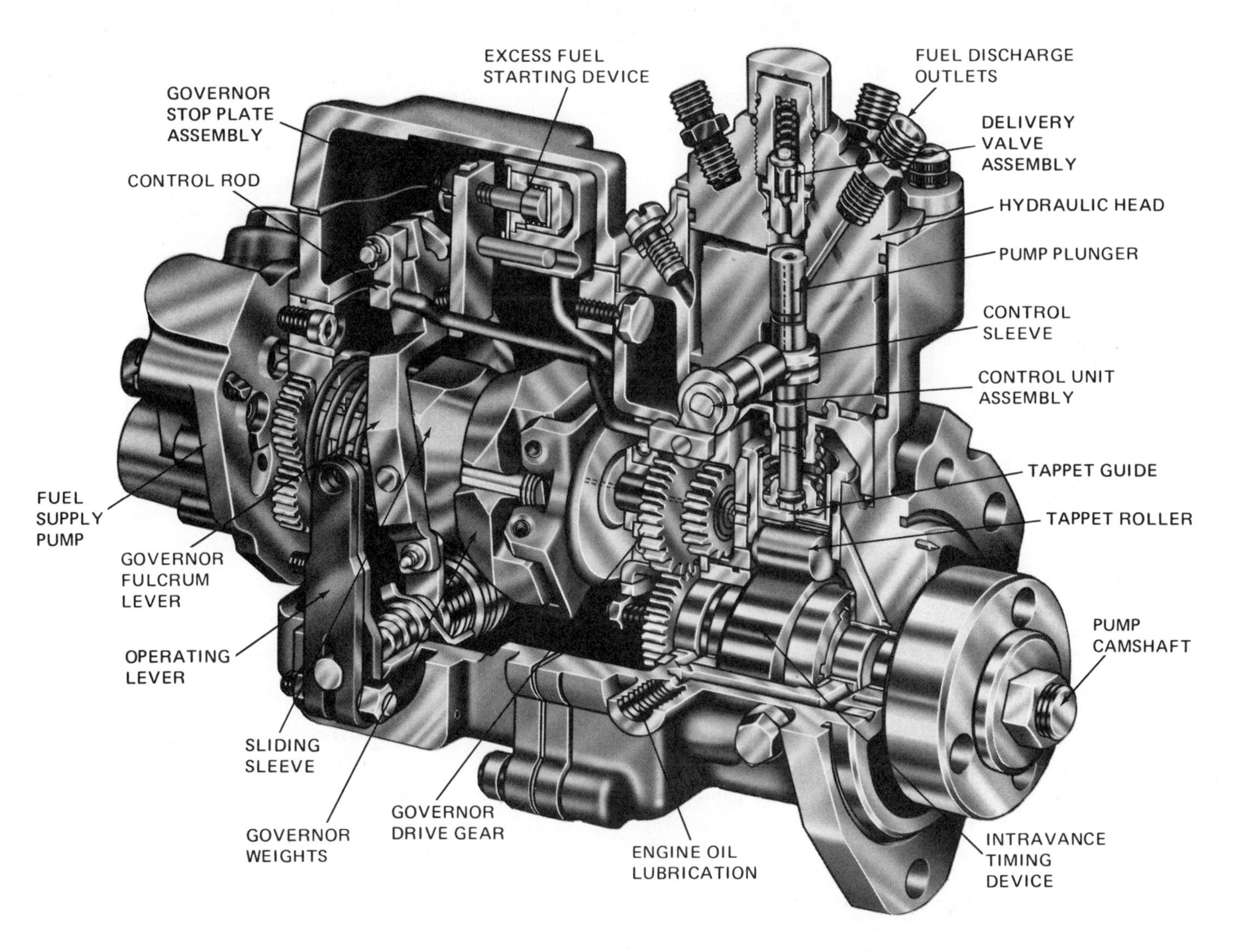

Fig. 8-16. Type PSJ American Bosch fuel injection pump. (American Bosch Arma Corp.)

stalling, and the maximum speed to prevent racing.

This governor is operated by a partial vacuum produced by a butterfly valve and venturi assembly, Fig. 8-15, located in the engine air intake manifold.

The principle underlying the operation of the pneumatic governor is that air passing through the tube tends to create a vacuum in another small tube entering it at right angles. The amount of vacuum created depends on the velocity of the air passing through the larger tube.

Diesel engines using a pneumatic governor are equipped with a venturi assembly, Fig. 8-15, in the air intake manifold. The venturi contains a small tunnel covering the opening to the vacuum tube and the butterfly valve controlling the air entering the engine. The valve is opened and closed by an operating lever. The maximum and minimum valve openings are limited by adjustable stops on the operating lever shaft.

The pneumatic governor attached to the end of the injection pump consists essentially of a housing divided into an air chamber by a flexible diaphragm which is directly connected to the injection pump control rack. The chamber nearer the pump is interconnected with the pump spray compartment and is therefore under atmospheric pressure. The other chamber is connected with vacuum tube of venturi. During engine operation, chamber is under partial vacuum.

To obtain the maximum speed the operator opens the butterfly valve by means of linkage connecting the operating lever to the foot accelerator. While the valve is open there is a large flow area through the venturi and the velocity of the air passing through the tunnel is comparatively small. This results in producing only slight vacuum in the line to the governor. It is not sufficient to draw the diaphragm and the pump control rack against the force of the main governor spring and the maximum fuel quantities are delivered by the pump. The maximum fuel quantity is controlled by means of an adjustable control rack stop mounted on the end of the pump opposite the governor.

To obtain medium speed, the butterfly valve is partly closed to reduce flow area through the venturi. If the engine load is suddenly increased, the speed will have a tendency to decrease. This results in less vacuum and the main governor spring forces the diaphragm and pump control rack toward greater fuel delivery, preventing any noticeable change in engine speed. Similarly, if the engine load is suddenly decreased, the speed will have a tendency to increase, but this results in more vacuum and the pump control rack will be drawn back into position of less fuel delivery.

To stop the engine, a separate shutoff arrangement is provided. The lever rotates the stop lever shaft bringing the shaft arm against the linkage pin and removing the pump central rack to a position of no fuel delivery.

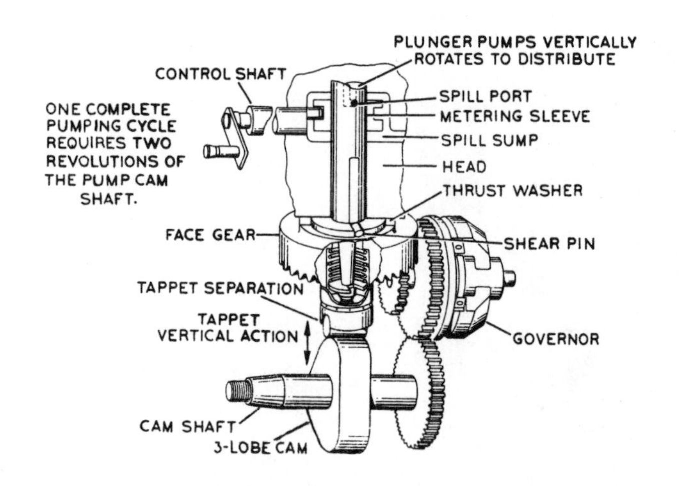

Fig. 8-17. Showing operation of pumping action of distributor type American Bosch PSJ injection pump. (American Bosch Arma Corp.)

AMERICAN BOSCH INJECTION PUMP (PSJ)

The American Bosch type PSJ fuel injection pump, Fig. 8-16, is a distributor type pump and differs from models described previously in that a single pump plunger is used for multicylinder engines. The single steel plunger reciprocates (moves back and forth) for pumping action, and rotates continuously for distribution of fuel to all discharge outlets. The constant reciprocating, rotating motion of single pumping element assures equal amounts of fuel volume to all cylinders.

In operation, the plunger makes one complete revolution for two revolutions of the pump camshaft, which is operating at engine crankshaft speed. The plunger rotates continuously by moving vertically through the pumping cycle. On an eight cylinder engine, the four lobe cam actuates the plunger eight times for two revolutions of the pump camshaft. Similarly, on a six cylinder engine the three lobe cam actuates the plunger six times for every two revolutions of the pump camshaft. This construction is shown in Fig. 8-17.

Fuel is injected only during the high velocity portion of the plunger stroke, and fuel metering is controlled by the variation of the position of the metering sleeve in relation to the fixed port closing position, which is the point at which the top of the plunger covers the intake ports.

Details of the fuel pumping principle are shown in Fig. 8-18. Fuel enters the pump from the supply system through the pump housing inlet connection. This action fills the sump area, and that portion of the head cavity between the top of the plunger and the bottom of the delivery valve, when the plunger is at the bottom of its stroke.

The tappet roller is held on the base circle of the cam by the plunger return spring during the filling sequence, shown at A, Fig. 8-18. As the rotating plunger moves upward in its stroke under cam action, it closes the two horizontal scallops which contain the inlet ports, trapping and placing the fuel under pressure. This pressure is increased until the spring-loaded delivery valve is opened as at B, Fig. 8-18.

As the plunger continues its upward stroke, fuel is forced through the delivery valve and is conveyed through the duct to the annulus (ring-shaped space) and distributing slot in the plunger. The vertical distributing slot on the plunger connects to the outlet duct, which is then registering as the plunger rotates. This is shown at C, Fig. 8-18.

The rotary and vertical motion of the plunger are so phased

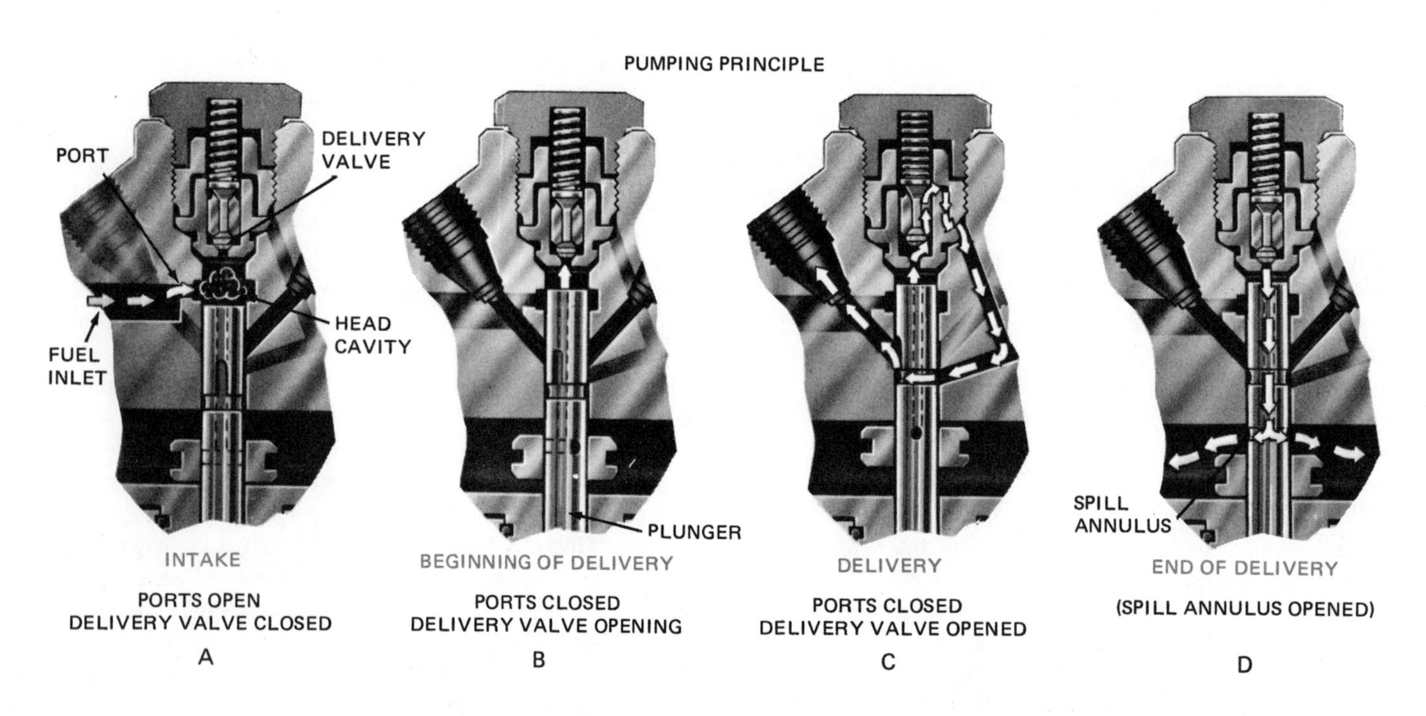

Fig. 8-18. Pumping principle of American Bosch type PSJ pump. (American Bosch Arma Corp.)

in relation to the outlet ducts that the vertical distributing slot overlaps only one outlet duct during each effective portion of each stroke. After sufficient upward movement of the plunger, its metering port passes the edge of the metering sleeve and the fuel under pressure escapes down the vertical hole in the center of the plunger into the sump surrounding the metering valve, which is now at supply pressure, D, Fig. 8-18. With the collapse of the pressure beneath it, the delivery valve then closes, during which action the piston portion of the valve blocks the passage before the valve reaches its seat. This reduces the pressure in the discharge system.

The quantity of fuel delivered per stroke by the American Bosch PSJ single plunger fuel injection pump is controlled by the variation of the position of the metering sleeve in relation to the fixed port closing position (the point at which the top of the plunger covers the intake ports). As the metering port on the plunger breaks over the top edge of the metering sleeve, pumping pressure is relieved down to the center hole of the plunger out into the sump surrounding the metering sleeve and fuel terminates despite the continued upward movement of the plunger.

When the metering (control) sleeve is lowered to its extreme point, Fig. 8-19, (no delivery) the spill annulus on the plunger is uncovered by the top edge of the sleeve before the upper end of the plunger can cover the intake ports. Under this condition, no pressure can be built up even after the intake ports are closed. Hence no fuel can be delivered. This is the shut off position.

If the metering sleeve is moved to the mid-position, the annulus on the plunger is uncovered by the sleeve later in the plunger stroke. The effective stroke of the plunger is longer and fuel is delivered, Fig. 8-19 (normal delivery). If the metering sleeve position is raised, the spill annulus on the plunger remains covered by the sleeve until relatively late in the plunger stroke. The effective stroke of the plunger is still longer and more fuel is delivered, Fig. 8-19 (maximum delivery).

The upward movement of the metering sleeve increases the quantity of fuel pumped per stroke and the downward movement decreases the quantity of fuel pumped per stroke.

The delivery valve assembly, Figs. 8-16 and 8-20, situated directly above the pumping chamber assists the injection function by preventing irregular loss of fuel from the delivery side of the system to the supply side, between pumping strokes. The valve assembly consists of the valve proper, containing a reaction piston and a tapered seat; the valve body, which receives the valve and has a corresponding tapered seat; and the valve spring secures delivery valve holder and cap screw.

After the plunger on its upward stroke covers the intake port, pressure is created. When this pressure overcomes the force of the spring holding the delivery valve on its tapered seat, the valve opens and fuel under pumping pressure flows through it and the distributor passing into the injection tubing as shown at A, Fig. 8-20. When plunger continuing its upward stroke opens spill annulus, there is a sudden drop in fuel pressure below delivery valve, and force of the valve spring acts to return the valve to its seat as shown at B, Fig. 8-20.

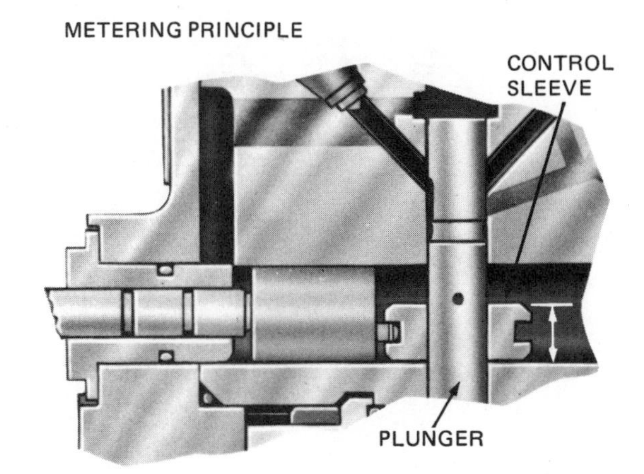

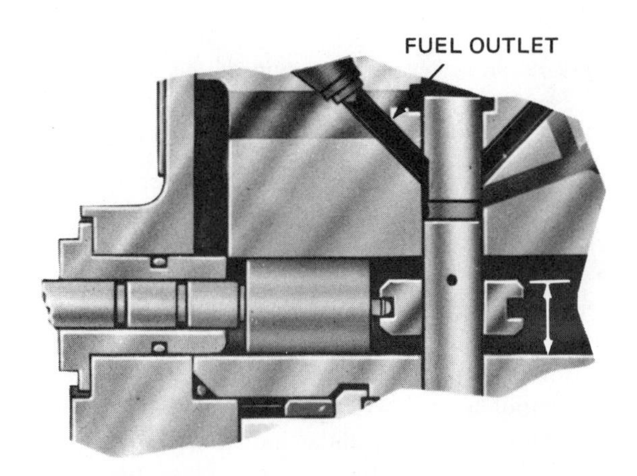

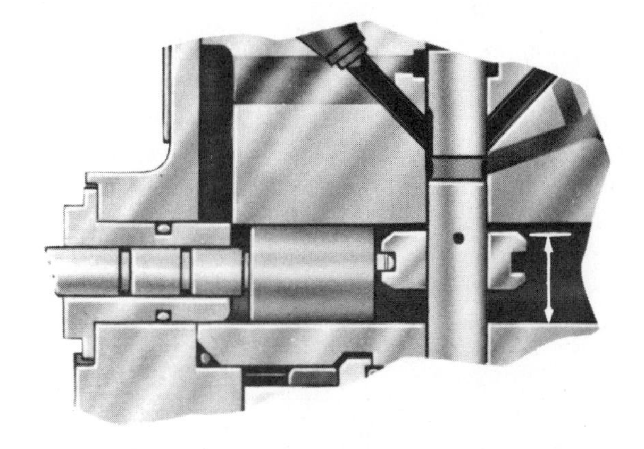

Fig. 8-19. Showing three positions of metering sleeve of American Bosch PSJ fuel injection pump. (American Bosch Arma Corp.)

After the valve starts down with its body, the lower edge of the retraction piston enters the valve bore and blocks the passage. The further downward movement of the valve retraction piston increases the volume on the high pressure side by the amount of the retraction piston movement, and consequently reduces the pressure in the line. This lowered

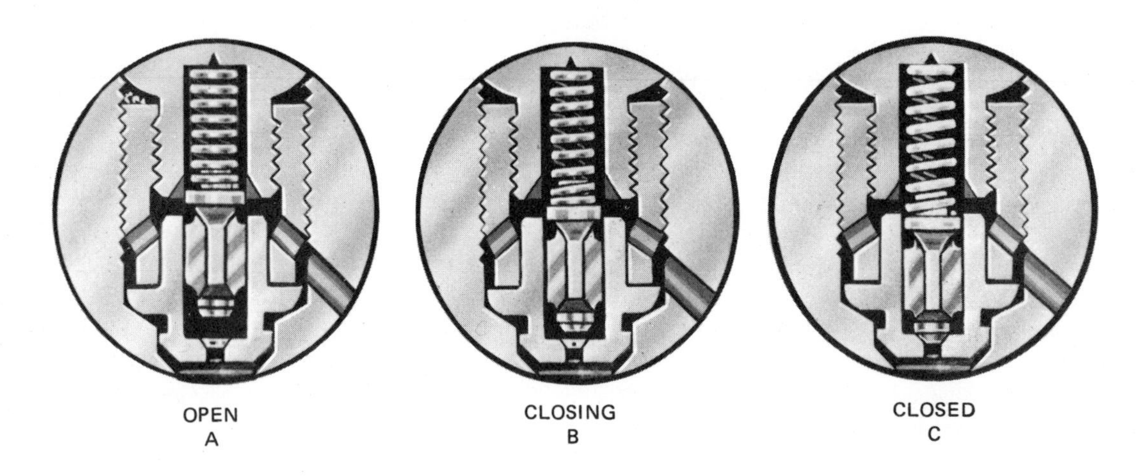

Fig. 8-20. Details of operation of delivery valves PSJ pump. (American Bosch Arma Corp.)

line pressure promotes rapid closing of the injection nozzle valve and reduces the effects of the hydraulic pressure waves that exist in the tubing between injections, minimizing the possibility of nozzle valve reopening prior to the next regular delivery cycle, as shown at C, Fig. 8-20.

PSJ VARIABLE SPEED GOVERNOR

The American Bosch PSJ variable speed governor, Fig. 8-16, is of the mechanical centrifugal type. It is assembled as a unit in the governor housing. The prime purpose is to serve as a means of presetting and maintaining within close limits, the engine speed within a nominal speed range irrespective of engine load. In addition, the governor controls engine idling speed to prevent stalling, and maximum speed to prevent over speeding.

The governor is gear driven from the injection pump camshaft. The governor weight shaft includes flyweights with their pivot pins mounted in a spider. A friction clutch is also included in the design. Adjustment for the degree of slippage is accomplished by means of shims. The clutch reduces vibrations in the pump camshaft. Movement of the flyweights moves a sliding sleeve, Fig. 8-16. The fulcrum lever bracket extends over the operating lever shaft, but is not directly connected to it. The torsion spring hub is firmly secured to the operating lever shaft and connected through the torsional spring to the fulcrum lever bracket. Spring tension tends to keep the tongue of the spring hub and the fulcrum lever bracket in line during operation of the governor. A movement of the foot pedal will rotate the spring hub and the fulcrum lever bracket, causing the fulcrum lever to turn about the pivot pin. This movement changes the position of the pump control unit, as the upper end of the fulcrum lever is connected through linkage to the control rod. The fuel quantity will be increased or decreased depending on the direction of operating lever movement.

INTRAVANCE TIMING DEVICE

The automatic internal timing device used with the PSJ American Bosch single plunger fuel injection pump is known as Intravance, Fig. 8-16. As the speed of the engine increases, ignition timing must be advanced to provide maximum performance. The entire mechanism is contained within the pump housing, and is an integral part of the injection pump, Fig. 8-21. It is operated by engine lube oil pressure and responds to engine speed only. The basic function of the Intravance is to sense engine speed, and set the beginning of the ignition to the specified ignition timing curve. Cut-in speed can vary from 800 to 1200 rpm. Current maximum mechani-

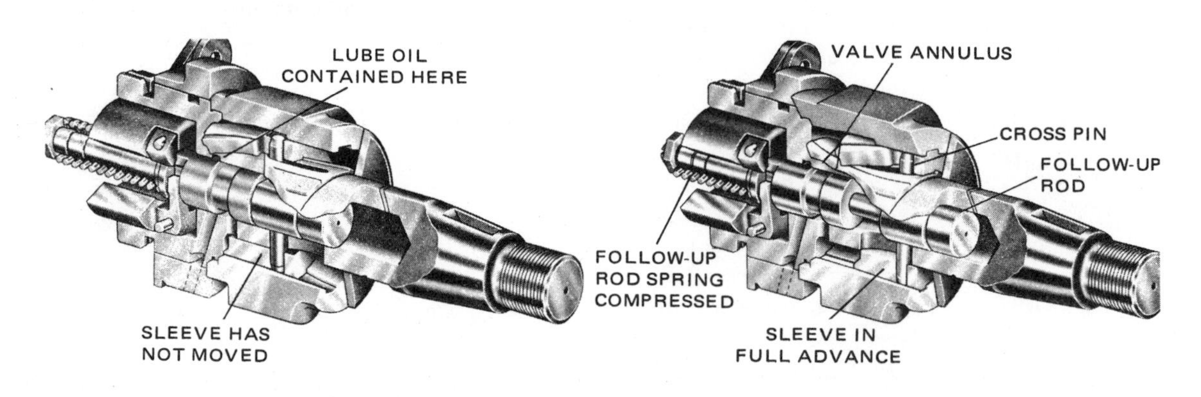

Fig. 8-21. The American Bosch Intravance is shown before cut-in speed at the left and in cut-in speed at the right. (American Bosch Arma Corp.)

cal advance is 12 1/2 deg.

In operation, lubricating oil is supplied through the pump housing to holes in the rear bearing. The bearing holes terminate in the circular groove in the shaft. Three holes in the shaft meet in the reservoir area in the interior of the shaft. While the engine rpm is less than cut-in speed, lube oil is contained in this area and cannot cross the metering edge of the control valve, Fig. 8-21. As long as the engine is operating in the range below cut-in speed, the Intravance weights cannot overcome the spring preload, and the control valve restricts the lube oil from any further movement from its recess in the inside diameter of the shaft. At cut-in speed, the weights have spline, this will cause rotation of the cam in a direction to advance beginning of injection up to 12 1/2 deg.

The followup rod, in moving to the right, has caused the spring to be compressed against the end of the control valve. This action forces the weight and valve back to a position where the metering edges are just closed and stops the flow of high-pressure oil to the pressure side of the sleeve. A decrease in engine speed would collapse the weights and the metering edge of the valve would stop the flow of lube oil. With no oil pressure behind the sleeve, the cam would retard due to the mechanical action of the cam and the spring on the followup rod. Intravance action is entirely independent of the governor.

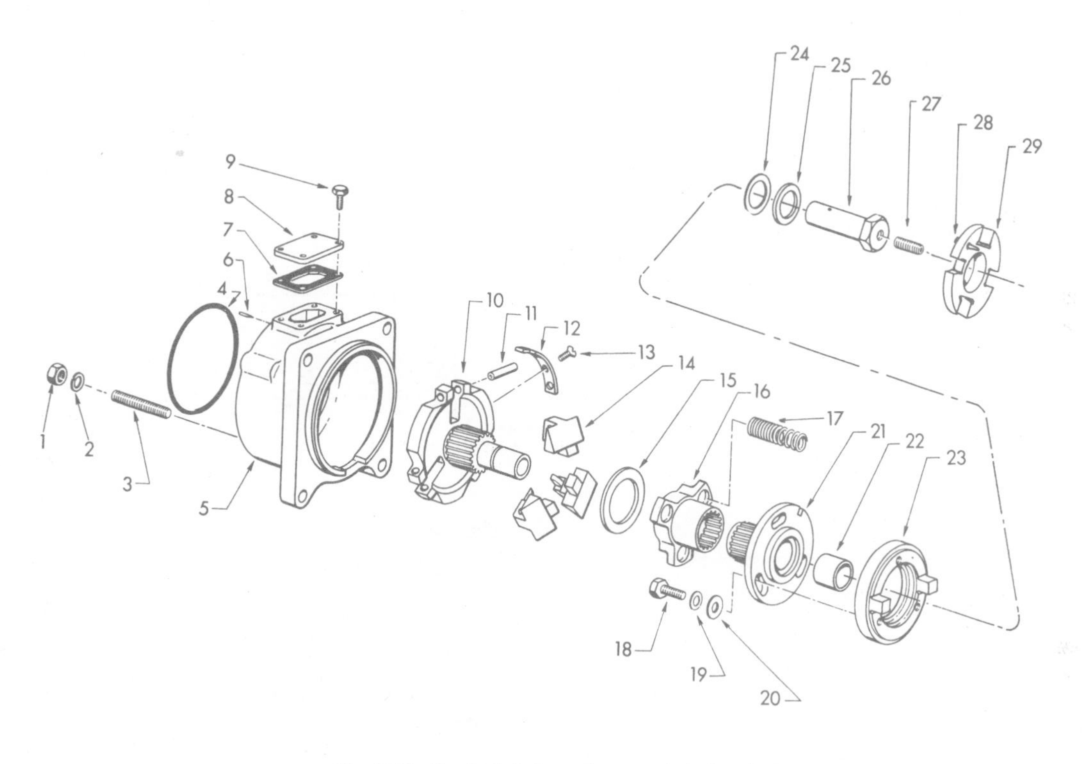

Fig. 8-22. Exploded view of external timing device.

1. NUT.
2. LOCKWASHER.
3. STUD.
4. GASKET.
5. HOUSING.
6. POINTER.
7. GASKET.
8. COVER, TIMING WINDOW.
9. SCREW.
10. SPIDER ASSEMBLY.
11. PIN, WEIGHT.
12. PLATE, TIMING.
13. SCREW.
14. WEIGHT, TIMING.
15. THRUST WASHER.
16. SLIDING GEAR.
17. TIMING SPRING.
18. SCREW.
19. LOCKWASHER.
20. WASHER.
21. HUB.
22. BUSHING.
23. COUPLING.
24. THRUST WASHER.
25. TIMING WASHER.
26. CAMSHAFT NUT.
27. SCREW.
28. DISC PIN.
29. DISC ASSEMBLY.

opened sufficiently to move the control valve slightly to the left, as shown in Fig. 8-21. This movement is enough to have the metering edge of the valve allow the high-pressure engine oil to enter the circular groove in the control valve (valve annulus). The engine oil continues through the two holes in the shaft to the pressure side of the splined sleeve. Then the splined sleeve acts as a piston and moves toward the right, carrying with it the cross pin and followup rod. The sleeve, pin and followup rod will move within the confines of the slot and the cam until full advance at a predetermined speed is reached. Then, depending on engine rotation and the angles of the

EXTERNAL TIMING DEVICE

Another method of advancing the timing of injection used by American Bosch is the external timing device, Fig. 8-22. The entire mechanism is enclosed by an aluminum housing. Its operation is as follows: As the engine commences to revolve the spider shaft, centrifugal force acting on the flyweights tends to open the weights from the collapsed position. The preloading of the three coil springs prevents their action until a specified engine speed is reached, at which point the forces of the spring and weights are balanced. As speed is further

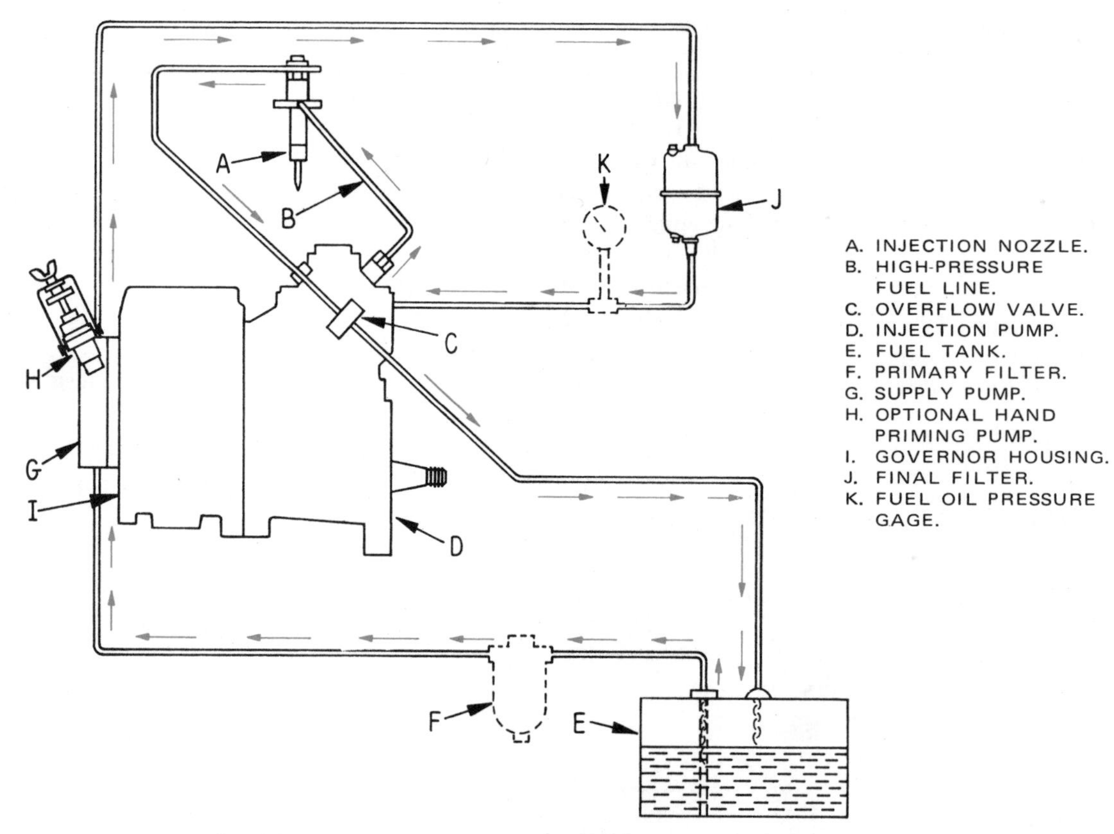

Fig. 8-23. Layout of PSJ fuel system. (American Bosch Arma Corp.)

increased, the flyweights are balanced. As speed is further increased, the flyweights spring out and the weight arms force the sliding gear toward the hub flange, overcoming the initial coil spring force, to a new position determined by the balancing spring forces. As a result of this longitudinal movement of the sliding gear, the injection pump shaft is rotated slightly in advance of the hub flange of the timing device by the action of the internal helical splines of the sliding gear and the external helical splines of both spider assembly and hub. As a result, the timing of the injection pump with relation to top dead center position of the engine piston, is advanced within a given range.

PSJ FUEL SUPPLY PUMP

The fuel supply pump used with the American Bosch fuel injection pump is of the gear type and is driven directly from the end of the governor shaft, Fig. 8-16. This is a positive displacement type pump. A built-in regulating valve prevents damage to injection equipment in the event of an obstruction in the fuel supply system. An overflow valve returns excess fuel to supply tank. The complete PSJ fuel system is shown in Fig. 8-23.

AMERICAN BOSCH NOZZLE AND HOLDER

The nozzle and holder assembly as used with the PSJ fuel injection pump is shown in Fig. 8-24, and is basically similar to the nozzle and holder used with the other American Bosch installations.

The nozzle and holder is designed to carry the high pressure fuel through the cylinder head to the combustion chamber and deliver it in the form of a fine spray through equal size holes in

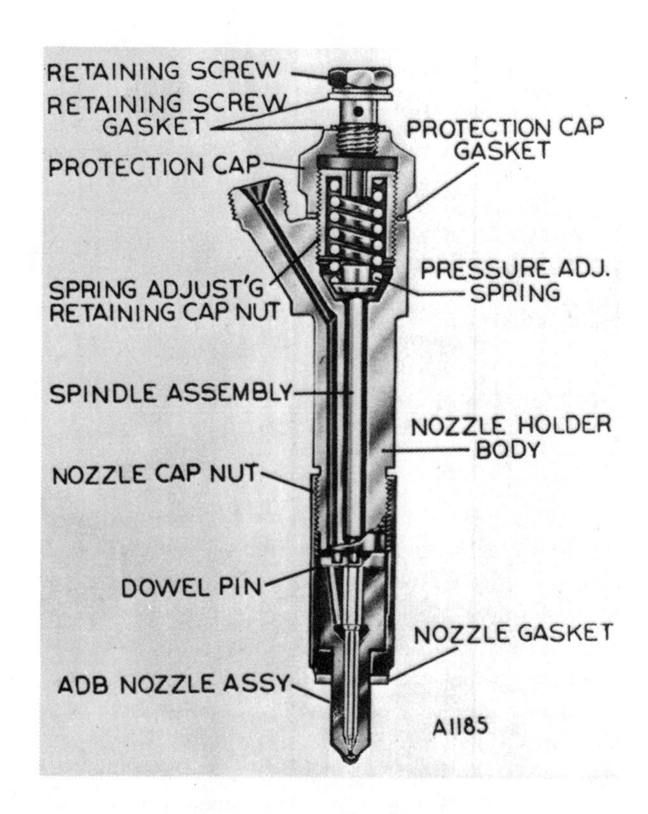

Fig. 8-24. Nozzle and holder. (American Bosch Arma Corp.)

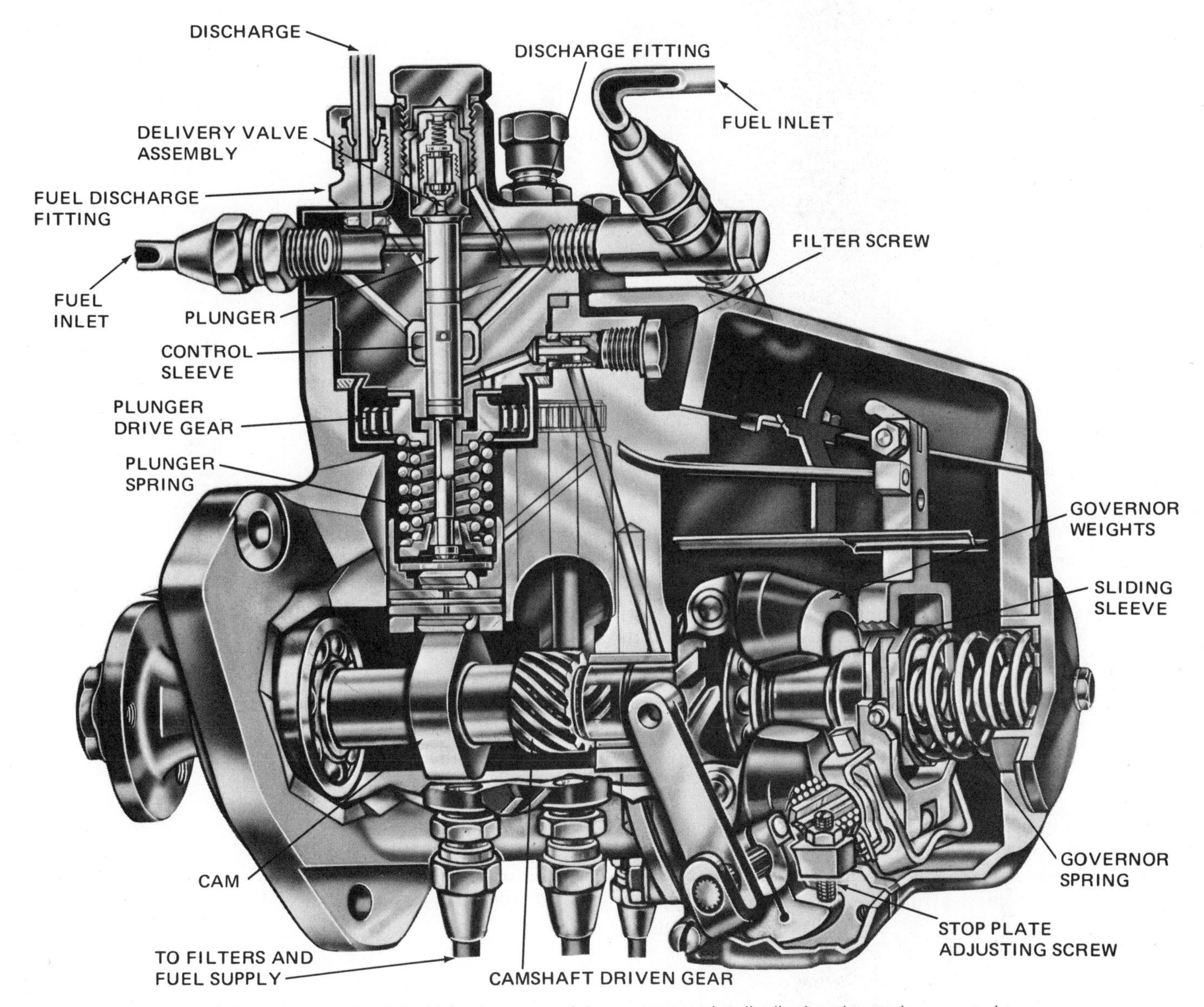

Fig. 8-25. Type PSB American Bosch fuel injection pump of the constant stroke, distributing plunger sleeve control type. (American Bosch Arma Corp.)

the nozzle tip. The holder consists of a forged body having a spindle and spring as well as high pressure fuel duct. The spindle bore and the spring chambers are utilized as a passage for leak-off fuel that bypasses and lubricates the nozzle valve.

This hole type nozzle consists of a body and valve. The valve spring loaded by the spindle and the spring is located in the nozzle holder. The spring is set to provide the specified nozzle opening pressure. As the plunger in the injection pump delivers fuel to the nozzle, the pressure in the line increases to a value sufficient to lift the nozzle valve against the spring load imposed upon it. At the point in the plunger stroke where the spill annulus on the plunger is uncovered by the metering sleeve causing delivery valve to close rapidly, the consequent reduction in line pressure enables the spring in the nozzle holder to seat the nozzle valve quickly, thereby terminating the injection of fuel into the combustion chamber for that particular stroke of the plunger.

The complete American Bosch PSJ fuel injection system is shown schematically in Fig. 8-23.

AMERICAN BOSCH TYPE PSB FUEL INJECTION PUMP

The American Bosch Type PSB Fuel Injection Pump is of the constant stroke, distributing plunger sleeve control type, Fig. 8-25. The plunger is actuated by a cam and tappet arrangement which also carries gearing for distribution functions.

The purpose of the pump is to deliver accurately metered quantities of fuel oil under high pressure to the spring nozzle to which the fuel is injected into the engine cylinders. This is at a definite timing relation to the engine firing cycle and within the required injection period. An integral governor of the mechanical centrifugal type is used to control fuel delivery as a function of engine speed. The governor is driven from an extension of the pump camshaft. The gear type fuel supply pump is mounted at the front of the injection pump and is driven from the distributor driven gear on the camshaft.

The operation of the American Bosch type PSB fuel

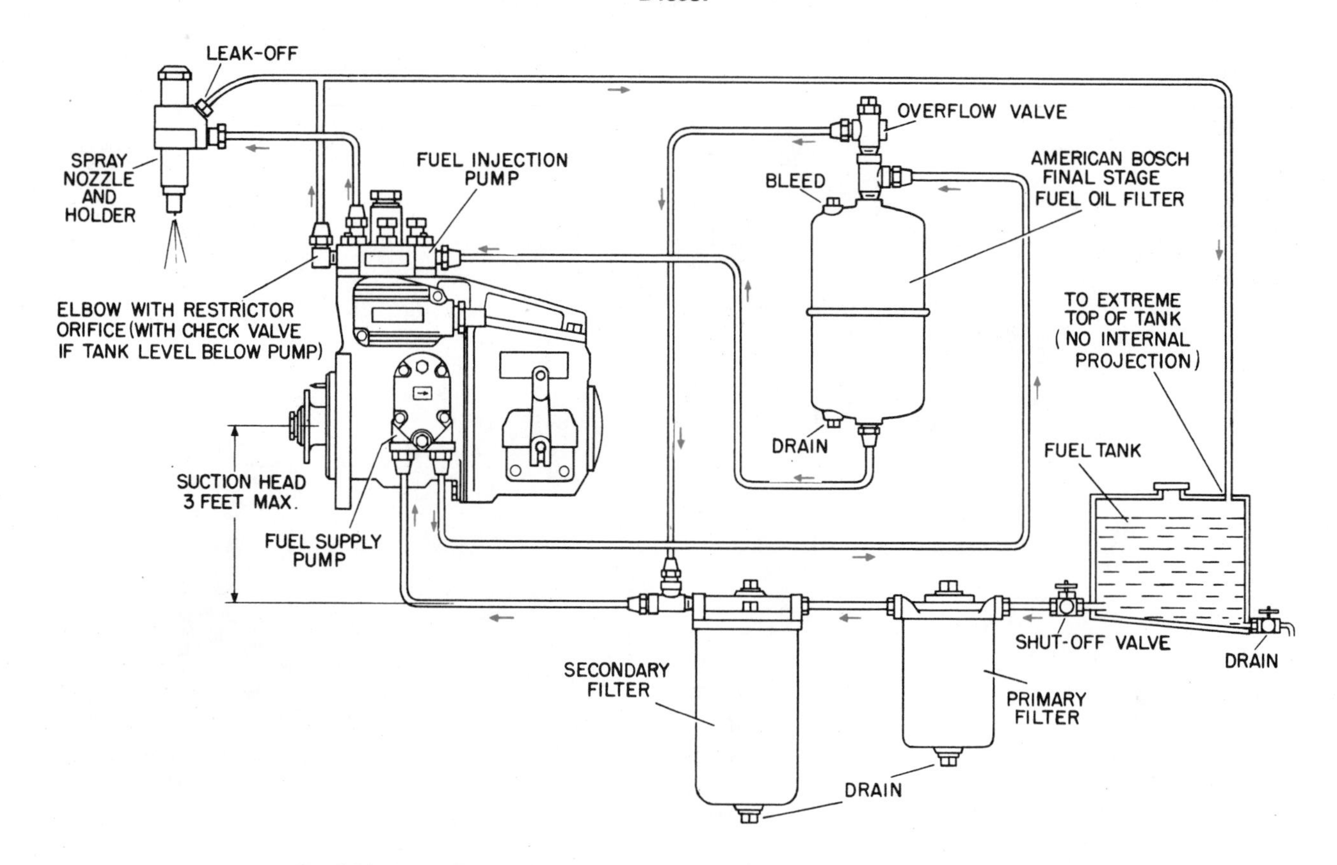

Fig. 8-26. Typical layout of injection system using type PSB fuel injection pump. (American Bosch Arma Corp.)

injection pump is as follows:

The complete system with fuel supply tank and filter is shown in Fig. 8-26. Fuel enters the fuel injection pump from the supply system, and fills that portion of the barrel cavity between the top of the plunger and the bottom of the delivery valve, when the plunger is at the bottom of its stroke, Fig. 8-27. As the rotating plunger moves upward in its stroke under cam action, it passes and closes the intake port, compressing the fuel and opening the spring-loaded delivery valve, as in B, Fig. 8-27. As the plunger continues its upward stroke, the fuel is forced through the delivery valve and is conveyed to the annulus in the plunger. The fuel then passes through the vertical distributing slot on the plunger to the outlet ducts with which the distributor slot is then registering, as shown at C, Fig. 8-27.

As the plunger continues its upward movement, its lower annulus passes the edge of the control sleeve and the fuel under pressure escapes down the vertical hole into the center of the plunger and into the sump surrounding the central sleeve which is at supply pressure, as shown at D, Fig. 8-27.

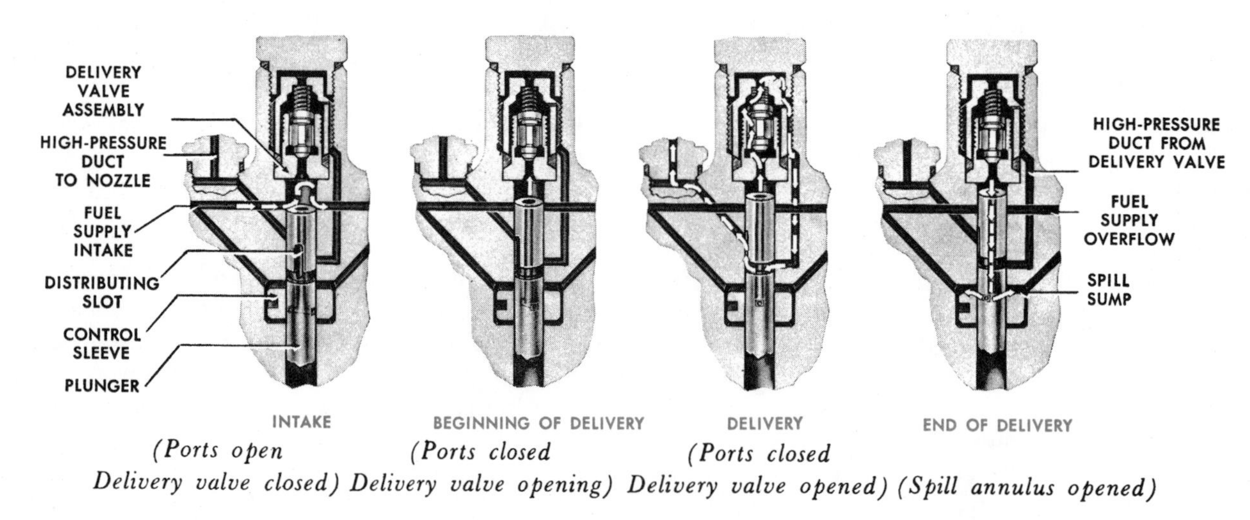

Fig. 8-27. Pumping sequence of the American Bosch type PSB fuel injection pump. (American Bosch Arma Corp.)

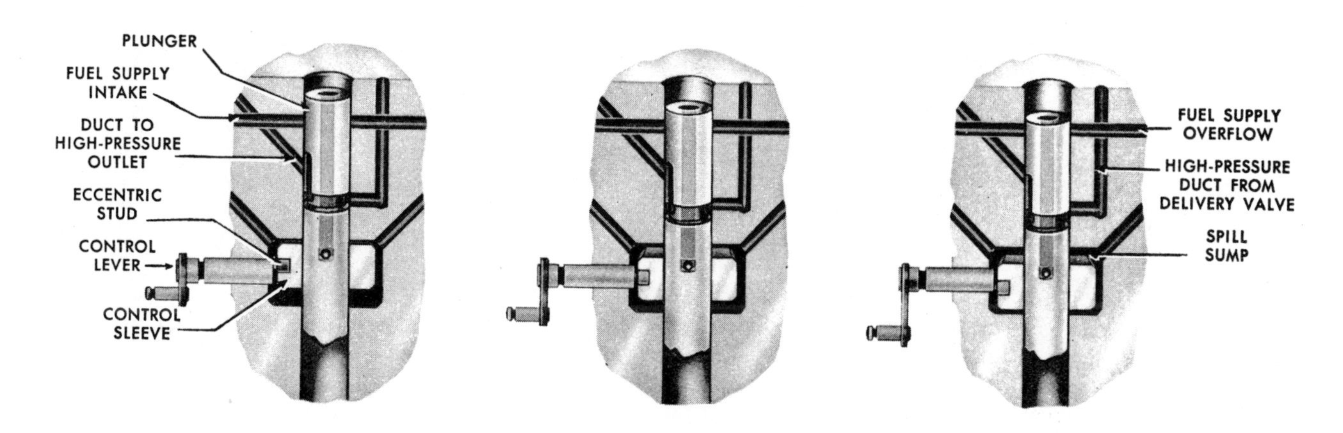

Fig. 8-28. Showing how quantity of fuel is controlled in PSB pump. (American Bosch Arma Corp.)

With the collapse of the pressure beneath it, the delivery valve closes. At the same time the piston portion of the delivery valve blocks the passage before the valve reaches its peak and thus reduces the pressure in the discharge system.

The quantity of fuel delivered per stroke is governed by variation of the position of the control sleeve in relation to the fixed port closing portion, Fig. 8-28. The closing position is the point at which the top of the plunger covers the intake port. As the spill annulus on the plunger breaks over the top edge of the control sleeve, pumping pressure is relieved down to the center hole of the plunger into the sump surrounding the control sleeve and full delivery terminates despite the continued upward movement of the plunger.

If the control sleeve is raised, the spill annulus on the plunger remains covered by the sleeve until relatively late in the plunger stroke. As a result, the effective stroke of the plunger is longer and more fuel is delivered, Fig. 8-28. If the control sleeve is lowered, the spill annulus on the plunger is uncovered by the sleeve relatively sooner, on the plunger stroke. Hence the effective stroke of the plunger is shorter and less fuel is delivered, as shown at B, Fig. 8-28.

When the control sleeve is lowered to its extreme point, the spill annulus on the plunger is uncovered by the top edge of the sleeve before the upper end of the plunger can cover the intake ports. Under that condition no pressure can be built up and no fuel can be delivered.

The delivery valve, Fig. 8-25, situated directly above the pumping chamber assists in metering the fuel, by preventing the irregular loss of fuel from the delivery side of the system to the supply side between pumping strokes.

INTERNAL SPRING GOVERNOR: The governor is of the variable speed mechanical centrifugal type with the weight assembly attached to the camshaft, Fig. 8-25. The governor action is accomplished through the flyweights acting against the movable sleeve backed up by springs floated in the opposite direction.

EXTERNAL SPRING GOVERNOR: This governor is designed for use on the American Bosch fuel injector when simplicity is the prime factor and neither close regulation nor adjustability of torque buildup are required.

This governor is of the mechanical centrifugal type and is attached to the fuel injector pump as an integral unit. It is driven on an extension of the pump camshaft. The governor action is accomplished through the flyweights acting against the movable sleeve which is loaded in the opposite direction by a spring-loaded fulcrum lever.

FUEL SUPPLY PUMP: The fuel supply pump designed for use for the American Bosch type PSB fuel injection pump is of the positive displacement gear type. This pump is mounted on the pad on the front side of the injection pump housing and is driven by a gear meshing with a gear on the camshaft. In some cases a separately mounted and driven diaphragm type pump may be used. Applied pressure at full load must be 6 to 8 psi.

MODEL 100 INJECTION PUMP

American Bosch model 100 fuel injection pump, Fig. 8-29, is a high-speed (up to 3200 rpm), variable timing, governor controlled, flange-mounted, high-pressure, single plunger injection pump. It is designed for over-the-road and off-highway vehicles. Model 100 is also used on marine and industrial applications. It is a multiple outlet unit.

The replaceable hydraulic head, Fig. 8-30, contains a delivery valve and a plunger which, in addition to being actuated axially by a multi-lobe cam, is continuously rotated to serve as a fuel distributor.

Fuel variation between cylinders does not need to be adjusted. Therefore, the only adjustments are for average fuel deliveries. Changes in fuel delivery are controlled by the vertical movement of the plunger metering sleeve. This sleeve is actuated by the control unit which, in turn, is operated by the control rod.

A centrifugal, mechanical type governor actuates the control rod. The governor controls the idle speed, maximum no-load speed and fuel delivery throughout the operating speed range for any given throttle position.

The fuel supply pump draws fuel from the supply tank, through a primary filter and supplies the fuel through a final stage filter to the hydraulic head, sump area, Figs. 8-29 and 8-30. Fuel pressure in the head sump area is controlled by the overflow valve assembly. The fuel supply pump contains an integral pressure relief valve which prevents fuel system damage in the event of downstream restriction.

An internal timing device (Intravance[R]) automatically advances or retards the beginning of fuel injection as engine speed requires, Fig. 8-21. Also, an internal excess fuel starting device provides increased fuel delivery at cranking speed.

BENDIX FUEL INJECTION SYSTEMS

The Bendix F series pumps, Fig. 8-31, are of the jerk pump type and utilize the principle of high-pressure metering by a helical edge on the plunger surface. Fuel quantity is controlled by altering the position of the lower plunger helix relative to the spill port. This is accomplished by rotating the pump plunger by means of a rack and gear sleeve, which engages the plunger. Timing of the ignition is controlled by the addition of an upper helix. The construction and operation of the Bendix Type F series pump is as follows: The plunger spring, Fig. 8-31, holds the lower spring plate, which in turn holds the plunger against the plunger follower. The plunger follower rides the engine tappet. The engine tappet moves the plunger up and is returned by the plunger spring, thus completing the pumping cycle. The barrel is stationary. The pump works on a constant stroke, variable output principle. The plunger stroke

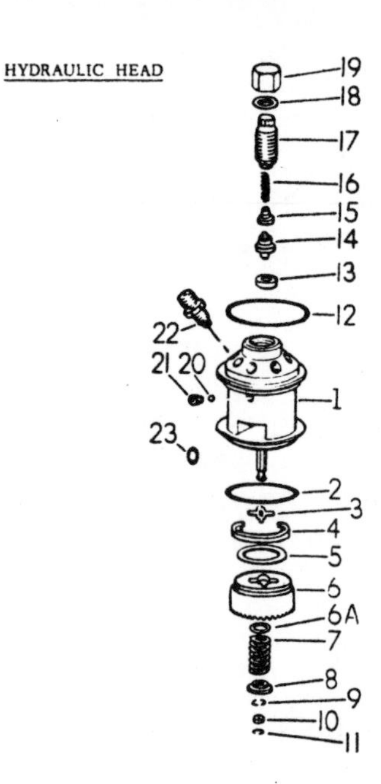

Fig. 8-30. Details of hydraulic head used with model 100 American Bosch injection pump. 1–Hydraulic head. 2–Gasket. 3–Plunger guide. 4–Gear retainer. 5–Thrust washer. 6–Face gear. 7–Spring. 8–Spring seat. 9–Retaining ring. 10–Plunger button. 11–Retaining ring. 12–O-ring. 13–Valve spacer. 14–Delivery valve assembly. 15–Valve spring guide. 16–Spring. 17–Delivery valve holder. 18–Gasket. 19–Cap nut. 20–Sealing ball. 21–Set screw. 22–Discharge fitting. 23–Spacer.

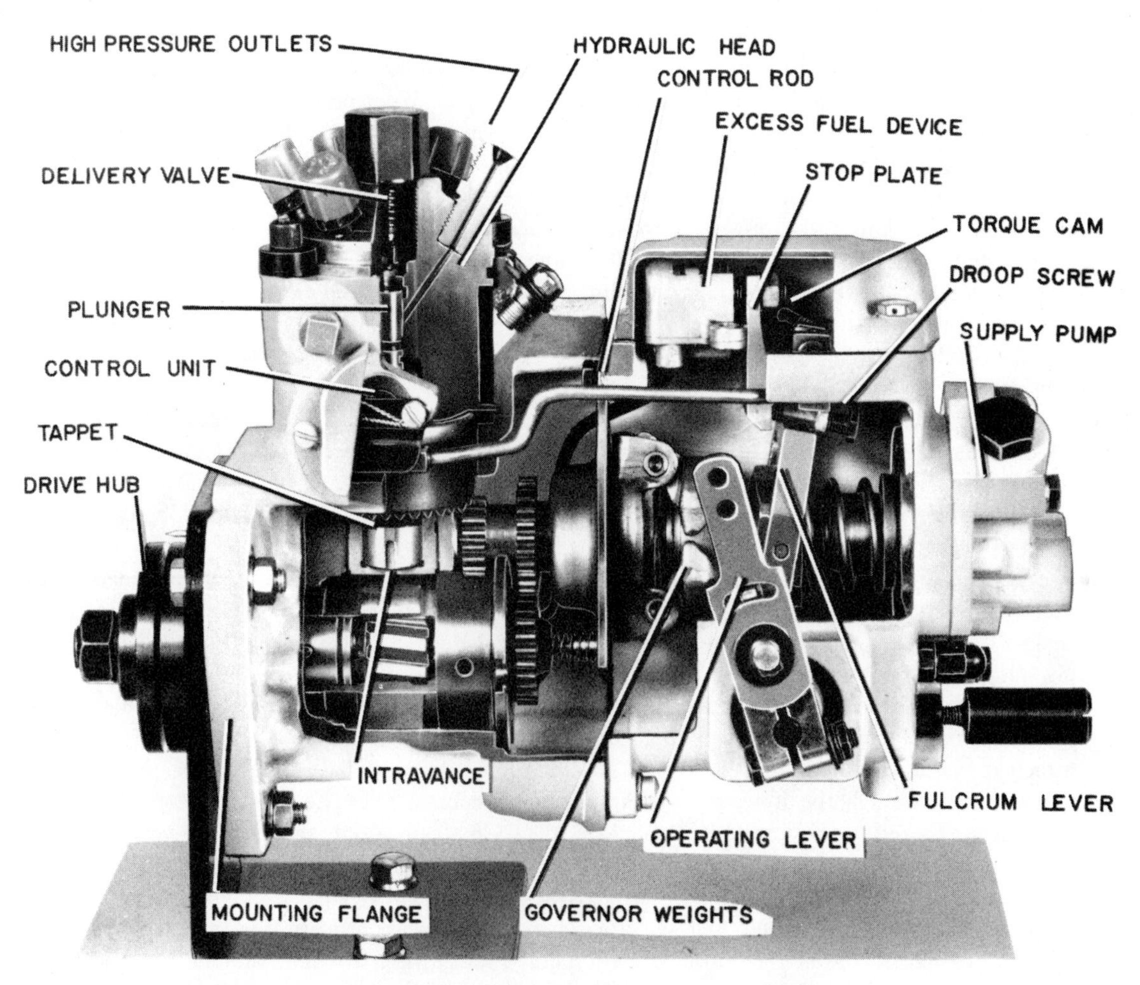

Fig. 8-29. Sectional view of American Bosch model 100 injection pump. (American Bosch Arma Corp.)

is constant in that it travels the same distance at all engine speeds. To vary fuel output of the pump, the plunger is rotated by means of its crossbar, the control sleeve, the control rack and throttle or governor linkages.

To pump fuel at high pressure it is necessary to bring it into a pressure chamber through an inlet, close the inlet, apply pressure for injection, terminate injection pressure and reopen inlet to admit more fuel. This fuel injection cycle is accomplished partially by location of inlet and spill ports in the barrel. It is further accomplished by a metering helix and a passage in the plunger that extends from end of plunger to the metering helix and side of plunger, Fig. 8-32. The metering helix, Fig. 8-32, is a sharp edge upper spiraling recess in the side of the plunger. Because of the form of the helix, rotation of the plunger will affect the amount of plunger rise required for it to uncover a spill port. The effective stroke then is the distance between the upper edge of the helix and the lower edge of spill port at the instant inlet ports close, Fig. 8-33. The rotation of the plunger and helix determines the duration of the injection by varying the distance the plunger must travel on the pumping stroke before the metering helix spills the fuel pressure. The complete pumping cycle is shown in Fig. 8-34. At A, in Fig. 8-34, the plunger is shown in its lowest position. The delivery valve is closed. The inlet ports are open. Because of transfer pump pressure, fuel enters the pressure chamber through these ports. This assumes that all air was bled from the pump during installation.

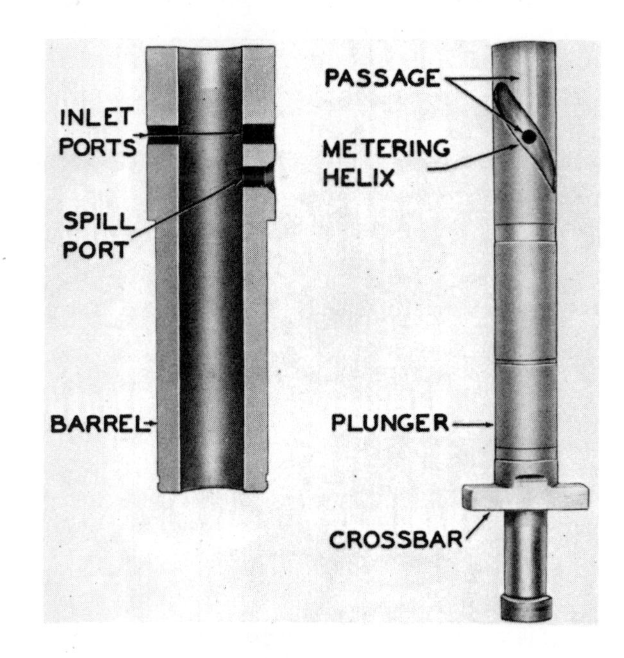

Fig. 8-32. Details of metering helix of Bendix series F injection pump. (Bendix Electrical Components Div.)

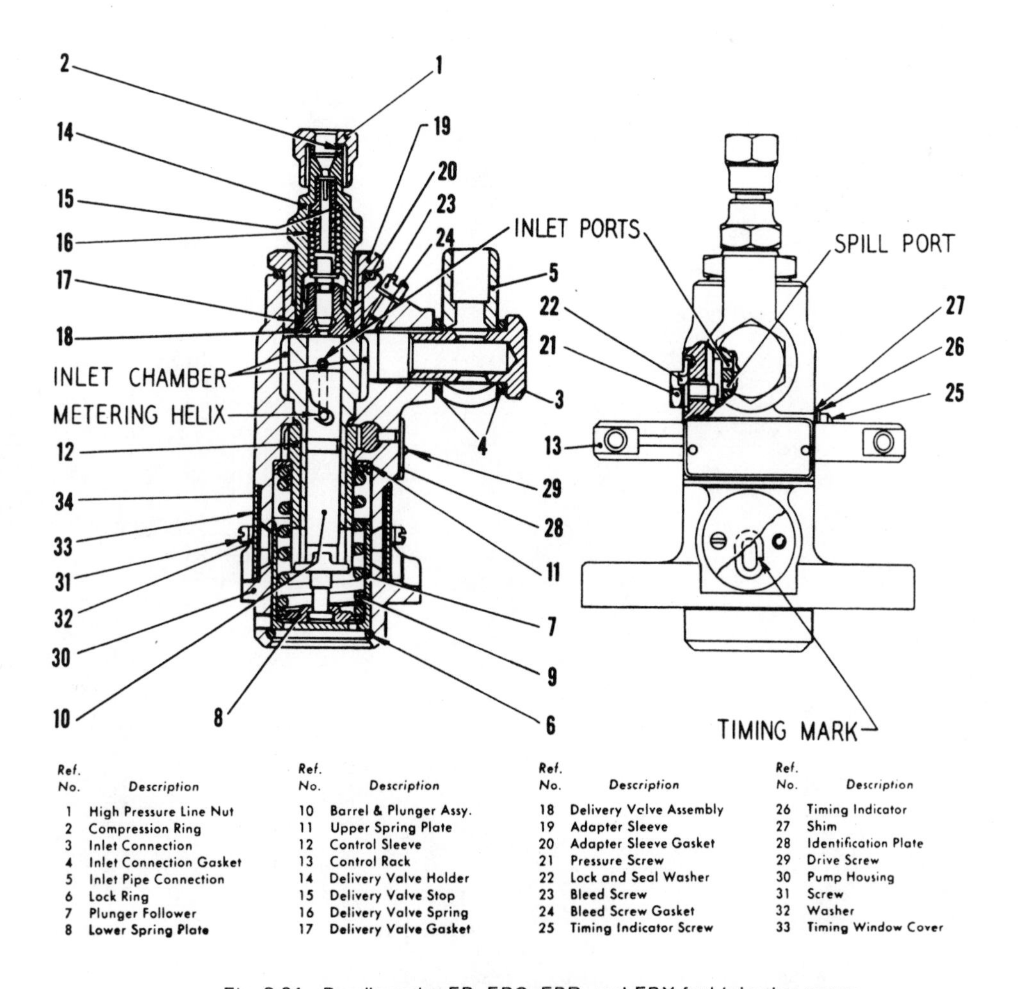

Ref. No.	Description	Ref. No.	Description	Ref. No.	Description	Ref. No.	Description
1	High Pressure Line Nut	10	Barrel & Plunger Assy.	18	Delivery Vclve Assembly	26	Timing Indicator
2	Compression Ring	11	Upper Spring Plate	19	Adapter Sleeve	27	Shim
3	Inlet Connection	12	Control Sleeve	20	Adapter Sleeve Gasket	28	Identification Plate
4	Inlet Connection Gasket	13	Control Rack	21	Pressure Screw	29	Drive Screw
5	Inlet Pipe Connection	14	Delivery Valve Holder	22	Lock and Seal Washer	30	Pump Housing
6	Lock Ring	15	Delivery Valve Stop	23	Bleed Screw	31	Screw
7	Plunger Follower	16	Delivery Valve Spring	24	Bleed Screw Gasket	32	Washer
8	Lower Spring Plate	17	Delivery Valve Gasket	25	Timing Indicator Screw	33	Timing Window Cover

Fig. 8-31. Bendix series FB, FBC, FBR, and FBX fuel injection pump. (Bendix Electrical Components Div.)

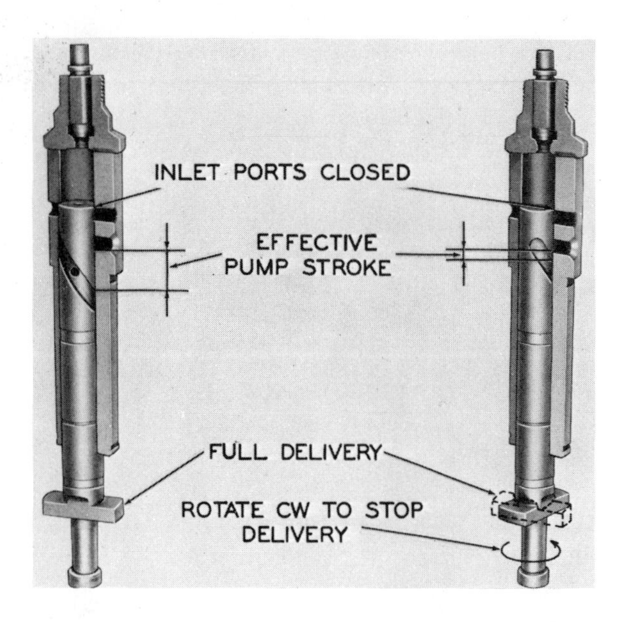

Fig. 8-33. Showing how plunger rotation changes effective stroke of Bendix series F injection pump. Rack setting is constant. (Bendix Electrical Components Div.)

In B, Fig. 8-34, the plunger has risen to where its end has closed the inlet ports. The pressure chamber is sealed. Pressure starts to build up and the delivery valve opens. High-pressure fuel is forced through the delivery valve and the high-pressure line. Injection of fuel into the engine begins the instant line pressure exceeds the pressure setting of the nozzle holder assembly.

Fuel injection will continue until the upper edge of the metering helix reaches the lower edge of spill port, as shown at C, Fig. 8-34. At this instant the high-pressure fuel in the pressure chamber, plunger center passage, and helix is released into inlet chamber. This stops injection because high-pressure line pressure drops and this allows the nozzle valve to close. The delivery valve assembly then closes. From this point to the top of the plunger stroke, the fuel in the pressure chamber spills into the inlet chamber. This means that the effective

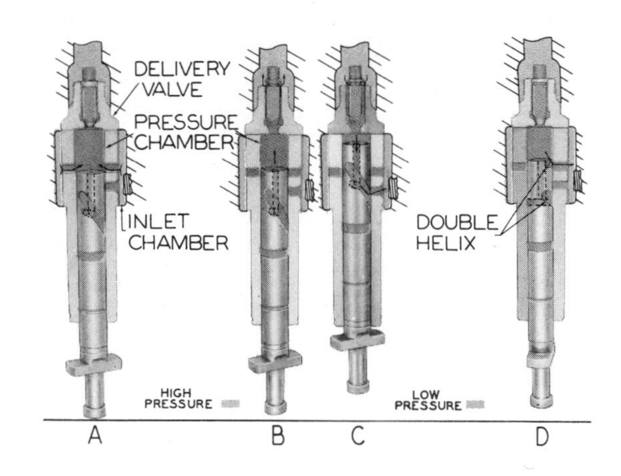

Fig. 8-34. Complete pumping cycle of Bendix FB pump. (Bendix Electrical Components Div.)

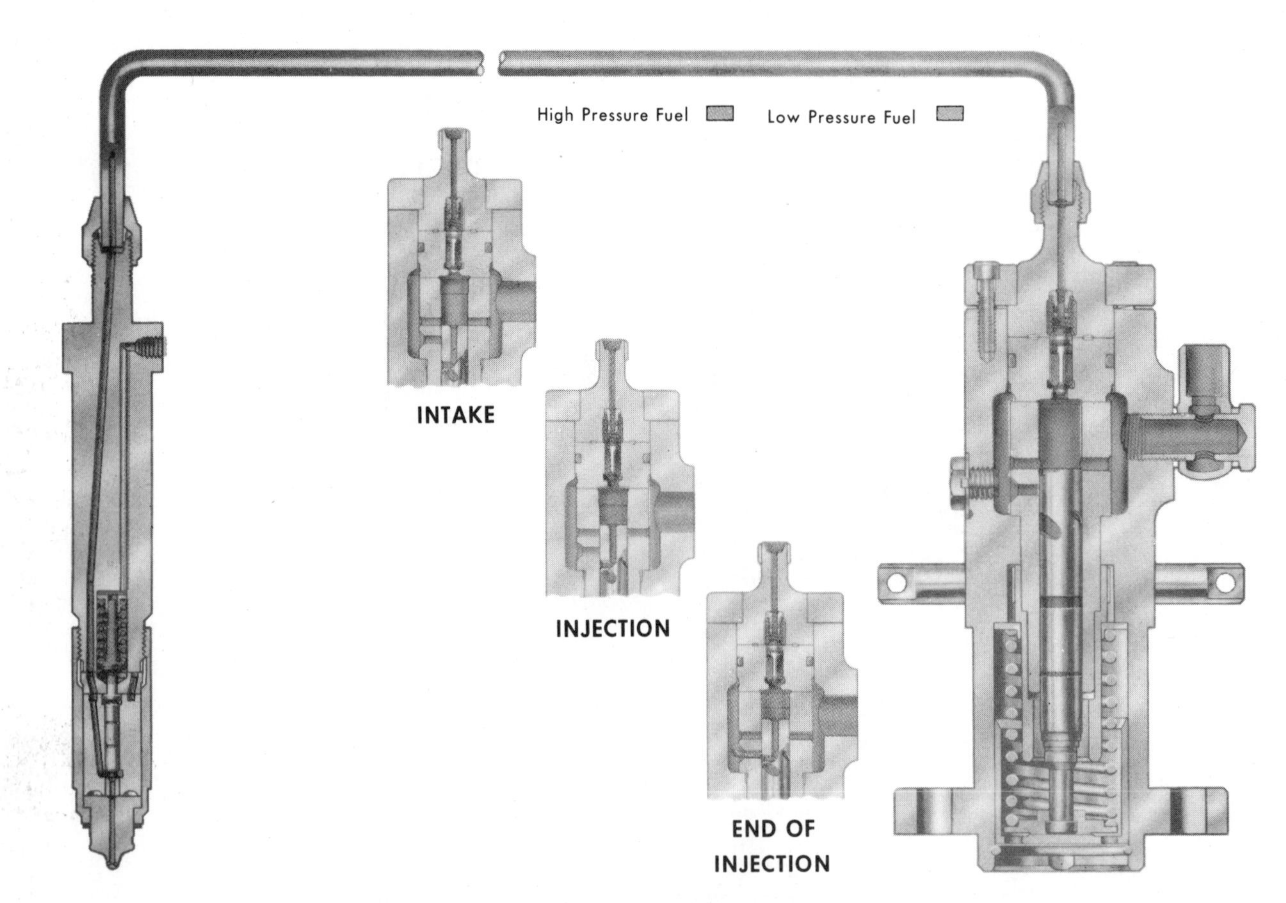

Fig. 8-35. High and low-pressure areas at various pumping stages. (Bendix Electrical Components Div.)

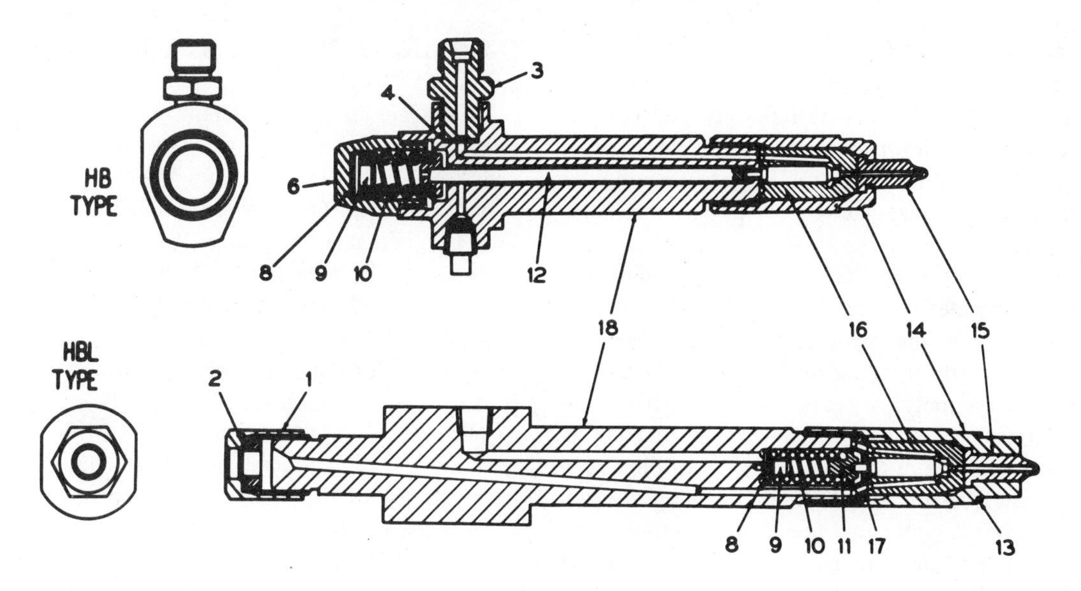

1. HIGH-PRESSURE LINE NUT.
2. COMPRESSION RING.
3. INLET NIPPLE.
4. GASKET.
5. LEAKAGE RETURN FITTING.
6. SPRING CAP.
7. SPRING CAP O RING GASKET.
8. SHIM.
9. PRESSURE SPRING SEAT.
10. NOZZLE SPRING.
11. SPRING BUTTON.
12. PRESSURE PIN.
13. GASKET (HOLDER TO ENGINE).
14. ASSEMBLY NUT.
15. SPRAY TIP.
16. NOZZLE VALVE ASSEMBLY.
17. STOP PLATE.
18. BODY.

Fig. 8-36. Sectional view of typical HB and HBL nozzle holders. (Bendix Electrical Components Div.)

stroke or duration of injection is the distance the plunger travels from inlet port closure to spill port opening. It can now be seen that by rotating the plunger, the helix will either arrive at the spill port sooner or later, depending on which way it is rotated.

Some plungers have a double helix. In this case the regular helix functions as already described. The second helix is at the top of the plunger directly above the engine idle end of the regular helix. This is shown in D, Fig. 8-34. Note that although end of plunger is well above inlet port, the top helix retards port closing. This helix has various configurations to meet various engine requirements. Its function is to provide a retard in timing of fuel injection at low engine speed.

The closed delivery valve, Fig. 8-31, maintains a column of fuel in a line between the pump and nozzle holder assembly. Precision lapped surfaces keep the high fuel pressure within the barrel and delivery valve chamber. Fig. 8-35 shows a typical diesel fuel injection installation which gives a comprehensive view of the pumping cycle as related to the complete system.

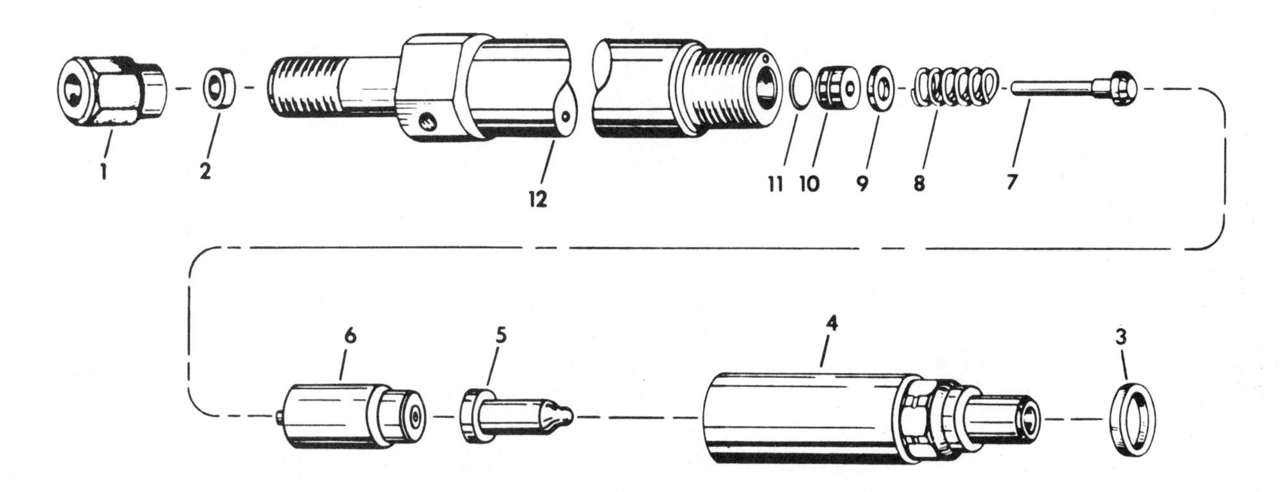

Fig. 8-37. Exploded view of Bendix H3LE nozzle and holder. 1—High pressure line nut. 2—High pressure compression spring. 3—Gasket holder. 4—Assembly nut. 5—Spray tip. 6—Nozzle valve. 7—Nozzle stop rod. 8—Nozzle spring. 9—Shim. 10—Needle stop guide. 11—Plate needle stop. 12—Body. (Bendix Electrical Components Div.)

BENDIX NOZZLE HOLDER

The general construction of the technical Bendix HB and HBL nozzle holder is as shown in Fig. 8-36. On this type of holder the high-pressure line connection is part of the holder body. Note they do not have inlet nipple parts nor bleeder screw parts.

The high-pressure line from the pump is attached to the nozzle holder body by means of a nut or inlet nipple. A drilled fuel passage extends from the high-pressure line entrance to a circular groove in the stop plate and/or to the circular groove in the nozzle valve assembly. A needle valve is seated by the nozzle spring pressure which is transferred by the spring button or pressure pin. Shim thickness determines the spring pressure. When fuel pressure builds up to opening pressure setting, it lifts the needle valve, and fuel is discharged through the spray tip into the engine cylinder. A small amount of fuel oil leaks past the needle, up through the needle well and to the spring well so all moving parts are lubricated. The oil then passes through a drilled passage to the leakage return fitting near the top of the body.

The body, stop plate, nozzle valve assembly and spray tip all have lapped (finally finished) surfaces where they join each other. These lapped surfaces are critical to the performance of the injector because they are the only means of containing the high fuel pressures. Details of Bendix type H3LE nozzle and holder are shown in Fig. 8-37.

BENDIX TYPE UO-G UNIT INJECTOR

The Bendix type UO-G unit injector, Fig. 8-38, is designed for use in small diesel engines and employs port and helix metering with a constant beginning and regulated end of injection. The plunger size ranges from 3 to 5.5 mm. in diameter. The operating stroke is 4.0 mm. A control rack and sleeve are provided to rotate the plunger and regulate the amount of fuel delivered.

CAV/LUCAS INJECTION SYSTEMS

There are several different types of CAV/Lucas fuel injection systems. The Minimec fuel injection pump, Figs. 8-39 and 8-40, is an in-line pump available in 2,3,4,5,6 and 8 cylinder models. Outputs are suitable for engines up to 1.5 liters (91.5 cu. in.) per cylinder.

A larger injection pump of this same type, called Majormec, is designed for use on diesel engines from 1.5 to 3.5 litres per cylinder. This pump is available in 4, 6 and 8 cylinder models. Two of the 6 cylinder models can be coupled for use on 12 cylinder engines. Also see page 78.

Minimec is a self-contained unit with a camshaft and governor assembly operating separate pumping elements for each cylinder of the engine. The pump casing consists of two main parts, an alloy cam box and governor housing, and a steel pump body. Located on an extension of the camshaft at the front of the pump is the governor which consists of a governor weight assembly located between the flanges of a channel sectioned back plate attached to the camshaft. When the camshaft revolves, the flanges of the back plate cause the weight assembly to revolve also.

The operation and construction of all the pumping elements is identical. Each plunger is operated by a separate cam on the camshaft. The cams are equally spaced on the camshaft in the engine firing order sequence. A pumping element is shown in the various stages of its operating cycle in Fig. 8-41.

At A the plunger is at the bottom of its stroke, and fuel under lift pump pressure fills the pumping element via the two ports in the barrel. At this stage the pressure in the barrel is not sufficient to lift the delivery valve off its seating against the action of the spring.

As the camshaft rotates, the plunger rises until position B is reached. At this point fuel can no longer enter or leave the element via the port in the barrel and further upward movement of the plunger compresses the fuel and begins to lift the delivery valve off its seat.

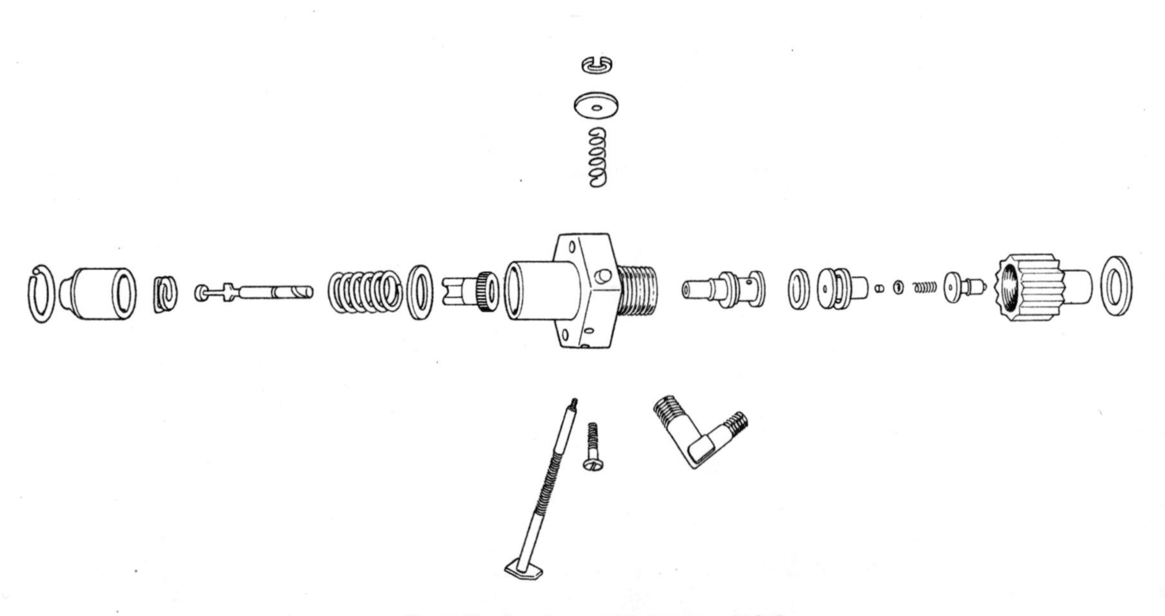

Fig. 8-38. Bendix unit injector type UO-G.

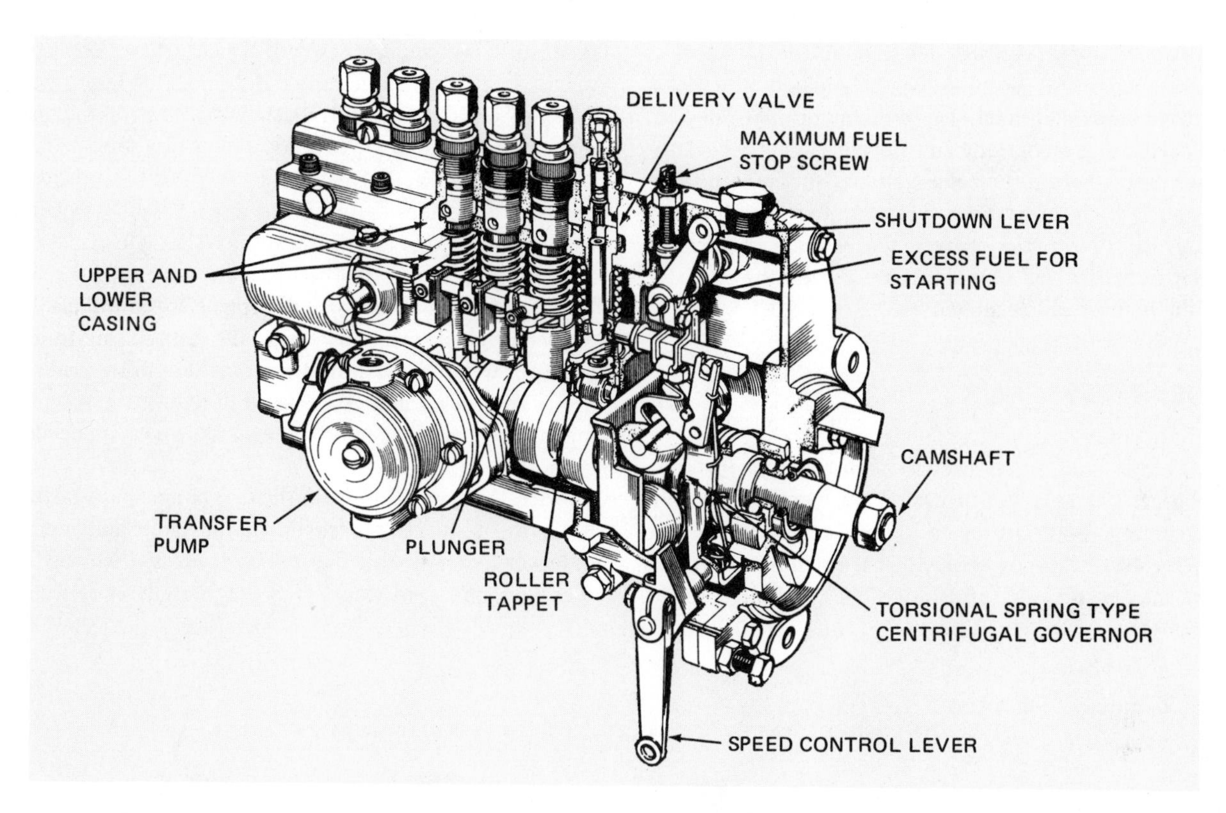

Fig. 8-39. Details of Minimec fuel injection pump. (CAV/Joseph Lucas, North America, Inc.)

When the fuel pressure is sufficient to lift the delivery valve completely off its seat and the piston clear of its guide position C, fuel passes along the pipe line to the injector. The fuel pressure developed by the plunger lifts the injector needle valve off its seat against the action of the spring and allows fuel in a highly atomized state to be sprayed into the cylinder.

Fuel continues to be injected until the plunger reaches position D when the upper edge of the helical groove has uncovered the lower edge of the spill port and allows fuel under high pressure to leak back down the center drilling in the plunger and out to the helical groove into the fuel galleries of the pump body. This reduces the fuel pressure in the barrel and the delivery valve spring assisted by the high pressure remaining in the pipe line causes the valve to close rapidly.

The piston entering its guide causes a slight but sharp reduction in the pressure of the fuel in the pipe line. The effect can be more clearly understood if it is compared with that of withdrawing its piston from its cylinder. As the piston is withdrawn the volume of the cylinder increases. If fuel or air, etc. are not allowed to enter or leave the cylinder, a reduction in pressure must result. The delivery valve also acts as a nonreturn valve and maintains pressure in the injector pipeline.

The sudden, if slight, reduction in pipeline pressure is sufficient to allow the injector needle valve to snap shut under the force of its spring. This prevents fuel dribbling from the injector, which would result in carbon buildup on the injector tip and low fuel delivery.

The plunger continues to rise completing its stroke, although the helical groove in the plunger prevents further fuel being delivered. The cam holds the plunger at the top of its stroke and only returns to the bottom when the corresponding engine cylinder is on the compression stroke again. This prevents the engine from running in the reverse direction in the event of a backfire.

To modify fuel delivery when the engine is operating at low speeds, a flat is ground on the piston portion of the delivery valve guide.

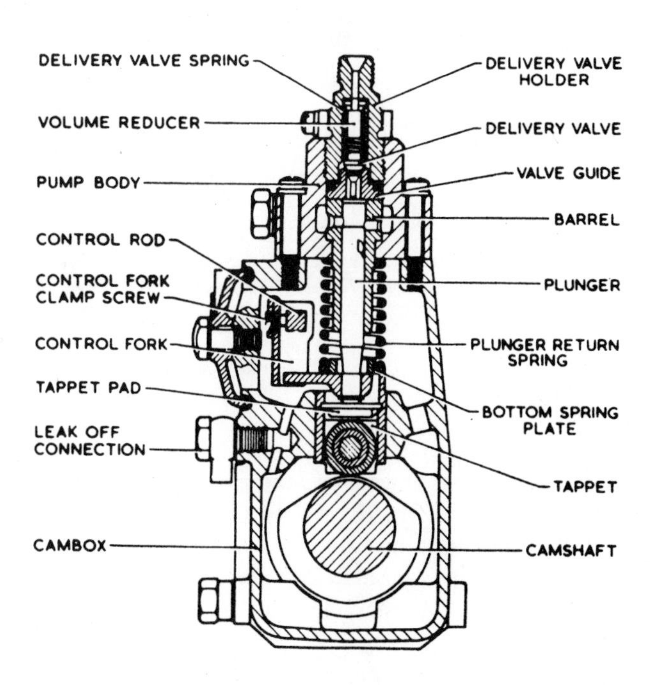

Fig. 8-40. Sectional view of CAV/Lucas Minimec injection pump.

The commencement of fuel delivery to the injectors is constant and occurs at position B. Immediately the rising plunger covers the inlet port. The amount of fuel injected, however, need not be constant and depends on the distance the plunger moves before the helical groove uncovers the spill port position D when pressurized fuel may flow back into the pump body. By rotating the plunger, the position of a helical groove relative to the spill port and consequently the effective stroke of the plunger can be altered.

CAV/LUCAS GOVERNOR

The governor used in connection with the fuel injection pump, Fig. 8-39, is a centrifugal type governor and is illustrated in Fig. 8-42. Its operation is as follows: As the camshaft revolves, the governor weight assembly is carried around by the backplate which is dowelled and bolted to the camshaft. Centrifugal force causes the governor weights to be thrown outward, but the angled slots in the weight carrier restrict the movement and convert part of the force to act along the axis of the camshaft. This force tends to move the weight carrier and hub away from the backplate against the action of the governor spring.

With the control lever, Fig. 8-42, in the idling position, the "dumbell" shaped roller is at the top of the ramp, and the governor spring-load is small. Under these conditions the weights are thrown outward and the rocking lever operated by the hub on the camshaft pushes the control rod backward, reducing the volume of fuel injected until a steady idling speed is obtained. If the engine speed decreases for any reason, the centrifugal force of the governor weights will also decrease and will be overcome by the governor spring-load. The rocking lever will pull the control rod forward increasing the volume of fuel injected. Consequently the engine speed will also increase. As the engine speed increases, the centrifugal force of the governor weights will again overcome the spring-load and cause the rocking lever and control rod to reduce the volume of fuel injected. Engine speed fluctuations are dampened by the governor.

With the control lever in the maximum speed position, Fig. 8-42, the roller is moved down the ramp and increases the governor spring-load. The spring-load is then sufficient to overcome the centrifugal force of the governor weights and move the weight carrier and hub along the camshaft. This motion is transmitted to the control rod by the rocking lever, which pulls the control rod forward, increasing the volume of fuel injected and also the engine speed. The maximum fuel position is reached when the stop control rod contacts the maximum fuel stop bell crank which is prevented from rotating by the maximum fuel stop screwed into the top of the governor housing.

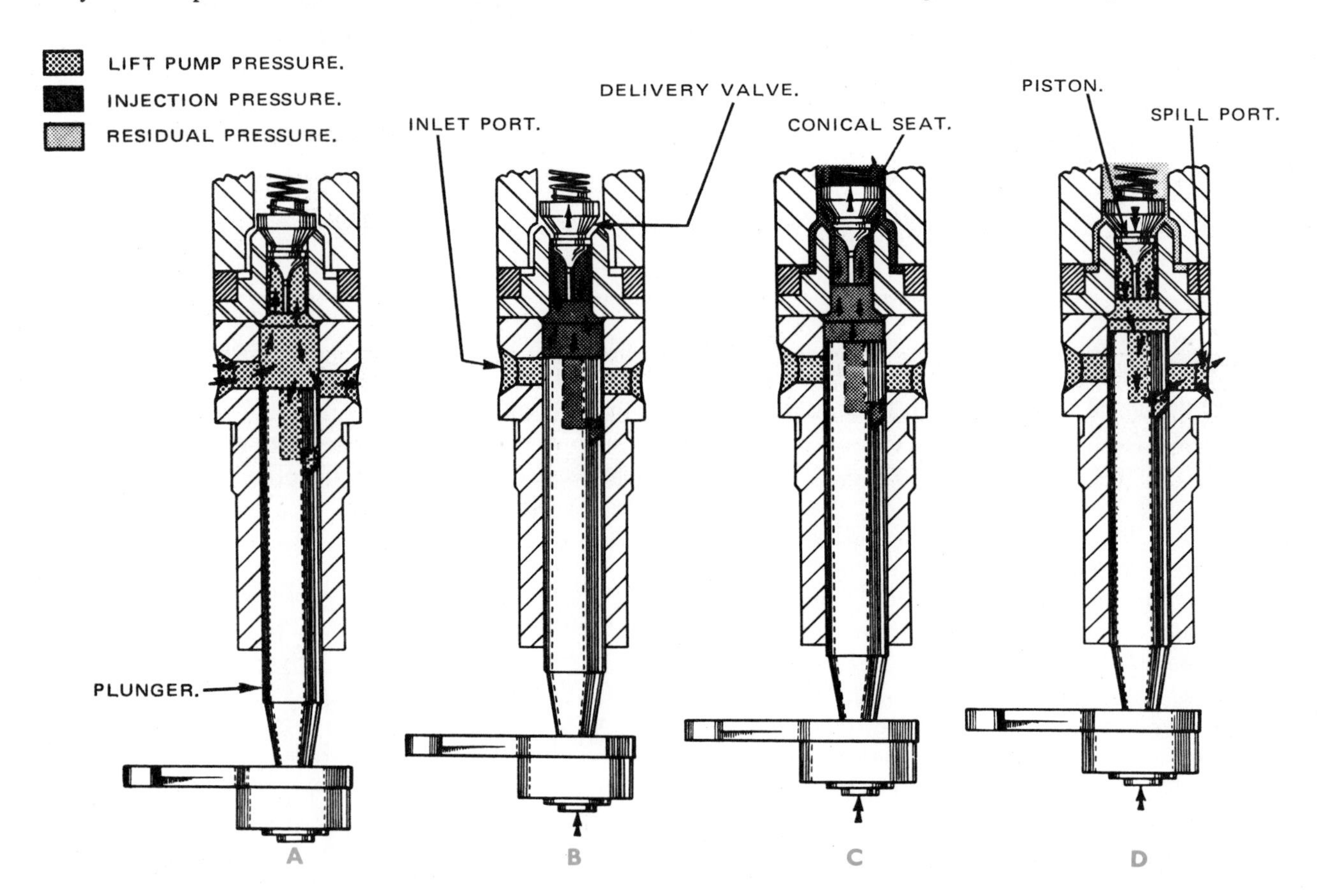

Fig. 8-41. Pumping element and delivery valve operation of Minimec fuel injection pump. (CAV/Joseph Lucas, North America, Inc.)

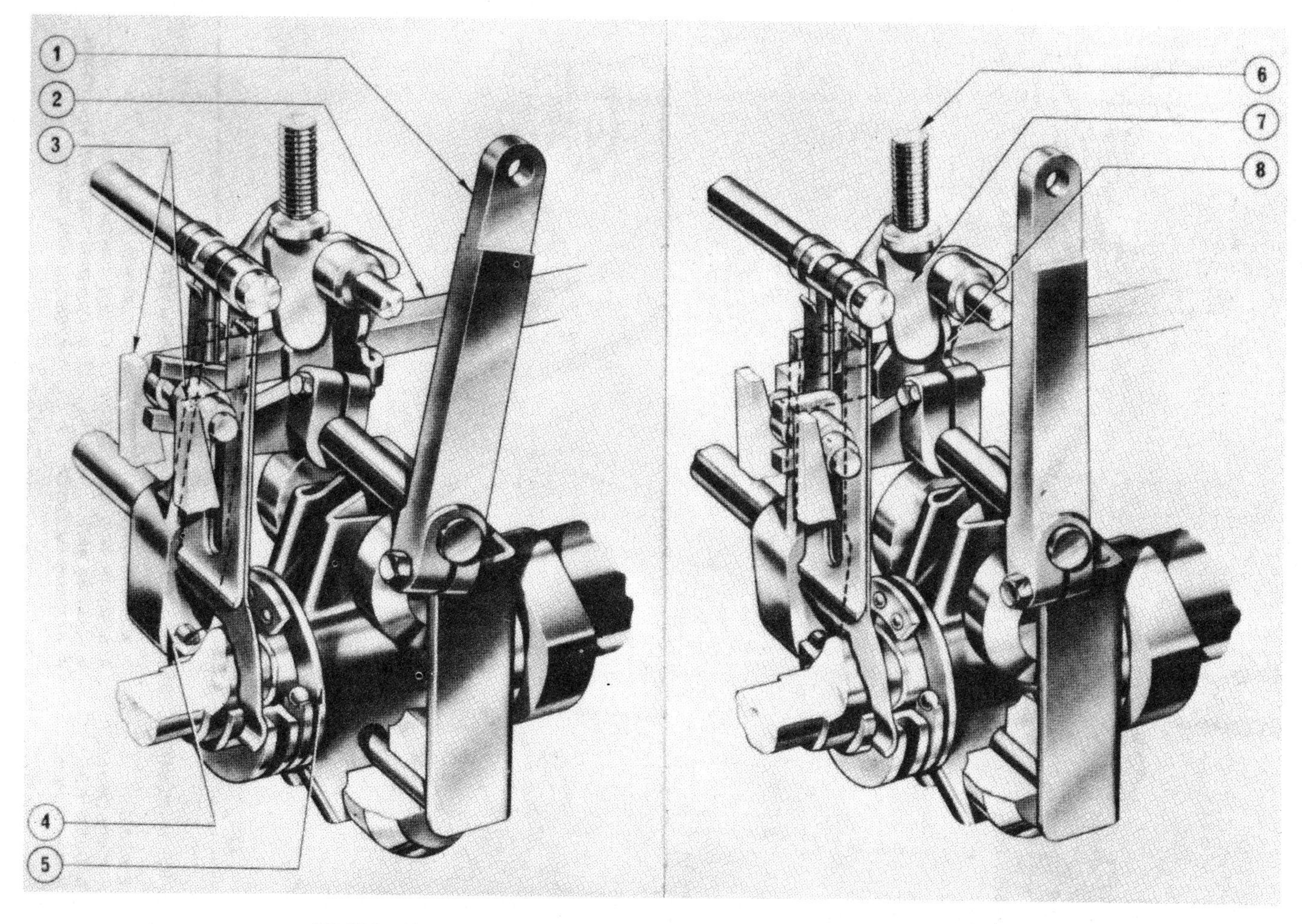

Fig. 8-42. Illustrating operation of governor. (CAV/Joseph Lucas, North America, Inc.)

1. CONTROL LEVER.
2. CONTROL ROD.
3. RAMP.
4. ROCKING LEVER.
5. HUB.
6. MAXIMUM STOP SCREW.
7. BELL CRANK.
8. CONTROL ROD STOP.

Should the engine speed continue to increase when the control rod has reached this position, the centrifugal force of the governor weights will overcome the spring-load and move the weight carrier and hub along the camshaft. The rocking lever will then push the control rod backward, reducing the volume of fuel delivered.

The engine speed is, at all times, proportional to the governor spring-load.

EXCESS FUEL DEVICE

An excess fuel device and stop control are incorporated in the top of the governor housing in some installations, Fig. 8-39. The excess fuel device provides the extra fuel required for starting. When the engine starts, its speed will rise until the centrifugal force of the governor weights overcome the governor spring force and cause the control rod to be pushed forward by the rocking lever. The spring will then return the maximum fuel stop bell crank to its original position.

The stop control consists of a sleeve, free to rotate on the excess fuel device spindle with a control lever on one end and an operating lever on the other. When the control lever is moved in a clockwise direction, the operating lever pushes the control rod forward by means of the stop, until the nodelivery position is reached. The engine will then stop.

The injector as used with the Minimec injector as installed on some Ford trucks, is shown in Fig. 8-43. This injector consists of a nozzle assembly, a nozzle holder assembly clamped together by the nozzle nut. The steel nozzle holder body incorporates lugs for clamping the injector into the cylinder head and contains a spring, spring seat, adjusting screw and spindle. The nozzle consists of a nozzle body with four spray holes and a needle valve.

Fuel from the injection pump enters the injector inlets adapter and passes through the injector body before reaching the needle valve seat, Fig. 8-43. The needle valve is held on its seat by the spring pressure acting through the spindle. The pressure of the fuel when the ports in the injection pump barrel are closed, causes the needle valve to open against the action of the injector spring. Fuel is then forced in a highly

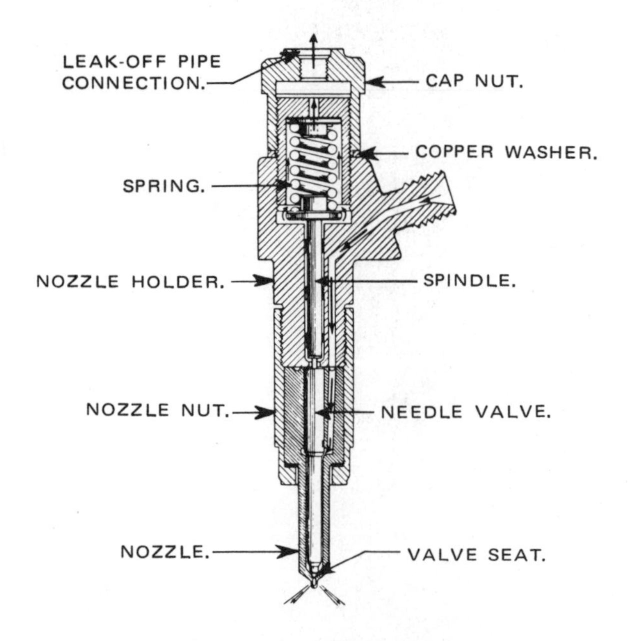

Fig. 8-43. Sectional view of CAV/Lucas Minimec injector.

atomized state through four holes in the nozzle tip. At the end of the injection the spring returns the needle valve to its seat. The spring pressure can be adjusted by the adjusting screw.

CAV/LUCAS INJECTION SYSTEMS

CAV/LUCAS produces both distributor type and plunger type fuel injection pumps. See page 74. Fig. 8-44 shows the distributor type DPA pump used on the Peugeot 504. The drive shaft, pumping and distributing rotor and sliding vein type transfer pump are an integral unit. The distributor is driven by the drive shaft, which couples the rotor to a drive hub located in the end of the pump housing.

The rotor, Fig. 8-45, consists of two parts, a pumping section and a distribution section. The latter is a close rotating fit in a stationary steel cylindrical body called the hydraulic head. The pumping section of the rotor is larger in diameter than the distributing section. It has a cylinder containing two opposed plain plungers. The plungers are operated by means of a stationary ring, two rollers and shoes which are carried in slots in the periphery (outside) of the rotor flange. Normally the cam ring has as many integral lobes as there are cylinders to be served.

As the rotor turns, one of the inlet ports in the rotor opens to the metering port in the head, Fig. 8-45, and fuel at metering pressure enters the rotor and separates the plungers. As the rotor continues to turn, the inlet port is closed and then with further rotation of the rotor, the distributor port

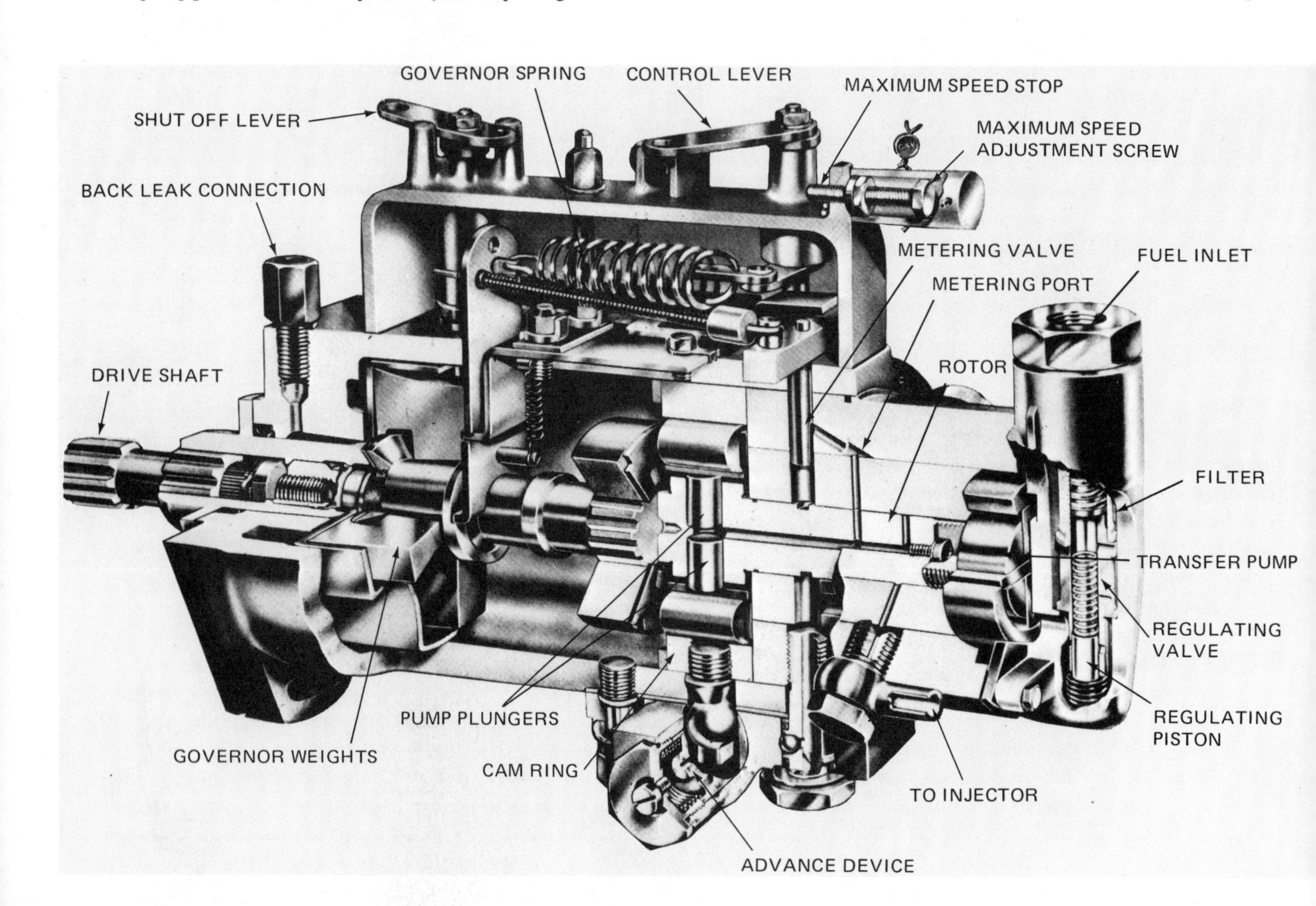

Fig. 8-44. Details of distributor type DPA injector pump with mechanical governor. (CAV/Joseph Lucas, Inc.)

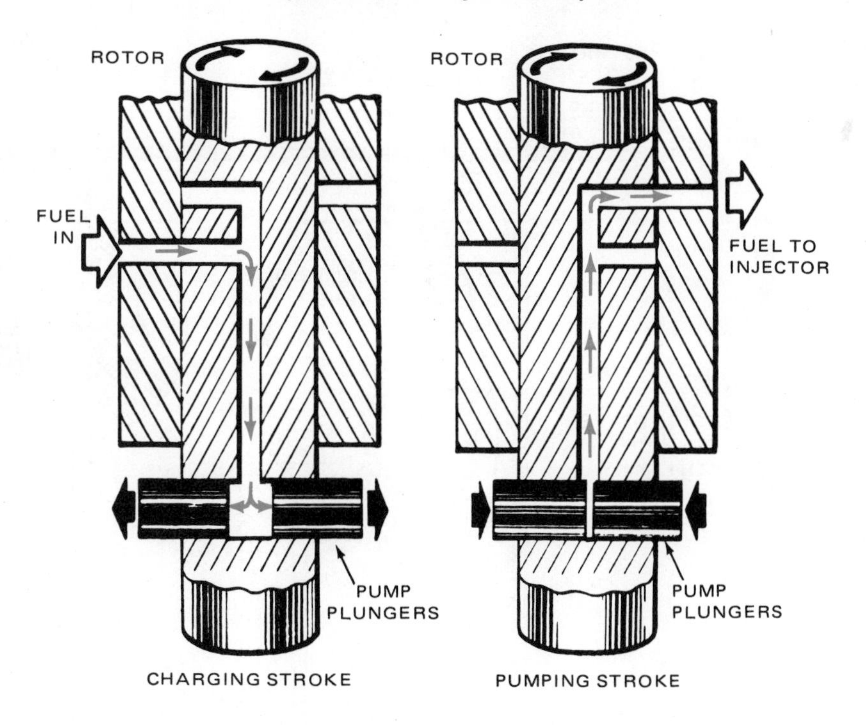

Fig. 8-45. The rotor of the CAV injector pump consists of two parts, a pumping section and a distributing section. (CAV/Joseph Lucas, Inc.)

aligns with one of the fuel outlets. Both rollers then contact opposing cam loads, Fig. 8-45, and the plungers are forced toward each other. This is the injection stroke. The fuels trapped between the plungers is forced back through the rotor and out to the injector.

Fuel displacement stops when the plungers reach the limit of inward travel imparted by the cam lobes. Shortly afterward the distributor port closes sealing off the line to the injector. As rotation of the rotor continues the cycle repeats itself.

CAV/LUCAS PRESSURE REGULATOR VALVE

The pressure regulating valve, Fig. 8-46, regulates transfer pressure, maintains the desired relationship between transfer pressure and the speed of pump rotation. In addition, it provides means of bypassing the transfer pump so the fuel passages in the hydraulic head can be primed when the pump is stationary. The regulating valve is shown in Fig. 8-44. Fig. 8-46 (left) shows a valve in static position. Since there is no

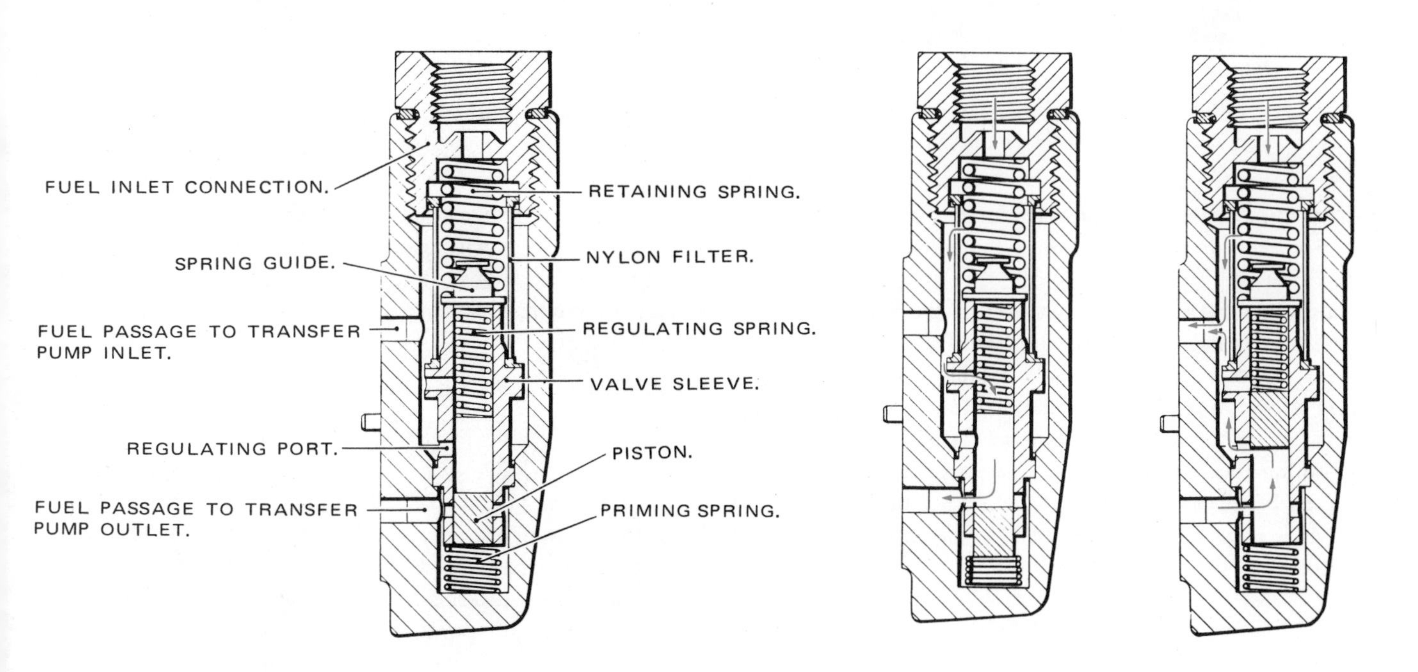

Fig. 8-46. Details of regulating valve of the CAV/Lucas injection pump.

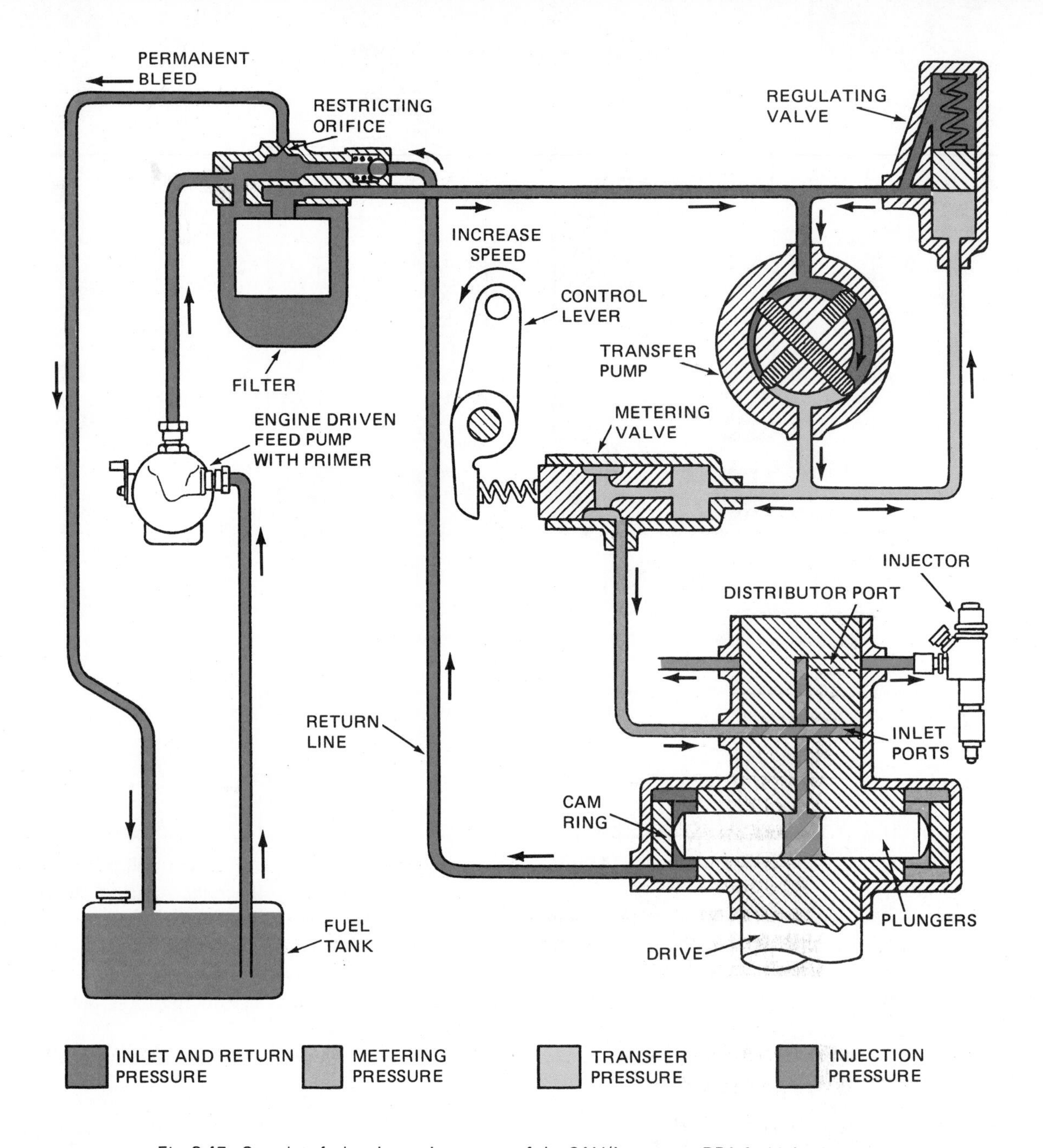

Fig. 8-47. Complete fuel and metering system of the CAV/Lucas type DPA fuel injection pump. (CAV/Joseph Lucas, Inc.)

fuel pressure within the end plate, neither spring is compressed. For hand priming the pressure caused by the hand primer across the transfer pump forces the piston valve down, compressing the spring and uncovering the primary port so that fuel bypasses the stationary transfer pump to fill the system by way of the hydraulic head. When the injection pump is operating pressure of the fuel from the transfer pump forces the piston valve in the sleeve back, Fig. 8-46 (right), uncovering the regulating port. The pressure on the piston is opposed by the regulating spring and a position of balance is reached. The delivery pressure of the transfer pump is controlled by the rating of the spring used.

CAV/LUCAS TRANSFER PUMP

The CAV/Lucas transfer pump, Figs. 8-44 and 8-47, is of the positive type. It has two vanes sliding inside an eccentric liner in the hydraulic head. The transfer pump rotor is carried in the end of the distributor rotor. The capacity of the transfer pump is considerably in excess of the requirements of the injection pump.

CAV/LUCAS METERING VALVE

A metering valve, Fig. 8-44, situated in the hydraulic head regulates the volume of fuel entering the rotor under the control of the governor, or of the hand throttle. The type of valve varies according to the type of governor fitted. With a hydraulic governor, a piston type valve is used. The valve is spring-loaded and controls the fuel according to its axial position. When a flyweight type governor is fitted, the valve is of a rotary type, with a slot cut in the periphery. The valve is rotated by the governor arm to regulate the entry of the fuel.

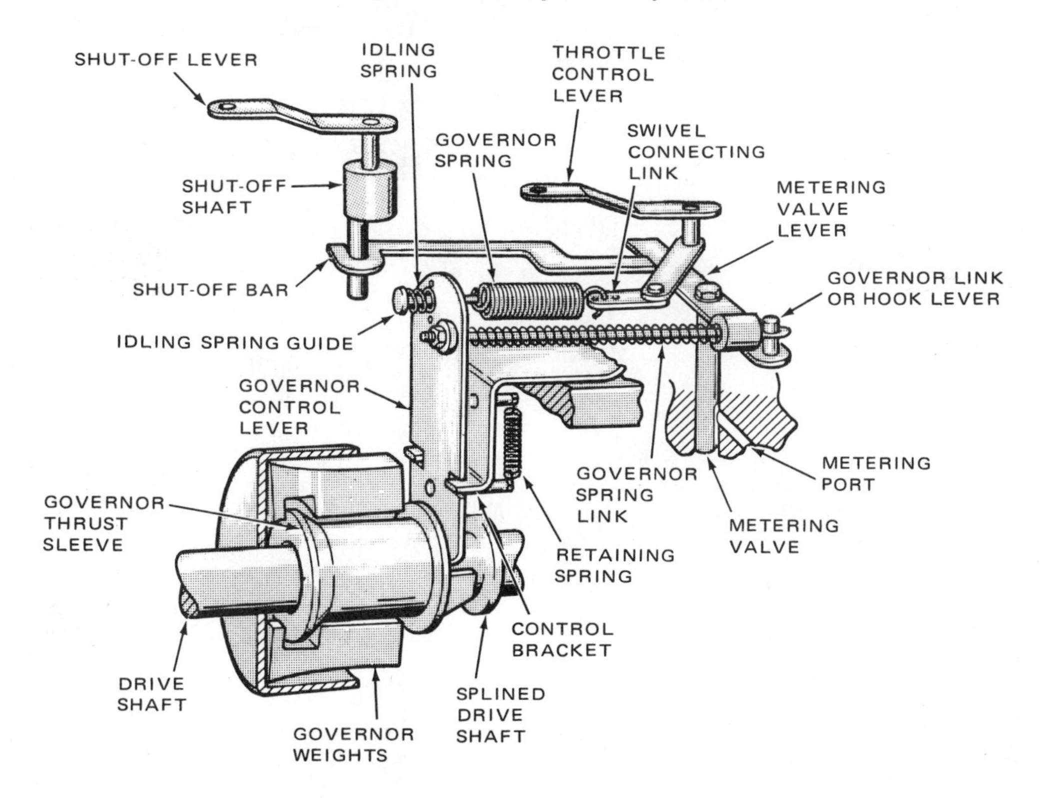

Fig. 8-48. Mechanical governor as used on the CAV/Lucas injection pump.

CAV/LUCAS FUEL METERING

Apart from small losses which occur during the injection stroke, the total volume of fuel introduced into the element is passed into the injector. Metering is effected by regulating the volume of fuel which enters the element at each charging stroke. The volume of the stroke is governed by two principal factors; the fuel pressure at the inlet port, and the time available for fuel to flow into the element while the inlet port in the rotor and the hydraulic head are in register. It is by controlling the pressure at the inlet port that metering is achieved.

Fuel oil enters the fuel injection pump at feed pressure and passes into the transfer pump, Fig. 8-47, which raises the pressure to what is known as transfer pressure.

Transfer pressure is proportional to engine speed and rises as the speed of rotation is increased. A predetermined relationship between transfer pressure and the speed of rotation is maintained by a regulating valve situated in the end plate of the pump.

Fuel at transfer pressure passes through the passage in the hydraulic head to the metering valve, which controls the flow of the fuel through the metering port. The effective area of the metering port is controlled by the metering valve, which is connected by suitable linkage to the accelerator pedal and the governor.

A pressure drop occurs as fuel passes through the metering orifice, reducing the fuel pressure to a level known as the metering pressure. The smaller the metering orifice, the greater will be the decrease in pressure and vice versa. Fuel at metering pressure passes into the inlet port to an obliquely (at a slant) drilled passage in the hydraulic head.

At idle speeds both transfer pressure and metering pressure are at their minimum values. The pressing of the accelerator moves the metering valve to a position where the effective area of the metering port is increased. This brings about an increase

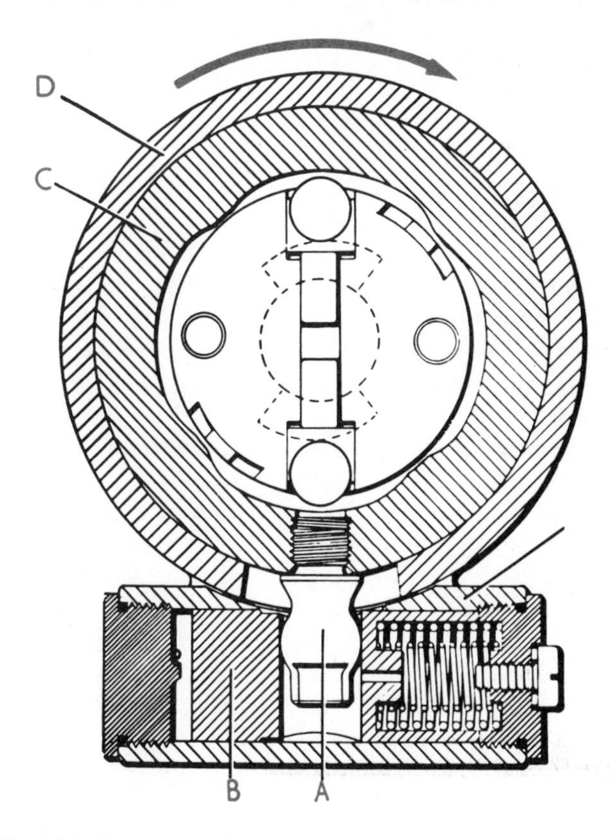

Fig. 8-49. Automatic speed advance mechanism on CAV/Lucas injector pump.

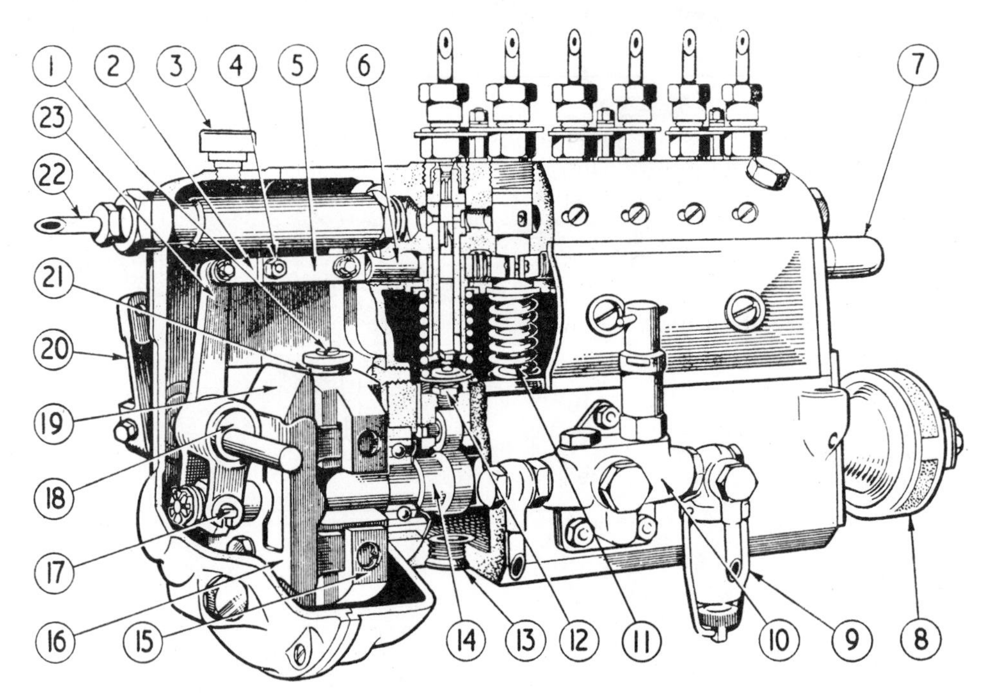

1. ADJUSTING NUT.
2. OUTER LINK FORK.
3. OIL LUBRICATOR.
4. SCREW FOR FORKS.
5. INNER LINK FORK.
6. CONTROL ROD.
7. CONTROL ROD STOP.
8. DRIVE COUPLING.
9. FILTER.
10. FEED PUMP.
11. SPRING.
12. TAPPET SCREW.
13. CLOSING PLUG.
14. CAMSHAFT.
15. FLYWEIGHTS.
16. BELL CRANKPIN RETAINING CAGE.
17. COUPLING PIN.
18. ECCENTRIC.
19. BELL CRANK LEVER.
20. CONTROL LEVER.
21. GOVERNOR SPRING.
22. FUEL INLET.
23. FLOATING LEVER.

Fig. 8-50. CAV/Lucas type BPE fuel injection pump.

in metering pressure and consequent increase in the quantity of fuel entering the pumping element at each charging stroke.

If the pedal is then released, the effective area of a metering orifice is reduced as engine speed is reduced as a result of the decreased fuel. Maximum fuel settings are made by limiting the maximum outward channel of the pumping plungers at a point where the desired fueling is obtained.

CAV/LUCAS GOVERNING

Both flyweight and hydraulic type governors are available on CAV/Lucas fuel injection systems. The mechanical type governor is shown in Fig. 8-48. Apart from the method of governing, the two types are similar in principle of operation. The fundamental working parts – hydraulic head, rotor, drive plate, adjusting plate, end plate and regulating valve are identical in both pumps.

The splined drive shaft, drive hub and quill shaft of the mechanically governed pump are replaced by a single splined drive shaft in the hydraulically governed pump. The metering valve differs as previously mentioned.

The mechanical governor: The DPA distributor type pump incorporating a mechanical governor is shown in Fig. 8-44. It will be noted that the flywheel assembly is carried on the splined drive shaft within the pump housing and that the governor control linkage is enclosed by a cover fitted on the upper face of the pump housing.

The governor weights, Fig. 8-48, are housed in pockets in the weight carrier, which is rigidly clamped between the end of the drive hub and a step on the drive shaft. When the pump is operating the drive hub, governor weight assembly and drive shaft rotate as a single unit. The weights are shaped so they pivot about one edge under the influence of centrifugal force. Such movement causes a thrust sleeve with which they are engaged to slide along the shaft. Movement of the thrust sleeve is transmitted to the metering valve by the pivoted governor arm and the spring-loaded hook lever.

Outward movement of the weights tends to close the metering orifice by rotating the metering valve, thus reducing the quantity of fuel reaching the engine cylinders at each injection.

The governor arm is spring-loaded by the governor spring and the tension of the spring acts in opposition to the centrifugal force, tending to oppose outward movement of the weights and to hold the metering valve in the maximum fuel position. Spring tension can be varied by moving the control lever to which the spring is connected. To increase tension, move the lever toward maximum speed setting.

When the control lever is moved to a position which calls for engine acceleration, increased spring tension is applied to the governor arm. This increased tension overcomes centrifugal force acting on the governor weights and the metering valve is rotated to maximum fuel position.

Engine speeds build up until the centrifugal force acting on the weights is sufficient to overcome the increased spring tension. The weights move outward and reduce the fueling by rotating the metering valve until the two opposed forces acting on the thrust sleeve are in balance.

TIMING INJECTION

The automatic speed advance mechanism is shown in Fig. 8-49. The piston B is free to slide in a cylinder machined in the body of the device E. Movement of the piston is transmitted by the cam ring C by the ball ended cam screw A causing the cam ring to rotate within the pump housing D.

Pressure exerted on the piston of the spring tends to hold the piston and the cam ring in the fully retarded position.

Fuel oil at transfer pressure enters the device through a fuel passage in the screw which secures the device to the pump housing. Transfer pressure acts upon the piston and tends to move the cam ring against the spring pressure.

Transfer pressure increases progressively as the engine speed is raised and the piston is moved along the cylinder to compress the spring and move the cam ring toward the fully retarded position. When the engine speed is decreased, transfer pressure falls off and the piston and cam ring are moved to the retarded position by spring pressure.

Any desired timing advance up to a maximum of 12 deg. (pump) is obtainable and the engine speed at which this is attained may be varied by installing weaker or stronger springs.

CAV/LUCAS BPE INJECTION PUMP

The CAV/Lucas BPE fuel injection pump is of the cam-operated spring return plunger type employing one pumping unit for each engine cylinder. It incorporates its own camshaft and tappet gear. This injection pump is illustrated in Fig. 8-50.

In this design the stroke of the pumping plunger is constant but that part of it which is actually pumping is variable, Fig. 8-51. By means of the helical edge which runs around the plunger (which itself can be rotated within the barrel) it is possible to make the point of cutoff earlier or later in the stroke as desired. The postion of the plunger stroke at which the helical edge will uncover the port is adjustable by rotating the plunger axially by means of a toothed quadrant.

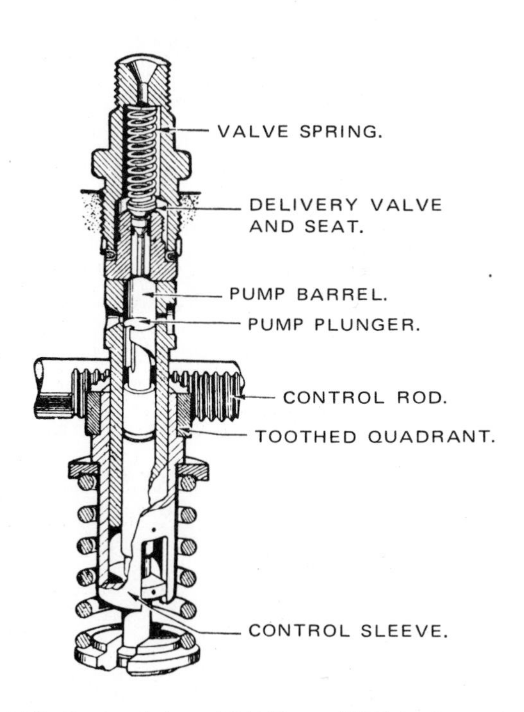

Fig. 8-51. Sectional view of CAV/Lucas BPE injection pump.

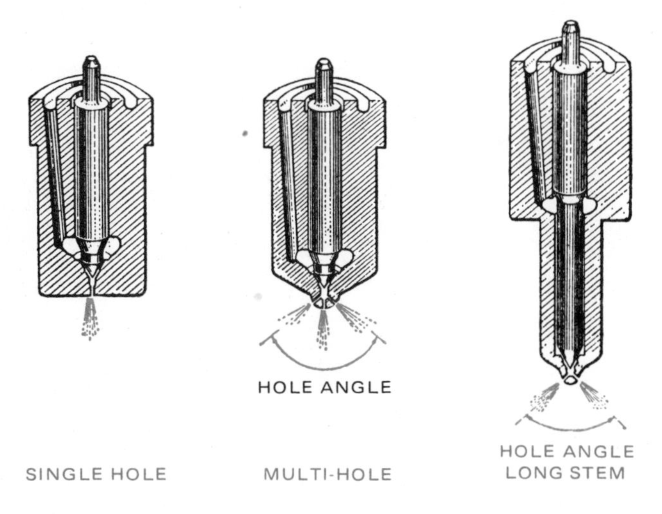

Fig. 8-52. Typical CAV/Lucas nozzles.

CAV/LUCAS INJECTION NOZZLES

The CAV/Lucas nozzles are of single hole, multiple hole and pintle design. All are of the closed type. The closed type nozzle is so-called because the nozzle is closed with a valve after each injection of fuel into the engine combustion chamber. These injection nozzles are shown in Fig. 8-52. The nozzle to use is dependent on the type of combustion chamber.

CAV/LUCAS DP15 PUMP

Internally generated fuel pressures are used to provide effective control of the various functions of late model CAV/Lucas DP15 fuel injection pumps, Fig. 8-53. These pressures are classed as transfer pressure, metering pressure, N^2 pressure and housing pressure. Prime pressure, then, is transfer pressure with metering, N^2 and housing pressures in that order of descending magnitude.

Pressure Generation and Regulation: Fuel is fed to the inlet of a rotor-driven, high-output pump, Figs. 8-53 and 8-54. This designated output pressure is controlled by a spring-loaded regulating valve that allows pressure to increase with speed. A portion of the fuel oil at transfer pressure is fed into the piston type metering valve that controls a variable aperature. At this point, a reduction in pressure occurs across the metering valve. This reduced pressure is termed "metering pressure."

At the end of the metering valve is a plate type valve that is spring-loaded by throttle action. Pressurized fuel flows continually through the variable gap between the plate valve and the metering valve to produce a controlled pressure drop, which is known as "N^2 pressure."

The term "N^2" is used because the pressure drop is proportional to the square of governor speed. N^2 pressure is unaffected by oil viscosity changes, but it is related to engine speed.

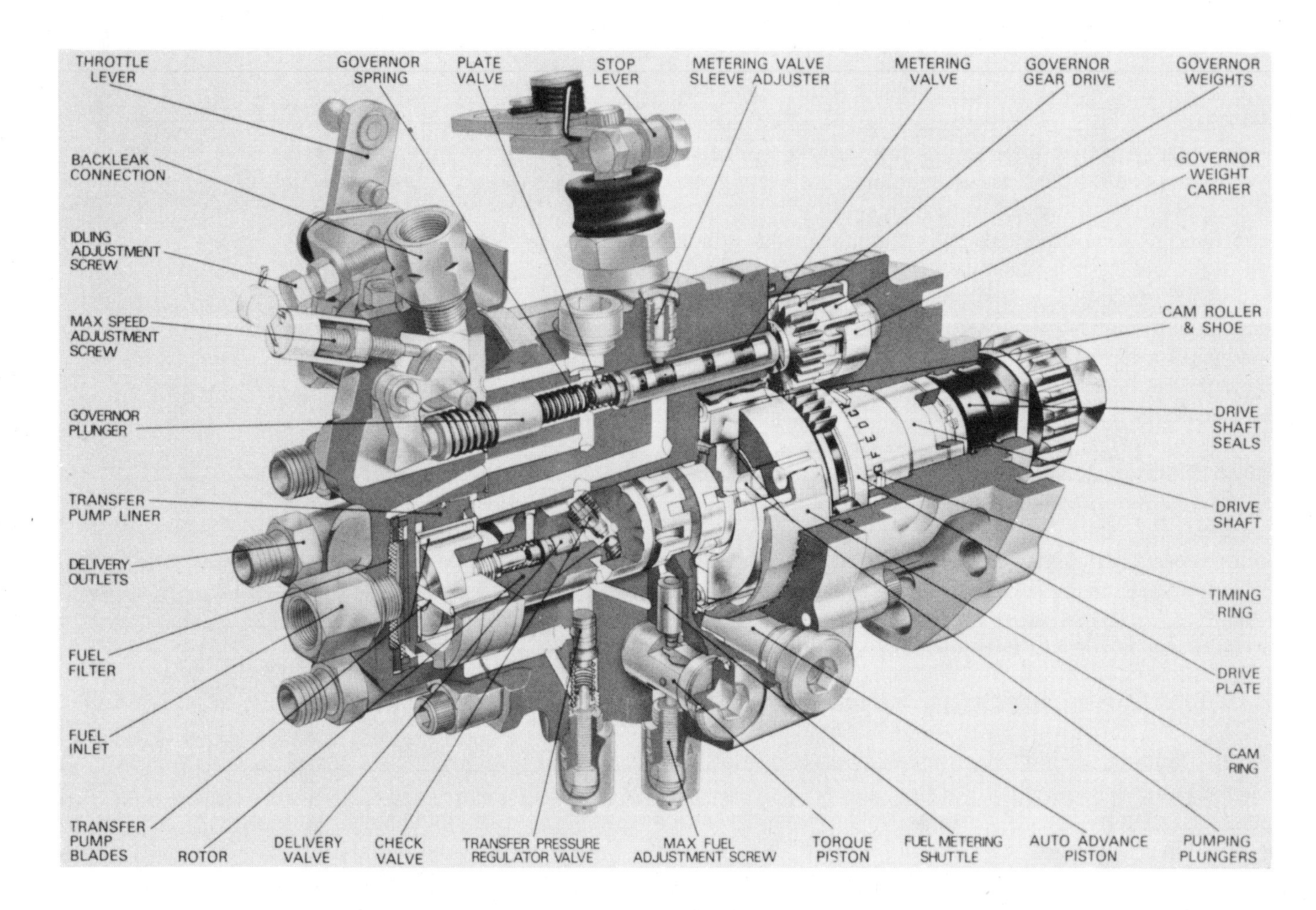

Fig. 8-53. Sectional view of CAV/Lucas model DP15 injection pump. (CAV/Joseph Lucas, Inc.)

Housing pressure is oil pressure generated within the pump housing by back leakage of oil from the operating pistons and internal valve spill. Pressure is maintained by the use of an outlet with a restricted orifice.

The hydraulic circuit for the DP15 fuel injection pump is shown in Fig. 8-54. Filtered fuel oil at feed pump pressure is fed to the inlet side of the transfer pump. Output from the transfer pump is maintained at transfer pressure by the regulating valve releasing oil to the feed pressure line.

Fuel for injection to the engine flows at transfer pressure to the metering valve, which produces metering pressure. Fuel at metered pressure flows past the fuel shut-off device to the second set of slots on the distributor rotor and on to the top of the right hand shuttle. At this point, fuel from the underside of the shuttle returns to housing pressure by way of slots in the first set of slots in the rotor. The downward movement of the shuttle due to this pressure differential is the shuttle charging stroke.

At the same time, fuel at transfer pressure flows to the underside of the left hand shuttle by way of aligned slots in the fourth set of slots on the distributor rotor. This pressure pushes the shuttle upwards and the fuel charge above the shuttle is pumped by way of the third set of slots in the rotor, to open the ball check valve and push apart the pumping plungers.

In the position shown in Fig. 8-54, the right hand shuttle is being charged while the left hand shuttle is discharging to the pumping plungers. As the rotor continues to rotate, the action of the shuttle alternates between filling and discharging, due to the slot spacing of the rotor and the hole alignment of the head. At the end of each shuttle discharge stroke, the ball of the ball check valve is pushed on to its seat by a small piston actuated by transfer pressure. This check valve isolates the bore of the rotor when the injection of fuel to the engine occurs.

Fuel injection to the engine occurs when the pumping plungers are forced inward by the action of the cam rollers on the cam lobes. When injection occurs, fuel at injection pressure forces open the delivery valve in the rotor bore and flows to the individual line delivery pressurizing valve and on to the specific injector.

The metering valve is connected to the governor weight assembly which imparts an axial movement caused by the pivoting of the governor weights due to variations in speed.

To increase engine speed, the speed control lever is rotated. This compresses the control spring, loads the plate valve and moves the metering valve toward the governor. This action provides increased fuel flow that results in a longer stroke of the shuttles and a corresponding increased stroke of the high pressure plungers.

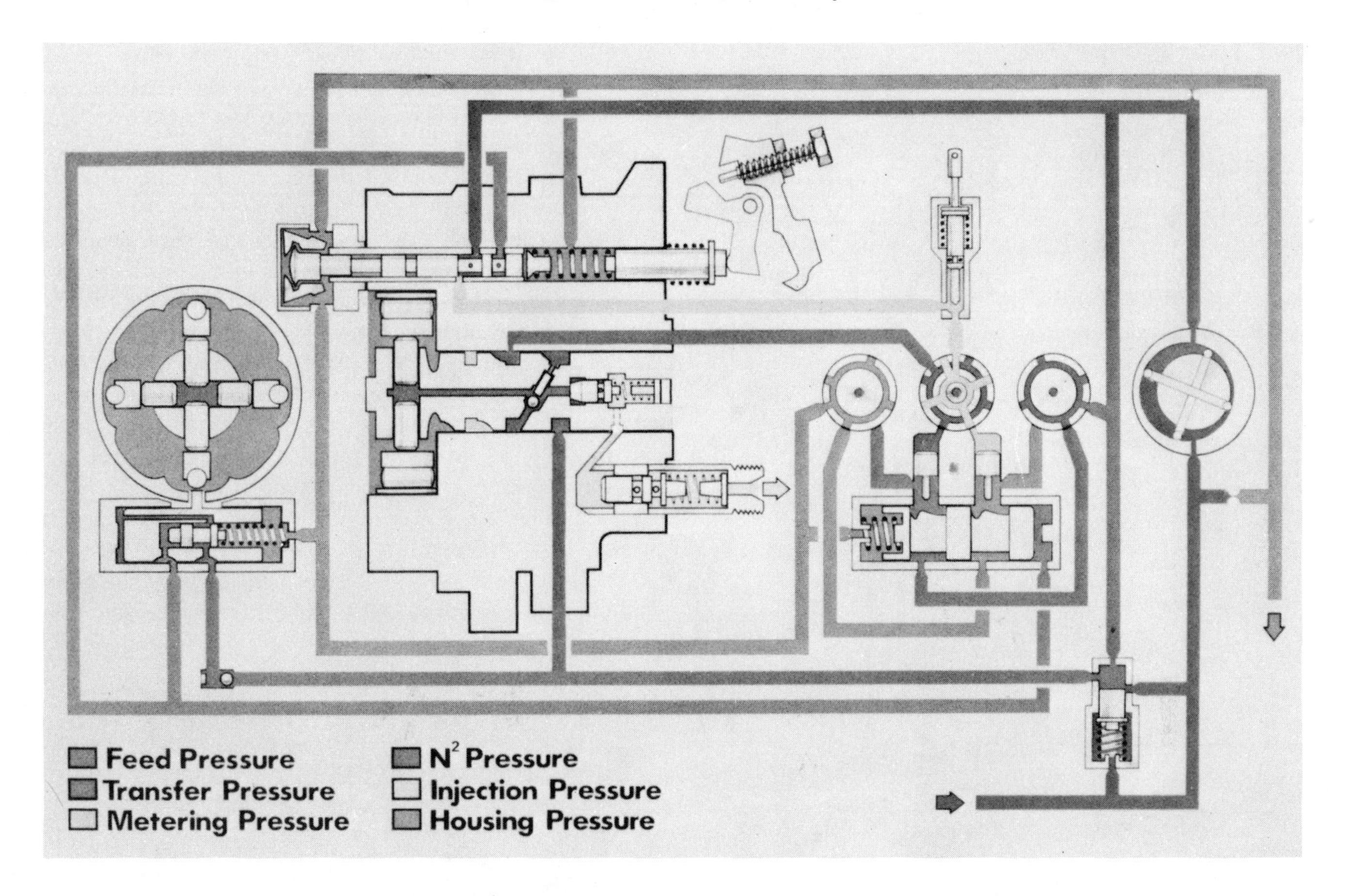

Fig. 8-54. Diagram of hydraulic circuit of DP15 injection pump. (CAV/Joseph Lucas, Inc.)

The outward limit of the shuttle stroke is determined by the position of the torque profiles. In Fig. 8-54, the profiles are shaped to give excess fuel (when selected) for starting the engine. Then, as speed increases, there is an increase in the maximum amount of fuel delivered. Movement of the torque piston changes the profile presented to the shuttle. This is achieved by N^2 pressure (which is speed sensitive) at one end of the torque piston, opposing the spring-loading of the piston.

N^2 pressure is also used to control the automatic advance device. A servo valve and its loading spring is located inside the advance piston and activated by fuel oil at N^2 pressure. As N^2 pressure builds up with increasing pump speed, the servo valve will move to allow oil at transfer pressure to flow past the land on the valve to the face of the advance piston. The piston will move against its spring loading and rotate the cam ring slightly to advance the injection of fuel. A ball type, non-return valve prevents the piston from being retarded during the point of injection. It also allows the cam ring to lock up at injection.

With decreasing speed, N^2 pressure falls. This causes the servo valve to move back and allow oil from the face of the advance piston to flow to the housing pressure side of the circuit. With that, the advance piston is moved back by its return spring.

For engine shutdown, the CAV/Lucas model DP15 pump incorporates a lever-operated, spring-loaded, shut-off valve. This valve cuts off the supply of metered fuel to the rotor, preventing fuel injection to the engine.

CAV/LUCAS INJECTORS

CAV injectors are of the modern closed type, incorporating a spring-loaded valve. They are responsive to the fuel pressures produced by the injection pump. When the injection fuel pressure is raised above a predetermined point, the valve is opened hydraulically by the fuel itself and stays open until the fuel pressure drops.

Two typical fuel injectors are shown in Fig. 8-55. The injector on the left is fitted with a conventional single hole nozzle. The injector on the right is a late type with a long stem nozzle, and the basic difference in the two injectors is in the valve design.

The nozzle valve has an elongated stem machined so that it is a clearance fit in the extended body of the nozzle. Extension of the valve stem in this way raises the close fitting part of the valve to a point in the cylinder head where it can be cooled. In this way, the valve is protected from high temperatures.

CUMMINS INJECTION SYSTEM

The Cummins injection system is used on Cummins engines which are used extensively in automotive trucks and also in the marine, stationary power plant and construction fields.

This injection system is known as the PT or pressure time system. It is based on the principle that the volume of liquid

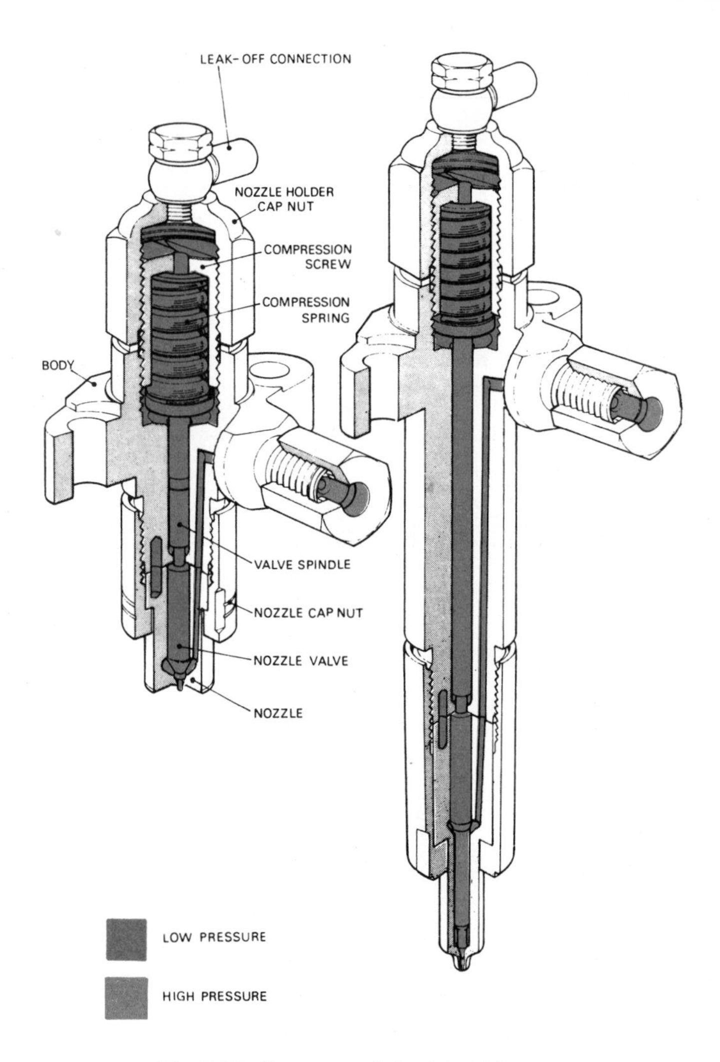

Fig. 8-55. Two types of diesel fuel injectors. (CAV/Joseph Lucas, Inc.)

flow is proportionate to the fluid pressure, the time allowed to flow and the size of the passage through which the liquid flows.

The complete PT fuel system consists of a fuel pump, Fig. 8-56, supply and drain lines and passages and the injectors. Originally, there were two types of PT fuel systems: one called PT type G, Fig. 8-57; the other is PT type R, Fig. 8-59. Type G is governor controlled. Type R is pressure regulated. Recently, type G was modified to minimize exhaust emissions and smoke. This new fuel injection pump is known as the PT type G with air fuel control fuel pump, Fig. 8-60.

The type PT-G fuel pump, Figs. 8-57 and 8-58, can be identified by the absence of the return line at the top of the fuel pump. The pump assembly is made of three main units.

1. Gear pump which draws fuel from the supply tank and forces it through the pump filter screen to the governor.
2. Governor controls the flow of fuel from the gear pump as well as the maximum and idle speeds.
3. Throttle which provides a manual control of fuel flow to the injector under all conditions in the operating range.

The PT type R fuel pump, Fig. 8-59, can be identified by the presence of the fuel return line from the top of the fuel pump housing to the supply tank.

The PT type R fuel pump is made of four main units:

1. Gear pump which draws fuel from the supply tank forcing it through the pump filter screen into the pressure regulator valve.
2. Pressure regulator which limits the pressure of the fuel to the injectors.
3. Throttle which provides a manual control of fuel flow to the injector under all conditions in the operating range.
4. Governor assembly which controls the flow of fuel from idle to maximum governor speed.

The gear pump and pulsation damper located at the rear of the fuel pump performs the same function on both the type G and type R fuel injection pumps. The pulsation damper mounted on the gear pump contains a steel diaphragm which absorbs pulsations and smooths fuel flow through the fuel system. From the gear pump, fuel flows through the filter screen.

The pressure regulator which is used only on the type R fuel pump functions as a by-pass valve to regulate fuel pressure

Fig. 8-56. Cummins PT fuel injector.

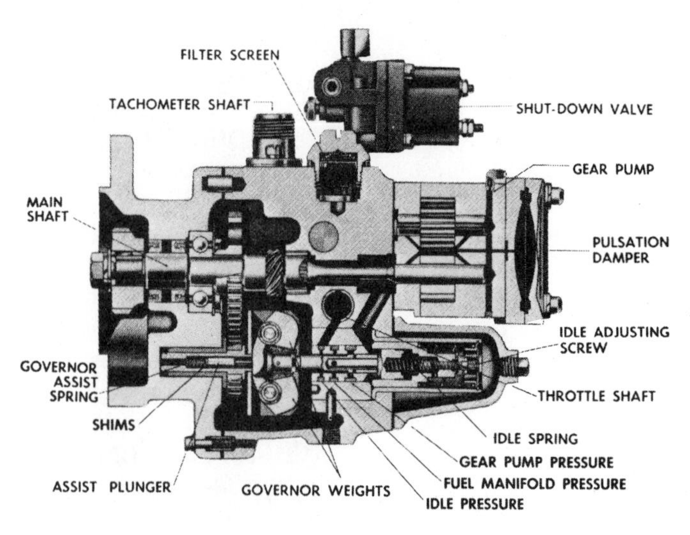

Fig. 8-57. Details of Cummins PT type G fuel pump.

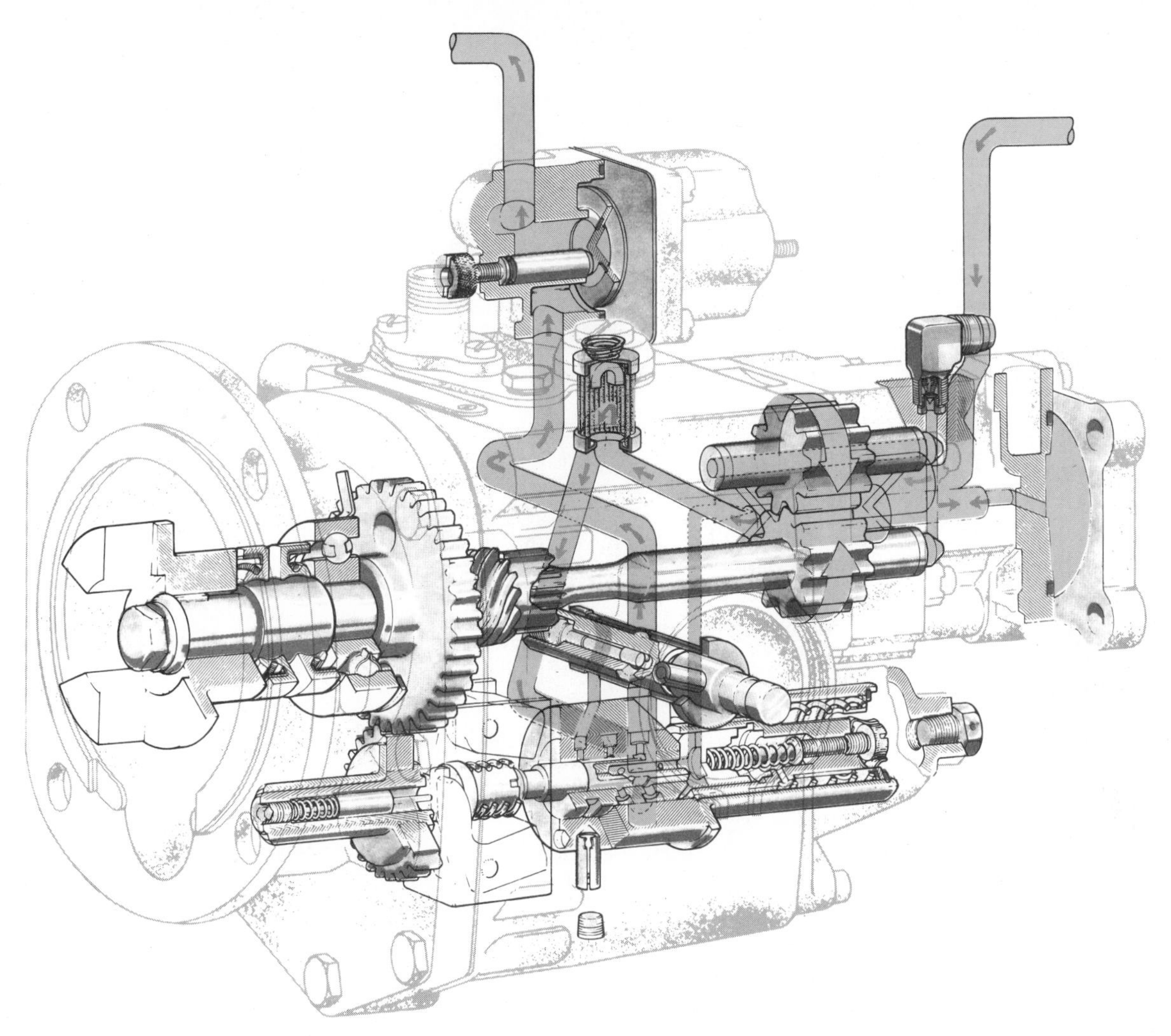

Fig. 8-58. Cross section of Cummins PT type G fuel pump and fuel flow.

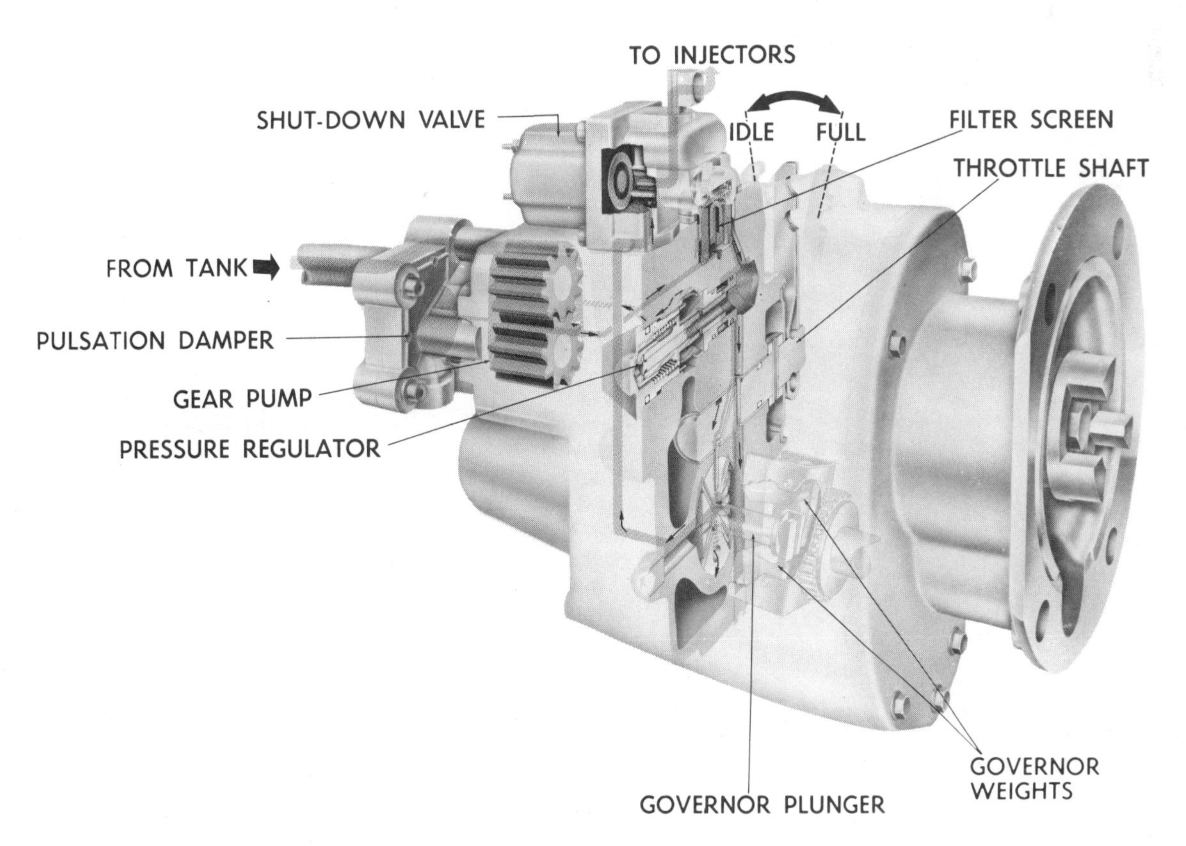

Fig. 8-59. Cummins PT type R fuel injection pump.

to the injectors. By-passed fuel flows back to the suction side of the gear pump, Fig. 8-59.

In both pumps the throttle provides a means of manually controlling engine speed above idle as required by varying conditions of speed and loads. In the type G pump, Fig. 8-57, fuel flows through the governor to the throttle shaft. At idle speed, fuel flows through the idle port in the governor barrel, past the throttle shaft. To operate above idle speed, fuel flows through the main governor barrel port to the throttling hole in the shaft.

In the type R pump, fuel flows past the pressure regulator to the throttle shaft. Under idling conditions, fuel passes around the shaft to the idle port in the governor barrel. For operation above idle speed, fuel passes through the throttling hole in the shaft and enters the governor barrel through the main fuel port.

The mechanical governor, Fig. 8-59, is the same on both type G and type R Cummins fuel pumps.

The governor which is of the centrifugal type is actuated by a system of springs and weights. The governor maintains sufficient fuel for idling with the throttle control in the idle position. In addition, the governor cuts off fuel to the injectors above rated rpm.

During operation between idle and maximum speed, fuel flows through the governor to the injector in accordance with engine requirements as controlled by the throttle and limited by the size of the idle spring plunger counterbore on the type G fuel pumps, and the pressure regulator on the type R fuel pumps.

When the engine reaches governed speed, the governor weights move the governor plungers so the fuel passages to the injector are shut off. At the same time, another passage opens and dumps the fuel back into the main pump body. In that way engine speed is controlled and limited regardless of the position of the throttle.

Fuel leaving the governor flows through the shutdown valve, inlet supply lines and onto the injectors.

CUMMINS ANEROID CONTROL

On some Cummins PT model G installations, an aneroid control is installed. The aneroid control is a fuel by-pass system that responds to engine air manifold pressure. The aneroid limits fuel manifold pressure when the air manifold pressure is below a preset value. When accelerating the turbocharged engine from speeds below normal operating range to approximately 1400 rpm, air manifold pressure is not sufficient for complete combustion of fuel unless fuel delivery is reduced to maintain a suitable air-fuel ratio. This is accomplished by the aneroid control.

CUMMINS AIR FUEL CONTROL FUEL PUMP

The PT (type G) Air Fuel Control (AFC) fuel pump, Fig. 8-60, is designed to supersede the PT (type G) fuel pump and aneroid combination on turbocharged engines. The air fuel control differs from the aneroid unit, which is basically an

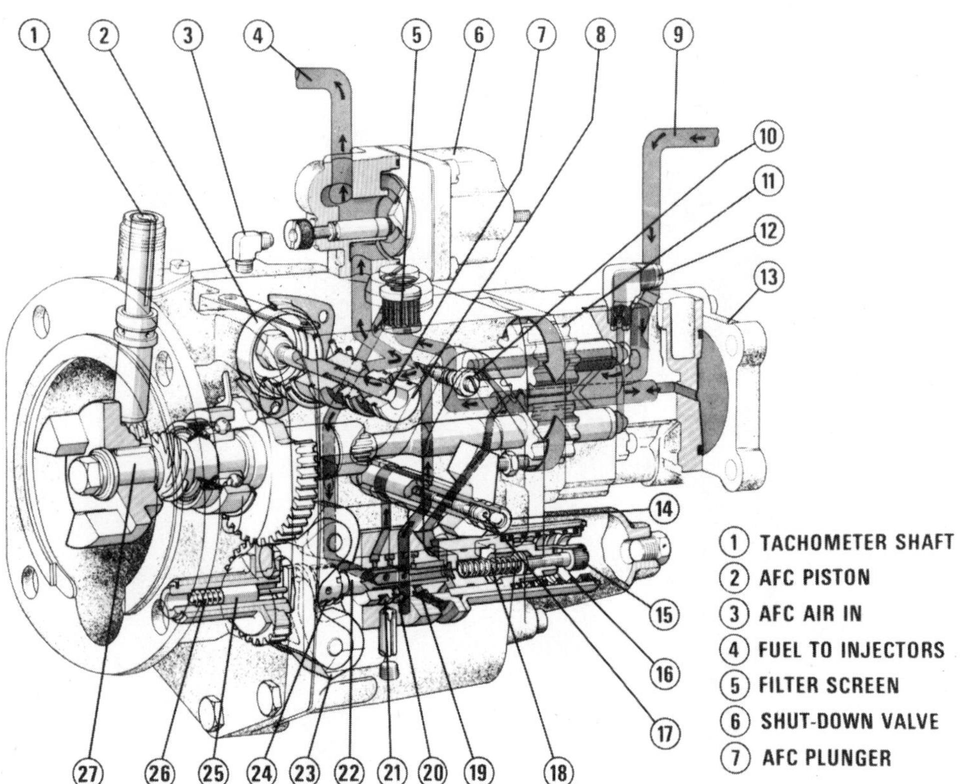

Fig. 8-60. Details of Cummins PT (type G) Air Fuel Control (AFC) fuel pump. (Cummins Engine Co., Inc.)

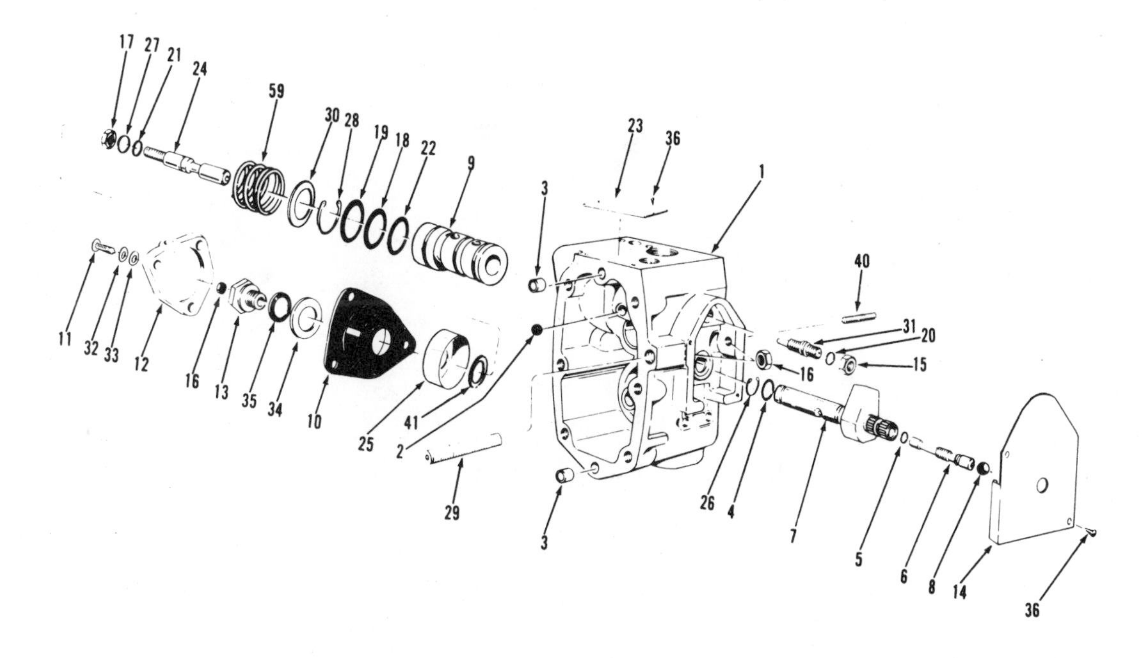

Fig. 8-61. Cummins PT (type G) air fuel control (AFC) fuel pump. 1—Fuel pump housing AFC (RH). 2—Ball plug. 3—Dowel. 4, 5—O-ring. 6—Fuel adjusting screw. 7—Throttle shaft. 8—Tamperproof ball. 9—AFC barrel. 10—AFC bellows. 11—Capscrew. 12—Fuel control cover. 13—Piston center bolt. 14—Throttle shaft cover. 15, 16—Nuts. 17- Lock nut. 18, 19, 20, 21, 22—O-rings. 23—Name plate. 24—Control plunger. 25—AFC piston. 26—Retaining ring. 27—Ring. 28—Snap ring. 29—Hex screw. 30—AFC spring. 31—Needle valve. 32—Washer. 33—Washer. 34—Retainer bearing washer. 35—Washer gasket. 36—Screw.

on-off, fuel by-pass device. Instead, the AFC functions as a fuel pressure and flow restrictor to provide the proper air-fuel delivery rate to the engine during acceleration.

Fuel pumps on engines not requiring the air fuel control feature will have a specially devised plug in the housing in place of the AFC barrel assembly. Therefore, the housing becomes standard for all engines. See Figs. 8-60 and 8-61.

There is an operational difference between the two PT (type G) pumps. In the PT (type G) fuel pump, the fuel passes directly from the throttle shaft through a passage to the shut-down valve. In the PT (type G) AFC pump, the fuel passes through the AFC unit after leaving the throttle shaft and before reaching the shut-down valve, Fig. 8-62.

The AFC control plunger is in the "no-air" position: 1. During engine start-up. 2. When the engine (turbocharger) speed and the resulting air manifold pressure is too low to overcome the pressure of the AFC spring. Fuel from the throttle shaft then flows through the no-air needle valve passage where the needle valve restricts the pressure and flow. The AFC plunger blocks fuel from passing through the AFC barrel. After passing the needle valve, the fuel flows to the shut-down valve.

As the turbocharger speed and intake air manifold pressure increases, the air pressure acting on the AFC bellows and piston overcomes the AFC spring force, causing the AFC plunger to move away from the AFC cover plate.

As the plunger moves, the plunger profile begins to allow fuel to flow through a drilling in the AFC barrel that is common to the drilling from the fuel pump throttle shaft. This allows fuel to bypass the no-air adjust needle valve and flow down the AFC barrel to a second drilling leading to the shut-down valve.

As the intake manifold air pressure increases, the AFC plunger uncovers more of the drillings until a minimum fuel restriction level is reached and intake manifold air pressure above the AFC unit holds the plunger in full fuel position. See Fig. 8-62.

CUMMINS INJECTORS

The Cummins fuel pump delivers fuel at moderate pressure to the injectors, and the injectors raise the pressure to produce

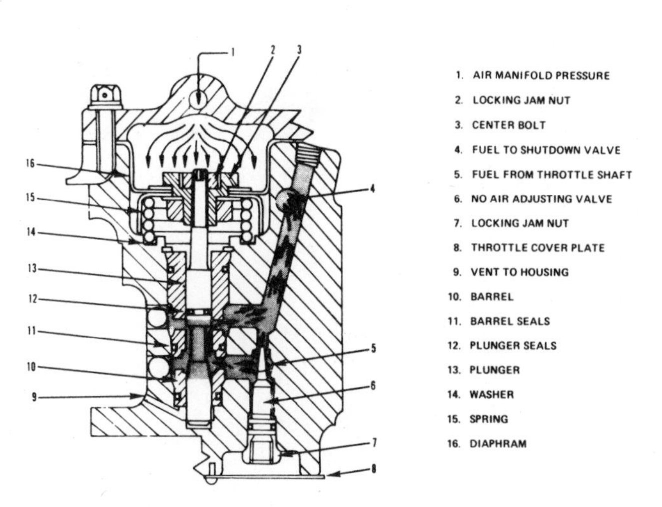

Fig. 8-62. AFC control plunger in full fuel position.

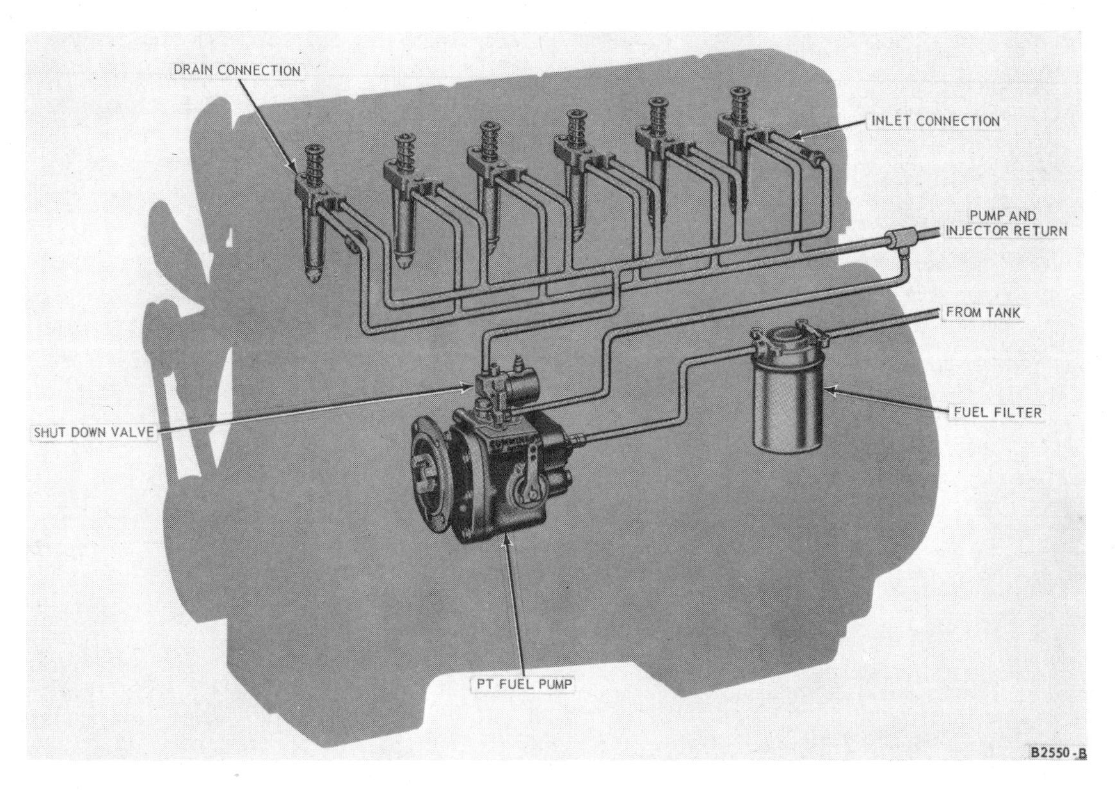

Fig. 8-63. Fuel flow diagram of Cummins pressure time injection system.

the desired spray, Fig. 8-63. In addition, the cam-operated injectors time the injection. The injectors combine the act of metering and injection.

There are two basic types of injectors, the flanged type, Fig. 8-64, and the cylindrical type, Fig. 8-65. In the flanged

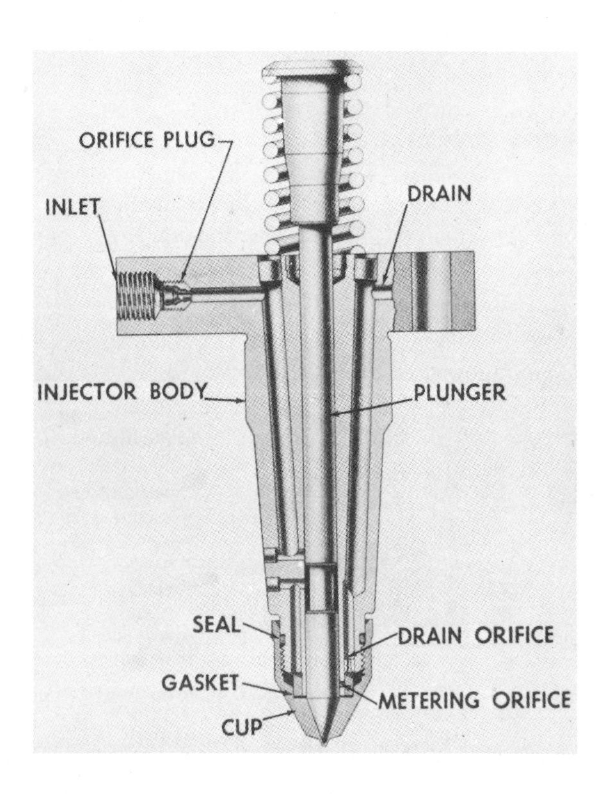

Fig. 8-64. Sectional view of Cummins flanged type injector.

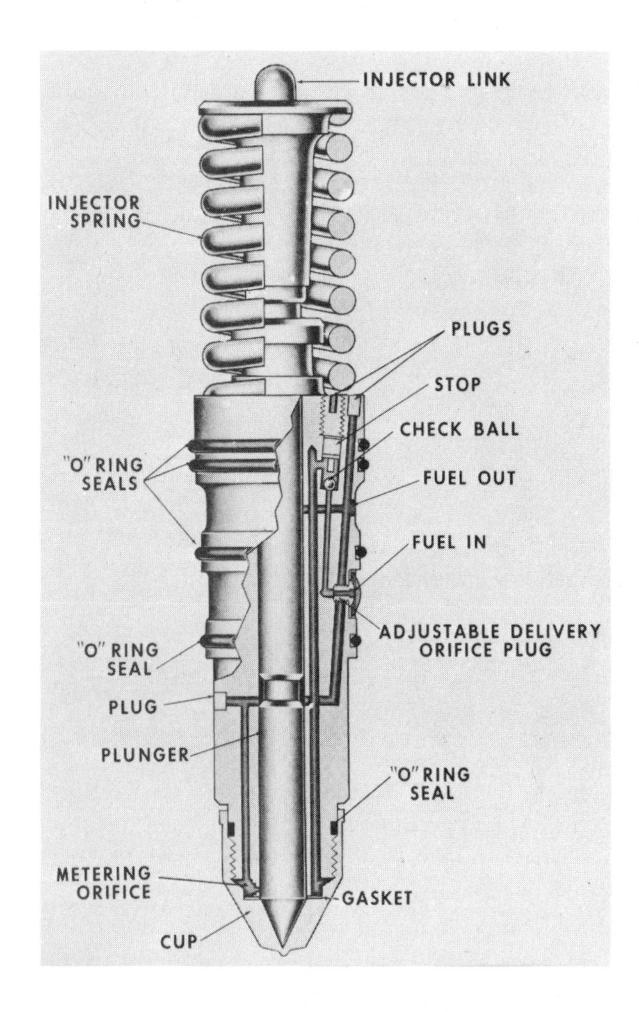

Fig. 8-65. Details of Cummins cylindrical injector.

injector, Fig. 8-64, fuel is supplied to and drained from the injector to external lines and connection, as shown in Fig. 8-63. From the inlet connection, fuel flows down the inlet passage of the injector, Fig. 8-66, around the plunger, between the body and the cup, up the drain passages to the drain connection and lines where it returns to the supply tank.

As the plunger rises, Fig. 8-66, the metering orifice is uncovered and part of the fuel is metered into the cup. At the same time, the rest of the fuel flows out of the drain orifice. The amount of fuel passing through the metering orifice and into the cup is controlled by the fuel pressure.

passage in the cylinder head at the front of the engine. A second drilling in the head is aligned with the upper injector radial groove to drain excess fuel. A fuel drain at the flywheel end of the engine allows return of the unused fuel to the fuel supply tank.

The fuel grooves around the injector are separated by O-rings, which seal against the cylinder head injector box, Fig. 8-67. This forms a leakproof passage between the injectors and the cylinder head injector bore surface.

The cylindrical injector contains a ball check valve. As the injector plunger moves downward to cover the fuel openings,

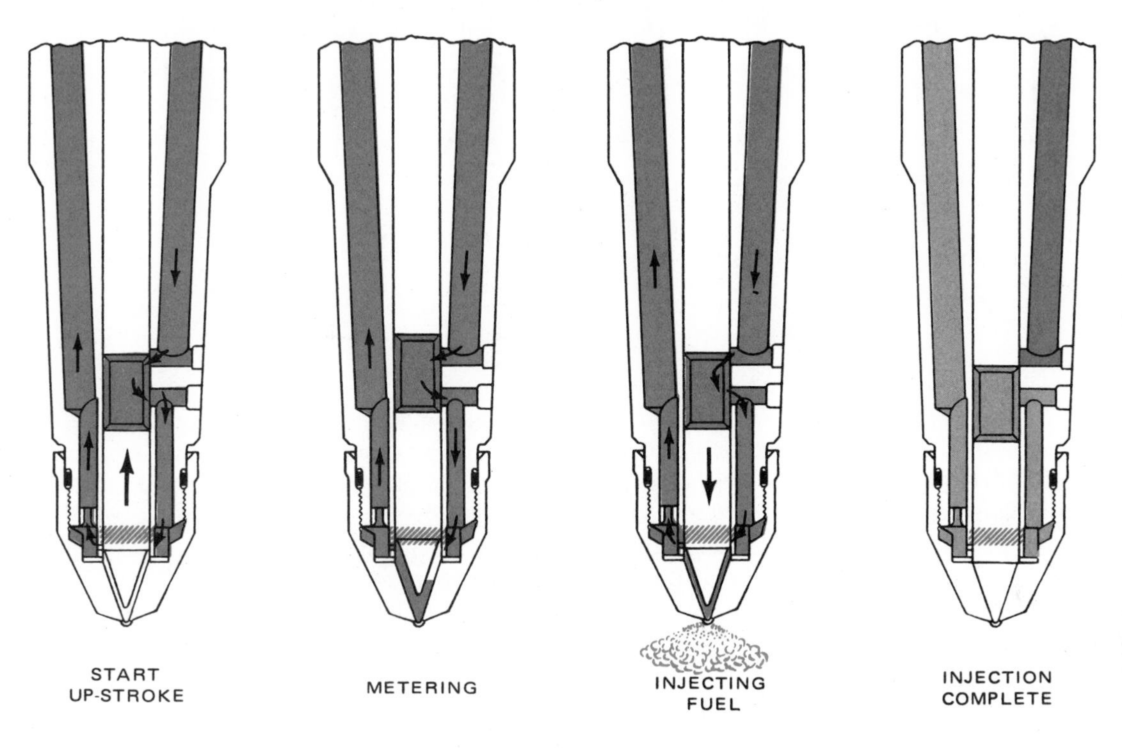

Fig. 8-66. Operation of Cummins flanged type injector.

During injection, Fig. 8-66, the plunger is forced downward through the action of the cam until the orifice is closed. Fuel in the cup is injected into the engine combustion chamber. While the plunger is seated, all fuel flows through the injector stops, Fig. 8-66. Injectors contain an adjustable orifice which meters fuel flow into the injector.

CUMMINS CYLINDRICAL INJECTOR

When cylindrical injectors, Fig. 8-67, are used, fuel supply and drain are accomplished through internal drill passages in the engine cylinder head. Such construction serves to maintain the fuel at an even temperature and viscosity. A radial groove, Fig. 8-65, around each injector mates with the drilled passages in the cylinder head and admits fuel through an orifice plug (which has been adjusted to size by burnishing on the test stand) in the injector body. A fine mesh screen at each inlet groove provides final fuel filtration.

Fuel flows from a connection on top of the fuel pump shut-down valve to the supply line into the lower drilled

an impulse pressure wave seats the ball and at the same time traps a positive amount of fuel in the injector cup for injections, Fig. 8-65. As a continuing downward plunger movement injects fuel into the combustion chamber, it also uncovers the drain opening, Fig. 8-67, and the ball rises from its seat. This allows free flow through the injector and out the drain which aids in cooling the injector.

EX-CELL-O FUEL SYSTEM

The Ex-Cell-O fuel system is used largely in marine service with the supply pump, governor and high-pressure injection pumps built into a single unit. However, it differs from the other designs in the operation of the plungers. Instead of being cam operated, the plungers move back and forth by means of a rotating swash plate. The surface of the swash plate is cut at an angle to its shaft. As the shaft with its swash plate are rotated the plungers are moved back and forth in their cylinders. The Ex-Cell-O injection pump is shown in Figs. 8-68 and 8-69.

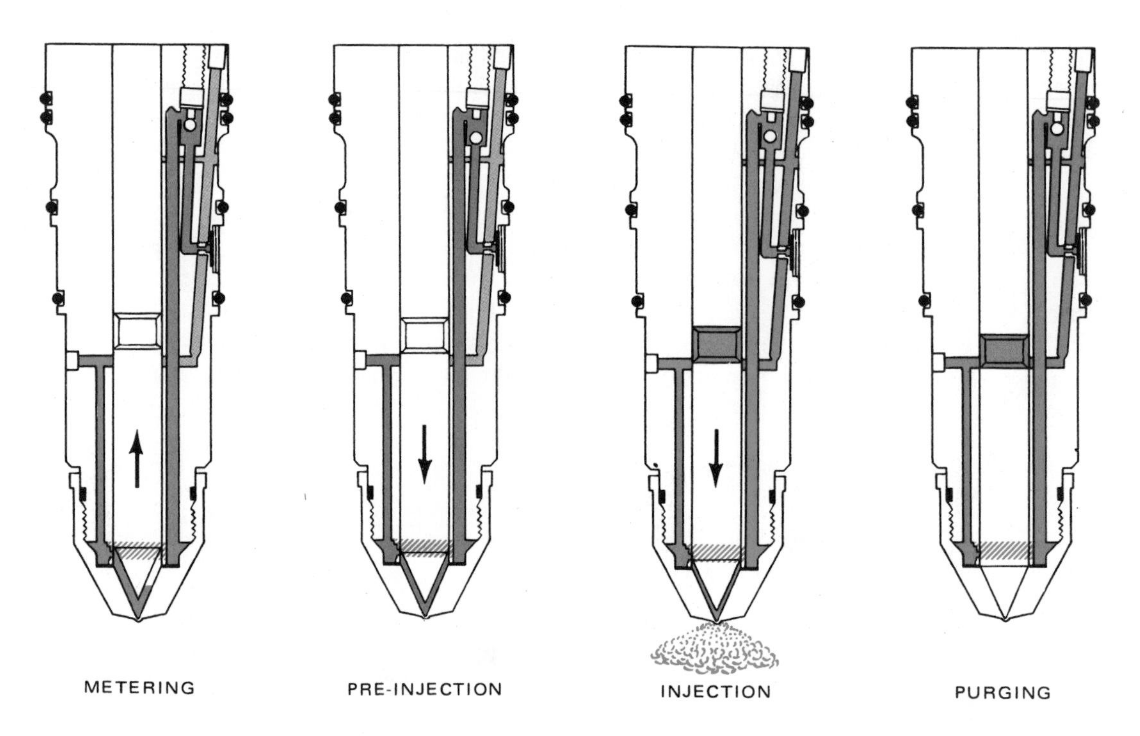

Fig. 8-67. Fuel injection cycle of Cummins cylindrical injector.

The supply pump, Fig. 8-68, is operated by an eccentric, Fig. 8-69, on the camshaft and sends oil under approximately 25 psi to the safety filter. As the swash plate is rotated, any point on the shoe plate will move back and forth. This movement, in turn, operates the injection plungers. The plungers force oil out to the spray nozzles where it is injected into the combustion chambers under high pressure.

The timing control and the throttle control regulate the amount of fuel that is injected at each plunger stroke, Fig. 8-69. The timing control also adjusts the time of injection in the engine operating cycle. It is manually operated by means of the remote control, Fig. 8-68.

The throttle control is operated entirely by the governor. An eccentric pin engages a stationary collar. When the collar is shifted, the rotor is simply moved lengthwise, increasing or decreasing the fuel setting of the rotor. The throttle control does not advance or retard the injection.

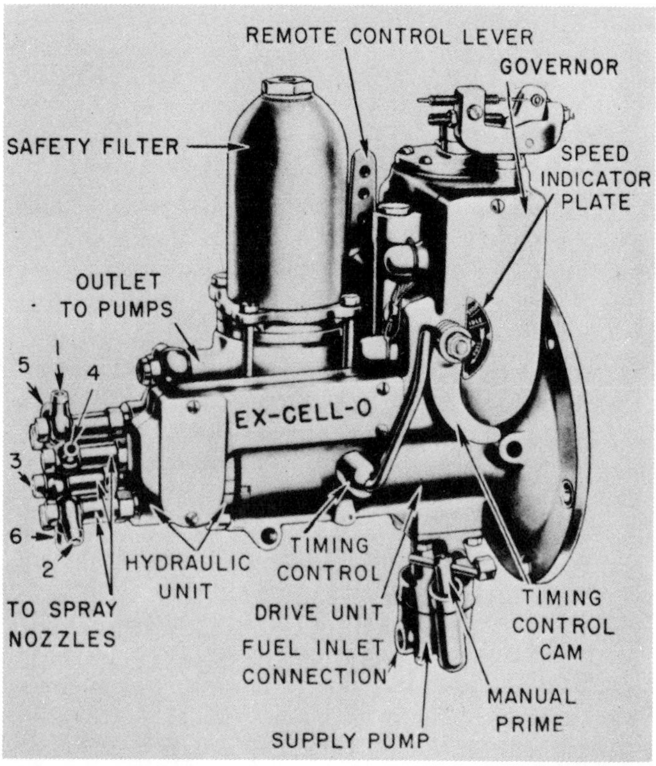

Fig. 8-68. External view of Ex-Cell-O fuel pump.

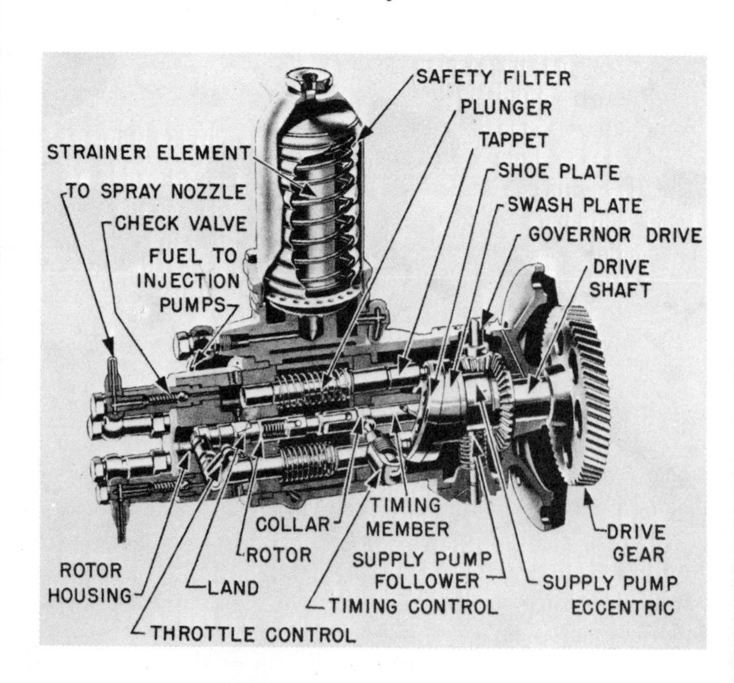

Fig. 8-69. Details of Ex-Cell-O fuel pump.

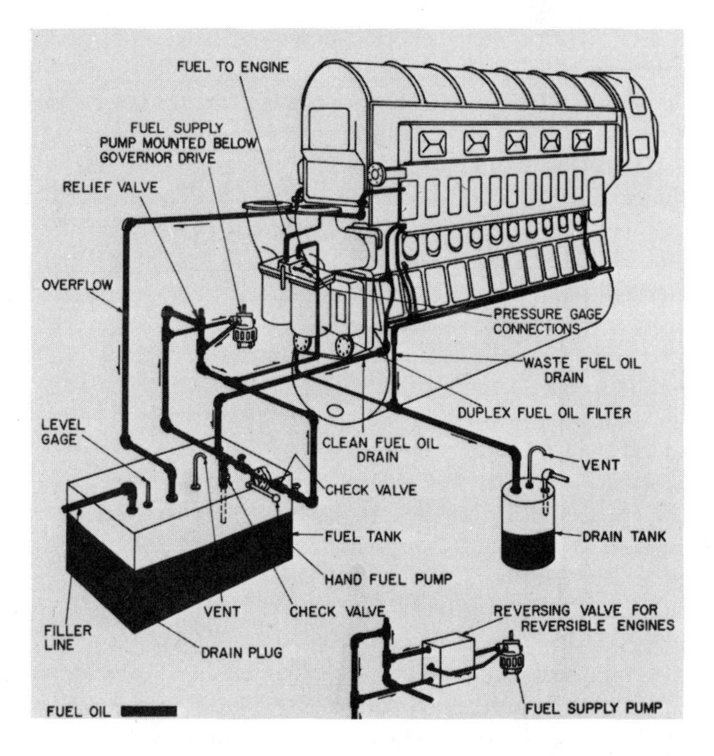

Fig. 8-70. Diagram of Fairbanks Morse oil piping. (Fairbanks Morse Power Systems Div., Colt Industries)

FAIRBANKS MORSE INJECTION SYSTEM

The complete fuel system as used on a Fairbanks Morse opposed piston diesel engine consists of a supply, ignition and drain systems, Fig. 8-70.

The supply system includes the supply tank, hand fuel pump, built-in supply tank on the engine, filter, gages and the necessary piping and fittings.

The hand operated pump when installed is used to fill the engine headers and deliver it through the filter to the engine inlet. A pressure of about 15 psi is built up and maintained in this fuel center.

A relief by-pass valve is provided in the piping between the pumps and the filter to protect the system in case of excessive clogging of the filter.

Excess fuel not used by the injection pumps returns to the supply tank to the clean fuel drain.

The fuel oil pump, Fig. 8-71, is of the positive displacement gear type and is mounted directly below the governor.

Fairbanks Morse marine engines are equipped with a fuel oil reversing valve. This valve is installed in the piping to maintain the direction of flow of oil when the engine direction of rotation is changed.

FUEL INJECTION SYSTEM

One pump and two nozzles are provided for each cylinder. Details of the pump are shown in Fig. 8-72. The injection pumps are cam operated. The camshaft, located in the upper crankshaft compartment, is chain driven from the upper crankshaft at the governor end of the crankshaft.

The injection pump, Fig. 8-73, receives fuel oil at low pressure from the fuel oil pump, meters it in correct amounts for injection, builds up a high pressure and delivers the fuel to the injection nozzle.

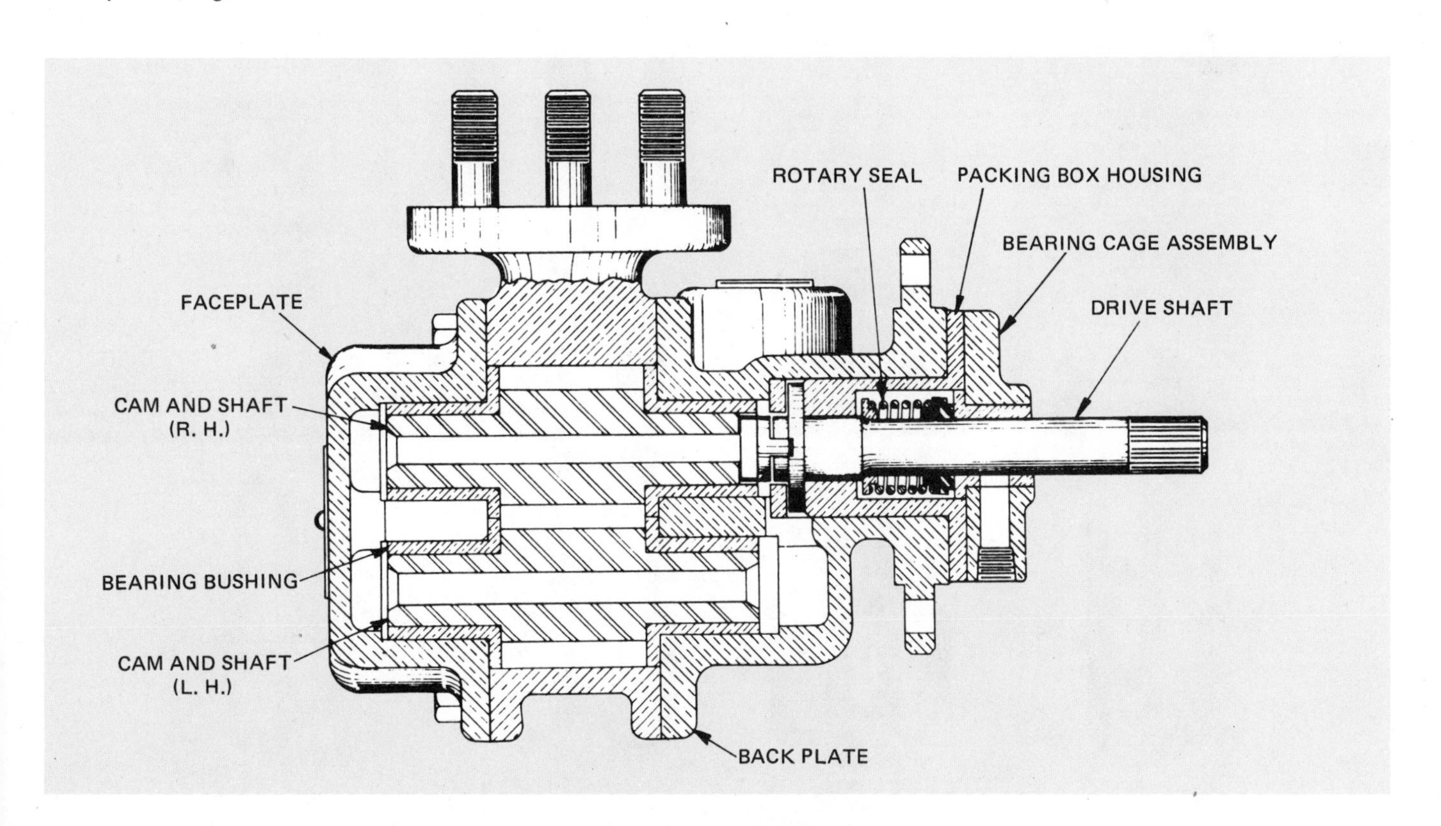

Fig. 8-71. Sectional view of fuel oil pump. (Fairbanks Morse Power Systems Div., Colt Industries)

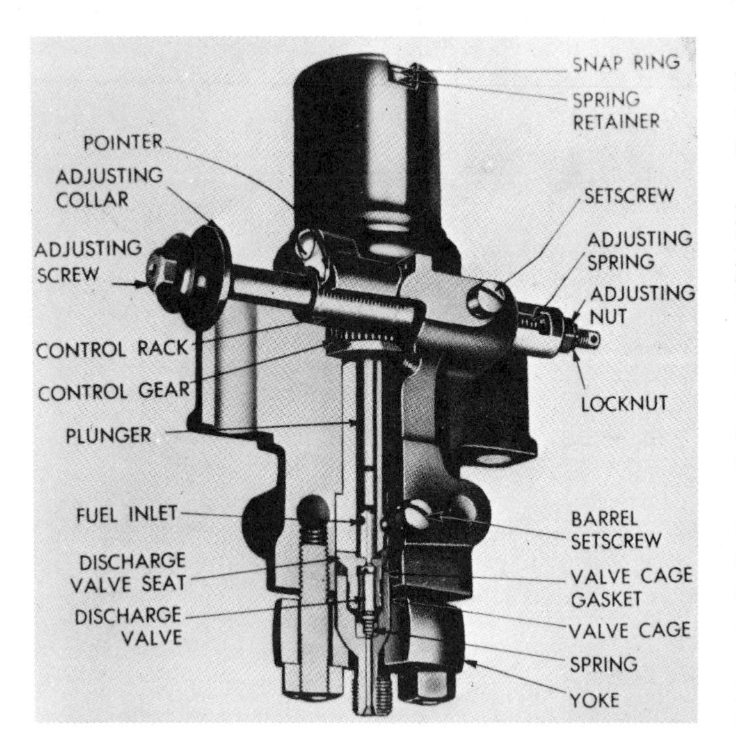

Fig. 8-72. Sectional view of injection pump. (Fairbanks Morse Power Systems Div., Colt Industries)

The injection pump plunger moves vertically in the pump barrel, delivering fuel through the discharge valve, Fig. 8-73, to the injection tubes and in turn to the injection nozzles which spray the fuel into the combustion chamber.

The plunger stroke is constant, therefore the area of plunger always admits an equal amount of fuel. The amount of fuel delivered to the nozzle depends on the position of the helix on the plunger relative to the port in the pump barrel. This position is controlled by the governor.

When the plunger is in its highest position, or on a low cam, Fig. 8-72, the fuel port is uncovered and the pump valve fills with fuel. The port is covered on the down stroke by the helix, and the fuel is delivered to the nozzle until a circular groove on the plunger uncovers the port.

As a result of the helix turning the plunger about one-quarter turn the plunger position is changed so the effective stroke varies from full-load to no fuel position, Fig. 8-73. The effective stroke is that part of the plunger stroke during which the port is covered and fuel is injected.

When the plunger is in "stop" position or "no fuel" position, the slot keeps the port uncovered during the entire stroke, thus bypassing all fuel. With the plunger turned so that the long edge of the helix edge coincides with the fuel inlet, the maximum amount of fuel is delivered.

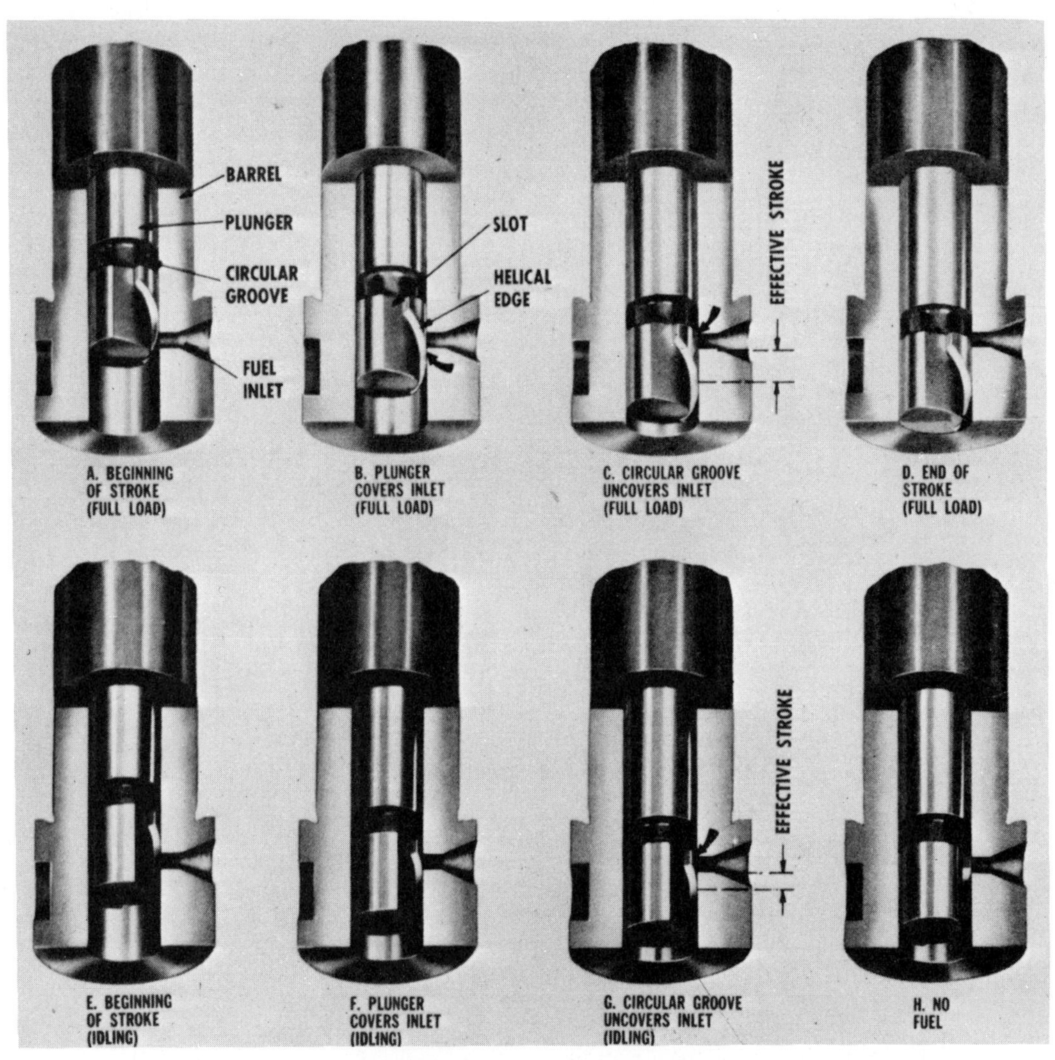

Fig. 8-73. Sequence of plunger positions. (Fairbanks Morse Power Systems Div., Colt Industries)

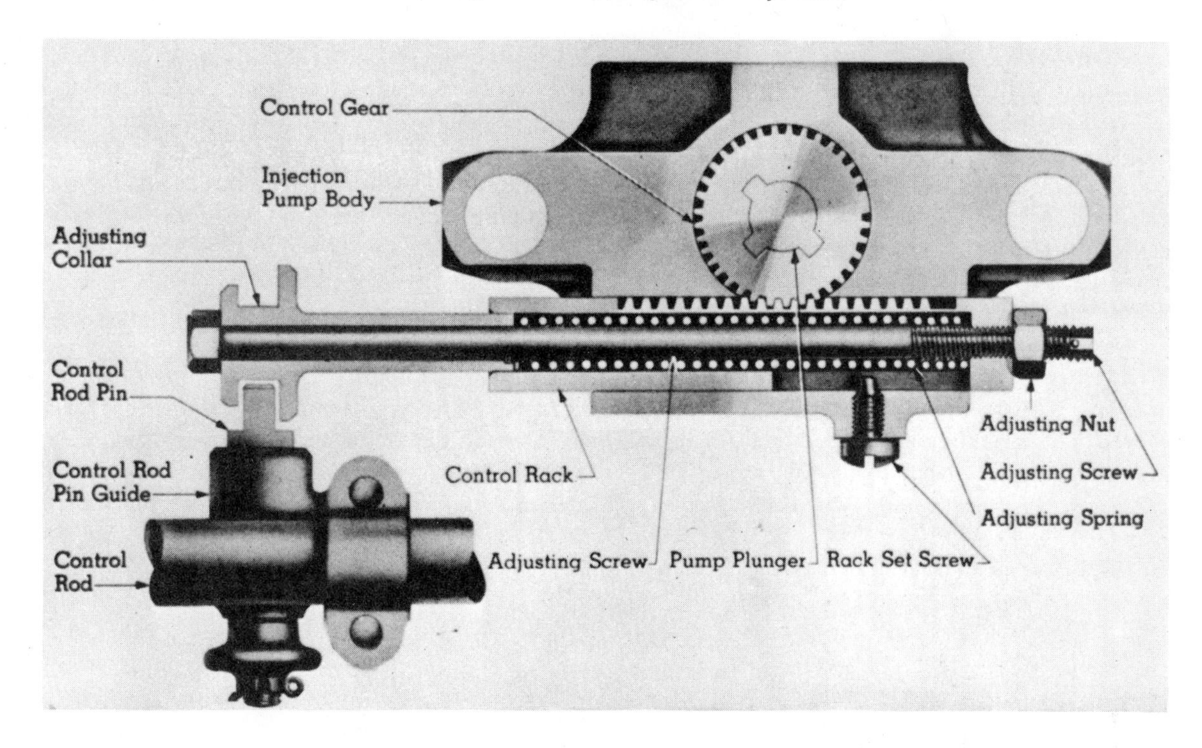

Fig. 8-74. Control mechanism of injection pump. (Fairbanks Morse Power Systems Div., Colt Industries)

The rotary position of the pump plunger is controlled by the governor by way of its linkage to the control rack of the pump and the control gear on the plunger, Fig. 8-74.

FAIRBANKS MORSE DUAL FUEL ENGINE INJECTION

The fuel supply system of the Fairbanks Morse dual fuel engine includes a duplex injection pump, Fig. 8-75.

The duplex injection pump receives fuel oil at low pressure from the supply pump, meters it, builds up a high pressure and delivers the fuel to the injection nozzle. The pump is of a two-plunger type. A pilot fuel plunger is connected to the main plunger and, therefore, has the same stroke.

The duplex pump is operated through a cam and tappet mechanism. The lower or pilot fuel pump delivers a constant amount of fuel at each stroke. The amount of fuel delivered by the pilot fuel pump is a small percentage of the full-load fuel oil operating requirement, due to the small cross-sectional area and the short stroke of the pump plunger. The pilot fuel pump delivery, during dual fuel operation, is used to initiate the combustion of the gas and air mixture. The fuel delivered by the pilot pump is not sufficient to sustain oil diesel operation.

Fig. 8-76 shows a pilot fuel pump plunger in successive positions of the downward stroke. At the top of its stroke fuel oil from the header flows through the inlet port, through the passages to the space below the plunger. As the plunger moves downward, fuel oil escapes from below the plunger until the port is closed. Further downward movement of the plunger builds up pressure on the trapped fuel until the spring-loaded discharge valve opens, releasing the high pressure oil into the

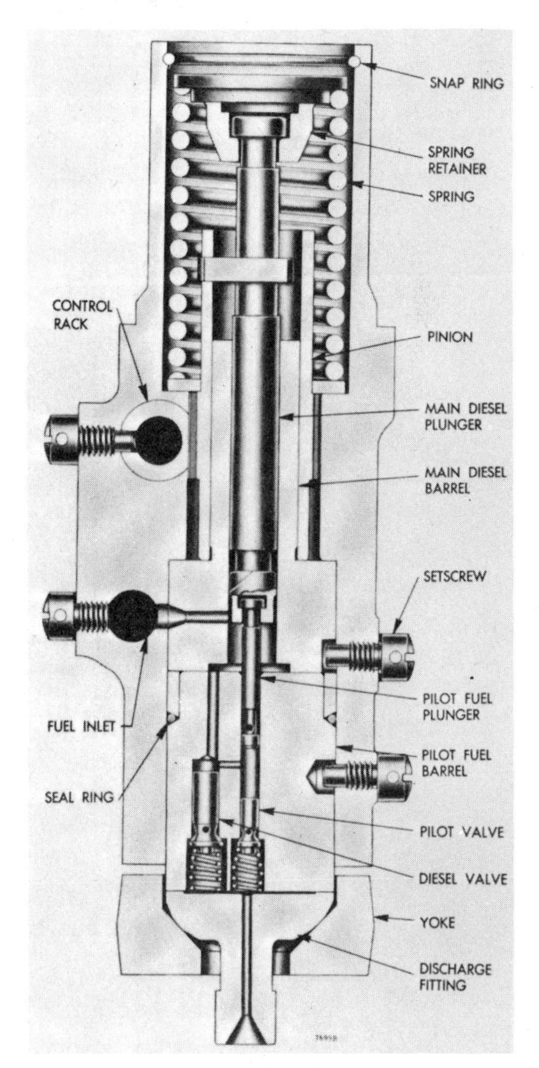

Fig. 8-75. Details of duplex injection pump used with dual fuel engines. (Fairbanks Morse Power Systems Div., Colt Industries)

injection tube. The injection continues until the relief in the plunger uncovers the inlet port at which time the trapped fuel is delivered through a hole in the plunger to the inlet passage. This condition exists until the end of the downward stroke.

The injection plunger moves vertically in the pump barrel, delivering fuel through the discharge valves and ignition tube to the injection nozzle and the combustion space. The injection pumps receive fuel from the engine header and delivers the fuel to the nozzles under high pressure.

As the plungers have a constant stroke the space in the barrel under the plungers always admits an equal amount of fuel from the supply system. The amount of fuel delivered to the nozzle depends on the position of the helix on the upper plunger relative to the port in the pump barrel. This position is controlled by the governor.

When the upper plunger is in its highest position, or on low cam, Fig. 8-75, the fuel ports are uncovered and the pump barrels fill with oil. The port is covered on the down stroke by the helix, and fuel is delivered to the nozzle until the top edge of the helix on the plunger uncovers the port.

Note in Fig. 8-76, that one-quarter turn changes the upper plunger position so the effective stroke varies from full-load to no-load position. The effective stroke is that part of the plunger stroke during which the port is covered and fuel is injected.

When the upper plunger is in "stop" or "no fuel" position, the slot lines up with the port during the entire stroke, thus bypassing all the fuel in the upper portion of the pump. The pilot fuel continues to pump fuel.

With the upper plunger turned so that the widest section of

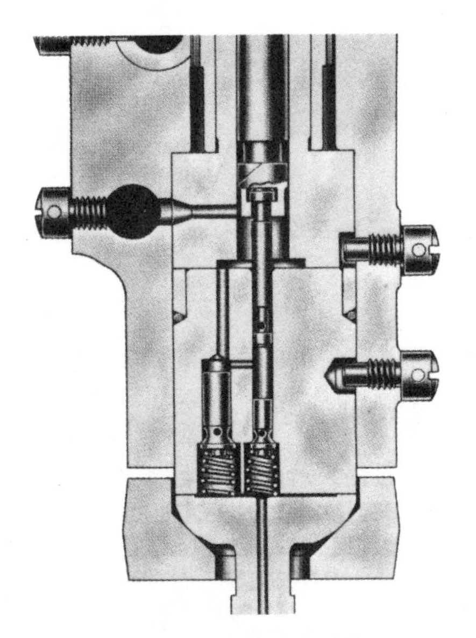

TOP OF STROKE-GAS OPERATION

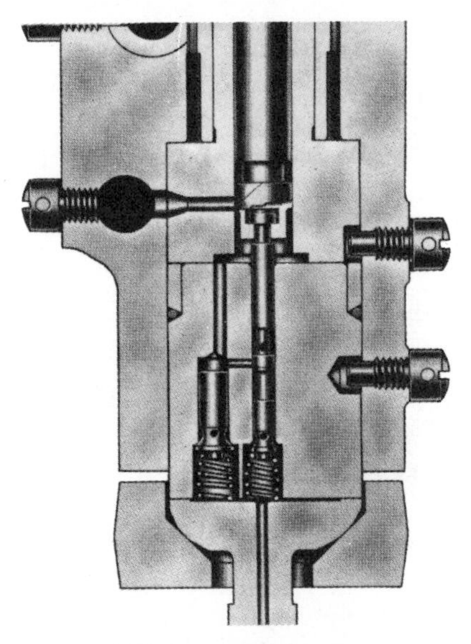

BEGINNING OF PILOT FUEL INJECTION-GAS OPERATION

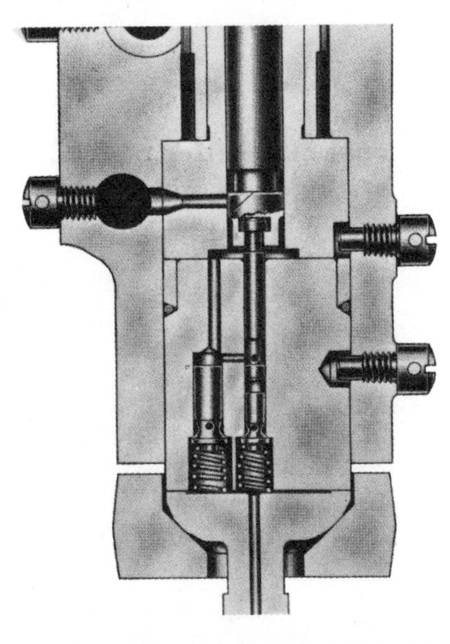

END OF PILOT FUEL INJECTION-GAS OPERATION

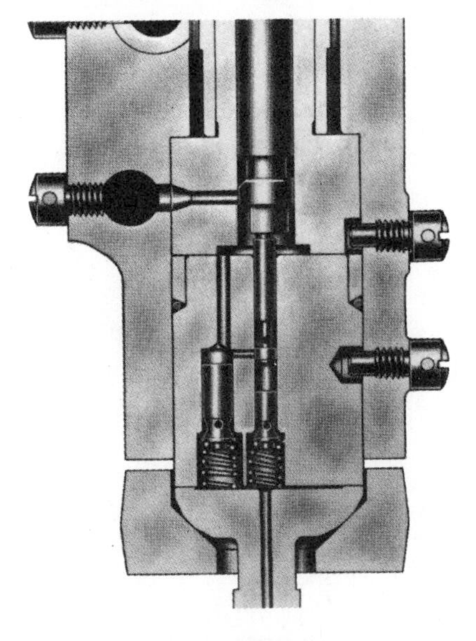

BEGINNING OF INJECTION-DIESEL OPERATION

Fig. 8-76. Pilot plunger positions. (Fairbanks Morse Power Systems Div., Colt Industries)

the helix coincides with the fuel inlet, the amount of fuel delivered will be maximum. On full-load, however, the effective stroke, Fig. 8-76, is only a portion of the maximum stroke.

The rotary position of the upper plunger, is controlled by the governor, through its linkage with the control rack of the pump. The control rack rotates the plunger through the plunger control gear, which is splined to the plunger, Fig. 8-74.

The control rack is connected at the adjusting collar, Fig. 8-74, to the control rod plunger. For an increase in fuel, the adjusting collar moves the control rack to the left. For a decrease in fuel the control rack is moved to the right. The adjusting collar is connected to the control rack through a spring. The spring-loaded slip connection is used so that should one or more pumps stick, the remaining pumps can still be controlled. Cutout of a duplex injection pump does not stop delivery of fuel by the pilot pump.

GAS ADMISSION VALVE

The gas admission valve of the Fairbanks Morse dual fuel engine, Fig. 8-77, is installed in each cylinder to admit gas from the gas header. The valves are actuated by a push rod and rocker through a tappet from gas cams on the control side camshaft. The valves located on the control side of the engine are installed in the left cylinder adapter as viewed from the control side of the engine.

GENERAL MOTORS UNIT INJECTORS

Three divisions of the General Motors Corporation build unit injectors for diesel engines. In these unit injectors, the injection pump, injector and spray valve form a single unit. A single unit is provided for each cylinder.

The Detroit Diesel Engine Division builds diesel engines for

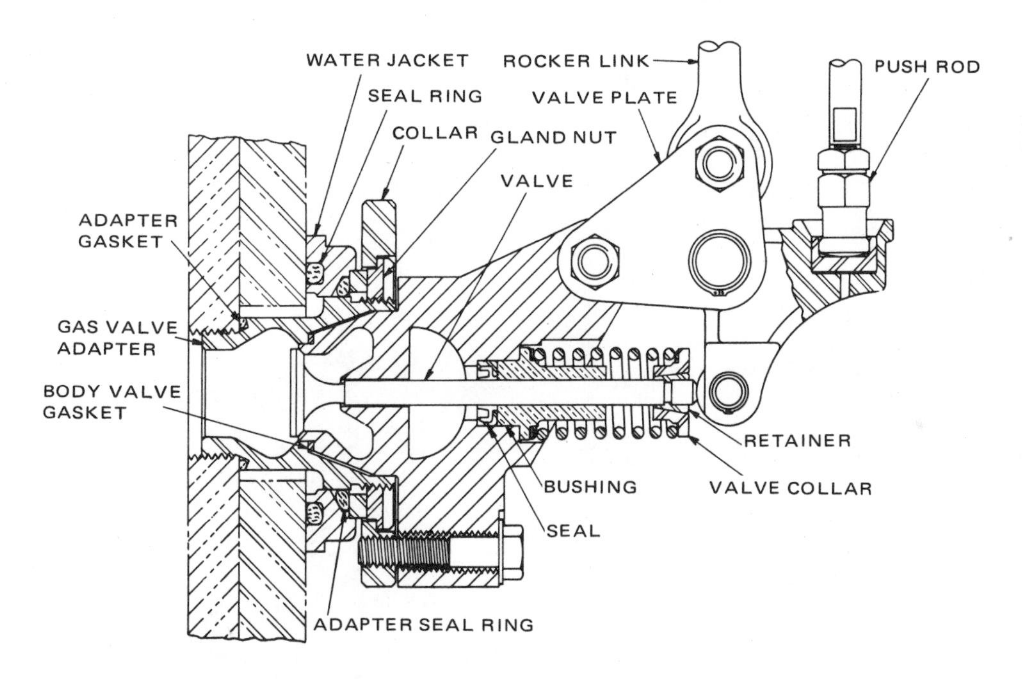

Fig. 8-77. Gas admission valves as used with dual fuel engines. (Fairbanks Morse Power Systems Div., Colt Industries)

The pilot fuel pumps deliver a constant amount of fuel oil on each stroke. The upper pump may discharge various amounts from no fuel to full fuel as determined by the position of the racks. These pumps discharge into the same injection tubes to separate discharge valves. Therefore a minimum fuel delivery by the pump will occur when the control rack is at O end rack. This will be the condition when operating an engine on gas.

The upper injection pump may be cut out or made inoperative by pulling the control rod plunger out of the control rack adjusting collar and moving the adjusting collar to the right, so the plunger and collar are no longer engaged. The pump will continue to pump pilot oil when the plunger and collar are disengaged.

the automotive field. The Electro-Motive Division produces engines primarily for locomotive applications, while the Cleveland Diesel Engine Division builds engines primarily for the marine field.

All three of the unit type injectors are quite similar in principle, but vary in size and also vary in arrangement of control mechanism and design of fuel supply. The description which follows pertains primarily to the injector produced by the Detroit Diesel Engine Division for use on General Motors Trucks and Coaches, such as the Series 51, 53, 71 and 110 engines, which are of the two-cycle type.

The General Motors unit injectors perform these functions:

1. Creates a high fuel pressure needed for efficient injection.

2. Meters and injects the amount of fuel required for varying engine loads.
3. Atomizes the fuel for mixing with the air in the combustion chamber.
4. Permits continuous fuel flow.

A typical fuel system used in GMC trucks and coaches is shown in Fig. 8-78. The fuel system includes fuel pump, fuel filters, fuel lines, fuel oil manifold and the fuel injectors.

Fuel is drawn from the fuel supply tank, through the primary filter by the action of the fuel supply pump. This is driven from the front end of the blower right rotor shaft. From the pump, fuel is forced through the secondary filter to

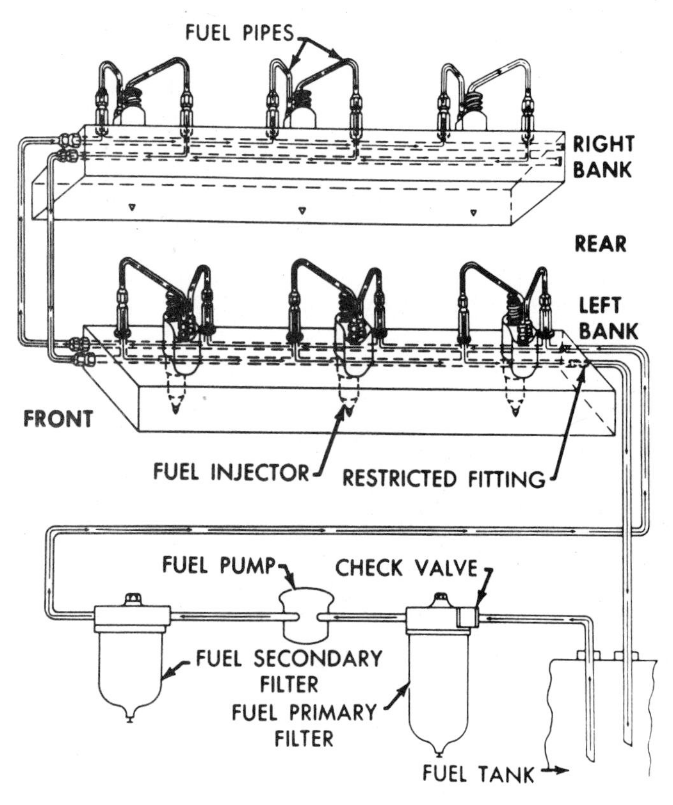

Fig. 8-78. Layout of General Motors fuel injection system for two-cycle engines. (Detroit Diesel Engine Div., General Motors Corp.)

the intake manifold, which is integral with the cylinder head. The intake manifold then supplies the individual injectors. Excess fuel is returned to the supply tank.

The installation of the injector in the cylinder head is shown in Fig. 8-79. This injector is of a unit type with the pump and nozzle forming a single unit. High-pressure fuel lines which characterize other systems are eliminated. Note in Fig. 8-79, the injector is operated through a cam push rod and rocker arm mechanism.

Two types of injectors are commonly used. One type, Fig. 8-80, is provided with crown type of valve. The other injector which has a needle valve is illustrated in Fig. 8-81. The exterior view of the GMC unit injector, Fig. 8-82, illustrates the method of identification. It must be remembered that injectors are especially calibrated for individual engines and are not interchangeable.

The operation of crown and needle valve injectors is similar

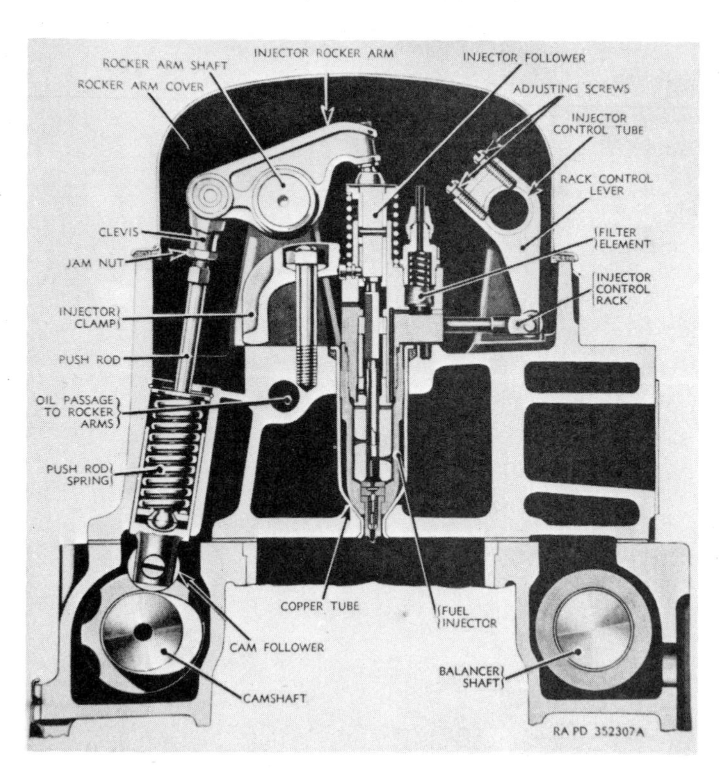

Fig. 8-79. GM unit fuel injector mounted in cylinder head.

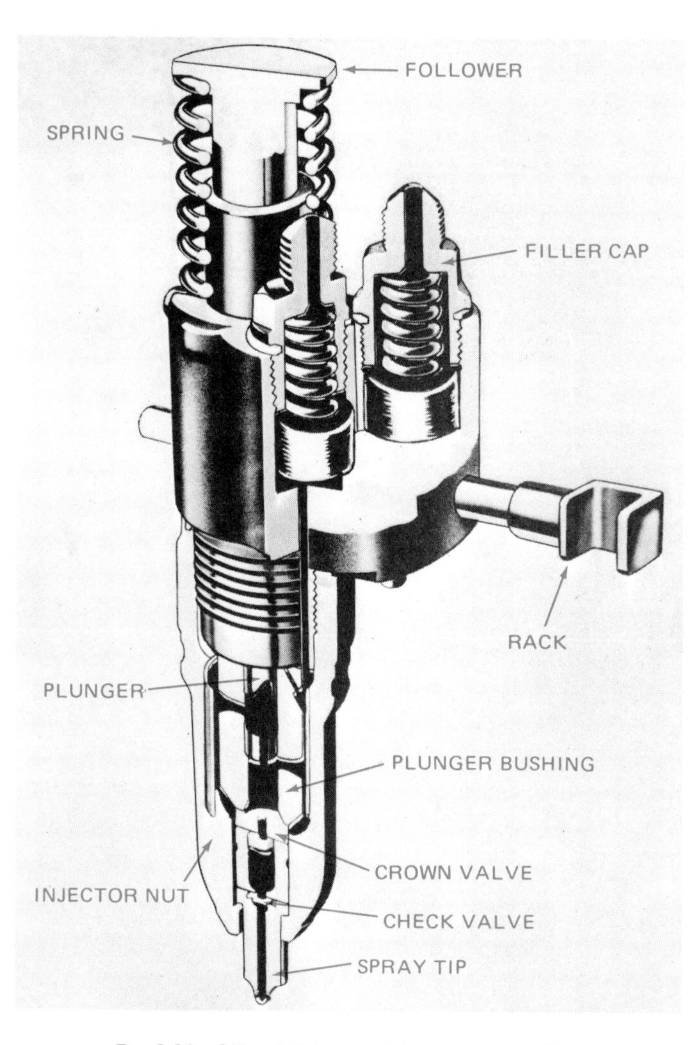

Fig. 8-80. GM unit injector with crown type valve.

except for the action of the needle and crown valves which occur at the time of injection. The following description of operation applies specifically to the needle valve type illustrated, Fig. 8-81.

Fuel under pressure enters the injector at the inlet side through a filter cap and filter element. From the filter element, fuel passes through a drilled passage into the supply

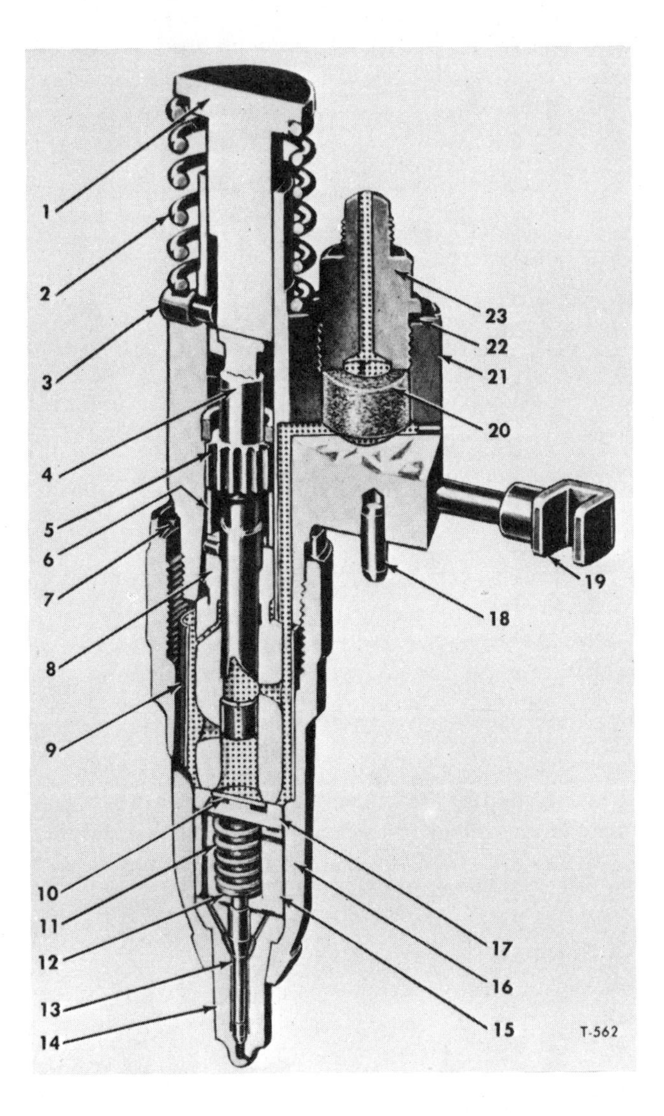

1. FOLLOWER.
2. FOLLOWER SPRING.
3. FOLLOWER STOP PIN.
4. PLUNGER.
5. GEAR.
6. GEAR RETAINER.
7. SEAL RING.
8. PLUNGER BUSHING.
9. SPILL DEFLECTOR.
10. CHECK VALVE.
11. VALVE SPRING.
12. SPRING SEAT.
13. NEEDLE VALVE.
14. SPRAY TIP.
15. SPRING CAGE.
16. BODY NUT.
17. CHECK VALVE CAGE.
18. DOWEL.
19. CONTROL RACK.
20. FUEL FILTER.
21. INJECTOR BODY.
22. GASKET.
23. FILTER CAP.

Fig. 8-81. GM unit injector with needle type valve.

chamber, between the plunger bushing and the spill deflector, in addition to the area under the injector plunger within the bushing. The plunger operates up and down in the bushing, the bore of which is open to the fuel supply by two funnel-shaped ports in the plunger bushing.

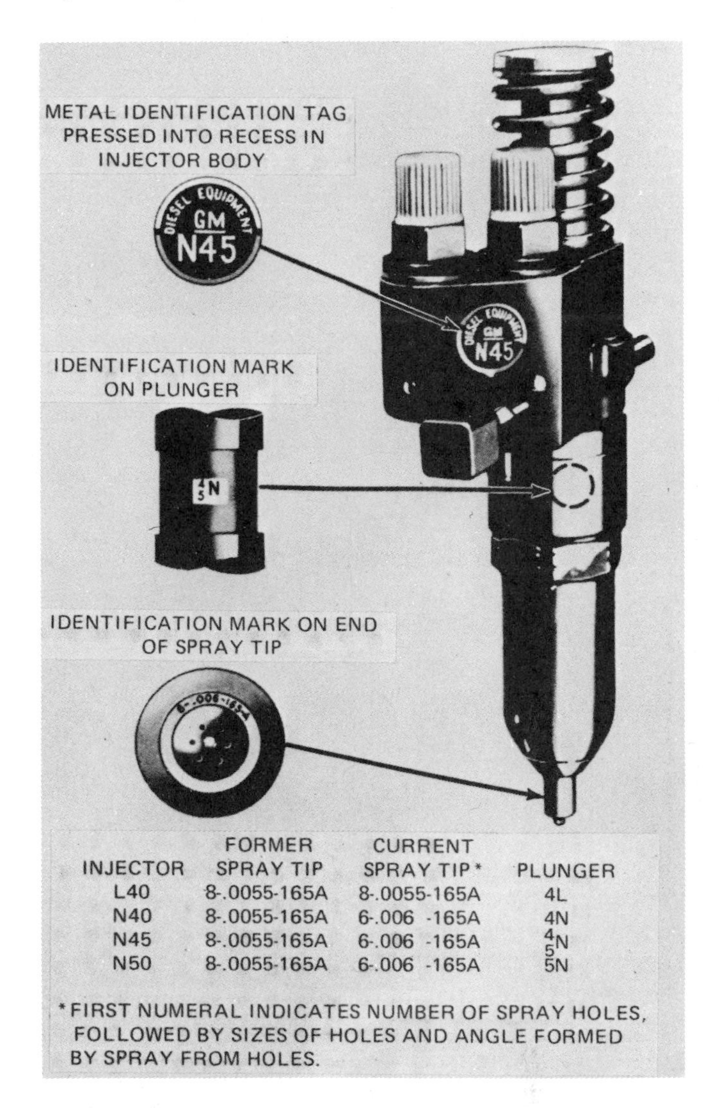

INJECTOR	FORMER SPRAY TIP	CURRENT SPRAY TIP*	PLUNGER
L40	8-.0055-165A	8-.0055-165A	4L
N40	8-.0055-165A	6-.006 -165A	4N
N45	8-.0055-165A	6-.006 -165A	4/5N
N50	8-.0055-165A	6-.006 -165A	5N

*FIRST NUMERAL INDICATES NUMBER OF SPRAY HOLES, FOLLOWED BY SIZES OF HOLES AND ANGLE FORMED BY SPRAY FROM HOLES.

Fig. 8-82. External view of GM unit injector.

The motion of the injector rocker arm, Fig. 8-79, is transmitted to the plunger, Fig. 8-81, by the follower which bears against the follower spring. In addition to the reciprocating motion, the plunger can be rotated during operation around its axis by the gear which meshes with the rack. For metering, the fuel and upper helix and the lower helix are machined in the lower part of the plunger. The relation of the helices to the two ports changes with the rotation of the plunger. As the plunger moves downward, under pressure of the injector rocker arm, a portion of that fuel trapped under the plunger is displaced into the supply chamber to the lower port until the port is closed off by the lower end of the plunger. A portion of the fuel trapped below the plunger is then forced up through a central passage of the plunger into the recess and into the supply chamber through the upper port until that port is closed off by the upper helix of the plunger. With the upper and lower ports both closed off, the remaining fuel under the plunger is subjected to increased pressure by the continued downward movement of the plunger.

When sufficient pressure is built up, it opens the flat, non-return check valve. Fuel in the check valve cage and spring cage passages, tip passages and tip fuel cavity is compressed

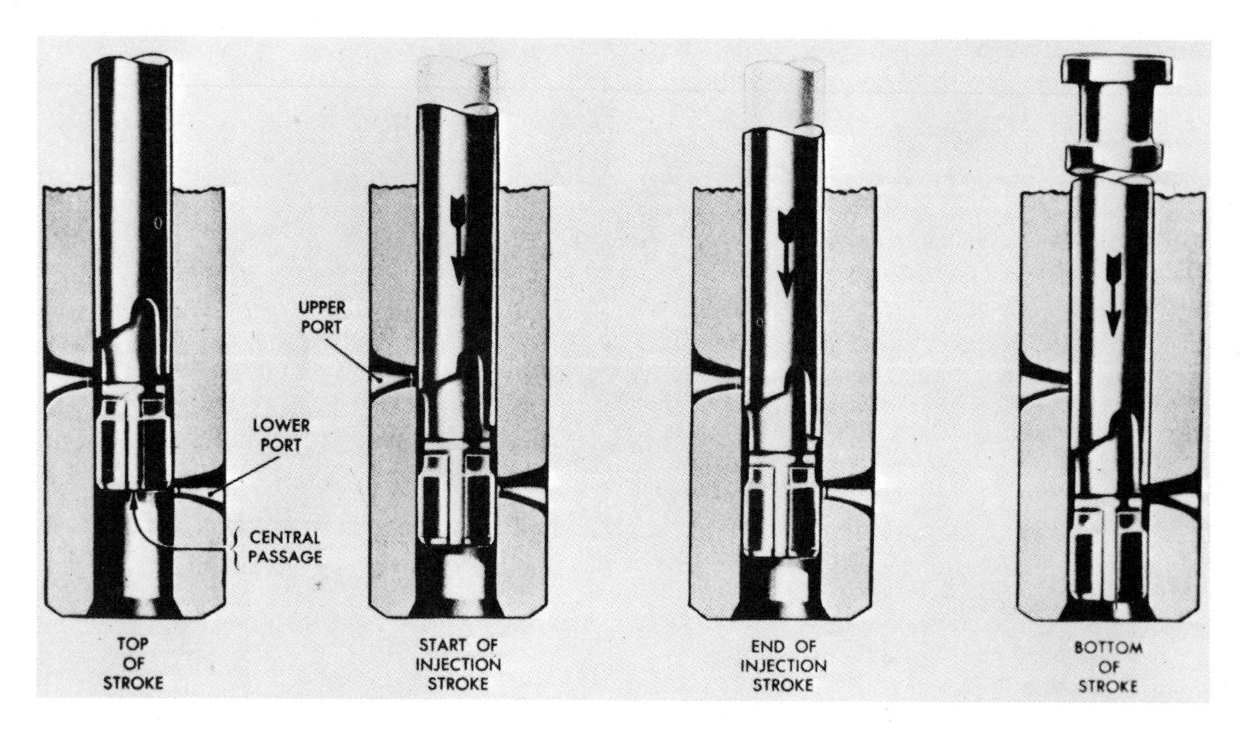

Fig. 8-83. Sequence of injector operation of GM unit injector by vertical travel of plunger.

until the pressure force acting upward on the needle valve is sufficient to open the valve against the downward force of the valve spring. As soon as the needle valve lifts off its seat, fuel is forced to the small orifices in the spray tip and is atomized into the combustion chamber.

When the lower land of the plunger uncovers the lower port in the bushing, the fuel pressure below the plunger is relieved and the valve spring closes the needle valve, ending injection.

A pressure relief passage has been provided in the spring cage to permit bleed off of fuel leaking past the needle pilot in the tip assembly.

A check valve, directly below the bushing, prevents leakage from the combustion chamber into the fuel injector in case a valve is accidentally held open by a small particle of dirt. The injector plunger is then returned to its original position by the injector follower spring. Fig. 8-83 shows the various phases of injector operation by the vertical travel of the injector plunger. On the return upward movement of the plunger, the high-

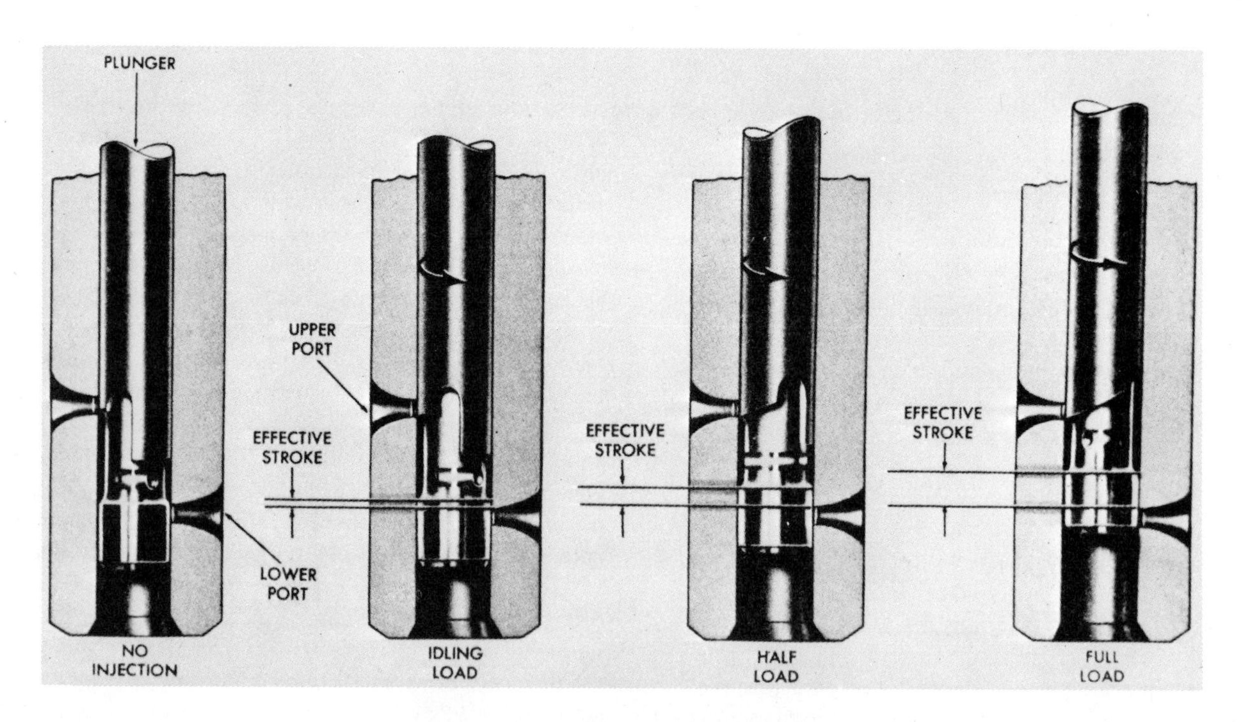

Fig. 8-84. Various plunger positions from no-load to full-load of GM unit injector.

pressure cylinder within the bushing is again filled with fuel oil through the ports. The constant circulation of fresh fuel through the injector renews the fuel supply in the chamber, helps cool the injector and also effectively removes all traces of air which might otherwise accumulate in the system and interfere with accurate metering of the fuel.

The fuel injector outlet opening, through which the excess fuel returns to the fuel return passage, and then back to the fuel tank, is directly adjacent to the inlet opening. This contains a filter element the same as the one on the fuel inlet side.

Changing the position of the helices by rotating the plunger, retards or advances the closing of the ports and the beginning and ending of the injection. At the same time, it increases or decreases the amount of fuel injected into the cylinder. Fig. 8-84 shows the various plunger positions from no-load to full-load. With the control rack pulled out all the way (no injection), the upper port is closed by the helix until after the power port is uncovered. Consequently, with the rack in this position, all of the fuel is forced back into supply chamber and no injection of fuel takes place. With the control rack pushed in (full injection), the upper port is closed shortly after the lower port has been covered, thus producing a maximum effective stroke and maximum injection. From this no injection position to full injection position (full rack movement), the contour of the upper helix advances the closing of the ports and the beginning of injection.

GENERAL MOTORS TORO-FLOW DIESEL

The General Motors Toro-Flow diesel engine is equipped with American Bosch Arma PSJ series fuel injection system described on page 62.

MURPHY INJECTION SYSTEM

The Murphy injection system which is of the cam operated injector type, Fig. 8-85, is used on Murphy diesel engines. These engines are used largely in the industrial, marine and electrical generating fields.

Fuel at relatively low pressure of approximately 20 psi enters the unit injector and flows to the annular chamber around the barrel. It then enters the plunger chamber through a port in the barrel when the plunger is on the up stroke, Fig. 8-85. The port is closed on the down stroke of the plunger and the fuel is forced to the dual check valves, which are of the flat seat type. Fuel enters and passes through the spray tip into the combustion chamber of the engine.

Fuel quantity is controlled by a rack and gear which rotates the plunger. The plunger has transverse helical slots connecting with actual radial slots to control the end of injection.

ONAN INJECTION SYSTEM

The Onan Corporation produces diesel operated electric power systems, Fig. 8-86. The Onan injection system uses a single plunger to distribute the fuel to the desired cylinder. The pump is actuated by the engine camshaft, which has a gear to drive the pump for rotary motion and an eccentric to transmit the reciprocating motion.

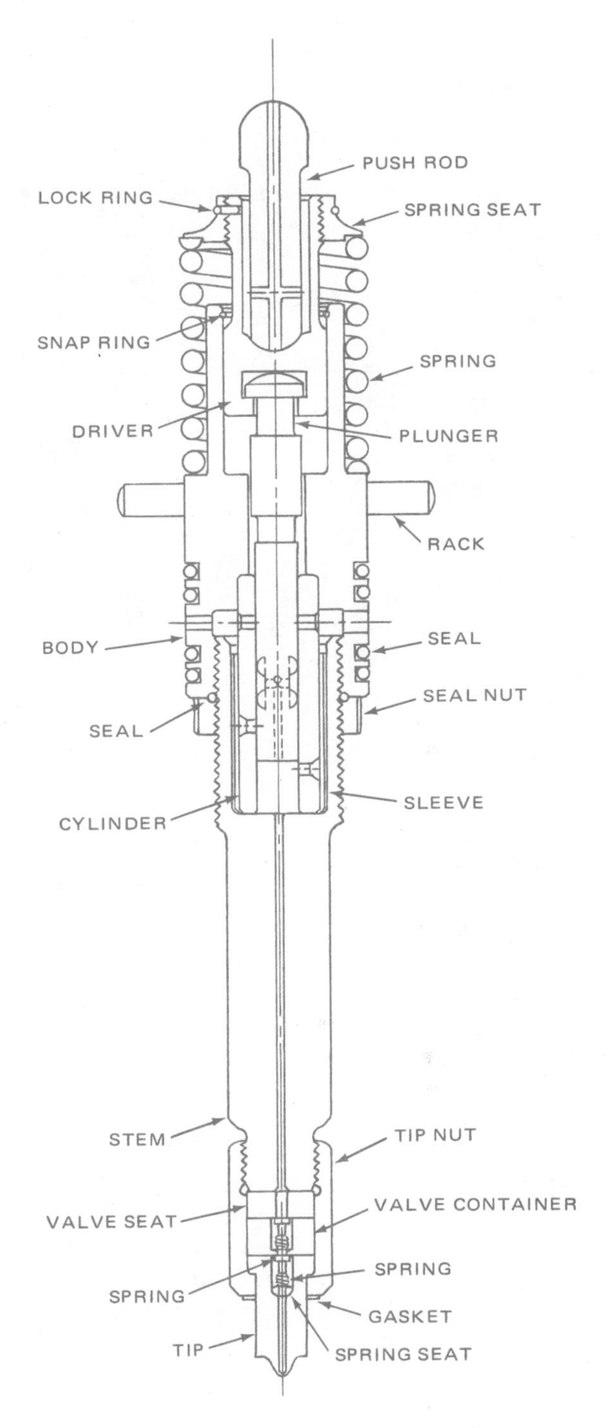

Fig. 8-85. Details of Murphy injector.

The pump consists basically of a ported plunger moving up and down with a sleeve around it to meter the quantity of fuel pumped, Fig. 8-87. The amount of fuel is controlled by the thickness of the sleeve, not by the length of the plunger stroke. Fuel pumping starts when the plunger port closes as it rises past the bottom of the sleeve. Pumping stops as the port is exposed above the sleeve.

By using a rotary motion, limiting the travel of the plunger,

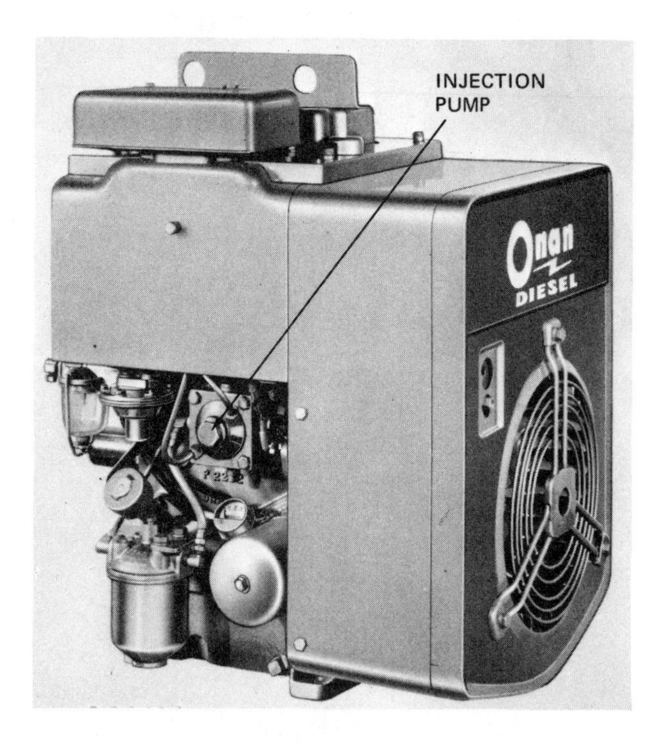

Fig. 8-86. Onan diesel engine generally is used in conjunction with electric power supply systems.

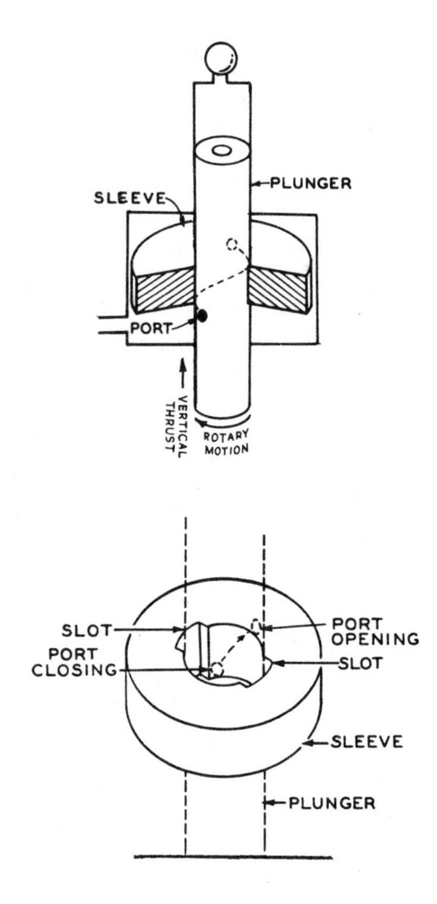

Fig. 8-88. Method of controlling fuel quantity in Onan injector.

and adding a vertical slot to the bore of the sleeve, fuel quantity can be controlled, Fig. 8-88.

The port can go above the sleeve but not below it. As the plunger starts up, its rotary motion closes the port on one edge of the slot and injection of fuel starts. Fuel injection continues until the port is exposed above the top of the sleeve. By turning the sleeve slightly so the slot closes the port fuel quantity can be controlled.

Sleeve rotation is controlled by a shaft with a small finger which engages a slot on the outer edge of the sleeve, as shown in Fig. 8-89. The shaft protrudes through a cover plate and a suitable throttle lever is attached to outer end of the shaft.

Note that the port opening and port closing are not dependent on each other. They are controlled by two different motions, so changing one will not change the other. The port opening is controlled only by the plunger lift and port closing is controlled only by the plunger rotation. It is evident that port closing and port opening can occur at the same time, resulting in no fuel injection. That method is used to stop the engine.

The method of distributing the metered fuel to the proper cylinder through a slot and several holes is shown in Fig. 8-90. The delivery valve at the top of the pumping chamber acts as a check valve but with a special purpose. It stops fuel from entering the pumping chamber from the high-pressure lines. It

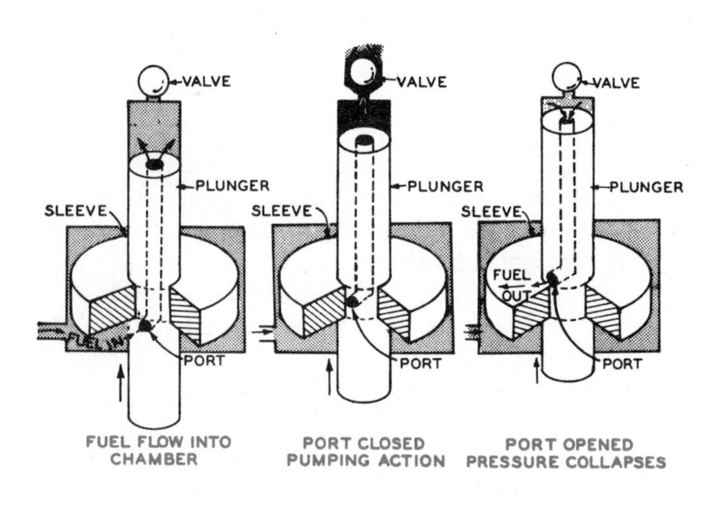

Fig. 8-87. Showing fuel flow in Onan injector.

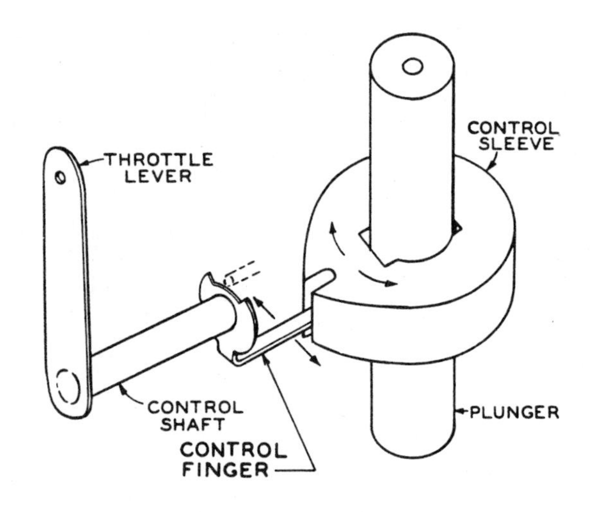

Fig. 8-89. Sleeve rotation is controlled by small finger. (Onan)

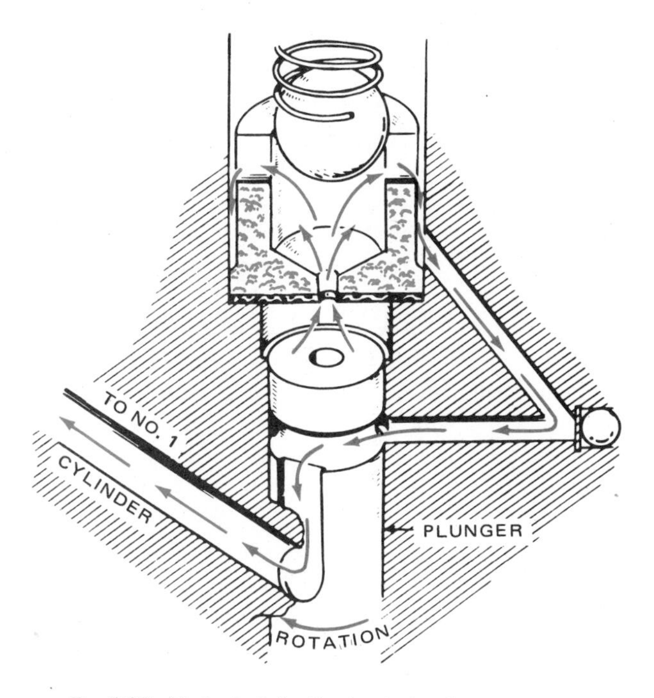

Fig. 8-90. Method of distributing fuel in Onan injector.

also tends to pull a little fuel from the lines at the end of the pumping cycle and thus shuts off the flow to the nozzle suddenly, as shown in Fig. 8-91.

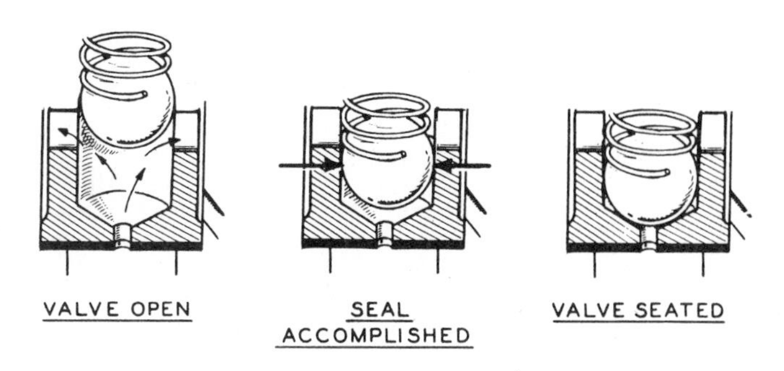

Fig. 8-91. Details of delivery valve in Onan pump.

ROOSA MASTER MODEL DB FUEL INJECTION PUMP

The Roosa Master fuel injection pump model DB is a single cylinder, opposed plunger, inlet metering, distributor type, Fig. 8-92. The rotating members of this fuel injection pump include the drive shaft, distributor rotor, which contains the plungers and mounts the governor and the transfer pump.

As shown in Fig. 8-92, the drive shaft engages the distributor rotor in the hydraulic head. The drive end of the rotor has a diametric bore containing two plungers.

The plungers are actuated toward each other simultaneous-

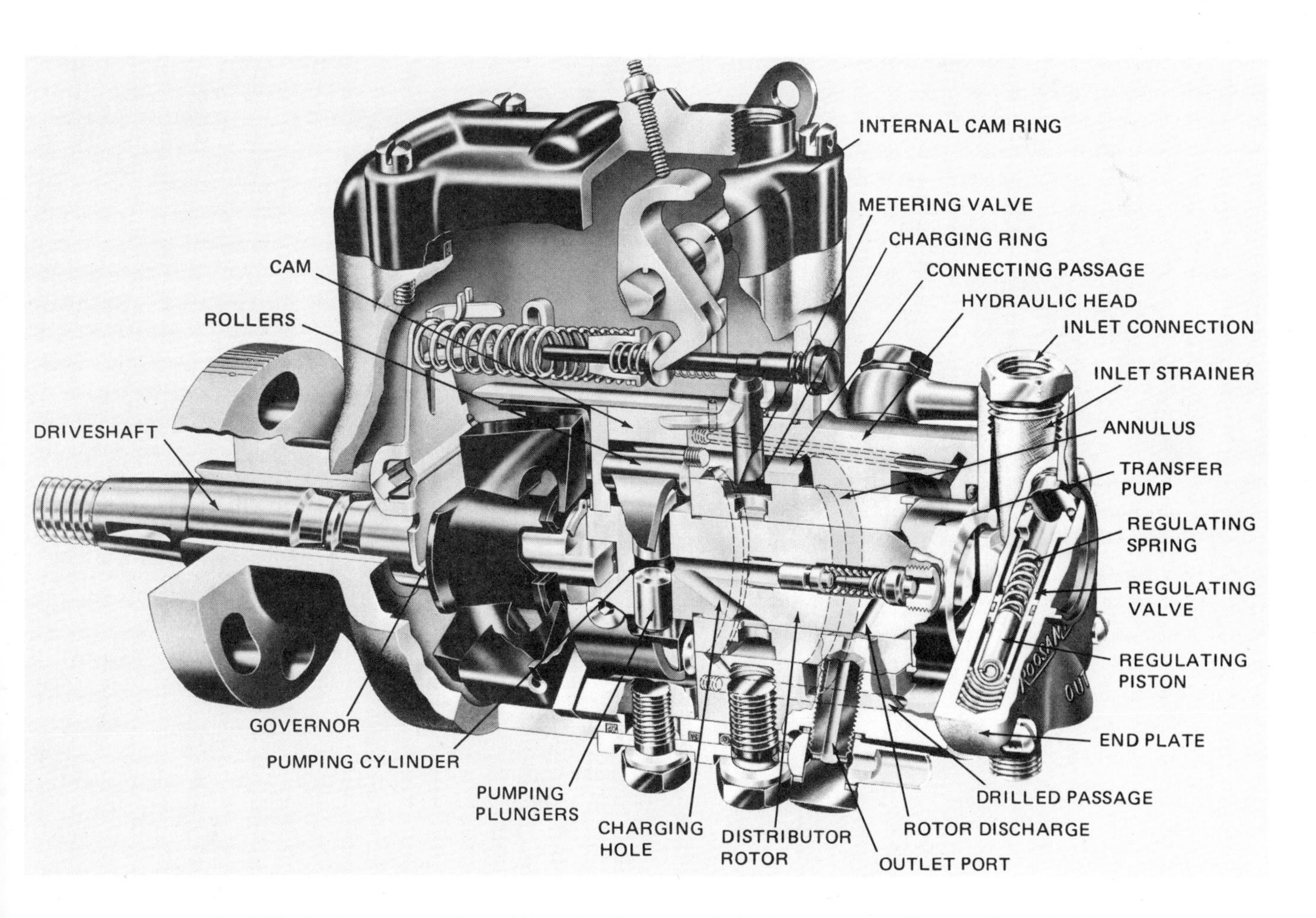

Fig. 8-92. Sectional view of Roosa Master distributor type fuel injection pump. (Standard Screw Co.)

ly by an internal cam ring through rollers and shoes, which are carried in guide slots in the flanged end of the rotor. Normally, there are as many lobes as there are cylinders to be served.

The transfer or supply pump in the opposite end of the rotor from the pumping cylinder, is of the positive displacement, vane type. It is covered by the end plate.

The model DB pump utilizes two types of distributor rotors to serve the requirements of different engine and combustion chamber types.

The simplest version incorporates a slot or channel which connects with the pumping cylinder by means of a single angled passage. The channel serves as a common passage for charging and discharging. The other type incorporates a single, angled passage for charging and an axial bore incorporating a delivery valve to serve all outlets for discharging. The hydraulic head contains the bore in which the rotor revolves, metering valve bore, charging ports and head outlets, to which are connected through appropriate fuel line connectors the injection pipes leading to the cylinders.

Covering the transfer pump, on the outer end of the hydraulic head, is the end plate. This assembly houses the fuel inlet connection, fuel strainer and transfer pump pressure regulating valve.

In operation, fuel is drawn from a supply tank into the pump through the inlet strainer by the vane type fuel transfer pump, Fig. 8-92. Since the transfer pump displacement greatly exceeds the injection requirements, a large percentage of fuel is bypassed through the regulating valve back to the inlet side. The flow thus bypassed increases with the speed and the regulating valve is designed so transfer pump pressure also increases with speed.

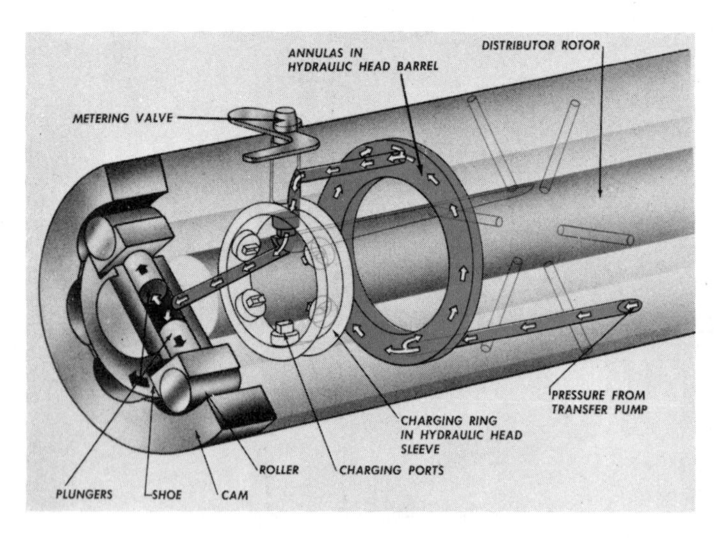

Fig. 8-93. Charge position of Roosa Master channel type rotor. (Standard Screw Co.)

Fuel, under transfer pump pressure, is forced through the drilled passage in the hydraulic head into the annulus. It then flows around the annulus to the top of the sleeve and through a connecting passage to the metering valve. The rotary position of the metering valve (controlled by the governor) regulates the flow of fuel into the charging ring which incorporates the charging ports.

As the rotor revolves, its single charging hole registers with one of the charging ports in the hydraulic head and fuel at transfer pump pressure flows through the angled passage to the pumping cylinder. The inflowing fuel forces the plungers outward a distance proportionate to the quantity to be injected on the following stroke.

If only a small amount of fuel is admitted into the pumping cylinder, as at idling, the plungers move out very little. As additional fuel is admitted, the plunger stroke increases to the maximum. It is limited by the leaf spring adjustment.

At this point (charging) of the cycle, the rollers are in the valley or relieved part of the cam between loads. The fuel is trapped in the cylinder for a slight interval after charging is complete.

The rotor charging port has passed out of registry with the head port and the rotor discharge port has not yet come into registry with the outlet port in the hydraulic head.

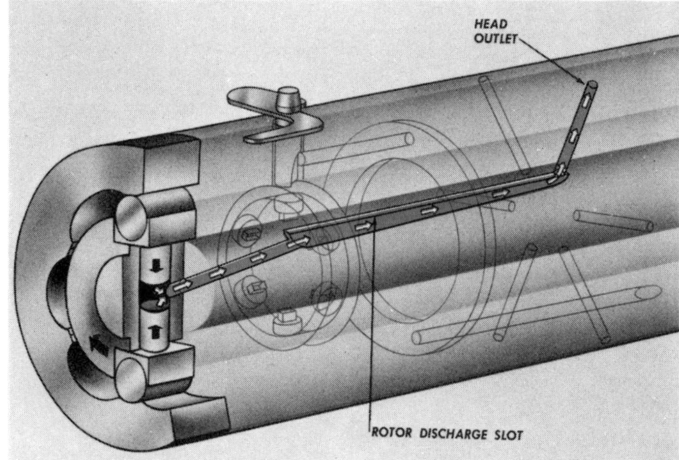

Fig. 8-94. Discharge position of Roosa Master channel type rotor. (Standard Screw Co.)

Further rotation of the rotor brings its discharge port into registry with an outlet port of the head at which point the rollers simultaneously contact the opposing cam loads and the plungers are forced toward each other. Fuel trapped between the plungers is forced from the pump through one of the outlet ports to an injection line.

As fuel at transfer pump pressure reaches the charging ring, slots on the rotor shank allow fuel and trapped air to bleed. This fuel fills the pump housing cavity and acts as a coolant as well as a lubricant, since it is allowed to return to the supply tank via the oil return connection in the pump housing cover. The return line also permits any air carried in the fuel, or originally contained in the pump to be carried out.

In addition an air bleed arrangement is incorporated in the hydraulic head which connects the outlet side of the transfer pump to the pump housing cavity. This allows air, which for any reason is carried into the end plate, to be bled back to the fuel tank via the return line.

CHANNEL TYPE ROTOR

The Roosa Master channel type rotor is illustrated in Figs. 8-93 and 8-94. These illustrations show the relationship of porting during the charging and discharging cycles.

As the rotor revolves, Fig. 8-93, the angled passage, leading from the slot or channel on the rotor shank registers with one of the charging ports in the hydraulic head. The fuel at transfer pump pressure then passes into the pumping cylinder forcing the plungers apart a distance proportionate to the amount of fuel required for injection on the following stroke. Only at full-load, will the plungers move to the most outward position which is controlled by the leaf spring setting (maximum fuel adjustment). This is the charging cycle. Note in Fig. 8-93 that while the channel is in registry with one of the charging ports, the shallow end of the channel is not in register as an outlet port. Note also that the rollers are between the cam lobes.

As the rotor continues to rotate, the channel passes out of registry with the charging port. For a brief interval the fuel in the rotor is trapped until the shallow end of the channel registers with one of the head outlets. As this registry takes place, both rollers contact the rise of cam lobes and are forced together, Fig. 8-94. This is the discharge or injection stroke. The fuel trapped between the plungers is forced through the angled passage, down the length of the channel and out the head outlet in registration, Fig. 8-94.

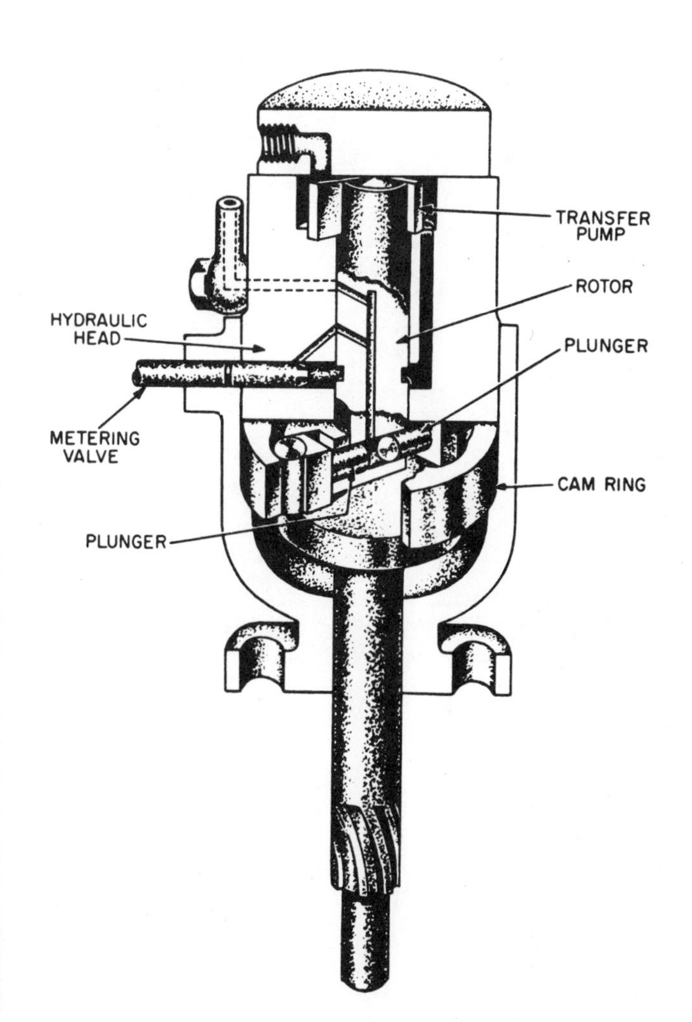

Fig. 8-95. Roosa Master fuel injection pump as used on some Gray marine engines in the U.S. Navy.

END PLATE OPERATION: The end plate in the Model DB Roosa Master injector pump is common to all models of the pump and varies only slightly between applications. It has three basic functions:

1. To provide passages for fuel, and cover and absorb end thrust of the transfer pump.
2. To house the pressure regulating valve.
3. To house the priming by-pass spring which permits fuel to bypass the transfer pump during hand priming.

The end plate and pressure regulating valve are shown in Fig. 8-92. During hand priming, the pressure differential across the transfer pump, caused by the hand primer, forces the piston down compressing a spring until the priming port is uncovered. Fuel then bypasses the stationary transfer pump to fill the system.

Fuel pressure forces the piston up the sleeve until the regulating port or ports are uncovered. Since the pressure on the piston is opposed by the regulating spring, the delivery pressure of the transfer pump is controlled by the spring rate and size and number of the regulating ports.

Another model of the Roosa Master fuel injector pump is shown in Fig. 8-95. This particular model is used to some extent by the U.S. Navy.

ROOSA MASTER PENCIL NOZZLE: The Roosa Master pencil nozzle, Fig. 8-96 has a body only 3/8 in. in diameter and is provided with an opening pressure and lift adjustment. Metered fuel flows through the inlet around the valve filling the nozzle body. When pressure overcomes the spring force, the valve lifts off its seat and fuel under high-pressure sprays through the holes. When fuel delivery ends and pressure drops, the spring returns the valve to its seat. During injection, fuel leaks past the valve guide, lubricating all moving parts.

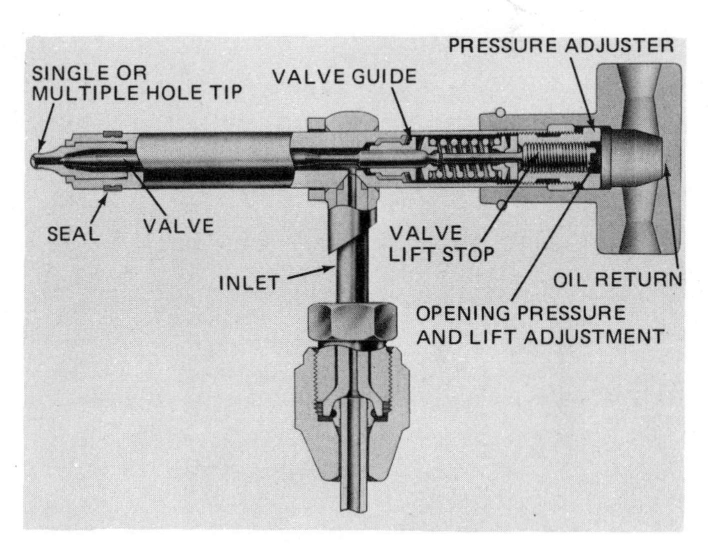

Fig. 8-96. Roosa Master pencil type injection nozzle. (Standard Screw Co.)

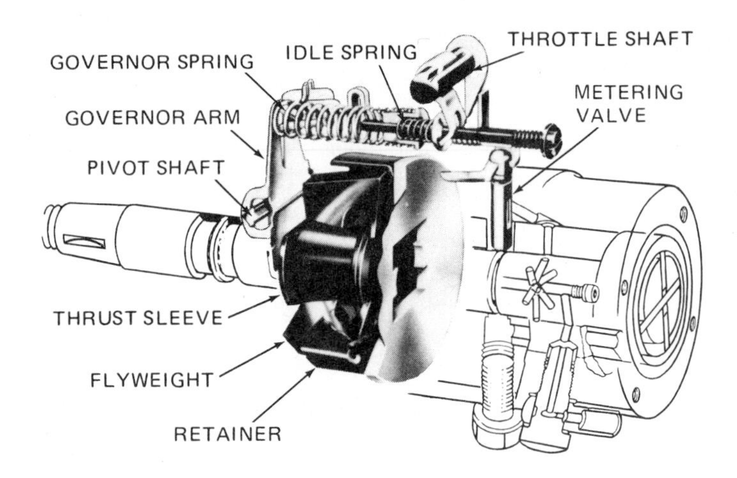

Fig. 8-97. Details of Roosa Master centrifugal governor.

ROOSA MASTER GOVERNOR

The governor used with the Roosa Master type DB injector, Fig. 8-97, is of the centrifugal type. In this design, the movement of the flyweights against the governor thrust sleeve rotates the metering valve. This rotation varies the registry of the metering valve slot with the passage of the rotor, thus controlling the flow of fuel to the engine. The force on the governor arm caused by the centrifugal action of the flyweights is balanced by the compression type governor spring, which is manually controlled by the throttle shaft linkage in regulating engine speed. A light idle spring is provided for more sensitive regulation at the low speed range. The limits of throttle travel are set by adjusting screws for proper idling and high speed positions.

ROOSA MASTER MODEL DM PUMP

The Roosa Master DM fuel injection pump, Fig. 8-98, is similar to the DP pump. However, model DM is designed for heavy-duty engines used in rugged construction and industrial applications. This pump is of the opposed plunger distributor type with inlet metering. A built-in governor is provided. The DM fuel injection pump is self-lubricating.

Working parts of the Roosa Master DM fuel pump are shown in Fig. 8-98, and the fuel flow plan is illustrated in Fig.

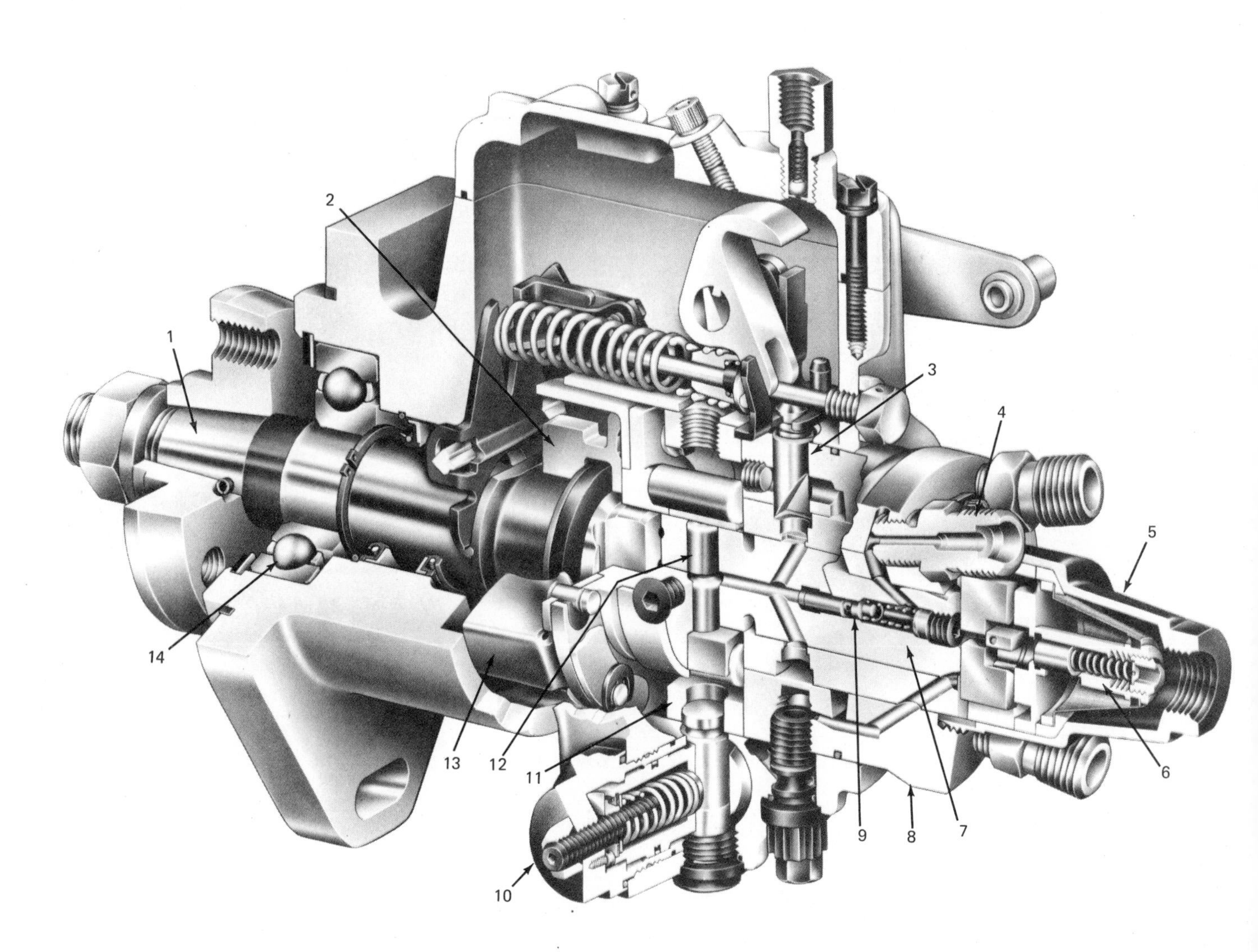

Fig. 8-98. Roosa Master model DM diesel fuel injection pump. 1—Drive shaft. 2—Governor weight. 3—Metering valve. 4—Discharge fitting. 5—Transfer pump. 6—Pressure regulator. 7—Distributor rotor. 8—Hydraulic head. 9—Delivery valve. 10—Advance. 11—Internal cam ring. 12—Pumping plungers. 13—Governor. 14—Drive shaft bearing.

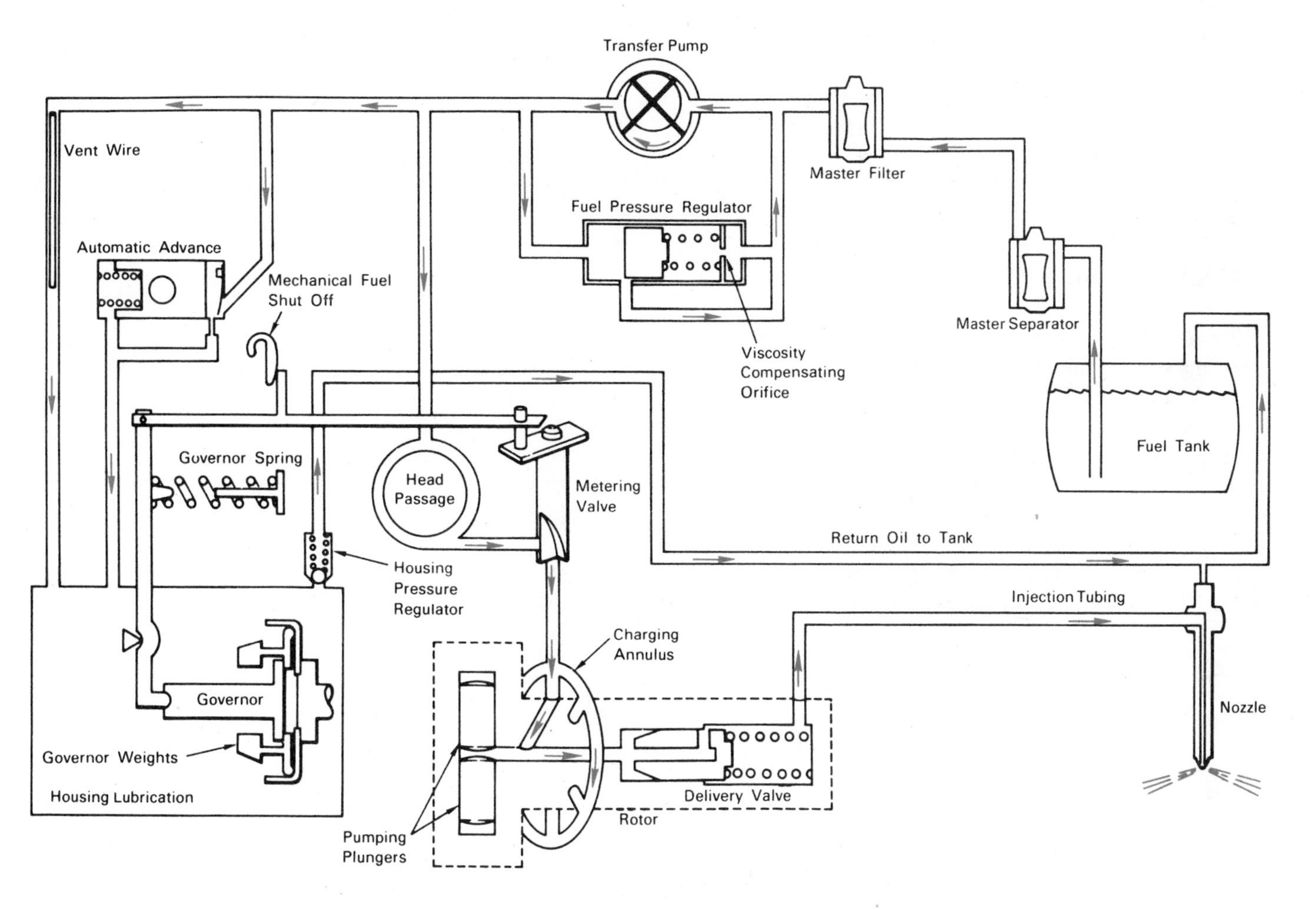

Fig. 8-99. Fuel flow diagram of Roosa Master model DM fuel injection pump. (Roosa Master Stanadyne/Hartford Div.)

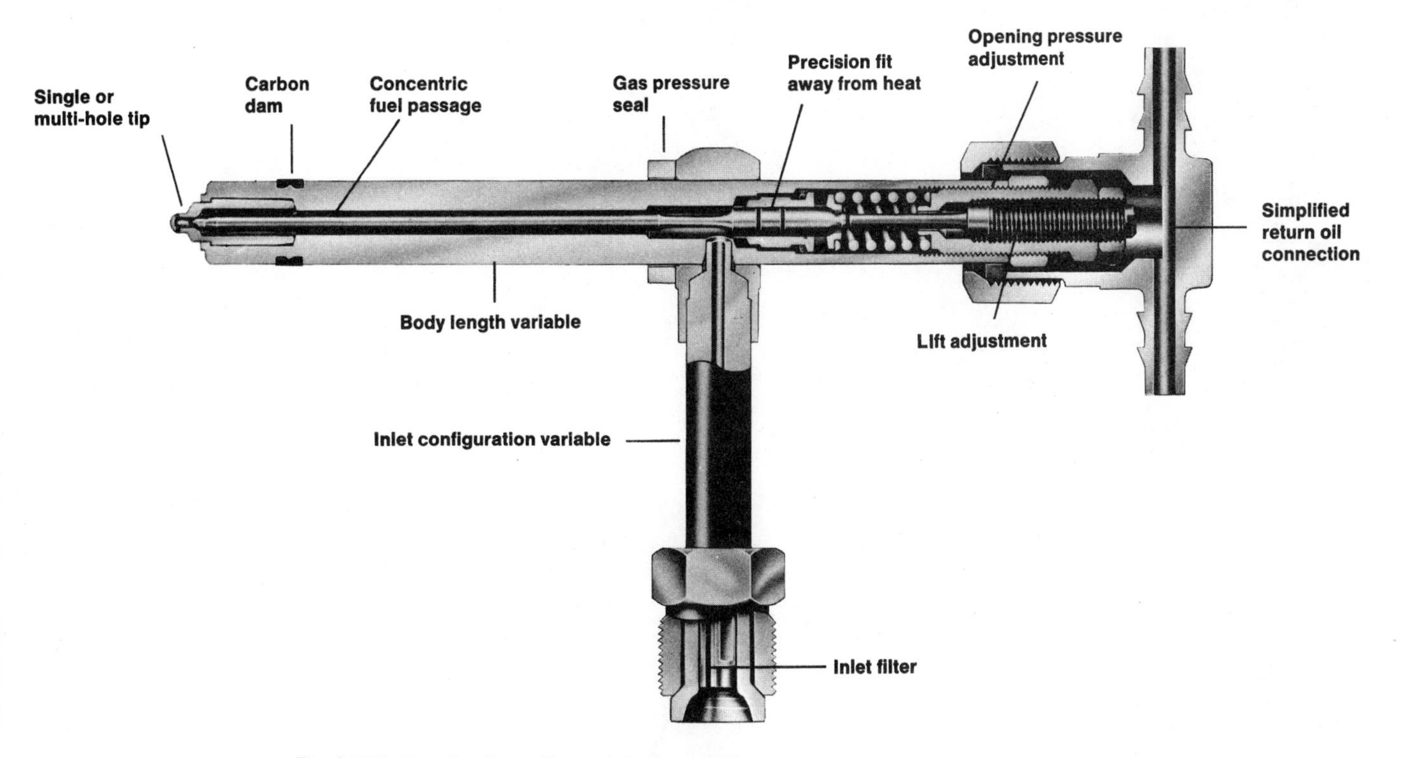

Fig. 8-100. Details of pencil type injection nozzle used in conjunction with DM fuel pump. (Roosa Master Stanadyne/Hartford Div.)

8-99. Note that the transfer pump draws fuel from the fuel tank through the master separator, then through the master filter. Next, the fuel is passed through separate headers to the head passage and charging annulus, and to the automatic advance and centrifugal governor. After passing through the head passage, metering valve, charging annulus and delivery valve, the fuel is passed to the injection nozzle.

ROOSA MASTER INJECTION NOZZLES

Details of the injection nozzle used in Roosa Master injection systems are shown in Fig. 8-100. This is a "pencil nozzle." It is principally a "hole type" nozzle designed for open combustion chamber diesel engines. The pencil nozzle operates by hydraulic differential pressures which can be adjusted by setting the desired opening. The nozzle opens inwardly from its seat in the tip. It is spring-loaded at the opposite end. The spring preload can be varied by means of the pressure adjusting screw. See Fig. 8-100. Valve lift is controlled by a separate screw located in the bore of the pressure adjust screw.

Fuel flow in the nozzle is through the inlet and around the valve to fill the nozzle body. When the fuel pressure overcomes the spring force, the valve lifts off its seat and fuel is injected into the combustion chamber through the nozzle holes.

When fuel delivery ends, the reduced pressure allows the spring to return the valve to its seat. With this type of nozzle, all moving parts are lubricated by the fuel oil.

CATERPILLAR FUEL INJECTION SYSTEMS

Caterpillar uses two basic injection systems, shown in Fig. 8-101. The precombustion chamber type differs from systems covered earlier in that the injection nozzle and precombustion chamber combine to form a single assembly which is screwed into the cylinder head. When necessary, the complete unit is replaceable.

The Caterpillar system provides individual pumps and injectors for each cylinder. The pump is placed at the side of the engine away from the heat of combustion. Because the

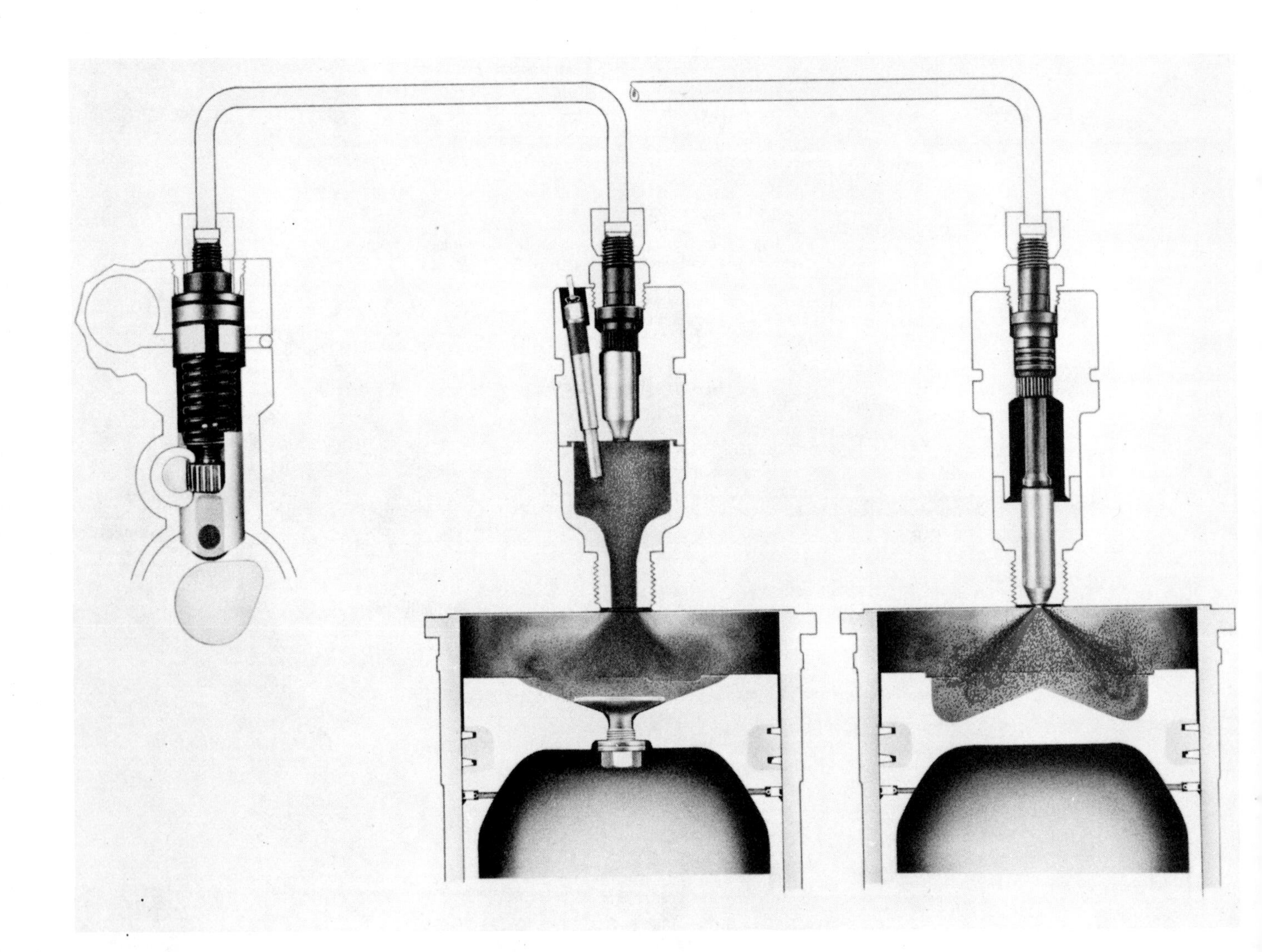

Fig. 8-101. Caterpillar fuel injection systems. Left. Fuel injection pump. Center. Precombustion chamber type. Right. Direct injection type. (Caterpillar Tractor Co.)

pump is separate from the injection nozzle and located in a cooler area, improved plunger life and performance is claimed.

Fuel is delivered from the fuel pump, Fig. 8-101 to the injector by the high-pressure line and connecting fittings to the valve body. It then flows through the space around the loosely fitted screw threads to the nozzle assembly. When the pressure reaches a predetermined point the poppet type valve opens and the fuel passes a single self-cleaning large diameter orifice, Fig. 8-101, into the precombustion chamber.

Fuel injection begins at pressures of approximately 1400 psi with maximum pressures of 4500 to 9000 psi which is considerably lower than pressure needed in direct injection.

The fuel injection valve is of the capsule type and is easily replaced. Because of the large diameter single orifice combined with the precombustion chamber, foul-free operation results on a wide range of fuels according to the manufacturer.

With the precombustion chamber design, Fig. 8-101, an electric glow plug aids in cold weather starting.

The complete system is shown in Fig. 8-102. This shows how fuel is drawn from the supply tank by the transfer pump. It then passes through the filter, to the manifold, to the injection pump and then to the injector.

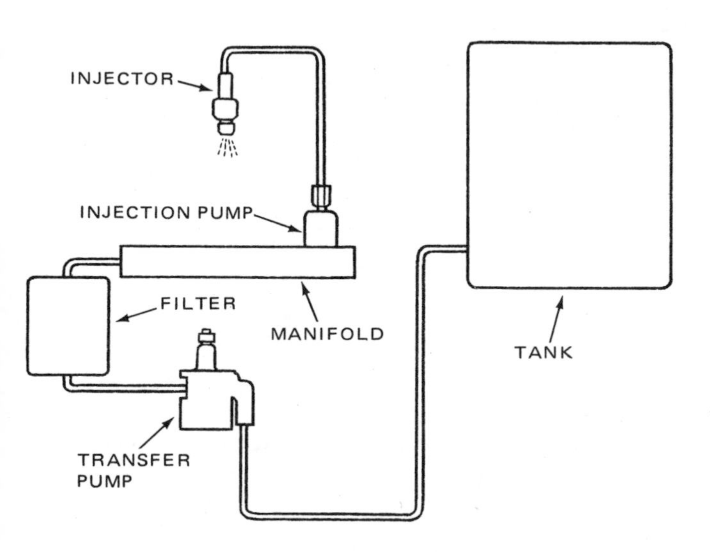

Fig. 8-102. Caterpillar injection system from supply tank to engine.

Injector pumps are interchangeable and no balancing or operating adjustments are necessary. Fuel pumps react to load changes through governor action.

The injector pump, Fig. 8-103, is cam operated and always makes a full stroke. The amount of fuel pumped per stroke is varied by turning the scrolled plunger of the pump. Rotation is achieved by governor action on the rack, which meshes with the gear segment of the pump plunger.

When the plunger is at the bottom of the stroke, fuel at approximately 30 psi enters the inlet port and flows out into the area above the plunger and down through the slot into the recess around the plunger. When the plunger moves upward, the inlet port is covered and fuel is trapped and forced back through the check valve. Fig. 8-103, shows how the helically cut plunger meters the fuel. In the shutoff position, the slot which connects the top of the plunger with the recess is in line with the port and no fuel is trapped and injected. In the idling position, the plunger is rotated so the narrow part of the plunger formed by the helix covers the port only for a short part of the stroke. This part is called the effective stroke. The effective stroke being short when idling, permits only a small amount of fuel to be injected.

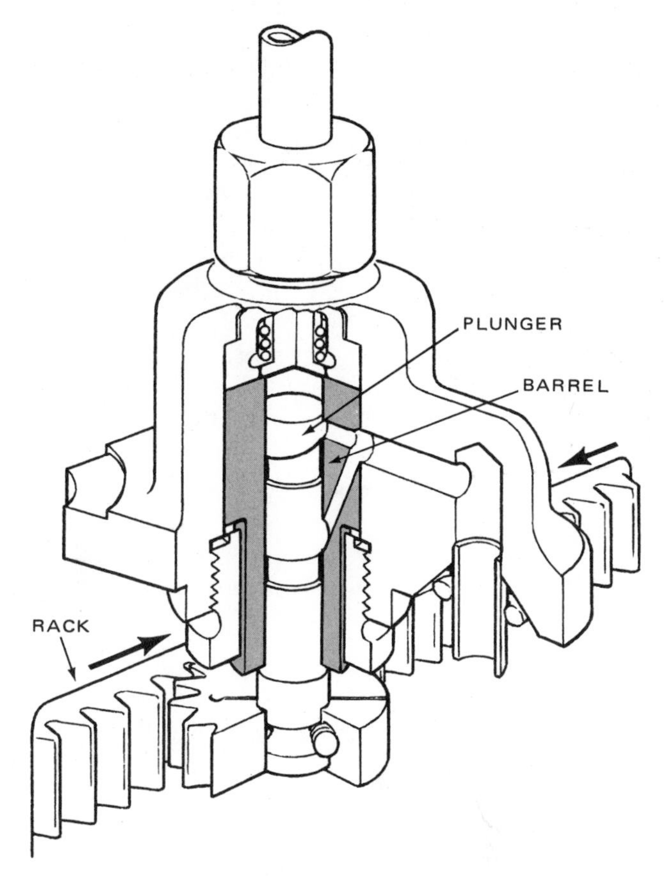

Fig. 8-103. Details of Caterpillar injection pump.

By rotating the plunger to the full-load position, a wider part of the plunger formed by the helix covers the port for a longer part of the stroke and more fuel is injected. Rotation of the plunger into intermediate position provides the correct amount of fuel to be injected to maintain a nearly constant engine speed during load variations.

The Caterpillar injection valve has only one large opening or orifice. This opening varies in size for different sizes of engines from .025 to .035 in. Because of its larger size it has a tendency to stay clean.

Injection starts into the precombustion chamber at 9 to 14 deg. before top dead center for all in-line engines up to 4.5 in. bore. On V-type engines, 4.5 in. bore, 5.4 in. bore, and larger engines have an automatic timing advance which advances ignition 1 deg. for every 100 rpm increase between 1400 and 2100 rpm.

CATERPILLAR SLEEVE METERING SYSTEM

The new Caterpillar fuel injection system is of the sleeve metering type, Fig. 8-104. It is designed specifically for use with the 3200 and 3300 series Caterpillar Tractor engines, Fig.

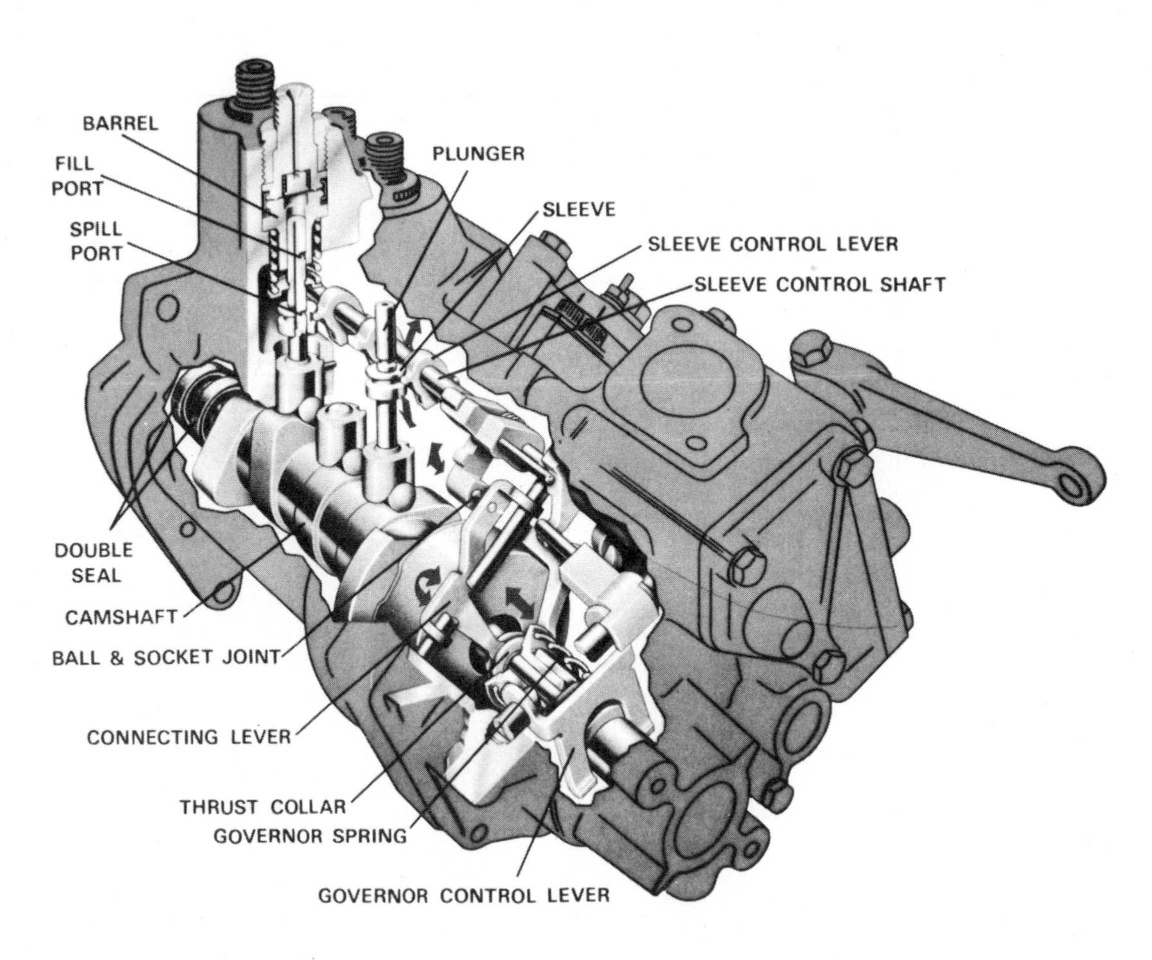

Fig. 8-104. Details of Caterpillar sleeve metering fuel injection pump and governor.

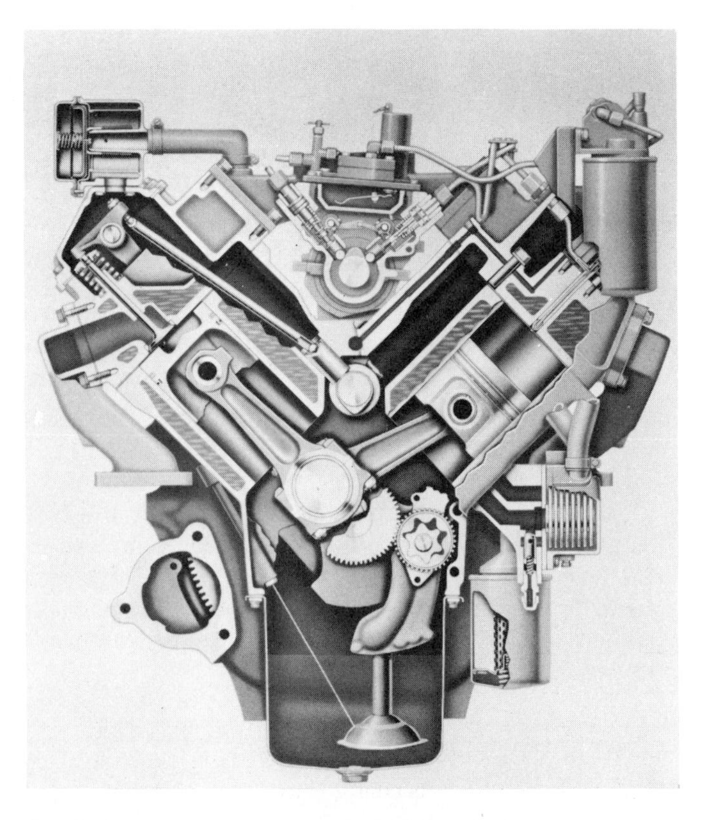

Fig. 8-105. Sectional view of Caterpillar direct injection engine, equipped with sleeve metering type of fuel injection.

8-105. The essentials of the new system include plunger A, in Fig. 8-106, that is machined with fill port E and spill port F. These ports are connected with a drilled passage extending downward from the top.

Sleeve B is free to slide up and down as positioned by the governor. Control rod C has finger type levers fitting into grooves in the sleeves to position them for metering the fuel. The upper part of the plunger rides in barrel assembly D.

There are no passages leading to the plunger fill ports. The entire housing is filled with fuel, immersing the plunger. Whenever the fill port is below the bottom edge of the barrel, it automatically fills with fuel from the pressurized reservoir.

The sequence of operations of the sleeve metering system is illustrated in four steps in Fig. 8-107:

1. The flow of fuel rushing into the fill port fills the pump chambers on the downward movement of the plunger, and the pump is charged with fuel.
2. At the correctly timed instant, the rotating cam lobe begins to push the plunger up again, closing the fill port as it passes upward into the barrel. Since the spill port is covered by the sleeve, the pump lines and fuel valves are pressurized and injection begins.
3. The injection continues as long as both fill and spill ports are completely closed by the barrel and sleeve.
4. Injection ends the moment the spill port edges above the sleeve, releasing the pressure in the plunger and, at the

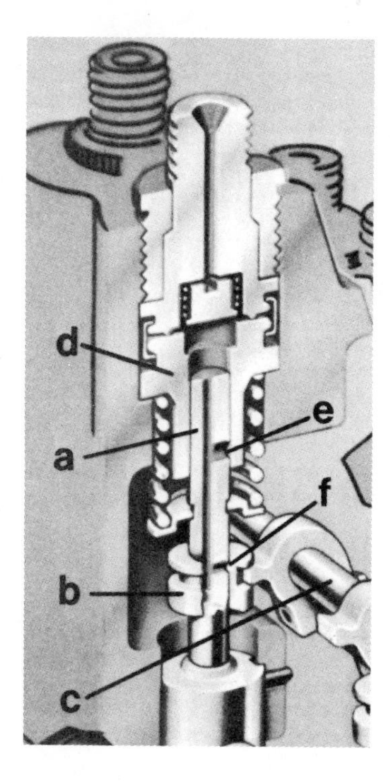

Fig. 8-106. Sectional view of Caterpillar fuel injection pump. A—Plunger. B—Sleeve. C—Control rod. D—Barrel assembly. E—Fill port. F—Spill port.

same time, letting fuel escape from the pump back into the housing.

To increase the amount of fuel injected, the sleeve is positioned higher on the plunger, which keeps the spill port closed for a longer period.

When more power is required, depressing the accelerator pedal will cause the governor control lever, Fig. 8-104, to move forward. This, in turn, compresses the governor spring and moves the thrust collar in the same direction. Movement of the thrust collar and the linkage rotates the sleeve control shaft and repositions the sleeves. When more power is called for, the sleeves lift and cover the spill ports for a longer time and more fuel is injected.

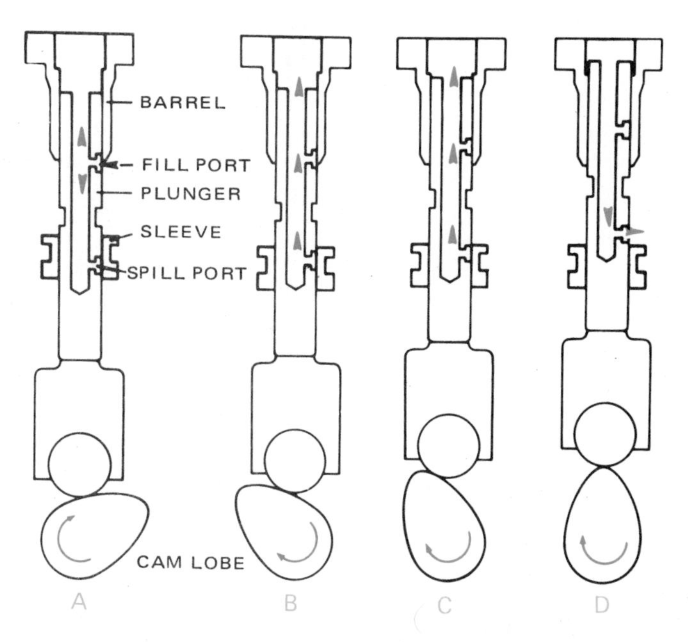

Fig. 8-107. Sequence of events in operation of sleeve metering fuel injection pump. (Caterpillar Tractor Co.)

WAUKESHA INJECTION SYSTEMS

Some Waukesha engines are equipped with American Bosch injection systems, others use a unit injector. The complete system consists of a supply tank, delivery pump, fuel filters, supply and return manifolds, pressure regulating valves and unit injectors, Fig. 8-108.

Fuel is pumped from the tank by a low-pressure delivery pump to several filters, Fig. 8-108. From the filters the fuel flows through the supply manifold to the various injectors. Excess fuel from the injectors flows through the return manifold to the return line back to the tank. Pressure regulating valves located in the right manifold supply line and the right return line are for the purpose of maintaining the desired supply pressure in the injectors.

The unit injector, Fig. 8-109, builds up the fuel pressure above that in the combustion chamber, to inject the fuel. It also meters the fuel accurately for the load requirements of the engine. In addition, it regulates the duration of injection and the time of injection with respect to piston travel for maximum combustion efficiency at various engine loads. The unit injector also atomizes the fuel and directs it into the combustion chamber.

The essential parts of the injector are a fuel supply circuit, follower assembly, plunger and bushing, rack and gear, and delivery valve assembly.

The continuous flow of fuel oil through the injector furnishes ample fuel for the engine requirements, cools injector parts, maintains uniform operating temperature and removes air from the injector.

The oil enters and leaves the injector through fine filter elements which prevent dirt from entering the unit while being handled. From the filter, however, the fuel passes through drilled passages in the body to the annular fuel chamber between the bushings and spill deflector. The spill deflector is a glass-hard, alloy steel sleeve which keeps the high pressure, high velocity by-pass fuel that flows out of the lower bushing port after injection from eroding the comparatively soft metal of the injector nut or body.

The follower assembly transmits the force from the rocker arm necessary to reciprocate the plunger and inject the fuel. At the same time it permits the plunger to rotate freely about its axis. The follower spring must exert the force necessary to return all reciprocating injector parts and the rocker arm assembly.

The stroke sequence of the injector is shown in Fig. 8-110. With the plunger at the top of the stroke, the bushing has been filled with fuel passing through the bushing ports from the annular fuel chamber. As plunger starts down, fuel is displaced back into the supply chamber through lower port, up through control passage and up through the upper port.

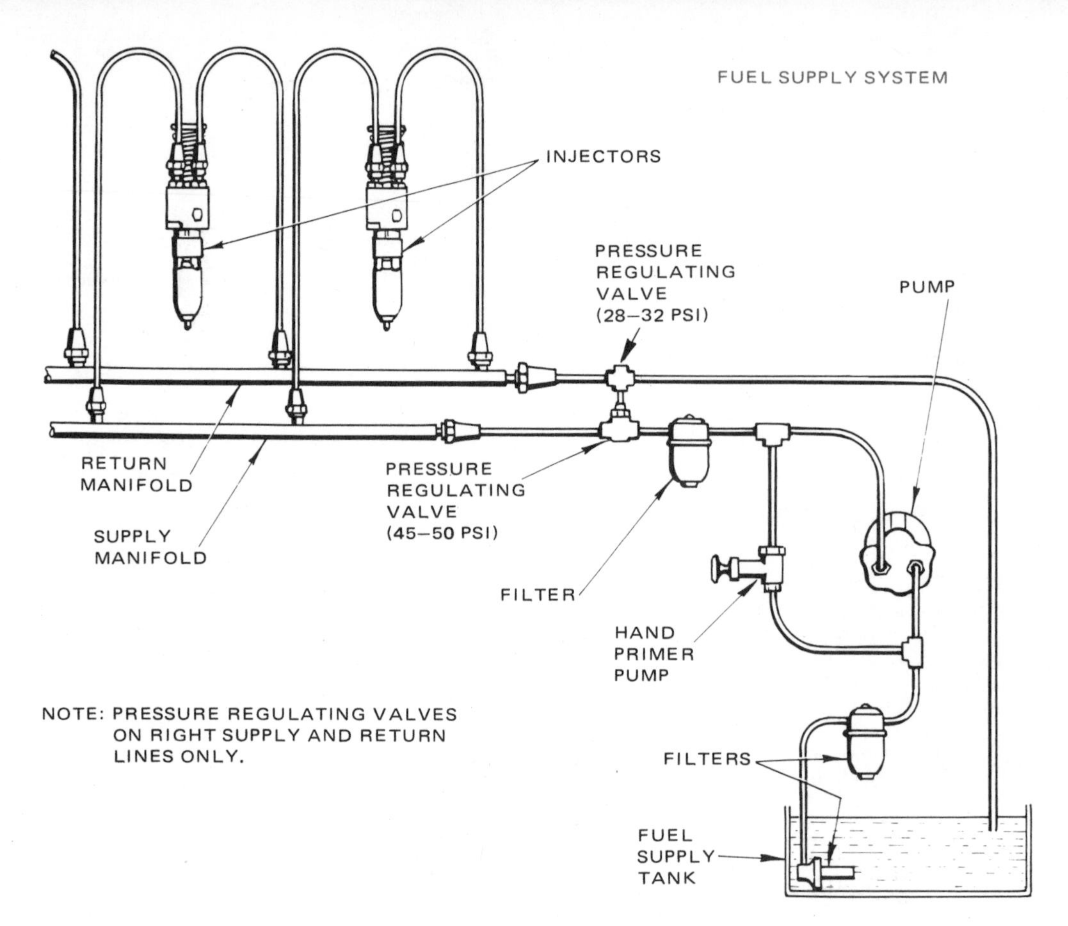

Fig. 8-108. Waukesha unit injection system.

When the lower port has been covered by the plunger, bypass continues through the upper port until it is covered by the plunger upper helix, at which point injection starts. Fuel is injected into the cylinder until further downward movement of the plunger causes the lower plunger helix to uncover the lower port. Bypass then occurs through the central passage of the plunger and lower port for the remainder of the downstroke. On the return stroke the bushing chamber is refilled with oil, first through the lower port, and then the upper port, and finally through the upper and lower ports together.

Concurrently with its reciprocating motion, the plunger may be rotated about its axis by the means of a rack meshed with a gear through which the plunger reciprocates. This plunger rotation controls the output of the injector by changing the relation of the plunger helices to the bushing port. When the rack is moved all the way out, no injection can occur during the downstroke of the plunger since the position of the helices is such that the lower port is uncovered before the upper port closes giving uninterrupted bypass. A certain amount of rack travel inward is used up by a "no fuel" travel for governoring purposes. This no fuel travel extends from the full out position to that point where the lower port opens and the upper port closes simultaneously. Further rack movement inward from this point gives an interval in the plunger stroke when both ports are closed and injection occurs. As the rack moves in, the quantity of fuel injected per stroke increases to a maximum at full rack travel.

The plunger helices have an injection timing function in addition to their metering function. The helices may be

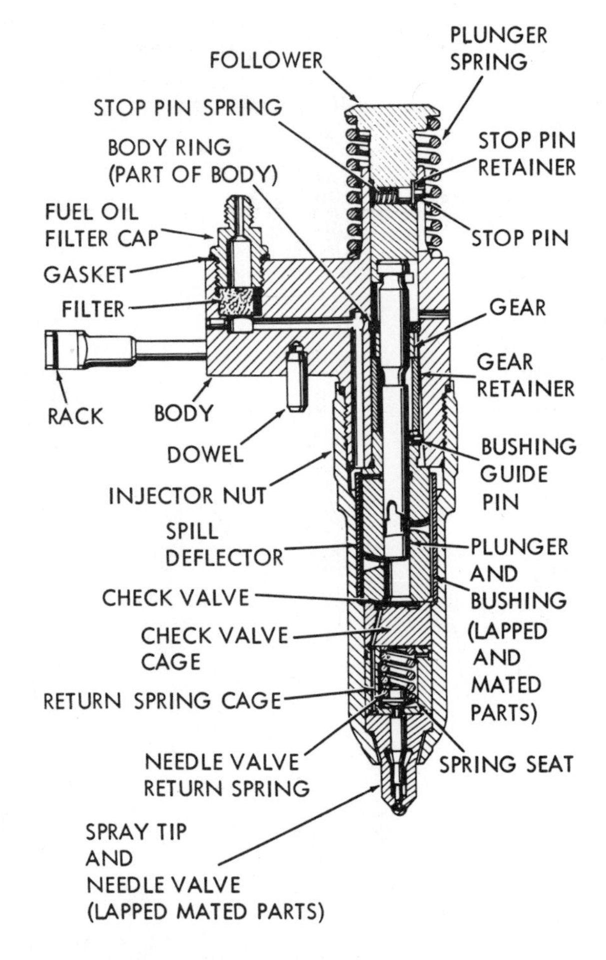

Fig. 8-109. Sectional view of Waukesha unit injector.

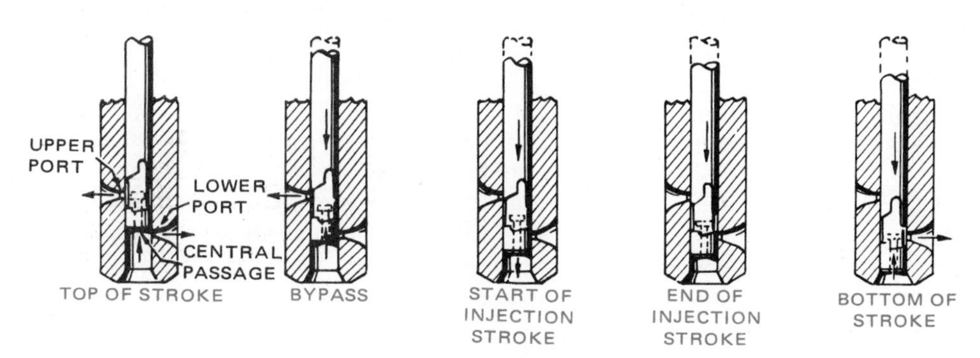

PHASE OF INJECTOR OPERATION BY
VERTICAL TRAVEL OF PLUNGER

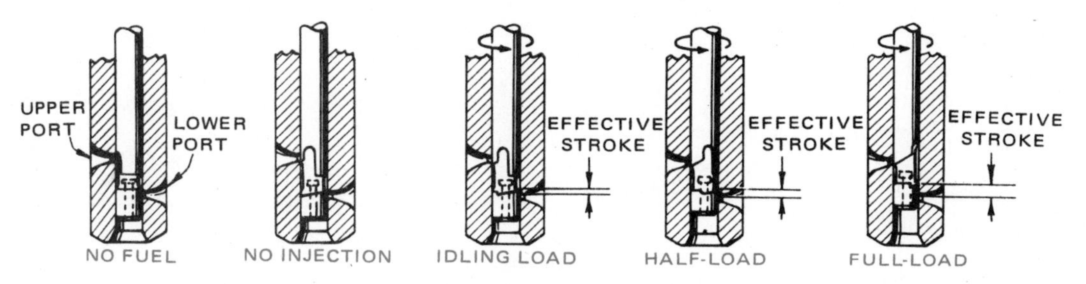

LOAD CONTROL FROM "FULL-LOAD" TO "NO LOAD" PRODUCED
BY ROTATING PLUNGER WITH CONTROL RACK

Fig. 8-110. Stroke sequence of Waukesha unit injector, showing load control from "full-load" to "no-load" produced by rotating plunger with control rack.

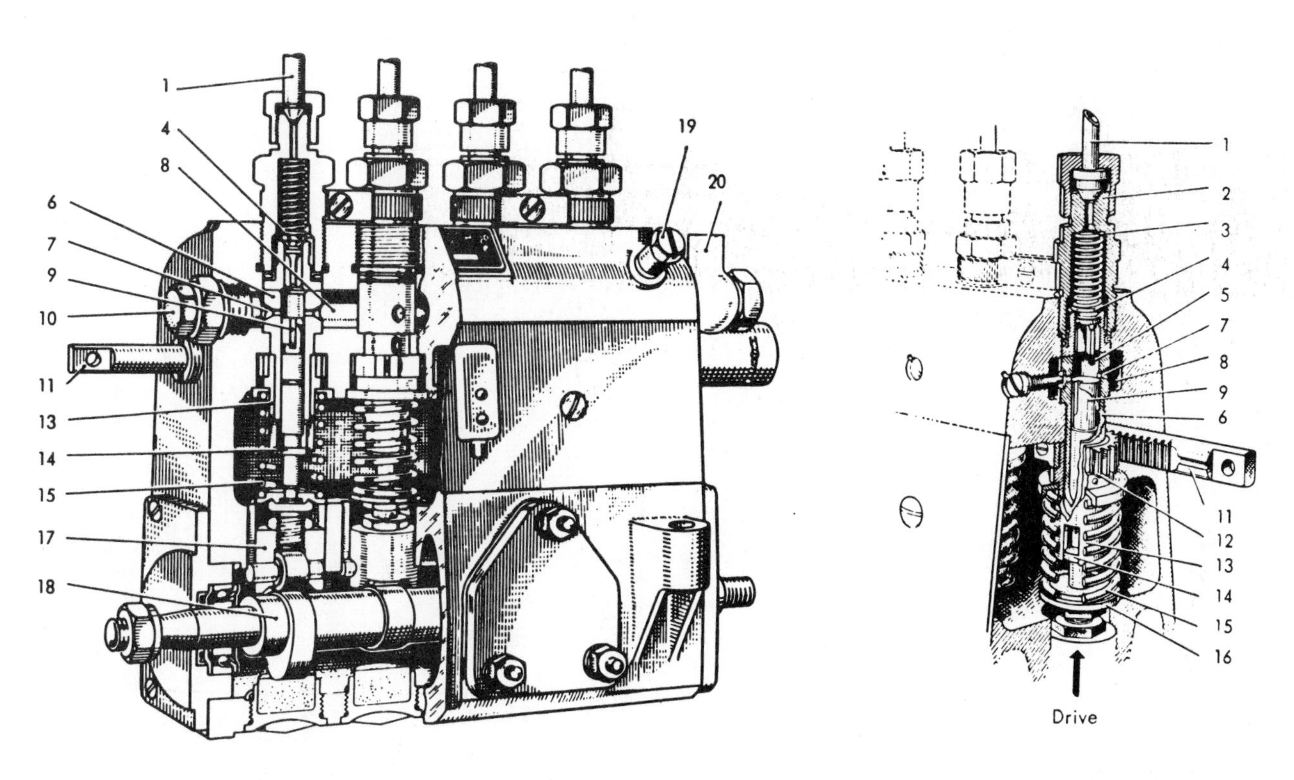

1. PRESSURE LINE.
2. PIPE CONNECTOR.
3. VALVE SPRING.
4. PRESSURE VALVE.
5. PRESSURE SPACE.
6. CYLINDER.
7. CONTROL PORT.
8. SUCTION SPACE.
9. PLUNGER.
10. SCREW PLUG.
11. CONTROL ROD.
12. PINION.
13. CONTROL SLEEVE.
14. PLUNGER LUG.
15. PLUNGER SPRING.
16. SPRING RETAINER.
17. ROLLER TAPPET.
18. CAMSHAFT.
19. BLEEDER SCREW.
20. FUEL INLET.

Fig. 8-111. Cross section of Robert Bosch PE injection pump used on some Mercedes-Benz engines. Left. Pump components. Right. Drive mechanism. (Daimler-Benz Aktien Gesellschaft)

machined to vary the time of the injections at various loads with respect to the piston position. Either/or both beginning and end of injection may be retarded, advanced or made constant with an increase of injector output depending on engine requirements.

During actual fuel injection, fuel passes from the bushing chamber through a needle delivery valve.

The needle valve consists of a closely guided, spring-loaded needle with a conical end which operates against a conical seat, Fig. 8-109. The cone angle of the valve is slightly larger than that of the seat to insure line contact. The valve opens at about 3000 psi and closes at about 1500 psi. The fuel flows from the bushing chamber through the drilled passages and annular grooves in the spring cage and seat to the valve chamber, from which it flows down to the spray tip and out of the spray nose into the combustion chamber of the engine.

An auxiliary valve keeps the high cylinder pressure from blowing back through the injector and air binding it in the event the spring-loaded valve is momentarily held open between injection cycles by a small dirt particle.

ROBERT BOSCH INJECTION SYSTEM

Multiple plunger injection pumps and single plunger distributor pumps are both produced by Robert Bosch GMBH, Germany.

The multiple plunger pump features an individual pumping element for each cylinder consisting of a barrel, a plunger with metering helix and delivery valve. The PE series are suitable for all engine applications. The PF series which are operated by the camshaft integral with the engine, are suitable for large engines.

The single plunger distributor pump has only a single plunger regardless of the number of cylinders. These pumps are known as the VA series. The reciprocating and rotating pumping plunger also serves to distribute the fuel to several outlets on the hydraulic head through individual delivery valves. In addition, the same main plunger being stepped, supplies fuel under pressure for the governing circuit. The fuel quantity, discharged from the pump is governed from one injection to another in a certain phase relationship with the main plunger.

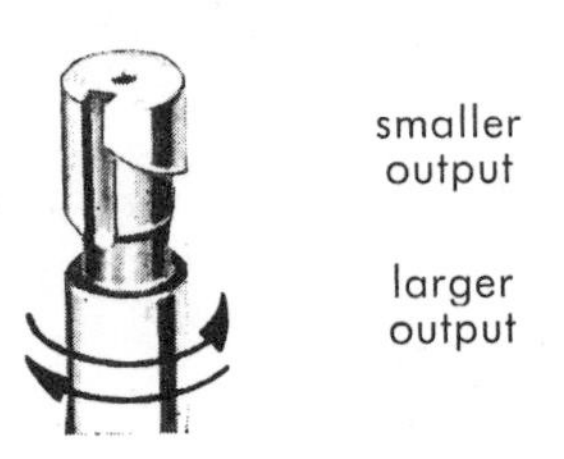

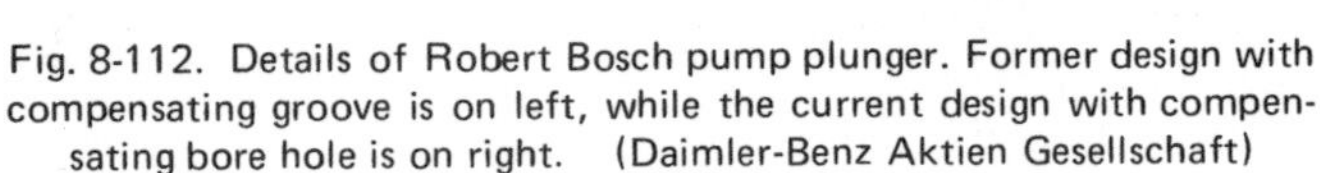
Fig. 8-112. Details of Robert Bosch pump plunger. Former design with compensating groove is on left, while the current design with compensating bore hole is on right. (Daimler-Benz Aktien Gesellschaft)

The injection pump as installed on current Mercedes-Benz cars is a PE series type. This pump contains one pump element consisting of cylinder and plunger for each engine cylinder, Fig. 8-111. The plunger is lapped in the cylinder and has a clearance of two to three thousandth of a millimeter, so it is sealed without special packing. Plunger and cylinder are interchangeable only as complete sets. An injection timing device is incorporated in the drive assembly of the pump so fuel injection is timed in relation to engine speed.

The discharge rate is changed by turning the pump plunger

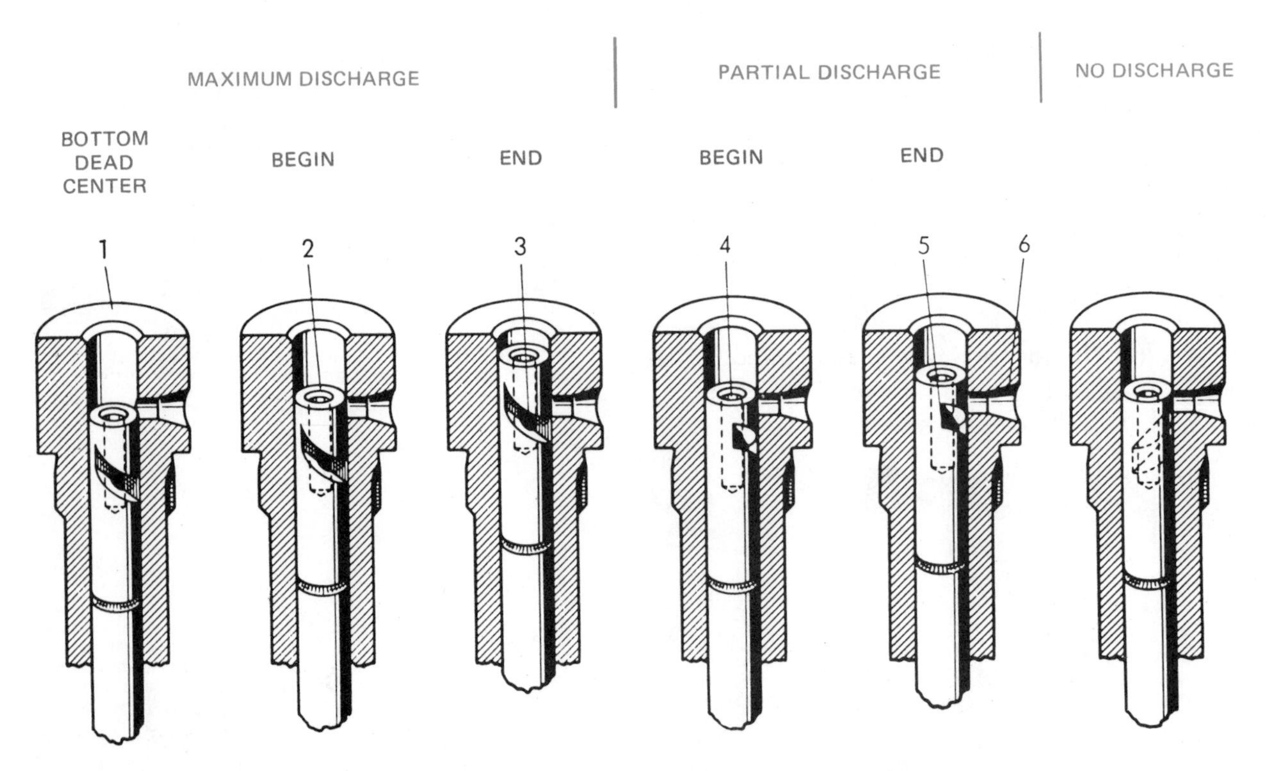

Fig. 8-113. Sequence of plunger positions. 1–Cylinder. 2–Plunger. 3–Sloped control edge. 4–Compensating hole. 5–Sloped groove. 6–Control port (inlet and return hole). (Daimler-Benz Aktien Gasellschaft)

(9), Fig. 8-111. A pinion (12) is clamped to the upper end of the control sleeve (13) which houses the pump cylinder. In the lower part of the control sleeve (13) are two longitudinal slots which guide the piston lugs. The control rod is geared to the pinion (12), Fig. 8-111. The pump plungers can be turned with the control rod; discharge rate of the pump can be infinitely varied from 0 to maximum.

The plunger has a sloped groove, of which the upper edge serves as the control edge. In the plunger is a longitudinal (compensating hole) which connects the pressure space (5), Fig. 8-111, with the sloped groove in the plunger, Fig. 8-112. The pump plungers of the former design have in place of the compensating hole a longitudinal groove in the plunger surface, which connects the pressure space (5) with the annular groove in the plunger.

The pump cylinder (6), Fig. 8-111, is provided with a control port (inlet and return hole) (7), which connects the pressure space (5) with the suction space (8), depending on the position of the plunger.

During the pressure stroke the pump plunger (9), Fig. 8-111, is lifted by a cam and during the suction stroke it is forced down again by the plunger spring (15). The stroke of the pump plunger cannot be varied. The suction space is filled with fuel constantly and is kept under pressure by the feed pump. If the pump plunger is at the bottom dead center the control port (6), Fig. 8-113, is opened and the pressure space is filled with fuel. During the upward motion the plunger closes the control port (6) and presses the fuel through the pressure valve into the pressure line. See Fig. 8-113.

The delivery ends as soon as the upper control edge has reached the control port, because the pressure is connected with the suction space by the compensating hole (4) in the plunger, Fig. 8-113, or in the case of the former units via the longitudinal groove, thus bringing about an equalization pressure.

The discharge rate is varied by turning the plunger. Fig. 8-113 shows that the plunger opens the control port sooner or later during the discharge depending on the amount the plunger has been turned. The output is largest if the plunger is turned full clockwise and the output is reduced if the plunger is turned counterclockwise. In order to reach zero output, the plunger must be turned fully counterclockwise. In that position, the upper control edge immediately opens the control port (6). In that way the pressure space (5) is continuously connected to the suction space (8) and there will be no discharge, Fig. 8-111.

ROBERT BOSCH INJECTION NOZZLE

The Bosch injection nozzle, Fig. 8-114, injects fuel delivered by the injection pump at a high pressure in a favorable spray pattern and at the proper moment into the combustion chamber. It distributes the fuel in such a way that a good combustible mixture is introduced in the combustion chamber. The nozzle is controlled by the fuel pressure. During the discharge stroke of the plunger, the pressure impulse is transferred through the injection lines, pressure passage of the nozzle holder, annular groove and inlet holes of nozzle holder insert, annular groove and pressure passages in the nozzle head until it reaches the pressure chamber in the injection nozzle, Fig. 8-114. If the discharge pressure becomes stronger than the tension force of the tension spring, the nozzle needle is lifted off its seat. Fuel is injected through the injection hole into the precombustion chamber of the engine and the main combustion chamber to produce a combustible mixture.

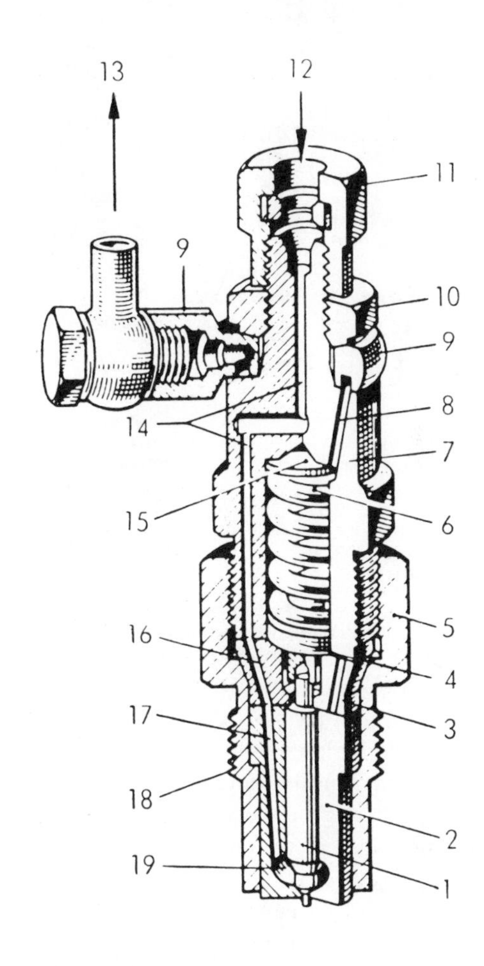

1. NOZZLE NEEDLE.
2. NOZZLE HEAD.
3. NOZZLE HOLDER INSERT.
4. PRESSURE BOLT.
5. CAP NUT.
6. TENSION SPRING.
7. NOZZLE HOLDER.
8. DRIP OIL PASSAGE.
9. ADAPTER.
10. HEX NUT.
11. CAP NUT TO INJECTION LINE.
12. FUEL INLET.
13. RETURN TO FUEL TANK.
14. PRESSURE PASSAGE TO NOZZLE HOLDER.
15. WASHER TO TENSION SPRING.
16. ANNULAR GROOVE AND INLET HOLES.
17. ANNULAR GROOVE AND PRESSURE PASSAGES.
18. MOUNTING THREAD.
19. PRESSURE CHAMBER.

Fig. 8-114. Robert Bosch nozzle holder and nozzle.

Drip oil reaching the nozzle holder (7), Fig. 8-114, has access to the annular groove of the drip oil connector (9), via passage (8) and flows back through the drip oil outlet (13) into the fuel tank. If, toward the end of the discharge stroke, the fuel pressure becomes weaker than the tension spring (6), the spring presses by way of the pressure bolt (4), the nozzle

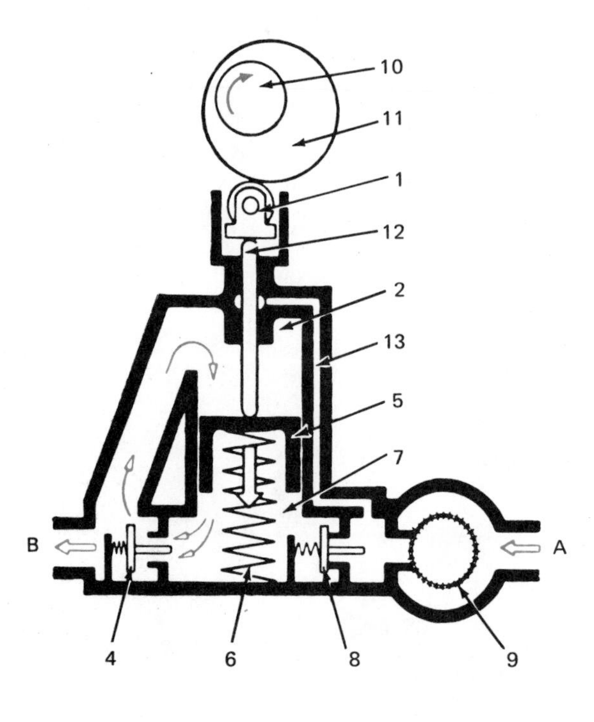

A. FUEL INLET.
B. FUEL OUTLET.

1. ROLLER TAPPET.
2. PRESSURE SPACE.
4. PRESSURE VALVE.
5. PLUNGER.
6. PLUNGER SPRING.
7. SUCTION SPACE.
8. SUCTION VALVE.
9. PREFILTER.
10. CAMSHAFT.
11. CAM.
12. ROD.

Fig. 8-115. Fuel feed pump in intermediate stroke position. (Robert Bosch)

needle (1) back against its seat; the injection is completed. With this the nozzle is closed until actuated again by the next discharge stroke. There must be no dribbling of oil.

The opening pressure of the nozzle can be adjusted by changing the initial load of the tension spring with the help of the washers (15), Fig. 8-114.

The Robert Bosch feed pump is shown in Figs. 8-115 and 8-116. This is a single acting plunger pump driven by the camshaft of the injection pump. The cam presses the plunger (5), outward against the plunger spring (6) by way of the roller tappet (1), and the pressure bolt (11), Fig. 8-116. In that way the fuel in the suction space (7) is discharged through the pressure valve (4) into the pressure space (2). This stroke is called the intermediate stroke. At the same time the plunger spring (6) is compressed. At the end of this stroke, the spring-loaded pressure valve (4) closes. The pressure bolt is not attached to the roller tappet or the plunger (5). As soon as the cam has reached its maximum lift, the pressure bolt and roller tappet are forced inward again in the direction of the injection pump by the pressure of the preloaded plunger spring (6). The result is that fuel is discharged from the pressure space through the filter into the injection pump. At the same time, fuel is drawn from the supply tank through the filter and suction valve (8) into the suction space (7) by the returning plunger (5). This stroke is called the discharge and suction stroke. The intermediate stroke serves only to discharge the fuel from the suction space into the pressure space and to tension the plunger spring.

A pneumatic type governor as used with some Mercedes-Benz engines is described on page 252.

ROBERT BOSCH DISTRIBUTOR PUMP

In addition to plunger type injection pumps, Robert Bosch produces a distributor type fuel pump, Fig. 8-117. Known as model VE, this pump contains an integrated supply or transfer pump, a timing device and a mechanical or electrical shut-off device. The governor is available in all speeds as well as in a combined version called a "road speed governor." This setup is preferable for passenger cars because, with it, the driver can achieve driveability similar to the performance level provided by a gasoline engine.

Fig. 8-117 shows a VE pump equipped with single delivery valves. In this application, they are designed as constant volume retraction valves. For additional control of fuel flow, these valves can be designed as torque control valves. It is also possible to install reverse flow throttle valves that integrate into the delivery valve holder. In this way, better control of hydrocarbon emissions is obtained.

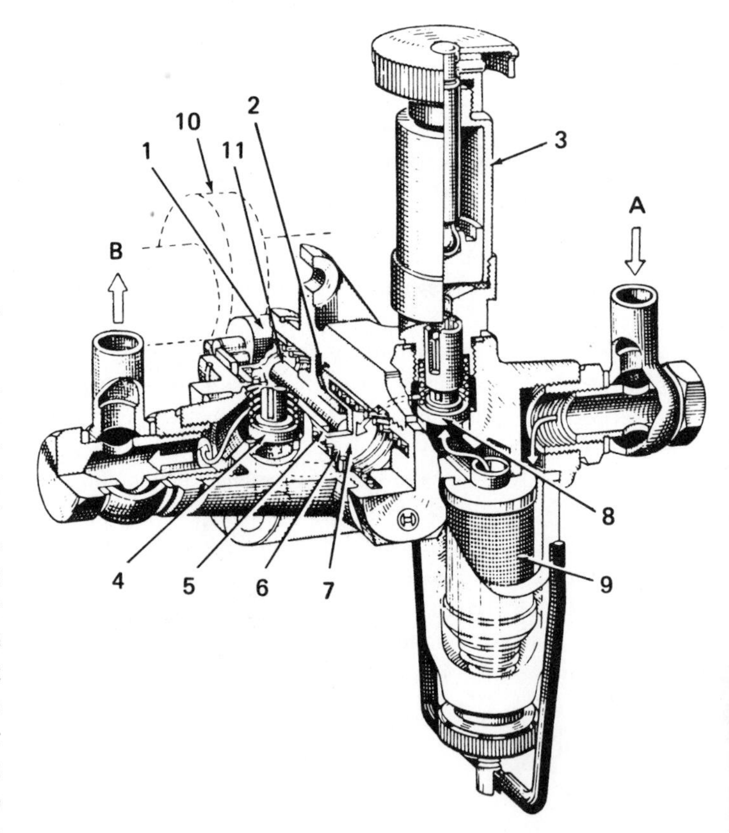

A. FUEL INLET.
B. FUEL OUTLET.

1. ROLLER TAPPET.
2. PRESSURE SPACE.
3. HAND PUMP.
4. PRESSURE VALVE.
5. PLUNGER.
6. PLUNGER SPRING.
7. SUCTION SPACE.
8. SUCTION VALVE.
9. PREFILTER.
10. CAM.
11. PRESSURE BOLT.

Fig. 8-116. Details of fuel feed pump. (Robert Bosch)

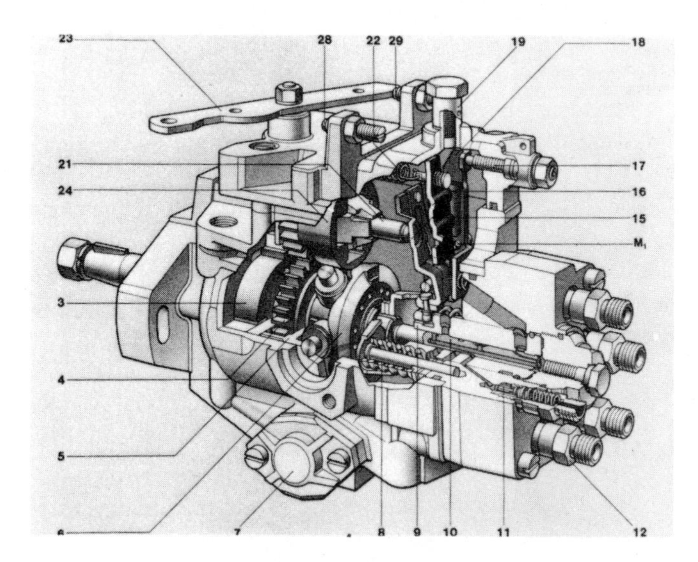

Fig. 8-117. Details of Robert Bosch model VE-F distributor type injection pump typical of Volkswagen Rabbit application: 3—Supply pump. 4—Governor drive. 5—Cam roller ring. 6—Cam plate. 7—Timing device. 8—Plunger return spring. 9—Regulating collar. 10—Distributor pump plunger. 11—Delivery valve. 12—Delivery valve holder. 15—Starting lever. 16—Tensioning lever. 17—Adjusting lever screw, full load. 18—Adjusting lever, full load delivery. 19—Excess flow valve. 20—Stop lever. 21—Sliding sleeve. 22—Governor spring. 23—Control lever. 24—Flyweight assembly. 28—Adjusting screw, rated speed. 29—Adjusting screw, idle speed. (Robert Bosch GmbH)

ROBERT BOSCH PUMP ON VOLKSWAGEN

The Robert Bosch VE-F distributor injection pump, Fig. 8-117, is used on the Volkswagen diesel engine. In operation, vane type pumping action draws fuel from the tank to supply the high pressure pump. Delivery pressure of the pump increases with the speed of the engine, although pressure is limited by the pressure regulator, valves and the restriction in the hollow core screw of the return line.

The high pressure pump, Fig. 8-118, increases the pressure and distributes fuel to the injector nozzles according to the firing order of the engine. A centrifugal governor regulates idle speed and governs the quantity of fuel injected at maximum speed. The ignition timing mechanism alters the start of fuel delivery in relation to engine speed.

Fig. 8-119 shows details of the injection nozzle. Fuel is pumped to the nozzle needle, which lifts from its seat when fuel pressure of 120-130 psi overcomes spring pressure. The nozzle needle is cooled and lubricated by the fuel. Excess fuel is returned to the supply tank.

FAIRBANKS MORSE INJECTION SYSTEMS

Fairbanks Morse of Colt Industries markets both distributor and in-line type fuel injection pumps and allied equipment.

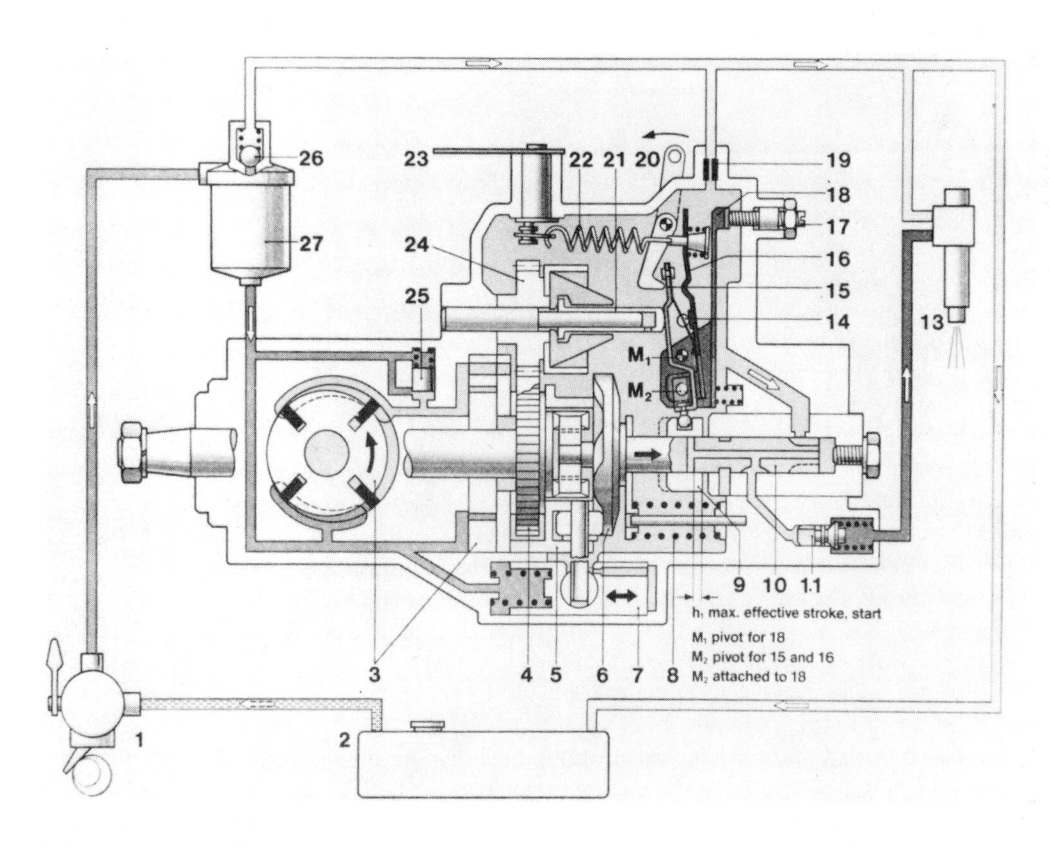

Fig. 8-118. Diagram of Robert Bosch model VE-F distributor type fuel injection pump: 1—Pre-supply pump. 2—Fuel tank. 3—Supply pump. 4—Governor drive. 5—Cam roller ring. 6—Cam plate. 7—Timing device. 8—Plunger return spring. 9—Regulating collar. 10—Distributor pump plunger. 11—Delivery valve. 13—Injection nozzle. 14—Tensioning lever stop. 15—Starting lever. 16—Tensioning lever. 17—Adjusting lever screw, full load. 18—Adjusting lever, full load delivery. 19—Excess flow valve. 20—Stop lever. 21—Sliding sleeve. 22—Governor spring. 23—Control lever. 24—Flyweight assembly. 25—Pressure regulating valve. 26—Overflow valve. 27—Fine filter. (Robert Bosch GmbH)

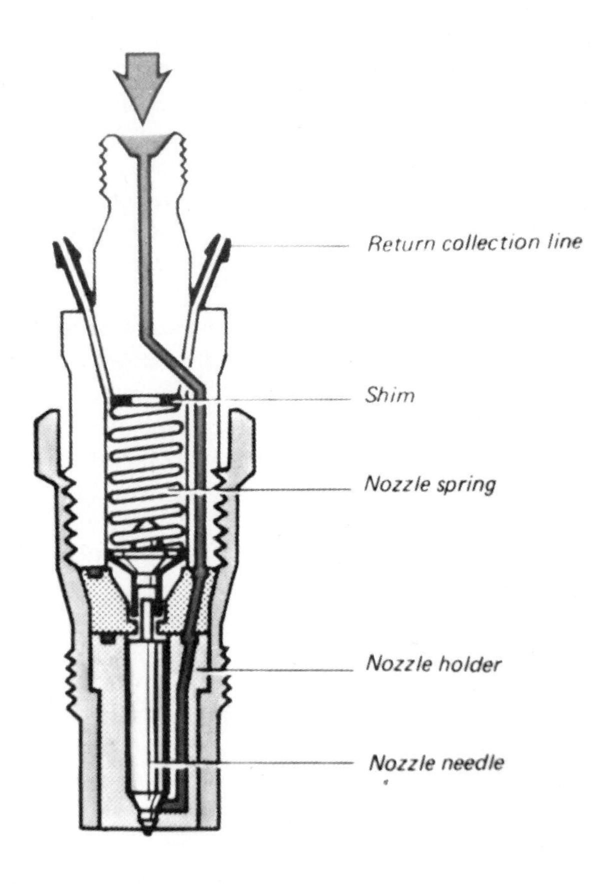

Fig. 8-119. Sectional view of the injection nozzle used with the VE pump. (Volkswagen of America, Inc.)

The series PRS pump, Fig. 8-120, is of the distributor type and is complete with governor, delivery curve shape control, excess fuel capability and automatic speed advance device. It is capable of operating on all fuels from gasoline to diesel in the multifuel form. In that form it requires engine oil lubrication. The low-pressure supply system consists of a supply pump which draws the fuel through the filters built into the flange portion of the pump. This maintains a constant pressure in the housing of the injection pump for charging the hydraulic head. A pressurizing valve maintains constant pressure and a permanent vent in the highest portion of the pump provides continuous circulation of fuel for cooling and lubrication at all times irrespective of whether fuel is being injected into the cylinders.

This recirculating system avoids the need of returning large quantities of fuel to the tank.

The fuel to be injected reaches the pumping chamber between the plungers of the hydraulic head through the interaction of filling and metering slots in the sleeve and rotor portions of the hydraulic head. A cam ring driven by the drive shaft, has internal profiles which operate the pumping plungers. The number and diameter of these plungers is determined by the injection characteristics of the engine. Start and end of injection is accomplished by the interaction of the metering slots and filling holes in a manner similar to in-line pumps. This gives a high-pressure cutoff at the start and end of injection. This results in a sharp definition of the injection period, which in turn results in improved fuel economy and reduced exhaust smoke.

Excess fuel is available for cold starting. This cuts out automatically, being dependent on the selection of the excess fuel springs in the governor arrangement. The excess fuel is achieved by a sharply defined beginning of injection, but there is no end of injection until the rollers pass over the nose of the cam, delivering additional fuel at a later timing in the engine cycle.

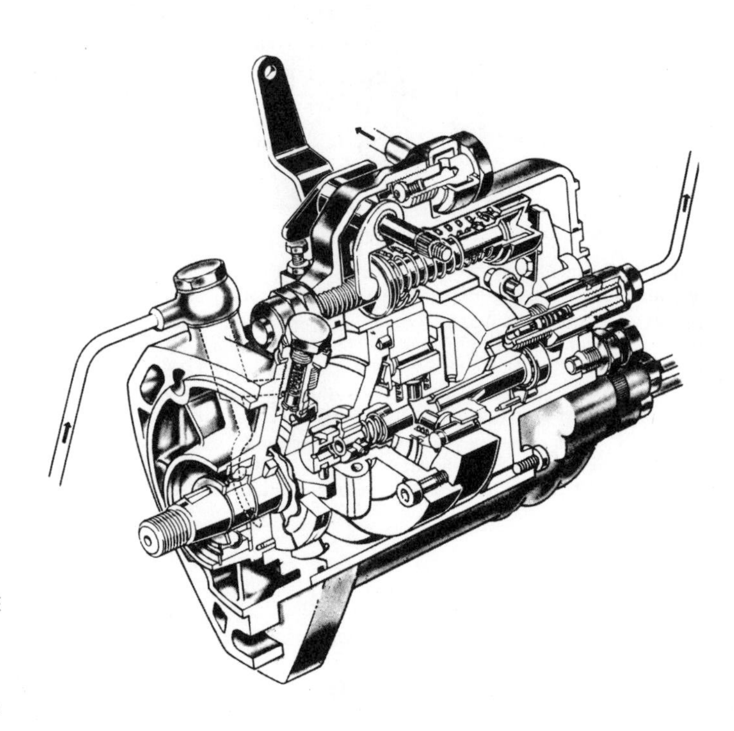

Fig. 8-120. Sectional view of S.I.G.M.A. fuel injection pump marketed by Fairbanks Morse, Colt Industries.

GM DIESEL ENGINE INJECTION

General Motors 5.7 litre diesel engine is similar in construction to Oldsmobile's 350 cu. in. gasoline engine. Originally introduced in Oldsmobile passenger cars, this 5.7 litre diesel engine now has applications in vehicles manufactured by all General Motors passenger car divisions, and also in Chevrolet trucks.

The fuel injection pump installed in the GM 5.7 litre diesel engine is a Roosa Master unit of the distributor type. It is classified by Roosa Master as "pencil type." Its main operating parts are shown in Fig. 8-121.

In studying the main operating parts of the pump, Fig. 8-121, the drive shaft engages the distributor rotor in the hydraulic head. The drive end of the rotor has two pumping plungers. The plungers are actuated toward each other at the same time by an internal cam ring through rollers and shoes which are carried in slots at the drive end of the rotor. The number of cam lobes equals the number of engine cylinders.

The transfer pump in the end cap (rear of rotor) is of positive displacement type. The end cap also houses the fuel inlet strainer and transfer pump pressure regulator.

The distributor rotor has two charging ports and a single axial bore with one discharge port to serve all head outlets to the injection lines.

The hydraulic head contains the rotor bore, the fuel metering valve bore, the charging ports and the head discharge fittings. The high pressure injection lines to the nozzles connect to the discharge fittings.

The pump has a mechanical governor capable of close regulation. The centrifugal force of the weights is transmitted through a sleeve to the governor arm through linkage to the fuel metering valve. The fuel metering valve can be closed to shut off fuel through linkage by an independently operated shut off lever.

The automatic advance is an hydraulic mechanism that advances or retards the beginning of fuel delivery from the pump.

Fuel flow principles for the Roosa Master pump, Fig. 8-122, are as follows: Fuel is supplied from the engine crankshaft operated fuel pump, through filters into the injection pump, through the inlet filter screen (1) to the vane type fuel transfer pump (2). Some fuel is by-passed through the pressure regulator assembly (3), back to the suction side.

Fuel under transfer pump pressure flows past the rotor retainers (4) into the connecting passage (5) in the head to the advance (6) and to the charging circuit (7). The fuel flows through a connecting passage to the fuel metering valve (8). The radial position of the fuel metering valve, controlled by the governor, regulates the flow of fuel into the charging ring (which has the charging ports).

As the rotor turns, Fig. 8-122, the two inlet passages (9) register with the charging ports in the hydraulic head, allowing fuel to flow into the pumping chamber. With further rotation, the inlet passages move out of registery and the discharge port of the rotor registers with one of the head outlets. While the discharge port is opened, the rollers (10) contact the cam lobes, forcing the plungers together. Fuel trapped between the plungers is then pressurized and delivered by the nozzle to the combustion chamber.

With the exception of the drive shaft bearing, the pump is self lubricated. In addition, an air vent passage in the hydraulic head connects the outlet side of the transfer pump with the pump housing. This allows air and some fuel to bleed back to the fuel tank through the return line. The by-passed fuel fills the housing and lubricates the internal parts.

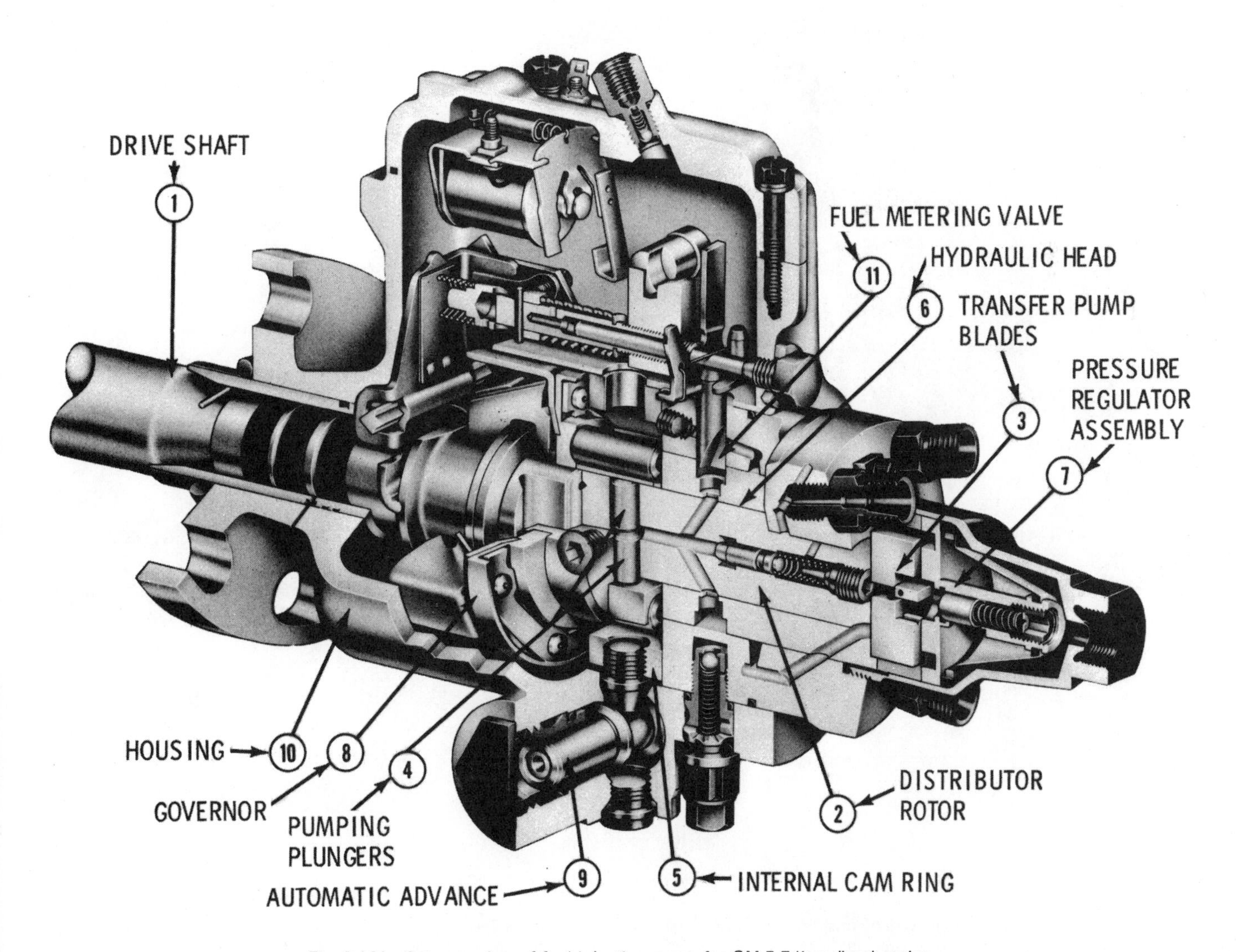

Fig. 8-121. Cutaway view of fuel injection pump for GM 5.7 litre diesel engine.

The positive displacement vane type transfer pump has a stationary liner and spring-loaded blades which are carried in slots in the rotor. Since the inside diameter of the liner is eccentric to the rotor axis, rotation causes the blades to move in the rotor slots. Blade movement changes the volume of fuel between the blade segments. As a result, transfer fuel output volume and pressure increases as pump speed increases.

The pressure regulator, Fig. 8-122, controls delivery pressure of the transfer pump. Fuel output from the discharge side of the transfer pump forces the piston in the regulator against the regulating spring. As the flow is increased, the spring is compressed until the edge of the regulating piston starts to uncover the pressure regulating slot. As a result, pressure increases with the speed.

The transfer pump works equally well with different grades of diesel fuel and varying temperatures. This is accomplished by a thin plate which has a sharp-edged orifice, and which is located in the spring adjusting plug.

The orifice allows fuel leakage past the piston to return to the inlet side of the pump. Flow through short orifice is almost the same when viscosity changes. The pressure exerted against the back side of the piston is determined by the leakage past designed clearance of the piston in the regulator bore and the pressure drop through the sharp-edged orifice.

With cold or thicker fluid, very little leakage occurs past the piston. The additional force on the back side of the piston from the thicker fuel is slight.

With hot or light fuels, leakage past the piston increases. Fuel pressure in the spring cavity also increases because flow through the orifice remains the same as with cold or thicker fuel. The increased fuel pressure assists the regulating spring and moves the piston, reducing regulating slot area. This resulting variation in piston position compensates for the leakage which would occur with the use of thin fuels. Therefore, design fuel pressures are able to be maintained over a broad range of fuel viscosity changes.

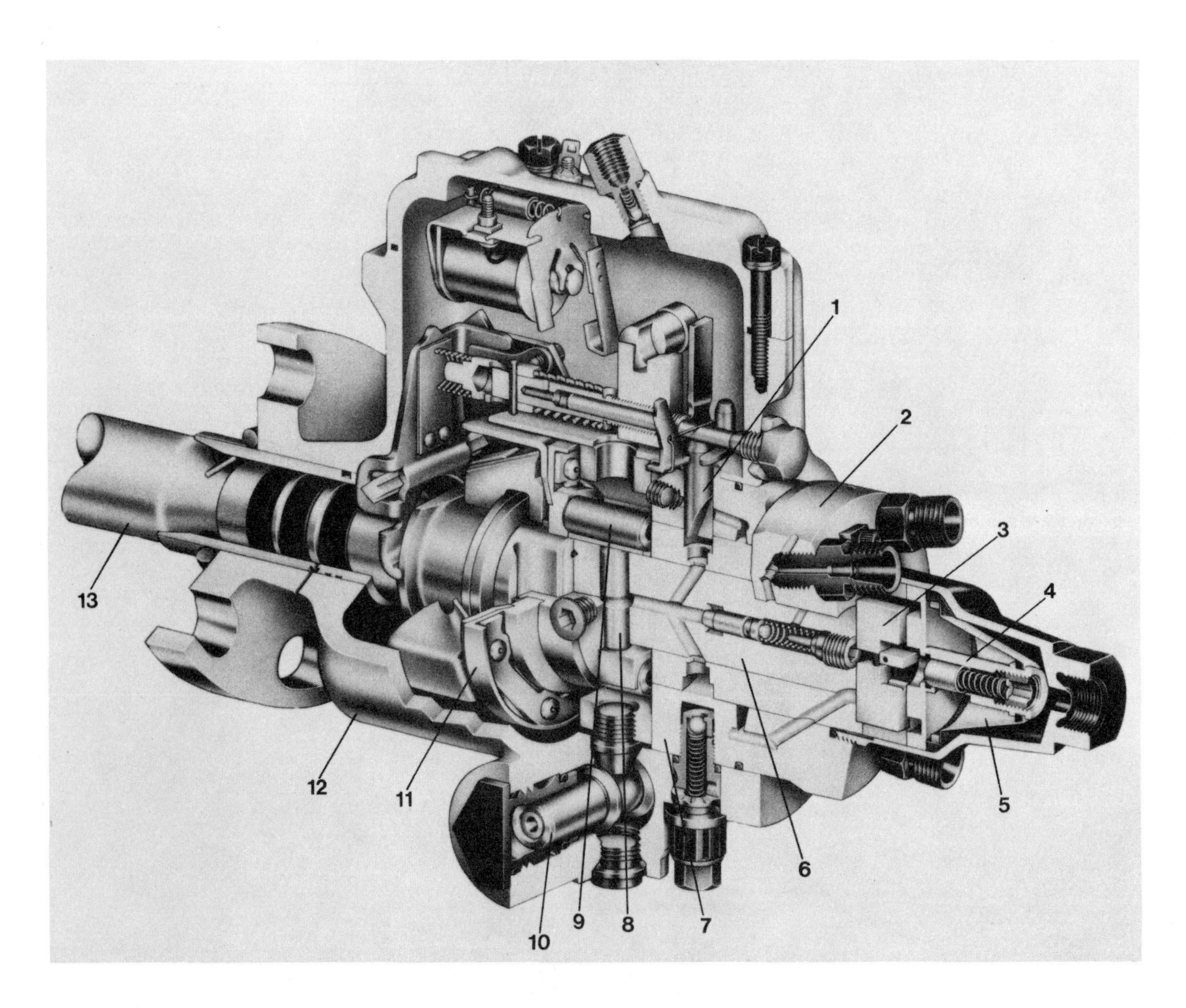

Fig. 8-122. Sectional view of diesel fuel injection pump of type installed on GM 5.7 L engine. 1—Fuel metering valve. 2—Hydraulic head. 3—Fuel transfer pump. 4—Pressure regulator assembly. 5—Fuel inlet filter screen. 6—Distributor rotor. 7—Internal cam ring. 8—Pumping plungers. 9—Rollers. 10—Automatic advance. 11—Governor. 12—Housing. 13—Drive shaft.

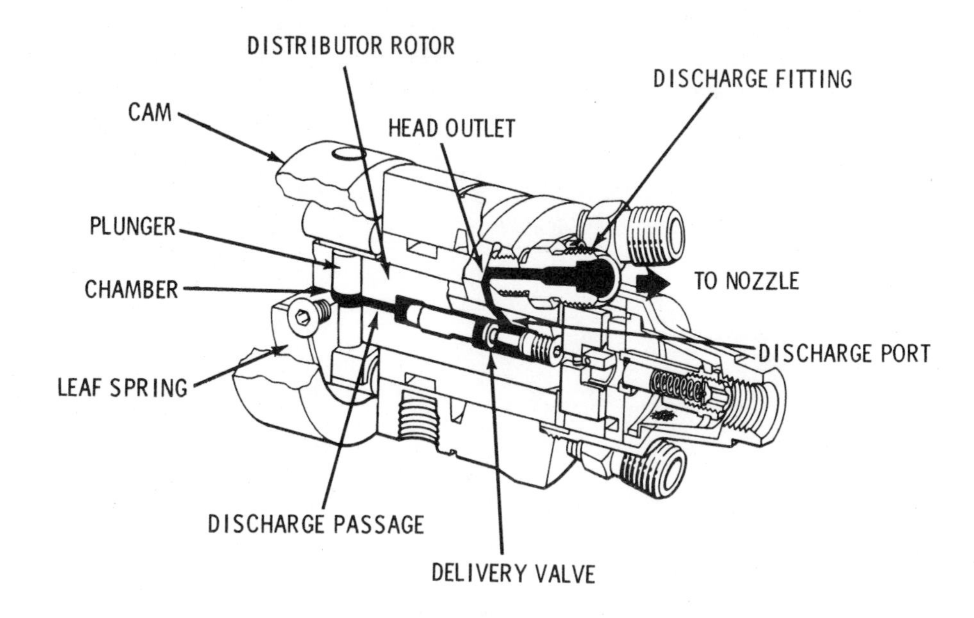

Fig. 8-123. Delivery valve and delivery cycle of fuel injection pump.

CHARGING CYCLE

As the rotor turns in the charging cycle, the two inlet passages in the rotor align with ports of the charging ring. Fuel under pressure from the transfer pump – controlled by the opening of the fuel metering valve – flows into the pumping chamber, forcing the plungers apart. The plungers move outward a distance equal to the amount of fuel required for injection on the following stroke. If only a small quantity of fuel is admitted into the pumping chamber, as at idling, the plungers move out a short distance. Maximum plunger travel and maximum fuel delivery are limited by a single leaf spring that contacts the edge of the roller shoes. Only when the engine is operating at full load will the plungers move to the most outward position. Note that while angled inlet passages in the rotor align with the ports in the charging ring, the rotor discharge port is not in alignment with a head outlet.

DISCHARGE CYCLE

As the rotor continues to turn, the inlet passages move out of alignment with the charging ports, Fig. 8-123. The rotor discharge port opens to one of the head outlets; the rollers contact the cam lobes; and injection begins. Further rotation of the rotor moves the rollers up the cam lobes, pushing the plungers inward. During this stroke, the fuel trapped between the plungers flows through the axial passage of the rotor and discharge port to the injection line. Delivery to the injection line continues until the rollers pass the innermost point of the cam lobe and begin to move outward. Then, the pressure in the axial passage is reduced, allowing the nozzle to close. This is the end of injection.

DELIVERY VALVE

The purpose of the delivery valve, Fig. 8-123, is to rapidly decrease injection line pressure (after injection) to a predetermined value lower than that of the nozzle closing pressure. The reduction in pressure causes the nozzle valve to return rapidly to its seat, causing sharp delivery cutoff which prevents improperly atomized fuel from entering the combustion chamber.

The delivery valve operates in a bore in the center of the distributor rotor. The valve requires no seat, only a stop to limit its travel.

Fuel under transfer pump pressure is discharged into a vent in the hydraulic head to form the fuel return circuit. Flow through the passage is restricted by a wire to prevent excessive return oil and undue pressure loss.

The mechanical governor serves the purpose of maintaining the desired engine speed within the operating range under various load settings. In operation, movement of centrifugal weights acts against the governor thrust sleeve, causing the governor arm and linkage hook to turn the fuel metering valve. Rotation of the valve varies the alignment between the fuel metering valve opening and the passage to the transfer pump, thereby controlling the quantity of fuel flowing to the plungers.

AUTOMATIC ADVANCE

The automatic advance, Fig. 8-121, is a direct acting hydraulic mechanism. It is powered by fuel pressure from the transfer pump, which rotates the cam and varies fuel delivery timing. The advance mechanism advances or retards start of fuel delivery in response to changes in engine speed.

A sectional view of the injection nozzle is shown in Fig. 8-124. It is principally a hole type nozzle operated by hydraulic differential pressures which can be adjusted by setting the desired opening. The spring preload can be varied by means of a pressure adjusting screw.

A schematic diagram of the complete diesel fuel injection system is shown in Fig. 8-125.

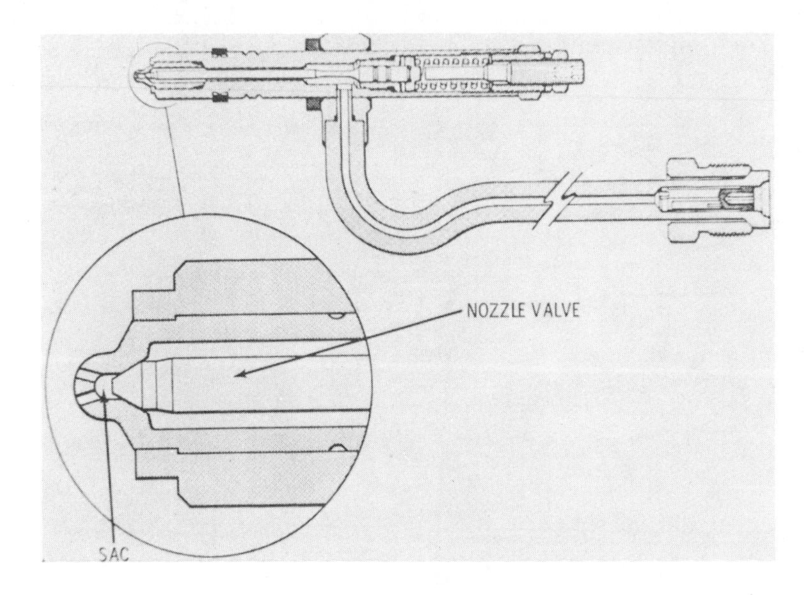

Fig. 8-124. Diesel fuel injection nozzle.

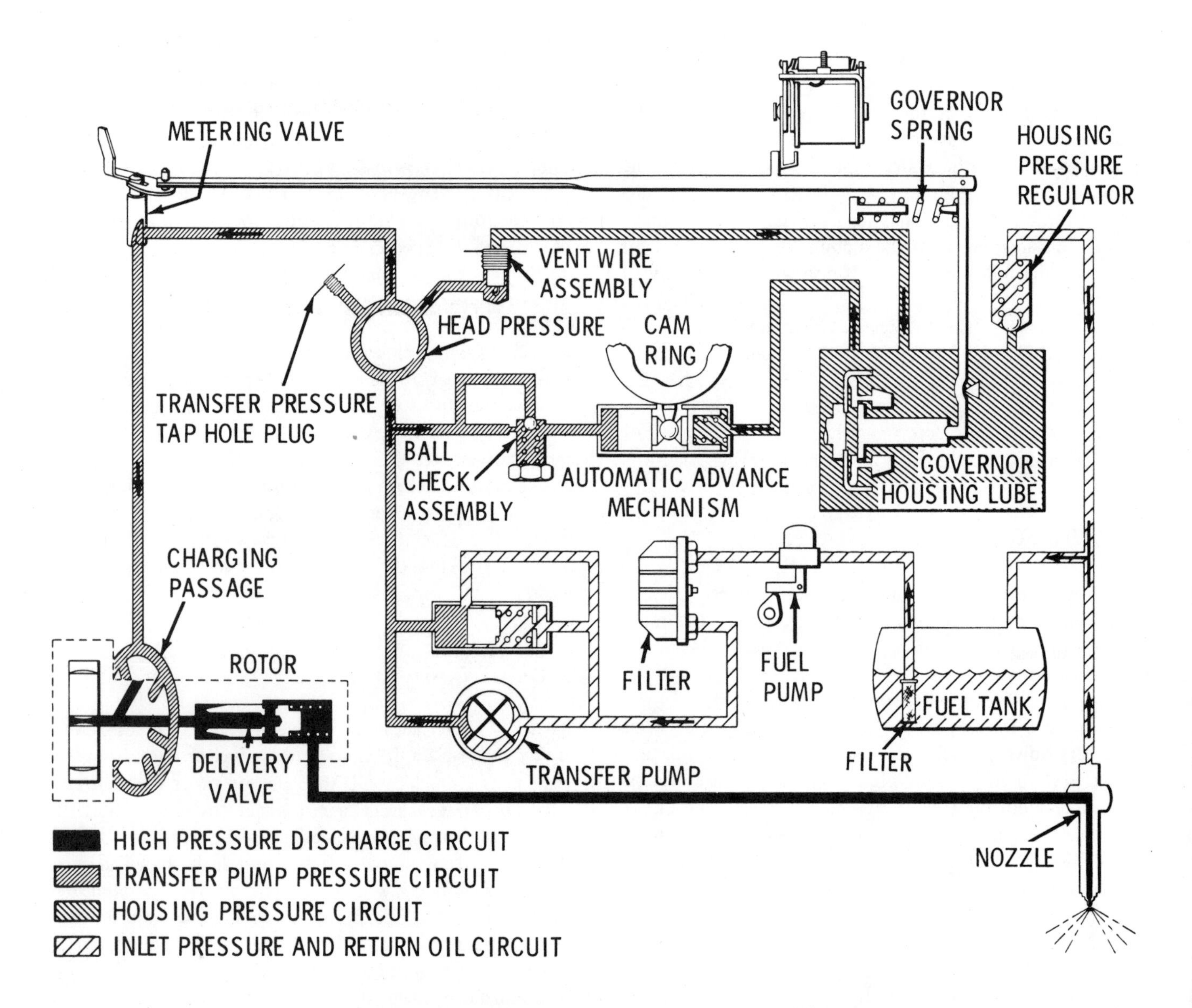

Fig. 8-125. Schematic of Roosa Master fuel injection system for 5.7 litre diesel engine.

REVIEW QUESTIONS – CHAPTER 8

1. The fuel injection system of a diesel performs more functions than the ____________ on the automobile.
2. At the rated load and speed some injection systems develop up to:
   a. 100 psi.
   b. 500 psi.
   c. 750 psi.
   d. 5000 psi.
3. With the diesel engine, it is necessary to _________ the quantity of fuel in accordance with the load on the engine.
4. Fuel should not be injected too early in the cycle since the temperature would be low and ____________ would be delayed.
5. Diesel fuel is not injected in a single quick spurt, but ___________ over a period of time.
6. The rate of injection varies with different ___________.
7. The two basic forms of injection are:
   a. Hydraulic.
   b. Mechanical.
   c. Air.
   d. Chemical.
8. The high-pressure pump of the common rail system discharges the fuel directly into a ____________.
9. In the mechanical injection pump there are five methods listed for controlling the quantity of fuel, they are:
   a. ________________________________.
   b. ________________________________.
   c. ________________________________.
   d. ________________________________.
   e. ________________________________.
10. What type of pump is the American Bosch type APE injection pump?
    a. Throttle inlet.
    b. Port control.
    c. Diaphragm opening.
    d. Throttle bypass.
11. In the APE American Bosch injection pump, the stroke is ________.
12. The fuel supply pump of the American Bosch injection pump is a:
    a. Diaphragm.
    b. Plunger.
    c. Centrifugal.
13. The pneumatic governor produced by American Bosch obtains the needed vacuum from which of the following:
    a. Air intake manifold.
    b. Separate vacuum pump.
    c. Plunger pump.
14. What type pump is the American Bosch type PSJ?
    a. Jerk pump.
    b. Distributor pump.
    c. Plunger.
15. The type of injection system produced by Bendix is ________.
16. To vary the output of the Bendix type F injection pump, the plunger is rotated by means of:
    a. ________________________________.
    b. ________________________________.
    c. ________________________________.
    d. ________________________________.
17. CAV produces the plunger type injection pump and the ___________ type.
18. In the CAV injection pump type DPA, what separates the plungers?
    a. Cam action.
    b. Centrifugal force.
    c. Fuel at metering pressure.
19. The CAV transfer pump is of the positive type. How many vanes does it have?
    a. Two.
    b. Four.
    c. Five.
    d. Seven.
20. The fuel volume of the CAV injection pump type DPA is governed by:
    a. Fuel pressure only.
    b. Time available for fuel flow.
    c. Length of stroke of plunger.
    d. Both time available and fuel pressure.
21. In the Cummins injection system, the letters "PT" stand for ____________.
22. There is a ____________ pipe line from the top of the fuel pump on the Cummins PT type R fuel pump.
23. The pressure regulator on Cummins PT type R pump functions as a ____________ valve to regulate the fuel pressure to the injectors.
24. In the type R Cummins pump, the fuel flows past the ___________ to the throttle shaft.
25. At what pressure does the Cummins fuel pump deliver fuel to the injectors?
    a. Low.
    b. Medium.
    c. High.
26. How many types of injectors does Cummins use?
    a. Two.
    b. Three.
    c. Four.
    d. Five.
    e. Six.
27. The Ex-Cell-O system is used primarily in the field of:
    a. Automotive power.
    b. Marine power.
    c. Electric generation.
    d. Portable equipment.
28. In the Ex-Cell-O system the plungers are operated by:
    a. Direct cam operation.
    b. Cam and push rod.
    c. Rotating swash plate.
    d. Hydraulically.
29. What type of mechanical fuel injection system is used by General Motors?

a. Distributor system.
b. Unit injector.
c. Common rail system.
d. Pump controlled.

30. The pump and nozzle of the General Motors system form a ____________ unit.
31. The Roosa Master model DB fuel injection pump is a:
    a. Single cylinder type.
    b. Double cylinder type.
    c. Four cylinder type.
32. What actuates plungers in the Roosa Master pump?
    a. Hydraulic pressure.
    b. Centrifugal force.
    c. Spring pressure.
    d. Cam ring through rollers.
33. In the Caterpillar fuel injection system, the injection nozzle and combustion chamber form ____________.
34. The Caterpillar system provides ____________ pumps and injectors for each cylinder.
35. Some Waukesha engines are equipped with American Bosch systems. What other type is used?
    a. Unit injectors.
    b. Distributor type injectors.
    c. Roosa Master injectors.
    d. Cummins injectors.
36. In the sleeve metering system, used in Caterpillar fuel injection, there are no passages leading to the fill port. True or False?
37. The American Bosch model 100 injection pump is a ____________ plunger pump.
38. The Roosa Master type DM fuel injection pump is designed particularly for small, high-speed diesel engines. True or False?
39. At what pressure does fuel flow to the underside of the left shuttle in a CAV DP15 fuel injection pump?
    a. Transfer pressure.
    b. Injection pressure.
    c. Output pressure.
40. The model VE Robert Bosch fuel injection pump is the ____________ type.
41. The fuel injection pump installed on the GM diesel engine is unit injection type. True or False?
42. The automatic advance in the GM diesel injection pump is a direct acting ____________ mechanism.

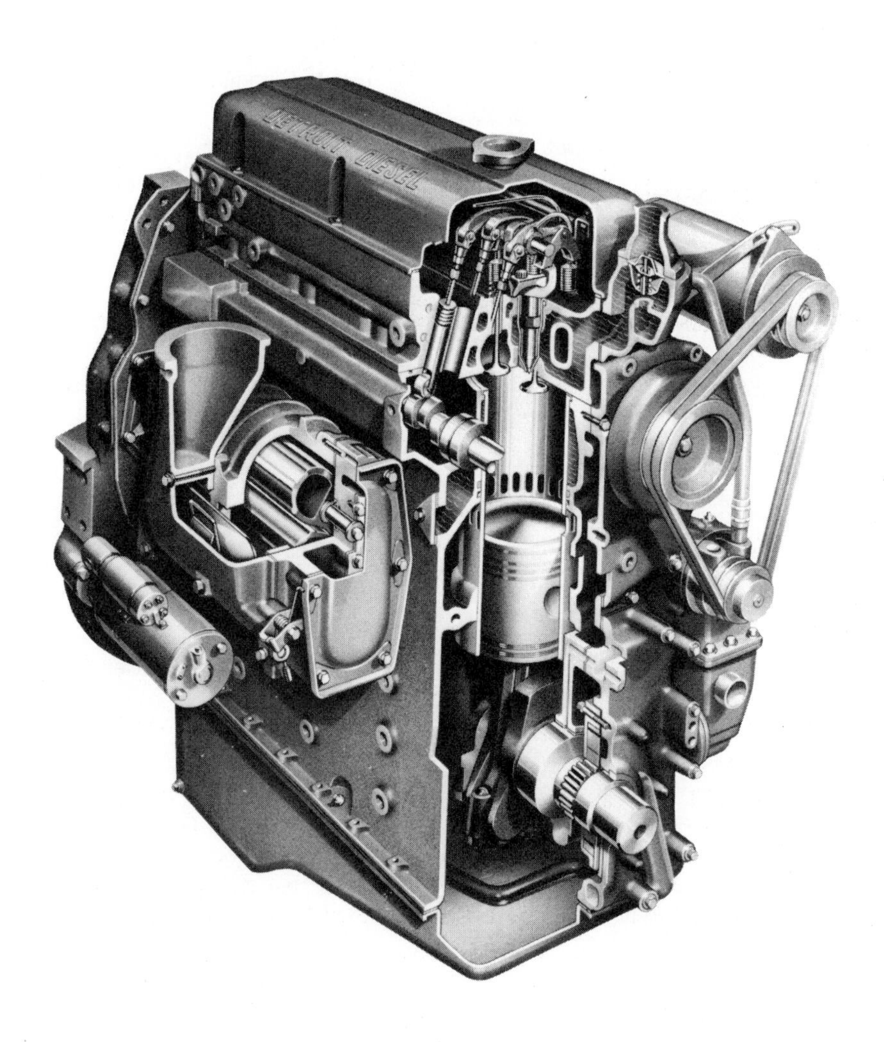

Sectional view of Detroit Diesel Allison Div. series 53 diesel engine.

Chapter 9
ATOMIZING FUEL

The continuing demand for increased horsepower per cu. in. of displacement has lead to extensive research into the design of combustion chambers, also atomization of the fuel and its penetration into the combustion chamber.

For maximum power and economy, it is essential that the engine be provided with fuel that is exactly timed and in correct quantities. Equally important is that the fuel be sprayed into the combustion chamber in such a manner that it is completely consumed, without producing smoke in the exhaust. This is a function of the injection nozzle. The design of the nozzle varies with each type and size of combustion chamber.

Nozzles are enclosed in holders which in turn are mounted in cylinder heads. A copper gasket is used between the injector and the cylinder head so that maximum heat will be transmitted to the engine coolant. This helps provide stable conditions for the injected fuel.

There are two major types of fuel injection nozzles, the open and the closed types. In the open type, no valve is provided to stop the flow of fuel from the nozzle. Instead the fuel supplied is controlled entirely by the fuel pump.

The closed type is provided with a valve (usually spring-loaded) near the exit orifice of the nozzle, which is controlled either mechanically or hydraulically.

An advantage of the open type nozzle is its freedom from becoming clogged with carbon particles or other solid material as the force of the spray keeps it clean. However, it is liable to dribble (fuel leakage). Consequently it is not used as extensively as the closed type nozzle.

While there is a possibility of the valve in the closed type becoming clogged with foreign material, this is largely overcome by adequate filtering of the fuel. This type has the decided advantage that it does not dribble. Dribble causes preignition. If fuel runs down and remains on the end of the nozzle carbon will form and cause after burning. In addition fuel will be wasted.

There are two basic types of the closed type injection nozzles, the pintle type and the hole type, Fig. 9-1. Each type has a valve and seat so the fuel lines are closed when no fuel is injected.

The valve of the pintle nozzle has an extension which protrudes through the hole in the bottom of the nozzle body. A hollow cone shaped spray is produced with this design. The nominal angle of the spray cone is usually between zero and 60 deg.; the angle depending on the type of combustion chamber. A pintle type nozzle generally opens at a lower pressure than the hole type because fuel flows more readily

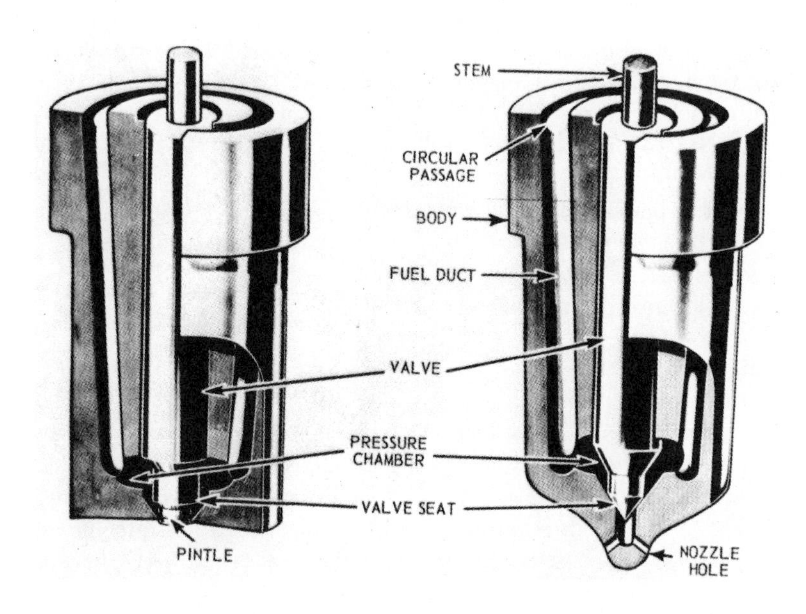

Fig. 9-1. Illustrating the difference between pintle, on left, and hole type nozzle, on right.

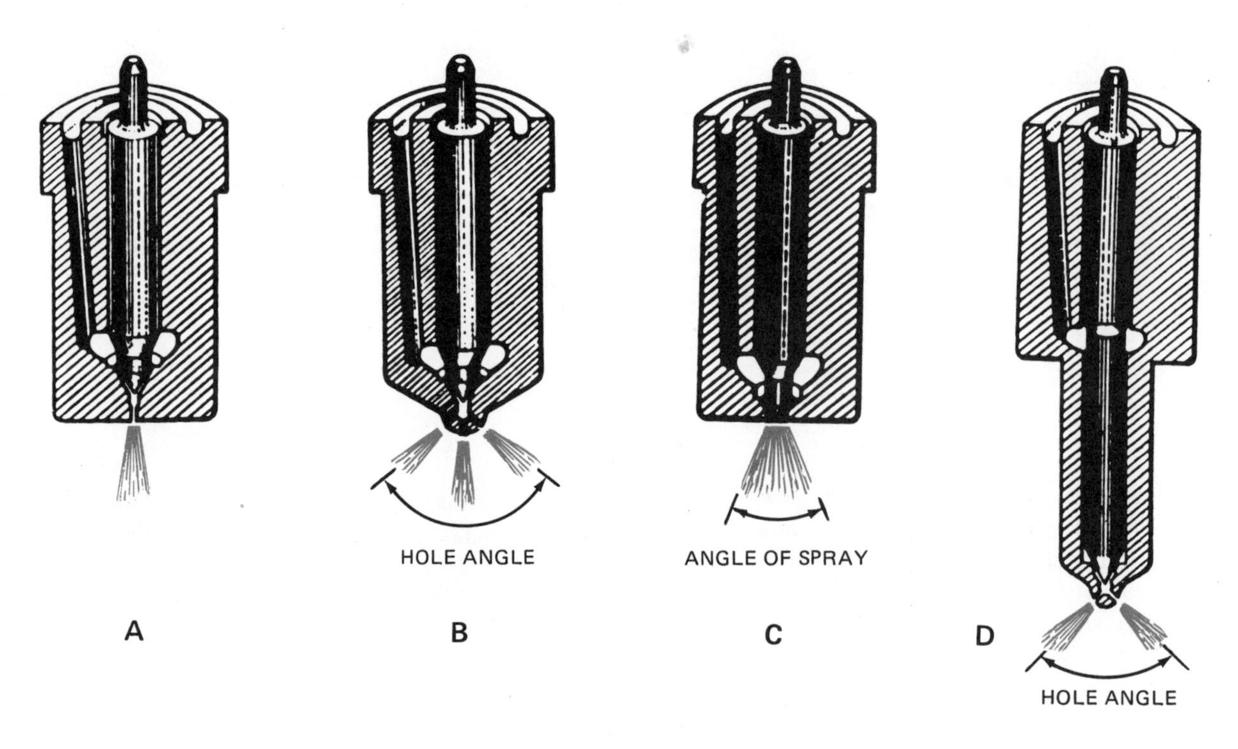

Fig. 9-2. Typical C.A.V. fuel injection nozzles. Note how angle of spray is affected by hole angle.

from the larger hole of the pintle type.

In addition to these major types of injection nozzles, there are annular ring, jet, fluted jet and the use of two impinging (close together) jets and the lip nozzle. These types are used only under special conditions.

The valves of the pintle nozzle have thin shanks or ends, the shape of which is designed to give the desired spray angle. The pin extends into the nozzle orifice so an annular spray (shape of ring) is formed. By varying the shape of the pin, the spray form ranging from a wide angle shape to a hollow cylindrical shape having a high penetration may be obtained. Some pintle nozzles are designed with a downward moving pintle in the direction of the flow of the fuel, while in other designs the pintle is lifted upward against the direction of the flow.

There are several variations of the hole type, which may be broadly classed as to the number of holes in the nozzle. Also the angle of the holes may be varied to provide wider or narrower spray forms, Fig. 9-2.

The diameter of the holes in both pintle and hole type nozzles is varied in accordance with the amount of fuel which is to be delivered. Fig. 9-1 shows the difference between the size of the spray from idling to full power on a Waukesha injection system.

As previously pointed out, the valves in the pintle type nozzle extend into the nozzle orifice so an annular space is formed. By varying the shape of the pins, different shaped jets may be formed. Research has shown that the penetration of the spray is dependent on the density and not the pressure in the combustion chamber. It has also been demonstrated that nozzles having a wide range of pintle angles will have different angles of spray during the early part of their injection, but approximately the same during latter stages. Also the cone angle of spray is the same as from plain, round hole orifices.

Another development of the pintle type nozzle is known as the C.A.V. Pentaux nozzle, Fig. 9-3. This nozzle has an auxiliary spray hole to assist easy starting under cold conditions. At engine starting speeds, the nozzle valve is not lifted sufficiently to clear the pintle hole, and the fuel is discharged through the auxiliary hole. At normal running speeds, when pressures in the fuel system are higher, the nozzle valve is withdrawn from the pintle hole allowing the bulk of the fuel to be discharged from it.

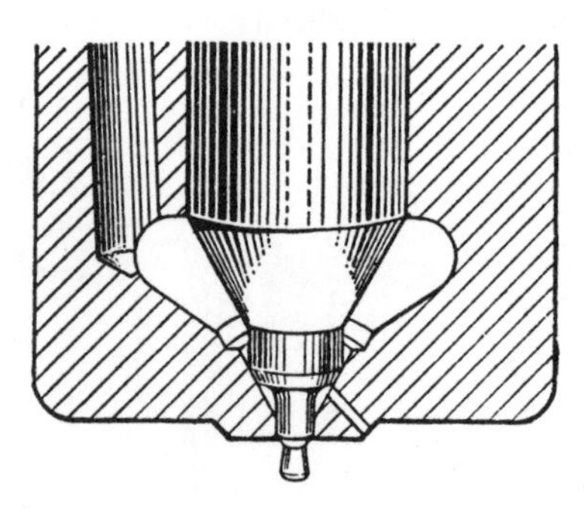

Fig. 9-3. Details of C.A.V. Pentaux fuel injection nozzle which is designed to aid in starting.

The pintle type nozzle is used largely in diesel engines with precombustion chambers, Fig. 9-4. The hole type nozzle is usually favored for direct injection engines, Fig. 9-5. The extent of penetration into the compressed charge of air and the form of injector fuel is dependent largely on length of nozzle orifice, diameter and number of orifice outlets, pattern or spray form, nature of fuel and pressure of the fuel. Another factor is the pressure wave set up in the system.

In general, the extent of penetration increases with the increasing ratio of the length of the orifice to its diameter, and best atomization occurs by decreasing the ratio of the orifice length to its diameter. Of course increasing the pressure within limits will also improve atomization.

Although atomization of the fuel is not as complete with the pintle type nozzle, penetration into the combustion chamber is greater. The multiple hole type nozzle provides good atomization but lacks penetration. Consideration of Figs. 9-4, 9-5, and 9-6 will show the installation of injector nozzles in different types of combustion chambers.

In the design and construction of fuel systems (including the fuel lines) it must be remembered that the action of the pump plunger also sets up pressure waves which results from

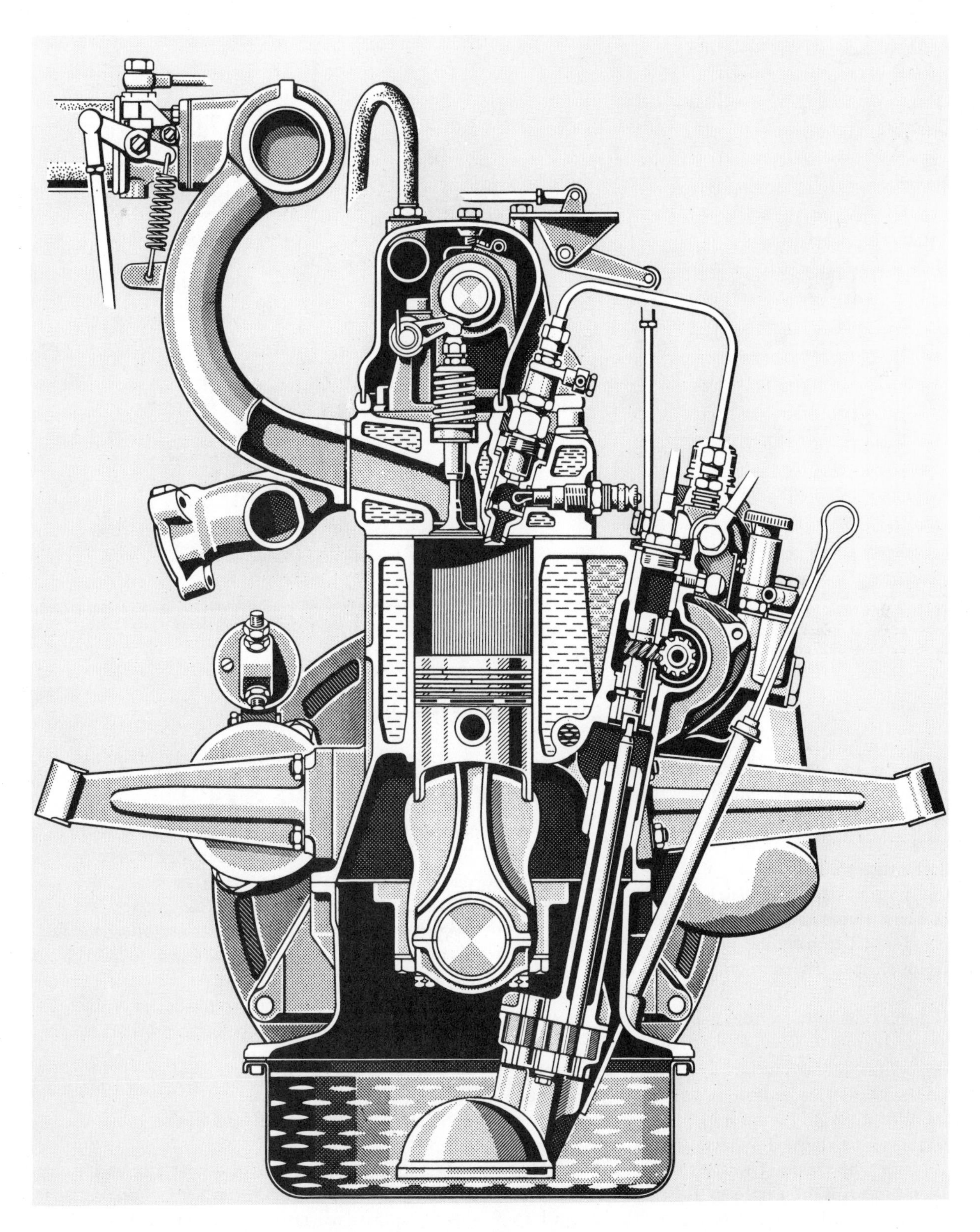

Fig. 9-4. Installation of fuel injector and glow plug in precombustion chamber of Mercedes-Benz 200D engine.

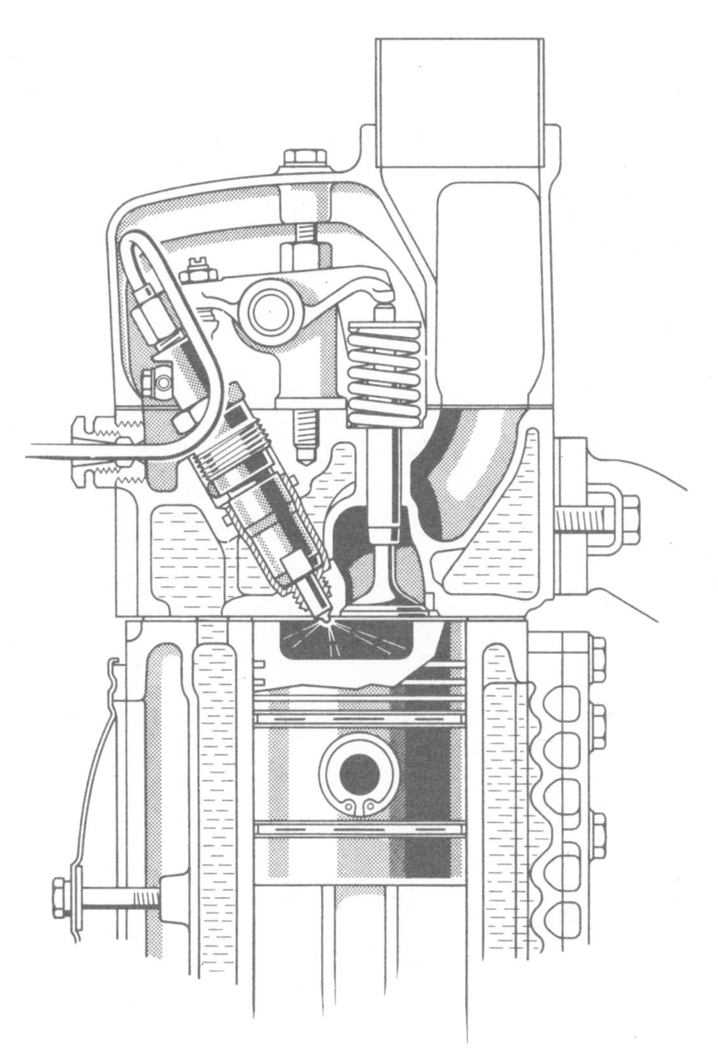

Fig. 9-5. Installation of injection nozzle in open combustion chamber of Mercedes-Benz OM 352 engine. Also note ample cooling around injector and valves.

the pressure buildup in the fuel oil. These pressure waves surge back and forth in the system with an action quite similar to that of a water hammer. Such pressure waves not only alter the characteristics of the injection system but also tend to cause cavities and erosion.

Most authorities are in agreement that fuel lines should be as short as possible and of equal length. However, Roosa Master engineers, Roosa and Hess, together with James Walker of the John Deere Company did not find this to be true on certain tractor engines. The same group pointed out in a paper presented before the Society of Automotive Engineers that improved atomization resulted from reducing the length of the hole in the nozzle from 0.027 in. to 0.010 in.

The form of the nozzle, Fig. 8-114, and the resulting spray, Fig. 9-7, as used by Mercedes-Benz in some of their passenger car engines, is of interest. The opening pressure of the nozzle of this system can be adjusted by changing the initial load of the tension spring by means of washers or shims, Fig. 8-114. The injection process is influenced by the shape of the nozzle needle and the throttle bore, and the nozzle head which provides pilot injection. The nozzle needle opens at first only a narrow annular gap through which only a little finely divided spray can pass. During further opening (caused by increased pressure), the cross section of the passage becomes larger. The main portion of the fuel is then injected toward the end of the needle (valve) stroke. The combustion and the engine performance is smoother because there is a slower increase of pressure in the combustion chamber.

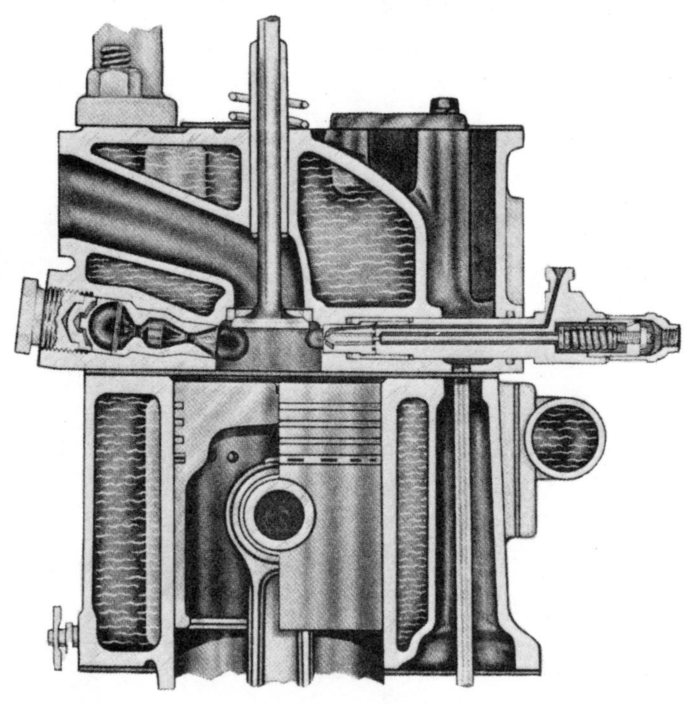

Fig. 9-6. Installation of injection nozzle in relation to energy cell in Minneapolis-Moline engine.

INJECTION NOZZLE COOLING

It is important to provide adequate cooling of the injector, otherwise fuel metering will vary, power will be reduced and carbon formation on the nozzle tip will increase. Nozzle tip temperatures usually should range between 480 and 570 deg. F. to prevent formation of carbon on the injector. Location of the injector in the cylinder head is a factor, as is the location of the injector pump. Some manufacturers place the pump away from the exhaust manifold. In addition, the fuel passages in the injector holder are positioned to provide maximum cooling for the injector, Fig. 9-8.

Special attention is given to the design of the cylinder head water jackets so adequate cooling is provided for the injector, Fig. 9-4.

DISTORTED SPRAY PATTERN

Distorted spray pattern of a nozzle or an injector may be indicated by such symptoms as a low firing pressure, loss of power, smoky exhaust and carbon deposits. Examples of distorted patterns are shown in Figs. 8-1 and 9-9.

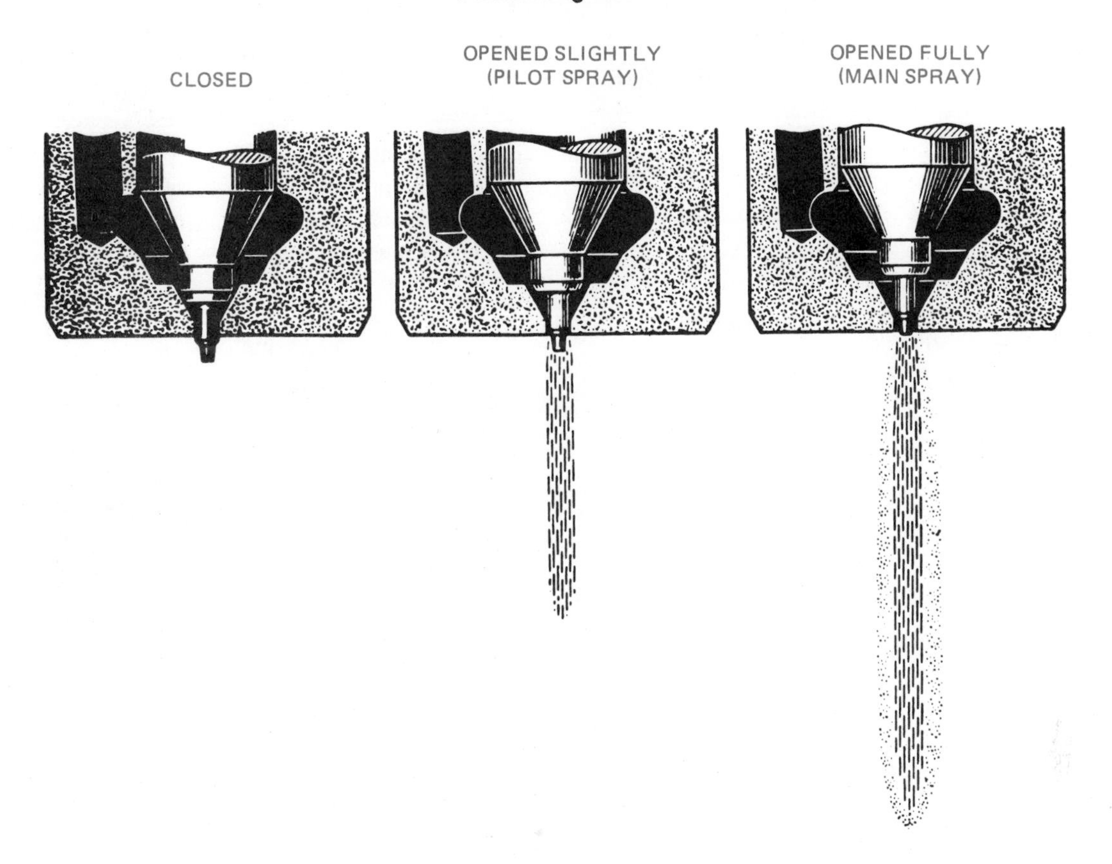

Fig. 9-7. Mercedes-Benz injector nozzle provides pilot injection.

Nozzles and injector tips are so designed that combustion should start before any appreciable quantity of fuel has struck the relatively cool surface of the combustion space. If the spray pattern is altered as the result of eroded valves, clogged orifices, broken pintle or other wear, combustion will be adversely affected.

Spray patterns can be checked by specialized equipment designed for the purpose.

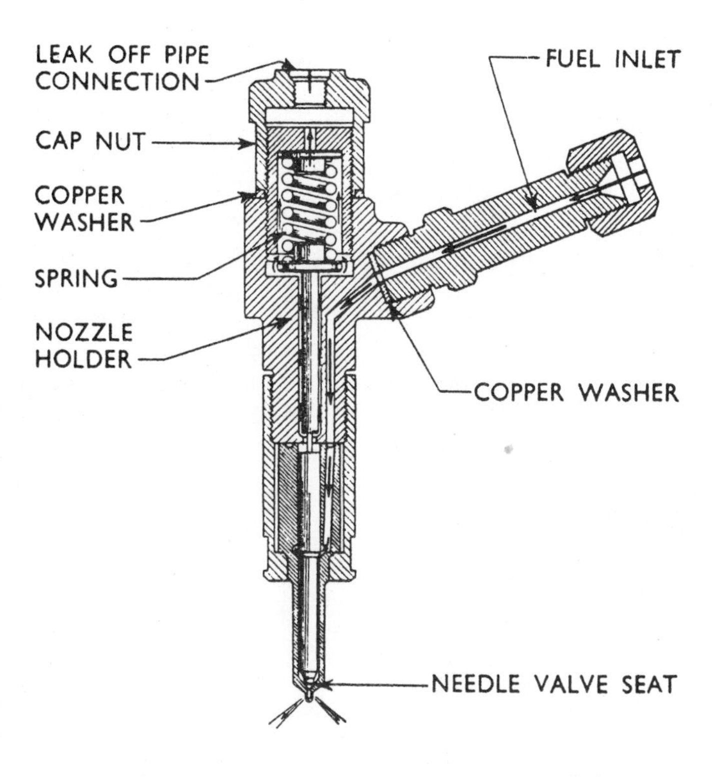

Fig. 9-8. Section through Simms injector as installed on some Ford engines.

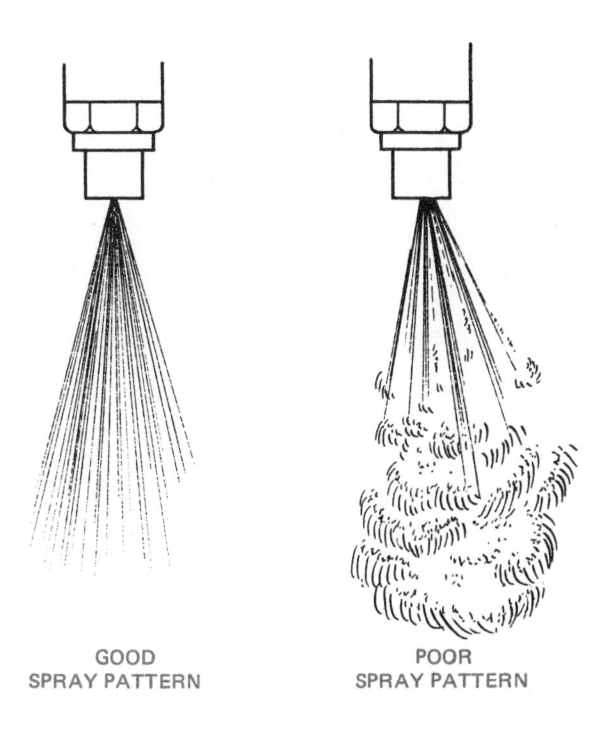

Fig. 9-9. Types of good and poor spray forms.

REVIEW QUESTIONS – CHAPTER 9

1. The type of nozzle most inclined to dribble is the _________ type.
2. Two basic styles of closed type injection nozzles are:
   a. ____________________.
   b. ____________________.
3. The pintle type nozzle opens at a _________ pressure than the hole type nozzle.
4. Injection nozzles used only under special conditions are:
   a. ____________________.
   b. ____________________.
   c. ____________________.
   d. ____________________.
5. The multiple hole nozzle provides good:
   a. Penetration.
   b. Injection.
   c. Atomization.
   d. Pressure.
6. Most authorities agree that fuel lines should be as short as possible and of _____________.
7. The opening pressure of the nozzle of the injector used on some Mercedes-Benz cars is _____________.
8. Distorted spray patterns of a nozzle or an injector may be indicated by such symptoms as:
   a. ____________________.
   b. ____________________.
   c. ____________________.
   d. ____________________.
9. Nozzle tip temperatures usually should range between _________ and _________ deg. F. to prevent formation of carbon on the injector.
10. The hole type nozzle is usually for ____________ engines.

Chapter 10

SCAVENGING, SUPERCHARGING, TURBOCHARGING

The diesel air intake system not only supplies clean air for combustion, but forces exhaust gases from the combustion chamber, which remain from the previous power stroke. This operation is known as scavenging. When exhaust gases remain in the cylinder they dilute the incoming charge of air and reduce the efficiency of the combustion. In addition, particularly in the two-cycle engine, incoming air provides cooling for the piston and combustion chamber (excessive temperatures reduce the volumetric efficiency).

The temperature of pistons, valves and combustion chambers varies with the frequency of the combustion cycle. For the same engine speed, two-cycle engine has twice as many combustion cycles per unit of time as a four-cycle engine and combustion chamber parts will have a higher temperature.

In engines with exhaust ports, the exhaust gases sweep across the piston head raising its temperature excessively. In addition, when the exhaust ports begin to open, hot gases come in contact with the side of the piston. This tends to burn the oil in the ring grooves, causing the rings to stick and in turn blow-by results.

Such temperatures are minimized by the scavenging air, increased thickness of the piston head, and by directing a jet of oil to the underside of the piston.

Scavenging must be accomplished in a relatively short portion of the operating cycle. In the two-stroke cycle engine the process takes place during the latter part of the down-stroke (expansion) and the early part of the up-stroke (compression), Fig. 10-1.

Note that in the two-stroke cycle, scavenging starts after the exhaust has opened and stops slightly after the exhaust has closed. The scavenging occurs through approximately 96 deg. of the cycle. Naturally the exact opening and closing of the

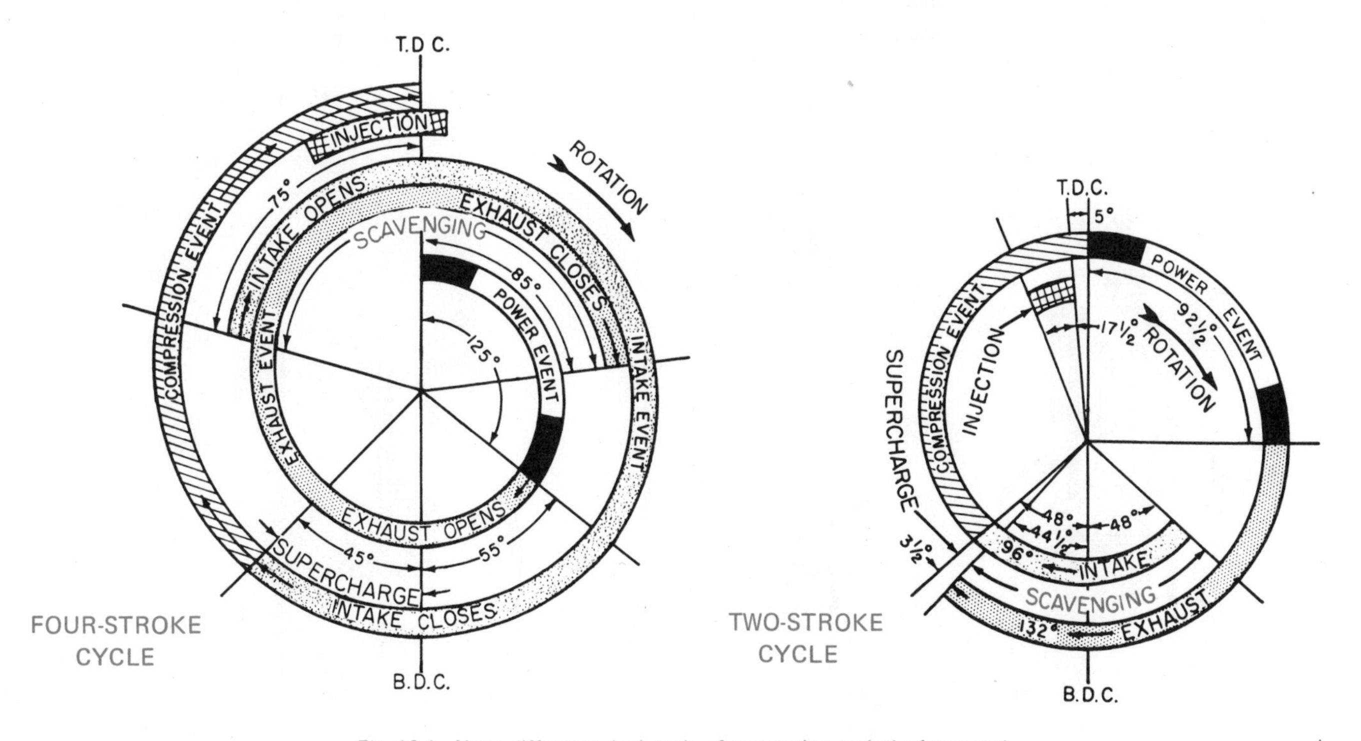

Fig. 10-1. Note difference in length of scavenging period of two and four-cycle diesel engines.

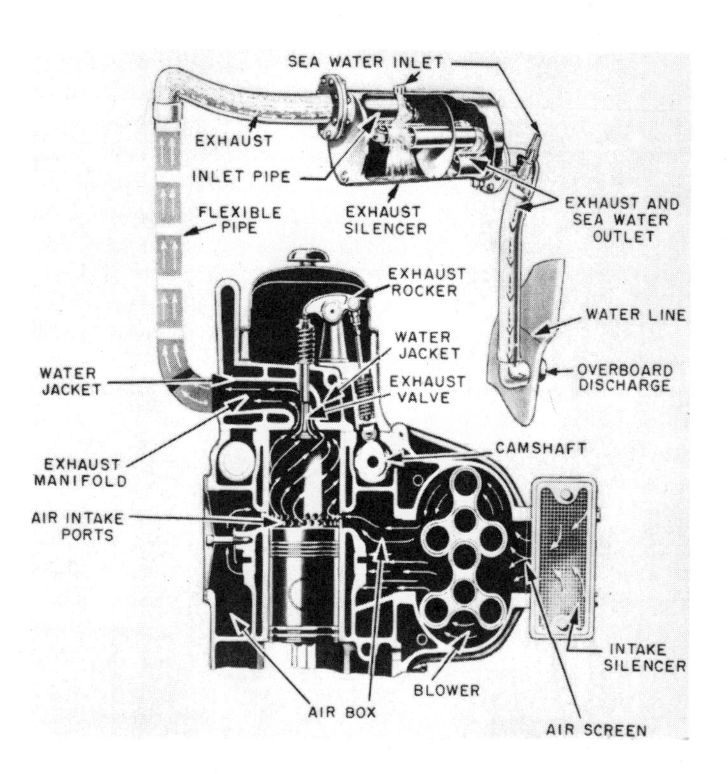

Fig. 10-2. Uniflow scavenging with blower on General Motors 71, two-stroke cycle engine.

ports will vary with different engines.

In the case of the four-stroke cycle the scavenging (air intake) occurs through 160 deg. of crankshaft rotation or almost twice as long as is the case of the two-stroke cycle. As a result, combustion chamber temperatures are reduced and a larger portion of the gases is swept from the cylinder.

Air may be supplied for scavenging as a result of the difference in air pressure between that existing in the combustion chamber as the piston moves down in the cylinder, and the normal atmospheric pressure existing at the air intake of the system. This is known as a naturally aspirated system. It is used extensively in motor vehicle engines of the four-cycle type, and some large industrial engines of the two-cycle type.

Due to the resistance to the flow of air through the manifold and valves, the air obtained by the naturally aspirated method is often insufficient to provide complete combustion of the fuel. To overcome that problem crankcase scavenging, pumps, superchargers, Fig. 10-2, and Turbochargers, Fig. 10-10, have been developed.

These devices compress the air and force it into the cylinder. More air is forced into the cylinder and in addition, burned gases which may have remained in the cylinder from the previous power stroke are forced out so only clean air remains. Such power scavenging is particularly necessary in the case of two-cycle engines. Details of scavenging pumps, supercharging and turbocharging units will be discussed later.

METHOD OF SCAVENGING

Scavenging of the cylinders must be so designed that a maximum of burned gases are swept from the cylinder.

The term scavenging is usually applied to two-cycle engines. However, in the case of four-cycle engines it is accomplished during the exhaust stroke of the cycle and during the initial period of the inlet stroke and is called supercharging.

There are several methods of scavenging in a two-stroke cycle diesel engine. These are generally known as cross flow, loop and uniflow scavenging.

In cross flow type scavenging inlet and exhaust ports are on opposite sides of the cylinder. When both ports are open, scavenging air enters through the inlet port, crosses top of the piston and leaves through the exhaust port. This is the simplest and probably the least effective method of scavenging as little of the exhaust gas in the upper part of the cylinder is removed and this will dilute the fresh air needed for efficient combustion.

To overcome difficulty experienced with cross flow

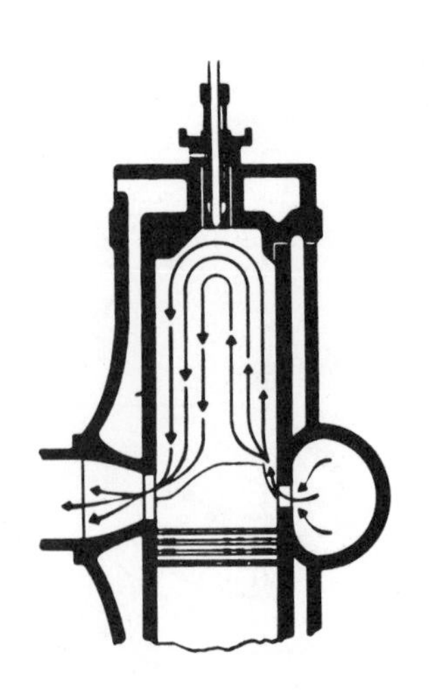
PORT DIRECT SCAVENGING

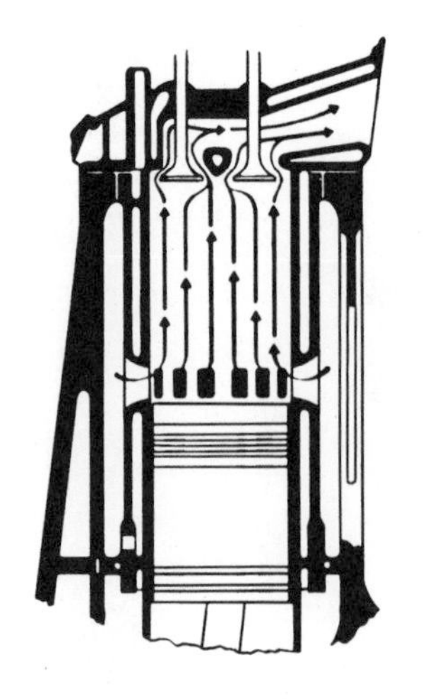
VALVE UNIFLOW SCAVENGING

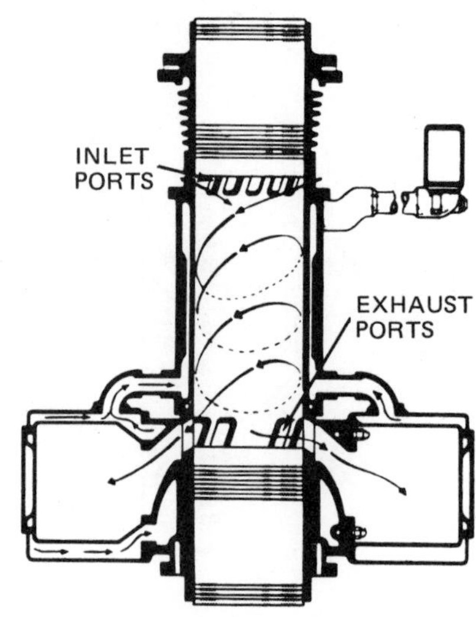

UNIFLOW PORT SCAVENGING

Fig. 10-3. Different methods of scavenging two-cycle engines.

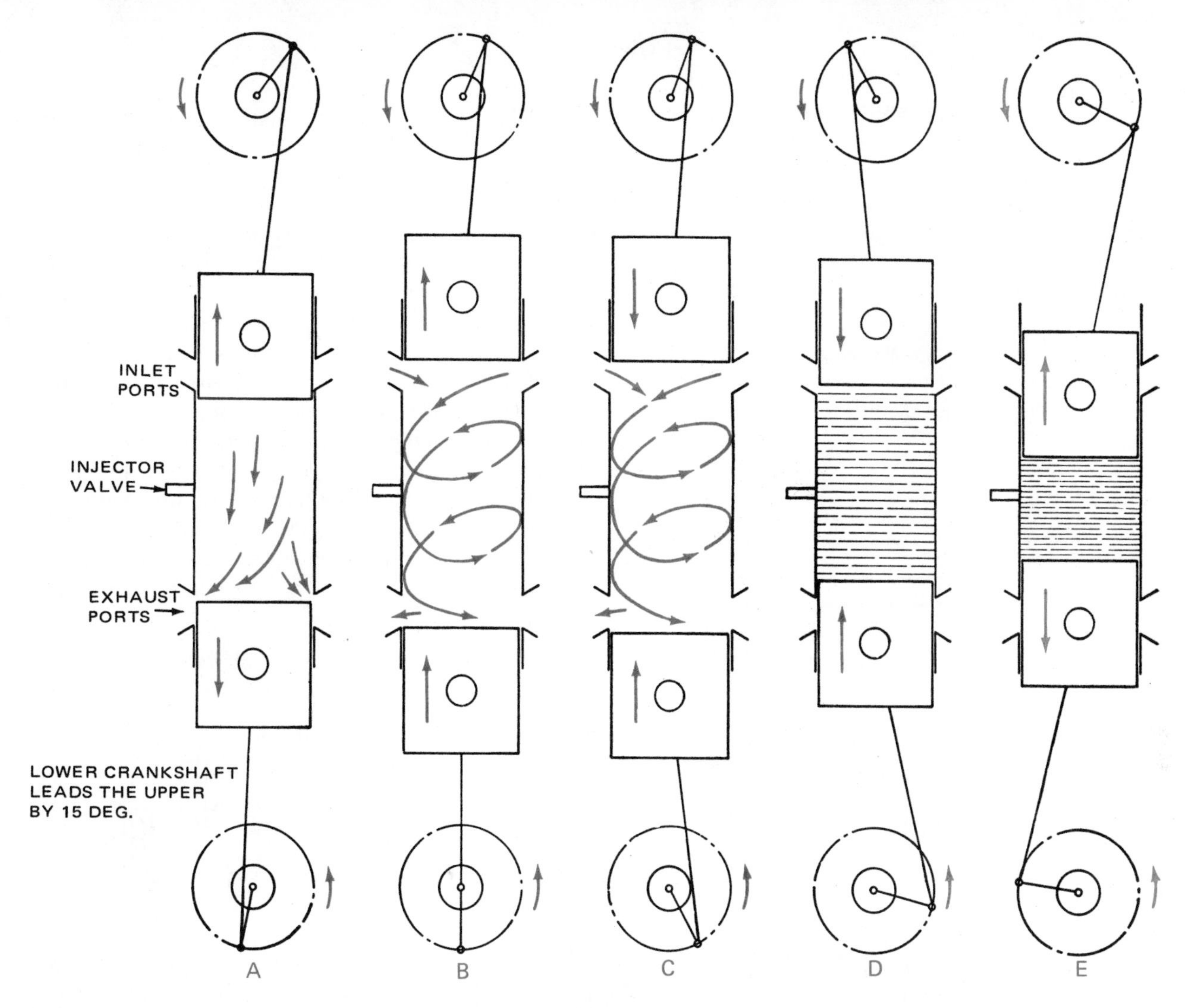

Fig. 10-4. Scavenging operation in opposed piston engine. (Fairbanks Morse)

scavenging, loop or port direct scavenging, Fig. 10-3, was introduced. In this design the end of the piston is contoured to direct incoming air upward to the top of the cylinder so it sweeps out the burned gases ahead of it.

In this design the ports are placed so the descending piston starts to uncover the exhaust port then the intake port. Pressure in the cylinder is materially reduced to a value below that existing in the intake system. The intake port is then uncovered and the scavenging air enters the cylinder as shown in Fig. 10-3.

In some designs the induction ports are provided with an air valve arranged to keep the burned gases from flowing into the inlet system. Such valves remain closed until the pressure within the cylinder has dropped below that of the scavenging air.

UNIFLOW SCAVENGING

In uniflow scavenging, Fig. 10-3, air enters the cylinder through the ports at the lower end of the piston stroke and passes upward and out through the exhaust valves located in the cylinder head.

In this system, when the piston has traveled 80 to 85 percent of the expansion stroke, cam-operated exhaust valves in the cylinder head are opened, Fig. 10-3. The exhaust gases are released and begin to escape from the cylinder. The piston continues to move downward. It uncovers the air inlet or scavenging ports, through which air enters the cylinder, forcing burned gases ahead of it. This continues until the piston on its upward stroke, covers the scavenging ports. After the ports are closed, the exhaust valves are closed.

OPPOSED PISTON ENGINE SCAVENGING

The cycle of operations including scavenging for an opposed piston engine (Fairbanks Morse) is shown in Figs. 10-3 and 10-4. Scavenging in this engine is of the uniflow or straight through type. The lower piston controls the exhaust ports; the upper piston, the scavenging or air inlet ports. To obtain the necessary preliminary release of exhaust gases by the exhaust ports ahead of the air inlet ports, the crank of the lower crankshaft is advanced in respect to the crank of the upper crankshaft. This condition is indicated at A, Fig. 10-4, where the exhaust port has been uncovered while the inlet port is still closed.

While the amount the lower crankshaft is advanced over the upper crankshaft varies, the amount usually ranges from 10 to 15 deg.

When pressure within the cylinder is reduced, the inlet ports are uncovered as shown at B, Fig. 10-4, and scavenging starts. After the exhaust ports are closed, additional admission of air takes place until the air inlet ports are closed by movement of the piston. Compression of the air charge is then started as shown at E, Fig. 10-4. Slightly before the pistons are closest together, fuel is injected and ignition takes place.

SCAVENGING METHODS

In the foregoing discussions of scavenging little mention has been made of methods providing air pressure to aid in forcing air through the cylinder. Two methods of providing such pressure are:

1. Compression of air in the crankcase.
2. By using a pump.

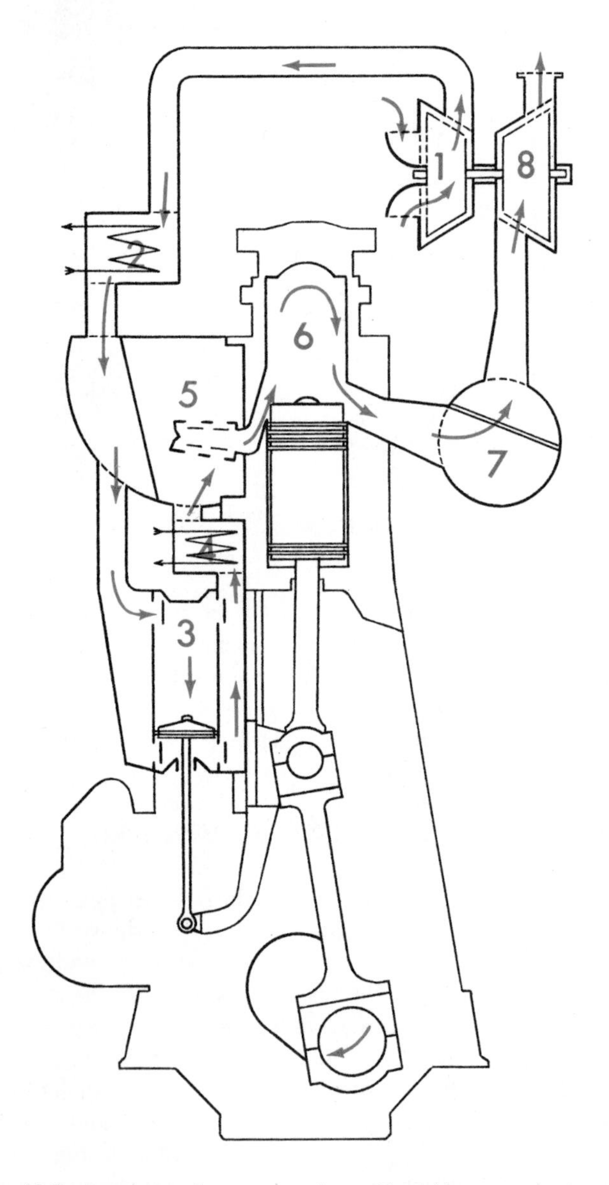

Fig. 10-5. A reciprocating pump and centrifugal blower are both used on this Fiat two-stroke single acting diesel engine.

1. CENTRIFUGAL BLOWER.
2. FIRST-STAGE COOLER.
3. RECIPROCATING SCAVENGING PUMP.
4. SECOND-STATE COOLER.
5. SCAVENGING AIR MANIFOLD.
6. MAIN CYLINDER.
7. EXHAUST GAS MANIFOLD.
8. EXHAUST GAS TURBINE.

CRANKCASE SCAVENGING

In crankcase scavenging on two-cycle engines as the piston moves upward, pressure in the crankcase is lowered until the check valve in the crankcase is opened. This permits air to enter. When the piston reaches the top of its stroke, injection takes place and the power stroke starts. The descending piston compresses the air in the crankcase closing the check valve. As the piston approaches lower center, the exhaust port is uncovered and exhaust gases start to pass out, lowering the pressure within the cylinder. Further movement of the piston uncovers the inlet port. This permits compressed air in the crankcase to pass into the cylinder. Note that the intake port is designed to direct air upward which with the contoured head of the piston scavenges the cylinder by the loop method. As the piston starts on its up or compression stroke, the cycle is repeated for crankcase scavenging in this example of a two-cycle diesel engine.

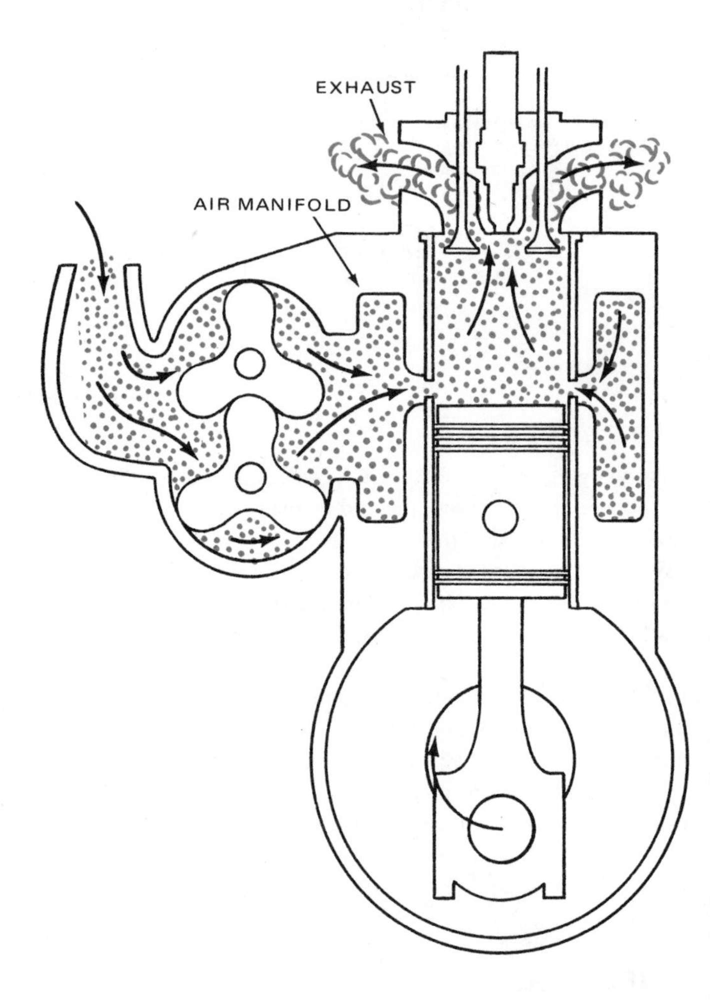

Fig. 10-6. Start of scavenging operation on two-cycle engine with blower. (General Motors Corp.)

In the two-stroke engine with crankcase scavenging, the displacement of the scavenging pump is the same as that of the working cylinder. Because of various losses through the ports, etc. and also because the clearance volume of the crankcase is large, two-stroke cycle engine scavenging is relatively low.

RECIPROCATING PUMP

The crankcase method of scavenging produces little pressure and consequently various types of pumps have been developed. One such pump is a reciprocating or piston type pump. This is of simple construction. One design is shown at 3, Fig. 10-5. This is a Fiat two-stroke single acting diesel engine with cross head, first and second stage cooler, scavenging air manifold, centrifugal blower and exhaust gas turbine. This system is known as a constant pressure system with an exhaust gas operated turbo blower in series with the reciprocating pump (one for each cylinder). The two stages of air cooling (after the turbo blowers and reciprocating pumps) are helpful when the unit is operated under tropical conditions.

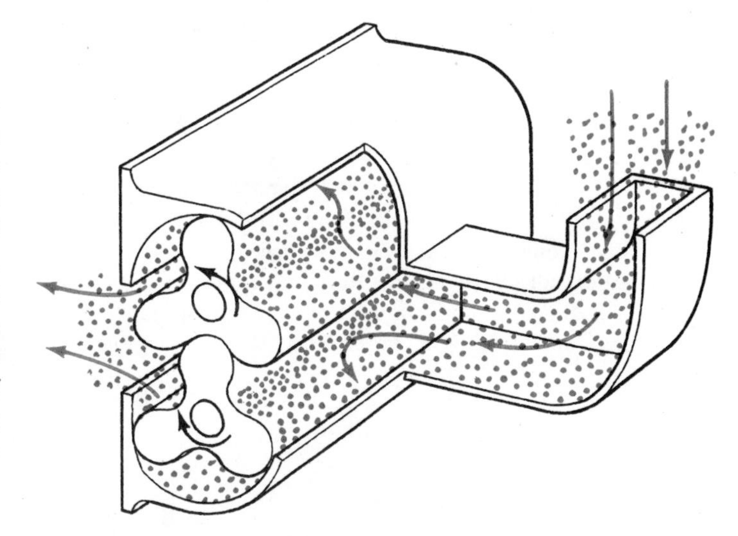

Fig. 10-8. Roots type blower with parallel lobes.

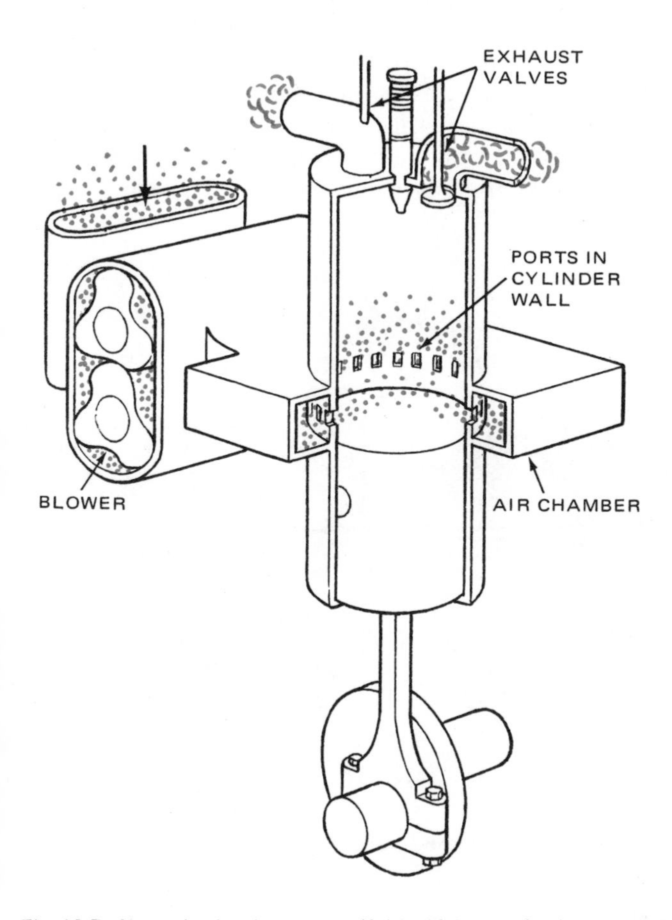

Fig. 10-7. Note air chamber or manifold which completely surrounds cylinder for better air distribution. (General Motors Corp.)

ROOTS TYPE BLOWER

The Roots type blower, Fig. 10-2, is used extensively for scavenging two-cycle engines. When the piston is at the bottom of its stroke just starting upward, Fig. 10-6, both intake and exhaust ports are open. Air is being blown in the ports which pushes exhaust gases left from the previous cycle out through the exhaust valves. When the pistons are about one quarter of the way up, the valves close. The ports are covered by the movement of the piston. Exhaust gases will have been expelled and the cylinder is full of fresh air. The rest of the stroke is an ordinary compression stroke at the end of which fuel is injected and combustion takes place.

In Fig. 10-6, it will be noted that air passes from the blower into an air manifold or chamber, Fig. 10-7, and enters the cylinder through ports which extend completely around the cylinder. This provides better distribution of air throughout the cylinder and improved scavenging results.

The Roots type of blower is essentially a gear pump with rotors of either two, three or four lobes each, enclosed in a suitable housing. This is classed as a positive displacement rotary blower. The rotors are so designed that they do not come into contact with each other or with the housing. Clearances are made as small as manufacturing methods permit and are approximately .005 in. Volumetric efficiency of the three lobe blower as installed on the General Motors type 71 engine decreases from slightly over 85 percent at 500 rpm (970 rpm blower speed) to about 83 percent at 2000 rpm engine speed.

One of the two rotors is direct driven through gears from the crankshaft and the shafts of the two rotors are in turn connected together through gearing. Fine adjustment of the angular relation of the lobes is obtained by means of splined shafts for the gears. The number of teeth in the gears are chosen so that a vernier adjustment is possible.

The lobes in the rotor may be made straight and parallel to the axis of rotation, Fig. 10-8, or twisted into a spiral form, Fig. 10-9. The latter design gives a more uniform and constant flow of air.

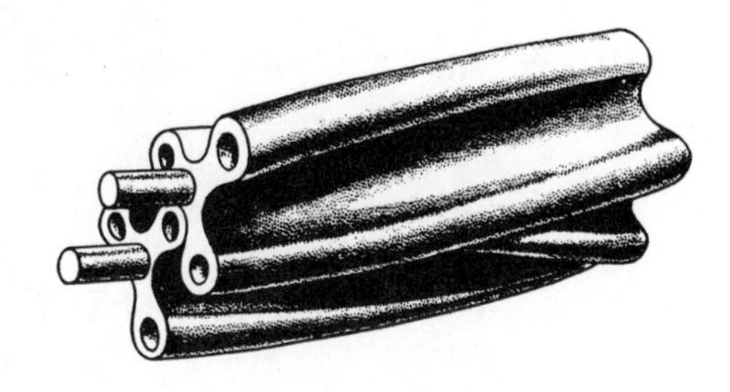

Fig. 10-9. Spiral lobes for Roots type blower designed to provide more uniform air flow.

The lobes of the rotor are made hollow to reduce the inertia. To stiffen the rotors internal ribs are used, or circular discs may be welded to the ends of the openings of the lobes.

One type Roots blower as installed in two-cycle engine was driven at 1.94 times crankshaft speed and had an air delivery of 943 cubic feet per minute at 2000 rpm. Scavenging pressure varied approximately as the square of the speed and the horsepower consumed as a cube of the speed.

A major advantage of the Roots type blower is that air delivery is almost directly proportionate of the speed. Normal operation ranges from 2000 to 6000 rpm.

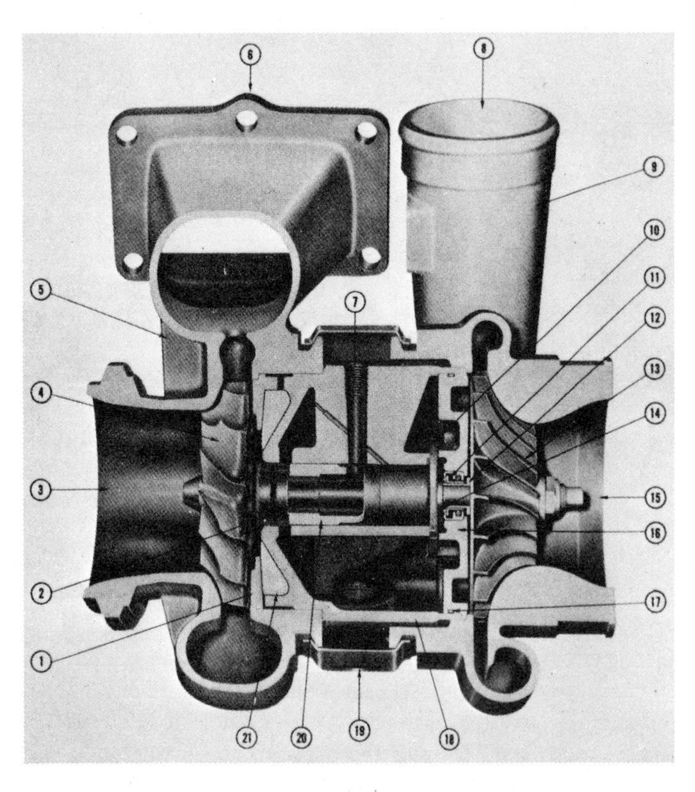

Fig. 10-10. Cummins type turbocharger with centrifugal type compressor.

1. HEAT SHIELD.
2. PISTON RING SEAL.
3. EXHAUST OUT.
4. TURBINE WHEEL.
5. TURBINE CASING.
6. EXHAUST IN.
7. OIL (COOLING) IN.
8. COMPRESSED AIR TO ENGINE.
9. CENTRIFUGAL COMPRESSOR CASING.
10. BEARING INSERT.
11. OIL SEAL.
12. SLEEVE.
13. COMPRESSOR WHEEL.
14. THRUST WASHER.
15. AIR IN.
16. SEAL PLATE.
17. SEAL RING.
18. BEARING HOUSING.
19. OIL OUT.
20. BEARING.
21. INSULATION PAD.

TURBOCHARGING AND SUPERCHARGING

Power that can be developed by an internal combustion engine is dependent to a considerable extent on type of fuel and how efficiently the fuel is burned. That in turn is dependent on the supply of air being adequate so complete combustion of the fuel is obtained.

By increasing the rate of flow of air to the engine more fuel can be burned efficiently and as a result power is increased. The rate of air flow can be increased by using blowers. The process is known as supercharging, and the equipment is called a supercharger or a turbocharger. In general when the blower is driven mechanically by an accessory shaft from the engine the device is called a supercharger. When the device is driven by the exhaust gases, it is called a turbocharger. In the diesel engine field most blowers are driven by exhaust gas and are called turbochargers.

Previously in this chapter the subject of scavenging was discussed. This is similar to supercharging as air is supplied to the cylinders, but it differs because with supercharging pressures are much greater. With scavenging the main purpose is to force out the burned gases. With supercharging the purpose is not only scavenging the cylinders of burned gases but also to supply air at a density greater than obtained by natural aspiration (movement of air by suction). Also, the term scavenging is applied only to two-cycle engines while supercharging is applied mainly to four-stroke cycle engines.

The major advantages of supercharging are: (1) Increased horsepower from an engine of given weights. (2) A supercharged engine usually costs less than a naturally aspirated engine of the same power. (3) Increased fuel economy.

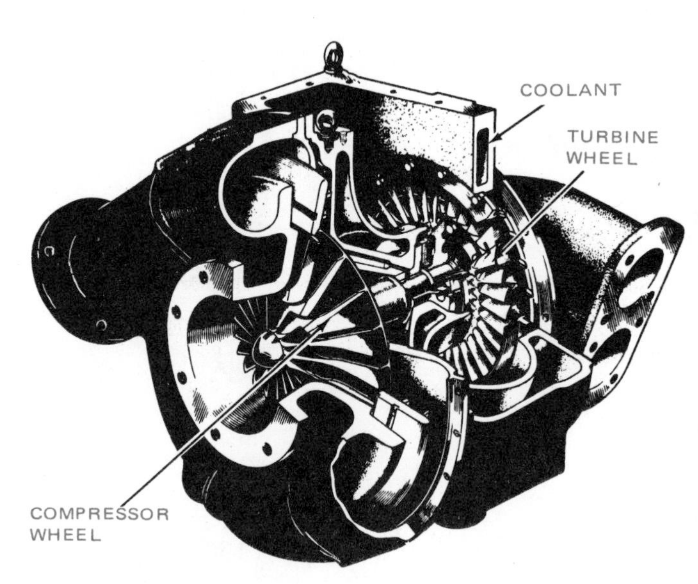

Fig. 10-11. Hispano-Suiza axial flow type turbocharger.

It has been found that the output of an engine can be increased about 50 percent by supercharging, without materially increasing bearing loads or heat stresses on parts such as pistons, rings and valves. With the use of intercoolers (these reduce temperature of supercharged air) the increase in developed power is considerably greater.

With normal supercharging ratios resulting in brake mean effective pressure (bmep) of approximately 150 psi the specific fuel consumption of the supercharged engine is usually less than that of the unsupercharged engine in the upper half of the load range. The brake mean effective pressure is an indication of the pressure needed for safe power output of an engine. It is determined by the average pressure on the pistons reduced by the amount needed to overcome loss to friction.

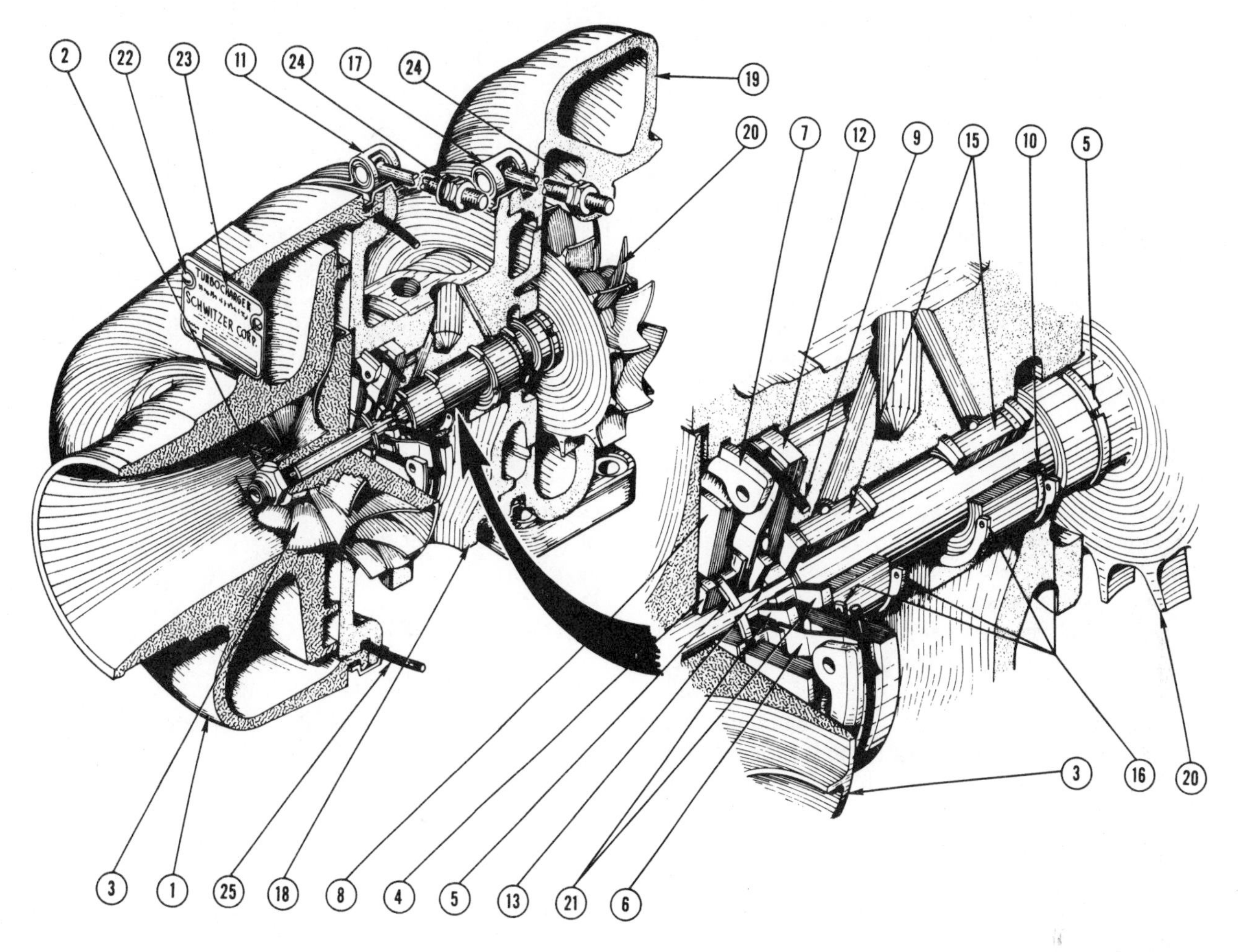

Fig. 10-12. Inward flow type turbocharger which weighs only 18 lbs. (Schwitzer Div., Wallace-Murray Corp.)

1. COMPRESSOR COVER.
2. COMPRESSOR LOCKNUT.
3. COMPRESSOR WHEEL.
4. FLINGER SLEEVE.
5. PISTON RING.
6. OIL DEFLECTOR.
7. SNAP RING.
8. COMPRESSOR INSERT.
9. O RING.
10. OIL CONTROL SLEEVE.
11. COMPRESSOR CLAMP BAND.
12. THRUST BEARING.
13. SPACER SLEEVE.
14. GROOVE PIN.
15. BEARING.
16. SNAP RING.
17. TURBINE CLAMP BAND.
18. BEARING HOUSING AND PIN ASSEMBLY.
19. TURBINE HOUSING.
20. SHAFT AND TURBINE WHEEL ASSEMBLY.
21. THRUST RING.
22. SELF TAPPING SCREW.
23. NAME PLATE.
24. CLAMP BAND LOCKNUT.
25. O RING.

CENTRIFUGAL TYPE BLOWER

The centrifugal type blower or compressor is of the nonpositive displacement type. It is used to provide pressures of 3 to 1 and over. At certain points of operation delivery increases substantially in direct proportion to the speed while the pressure increases as the square of the speed.

The centrifugal type supercharger is usually operated at speeds ranging up to 50,000 rpm. In some designs the centrifugal type blower is mechanically driven through gears from the crankcase. However, in most cases it is driven by pressure of hot exhaust gases at a saving in power.

Centrifugal compressors or blowers are shown in Figs. 10-10 and 10-11. Fig. 10-10 shows a Cummins unit, Fig. 10-11, a Hispano-Suiza unit. These are both driven by turbines powered by the exhaust and are classed as turbochargers. Centrifugal compressors include diffusion blades designed to convert the velocity of the air into static pressure. This is usually accomplished by increasing the radius of the air spin and by increasing the area.

While some centrifugal compressors have been driven directly through gearing from the engine crankshaft the current practice is to use energy of the exhaust gas for that purpose.

The centrifugal compressor provides air at pressure ratios up to 3 to 1. In operation, air enters blower at the center and passes from the impeller to a stationary diffuser form in the housing. In the diffuser much of its kinetic energy (energy developed by a moving body as a result of its motion) is converted to increased pressure. The air is then discharged from the perimeter of the housing.

Design of diffuser, Fig. 10-14, varies but in general consists of an open passage which increases in cross-sectional area which reduces the velocity of the air. Another design uses diffuser blades to perform the same function. In such designs an annular space contains curved blades and outside the diffuser there is a volute which receives air set in motion by the impeller and conducts it to the outlet of the housing.

The gear-driven rotary blower turns on a fixed ratio to that of the crankshaft and a supply of air, which is approximately constant, is supplied to the engine. As a result the engine will be capable of supplying a high mep (mean effective pressure) at all speeds. Where there is little space available and high output is necessary over a wide range of speed, a gear-driven supercharger is often preferred.

TURBOCHARGING

A turbocharger is a device consisting of an engine exhaust operated turbine which drives a centrifugal compressor, Figs. 10-10 and 10-11. It is used to supply air for scavenging and combustion. It is driven at speeds up to 100,000 rpm.

Through constant improvements in design and materials, the weight of the turbocharger has been constantly reduced. See Figs. 10-12 and 10-13.

Turbochargers operate not only under conditions of extreme heat but also under extreme variations in temperature. On the exhaust or input side of the turbocharger temperatures are in the 1500 deg. F. area. On the output side, the air compressed by the compressor is at the 300 deg. F. level.

Since exhaust gases are used to drive the turbine, use is made of what otherwise would be wasted energy. In the case of the engine driven supercharger, tests show the 8 to 10 percent of the total power developed is used for driving the supercharger. In the case of the turbocharger losses that may result from back pressure are usually of slight consideration.

Increased power resulting from the installation of a turbocharger will vary with different engines and characteristics of the turbochargers. However, on one four-stroke engine, brake horsepower was increased 37.5 percent over the naturally aspirated engine. On another engine, the increase was 56.0 percent. Such increase is largely due to: (1) More complete scavenging. (2) Adequate air to support complete combustion. (3) Improved mechanical efficiency of the engine. There is no appreciable increase in friction hp with the increase in output. Turbocharged systems usually include cooling devices to maintain the temperature of the scavenging air at a value which will produce the greatest power. See Fig. 10-19.

Since operation of the turbocharger is dependent on the energy received from exhaust gases, the speed tends to increase as the load on the engine increases. In other words, the speed is more dependent on engine load than on engine speed. If engine speed remains constant but load is increased, the speed of the turbocharger will increase and the air delivery will also increase.

TURBINE CONSTRUCTION

Depending on how the energy of the exhaust gases is used in the turbine, there are three different classifications of turbine construction:

1. Reaction.
2. Impulse.
3. Mixed flow.

In the reaction type turbine the rotor buckets have a smaller discharge area than they have an entrance area. This results in an increase in velocity, but a decrease in pressure, as exhaust gases flow through the buckets.

In the impulse turbine all the pressure of the gases entering the turbine is stepped up to the highest possible velocity by

Fig. 10-13. Showing development of Schwitzer turbocharger. The larger unit which weighs 60 lbs. and rotates at 48,000 rpm has same output as smaller unit which weighs only 18 lbs. and rotates at 100,000 rpm.

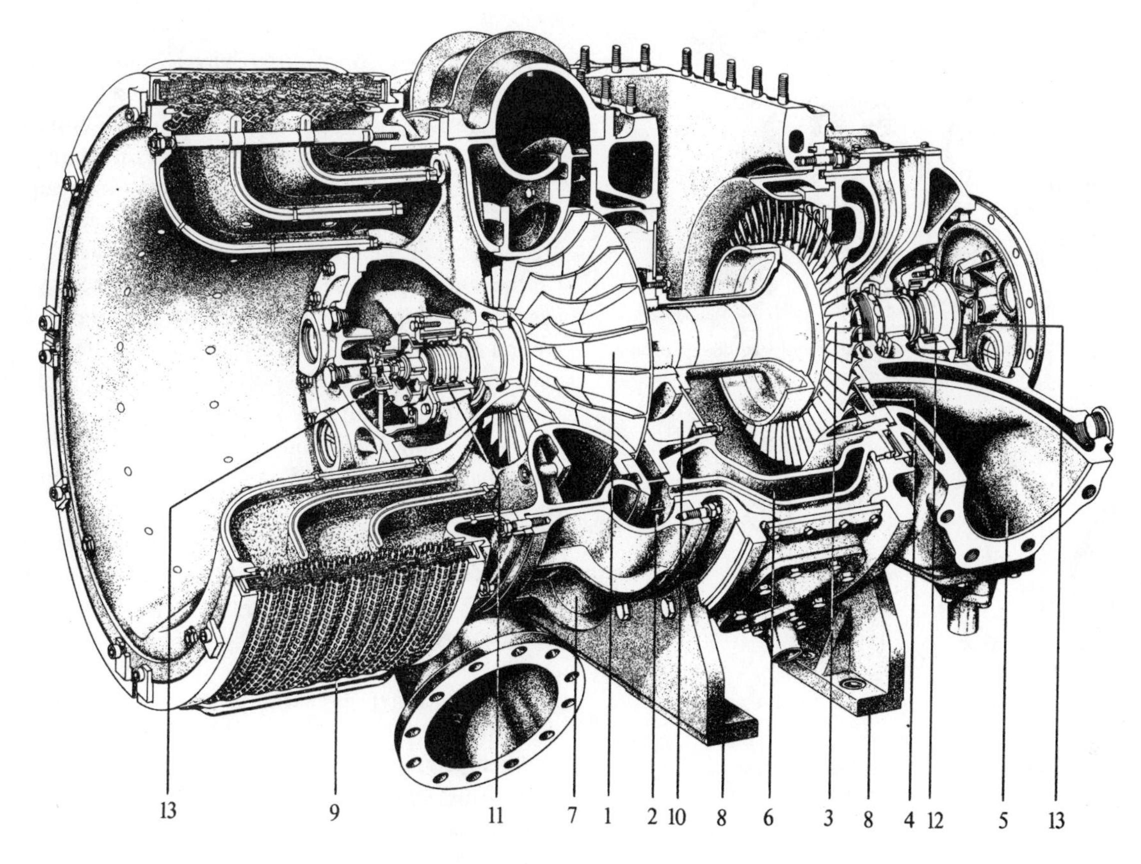

Fig. 10-14. Cutaway view of Brown Boveri turbocharger.

1 Compressor wheel.
2 Diffuser.
3 Turbine.
4 Nozzle ring.
5 Gas inlet casing, water-cooled.
6 Gas outlet casing, water-cooled.
7 Compressor casing.
8 Feet.
9 Filter-silencer.
10 Partition wall (heat insulating).
11 Ball bearing assembly, journal and thrust bearing.
12 Roller bearing assembly, journal bearing.
13 Lubricating oil pumps.

the contour of the guide vanes. The gases are then directed against the rotor buckets. The turbine rotor is driven by the impulse due to reversing the direction of the flow through the buckets.

In the mixed flow turbine, gases are expanded partly in the guide vanes increasing the velocity which in turn causes an impulse on the rotor. With the decreasing area of the rotor buckets, there is further increase in velocity and decrease in pressure. The increase in velocity causes a reaction opposite to the direction of gas flow aiding the rotation of the turbine rotor.

The turbines in turbochargers are mostly of the mixed flow type. The degree of impulse or reaction principle varies with the different manufacturers.

As a result of friction and outlet losses, the efficiency of the impulse turbine is usually greater than that of the reaction turbine.

GAS FLOW IN TURBINES

Designs of turbines vary also in the direction in which the exhaust gases pass through the unit. The direction of gas flow may be either: (1) Axial. (2) Inward. (3) Mixed.

The Cummins turbocharger, Fig. 10-10, and the Schweitzer, Fig. 10-12, are of the inward flow design; the Hispano-Suiza unit, Fig. 10-11, is an example of axial flow.

In the inward flow turbine exhaust gases enter the housing which surrounds the unit (6) Fig. 10-10. The gases then pass tangentially (at an angle) against the rotor blades in the direction of rotation, leaving at the center of the rotor as shown at (3) Fig. 10-10.

In the axial flow turbocharger, Fig. 10-11, exhaust gases enter along the axis of the rotor shaft. They pass through the nozzle guide vanes of the nozzle ring, then through the blades of the rotor.

Another example of the inward flow turbocharger is shown in Fig. 10-12. This is a Schwitzer D Series Unit used extensively in automotive type vehicles. On V type engines, it is often installed in tandem (one back of another).

In general, axial flow and mixed flow type turbochargers are installed in low and medium speed diesel engines. In the high speed diesel engine field the inward flow turbine, Figs. 10-10 and 10-12, is used extensively.

A combination of the high speed engine driven supercharger and turbocharger has a capability of producing unusually high torque at low engine speed. Such a combination helps to eliminate smoke during periods of acceleration. The supercharger gear driven by the engine supplies sufficient air required by the engine at lower speeds and loads, when there is usually insufficient exhaust gas energy. As engine speed increases, the turbocharger automatically assumes the supercharging of the engine and relieves the horsepower required to

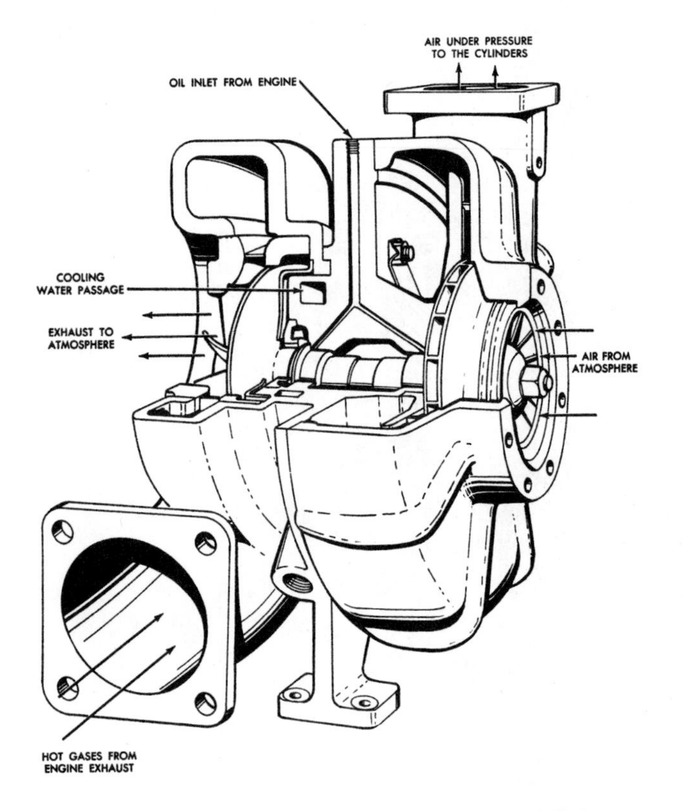

Fig. 10-15. Details of Elliot turbocharger as installed on Waukesha engine.

drive the supercharger.

The Fiat two-stroke single acting diesel engine utilizes a piston type compressor and also a turbocharger. See Fig. 10-5.

COOLING TURBOCHARGER

With the temperature of the exhaust gases, which operate the turbocharger, in the vicinity of 1350 deg., it is essential that provision be made for cooling the unit. In Fig. 10-11, the water jackets are illustrated. In Fig. 10-10, oil for lubrication and cooling enters at (7), and leaves the unit at (19).

In the Brown Boveri turbocharger, combination lubricating and cooling is obtained as follows: bearings (12) and (13), Fig. 10-14, are fitted at each end of the rotor. Lubrication of the bearings is independent of that of the engine. Lubrication discs (14) are fitted on each end of the shaft which supplies oil to the bearings. For larger machines, lubricating oil pumps are used instead of the discs. The bearing casing acts as an oil tank. On the compressor side, lubricating oil is cooled by air blowing over the bearing casing; on the turbo side, water is in a jacket around the casing. Fig. 10-14 also shows the heat insulating partition wall designed to prevent transmission of heat in the turbine into the compression area.

The cooling provisions are also shown in the Elliot turbocharger, Fig. 10-15, as installed on some Waukesha engines. This turbocharger is water-cooled by water drawn from the engine cooling system, which is passed through the intermediate casing water jacket at all times while the engine is operating. The cooling water is circulated through a cast-in-jacket in area adjacent to turbine end bearing. The water inlet connection is at an angle at the bottom of the intermediate casing. The water discharge is at the top. Water circulating through the turbocharger should be regulated at such a rate that temperature rise does not exceed 30 deg. at full engine load. This will restrict turbine distortion to a reasonable amount. Discharge temperature should not exceed 200 deg. F.

Because of large variations in temperature throughout the supercharger, different materials each with a different coefficient of expansion, close clearances between rotating parts and housings, together with high rotational speeds, it is essential that adequate cooling of the supercharger be provided and maintained. Should the cooling system fail or be inadequate the resulting expansion of various parts would quickly ruin the unit. An air-cooled turbine which eliminates water connections to the turbine and interconnecting water piping to the engine cooling system is produced by the Airesearch Div. of Garrett Corp. As a result, overall weight is reduced and space saved. The Garrett turbocharger features an internal subdivision, Fig.

Fig. 10-16. Control system of Garrett turbocharger. (Garrett Airesearch, Industrial Div.)

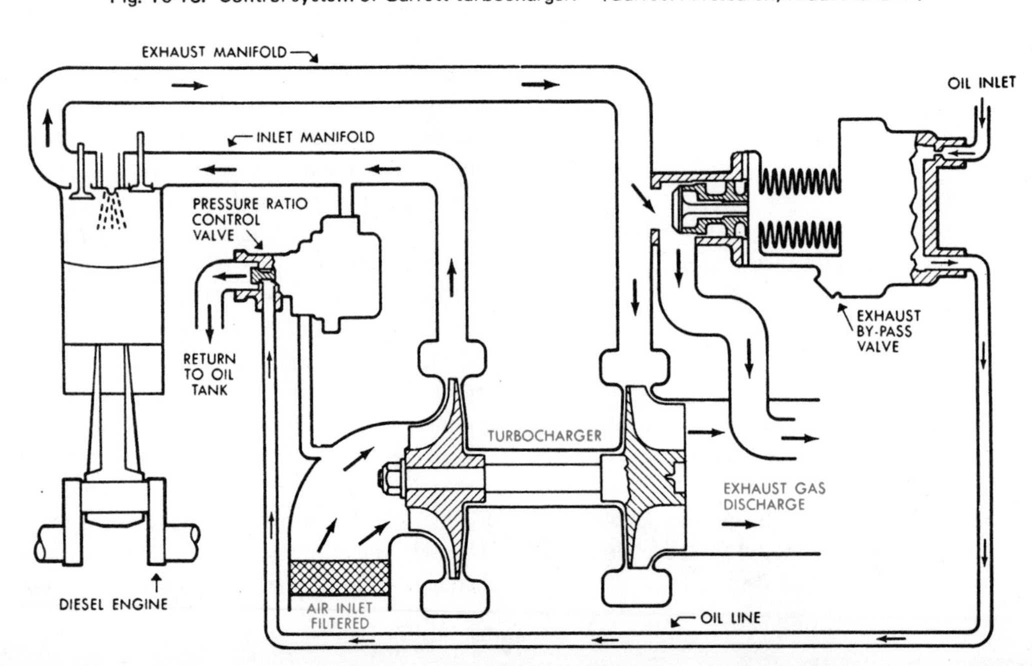

10-16, that directs exhaust gases through different admission passages into the turbine wheel. It is claimed as a result the exhaust pulsations (as received from the engine) are more efficiently utilized and the amount of useful supercharge power for engine acceleration is increased.

SUPERCHARGING PERFORMANCE

Brake horsepower and fuel economy are both increased by means of supercharging. In the case of two-cycle engines, the increase in horsepower can be as much as 35 to 50 percent. Up to 100 percent and over may be obtained with a four-cycle engine, over the naturally aspirated engine.

Fig. 10-17 shows the curves of a typical General Motors 12 cylinder engine. These curves reveal that with the installation of a supercharger there is an increase of 200 hp and over 600 ft. lb. of torque. Fuel per brake hp hour consumption increases approximately .025 at 1600 rpm. It will also be noted that the supercharged engine attains maximum torque at a lower speed than the naturally aspirated engine.

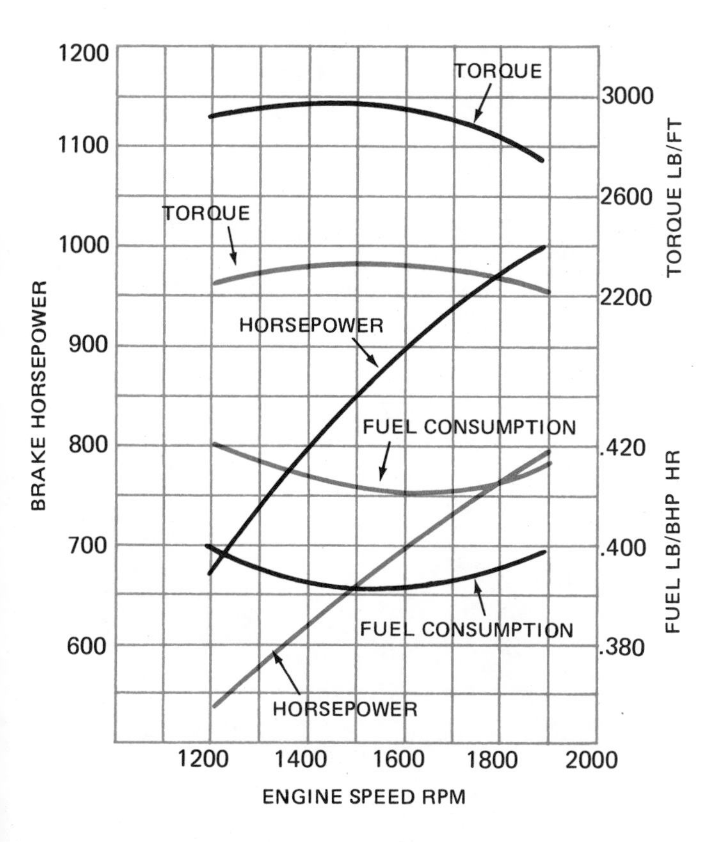

Fig. 10-17. Comparison of power developed by nonsupercharged (red curves) and supercharged (black curves) engines.

Fig. 10-18 shows the hp curves of a typical truck engine. It will be noted that at 2100 rpm of the naturally aspirated engine, the hp is 240 hp, while that of the supercharged engine is 320 hp. When aftercooling is added brake hp is increased to nearly 380 hp.

Improvement in torque and fuel consumption per brake hp hour are also impressive. At 2100 rpm fuel consumption for

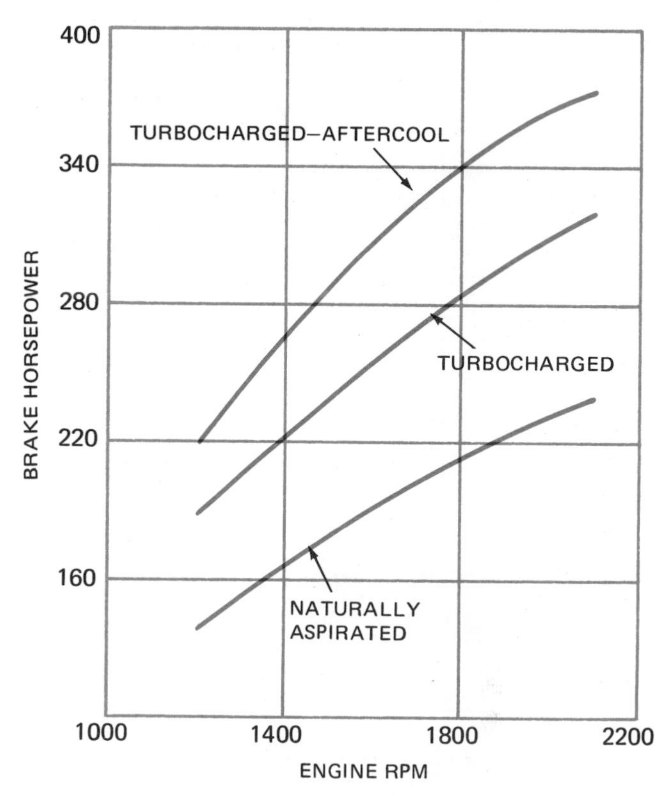

Fig. 10-18. Performance curves of typical truck engine showing bhp with naturally aspirated engine, turbocharged and turbocharged with after cooling.

that particular engine naturally aspirated was .425 lb. per bhp hour and the supercharger was .38 lb. per bhp hour. Only a slight improvement was attained by aftercooling.

Torque improvement for this same engine was 675 for the naturally aspirated engine, 875 for the supercharged engine, and 1025 ft. lb. for the supercharged and aftercooled engine.

DUAL BLOWER INSTALLATION

Power developed by an engine is dependent to a considerable extent on how much fuel it can burn. That in turn, depends on how much air can be supplied to the cylinders. In some designs it has been found advantageous to install two blowers in series or parallel. The method of connecting the blowers is dependent on the amount of air required and the type of equipment. In some installations such as the Cooper-Bessemer, a turbocharger is used for the initial stage of compression. The air is further compressed by a gear-driven centrifugal blower. The air passes to an aftercooler, then to the engine. The installation on a Fairbanks Morse engine first uses a gear-driven rotary blower. The air is then further compressed in the turbocharger. After passing through the intercooler it enters the air manifold.

The Schwitzer supercharger-turbocharger combination as installed on some trucks, is designed to produce high torque at low engine speeds, and reduce smoke during acceleration. The supercharger, gear-driven by the engine, supplies additional air required at low engine speeds when little exhaust gas energy is available. The turbocharger takes over as engine speed and load

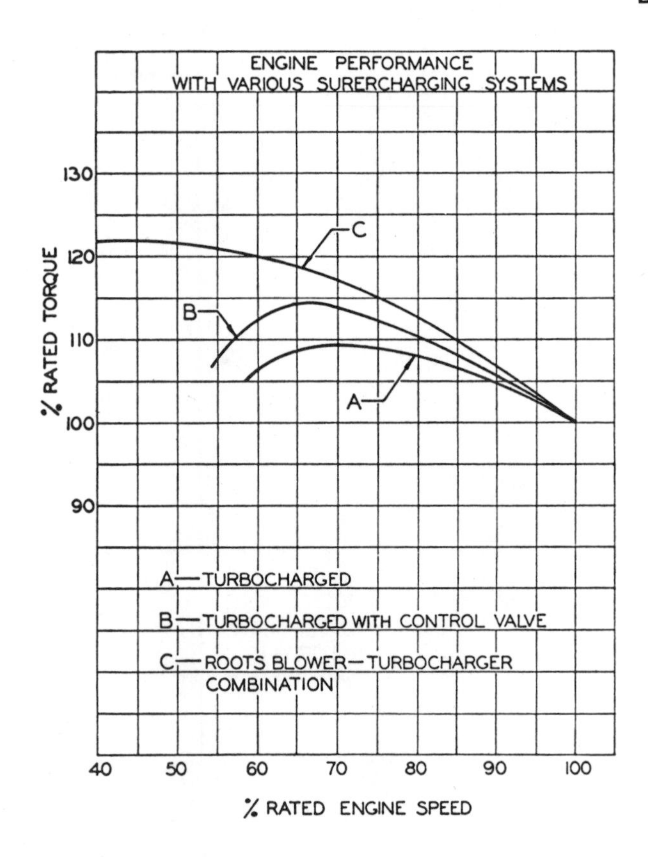

Fig. 10-19. Torque curves produced by different types of supercharging systems produced by Schwitzer.
(Schwitzer Div. Wallace-Murray Corp.)

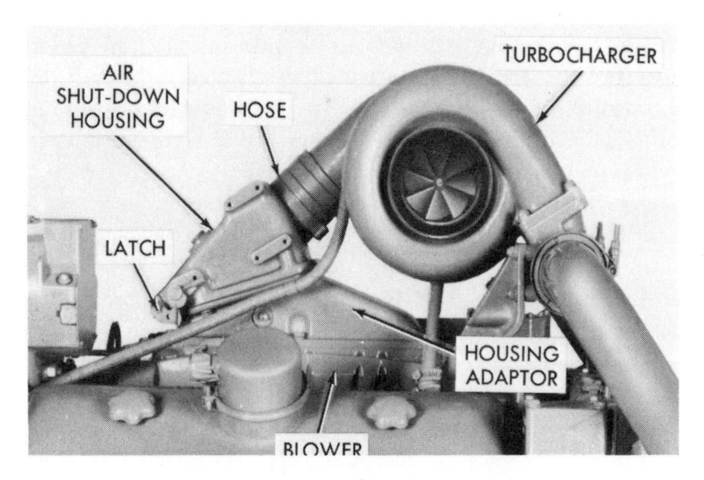

Fig. 10-20. Installation of a Roots type blower and a turbocharger on a series 92 Detroit Diesel Allison engine.

increase thereby reducing the hp required to drive the blower.

Fig. 10-19 shows the power produced by an engine with a turbocharger and control valve, and with a Roots blower turbocharger combination. The control valve is a device which bypasses the exhaust gas around the turbine internally and enters the exhaust stream ahead of the exhaust stack connection in the turbine housing.

TURBOCHARGED DETROIT DIESEL

The installation of both a blower and a turbocharger on a Detroit Diesel Allison series 92 engine is shown in Fig. 10-20. The Roots type blower is mounted between the cylinders of the V type engine. The turbocharger is placed above the blower.

The 92 series engine is built in V-6, V-8 and V-12 models. An in-line six also is available. The V-8 engine develops 360 hp at 2100 rpm without the turbocharger. With the turbocharger, power is stepped up to 430 hp at the same speed. Torque and fuel economy are also improved.

Another feature of this engine is the air-to-water heat exchanger mounted directly below the Roots blower. This unit aids in reducing exhaust emissions. The cross head piston, Fig. 12-22, has the dome and skirt of malleable iron and plated skirt. These features also contribute to improved cooling.

Exhaust smoke and emissions from this turbocharged engine have met Federal standards without difficulty in initial tests. Exhaust smoke under acceleration was 10.67 compared to 20 for the Federal requirements. Exhaust emissions (BSCO) were 4.98 compared to the Federal standard of 40. The installation of both the blower and a turbocharger on a series 149 Detroit Diesel Allison engine is shown in Fig. 10-21.

INTERCOOLERS

As air is compressed, its temperature will rise. For most efficient combustion the temperature must be relatively low and maintained within certain limits. To reduce the tempera-

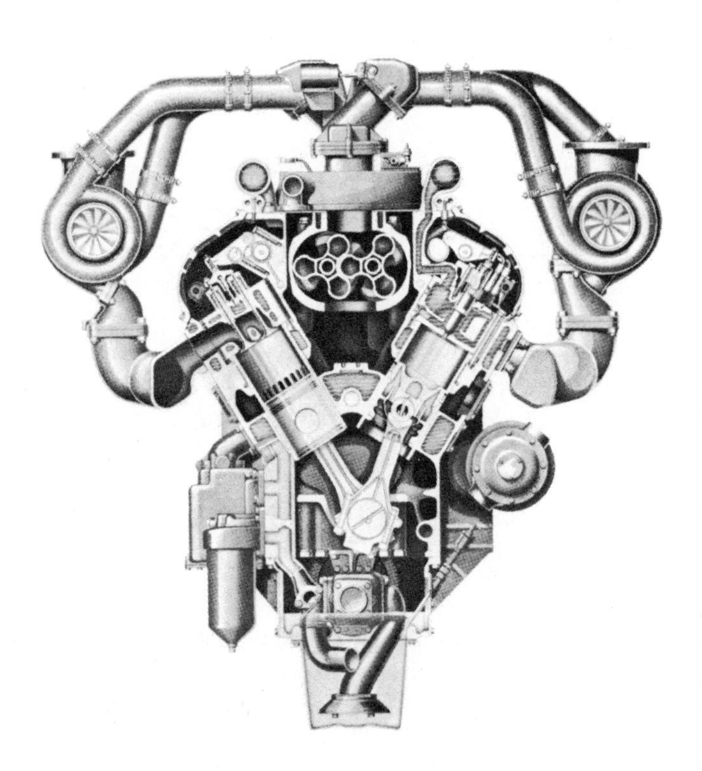

Fig. 10-21. Sectional view of the installation of a blower and turbocharger on a series 149 Detroit Diesel Allison engine.

ture of the air after it has been compressed by the supercharger an intercooler (also known as an aftercooler) may be installed between the supercharger and the engine, Fig. 10-22. By reducing the temperature of the air entering the engine, its density is increased and thus a greater amount of air by weight enters the cylinder. As a result, combustion efficiency is improved without increasing fuel consumption, Fig. 10-17.

INTERCOOLER DESIGN

Some designs of intercoolers follow closely the design of the familiar automobile radiator. An industrial type of intercooler is shown in Fig. 10-23. In general, the low-pressure type is made for combustion and scavenging air blowers operating at pressures of 3 psi on two-cycle engines. High-pressure coolers are used for four-cycle supercharged diesel engines. They provide, at reduced temperatures, air pressures from 4 psi to 30 psi for both scavenging and combustion.

When air entering the supercharger is approximately 90 deg. F. and is compressed to approximately twice the pressure of the atmosphere, its temperature will be increased to approximately 300 deg. F. The conventional aftercooler using water jacket coolant will bring the temperature down to approximately 200 deg. F. As a result of this reduction in temperature, more and cooler air enters the engine which operates at a lower temperature. In addition, peak cylinder pressures and exhaust gas temperatures are lower.

The Nordberg series 21 engines have an interesting combination of first and second stage intercoolers, two turbochargers and a centifugal scavenging blower, Fig. 10-24. In this system, air is drawn through the filter and is compressed in a motor-driven scavenging blower. It then passes through an intercooler where the heat of compression is removed. Air then enters the turbochargers. The turbochargers compress the scavenging air and discharge it through the second stage intercooler into the scavenging header.

Fig. 10-23. Type of air cooler with two separate tube bundles with one block for easier cleaning. (GEA Airexchangers, Inc.)

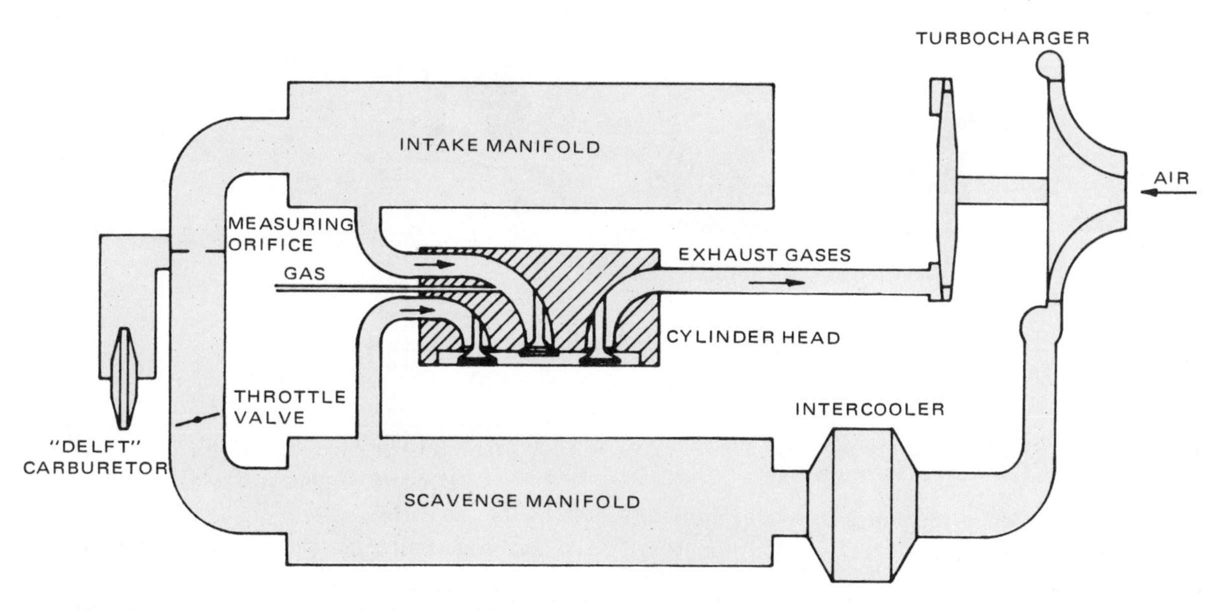

Fig. 10-22. Installation of intercooler on diesel gas engine. (N.V. Motorenfabriek Thomassen)

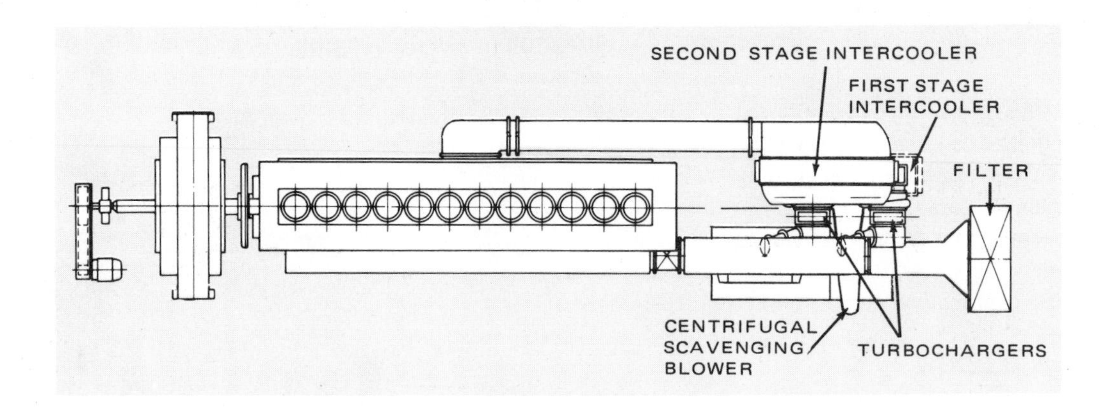

Fig. 10-24. Twelve cylinder power plant generating unit with motor driven centifugal blower, two turbochargers and first and second stage intercoolers. (Nordberg Mfg. Co.)

MERCEDES-BENZ 300 SD TURBOCHARGER

Mercedes-Benz engineers found that conventional turbochargers used on heavy-duty diesel engines have relatively large housings if they were designed to provide boost pressures required in the upper speed range. If these conventional turbochargers were used on passenger car diesel engines having a wider speed range, the boost pressure at the lower speed range would not be satisfactory.

The turbocharger adopted by Mercedes-Benz, Fig. 10-25, is the result of collaboration of Mercedes-Benz engineers and the Garrett AiResearch Industrial Division. It has improved compressor efficiency in a wide flow range. At engine speeds above 1,600 rpm, compressor efficiency is between 65 and 74 percent.

This turbocharger is about one-half conventional size. It has low weight of rotating masses and small diameter rotating blades. Maximum boost is attained at an engine speed of 2,000 rpm. At higher speeds, boost is maintained at a constant pressure level by the exhaust gas turbine wastegate valve.

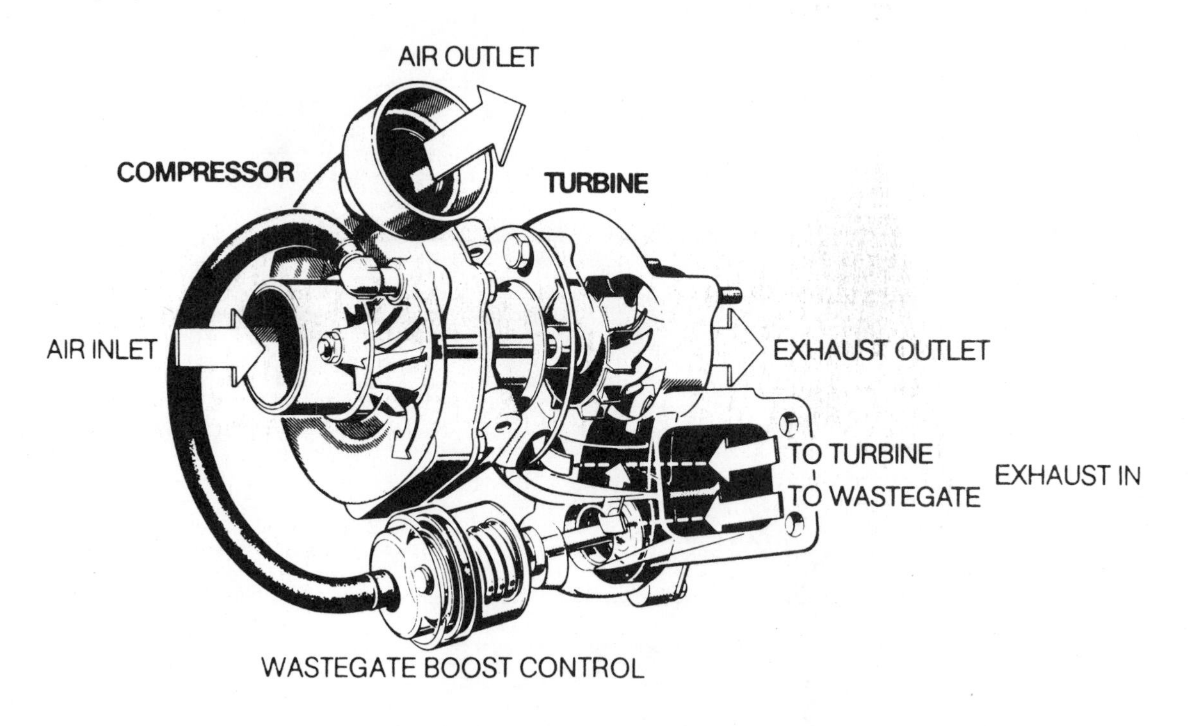

Fig. 10-25. Garret turbocharger of type installed on Mercedes-Benz diesel.

REVIEW QUESTIONS – CHAPTER 10

1. The purpose of the air system of a diesel engine is to supply clean air for purposes of combustion and ____________.
2. The air is also designed to _________ the combustion chamber.
3. At the same engine speed, the two-cycle engine has ______ as many combustion cycles per unit of time as a four-cycle engine.
4. The engine that has the longer scavenging period is the ________________.
5. In a naturally aspirated engine, the air is supplied by:
   a. A supercharger.

b. A centrifugal blower.
c. A vacuum chamber.
d. An injection tube.

6. The term "scavenging the cylinders" of a diesel engine is usually applied to:
   a. A two-stroke cycle engine.
   b. A four-stroke cycle engine.
7. In loop scavenging, the top of the piston is:
   a. Flat
   b. Contoured.
   c. Slanted.
   d. Depressed.
8. In uniflow scavenging how does the scavenging air move?
9. In the crankcase method of scavenging, the method used to produce the air pressure is by a:
   a. Supercharger.
   b. Centrifugal pump.
   c. Movement of engine piston.
   d. Natural aspirator.
10. The Roots type blower is usually used on a ____________ cycle engine.
11. The rotors of a Roots type blower are always ________ to the axis of rotation.
12. The scavenging pressure of a Roots type blower varies:
    a. Directly as the speed.
    b. As the square of the speed.
    c. As the cube of the speed.
13. The method which supplies the greater pressure is _________.
14. For the same power developed, the system that costs less is:
    a. Naturally aspirated.
    b. Supercharged.
    c. Centrifugal pump.
15. The type of engine that provides the greater fuel economy is the_____________.
16. For the same weight, the type of engine that develops the greater horsepower is _____________.
17. The centrifugal type supercharger is normally operated at speeds ranging up to:
    a. 5,000 rpm.
    b. 15,000 rpm.
    c. 20,000 rpm.
    d. 25,000 rpm.
    e. 50,000 rpm.
18. A diffuser converts air pressure from the supercharger into ____________pressure.
19. As air from the supercharger enters the diffuser, the cross-sectional area of the diffuser_____________.
20. A turbocharger is driven at speeds:
    a. Up to 10,000 rpm.
    b. Up to 20,000 rpm.
    c. Up to 50,000 rpm.
    d. Up to 100,000 rpm.
21. In the case of an engine driven supercharger, the percent of the developed power used to drive the supercharger is:
    a. 2 percent.
    b. 5 percent.
    c. 8 percent.
22. The speed of a turbocharger tends to ____________with the load on the engine.
23. Three reasons for the increase in power due to a turbocharger are:
    a. ________________________________.
    b. ________________________________.
    c. ________________________________.
24. In the reaction type turbine, the rotor buckets have a much ____________discharge area than they have an entrance area.
25. The turbines in turbochargers are generally which of the following type:
    a. Reaction.
    b. Impulse.
    c. Mixed flow.
26. The type of turbine which has the greater efficiency is the:
    a. Impulse.
    b. Reaction.
    c. Activated.
27. The type of turbine usually used in the high speed diesel field is:
    a. Axial flow.
    b. Inward flow.
    c. Mixed flow.
28. The approximate temperature of the gases entering the turbocharger is:
    a. 212 deg.
    b. 550 deg.
    c. 1000 deg.
    d. 1350 deg.
    e. 1500 deg.
29. Temperature throughout a turbocharger varies_________.
30. In the case of a two-cycle engine, the horsepower developed as the result of a supercharger can be as much as ___________.
31. The percentage increase in power that can be expected from the installation of a supercharger on a four-cycle engine is:
    a. Up to 25 percent.
    b. Up to 35 percent.
    c. Up to 50 percent.
    d. Up to 75 percent.
    e. Up to 100 percent.
32. In some designs it has been found to be advantageous to install _________ superchargers.
33. For efficient combustion the temperature of the incoming air should be relatively_____________.
34. An intercooler will reduce the temperature of the air to about__________deg. The coolant being obtained from the water jacket.
35. Some Detroit Diesel Allison engines are equipped with both a blower and a turbocharger. True or False?

Chapter 11

CONSTRUCTION AND BASIC DESIGN

In this chapter we will study variations in diesel engine design.

As pointed out previously, the vertical in-line engine is especially popular in the commercial vehicle and industrial field. Fig. 11-1 shows a six cylinder 330 cu. in. Ford diesel designed for industrial service, with parts identified. Fig. 11-2 shows an external view of the same engine.

Note that the fuel injection pump on this engine is operated from a gear-driven accessory shaft. Exhaust and intake valves are operated by means of push rods and rocker arms. Valve lifters are a mechanical type. Pistons are provided with four piston rings. Cylinders are of the wet sleeve type, with full length water jackets. The combustion chamber is of the open type. It is formed in the piston crown with direct fuel injection. Note that an injector leak-off pipe is provided and the fuel oil filter is mounted on the side of the engine, above and close to the fuel injection pump. Fuel lines from the injector pump to the injectors are relatively short. All are approximately the same length.

In the truck field, the six cylinder in-line engine is used extensively. The Cummins C-180 six cylinder in-line engine, Fig. 11-3, having a displacement of 464 cu. in. is a supercharged model, which is used in Ford C8000, F8000, and T8000 trucks. This engine develops 180 hp at 2500 rpm. It has a compression ratio of 14.5 to 1. When naturally aspirated, it develops 160 hp at 2500 rpm.

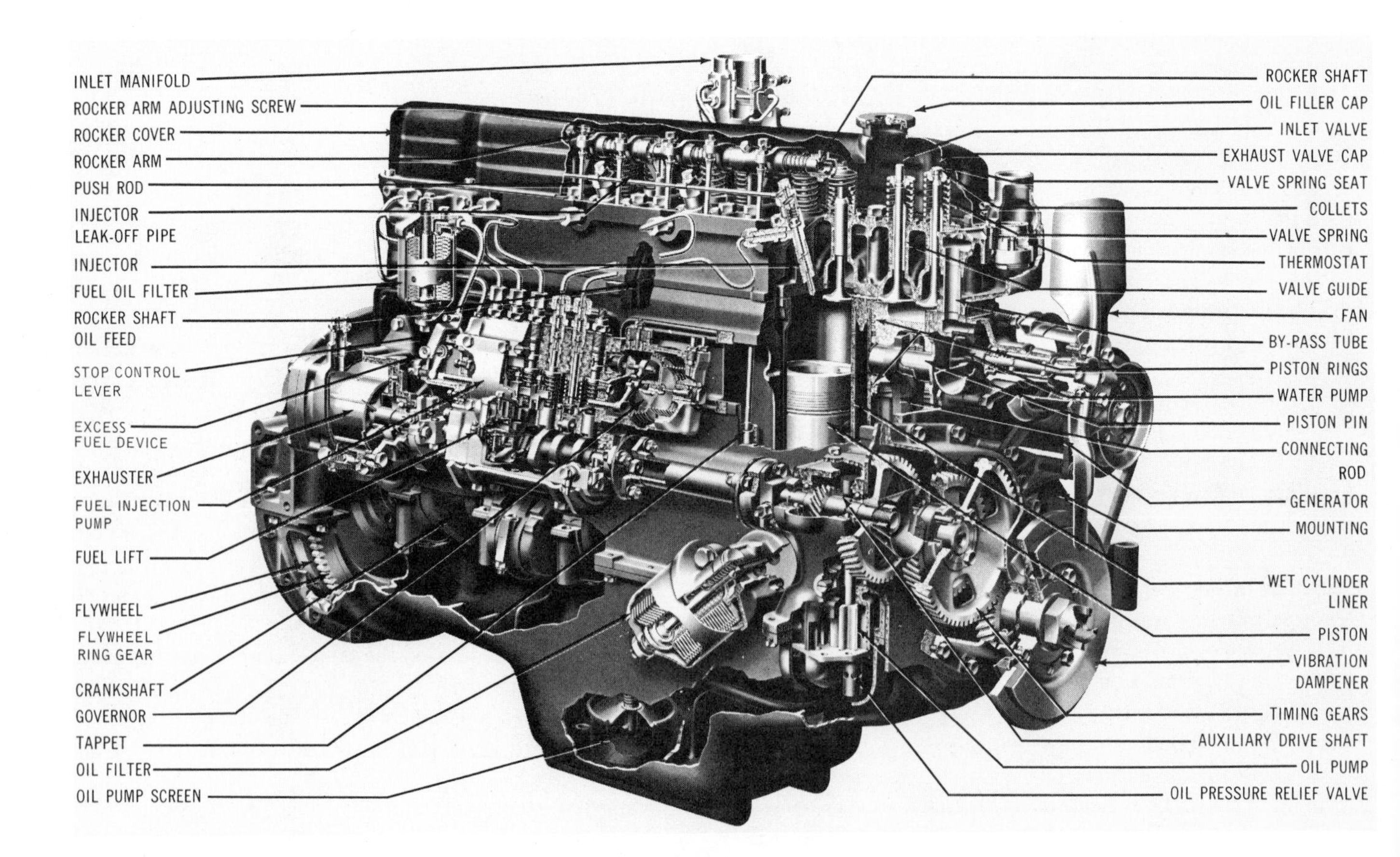

Fig. 11-1. Details of Ford 330 cu. in. six cylinder in-line engine. (Ford Motor Co.)

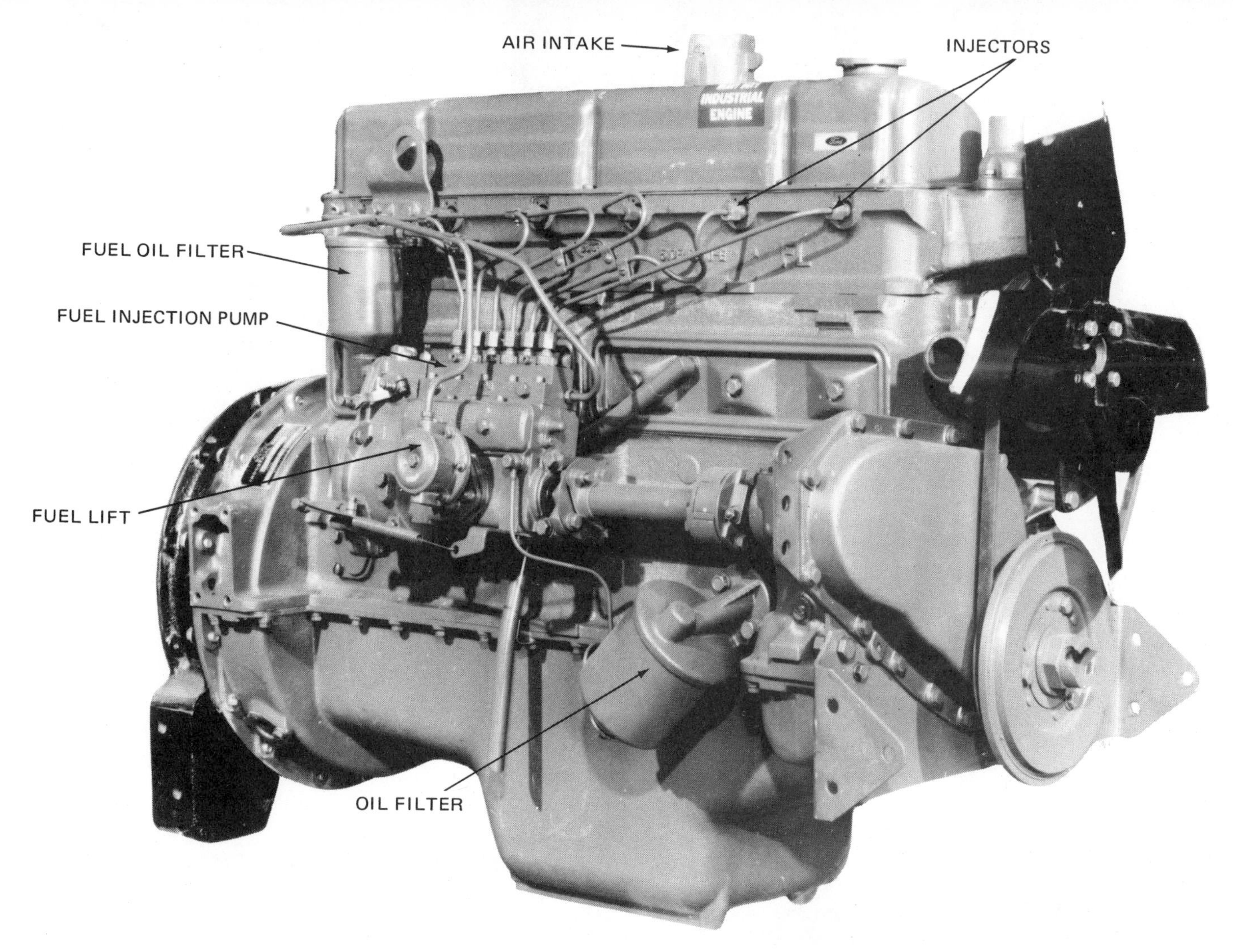

Fig. 11-2. Exterior view of Ford 330 cu. in. industrial engine. (Ford Motor Co.)

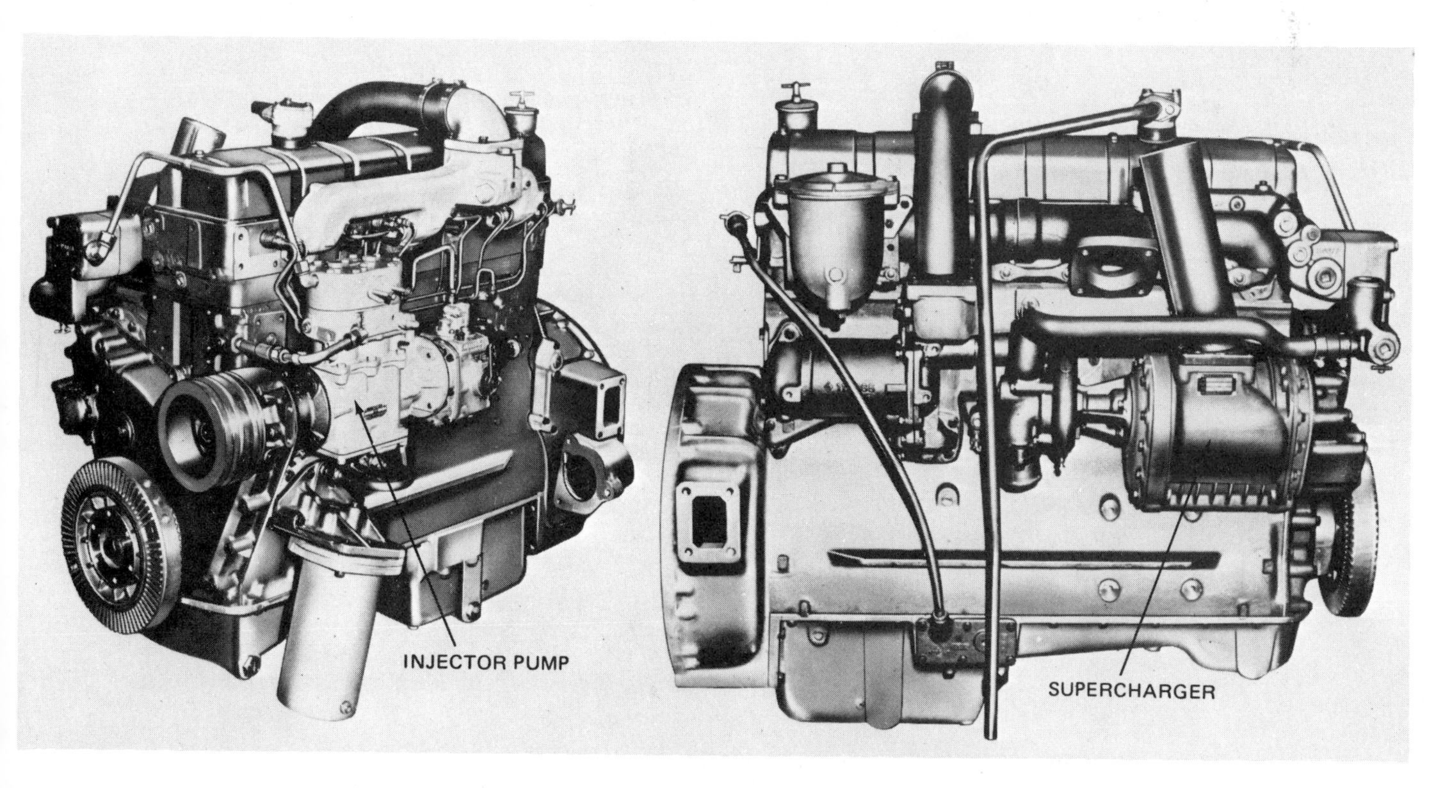

Fig. 11-3. Left and right sides of Cummins diesel engine as installed in some Ford trucks. Note location of supercharger and injector pump locations. (Ford Motor Co.)

Fig. 11-4. Eight cylinder in-line diesel engine designed for marine propulsion and electric generation. Note turbocharger at front end and removable panels permitting removal of connecting rods for service. Long tie rods are used to prevent high tensile stresses in the cast iron structure. (Stork-Werkspoor, Holland)

Fig. 11-5. The International 9.0 litre V-8 diesel engine is standard for Cargostar cab-over models and for S-Series conventional cab medium-duty diesel trucks. This mid-range diesel engine is offered at 165 hp and 180 hp ratings at 2800 rpm for applications up to 50,000 lbs. gross combined weight. (IH Truck Group of International Harvester)

In-line diesel engines are used for ship propulsion and electrical generation. The Stork-Werkspoor diesel, Fig. 11-4, is classed as a four-stroke, medium speed, heavy fuel, trunk piston engine, turbocharged with air cooling. The bed plate, which is U-shaped, supports the cast iron block. Crankpins are of large diameter, resulting in low bearing loads.

The Werkspoor engine has an initial design rating of 500 bhp per cylinder at 500 rpm. It is available in 6, 8 and 9 cylinder models.

INTERNATIONAL 9.0 LITRE V-8

International Harvester's 9.0 L V-8 is a four-stroke cycle, naturally aspirated, mid-range diesel engine weighing 1270 lbs. See Fig. 11-5. It is available at two different horsepower ratings. The 165 hp engine produces 366 lb. ft. of torque at 1200 rpm. The 180 hp engine produces 401 lb. ft.

The International 9.0 L diesel engine has a bore of 4.510 in. (114.6 mm) and a stroke of 4.3125 in. (109.5 mm). It has a compression ratio of 19.1 to 1. It is a non-sleeved engine with fully counterbalanced, forged steel crankshaft, three-ring aluminum alloy pistons, steel-backed aluminum main and connecting rod bearings, replaceable valve seat inserts and guides, carbide wafer-faced tappets.

The fuel injection pump is an in-line, multiple plunger Robert Bosch unit with a mechanical flyball, all-speed governor. The fuel injection nozzles are mounted from the outside of the cylinder head cover for ease of servicing. See Fig. 11-5.

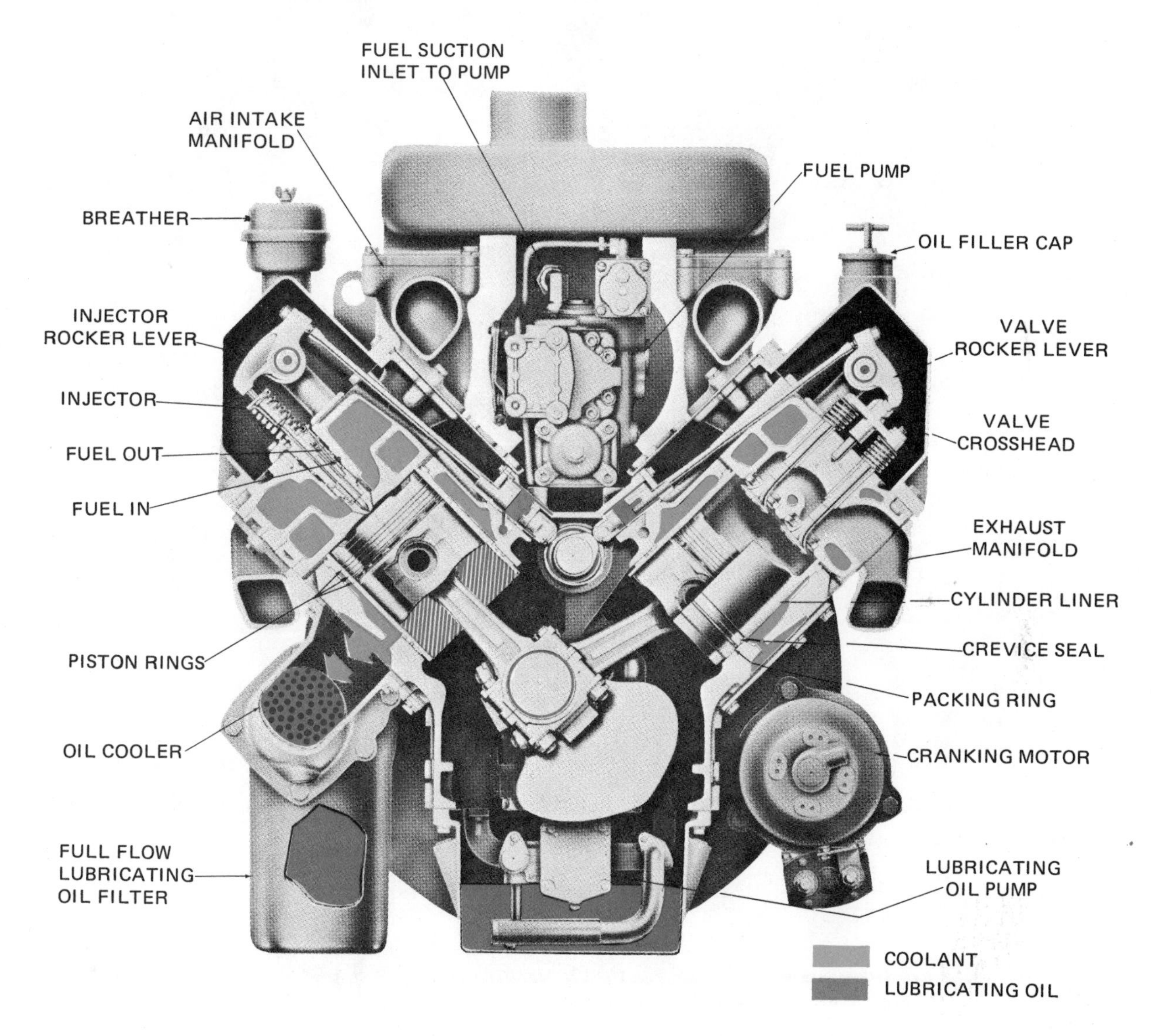

Fig. 11-6. Sectional view of Cummins V-8 diesel engine showing the large capacity water jackets, oil cooler and oiling system. (Cummins Engine Co.)

V-TYPE ENGINE

In the automotive field, for higher power, the V-8 type engine is virtually standard. Among the many reasons for the popularity of the V-type engine is its compact design resulting in lower space requirements, improved distribution of air to the cylinders, casting less liable to distortion than an in-line eight, and reduced torsional vibration because of the shorter crankshaft.

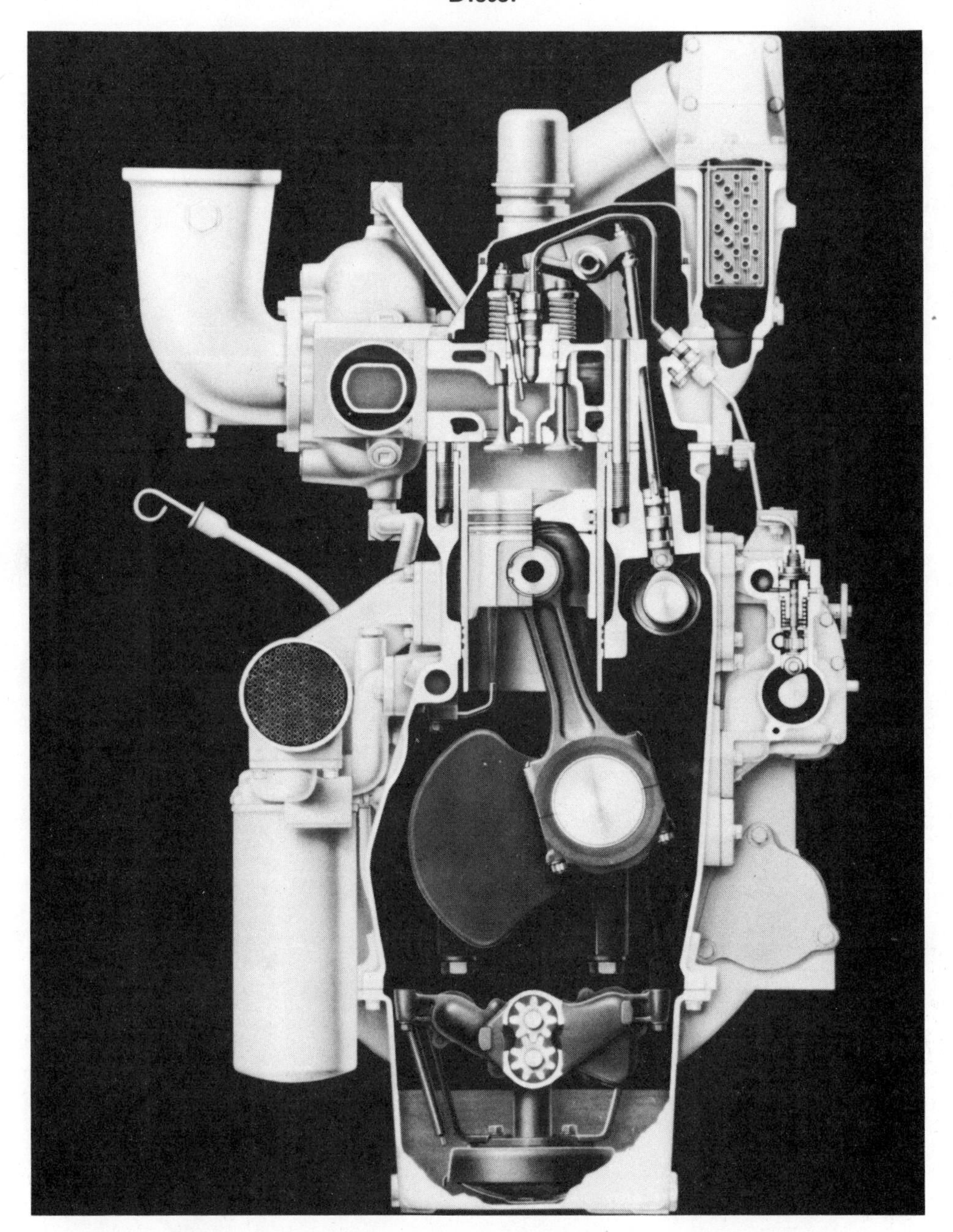

Fig. 11-7. Sectional view of in-line, six cylinder model 3406 turbocharged diesel engine. Note the injection pump, injector and precombustion chamber. This engine is used primarily in truck and bus applications. It develops up to 375 hp. (Caterpillar Tractor Co.)

Fig. 11-6 shows the flow of coolant and lubricants in the Cummins V-8 series 185 engine. Note the water jacketing in the cylinder head and around the wet type cylinder sleeve. A cooler is provided for the lubricating oil. Fuel is injected directly into the cylinders with the piston head forming the combustion chamber. For increased stiffness of the cylinder block, the sides of the crankcase are carried well below the center lines of the crankshaft. Note the location of the oil cooler and the full flow oil filter.

CATERPILLAR 3406 DIESEL

The model 3406 Caterpillar diesel engine is a six cylinder model with the usual Caterpillar precombustion chamber, Fig. 11-7. The cylinder bore is 5.4 in. The stroke is 6.5 in. Displacement is 893 cu. in.

The Caterpillar 3406 is designed specifically for truck and bus installations. This turbocharged diesel engine develops from 280 to 375 hp at 2100 rpm, depending on accessory equipment. The pistons are made of aluminum alloy. The full length cylinder sleeves are forged boron steel, hardened and shot peened.

Model 3406 has four valves per cylinder, operated by push rods and rocker arms. The valve seats are replaceable inserts. The inlet seats are stainless steel; exhaust seats are nickel-base alloy. The valves are faced with chrome-tungsten alloy, and the valve rotators provide increased valve life.

Turbocharging with after cooling provides the Caterpillar 3406 engine with two advantages, increased power and lower exhaust temperatures.

The Waukesha model L-1616D series is a V-type engine used extensively in the marine field, Fig. 11-8. This is a 12 cylinder four-stroke cycle engine with overhead valve and open combustion chamber. Displacement is 1616 cu. in. The naturally aspirated model develops 508 hp at 2400 rpm. The turbocharged model develops 719 hp at 2200 rpm. The crankcase is a Y-shaped iron alloy casting with 12 wet-type cylinder sleeves. At the upper end of the sleeves, seating is provided by the cylinder head gasket which seals the mating surfaces between the flanges of the sleeve and the crankcase deck. At the lower end of the sleeve, three rubber seal rings are provided for each sleeve. Each main bearing is supported, in located at the front end of the crankcase.

Cylinder heads are cast in pairs. Each head carries two complete rocker arm assemblies which operate four valves for each cylinder. Each set of valves operates off one rocker arm by means of a valve actuator, which opens two valves at the same time. Valve seats are of Stellite (nonferrous metal containing cobalt, tungsten, chromium, carbon). Contrary to current automotive gasoline engine practice, valve guides are of the replaceable type.

Pistons are aluminum, cam ground, tapered and of the full skirt type. Two compression, a scraper and a single oil ring are used.

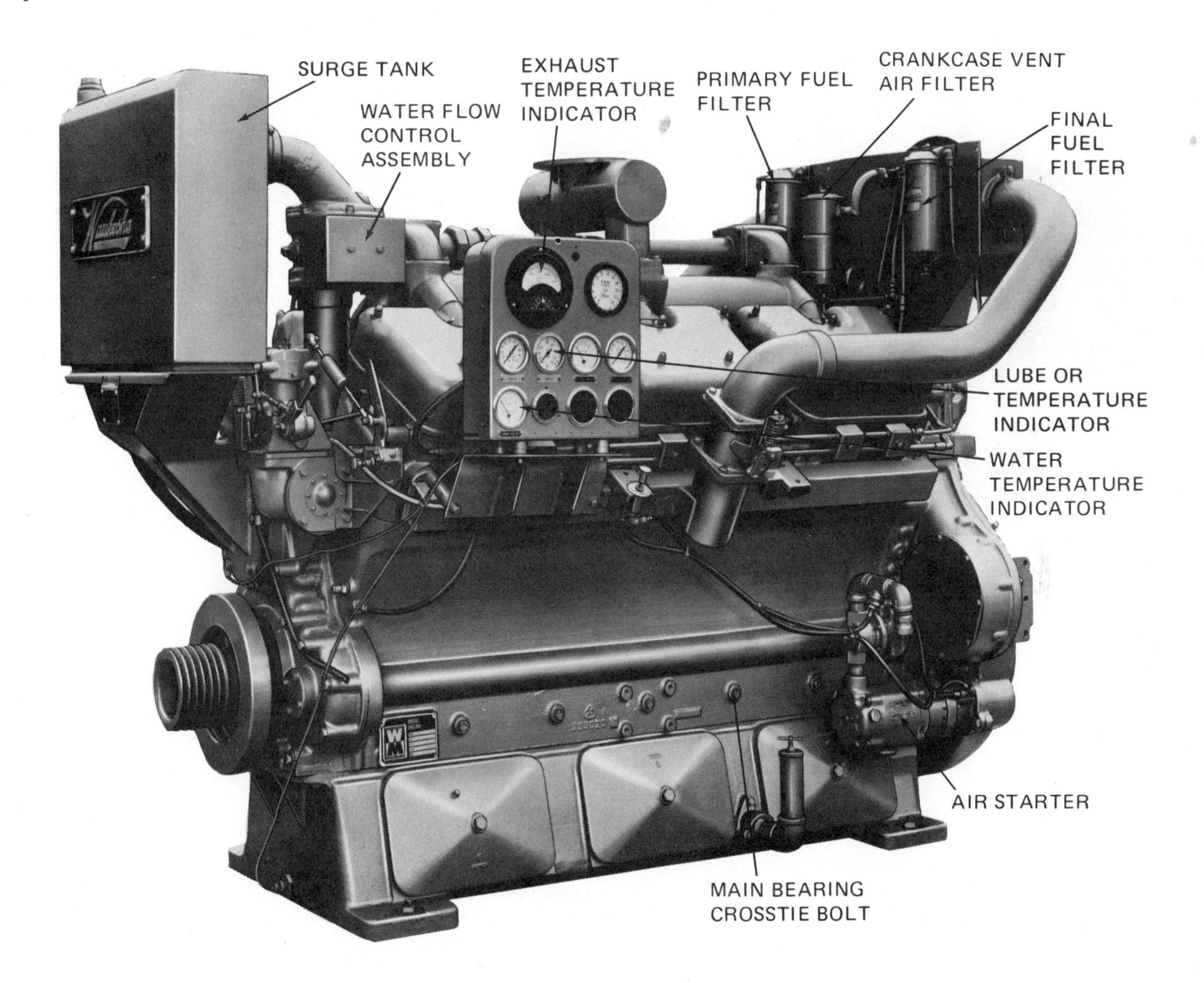

Fig. 11-8. This 1616 cu. in. 12 cylinder industrial engine operating on the four-stroke cycle, develops 719 hp at 2200 rpm. Note instrumentation. Model shown is used largely in the marine field. Major parts are interchangeable between right and left banks. (Waukesha Motor Co.)

addition to two Marsden locknuts, by two place tie bolts to side of the crankcase to secure bearing cap rigidly, Fig. 11-8.

The rear of the crankcase provides a housing for the crankshaft drive gear, cam gear, idler gear and water and oil pump gears. The camshaft driven gears which supply power for the governor drive and front end accessory drive gears, are

Connecting rods are drop forged and rifle drilled to provide pressure lubrication to the piston pins; also cooling for the piston. To reduce stress on connecting rod bolts, the connecting rod bearing cap is machined at an angle rather than at right angles to the center line of the rod. A dowel is provided to insure accurate location of the cap and rod.

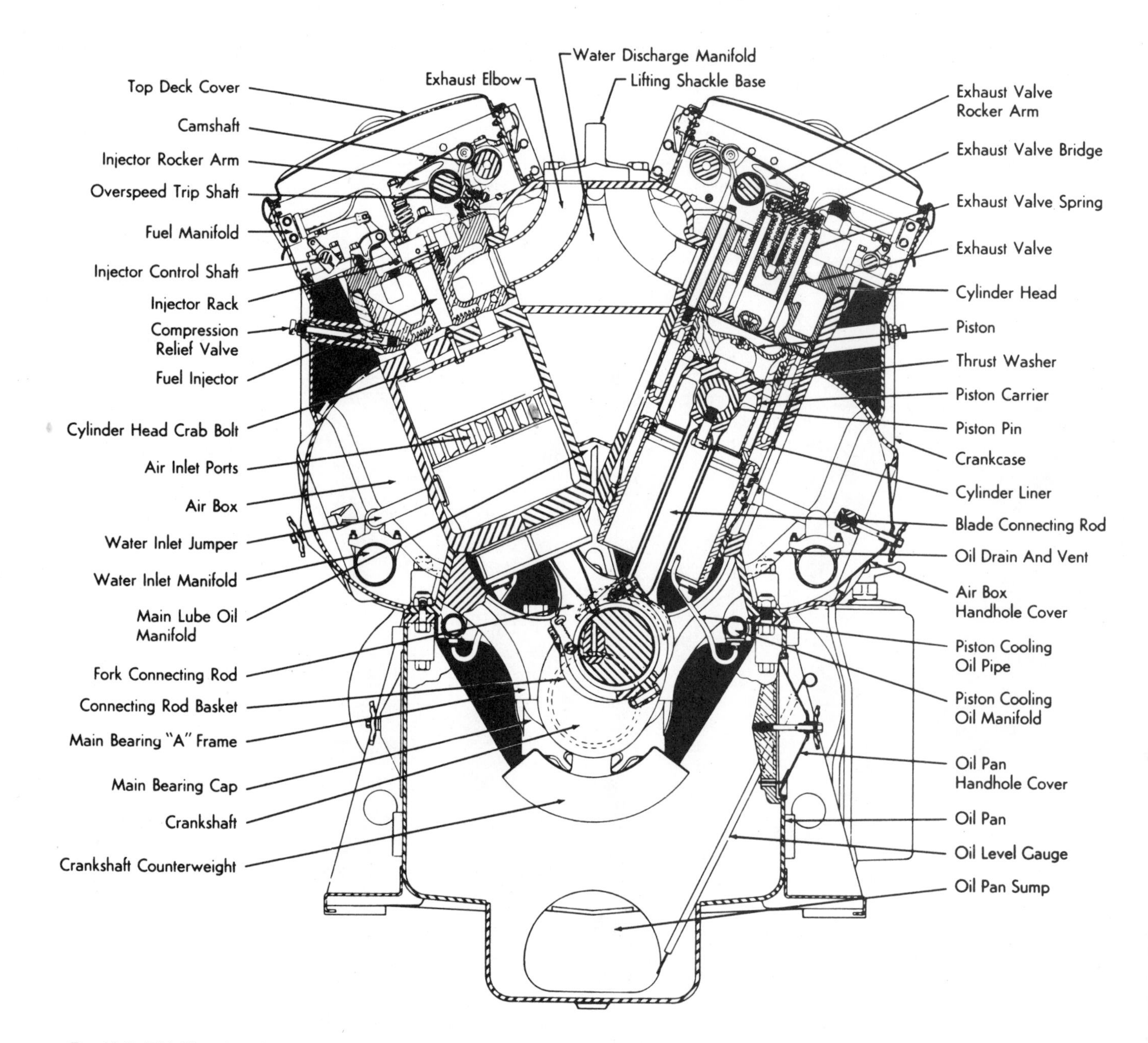

Fig. 11-9. This V-type engine is designed for railroad locomotives, electric generating plants, industrial power units and marine applications. Air intake is through ports in the cylinder, while exhaust is through four valves in the cylinder head. The cylinders of this two-cycle engine are at an angle of 45 deg. Each cylinder has a displacement of 645 cu. in. (Electro-Motive Div., General Motors Corp.)

The crankshaft is carried in seven steel backed copper lead alloy main bearings. Ample gages are provided, including a marine gear oil temperature indicator and an exhaust temperature indicator, Fig. 11-8. Note that compressed air starting is used, as is usual on large industrial and marine installations.

The V-type engine is also used extensively in the locomotive field. Figs. 11-9 and 11-10 shows Model 645 two-cycle V-type engine built by the Electro-Motive Div., General Motors. This is available in 6, 8, 12, 16 and 20 cylinder models. Cylinder bore and stroke are 9 1/6 in. and 10 in. respectively. In its various forms and sizes it produces a hp range of 750, 1000, 1500, 2000, 3000, and 3600 traction. Rated speed of these engines is 900 rpm.

Compression ratio is 14.5 to 1 on turbocharged engines; 16 to 1 on normally aspirated engines. The Model 645 is available in 8, 12 and 16 cylinders, Roots blower and 12, 16 and 20 cylinders turbocharged models. Air inlet ports are in the cylinder liner, while four exhaust valves are in the cylinder head, Figs. 11-9 and 10-2. Fuel injectors are General Motors unit injectors of the needle valve type. Crankshaft main bearings diameter 7 1/2 in., crankpin 6 1/2 in. while the piston pin diameter is 3.68 in. An end view of the model 645 engine is shown in Fig. 11-10.

Starting is by air motor or electric motor. In addition to being used in the locomotive field, it is also found in marine and industrial installations.

Fig. 11-10. End view of engine shown in Fig. 11-9, showing turbocharger at top. (Electro-Motive Div., General Motors Corp.)

HORIZONTAL ENGINES

When overhead clearance is at a premium, horizontal engines are often used. A Cummins engine of this type is shown in Fig. 11-11. This is a six cylinder model with four valves per cylinder. It is available with supercharger, turbocharger, as well as natural aspiration. Power ranges up to 325 hp at 2100 rpm for the 743 cu. in. engine.

RADIAL ENGINES

Compactness is a major advantage of the radial type engine. In approximately one-half the space, the radial engine delivers about the same hp as an equally rated in-line engine. Fig. 11-12 shows a Nordberg radial engine. Such engines are used to generate electricity for the electrolytic reduction of aluminum, produce power in central power stations, drive pumps in flood control projects, and in sewage disposal plants.

The crankshaft of these engines is in a vertical plane with the single crank at the top. Cylinders which are equally spaced radially about the shaft in a horizontal plane, are bolted to the frame. All connecting rods are identical. There is no master connecting rod as in the case in some types of radial engines.

These engines operate on the two-cycle principle. Ports are provided for scavenging and exhaust. There are no valves of the poppet type. Scavenging air is supplied by a separate blower.

Firing order is consecutive around the circle. Balance is achieved by actual convergence of combustion pressures and inertia forces on the crankshaft axis. Inner ends of the connecting rods are secured by knuckle pins spaced around a master bearing (crankpin bearing). Reciprocating motion of each piston is translated into rotary motion of the crankshaft with the connecting rods, and the master bearing, which is restrained from rotating by a simple constrained linkage assembly. This makes the master bearing gyrate bodily with the crankpin. This linkage does not restrain crank motion of the engine mechanism.

Two directly opposed connecting rods, Fig. 11-13, are rigidly connected to special longer knuckle pins which extend above the master bearing and form two cranks. A restraining link connects the two cranks together. A constrained linkage is

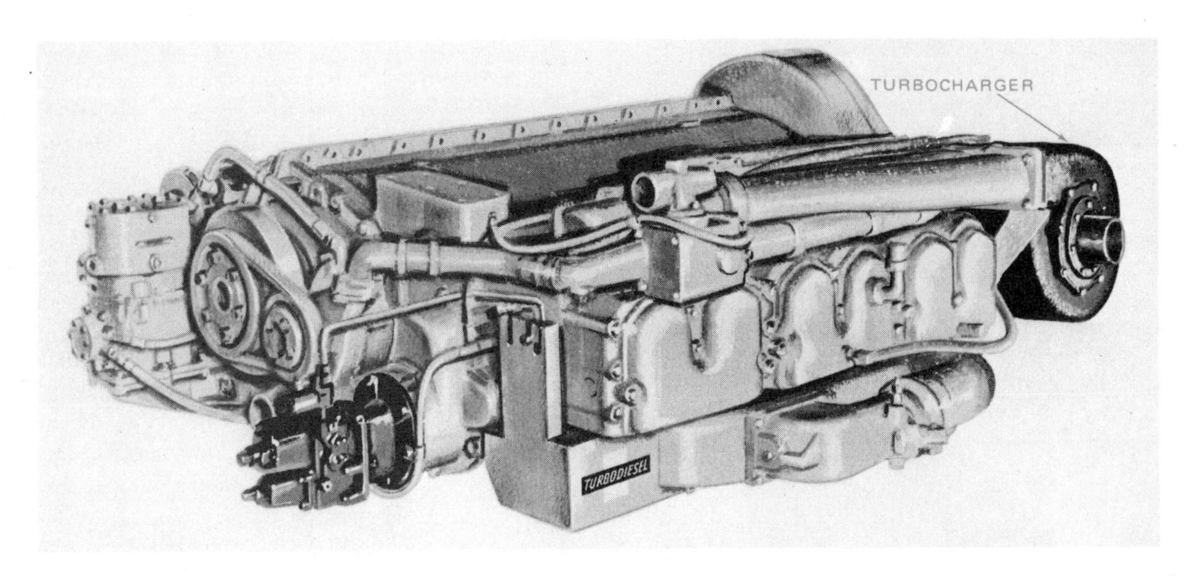

Fig. 11-11. When vertical space is at a premium, cylinders are placed horizontally. This is a six cylinder opposed engine of 743 cu. in. displacement. Turbocharged it develops 202 hp at 2100 rpm. Four valves per cylinder are provided. (Cummins Engine Co.)

Fig. 11-12. Nordberg radial engines are installed with the axis of the crankshaft in a vertical position, thus saving space. These are two-cycle engines in both naturally aspirated and supercharged models. Used extensively in generation of electricity and other industrial purposes. This is a twelve cylinder model with 14 in. bore and 16 in. stroke. Nonsupercharged, it develops 2125 hp. (Nordberg Mfg. Co.)

formed by the two opposite pistons, corresponding connecting rods, special knuckle pins and cranks all connected to the restraining link. This linkage permits the main crankshaft pin with its master bearing to revolve freely, yet it prevents the master bearing from rotating. Each knuckle pin describes an identical circle resulting in the same motion for each connecting rod and piston and attaining inherent balance.

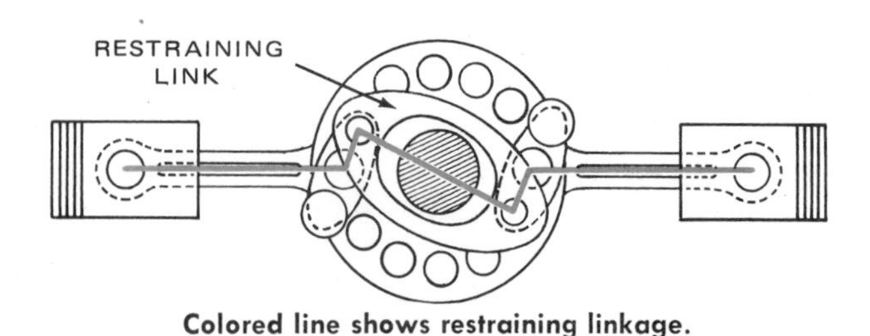

Colored line shows restraining linkage.

Fig. 11-13. Details of restraining link of Nordberg radial engine.

TWIN BANK CYLINDERS

Twin bank or parallel cylinders is another design used in some large displacement diesel engines. In the twelve cylinder design, Fig. 11-14, built by Sulzer (Switzerland) the combustion chamber is of the open type. There are two valves per cylinder (one intake and one exhaust). The two crankshafts which are geared together are synchronized and have a viscous (tendency of oil to resist flowing) type vibration damper.

For fuel injection there is one individual pump for each cylinder incorporating injection timing control by double helix plungers.

The cooling system pump is a separate electrically driven unit. A single water circuit for engine jackets, oil cooler and charge air cooler is provided.

OPPOSED PISTON ENGINE

An example of the V-type opposed piston engine is shown in Fig. 11-15. This is a two-cycle V-type engine produced by Fairbanks Morse. Combustion chambers are formed between pistons in the upper and lower cylinders. Openings for fuel injection nozzles, air start and cylinder release valves are in the upper cylinder. The upper piston is smaller and controls the exhaust gases. This has only one-half the speed of the lower piston. The upper and lower crankshafts are connected through a gear train. They are timed to open the exhaust ports in advance of the air ports for efficient scavenging. This also

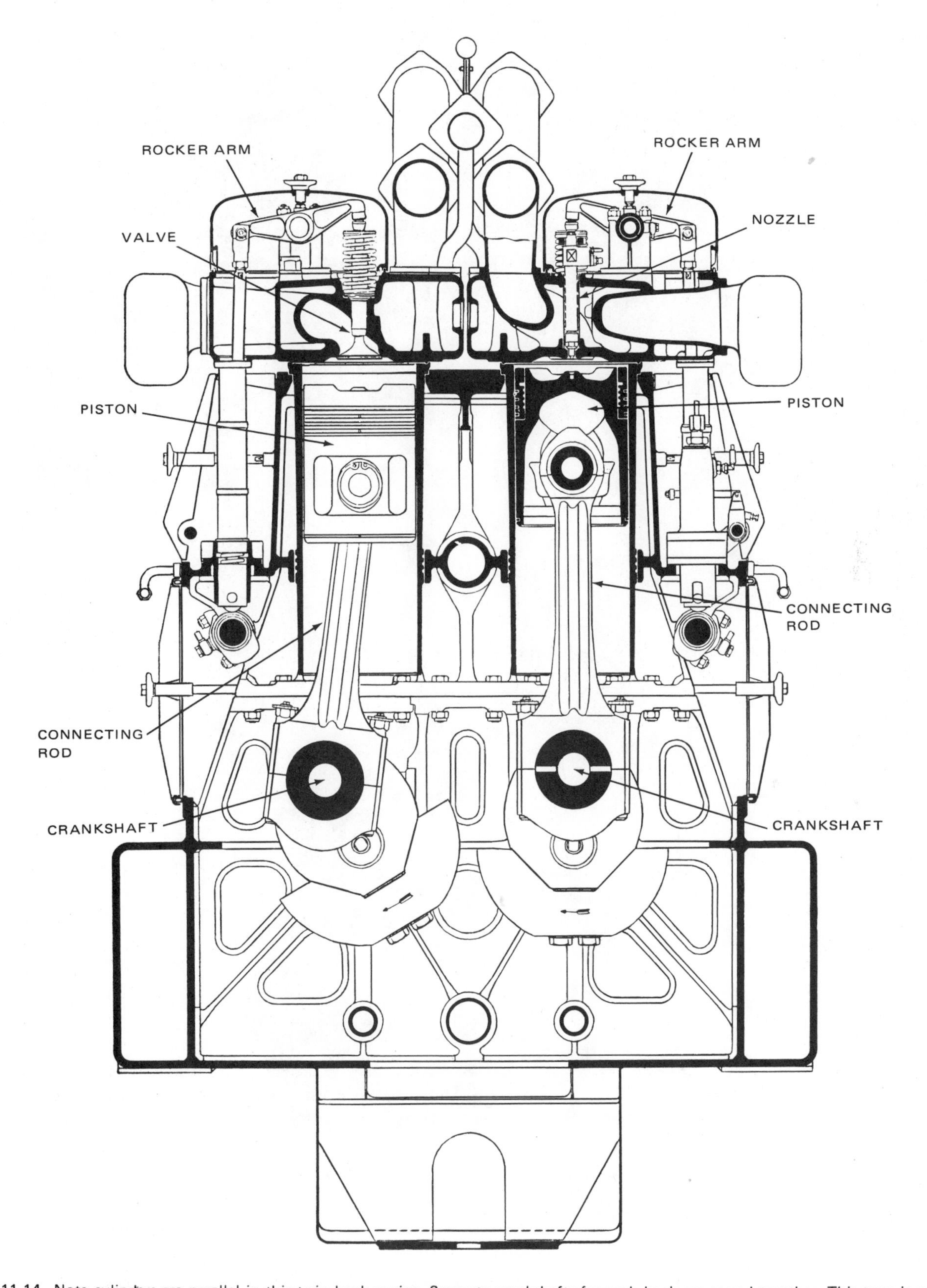

Fig. 11-14. Note cylinders are parallel in this twin bank engine. Separate crankshafts for each bank are geared together. This type is used in railway locomotives and develops 2750 hp at 800 rpm. Engine is of the direct injection type. (Sulzer Bros. Ltd., Switzerland)

sets the closing timing for charging the cylinders at turbocharged pressure.

Connecting rods for the V-type engines are fork and blade construction for the lower pistons, while the upper connecting rods are conventional side by side construction.

Model 38A20 two-cycle V-type engines develop 1000 to 1250 hp per cylinder at 400 to 450 rpm. They are produced in 12 and 18 cylinder models. Similar design is available in 6 and

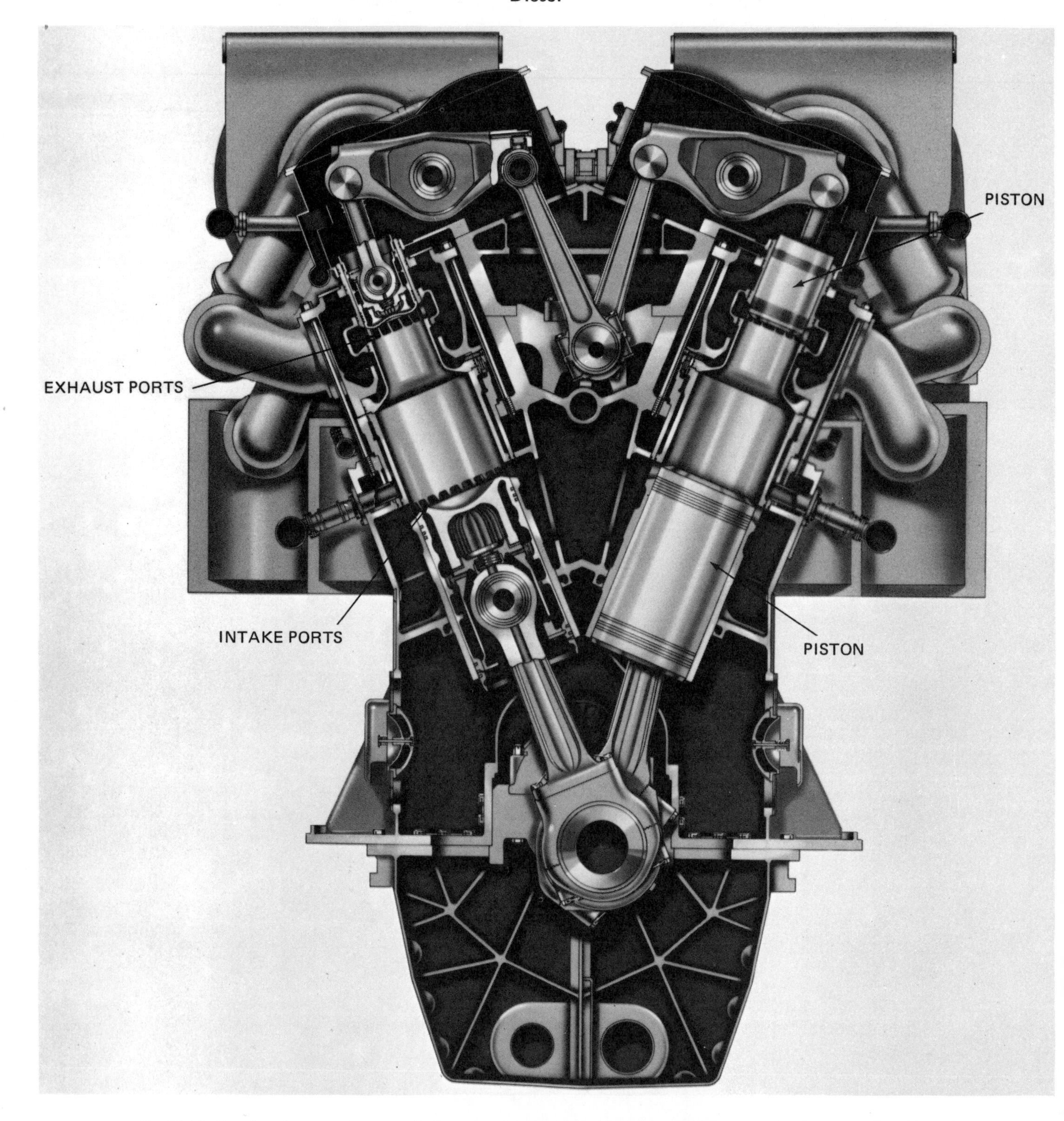

Fig. 11-15. Details of two-cycle opposed piston type engine. Note location of exhaust and intake ports. This V-type engine develops 1000/1250 hp per cylinder at 400/500 rpm. It is turbocharged and is capable of burning gas or oil. (Fairbanks Morse Power Systems Div., Colt Industries)

9 cylinders in-line engines. The 38A20 engine is designed for both marine and stationary service.

Another type of opposed piston engine, also produced by Fairbanks Morse, is shown in Fig. 11-16. This is a vertical type engine designed to burn either liquid or gaseous fuel as the primary fuel. One method is to operate as a conventional diesel engine on fuel oil; the other is to operate as a dual fuel engine, burning primarily a gaseous fuel, plus a small quantity of liquid fuel. Another variation of this engine burns gas only, it has an electric ignition to initiate combustion.

In this type engine, air from the positive displacement type blower is introduced into the cylinder and compressed between two pistons which operate vertically toward each other and in each cylinder.

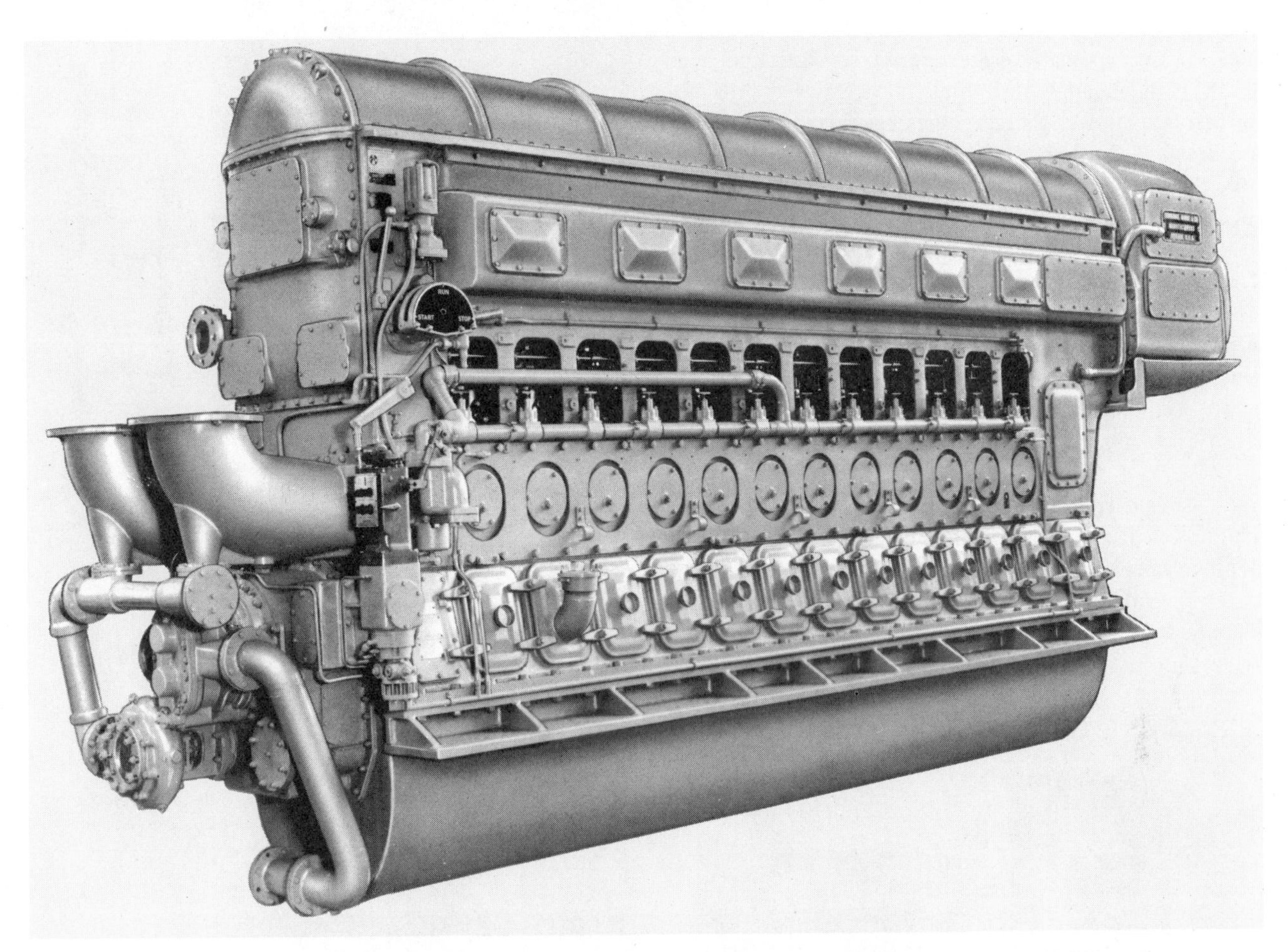

Fig. 11-16. External view of opposed piston engine. The model shown develops 1365-1720 KW at 720-900 rpm. (Fairbanks Morse Power Systems Div., Colt Industries)

The upper and lower pistons drive separate crankshafts which are interconnected by a vertical drive, Fig. 11-17. Fresh air is admitted to the cylinder and exhaust gas is expelled by the piston uncovering and covering the ports at the upper and lower end of the cylinder respectively. Fig. 10-4 shows a lower crankshaft past outer dead center and the upper crankshaft on outer dead center.

The combustion spaces formed between the recessed heads of the two pistons of the crankshaft approach inner dead center. When operating on fuel oil, injection begins at approximately 9 deg. before the lower piston reaches inner dead center.

In the case of the dual fuel engine, gas admission begins shortly after the lower piston covers the exhaust ports and

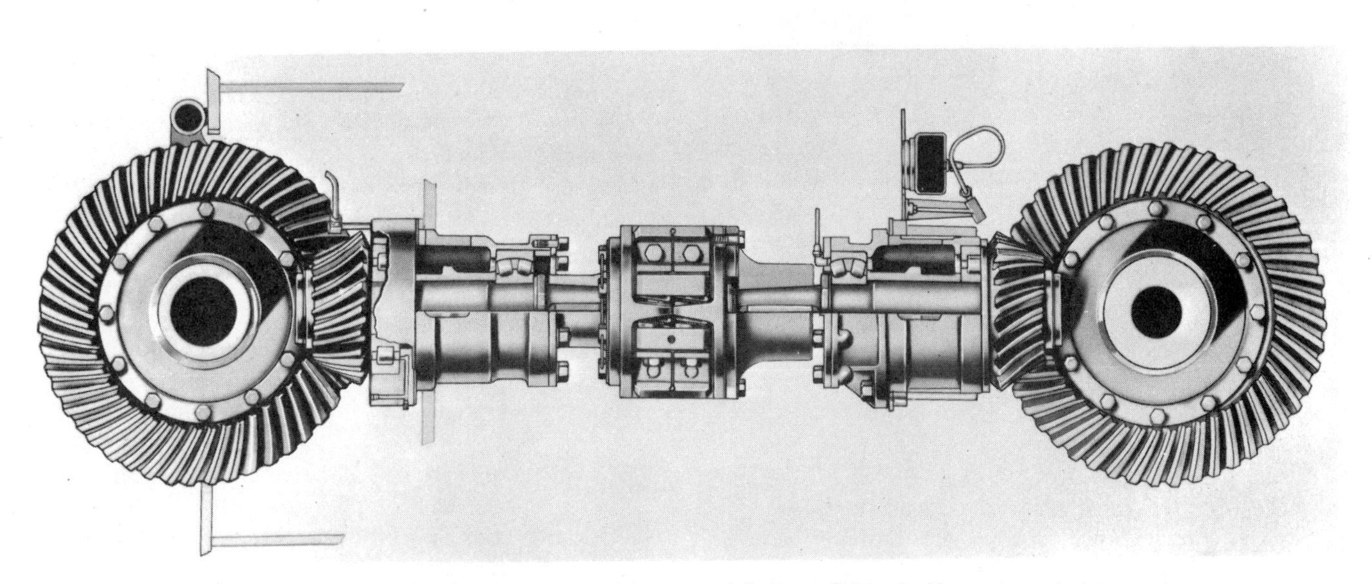

Fig. 11-17. Method of connecting upper and lower crankshafts on Fairbanks Morse opposed piston engine.

continues for approximately one quarter revolution of the crankshaft. A small amount of fuel oil is injected into the cylinder before the lower piston reaches inner dead center.

When designed for operation as a spark ignition engine, the cycle of operation is similar to the oil diesel cycle with the spark occurring approximately 5 deg. after inner dead center.

SMALL AIR-COOLED ENGINES

For portable power application, driving small electric generators, air compressors, and water pumps, small diesel engines are used extensively. Such engines in general, range up to 30 hp. They are available in one, two and four cylinder models.

In most cases small diesel engines operate on the four-cycle principle; however, a two-cycle uniflow engine with its piston ported intake valve and poppet exhaust valve is also used.

Figs. 11-18 and 11-19 illustrate the Onan air-cooled engine with electric starter and cooling fans mounted on the end of the crankshaft. Being air-cooled, there is a saving in weight which results in improved portability. Glow plugs are provided for ease of starting.

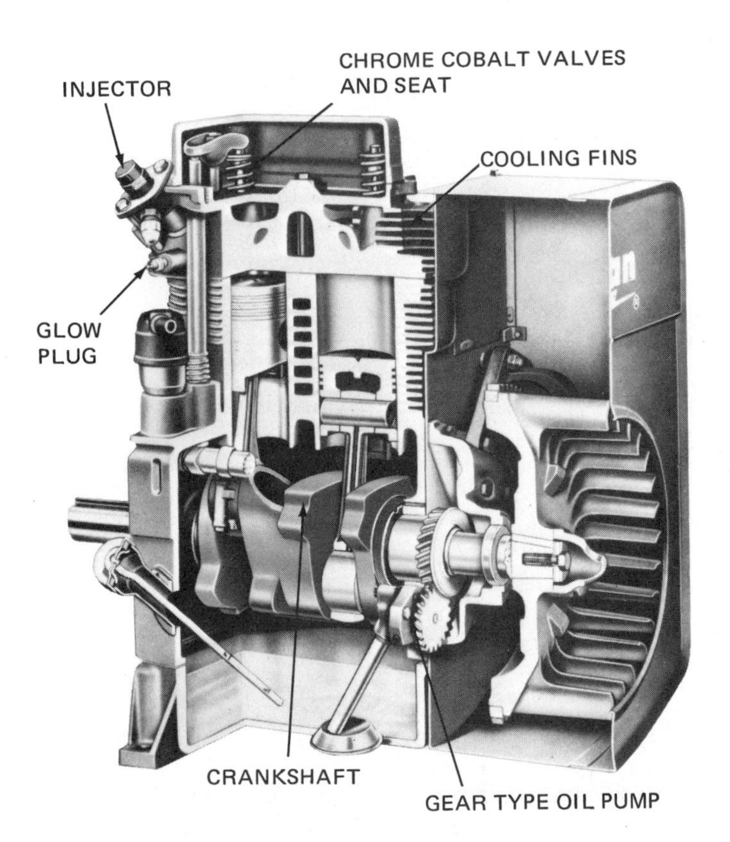

Fig. 11-18. Sectional view of two cylinder Onan air-cooled engine. Note cooling fins on cylinders and fan blades in flywheel. (Onan)

The single cylinder models develop 7.2 hp at 2400 rpm, the two cylinder models develop 14.6 hp at 2400 rpm. The four cylinder models produce 27.5 hp at 2400 rpm. Displacement is 30 cu. in. per cylinder. The power-to-weight ratio for the three models is 31.3 lb. per hp, 18.4 lb. per hp, and 15.9 lb. per hp respectively.

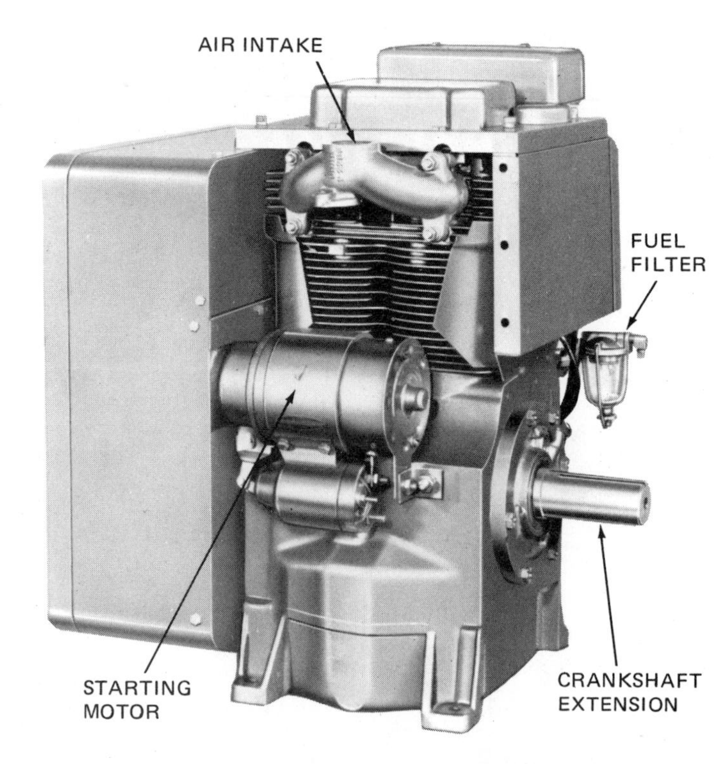

Fig. 11-19. External view of Onan two cylinder air-cooled engine. This is a four-cycle type with American Bosch injection.

NAPIER DELTIC

The Napier Deltic engine produced by Paxman Engine Div. of the English Electric Diesel Ltd., Colchester, England is an unusual design. This two-stroke cycle, opposed piston engine derives the name "Deltic" from the triangular arrangement of the three cylinder blocks which form an inverted delta symbol. It is used extensively on high-speed naval craft, mine sweepers, coastal ships, motor yachts, rail traction and in general industrial service.

The Model CT-18-42K, Figs. 11-20 and 11-21, is a turbo blown eighteen cylinder engine having charge air-coolers which reduce the temperature of the air entering the cylinders and lower the piston temperature. As a result more fuel can be burned with attendant increased output.

The basic Deltic engine comprises three cylinder blocks and three crankcases tied together by a system of steel bolts of high tensile strength, which pass through the cylinder blocks and crankcases. Short crankshafts are used with each crankpin carrying one inlet and one exhaust piston, Fig. 11-20, on plain and forked rods respectively.

Torsional vibration is eliminated by means of viscous dampers on each crankshaft and quill shafts in tune with the dampers.

The air inlet manifolds, coolant jackets and exhaust ports are cast integrally with each of the three cylinder blocks. The replaceable cylinder liners are located against machined shoulders formed by the inlet end of the cylinder block.

The crankshafts have hollow crankpins, are nitrided (hard surfaced) all over and the bearing surfaces are finished by lapping. The two top crankshafts which are identical, rotate in

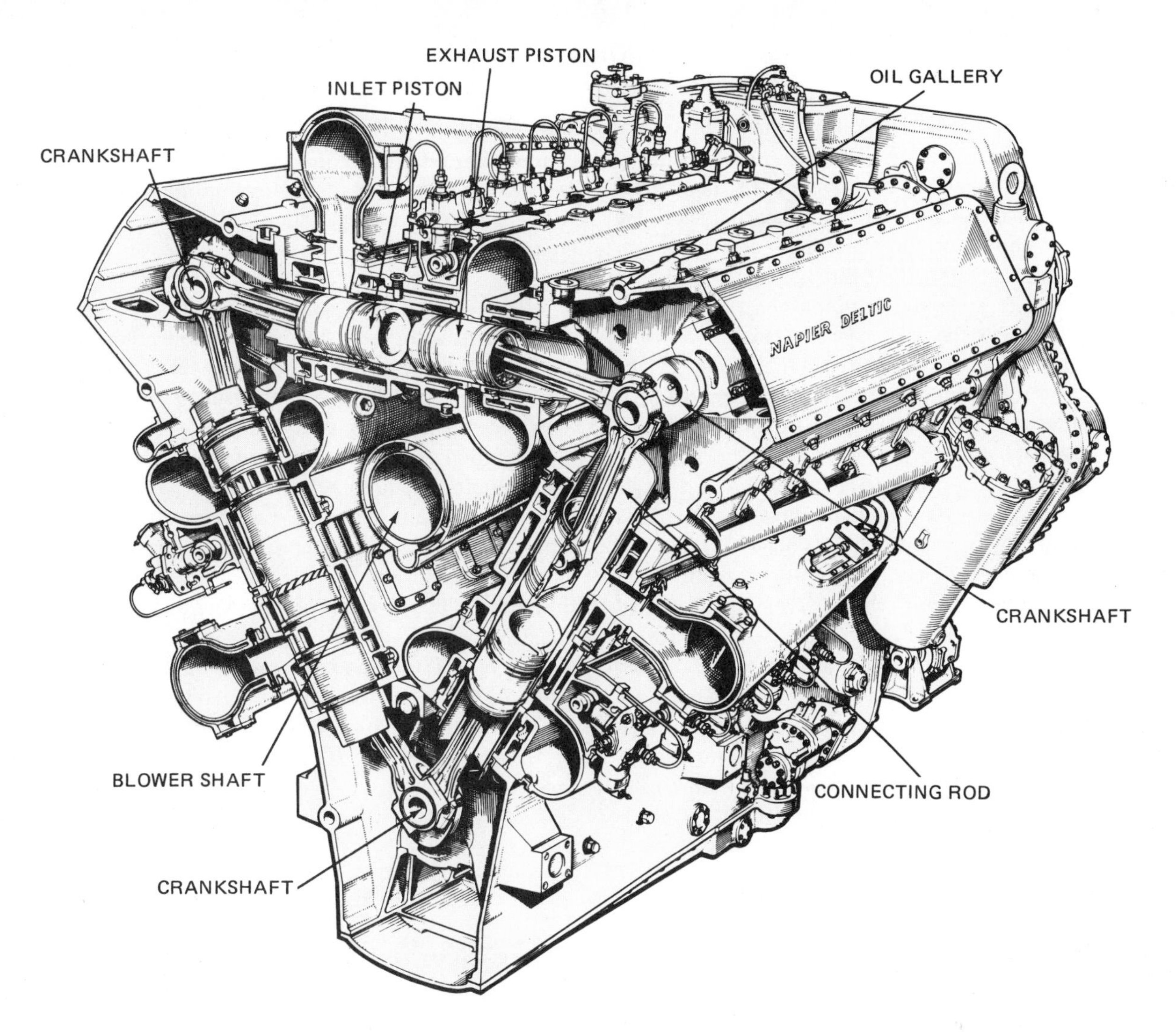

Fig. 11-20. Cross section through No. 2 cylinders of British Napier Deltic engine, produced by Paxman Engine Div. of English Electric Diesels Ltd., Colchester, England.

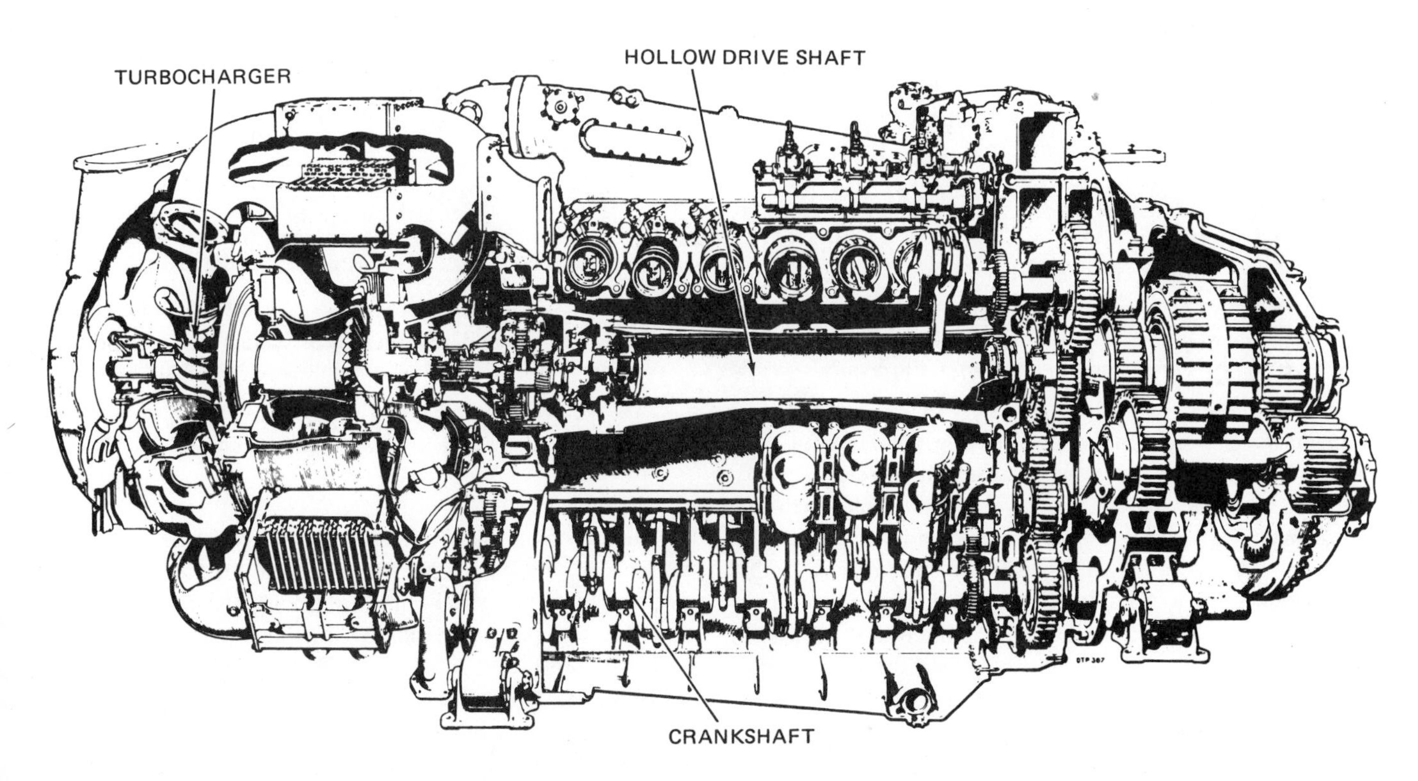

Fig. 11-21. Horizontal section of Napier Deltic engine also shown in Fig. 11-20. Note hollow drive shaft and turbocharger.

a clockwise direction when viewed from the front of the engine. The bottom crankshaft has opposite handed throws that rotate in a counterclockwise direction so the correct phase relationship of the pistons may be obtained. Pistons are fitted with three compression rings and three oil rings. Power is transmitted from each crankshaft through what the British call "quill" shafts and gearing to the output shaft. The output shaft is connected with the turbocharger transmission by a hollow drive shaft which passes through the center of the triangle formed by the cylinder blocks, Figs. 11-20 and 11-21.

The turboblower of the Deltic type CT-18-42K engine is mounted at the forward end of the engine. The assembly consists of a single sided centrifugal compressor and a single stage axial flow turbine with four light alloy housings. The housings are cleaned by passage of engine coolant through the jackets formed in the housing. The rotor assembly is a turboblower consisting of two shafts, shrunk and bolted together. The compressor end of the shaft is hollow and is splined and shouldered together to accommodate the impeller and inducer. The turbine half of the shaft and turbine disk are formed integrally. A heat resistant steel wire passes through each turbine blade to reduce blade flutter. The rotor shaft is driven at a ratio of 5.153 to 1, to the crankshaft speed. Engine speed of this CT-18-42K designed for high-speed naval craft, is 2100 rpm at maximum power of 3700 bhp. Continuous bhp is 2750 at 1800 rpm.

Each cylinder is provided with a single injector supplied by an individual constant stroke variable delivery fuel injection pump. The six injection pumps on each camshaft housing are supplied with fuel from a common pressurized fuel gallery cast in the camshaft housing. An engine driven fuel circulating pump supplies fuel to the three camshaft housing supply galleries in excess of engine requirements. Excess fuel is returned to the supply tank through a pressurized valve. Air locks in the fuel system are eliminated.

Cooling of the Napier Deltic engine is effected by a closed circuit circulating system. The heat transfer qualities of the engine lubricating oil are also a factor.

The engine coolant and lubricating oil are both cooled by water through combined or separate heat exchanges. Thermostatic valves control the flow to the heat exchanger and employ bypass circuits to maintain efficient operating temperature of the engine.

VARIABLE COMPRESSION RATIO ENGINE

Engineers are constantly investigating methods of improving the diesel engine. Much research is directed toward increasing the developed horsepower per cubic inch of displacement, improving fuel economy, easier starting and reduction of weight. A noteworthy accomplishment is the development of the variable compression ratio engine. This research started in 1952 by the British Internal Combustion Engine Research Association. Recently the Continental Aviation and Engineering Corp. obtained the patent rights to manufacture this engine in the United States.

Research has shown that a diesel engine with direct injection can be started without starting aids at zero degrees F. with about 16 to 1 compression ratio. While at -25 deg. F. a compression ratio of 19 to 1 is needed. Once started an engine will idle satisfactorily with a compression ratio of 12 to 1.

Comparison with a gasoline engine where starting aids are required even with a compression ratio of 25 to 1, is of interest.

With a diesel using a compression ratio of 19 to 1, starting will be satisfactory but the peak cylinder pressures become exceedingly high as the load on the engine is increased. Such high pressures would necessitate using a very heavy structure to provide the necessary strength.

A solution to the problem is an engine which will provide a high compression ratio for starting, and a lower compression ratio as the load is increased. Facts concerning such an engine were presented in papers presented to the Society of Automotive Engineers by W. A. Wallace, John C. Basiletti and Edward F. Blackburne, of the Continental Aviation and Engineering Corp.

The variable compression ratio (VCR) is produced by a piston of special design, which operates similar in a degree to the familiar hydraulic valve lifter. See Fig. 11-22.

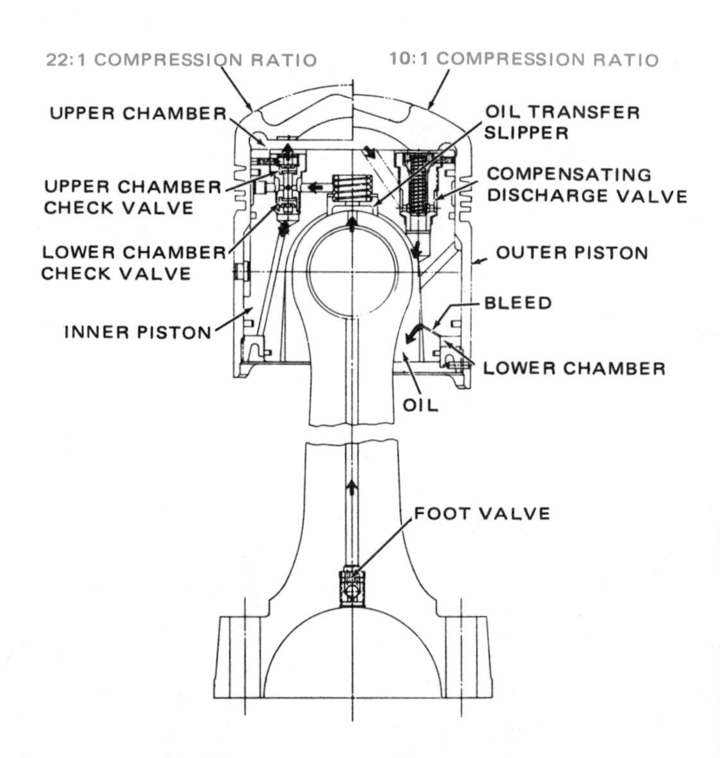

Fig. 11-22. Details of variable compression ratio piston. (Continental Aviation and Engineering Corp.)

The piston consists of two basic pistons, one within the other. The outer piston is free to slide up and down in relation to the inner piston within limits of 0.2 in. This movement is in response to pressure of combustion and under restraint of hydraulic pressure of engine oil. The movement of the outer piston in relation to the inner piston provides the VCR. There is no change in engine displacement.

The complete VCR system includes the connecting rod which is designed to allow engine oil to be carried from the

Fig. 11-23. Continental engine equipped with variable compression ratio pistons produced 1.32 horsepower per cu. in. displacement. (Continental Aviation and Engineering Corp.)

connecting rod bearing through a slipper, Fig. 11-22, to the inner piston.

The outer piston which is of cast iron, may be compared to the normal piston. The combustion chamber is formed in the crown of the piston. The outer piston also carries the piston rings. The inner piston is an aluminum forging which contains the valves and hydraulic seal rings. As shown in Fig. 11-22, the inner piston is connected to the piston pin in the conventional manner. The travel of the outer piston in relation to the inner piston is limited by the piston retaining ring which provides a mechanical stop for the upward motion of the piston.

The hydraulic system which controls the movement of the outer piston includes an upper chamber consisting of cylinder clearance space between the two pistons and a lower annular chamber. Spring-loaded valves in the inner piston control the flow of engine oil to the chambers.

The lower and the upper chambers are filled with lubricating oil through the check valves which are supplied with oil from the connecting rod by way of the oil transfer slipper. The upper chamber is supplied with a compensating discharge valve, and the lower chamber with a bleed which controls the amount of movement between the two pistons on exhaust and intake strokes. Oil from the bleed is discharged directly into the crankcase.

As engine power is increased, the compression pressure on top of the outer piston is also increased. When the pressure exceeds the setting of the compensating discharge valve, a small amount of oil leaks into the crankcase, decreasing the volume of the upper chamber. This permits the upper piston to move down in relation to the inner piston. This in turn increases the clearance volume between the outer piston and the cylinder combustion dome with a corresponding decrease in compression ratio. The process continues until the upper piston contacts the inner piston and there is then no further decrease in compression ratio.

During the latter part of each upward stroke of the piston and early part of each downward stroke, the inertia of the oil acting upward opens the upper and lower check valves into the upper and lower chambers. At the same time, the inertia of the outer piston tends to raise the outer piston in relation to the inner piston. Simultaneously, the volume of the lower chamber is reduced and oil is forced out through the bleed hole. The lower chamber acts as a dashpot preventing outer piston from moving more than a predetermined amount in relation to inner piston. The limit of upward movement is approximately 0.005 in. This varies with different installations.

If combustion pressure is suddenly reduced, the piston will return immediately to the position of higher compression ratio. This piston changes from its lowest to its highest compression ratio in 50 to 60 cycles regardless of engine speed. The foot valve in the connecting rod is designed to prevent reverse flow of oil due to inertia.

Early tests made by Continental Aviation and Engineering Corp. engineers and reported in the transactions of the Society of Automotive Engineers showed an increase of 0.5 to 0.75 bhp/cu. in. In more recent tests comparison was made with a Continental AVDS-1100-1 engine of 1120 cu. in. displacement and a similar engine fitted with variable compression ratio (VCR) pistons, Fig. 11-23. The gross horsepower per cu. in. of displacement for the conventional engine was 0.49 and for the VCR engine 1.32, Fig. 11-24. Pounds per gross horsepower was 5.7 compared to 2.3. Gross brake horsepower was 550 compared to 1475 hp for the VCR engine or an increase of 167 percent. Maximum torque was 1075 at 2000 rpm for the conventional engine and 2800 lb. ft. at 2200 rpm for the VCR engine.

The brake horsepower per cu. in. is particularly interesting. Passenger car gasoline engines until recently have not approached one hp/cu.in. For example the current 427 cu. in. engine in the Corvette develops 1.019 hp/cu.in. (highest in the list) while the Continental VCR diesel produces 1.32 hp/cu.in.

The engineers point out that the VCR design provides an answer to the problem of piston cooling as well as improved starting, better idling and improved endurance.

Pistons are materially heavier for the VCR design. The total increase in weight for the piston and connecting rod is in excess of four pounds.

Tests showed that oil viscosity had no appreciable effect on the operation of the piston. With a compression ratio of 21 to 1 the engine started successfully in normal cold climates without the use of starting aids. To permit successful starting in Arctic temperatures below -25 deg. F. a simple manifold heater provided starting ability to below -45 deg. F. The engineers also pointed out the small difference in starting ability temperatures of normal diesel fuel and Army CITE fuel. The cetane numbers for these fuels is 35 for the CITE (normal, no improver) and 42 for Arctic diesel fuel.

With 80 octane gasoline as fuel and 21 to 1 compression ratio, the low temperature limit for unaided starting was +17 deg. F. In other cold starting tests it was found that with a manifold type heater it required 30 seconds to start when Arctic diesel fuel of 40 octane was used. When the fuel was 35 cetane CITE fuel it required 40 seconds. When the fuel was 86/93 octane gasoline the engine started in 44 seconds.

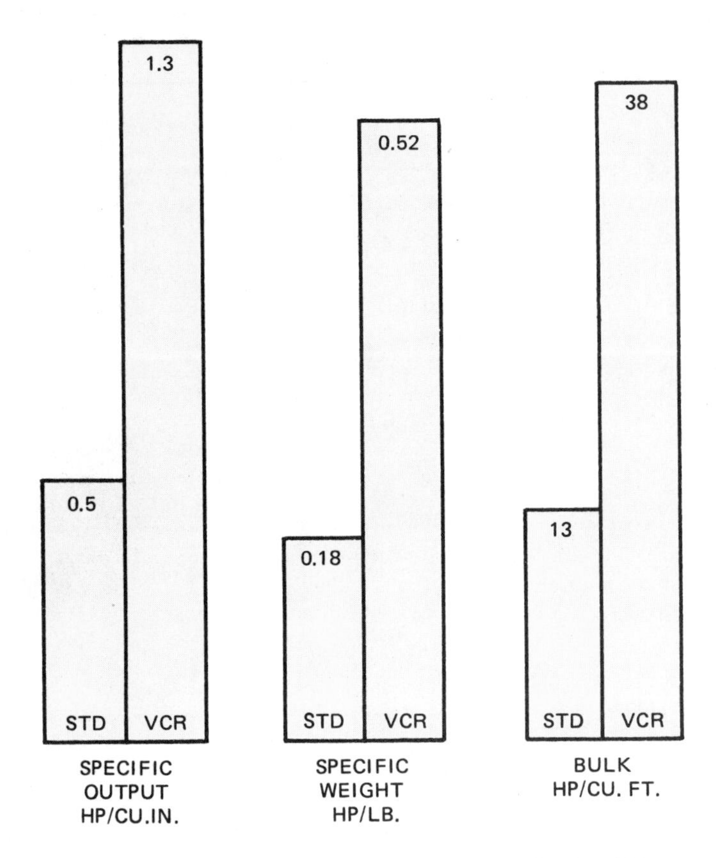

Fig. 11-24. Comparison of output of conventional and variable compression ratio engines of same size. (Continental Aviation and Engineering Corp.)

FIVE CYLINDER DIESEL ENGINE

The Mercedes-Benz five cylinder diesel engine, Fig. 11-25, is used to power the 300D model. It has a displacement of 183 cu. in. (3005 cc) and develops 77 hp at 4000 rpm.

Crank throws of this five cylinder engine are 72 deg. apart. Consequently, primary and secondary inertia forces are zero,

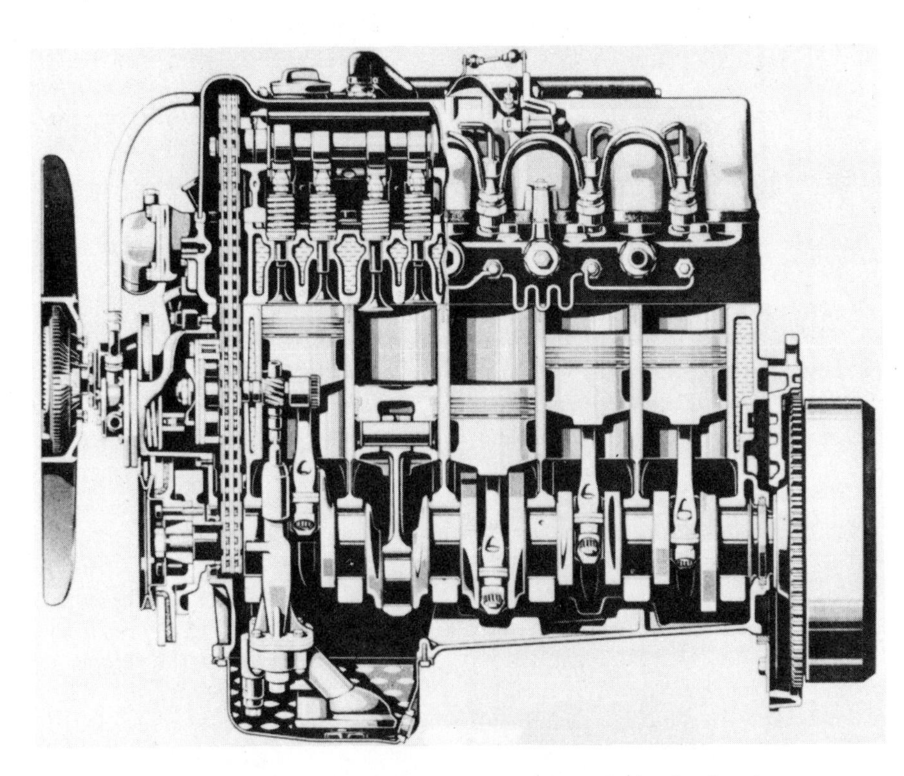

Fig. 11-25. Sectional view of Mercedes-Benz five cylinder diesel engine.

the same as in a six cylinder engine. However, there are unbalanced couples. The unbalanced couples (same as inertia forces) rest on suspension points of the engine and cause vibrations around the vertical and transverse axis.

In this Mercedes-Benz five cylinder engine, the secondary inertia forces tend to induce bending vibrations in the cylinder block. This necessitates the use of two shock absorbers parallel to the front mounts of the engine, in order to limit vehicle vibration when starting and stopping the vehicle.

The overhead camshaft is chain driven, and a precombustion chamber with glow plug is used. A Robert Bosch fuel injection pump is mounted on the left side of the engine. This is a five plunger, in-line unit with a mechanical governor. The fuel pump is attached to the injection pump and is driven by an extra cam on the camshaft of the injection pump.

The injection volume is controlled by an upper and lower helix in the plungers. These units control the beginning and end of the injection, depending on the desired load. Injection timing is a function of engine speed and is controlled by the automatic injection timer (operated centrifugally). The mechanical governor controls the delivery of the pump and operates in the idle and maximum speed ranges. It matches the delivered fuel quantities to the air flow characteristics at high engine speed. The driver selects the amount of fuel in the range between idle and maximum with the accelerator pedal.

At low ambient temperatures, and with a cold engine, an additional quantity of fuel is released by the injection pump upon depressing the accelerator pedal once. The glow plugs remain energized until the engine has started and the starting key is turned to the second position. An "additional time relay" ends the glow plug operation if after 150 seconds a start has not occurred.

Exhaust emissions of the Mercedes-Benz diesel are said to be below government standards. Even HC and CO requirements can be met. Further reduction of NO_x is said to be difficult, since catalysts are not effective due to the constant presence of oxygen in the exhaust.

TURBOCHARGED FIVE CYLINDER ENGINE

The Mercedes-Benz turbocharged five cylinder diesel engine, Fig. 11-26, powers the 300SD model. This engine is of precombustion chamber design, which the engineers feel is most suitable for use with a turbocharger.

The turbocharged engine is based on the naturally aspirated five cylinder engine, Fig. 11-25. Both engines have a compression ratio of 25.1 to 1. Fuel consumption by the turbocharged engine is materially lower.

Main bearings are identical, but rod bearings were redesigned for the turbocharged version to withstand the higher

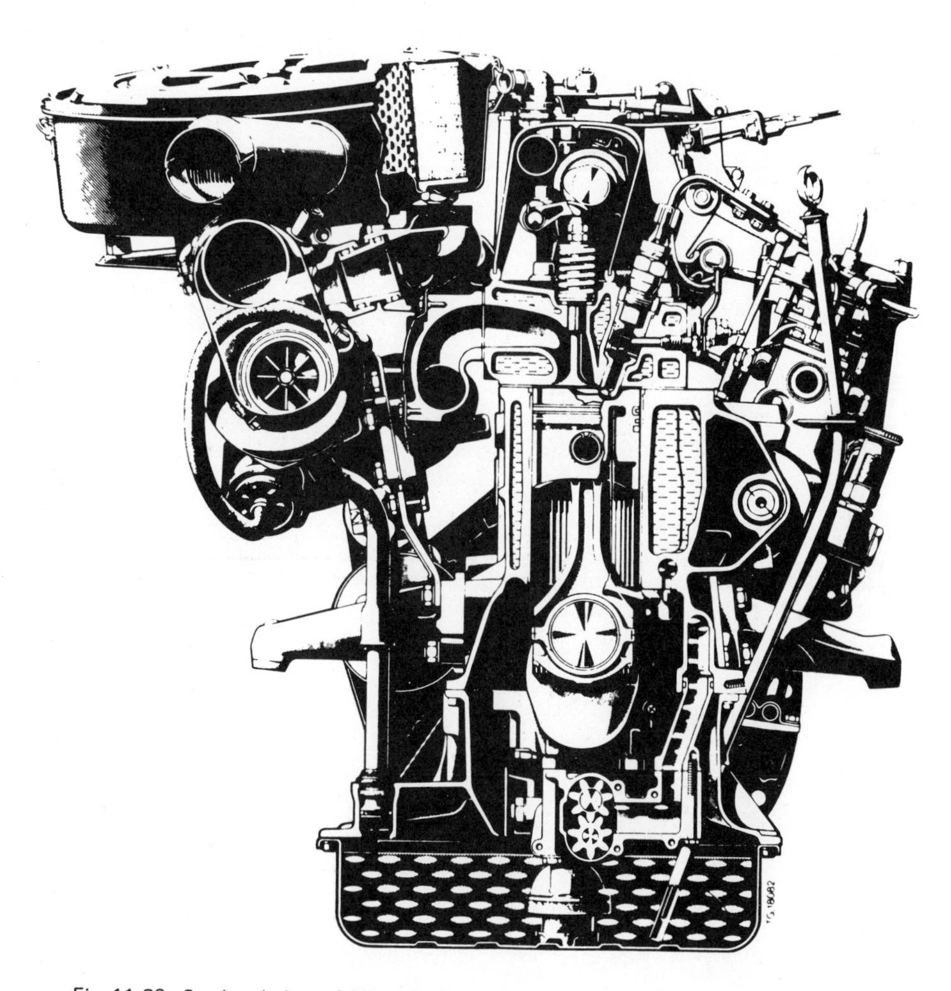

Fig. 11-26. Sectional view of Mercedes-Benz turbocharged, five cylinder diesel engine.

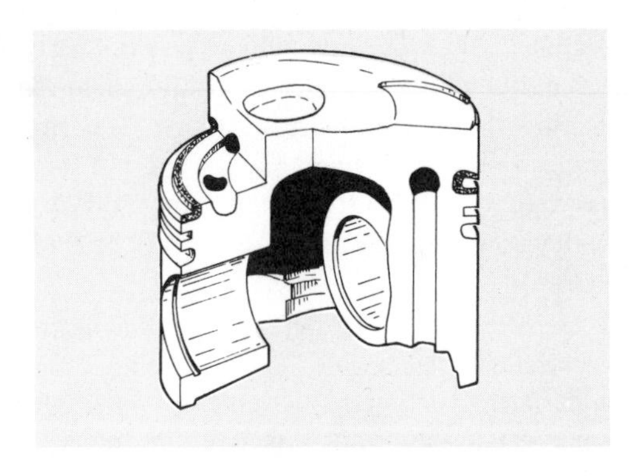

Fig. 11-27. Note integral cooling channel and cast-in ring carrier on this special piston design for the Mercedes-Benz turbocharged five cylinder diesel engine.

loads. The pistons were strengthened, have a built-in support for the upper ring and are provided with an oil pipe, Fig. 11-27. Fixed oil jets mounted in the crankcase, Fig. 11-28, direct the engine oil upward into the interior of the pistons to cool and lubricate the area.

For added strength, the diameter of the piston pin bearings in the pistons was increased 5/64 in. (2 mm). To further withstand higher combustion chamber temperatures and to limit wear, valve head thickness was increased at the edges.

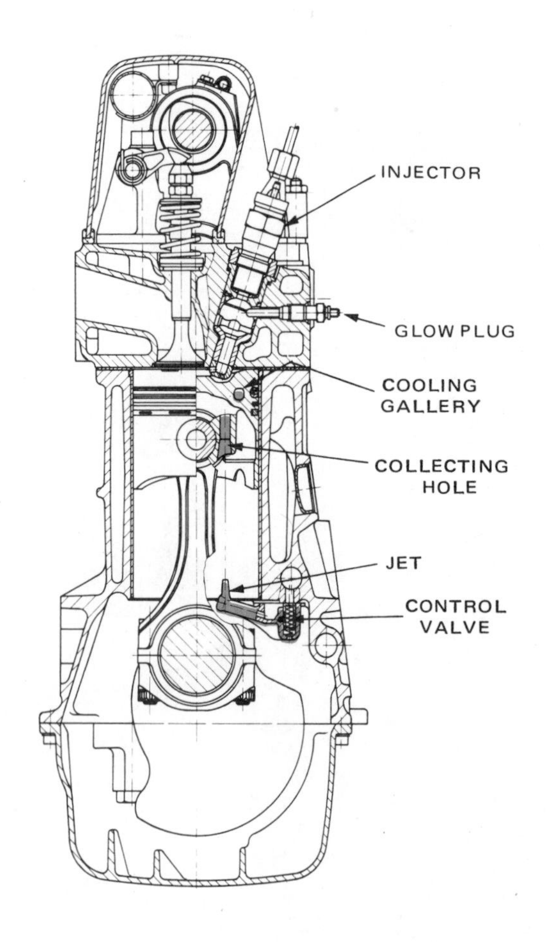

Fig. 11-28. In Mercedes-Benz five cylinder engine, a jet sends a stream of oil upward into a collecting hole that carries the lubricating oil into a cooling gallery.

According to Mercedes-Benz engineers, increased thermal loads due to the use of a turbocharger are compensated for by improved dissipation of heat into the cylinder head. Special design of the combustion prechamber insert results in a temperature of 1652 deg. F. (900 C) at the lower area of the chamber (about the same as encountered with naturally aspirated engines). The turbocharger used with the five cylinder diesel is of special design. See Fig. 10-25.

Fig. 11-29 shows the increase in power resulting from the installation of the turbocharger. Note that brake horsepower (bhp) peaks at 110 at about 4,000 rpm. Fuel consumption by the turbocharged engine was materially lower than by the naturally aspirated unit. Hydrocarbon and carbon monoxide emissions were reduced. Oxides of nitrogen were slightly higher with turbocharger.

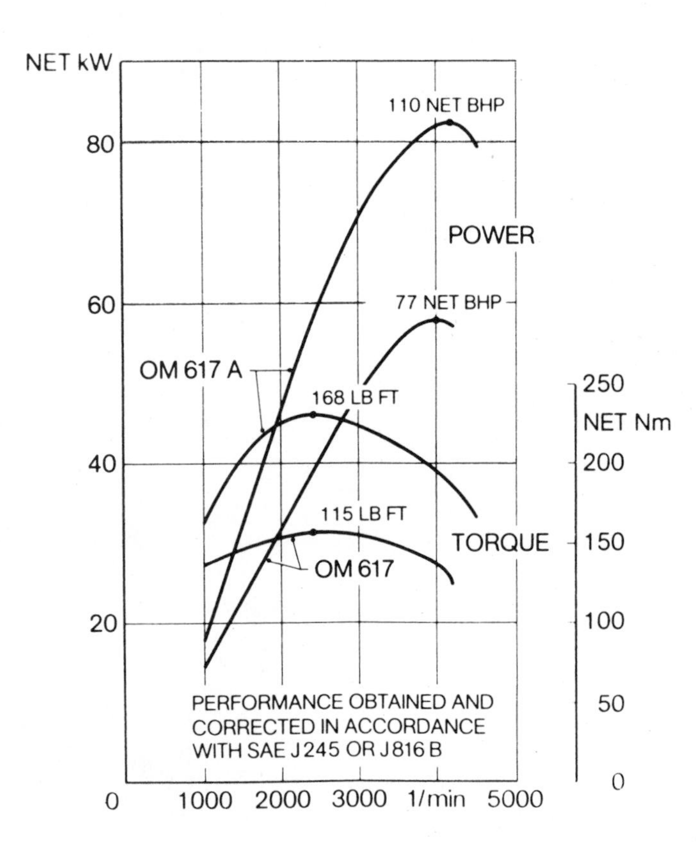

Fig. 11-29. Power and torque curves compare performance of the Mercedes-Benz five cylinder diesel engines: OM 617-Naturally aspirated; OM 617 A-Turbocharged. Metric units include: kW-Kilowatts; Nm-Newton-metres.

GM DIESEL ENGINES BY OLDSMOBILE

The General Motors 5.7 litre V-8 diesel engine installed in Oldsmobiles and Cadillac Sevilles develops 125 hp at 3,600 rpm. See page 167. It features a variation of the precombustion chamber, Fig. 11-30. The injectors are Roosa Master model D-82 described in Chapter 8.

Bore of this GM V-8 is 4.057 in. (103.0 mm) and the stroke is 3.385 in. (86.0 mm), which is the same as the 350 cu. in.

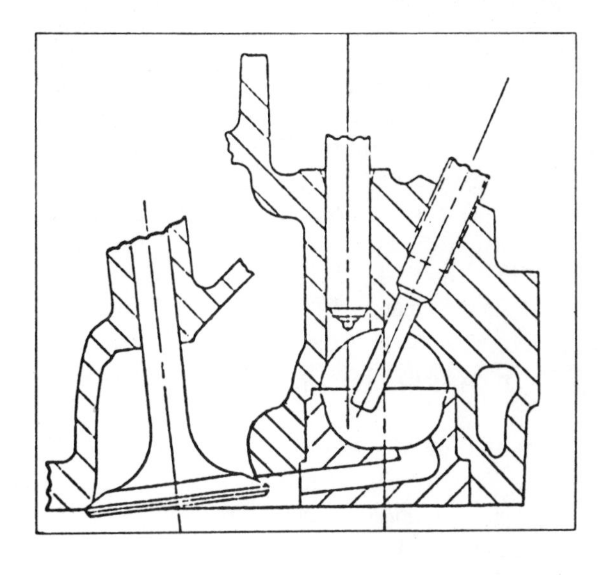

Fig. 11-30. GM 5.7 L diesel engine precombustion (swirl) chamber.

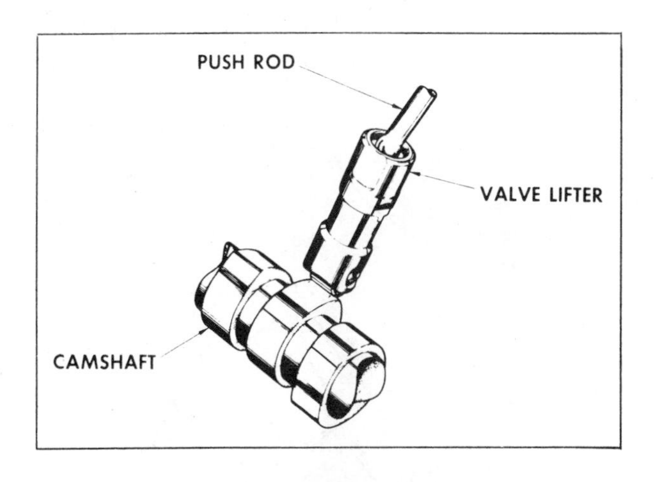

Fig. 11-32. Roller hydraulic valve lifters are used on later GM 5.7 L diesel engines. Rolling action of needle bearings on camshaft lobes reduces friction and allows the oil and oil filter change intervals to be extended to 5,000 miles.

(5.7 litre) gasoline powered engine. Many other dimensions remain the same but the cylinder heads, of course, are different. Also, the diameter of the piston pin was increased about 7/64 in. (3 mm). Piston weight was stepped up 155 grams. These changes are directly related to the increase in bearing loads which, in turn, resulted from the higher compression of the diesel. The compression ratio is 23 to 1. Horsepower and torque curves are shown in Fig. 11-31.

Later GM 5.7 litre diesel engines feature several engineering improvements: Roller hydraulic lifters, Fig. 11-32, replace the flat follower type. A new, modulated exhaust gas recirculation valve and internal EGR system are used. A large capacity water separation system located in the fuel tank also includes an instrument panel warning light.

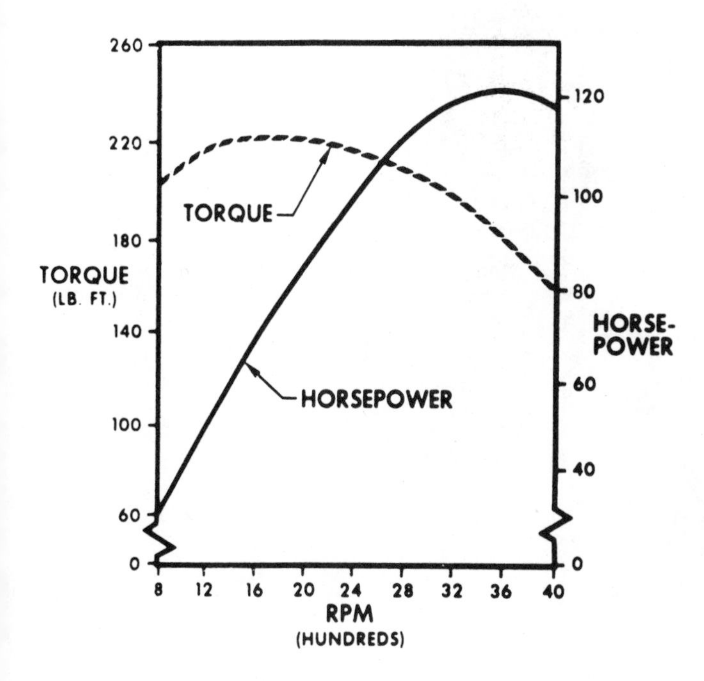

Fig. 11-31. Performance curves of GM 5.7 L diesel engine.

VOLKSWAGEN DIESEL ENGINE

Volkswagen's in-line, four cylinder diesel engine, Fig. 11-33, has a 1.5 L displacement. It develops 50 hp (DIN) at

Fig. 11-33. Volkswagen's 1.5 L (89.7 cu. in.) diesel engine.

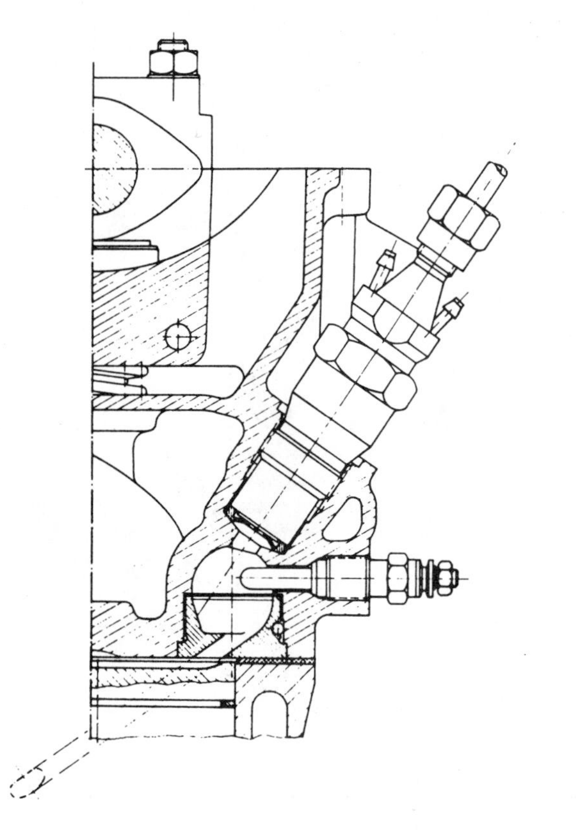

Fig. 11-34. Close-up of the precombustion chamber and glow plug in a VW Rabbit diesel engine. (Volkswagen of America, Inc.)

5000 rpm. Compression ratio is 23 to 1. Bore and stroke are 76.5 mm x 80 mm (3 1/64 in. x 3 5/32 in.). A Robert Bosch distributor type fuel injection pump injects the fuel into the swirl or turbulence chamber, Fig. 11-34. A glow plug is provided for easy starting. See Fig. 12-51.

The VW diesel is water cooled. The water jackets extend the full length of the cylinders, keeping cylinder distortion to a minimum. To conserve space, the engine is mounted transversely in the chassis.

PEUGEOT DIESEL ENGINE

The Peugeot model XD and XDP four cylinder diesel engine, Fig. 11-35, develops 58.5 hp with cooling fan engaged and 65 hp with the fan disengaged. Displacement is 2112 cm^3 (128 cu. in.) with a bore of 90 mm (3.5 in.) and a stroke of 83 mm (3.47 in.). The compression ratio is 22.2 to 1.

The overhead valves are operated by push rods. The combustion chamber is described as a Ricardo Comet V type, more generally known as a swirl or precombustion type. A glow plug is provided for easy starting.

The engine is placed at an angle of 22 deg. in the chassis. An unusual feature is the use of cooling fins on the oil pan. Acceleration of 22 seconds for 400 meters (437 yards) from a standing start is claimed. Fuel injection by Robert Bosch is the EP/VM model with a deferred injection device.

Fig. 11-35. Sectional view of Peugeot 2112 cm^3 diesel engine.

REVIEW QUESTIONS – CHAPTER 11

1. The fuel injection pump is driven on the Ford six cylinder 330 cu. in. engine by the:
   a. Gear driven accessory shaft.
   b. Cog type belt.
   c. Extension of the crankshaft.
2. The valves are operated in the Cummins C-180 engine as used in some Ford models by:
   a. Push rods.
   b. Overhead camshaft.
   c. Rotating lobe shaft.
3. The type injection pump used by the International model UDT engine is the:
   a. Rotary type.
   b. Unit injector.
   c. Multiple plunger type.
4. The cylinder heads are cast for the Waukesha model L-1616D, twelve cylinder engine:
   a. Individually.
   b. In pairs.
   c. One for each bank.
   d. In double sections.
5. The Electro Motive Div. of General Motors has a model 645 engine. This is a ____________ cycle engine.
6. When overhead clearance is limited, the type of engine that should be considered is:
   a. V-type.
   b. In-line.
   c. Horizontal.
   d. Radial.
7. The major advantage of the radial type engine is ____________.
8. In the Fairbanks Morse opposed piston engine, the fuel injection nozzles are located in:
   a. Lower cylinder.
   b. Upper cylinder.
   c. At mid point.
9. The major characteristic of the Napier Deltic engine is ____________.
10. What is the angle between the crank throws of the Mercedes-Benz five cylinder engine?
11. What make of fuel injection pump is used on the VW Rabbit diesel engine?
12. The Peugeot diesel engine is of the direct injection type. True or False?

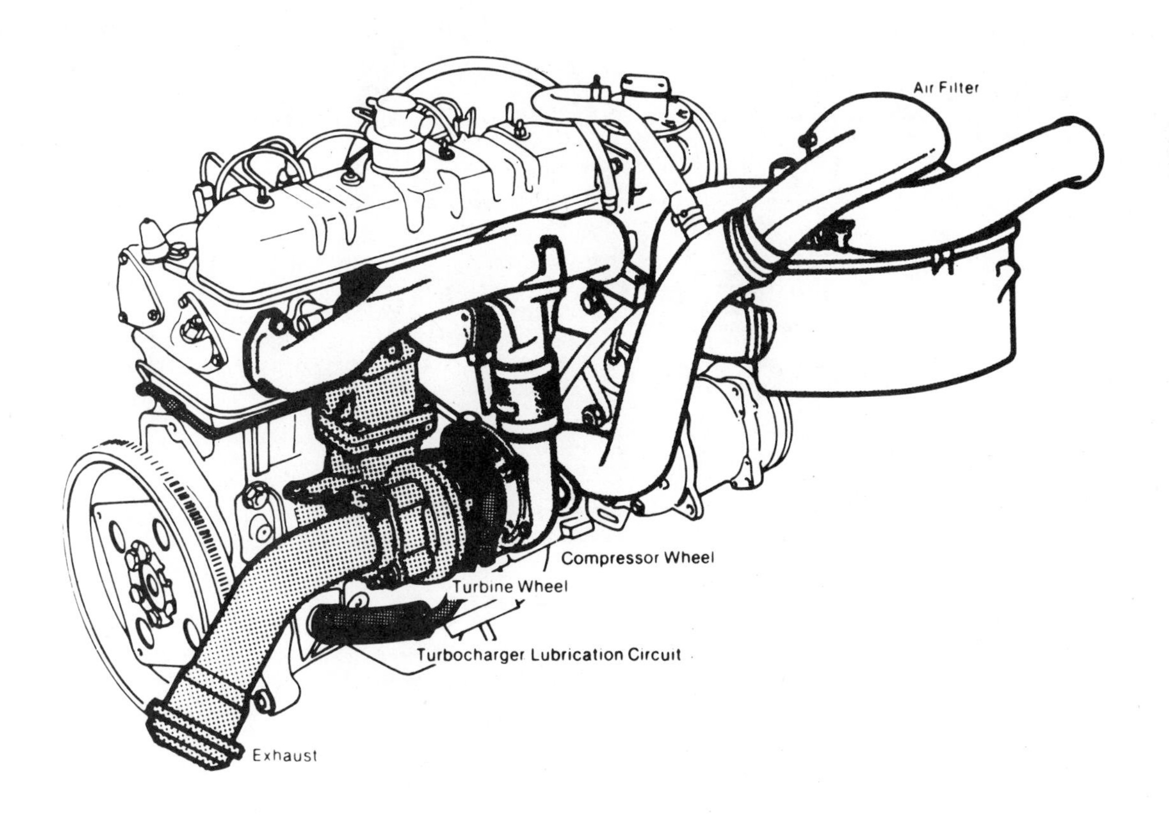

Shaded area indicates location of turbocharger as installed on Peugeot XD2PT diesel engine. This 2.3 litre (140 cu. in.), four-stroke cycle, four cylinder engine develops 80 hp.

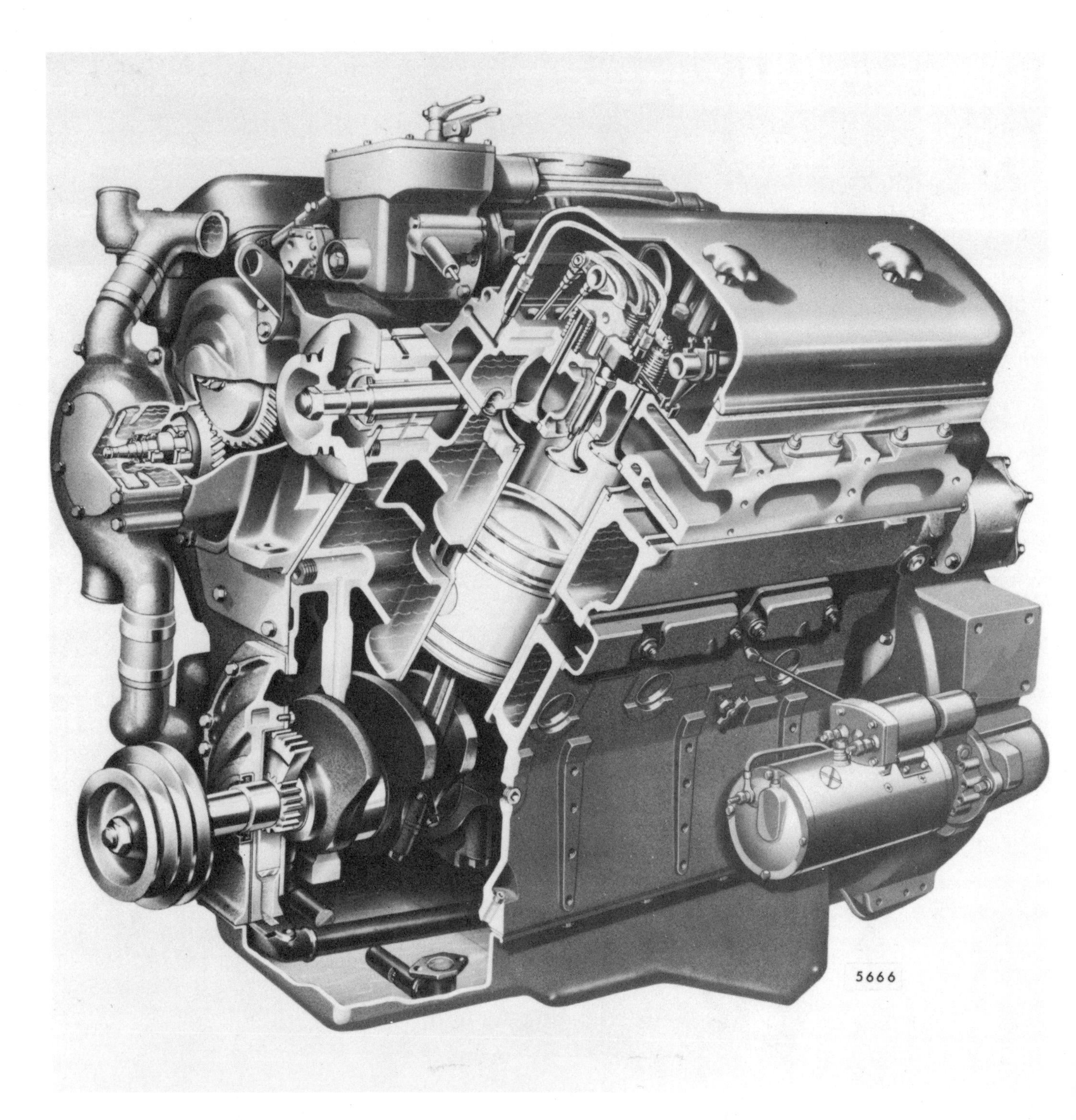

Cutaway view of Detroit Diesel Allison series 71 V-type engine. This is a two cycle power plant with unit injectors. Most models have crosshead pistons with separate crown and skirt sections to improve piston pin bearing life. The 71 V-8 truck model develops 318 hp at 2100 rpm.

Chapter 12
DETAILS OF ENGINE PARTS

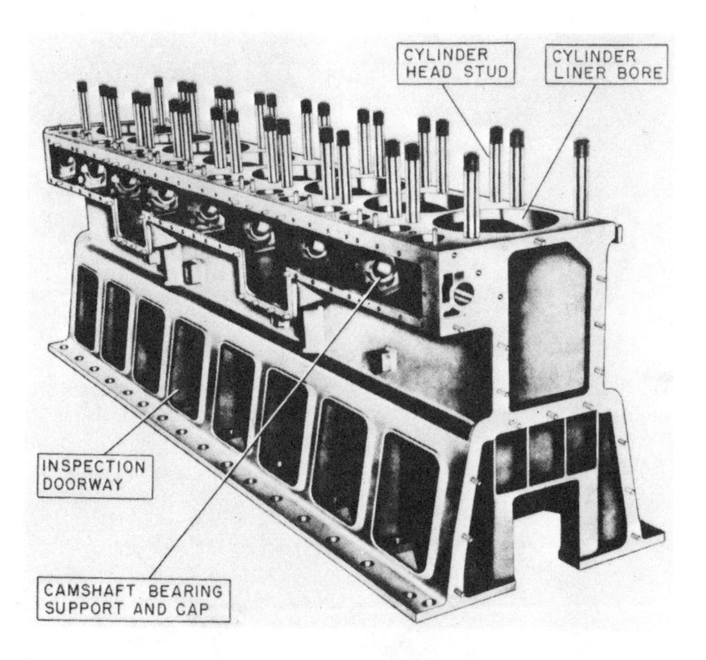

Fig. 12-1. Large cylinder block for industrial engine constructed of welded steel forgings and steel plate.
(Cooper Bessemer Div., Cooper Industries)

This chapter deals with individual engine parts such as cylinder blocks, cylinder heads, crankshafts, connecting rods, pistons, pins and rings. Particular emphasis will be placed where the design and construction differs materially from the same parts used in the gasoline spark ignition engine.

Major differences in construction result directly from the much greater compression pressures, higher heat and also the fact that in many instances the displacement is much greater than that encountered in the automotive field. In addition, many diesels are in continuous operation and are only stopped when repairs become necessary.

Fig. 12-1 illustrates a large Cooper-Bessemer cylinder block which is designed for marine work. This eight cylinder unit is constructed of welded steel forgings and steel plate. This type of block is secured to a separate engine base. When the two parts are bolted together they form the frame for the main bearings which carry the crankshaft. Note that the camshaft bearing supports, consisting of four transverse members are an integral part of the cylinder block. Pads welded to the exterior of the block are machined to carry engine parts and accessories. Each cylinder liner and cylinder head has its own water jacket.

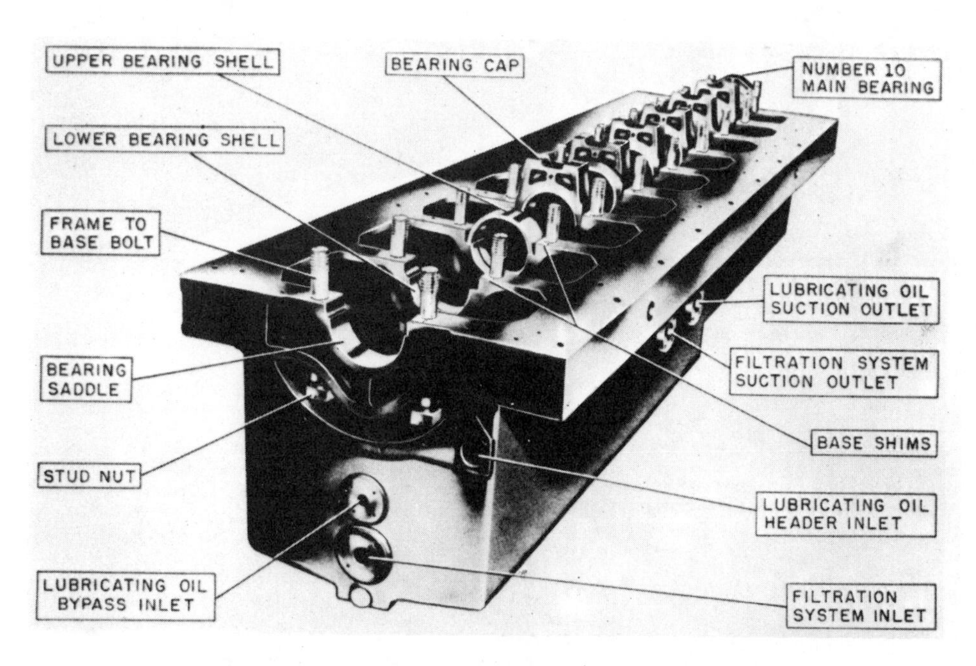

Fig. 12-2. Engine base for Cooper Bessemer engine shown in Fig. 12-1.

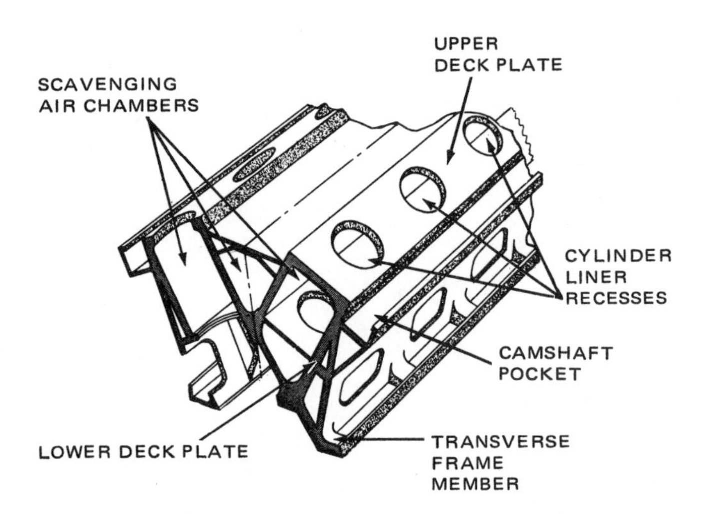

Fig. 12-3. Large V-type engines are often constructed of forgings and steel plates welded together.

The engine base for this engine is shown in Fig. 12-2. The base supports the main bearings. This also serves as the oil pan, which is combined with the bed plate.

In the automotive field, the cylinder block and crankcase form a single casting. In the industrial and marine field it is not unusual to have the cylinder block constructed of forgings and steel plates which are welded together, Fig. 12-3. In this V-type construction, the upper and lower deck plates on each side of the V are bored to receive the cylinder liners. The space

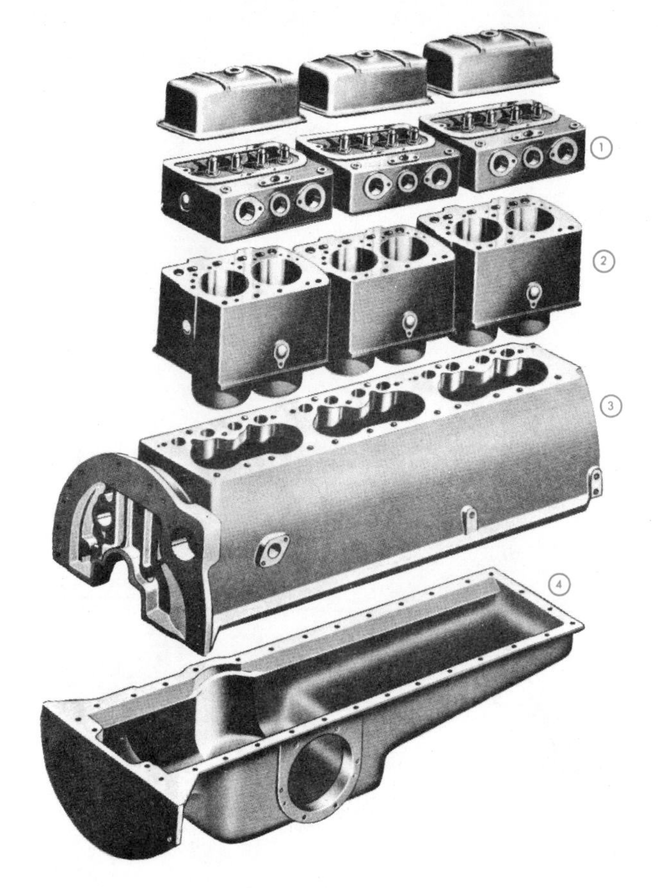

Fig. 12-4. Parts of a large tractor engine. Note that cylinder heads and cylinder blocks are cast in pairs. 1—Cylinder heads. 2—Cylinder blocks. 3—Upper crankcase. 4—Oil pan. (Minneapolis-Moline)

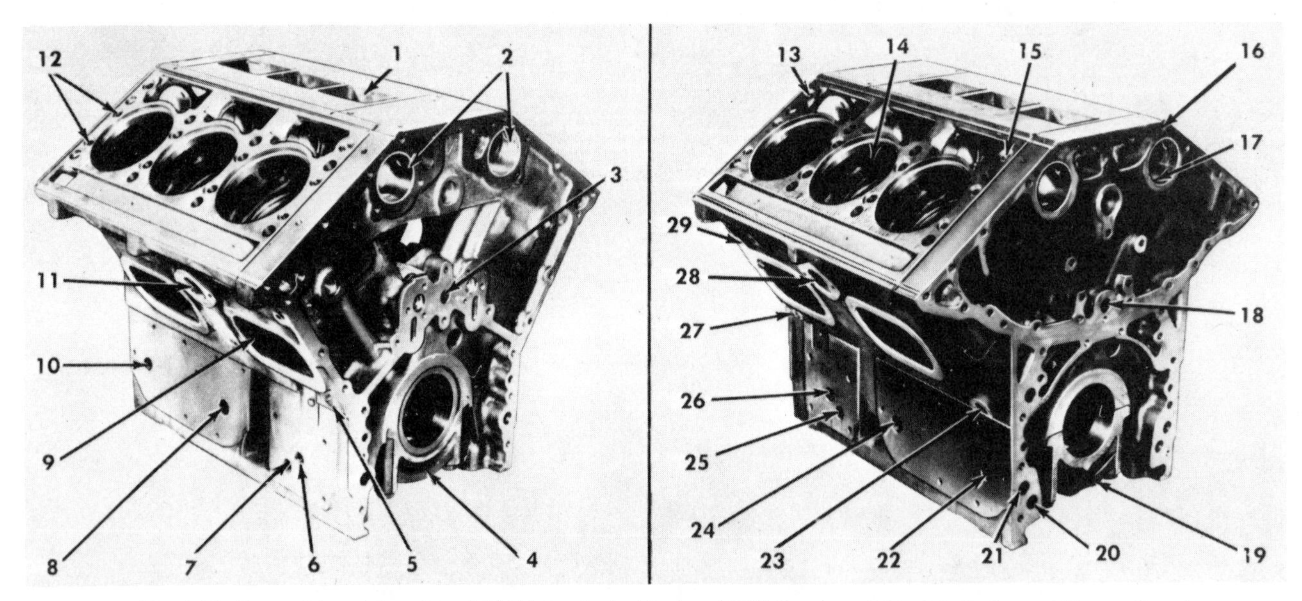

Fig. 12-5. Front side and rear side of GM V-6 cylinder block. (GMC Truck and Coach Div., General Motors Corp.)

1. AIR BOX.
2. CAMSHAFT BORE.
3. OIL GALLERY – MAIN.
4. CAP – MAIN BEARING (REAR).
5. OIL PRESSURE TAKE-OFF OPENING.
6. OIL PASSAGE FROM FILTER TO COOLER.
7. OIL PASSAGE TO FILTER FROM PUMP.
8. OIL PASSAGE TO OIL COOLER.
9. AIR BOX.
10. OIL PASSAGE FROM OIL COOLER.
11. WATER PASSAGE FROM OIL COOLER TO BLOCK.
12. WATER PASSAGE TO CYLINDER HEAD.
13. OIL PASSAGE TO CYLINDER HEAD.
14. BORE FOR CYLINDER LINER.
15. OIL PASSAGE TO CYLINDER HEAD.
16. CYLINDER BLOCK.
17. OIL GALLERY TO CAMSHAFT.
18. OIL GALLERY – MAIN.
19. CAP – MAIN BEARING (FRONT).
20. OIL PASSAGE FROM OIL PUMP TO FILTER.
21. OIL PASSAGE TO MAIN GALLERY VIA FRONT COVER.
22. OIL PASSAGE FROM OIL COOLER.
23. OIL PRESSURE TAKE-OFF OPENING.
24. OIL PASSAGE TO OIL COOLER.
25. OIL PASSAGE TO FILTER FROM PUMP.
26. OIL PASSAGE FROM FILTER TO OIL COOLER.
27. OIL PRESSURE TAKE-OFF OPENING.
28. WATER PASSAGE FROM OIL COOLER TO BLOCK.
29. OIL PRESSURE TAKE-OFF OPENING.

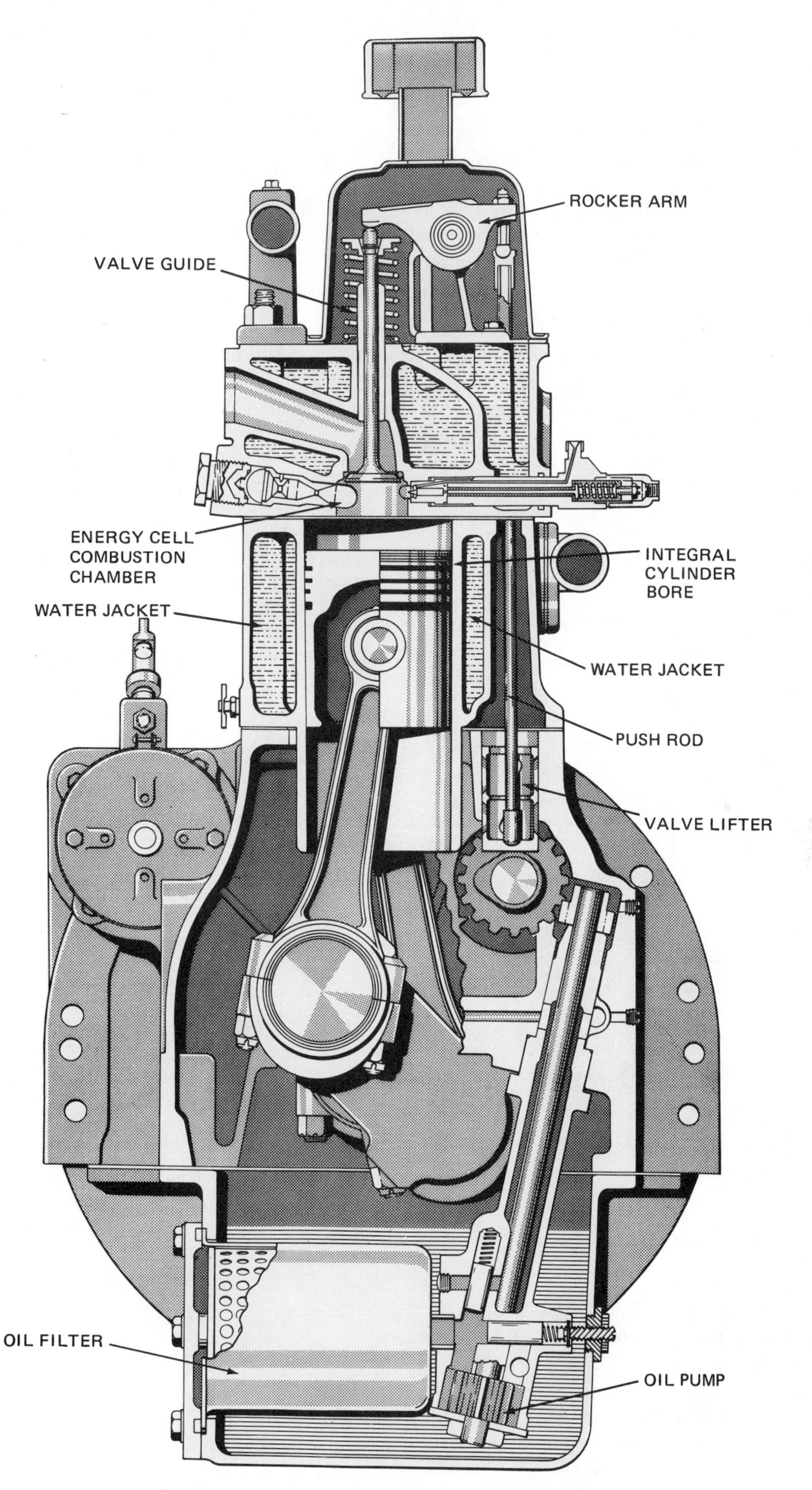

Fig. 12-6. Integral type of cylinder bore. (Minneapolis-Moline)

between the decks and the space between the two banks of cylinders form the scavenging air chambers. In some cylinder blocks of this type, the liner bore in the lower deck plate is made with a groove that serves as a cooling water inlet for the liner or sleeve. Some V-type cylinder blocks are constructed with mounting pads for the main bearing seats as integral parts of the forged transverse members at the bottom of the block.

While the usual practice is to have the cylinders form a single unit, the Minneapolis Moline tractor engine has the cylinders cast in pairs (2), Fig. 12-4. Similarly the cylinder heads (1), Fig. 12-4, are in pairs. The crankcase is a single unit. An unusual feature is the oil pan, a casting and which provides added strength for the assembly.

An automotive V-6 cylinder block is shown in Fig. 12-5. The cylinder block and crankcase form the main structural part of the engine. This one-piece casting of alloy cast iron, is a product of General Motors. Cylinder bores are counterbored. Each is fitted with an insert to support the cylinder liners into which a number of air inlet ports are drilled. Hole plates are provided for access to the air chamber, and for inspection of the cylinder wall, pistons and rings, and the air intake ports in the cylinder walls.

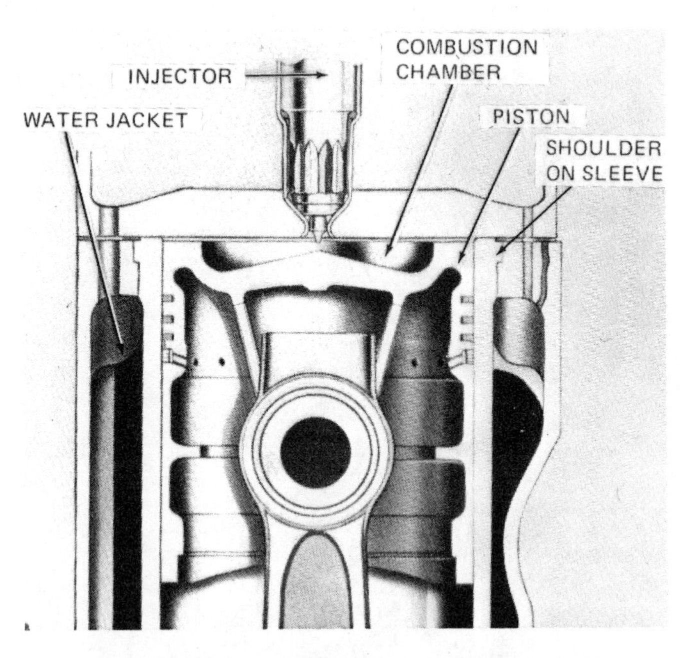

Fig. 12-7. Note shoulder on sleeve of wet-type cylinder liner. (GMC Truck and Coach Div., General Motors Corp.)

CYLINDER LINER

The barrel or bore in which the piston moves may be either integral with the main cylinder blocks, Fig. 12-6, or, it may be a separate liner, Fig. 12-7, which is also known as the sleeve. The construction first mentioned has the advantage of lower cost but the disadvantage of not being replaceable. Most diesel engines have replaceable sleeves. When cylinders become worn, it is a relatively simple matter to remove the worn sleeve and install a new one. This restores the cylinder bore to its original condition.

Fig. 12-8. Showing air ports in cylinder sleeve for two-cycle engine. (GMC Truck and Coach Div., General Motors Corp.)

When the cylinder bore is an integral part of the cylinder block, and the cylinder becomes worn, reconditioning is necessary. In some cases the machining can be done without removing the engine from the installed position by means of portable conditioning equipment. In the case of large diesel engines, it is necessary to take the cylinder block to the machine shop, which is a costly operation.

Another advantage of the cylinder sleeve design is that the sleeves can be manufactured of special alloys, while the cylinder block can be produced of lower cost material.

Fig. 12-7 shows a cylinder sleeve which is formed with a shoulder at the top which prevents vertical movement of the sleeve after the cylinder head is bolted in position. The outer area of the sleeve forms a wall of the water jacket and is called

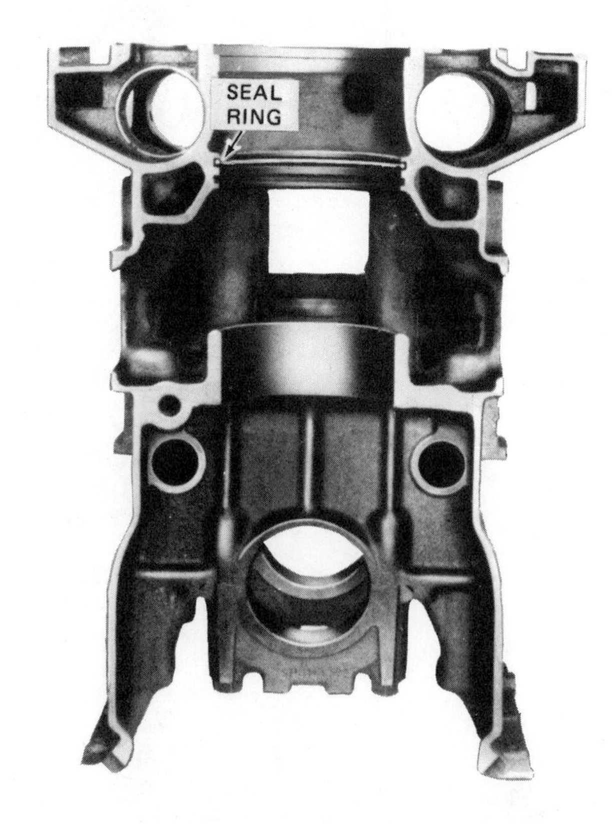

Fig. 12-9. Note location of cylinder liner seal ring in cylinder block bore. (GMC Truck and Coach Div., General Motors Corp.)

a wet-type sleeve. Most diesel engines use this design as it has the advantage of easy replacement. However, some engines are provided with dry-type sleeves. In such designs the cylinder blocks are formed with complete water jackets which are bored out for the cylinder sleeve. The sleeve is then a press fit in the cylinder bore.

When cylinder sleeves are used in two-cycle diesel engines such as those made by General Motors, the cylinder sleeves (or liners) are provided with ports for admission of air to the cylinder, Fig. 12-8. These wet liners are mounted in the cylinder block. A sealing ring is provided at the upper end to seal the coolant, Fig. 12-9. These are a slip fit in the cylinder block and are inserted from the top. The flange of each liner rests on a counterbore and seal ring at the top of the cylinder block.

To seal the top of the liner, a compression gasket is placed on the top of the liner as shown in Fig. 12-10. The compression gasket is then compressed (sealing the joint) when the cylinder head is installed. Without this gasket the compression in the combustion chamber would tend to leak into the water jacket.

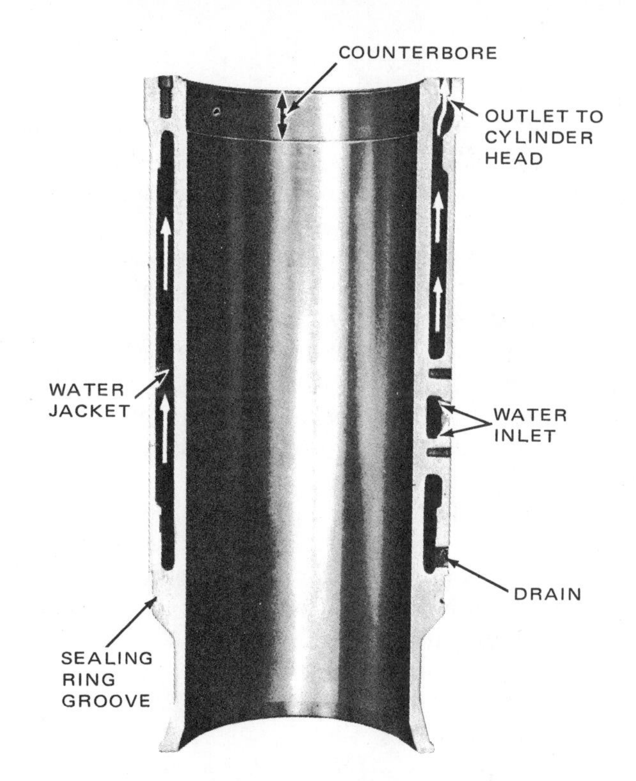

Fig. 12-11. Detail of wet-type cylinder sleeve with integral water jacket.

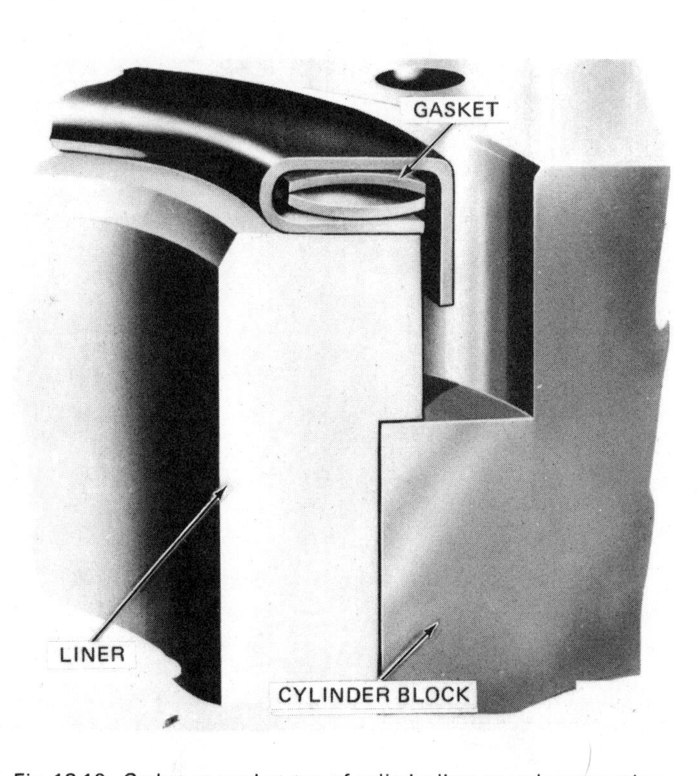

Fig. 12-10. Gasket as used at top of cylinder liner to seal compression. (GMC Truck and Coach Div., General Motors Corp.)

Another type of cylinder sleeve, Fig. 12-11, is formed with an integral coolant jacket. Such jackets may be shrunk on or cast with the sleeve, as shown in Fig. 12-11. Water is admitted into the lower section of the jacket and leaves through the top. Such designs are used extensively on large industrial type diesel engines.

In addition to the retaining shoulder at the top of the cylinder sleeve, most wet sleeves are formed with grooves at the lower end in which sealing rings or gaskets are installed, Fig. 12-12. These are designed to prevent leakage of coolant

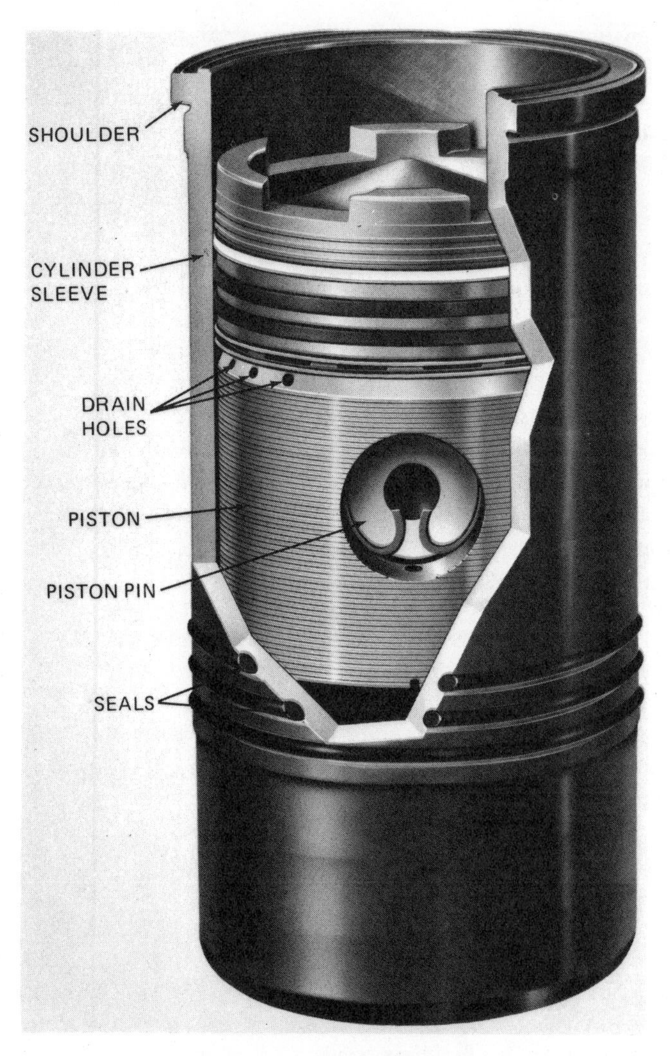

Fig. 12-12. Note sealing rings at lower end of wet-type cylinder sleeve as installed on International diesel engines. (Sealed Power Corp.)

from the water jacket. Usually two sealing rings are provided and these are of the familiar Neoprene O type. Fig. 12-12 shows a cylinder sleeve with piston produced by the Sealed Power Corporation, as a replacement for International Harvester trucks. Note the O rings at the bottom of the sleeve. The sleeves have a cross hatched finish on the inner diameter permitting rings to seat quickly.

Cylinder sleeves or liners, as they are also called, are usually made of an alloy of cast iron. Alloys such as chromium, nickel and molybdenum are being used. In some cases the sleeves are sand cast, in other cases they are centrifugal cast. Chrome plating is used extensively to reduce wear.

Operating conditions of cylinder sleeves are severe. Not only must they withstand high pressures of diesel operation and heavy thrust of the connecting rods, but also the high temperatures. Such operating temperatures on a two-cycle industrial engine may range from 1100 deg. F. at the top of the liner, where the cooling conditions are poor, to 400 deg. F. only a short distance below the top. At the lower end of the sleeve the temperature is still less.

PISTONS

With compression pressures reaching 900 psi and exhaust temperatures approximating 1000 deg. F., diesel engine pistons are subjected to severe operating conditions. The pistons must be designed to withstand such conditions. Consequently they are heavier than the familiar gasoline engine piston and have special provisions for cooling.

Aluminum and cast iron are generally used in the manufacture of diesel pistons. In some instances, large pistons are assembled which have a crown made of a steel forging or cast steel. Another design provides a cast iron or heat-treated steel forging insert for the piston pin insert. Inserts made of NiResist (a nickel alloy), other alloys, and cast iron are also used as inserts for top ring grooves in aluminum alloy pistons as a reinforcement and wear reducing feature. Another way to control expansion is the installation of alloy steel braces placed across the piston pin bosses. Steel rings are also used in the piston crown for added strength.

Aluminum has several advantages as piston material. It is lighter than cast iron, weighing .097 lb. per cu. in. and is almost one-third the weight of cast iron which weighs .284 lb. per cu. in. In addition, it has excellent heat conductivity . . . nearly twice as much as cast iron.

The strength of aluminum decreases faster than cast iron as its temperature increases, and it is necessary to use heavier sections to obtain the strength needed to carry heavy loads.

The lighter weight of aluminum is important from the standpoint of vibration, and the load on the bearings and crankshaft.

While the wear of aluminum pistons in some installations may be greater than cast iron, this is offset by the fact that the aluminum pistons, because of the greater heat conductivity, are less subject to scoring.

Brief mention has been made of various type of inserts which are used in pistons. Some inserts are used to reduce the wear of the ring groove, while other types of inserts are used to control expansion of aluminum pistons, Fig. 12-13.

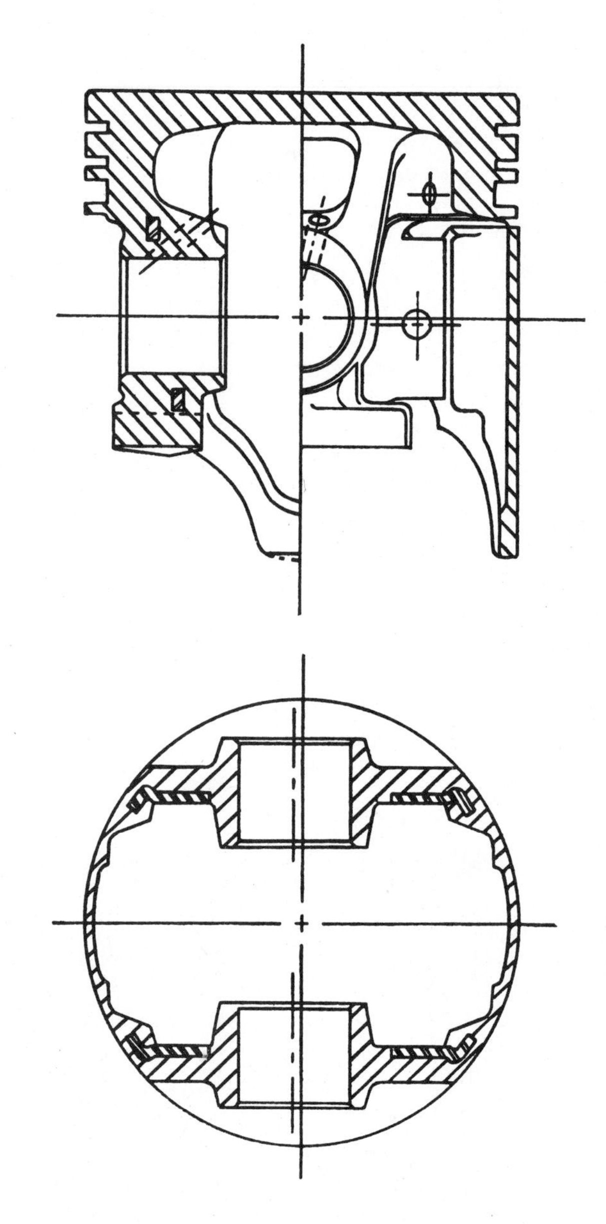

Fig. 12-13. Cross section of Bohn Aluminum open slipper-type piston, indicating struts within casting to control expansion. (Bohn Aluminum and Brass Corp., Div. Universal American Corp.)

Aluminum melts at a temperature of 1260 deg. F., cast iron at 2768 deg. F. The temperature at the top of an aluminum piston can be such that the metal approaches a malleable condition. Consequently, the force of combustion tends to crush the upper compression ring groove and nullify the effectiveness of the compression ring.

This problem is met by some designers by the installation of an insert of cast iron or other metal in the ring groove. Such inserts or rings are bonded to the aluminum piston by the Al-fin method or other special process.

NiResist, a nickel-bearing iron alloy is frequently used for ring inserts, Fig. 12-14. This material expands the same as aluminum. It provides strength and wear characteristics needed, and prevents the upper ring groove from being crushed.

Special heat-treated spring steel bands are also used in the head or crown of the piston or in the top ring groove, Fig. 12-15. In either case it acts as a heat dam. When built into the ring groove it provides a stronger surface for the ring so ring groove wear is reduced, Fig. 12-13.

Some diesel engine pistons have flat crowns. Others are formed to provide the combustion chamber with turbulence during the combustion process. For example, the M combustion chamber system has a spherical chamber which is formed in the crown of the piston, Fig. 5-7. The Mercedes Benz, which has a precombustion chamber, uses a flat top piston. Some pistons are provided with cast-in cooling passages behind the compression rings, Fig. 12-14.

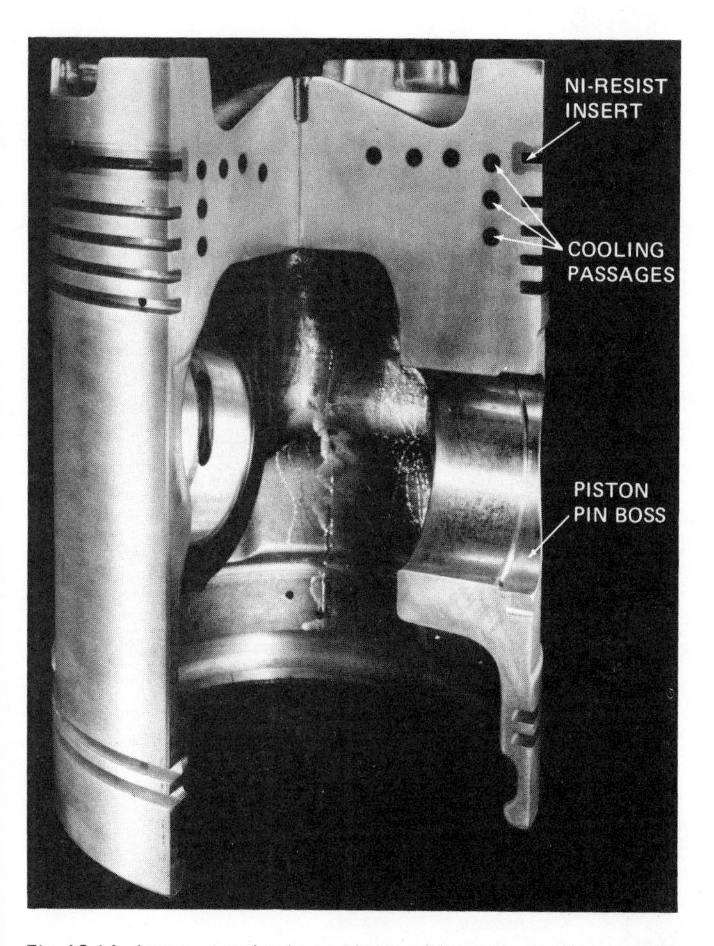

Fig. 12-14. Large cast aluminum piston with cast-in cooling passages behind compression ring grooves. Note also NiResist insert for upper compression ring. (Aluminum Co. of America)

A different type piston is used in the Series 149 engine, Fig. 12-16. In a discussion of the engine at an SAE meeting, Engineers of the Detroit Diesel Allison Division of General Motors Corp. stated that studies indicated that ring belt cooling by means of a control oil spray from the connecting rod was marginal for a piston of size used in the 149 engine. In addition, the interrupted flow caused by the variable inertia forces of a reciprocating assembly, particularly at higher engine speed, seriously reduced the quantity of oil available for piston cooling. For those reasons, a floating skirt piston design, Fig. 12-16, was chosen which has these features: (1) A continuous supply of entrapped oil to cool the ring belt and

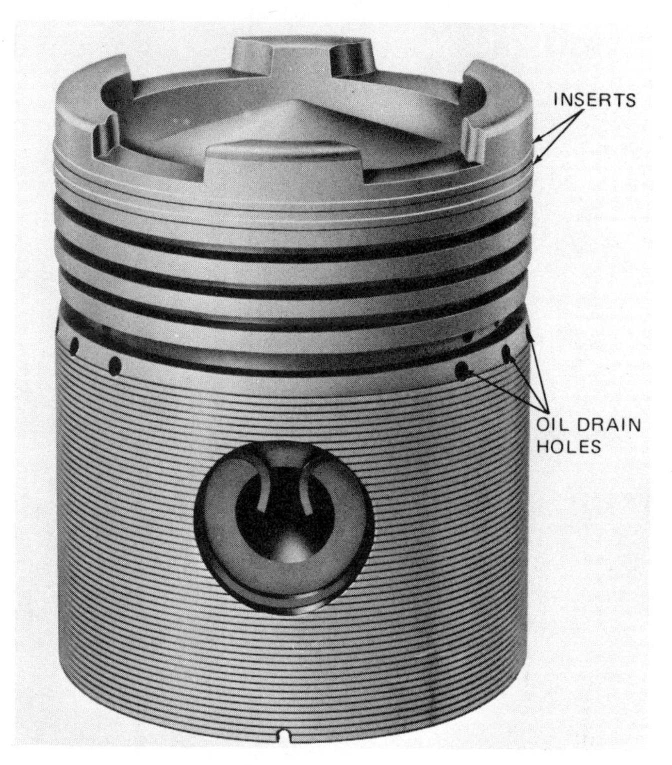

Fig. 12-15. One type of piston for open combustion chamber. Note inserts in piston crown and oil drain holes at top of skirt. (Sealed Power Corp.)

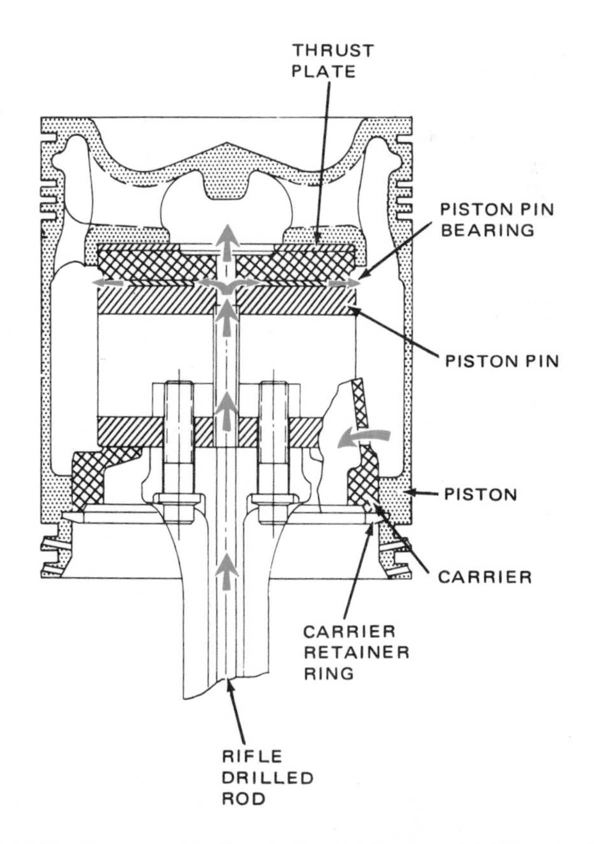

Fig. 12-16. Piston used in Detroit Diesel Allison model 149 engine has a continuous supply of entrapped oil to cool ring belt and combustion bowl, heavy ribbing on the underside of the crown and the combustion load is transferred directly from crown to piston pin. It also has increased piston pin bearing area and pressure lubrication to pin area. (Detroit Diesel Allison Div., General Motors Corp.)

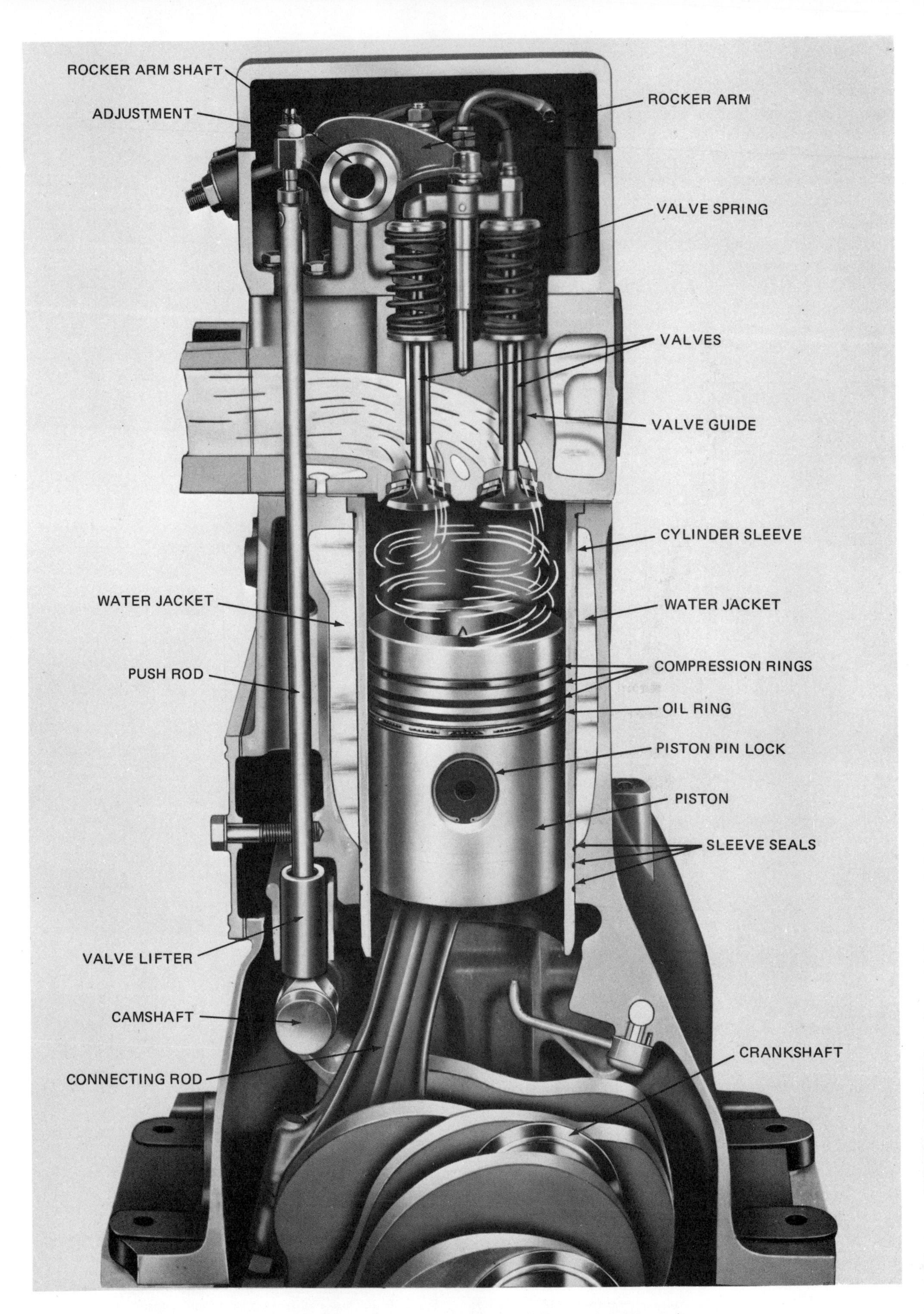

Fig. 12-17. Example of trunk type piston. (Allis-Chalmers Mfg. Co.)

the combustion bowl. (2) Increased heat transfer from the piston to the cooling oil provided by ribbing the underside of the piston crown. The size, quantity and location of these ribs can be varied to match various engine operating conditions. (3) The combustion load is transferred directly from the crown to the pin. This, together with the symmetrical skirt design, maintains piston roundness for conformity under various conditions of load and temperature. (4) Greatly increases piston pin bearing area. (5) Pressure lubrication for the piston pin.

Fig. 12-16 shows the outer piston and inner carrier, both of iron, with a steel backed bronze thrust plate. Oil is trapped in the chamber between the piston and the carrier to insure a constant scrubbing action of all inner surfaces. This provides a significant cooling effect. The piston is fed through the rifle drilled connecting rod, which directs the main oil flow to the piston dome. In a 700 hp engine, tests indicated piston temperatures were reduced from 600 to 265 deg. F.

TRUNK TYPE PISTONS

A piston which is unusual in construction is the trunk type. In this design, the piston is of unit construction and the skirt is made of sufficient length so it can take side thrust without scoring. One such example is shown in Fig. 12-17. The upper end of the connecting rod is connected directly to the piston by means of a pin which passes through the bosses on each side of the piston, Fig. 12-14. The piston can oscillate on the pin as it moves up and down in the cylinder.

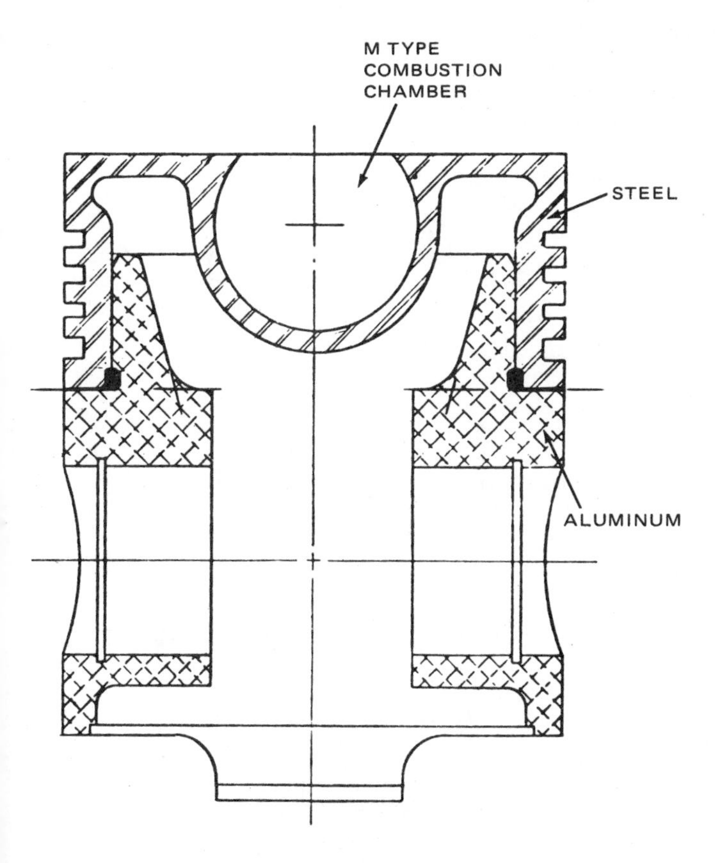

Fig. 12-18. Details of composite aluminum and steel piston design. Note M-type combustion chamber in crown of piston.

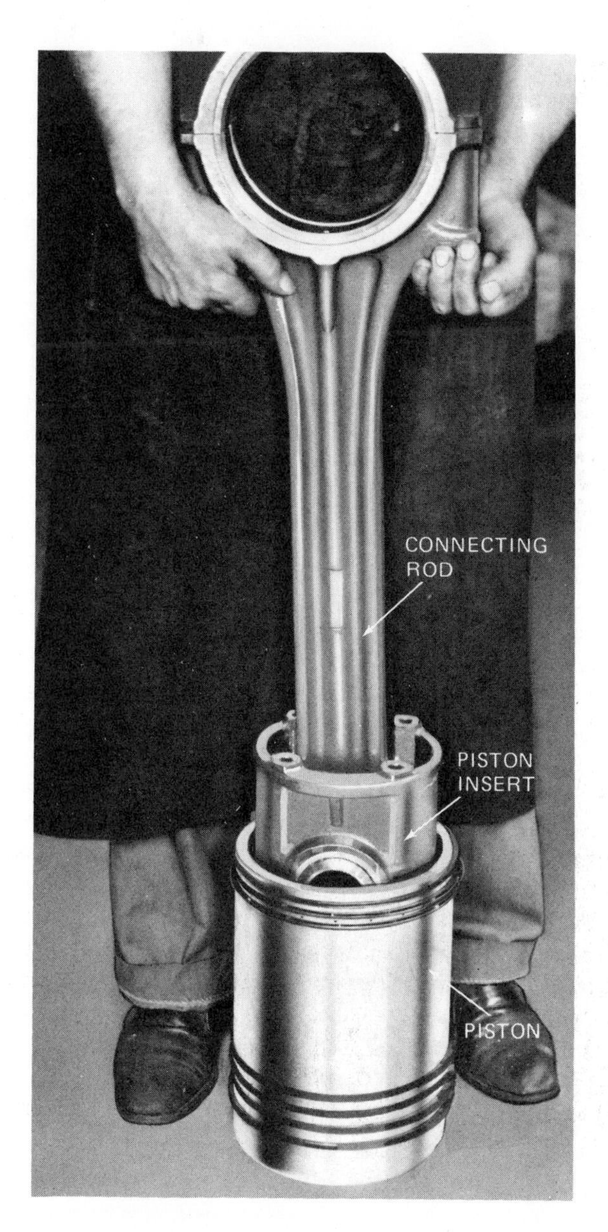

Fig. 12-19. Large composite type piston used in 8 1/8 in. bore and 10 in. stroke engine. (Fairbanks Morse Power Systems Div., Colt Industries)

COMPOSITE PISTONS

As has been pointed out, the piston must be strong enough to withstand the heavy force of expanding combustion gases; also be able to take side thrust without undue wear. In addition, the piston must be kept light in order to reduce bearing loads to a minimum.

One method of meeting those requirements is to make the piston crown of a metal such as malleable iron or steel (these are strong enough to take combustion forces) and use aluminum because of its lightness and ability to conduct heat, for the piston skirt. Originally the problem was to find a satisfactory method of joining the two metals. Recent advances in metallurgy have overcome this problem. A composite type piston is shown in Fig. 12-18. In this piston, steel is used for the piston crown and aluminum for the skirt. The two dissimilar metals are joined by electron welding.

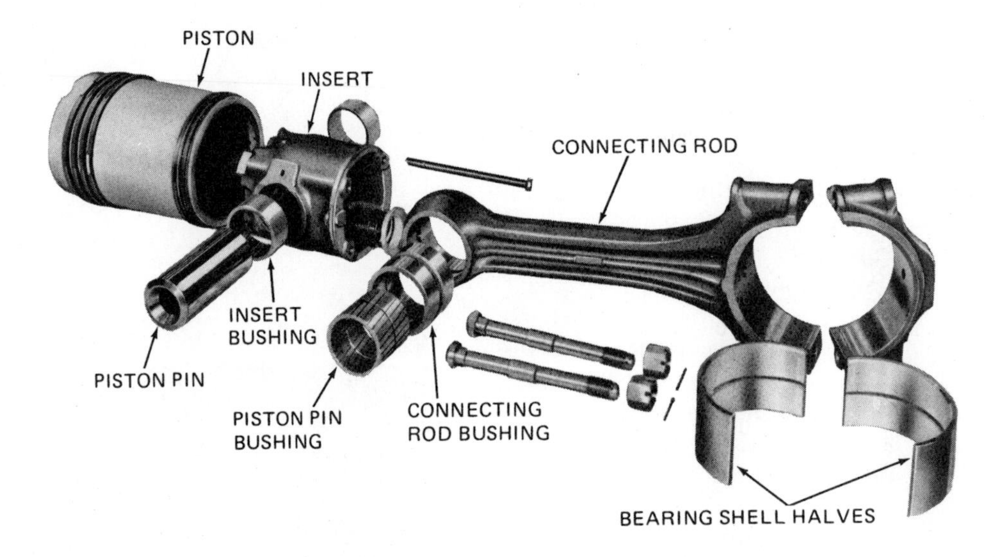

Fig. 12-20. Exploded view of Fairbanks Morse piston and connecting rod assembly.

Another type of composite piston is used by Fairbanks Morse in one of their 8 1/8 in. bore and 10 in. stroke engines, Figs. 12-19 and 12-20. The outer piston carries the piston rings and includes both piston crown and piston skirt. The inner piston or insert provides the mounting for the piston pin. The two parts of the piston are bolted together and form an enclosure for the piston cooling oil. The cooling feature of this piston design will be discussed later.

CROSSHEAD PISTON

In long stroke, large bore engines, the side thrust of the connecting rod reaches extremely high values. For a piston of normal size this would result in excessively high unit loading. The load per square inch can be reduced by using a longer piston skirt. However, this solution results in excessive piston weight and attendant inertia loads on the engine bearings. One method of solving the problem used on larger engines, is to use a crosshead. In such a design, the piston is connected to a vertical member, the lower end of which is attached to a sliding member called a crosshead, Fig. 12-21.

A crosshead moves up and down in guides and is connected on the lower end to the connecting rod. It will be noted in Fig. 12-21, the crosshead takes the side thrust and the piston has only vertical reciprocating motion.

In a piston designed by the Detroit Diesel Allison Div. of General Motors Corp., and called the "Crosshead," the crown of the piston absorbs the thrust of combustion and is attached to the skirt in such a manner the skirt takes all of the side thrust. As shown at A in Fig. 12-22, the piston is made in two parts. The piston crown carries the compression rings and a prefinished piston pin bearing is inserted in the lower part of the piston crown which is made of malleable iron. To help cool and transmit the combustion load, eight struts reach from the piston crown to the piston pin carrier. These struts, which are of the cooling fin type, distribute gas pressure loads to the slipper bearing and through the piston pin to the connecting rod assembly.

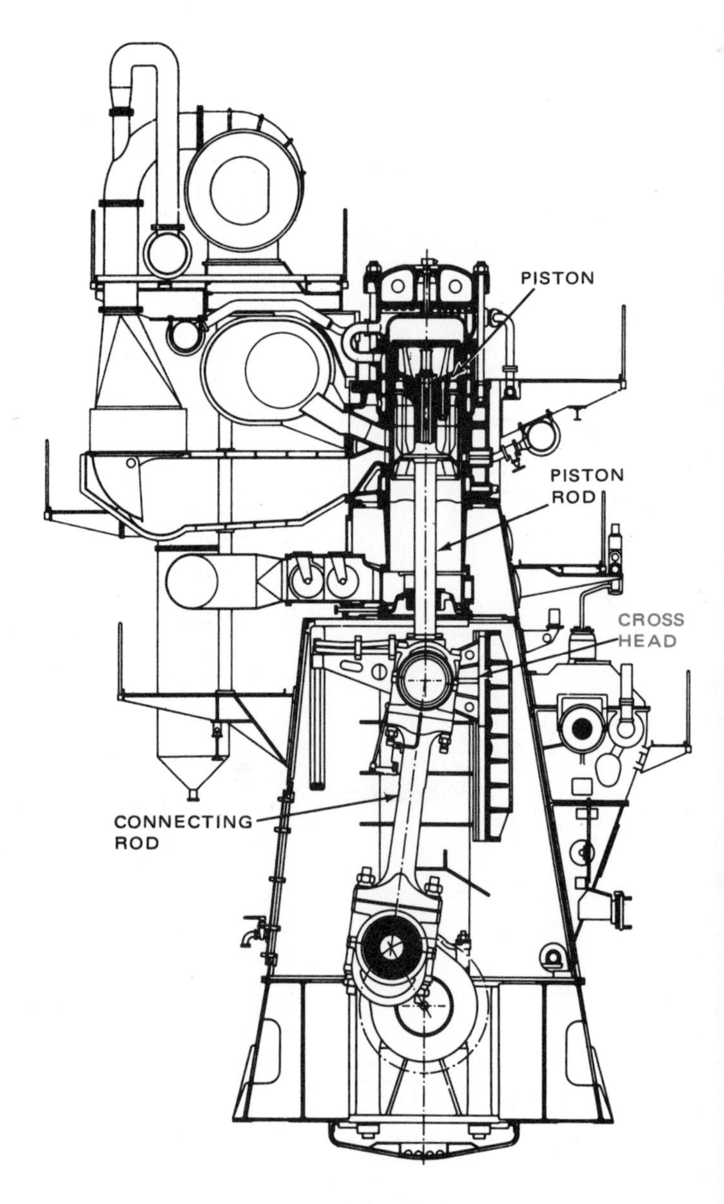

Fig. 12-21. M.A.N. engine with crosshead.
(MAN Maschinenfabrik)

The skirt of the Detroit Diesel crosshead piston is free from the crown and is held in place by the ends of the connecting rod pin. The skirt which carries the side thrust from the connecting rod is made of malleable iron, tin plated to reduce scoring tendencies.

The skirt also forms the outside of the oil well which carries cooling oil supplied from the rifle drilled connecting rod. This well supplies what is termed as "cocktail shaker" cooling for the piston crown. A standpipe in the well returns excess oil to the crankcase. The top of the connecting rod forms a saddle into which the piston pin is bolted, so there is no problem of lateral movement of the pin.

The Sulzer piston, shown at B in Fig. 12-22, slowly rotates around its longitudinal axis. At each stroke, a new, oil-wetted portion of the piston is in contact with the pressure side of the cylinder liner. The piston rings also rotate, so local overheating resulting from blow-by at the ring gap is avoided.

PISTON COOLING

Temperature of combustion in a diesel is in excess of 3000 deg. F. The temperature is dependent on type of fuel, compression and characteristics of the combustion chamber. Consequently the temperature of the piston will also vary. In the case of a cast iron piston, the temperature may be in excess of 1200 deg. F. An aluminum piston will have a temperature ranging in the 600 deg. F. area. It is essential that pistons be designed so they keep as cool as possible.

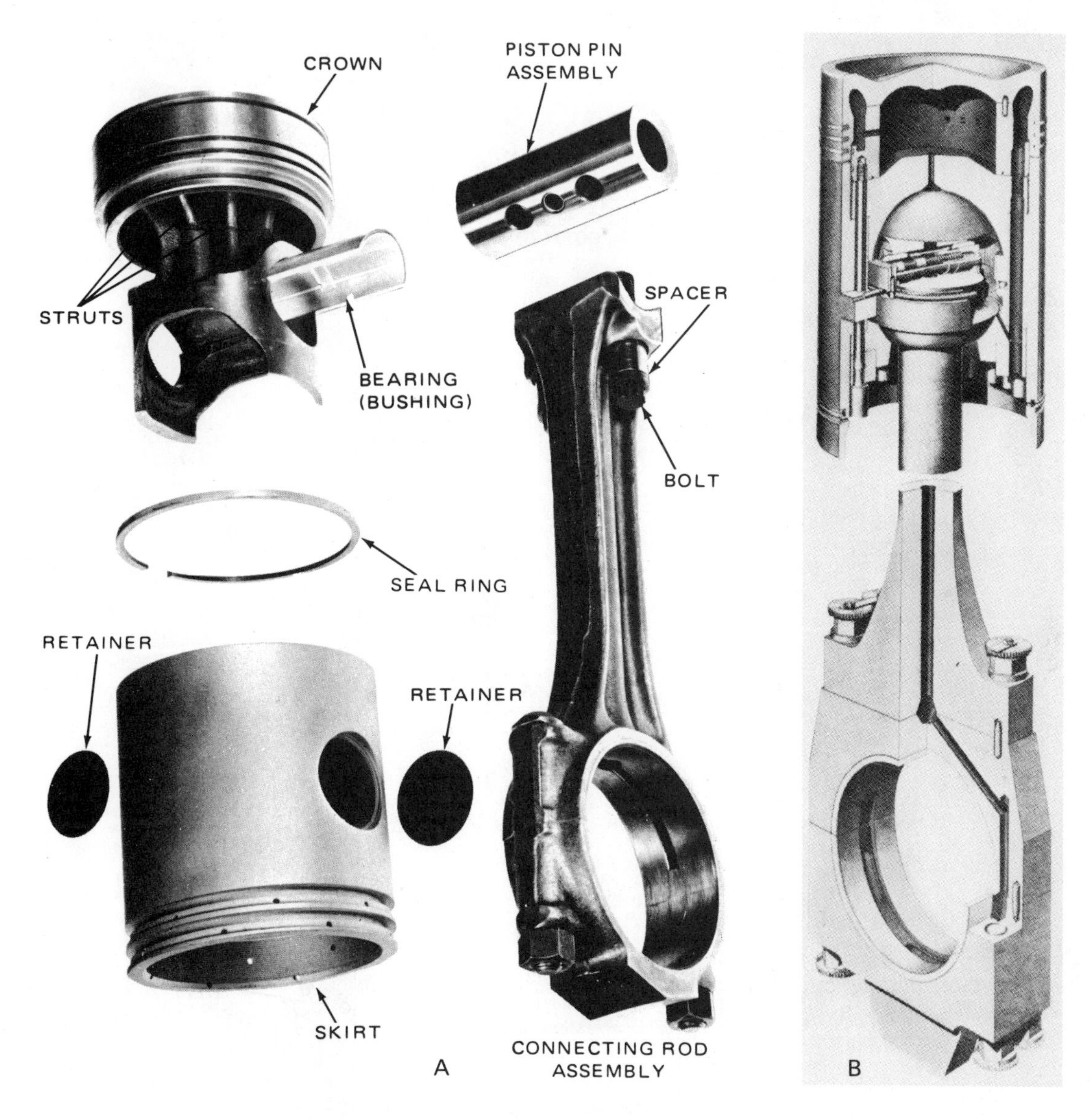

Fig. 12-22. A—Construction of Crosshead piston permits the crown and the skirt to rock on piston axis independent of each other. (Detroit Diesel Allison Div., General Motors Corp.) B—Sulzer piston is symmetrical, oil-cooled and, along with reciprocating motion, it has a slow rotating motion around its longitudinal axis. (Sulzer Bros. Ltd.)

Special provision is made to cool the pistons. Water which was used in a few designs in the past, has been replaced by oil.

Three general methods of supplemental cooling which are in general use are: (1) Circulation. (2) Shaker. (3) Spray.

In the circulation method, oil is circulated through grooves behind the ring belt. This method is used in some large diesel engines such as the Alco locomotive engine. In this system oil is supplied by way of the connecting rod which is rifle drilled. It then passes through the hollow piston pin. From there it passes to circular grooves machined in the piston head. In another design oil is circulated through coils of tubing cast in the piston head.

In the shaker method, a compartment in the piston head is supplied with oil from the rifle drilled connecting rod. The motion of the piston shakes the oil so it is spilled into channels

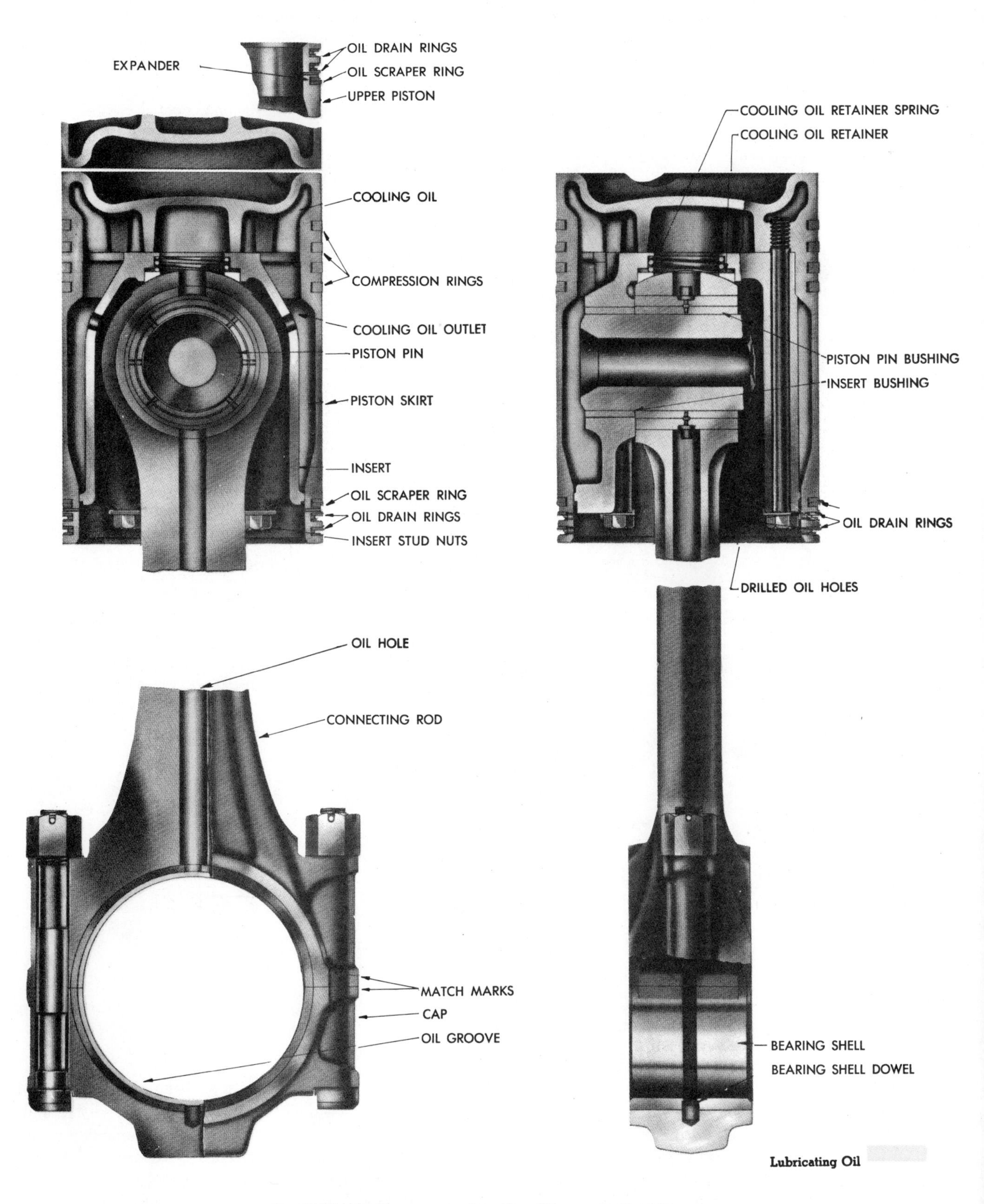

Fig. 12-23. Showing passage of cooling oil in composite piston. (Fairbanks Morse Power Systems Div., Colt Industries)

or pipes which return it to the crankcase. Fairbanks Morse and Cooper Bessemer use variations of this design in some of their engines.

In the Fairbanks Morse structural design, Fig. 12-23, with composite pistons, an enclosure for oil is formed between the inner piston (insert) and the outer piston. Lubricating oil flows through the connecting rod, around the outer annular groove in the rod bushing into the piston, and is then discharged. The piston retainer acts as a seal for the cooling oil in the piston crown.

Spray cooling of pistons is probably the most frequently used method of cooling pistons in diesel engines. One such design is found in the Murphy diesel, Fig. 12-24. In this design, oil from the connecting rod is sprayed through a jet against the underside of the piston head. To assist in cooling and also provide additional strength for the piston head, the underside of the piston head is finned. These fins provide additional cooling area as well as strength.

PISTON COATING

To reduce the possibility of piston scoring, particularly during the break-in period, pistons, both cast iron and aluminum are frequently given a special surface coat of tin plating. Some designs are provided with tin plated skirts and chrome plating for the upper ring lands.

Anodizing is used on some aluminum pistons of relatively small diameter. It is also used on compression ring grooves. Tin plating is used on the oil ring grooves and piston skirt.

Another idea used in aluminum pistons of small diameter, is to knurl the piston skirt. The fine grooves formed in the knurling process provide tiny oil reservoirs and improve lubrication.

PISTON DIMENSIONS

The diameter of the piston is determined to a considerable extent by the power to be developed. Fig. 12-19 shows an 8 1/8 in. piston used in the Fairbanks Morse 38D 1/8 diesel. In the illustration, the piston insert and connecting rod are being withdrawn from the piston.

The length of the piston skirt in relation to the piston diameter varies in different engines. Longer piston skirts provide a greater area to distribute the side thrust of the connecting rod. Quieter operation and reduced cylinder wear result. Increasing the length of the piston stroke increases the weight of the piston, and consequently the inertia forces are also stepped up, which further increases the load on the connecting rod bearings and crankshaft.

When pistons are connected directly to the upper end of the connecting rod, which is in turn connected with the crankshaft, the piston is known as trunk type piston, Fig. 12-17, this takes the thrust of the connecting rod. In another construction, the piston is connected to a piston rod which has only vertical motion. The lower end of the piston rod is connected to a crosshead which in turn is attached to the upper end of the connecting rod. The crosshead takes the side thrust of the connecting rod. Such an engine is known as a crosshead engine, Fig. 12-21. Such a design is used primarily on large slow speed engines.

PISTON RINGS

There are many different designs of piston rings, but there are two main classifications: compression rings and oil rings. As the names imply, compression rings are designed primarily to seal compression so it does not pass the pistons. Oil rings are designed to control the amount of oil on the cylinder walls so that there is enough for lubrication, but not such an excess it would reach the combustion chamber.

Both types of rings, particularly the compression rings, transmit heat from the piston to the cylinder wall or cylinder sleeve.

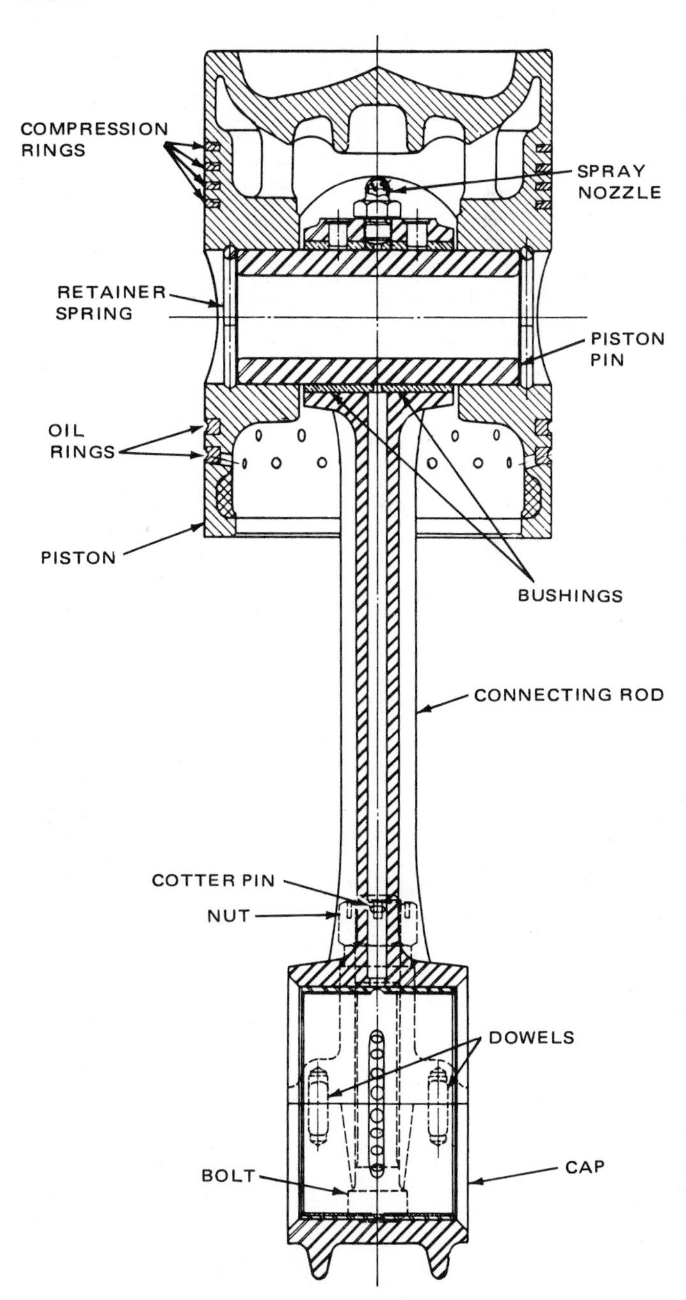

Fig. 12-24. The underside of the piston crown is ribbed and is cooled by a spray of oil from the rifle drilled connecting rod. (Murphy Diesel Co.)

Compression rings, Fig. 12-24, are placed at the top of the piston. In the illustration four compression rings are indicated. The installation is on a large industrial Murphy diesel engine. There are many pistons with only one or two compression rings. On some installations as many as six are used.

Piston rings are designed so the diameter before installation, is slightly larger than that of the cylinder bore. When installed and the engine is at operating temperature, ends of the rings will not contact each other.

Most manufacturers lay great stress on the ring exerting even pressure against the cylinder wall for the full circumference. When pressure varies, points of largest pressure will receive maximum wear. Other areas may not contact the wall and combustion leakage will occur.

Some manufacturers emphasize the importance of the type of ring gap, Fig. 12-25, while others contend it has little significance.

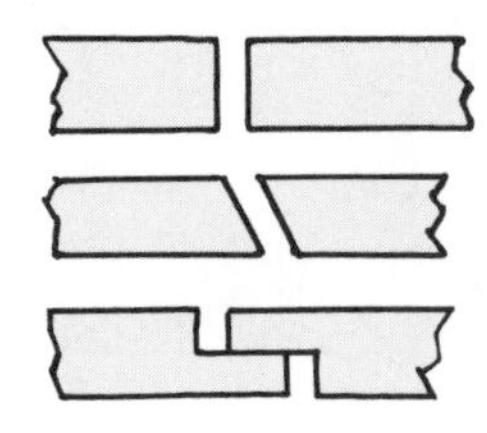

Fig. 12-25. Three common types of piston ring gaps.

Cast iron is the conventional material used for piston rings, though narrow steel bands are also used. In the case of oil control rings, the two steel runners will be separated by a crimped expander which, in addition to exerting radial pressure, also serves to keep the steel runners pressing against the upper and lower walls of the ring groove, Fig. 12-26. This

Fig. 12-26. Stainless steel oil ring with expander. (Sealed Power Corp.)

oil control ring produced by the Sealed Power Corp. is of stainless steel. A compression ring produced by that same company is shown in Fig. 12-27. Here the basic ring casting is machined with a shallow channel in the face. Molybdenum metal is sprayed into the channel. The outer surface is then ground to finish size. Faster seating and extra lubrication is claimed for this compression ring which is especially designed for use in the top groove.

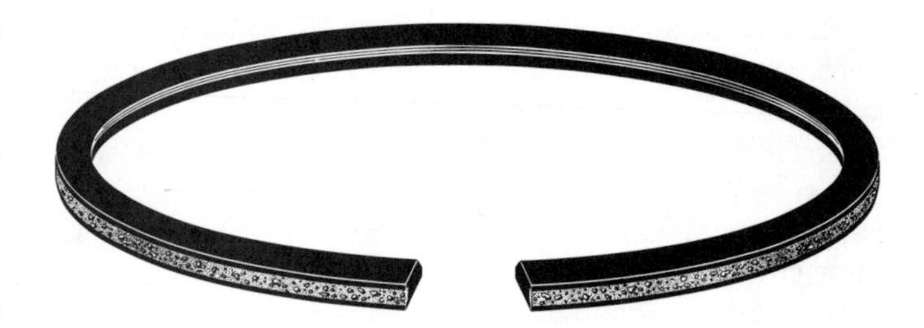

Fig. 12-27. Molybdenum filled top compression ring. (Sealed Power Corp.)

The basic compression ring is usually a rectangular section, Fig. 12-28. However, to provide quicker seating the face of the ring is often tapered approximately one degree. This provides initially a line contact with the cylinder wall, which (because of high unit pressure) quickly seats, assuring improved sealing against compression losses.

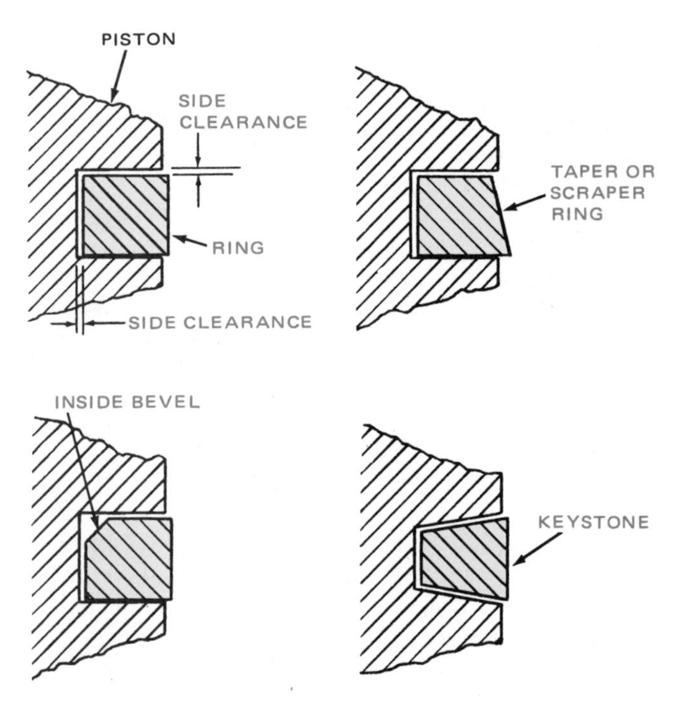

Fig. 12-28. Four common types of compression rings.

The compression ring with an inside bevel, Fig. 12-28, utilizes gas pressure to force the bottom edge of the ring to contact the cylinder wall with high pressure. This helps to seal against compression losses.

Several advantages are claimed for the keystone form of compression ring. Because of its wedge or keystone shape, the forces of compression cause a certain degree of motion between the ring and the piston. The action is claimed to free the ring of carbon accumulation, thereby helping prevent sticking.

Compression rings are often chrome plated or molybdenum coated to provide increased life and greater freedom from scoring.

As previously pointed out, oil rings are designed to prevent excess oil from reaching the combustion chamber and at the same time maintain adequate lubrication for the piston and compression rings.

Oil rings are placed immediately below the compression rings. In some cases one or more additional rings is placed below the piston pin, Fig. 12-24.

There are a large number of oil control rings; many are of two or more segments; usually they have some form of expander, Figs. 12-29 and 12-30. Some designs have slots or holes through the ring. Excess oil from the cylinder walls passes through such holes and then through drain holes in the back of the ring groove in the piston. Excess oil is returned to the crankcase. Some oil rings also act as scraper rings which scrape excess oil from the cylinder walls into a beveled groove in the piston, provided with drain holes, Fig. 12-31.

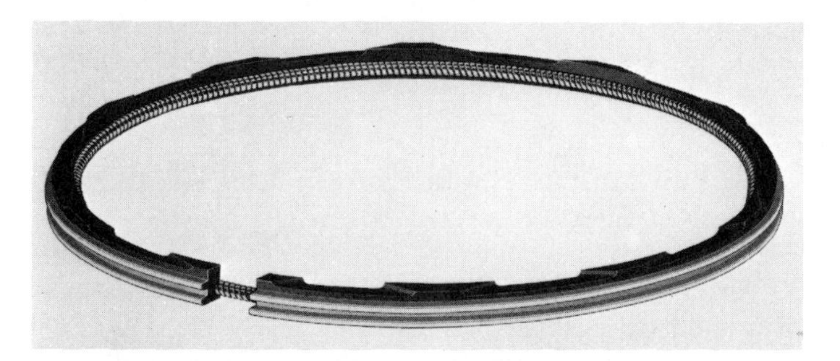

Fig. 12-29. Heavy-duty oil ring with coiled spring expander. Note oil does not drain through spring. (Sealed Power Corp.)

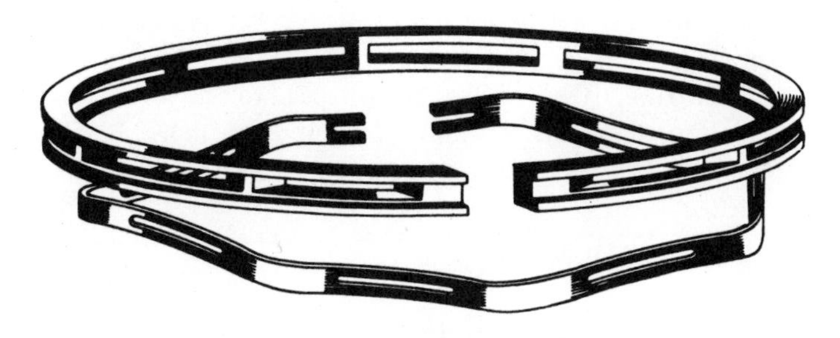

Fig. 12-30. One-piece oil ring with expander.

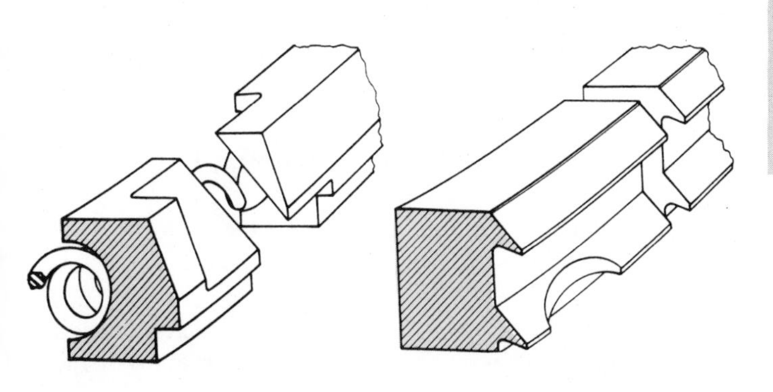

Fig. 12-31. Left. Conformable scraper seal oil ring with beveled edge and step seal joint. Right. Double hook scraper oil ring. (Koppers Co., Inc.)

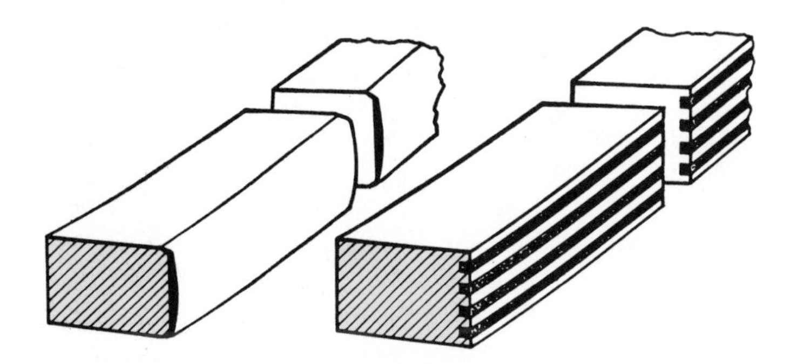

Fig. 12-32. Left. Crowned, chromium-plated compression ring. Right. K-filled compression ring. (Koppers Co., Inc.)

It is important the shape of the ring conform to the shape of the cylinder. Theoretically this is circular, but operating conditions and variation in coolant temperature frequently cause the shape of a cylinder to vary from the true circle. Piston rings are, as a result, often designed with a degree of flexibility so they will conform to such changes.

To make rings better conform to the shape of the cylinders, they are often made in two (or more) pieces with a special expander type ring being used to the pressure required to force the ring against the cylinder wall. The basic rings are usually of cast iron. Being of thin section they will readily conform to variations in the shape of the cylinder. Compression rings usually are made of cast iron in one piece. Some are molybdenum coated; others are chromium plated, Fig. 12-32. K-filled compression rings are used in certain industrial engines. With this type of compression ring, the facegrooves are filled with a porous, scuff-resistant material to aid in lubricating the upper cylinder wall. See Fig. 12-32.

VALVES AND VALVE SEATS

Because of the extreme operating temperatures, excessive gas pressures, corrosive quality of fuel, and hammering due to the high pressure of the valve springs, it is essential that diesel engine valves be constructed of special alloys. The effects of the temperature variation are particularly severe.

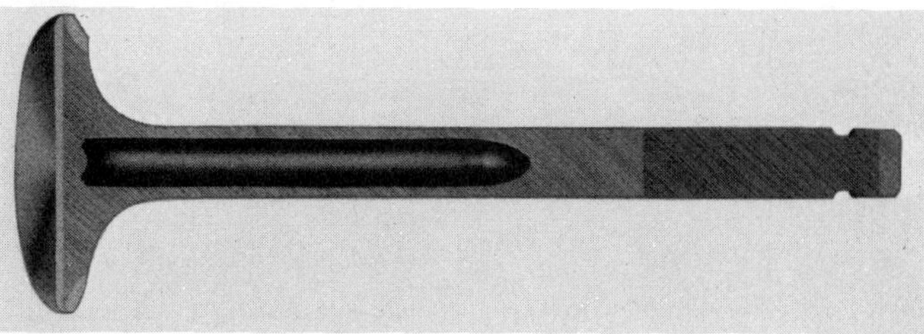

Fig. 12-33. Hollow stemmed valve, filled with sodium which assists in dissipating the heat. (Sealed Power Corp.)

In regard to gas pressure, this can easily exceed 2800 lbs. on a 2 in. diameter valve head. On a 2 1/2 in. valve, this would reach 4350 lbs. Naturally, valves have to be designed to withstand such pressures.

To withstand high temperatures, many diesel exhaust valves

are made of special alloy steel having a high resistance to heat. The edge (margin) of valves for diesel engines also is much thicker than is customary for valves in gasoline engines.

Some diesel engine valves have hollow stems partly filled with mineral salts, which help dissipate the heat, Fig. 12-33. Engineers' tests show that over half the heat leaves the valve through the valve face. Fig. 12-34 shows heat distribution.

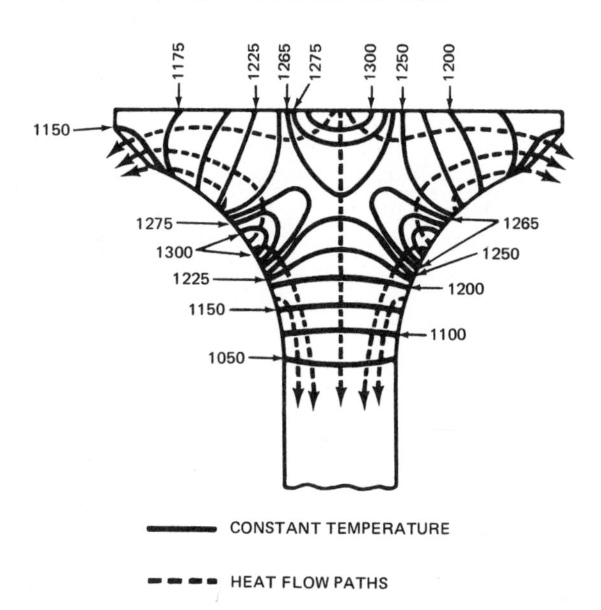

Fig. 12-34. Heat flow in exhaust valve. Note temperatures range from 1050 to 1275 deg. F. (TRW Thompson Products, TRW, Inc.)

Since a large portion of the heat from the valves is dissipated to the valve head and seat, it is essential that particular attention be paid to the width of the seat. While a wide seat has the advantage of being able to conduct a greater amount of heat, it has the disadvantage of holding greater amounts of carbon which will cause leakage. On the other hand, narrow valve seats will more quickly provide a pressure resistant seal. Compromise is, therefore, necessary. A seat width of approximately 1/8 in. is customary for many high-speed engines. General Motors 71 Series engine specifications call for a seat width of 1/16 in. to 3/32 in. for both two valve cylinder heads and the four valve cylinder heads.

Valves are often of two-piece construction, the valve head and neck being one portion, and the stem the other. The best material for each portion can be selected. Stems are often made of alloy steel which operates in a relatively cool area and provides good bearing qualities for reciprocating motion in the valve guide. The valve head, as previously indicated, must withstand the high temperatures of exhaust gases, high pressures, pounding and the corrosive action of burned fuel, and is made of special alloy steel to withstand such conditions.

Valve seat inserts are provided in many engines so seat life is extended. They also have the advantage of being replaceable. Several different kinds of material are used for valve seat inserts, Fig. 12-35, including special alloys of cast iron for

Fig. 12-35. Alloyed steel valve seats are used to withstand hammering of hot valves.

intake valve seats; for exhaust valve seats stellite and hardened chrome vanadium steel are used. Such inserts are installed with an interference fit.

The angle of the valve face and valve seat varies with different engine designs. Seat angles are usually either 30 or 45 deg. While many valves and seats have the same angle (i.e. either 30 or 45 deg.), some valves have a face angle one-half degree less than the angle of the seat. This is done to provide quicker seating of the valve when the engine is placed in service. In the design of automotive diesel engines some manufacturers favor 30 deg. valve seats while others use 45 deg. The GMC Toro engine has 45 deg. valves, while Diamond Reo, Cummins, Ford and Detroit Diesel have 30 deg. valves.

VALVE GUIDES

Replaceable valve guides, Fig. 12-17, are provided for most engines. These not only provide a guide and bearing for the valve stems, but also aid in conducting heat from the stem to the water jacket which surrounds the guide.

A long valve guide is desirable from the standpoint of providing adequate bearing area for the valve stem and to conduct heat away from the valve, but a long guide is apt to cause valve sticking. This is particularly true in the case of exhaust valve guides when they are extended into the port toward the valve head. For that reason many exhaust valve guides are cut off flush with the head boss in the port. Counterboring of the guide is also used as an aid against sticking.

ROTATORS AND SPRING RETAINERS

In the valve mechanism, Fig. 12-36, the valve does not rotate; instead, the valve face strikes the valve seat in the same area.

Fig. 12-36 shows the split cone method of retaining the valve spring. Another method is to use a horseshoe shaped retainer.

Under overload operating conditions, valve head temperatures will often reach the 1600 deg. F. range. As temperatures rise, valve deterioration increases. Such deterioration results from combustion products forming on the valve face in nonuniform patterns. This permits blowby gases to pass through and raise the temperature of the valve head. Early valve failure results. Furthermore, the temperature around the edge of the exhaust valve varies.

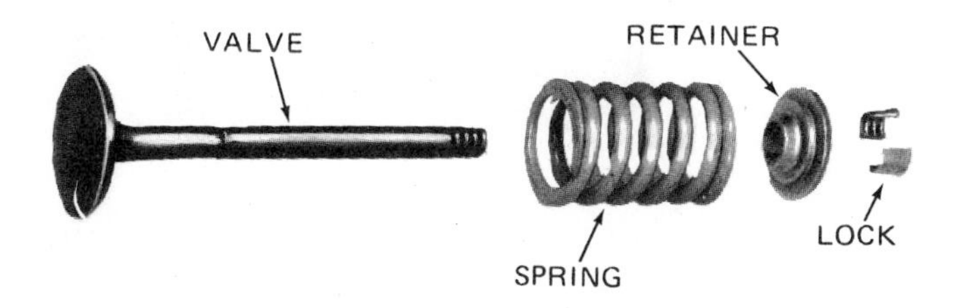

Fig. 12-36. Typical valve and spring assembly.

These two conditions contribute to valve burning. This is further augmented by the fact that valve guide wear is normally uneven. Rotating the valve will wipe off combustion accumulations, produce more uniform temperatures around the circumference of the valve head and result in even distribution of wear in the valve stem and valve stem guide.

There are several different designs of valve rotators. The type used on some General Motors 71 engines is shown in Fig. 12-37. The rotator is used in conjunction with a special valve spring cap, Fig. 12-37. The valve rotator includes six spring-loaded steel balls in individual grooves. The grooves are tapered so the balls aid rotation in one direction only, on each downward stroke of the exhaust valve. When the valve closes, rotation in the opposite direction is prevented since the narrow ends of the grooves restrict movement of the steel balls. The deflector indicated in Fig. 12-37 is designed to deflect hot gases away from the valve stem and guide, reducing the formation of carbon.

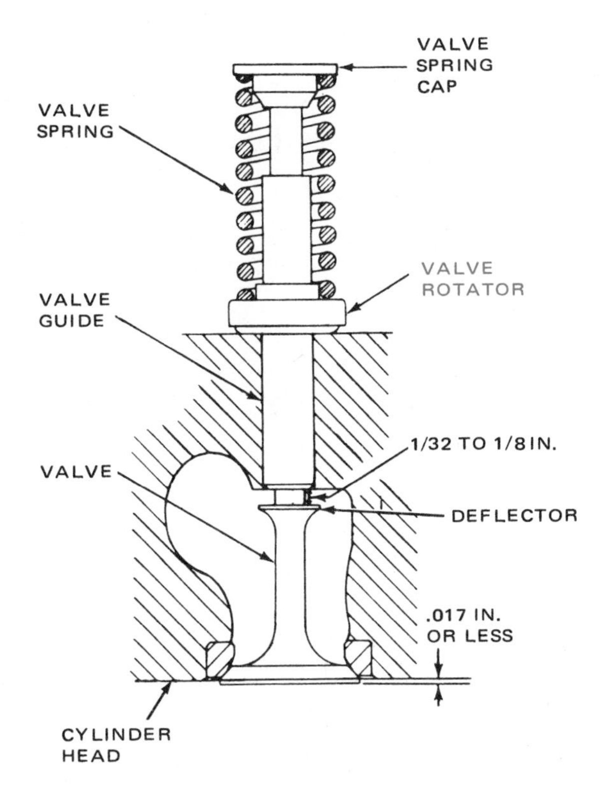

Fig. 12-37. Valve rotator as installed on some engines. (Detroit Diesel Engine Div., General Motors Corp.)

The type used on some Ford industrial engines is shown in Fig. 12-38. This is known as a nonpositive release type rotator. During each valve cycle the rocker arm forces the collets or locks away from the shoulder on the valve stem, momentarily releasing the load from the valve, permitting rotation.

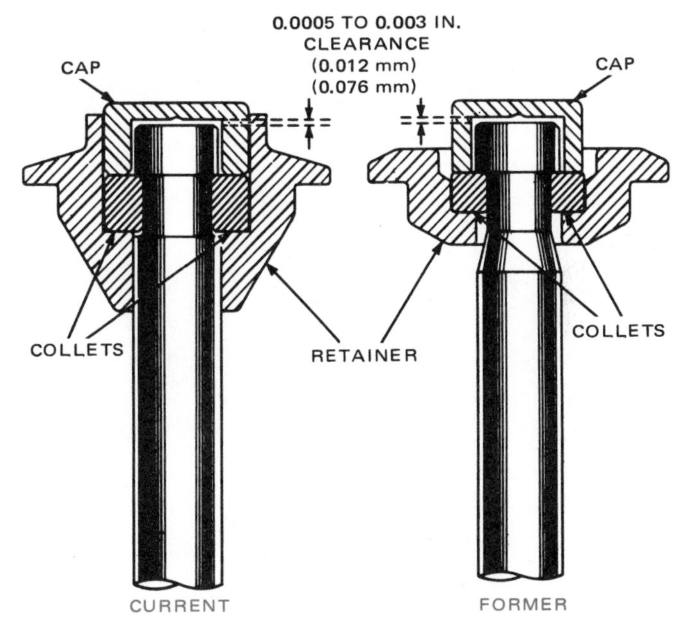

Fig. 12-38. Exhaust valve and rotator parts used on some Ford Industrial engines. (Ford Motor Co.)

A Sealed Power valve and rotator is shown in Fig. 12-39. The operation of this rotator is similar to that described for the General Motors design.

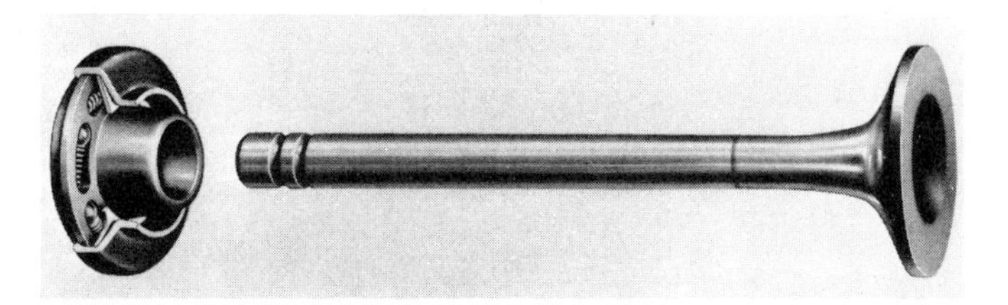

Fig. 12-39. Valve with rotator. Note steel balls on ramp with spring. (Sealed Power Corp.)

VALVE CAGES

To facilitate servicing, valves in large engines are often placed in removable cages, Fig. 12-40. Some cages are provided with water jackets for improved cooling of the valve area.

Considerable time is saved when servicing such valves as the

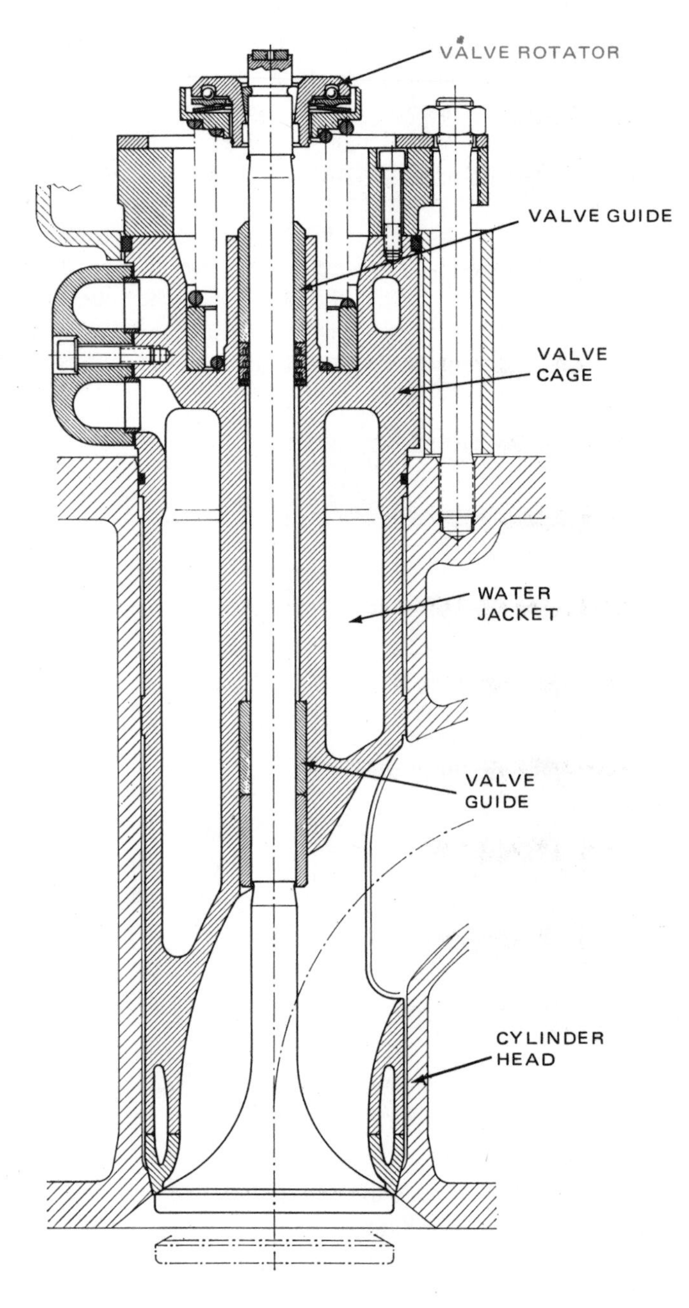

Fig. 12-40. Valve with rotator mounted in valve cage. (MAN Maschinenfabrik AG)

individual cages can be quickly removed and standby units installed.

CAM FOLLOWERS

Cam followers, or valve lifters as they are also called, are the part of the valve operating mechanism that changes the rotary motion of the camshaft to reciprocating motion to open the valve. The cam followers ride the cam and are raised by the section of the cam as the camshaft rotates.

Three types of cam followers or valve lifters are shown in Fig. 12-41. The mushroom type follower on the left is the simplest type of lifter. The roller lifter shown at the center of the illustration has the advantage of reduced friction. A rocker mounted roller lifter is shown on the right of Fig. 12-41. This

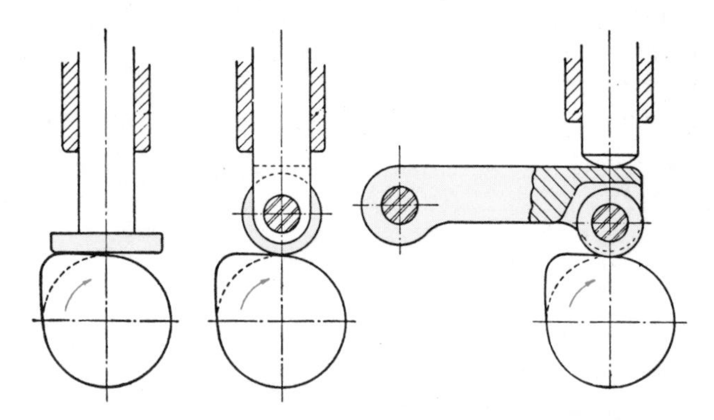

Fig. 12-41. Different types of cam followers. Left. Mushroom type. Center. Roller type. Right. Hinged roller type.

design has the advantage of not producing side thrust. Side thrust can become excessive with some types of high lift cams.

Followers are usually made of carburized low carbon steel. In some cases the followers are given further surface treatment, such as chrome plate, to provide a more durable working surface.

Hydraulic valve lifters are often installed instead of mechanical type lifters, to provide automatic adjustment of the clearance which must be provided to compensate for expansion and contraction of the parts due to changes in temperature. A typical hydraulic lifter is shown in Fig. 12-42.

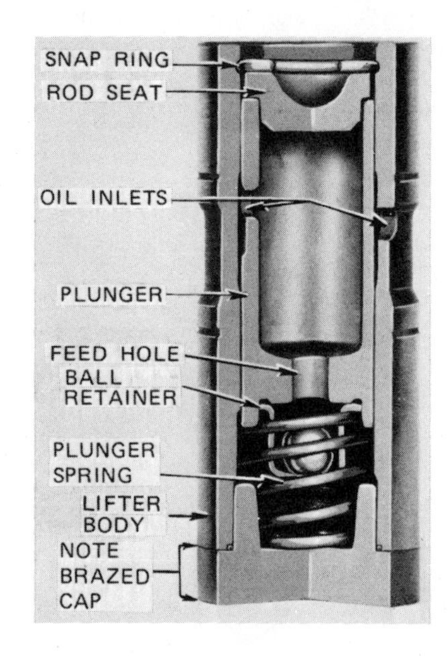

Fig. 12-42. Details of hydraulic type valve lifter.

When the hydraulic lifter is riding on the low point of the cam, the plunger spring keeps the plunger and push rod seat in contact with the push rod. When the lifter body is raised by the cam lobe, the ball check valve closes off the transfer of oil from the reservoir below the plunger. The plunger and lifter body then rise as a unit, pushing up the push rod and opening the engine valve. As the lifter body descends, the plunger follows until the valve closes. The lifter body continues to follow the cam to its lowest point, but the plunger spring keeps the plunger in contact with the push rod. The ball check valve will then move off its seat and the lifter reservoir will remain full.

ROCKER ARMS

When the camshaft is in the crankcase and the valves are overhead, a push rod is used to transmit motion of the cam and lifter to a rocker arm on the cylinder head, Figs. 12-17 and 12-43. The motion of the rocker arm then opens the valve.

In the case of overhead camshafts, the cams may operate directly on the followers, or in some designs the cams first operate on a rocker lever, Fig. 12-44.

Adjustment for clearance is provided on the push rod end of the rocker arm as shown in Figs. 12-43 and 12-44.

Pairs of valves may also be operated simultaneously from a single push rod by means of a bridge or crosshead, Fig. 12-45. Still another method is shown in Fig. 12-46. It will be noted that the rocker arm actuated by the push rod, in addition to depressing one valve also operates another rocker arm. This in turn opens the second valve.

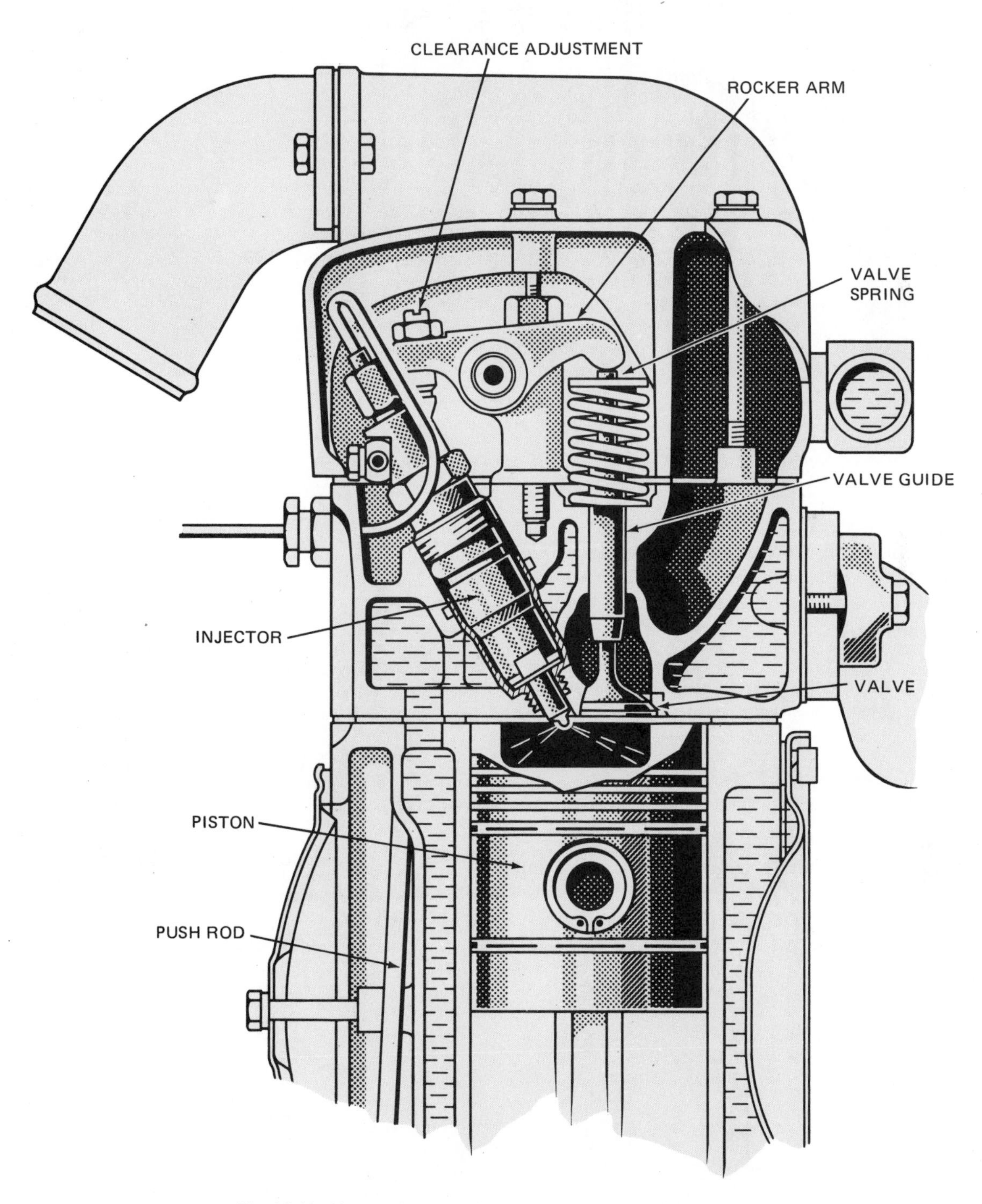

Fig. 12-43. Note push rod, rocker arm and clearance adjustment. (Daimler Benz Aktiengesellschaft)

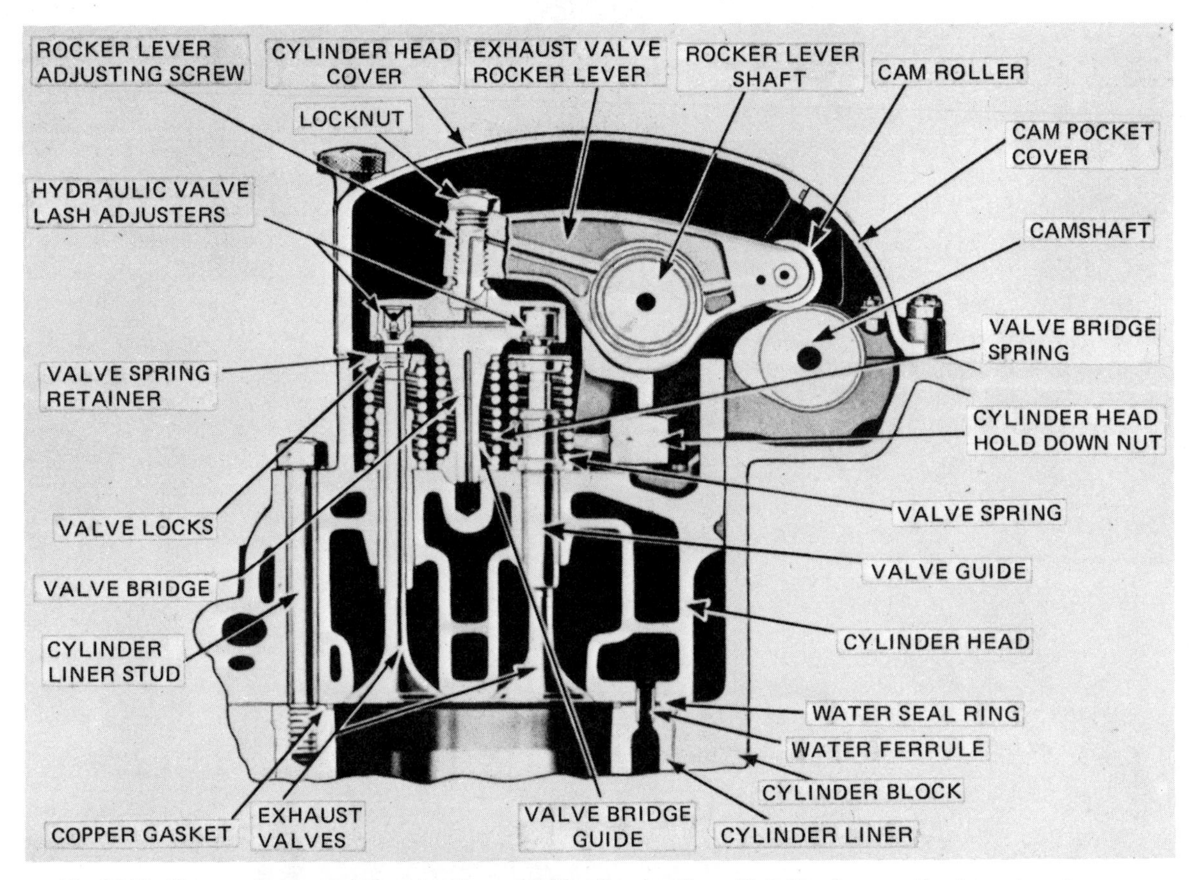

Fig. 12-44. Note overhead camshaft and rocker arm with roller, together with bridge for operating two valves at once, on this GM 278 diesel engine. (Detroit Diesel Engine Div., General Motors Corp.)

The Nordberg dual fuel engine, Fig. 12-47, has a rocker arm of unusual construction. A roller lever is operated by the engine cam. The push rod is pinned to the rocker lever. The push rod in turn operates a rocker arm.

VALVE SPRINGS

In order to close the engine valves after being opened by the action of the cam, coil springs are provided, Figs. 12-36

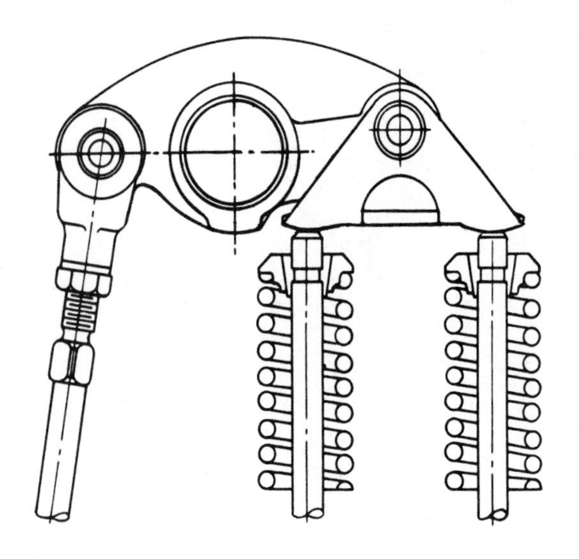

Fig. 12-45. Details of push rod, rocker arm and bridge for operation of two valves as used on some General Motors engines.

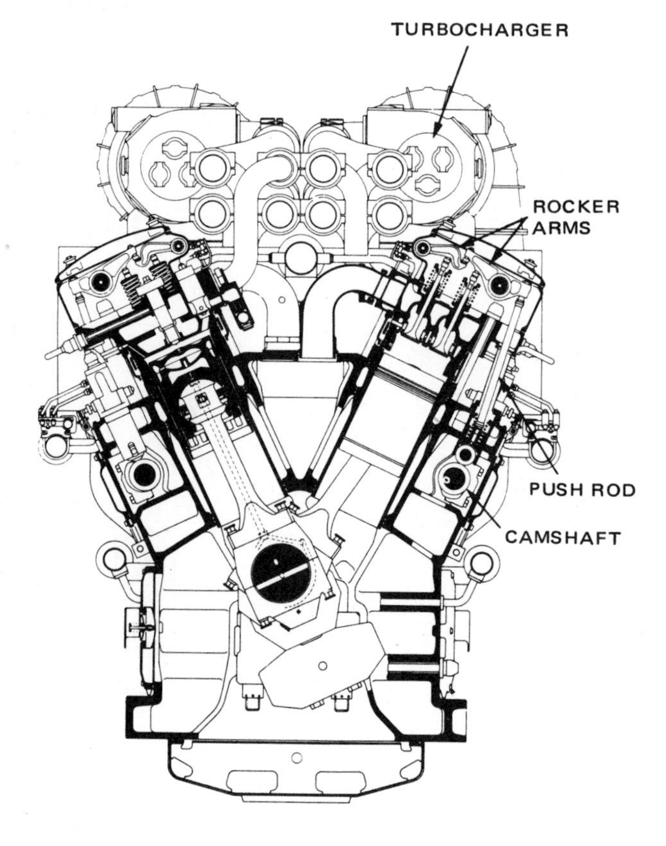

Fig. 12-46. Note how single push rod and rocker arm operates second hinged rocker arm on this Sulzer turbocharged locomotive diesel engine. (Sulzer Bros. Ltd.)

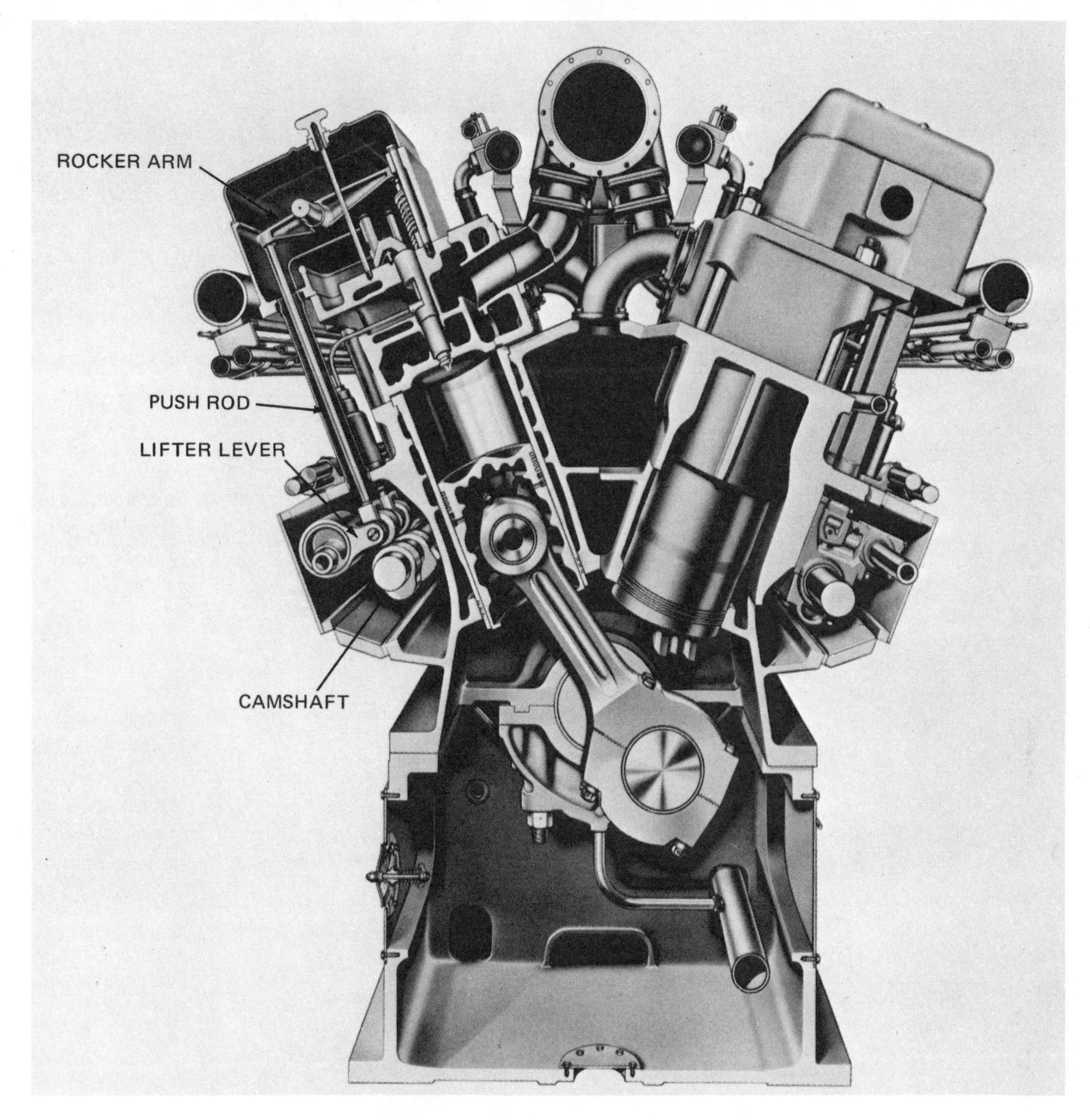

Fig. 12-47. Nordberg dual fuel engine with push rods pinned to roller type lifter levers. (Nordberg Mfg. Co.)

and 12-37. The valve springs are made of special alloy steel designed to withstand heat, corrosion and continued flexing. To reduce valve flutter that may occur (particularly at high speeds), it is customary to wind the coils closer at one end of the spring than the other. Spring dampers are used for the same purpose. Tapered springs or two springs, one within the other, are also used to dampen the flutter.

CAMSHAFTS

Camshafts usually are forgings but, in some cases, they are castings. Fig. 12-48 illustrates a pair of camshafts. The purpose of the camshaft is to open the valves. Cams usually contact valve lifters which, in turn, open the valves by means of push rods and rocker arms, Fig. 12-47. The cams are mounted in the crankcase and driven directly through gearing, Fig. 12-49.

In some engines, however, the camshafts are mounted above the cylinder head. Each cam contacts a follower which, in turn, contacts the valve stem. See VW Rabbit engine in Fig. 12-51. This design eliminates the customary push rods and rocker arms. Also note that a cogged belt (or long link chain) is used to drive the camshaft. In other installations, a series of gears may be used to drive the overhead camshaft. See gears 3, 4, 7 and 1 in Fig. 12-50.

To eliminate side thrust, some engines use a roller follower between the cam and the valve stem.

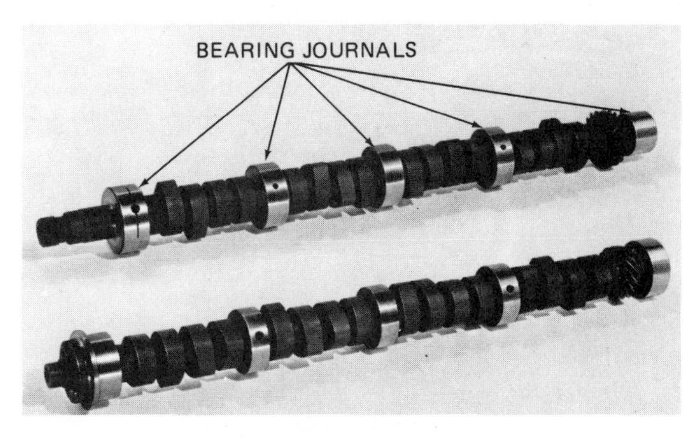

Fig. 12-48. Two camshafts as used in V-type engine. (Sealed Power Corp.)

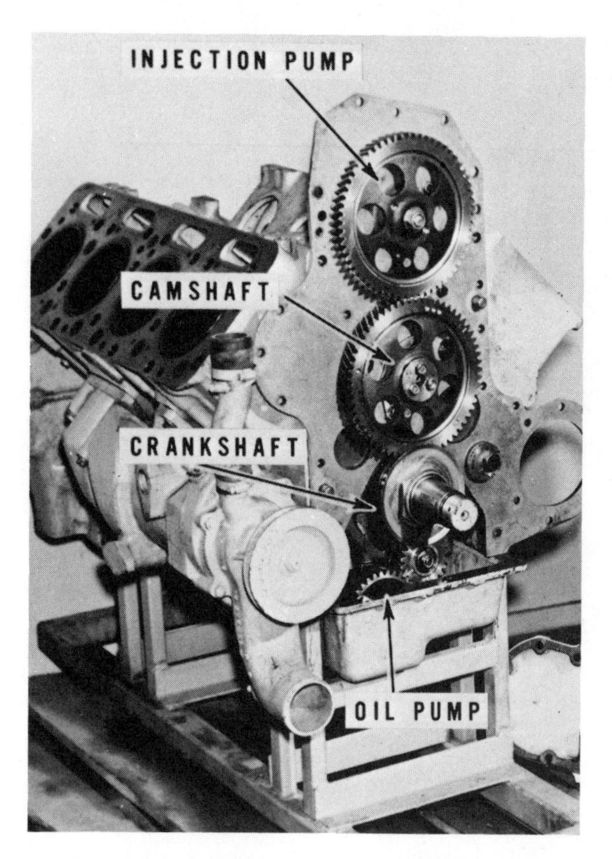

Fig. 12-49. Gear train on International V-8 engine. (International Harvester Co.)

Caterpillar uses a dual overhead camshaft on one of their engines. One camshaft operates two intake valves per cylinder, while the other operates two exhaust valves per cylinder.

Camshafts are usually mounted in sleeve type bearings.

The contour of the cams is dependent on the timing desired. In small and medium size engines the cams are forged or cast integral with the shaft. On larger engines the shaft may be built up in sections, or individual cams may be assembled on the shaft.

CAMSHAFT DRIVES

The type of drive used for the camshaft and accessories in diesel engines depends largely on the location of the camshaft and the equipment to be driven. When the camshaft is in the crankcase, a helical gear train layout generally is used. See Figs. 12-49 and 12-50.

When the camshaft is overhead, a long chain is often used to drive it. Many small engines, such as the Volkswagen Rabbit engine shown in Fig. 12-51, use a cogged belt made of a synthetic material.

CRANKSHAFTS

The crankshaft, as its name implies, is made of a series of cranks, one for each cylinder in the case of in-line engines. In

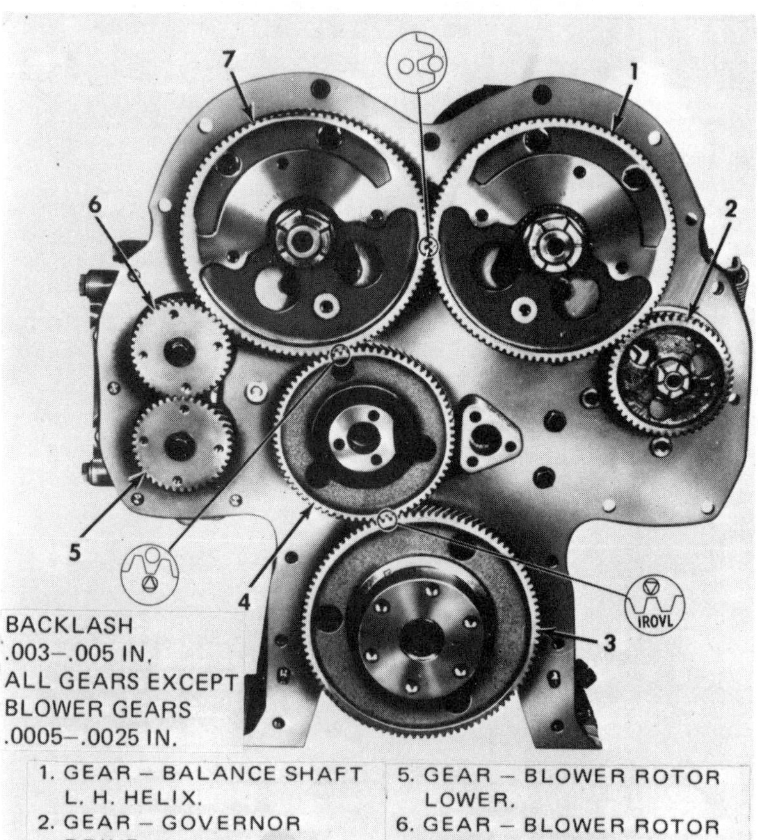

Fig. 12-50. Gear train and timing marks on 53 series three cylinder RA engine. (GMC Truck and Coach Div., General Motors Corp.)

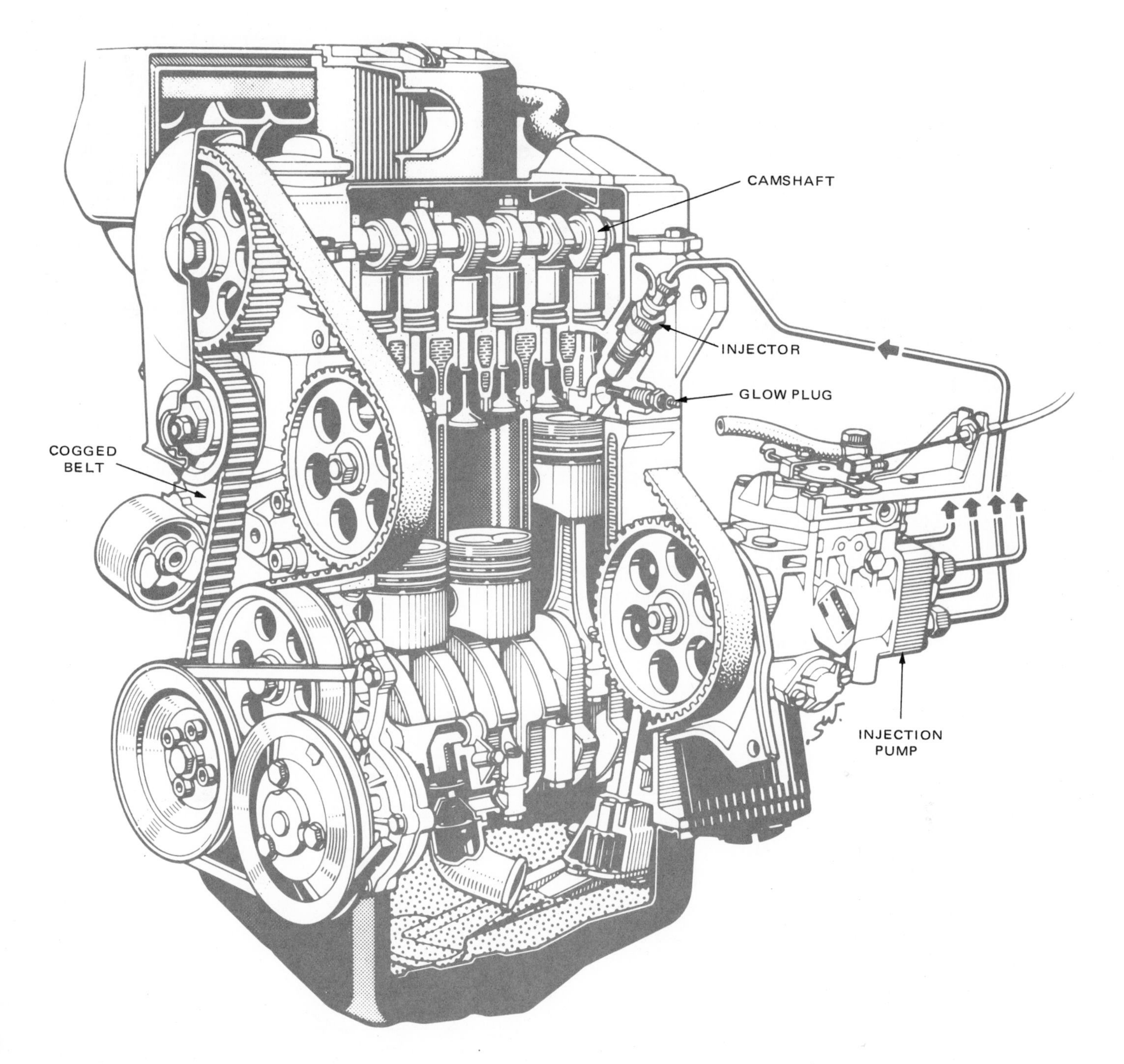

Fig. 12-51. A cogged belt is used to drive the overhead camshaft on the Volkswagen Rabbit engine. (Volkswagen of America, Inc.)

the case of V-type engines, each crank will serve a pair of cylinders. The function of the crankshaft is to convert the reciprocating motion of the piston and its connecting rod, into rotating motion, which is needed to drive generators, pumps, ships propellers and the wheels of automobiles and trucks.

Fig. 12-52 shows a seven main bearing crankshaft as installed in an Allis Chalmers six cylinder in-line engine. Fig. 12-53 shows a Fairbanks Morse crankshaft as used in a two-cycle opposed piston engine, Model 38D 818.

The throw of the crankshaft is equal to the stroke of the engine, (i.e. the distance the piston moves from the top of the stroke to the bottom of the stroke).

The crank arrangements of two-stroke cycle engines are

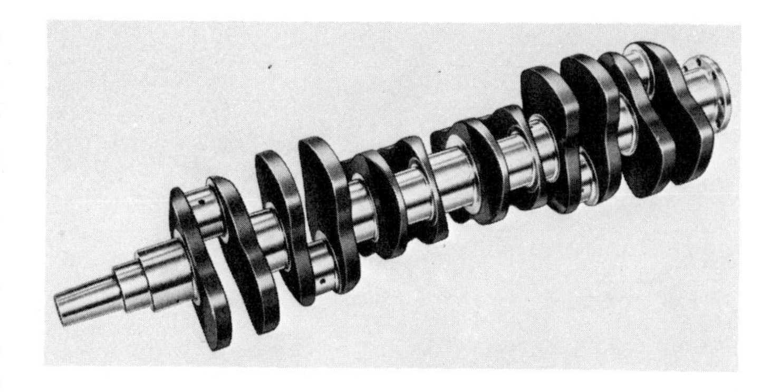

Fig. 12-52. Seven main bearing, six cylinder crankshaft as used in tractor engine. (Allis Chalmers Mfg. Co.)

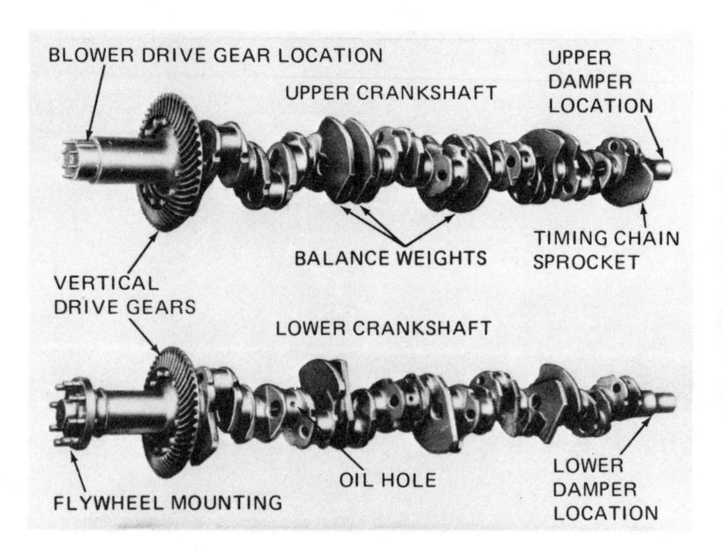

Fig. 12-53. Upper and lower crankshafts in opposed piston engine. (Fairbanks Morse Power Systems Div., Colt Industries)

different than those of a four-stroke cycle engine of the same number of cylinders. For example, the usual arrangement of the crank throws of a four-stroke cycle six cylinder in-line engine is to have the throws 120 deg. apart, using a firing order 1-5-3-6-2-4, Fig. 12-54. To use that crankshaft in a two-cycle engine, it will be necessary to have two cylinders fire at the same time. This would largely nullify the advantage of multicylinder construction.

The crankshaft of a six cylinder two-cycle in-line engine will be arranged with the crank throws 60 deg. apart, Fig.

Fig. 12-56. Balancing a crankshaft on an electronic balancer.

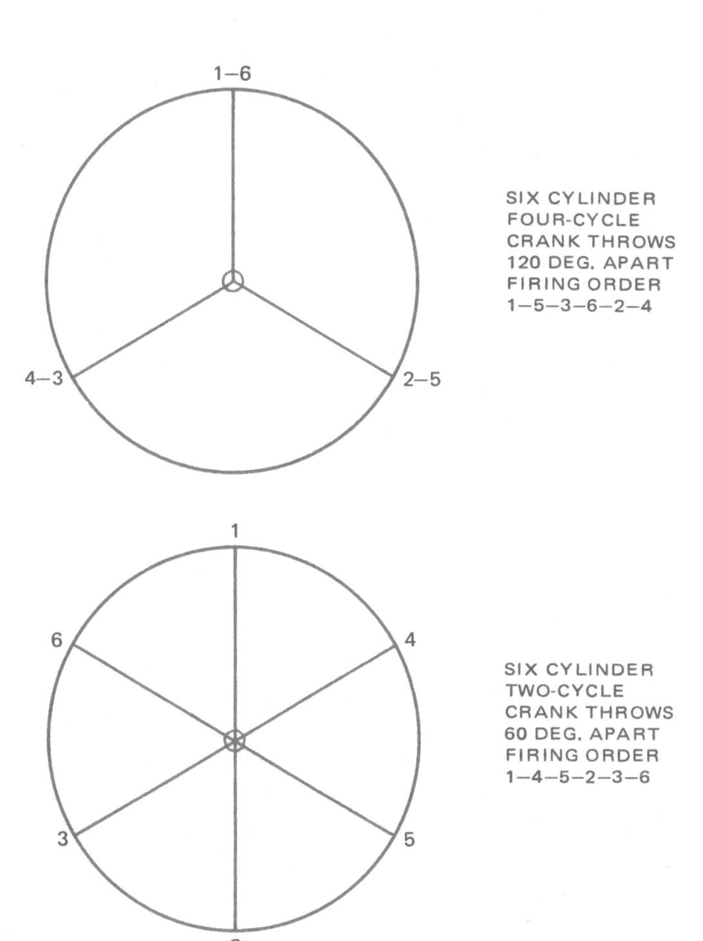

Fig. 12-54. Top. Location of throws on a four-cycle crankshaft.
Fig. 12-55. Bottom. Location of throws on a two-cycle crankshaft.

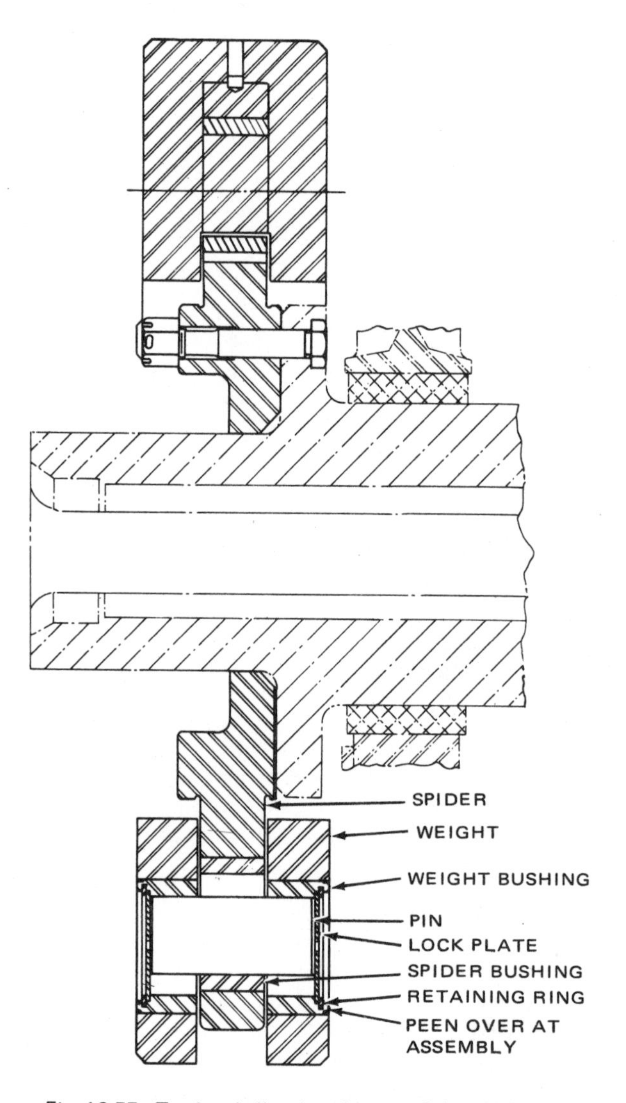

Fig. 12-57. Torsional vibration damper of the pendulum type. (Fairbanks Morse Power Systems Div., Colt Industries Inc.)

12-55. This crankshaft is used with a firing order of 1-4-5-2-3-6, or 1-6-2-4-3-5 is also possible. In the case of the five cylinder, four-cycle engine (Mercedes-Benz), the crank throws are 72 deg. apart. The firing order is 1-2-4-5-3, with a firing interval of 144 deg.

Any rotating mass has to be balanced to keep vibration to a minimum. Consider that the crankshaft on a single cylinder engine has a single crank throw. To attain balance, an equal weight must be placed opposite the crank throw.

A state of unbalance causes a centrifugal force to be formed which increases with the speed of rotation. Centrifugal force causes vibration and increases the load on the main bearings. The centrifugal force in the case of the crankshaft is counterbalanced by weights placed opposite the crankpin. Such weights create an equal and opposite force so the two forces counteract each other. When calculating the weight required to obtain balance, it is also necessary to include a portion of the reciprocating weights of the piston and connecting rods.

Balancing crankshafts is a precision operation and can be done to a fraction of a gram with precision balancing equipment, Fig. 12-56.

The forces considered so far are known as inertia forces. In the four-cycle engines of more than four cylinders, the crank throws can be arranged so that the inertia forces will balance each other.

In addition to the inertia forces there are forces set up that are due to explosive forces acting on the piston and in turn on the crankshaft. These tend to twist the crankshaft and cause what is known as torsional vibration. In general, the longer the crankshaft the greater the torsional vibration. Torsional vibration can be minimized by the addition of vibration dampers, Fig. 12-51, which are attached to the front end of the crankshaft.

The torsional damper, Fig. 12-57, is a Fairbanks Morse unit, and is installed on their four and five cylinder engines. It consists of a spider fitted with slotted weights which are free to move in and out on weight pins.

Another type of damper consists of a flywheel ring which is bonded to the hub of a thick layer of rubber.

Fig. 12-58. I-beam type connecting rod. (International Harvester Co.)

CONNECTING RODS AND PINS

The connecting rod and piston pin are connecting links between the piston and the crankshaft and change reciprocating motion of the piston to rotating motion at the crankshaft, Fig. 12-6.

Connecting rods are in most cases of I-beam section; a few are of tubular section. Connecting rods are steel forgings. A typical automotive connecting rod is shown in Fig. 12-58.

The Waukesha Diesel L-1616-D uses I-beam section rods, but the big end bearing is split at an angle of 45 deg., Fig. 12-59. This reduces the tensile stresses on the bolts. Note the serrated surfaces of the bearing cap and mating surface of the rod. This design strengthens the structure and insures against movement of the two surfaces. Another method of maintaining proper alignment between the cap and the rod is by means of dowel pins.

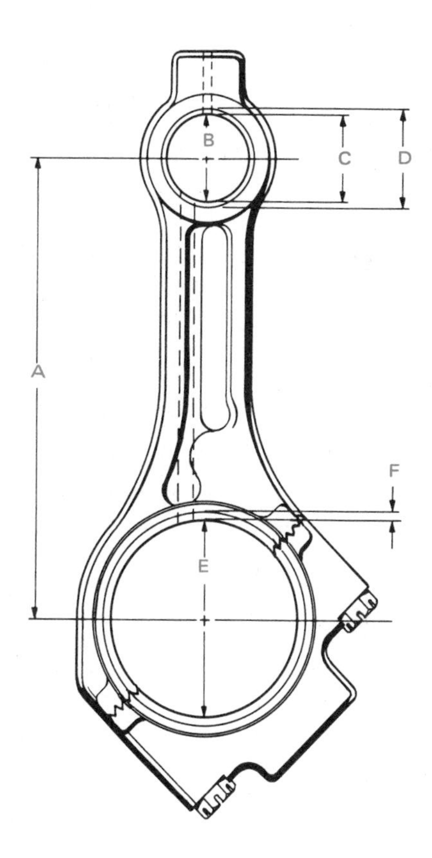

Fig. 12-59. Note serrated surface of parting halves of rod and cap to maintain alignment, also the joint between the cap and the rod is at an angle. (Waukesha Motor Co.)

Still another method of joining the cap to the rod is found on the Murphy diesel connecting rod, Fig. 12-60. This type of connecting rod is used on the Murphy engine up to approximately 400 hp. Here the rod bearing bolts are at right angles to the center line of the rod, so all tensile loading on the bolts is removed. They serve therefore only as a clamping device. Fig. 12-60 clearly shows the oil passage through the rod which carries oil to the piston pin and to the spray nozzle which sprays the underside of the piston with oil, keeping its temperature down.

Because of their direct relationship to balance of crankshaft and reciprocating parts, the manufacturing tolerance of connecting rods is of interest. In the Waukesha rod, shown in Fig. 12-59, the following tolerances are observed:

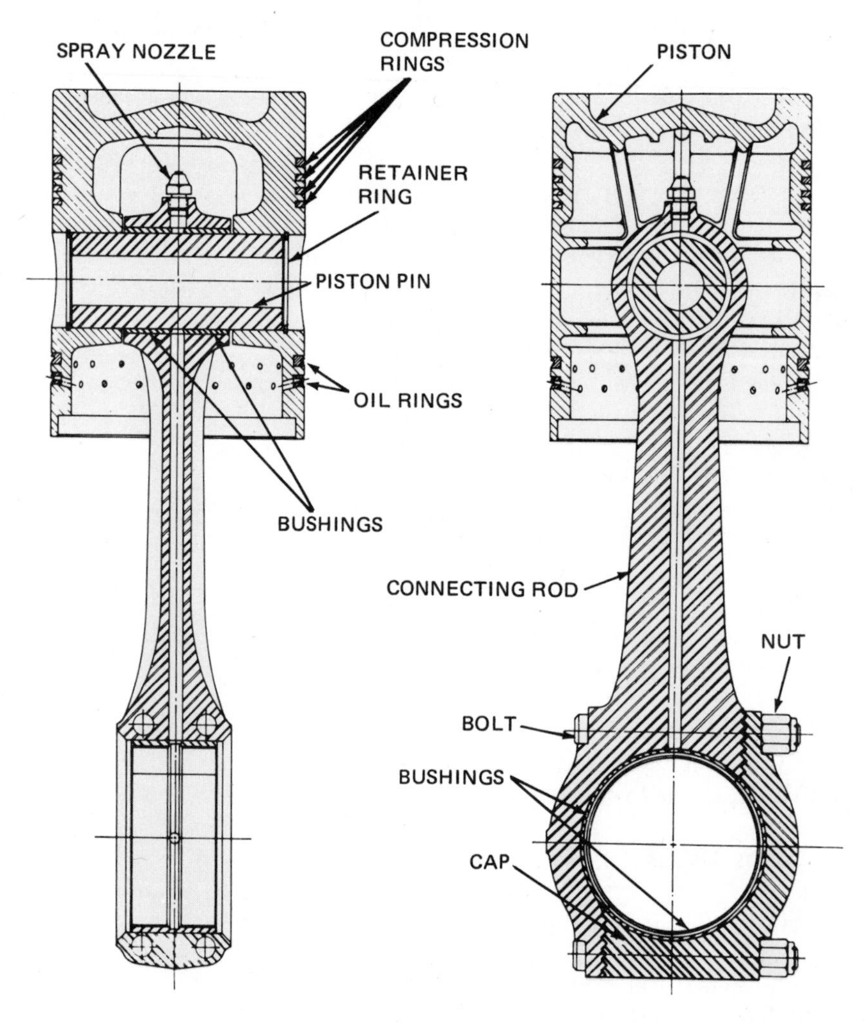

Fig. 12-60. Connecting rod bolts are at right angles to center line of connecting rod to reduce interference with crankcase walls. Note also staggered surfaces of cap and the serrated surfaces. (Murphy Diesel Co.)

A. Rod Length Center to Center	10.501 –10.502
B. Rod Small End Final Size	2.495 – 2.2500
C. Bushing Bore Diameter	2.0620– 2.0625
D. Bushing Press in Rod Diamond Bored	.002 – .004
E. Rod Large End Finish Size	4.5630– 4.5635
F. Bearing Running Clearance	.005 – .0020
G. Weight Variation	1/2 oz.

In some V-type engines connecting rods from opposite banks of cylinders are mounted side by side on the same crankshaft journal. Another construction is known as the fork and blade type, Fig. 12-61. This type is used in some Electromotive two-cycle engines. One rod is known as the forked rod which straddles the blade rod. There are other variations of fork and rod construction. In one of these, the blade rod has no bearing cap, but rides on the outside of the bearing shell and is held in place by flanges on the forked rod bearing cap.

Articulated construction of connecting rods is shown in Fig. 12-61. The main rod is of conventional construction except for a boss or projection with an eye to which the other rod is connected.

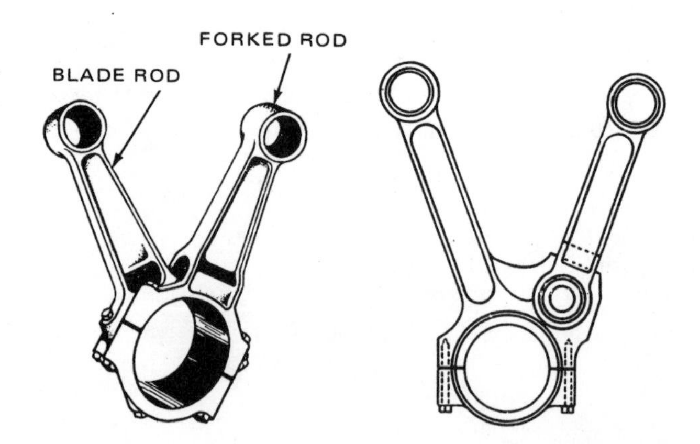

Fig. 12-61. Types of connecting rods used in V-type engines. Left. Fork and blade type. Right. Articulated rods. The usual construction is to place the rods side by side on the crankpin.

PISTON PINS

Piston pins, also called wrist pins, form the connection between the upper end of the connecting rod and the piston. The design is such that the piston can swing back and forth (oscillate) in a small arc about the pin and the rod.

Piston pins are usually of tubular construction to keep the weight to a minimum. The diameter is determined by the load it must carry. The load includes combustion pressure and inertia loads resulting from the weight of the piston and the speed. The diameter of the pin is usually larger for diesel engines in proportion to the bore, than gasoline engines, because of higher combustion pressures. Thickness must be adequate to provide sufficient strength to withstand the loads, otherwise the pin will deform. Piston pins are always case-hardened and ground to finish size which is held to within extremely close limits.

There are three main methods of mounting the pin in the piston and connecting rod. The problem is difficult as the assembly is subject to large inertia loads, rapid acceleration, and deceleration. The weight must be kept to a minimum and the pin must be kept from moving sideways, otherwise it will score the cylinder wall.

In the Detroit Diesel crosshead piston, a prefinished heavy-duty type piston pin bearing is inserted into a broached slot in the lower portion of the crown to serve as the load carrying wear surface on the piston pin, Fig. 12-22. The top of the connecting rod is precision machined to form a saddle into which the piston pin is bolted in this Detroit Diesel crosshead piston.

If the pin is fastened in the piston with a small screw, Fig. 12-62, the assembly will be lighter than when a bolt is used in the connecting rod. However, if the pin is clamped in the rod it will have double the bearing surface as there will be a bearing in each piston boss. If the pin floats in both piston and rod, it will have the maximum bearing surface and the lightest weight.

Piston pins are lubricated by pressure, receiving oil from the rifle drilled connecting rod. In some small engines, the oil is supplied by splash or throw-off from the big end of the connecting rod.

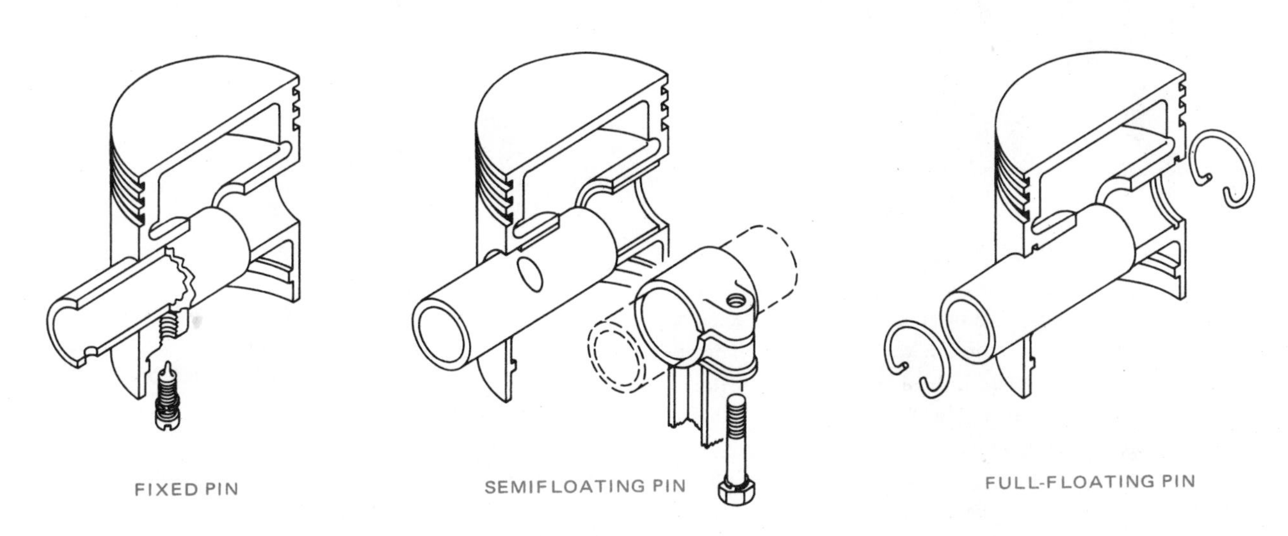

Fig. 12-62. Methods of retaining the piston pin to prevent side motion.

The three methods of mounting the pin at the piston which have been used extensively, Fig. 12-62, are:

1. Pins anchored in the piston with the upper end of the connecting rod oscillating on the pin.
2. Pins clamped in the rod, with the pin oscillating in the piston.
3. Pins which are full floating in both rod and piston with lock rings at each end of the pin to prevent sideway movement of the pin.

As the piston pins have oscillating motion in the bearing surface and not continuous rotating motion, as is the case of the crankshaft, the steel piston pin can bear directly in an aluminum piston, or in bronze bushings in either cast iron or aluminum pistons.

In some cases needle bearings have been used on piston pins. Needle bearings naturally require hardened inner and outer races. As the piston pins are always case-hardened, they serve as the inner races. For the outer race a hardened bushing is used.

PISTON RODS

The piston rod is of either I-beam or round tubular section which connects the piston to the crosshead on engines of such design, Fig. 12-21. Its motion is straight reciprocating and it contains passages for lubricating and cooling oil. The crosshead transmits the motion from the piston rod to the connecting rod, which in turn is connected to the crankshaft. The crosshead in such designs takes the side thrust usually taken by the piston in conventional construction.

The crosshead construction is on large diesel engines used in Marine and Industrial service. The piston rod and crosshead increase the weight of reciprocating parts and consequently bearing loads are also greater.

BEARINGS

Main and connecting rod bearings used in engines of recent manufacture are of the precision type consisting of a hard metal backing and a facing of bearing metal, Fig. 12-63. The backing material may be made of bronze or steel. Occasionally cast iron is used as backing on large engine bearings. Steel is the most common material. Such bearings are often referred to as steel backed bearings, or precision bearings.

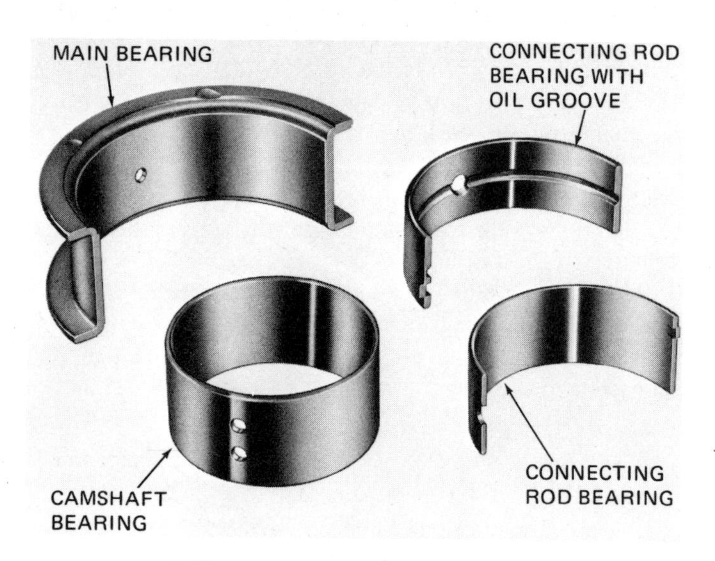

Fig. 12-63. Types of slip-in and sleeve type engine bearings.

There are many conditions that have to be satisfied by the materials used for main and connecting rod bearings. The bearing surface should be capable of absorbing tiny particles of material which result from normal wear of parts. This is known as embedability. Bearings should be able to conform to small variations in the surface of the bearing journal. Bearing must be corrosion resistant and be able to withstand normal operating temperatures. Bearing material must be strong enough to withstand loads imposed. This condition is known as fatigue strength.

One of the first materials to be used as a bearing surface for main and rod bearings is babbitt, an alloy of tin, with copper and antimony. This material is still used. It is backed by either steel or bronze. A later type of babbitt, in which lead is substituted for the tin, is known as lead-base babbitt. Steel is usually used for backing such bearings which are generally used where loads are relatively light, not exceeding 1200 psi. The fatigue strength of this material is relatively low. The percentage of the various ingredients varies with different manufacturers.

Loads up to 7000 psi can be accommodated by steel backed copper lined bearings with a coating of tin-base babbitt. Such babbitt is usually about 0.005 in. thick. If the babbitt fails the crankshaft can operate on the bronze without damage to the shaft.

Good corrosion resistance is claimed for solid aluminum bearings with an overlay of babbitt. Unit loads up to 6000 psi may be carried.

Another type of bearing is centrifugally cast solid bronze. Such bearings are frequently used for highly loaded piston pins and crosshead bushings.

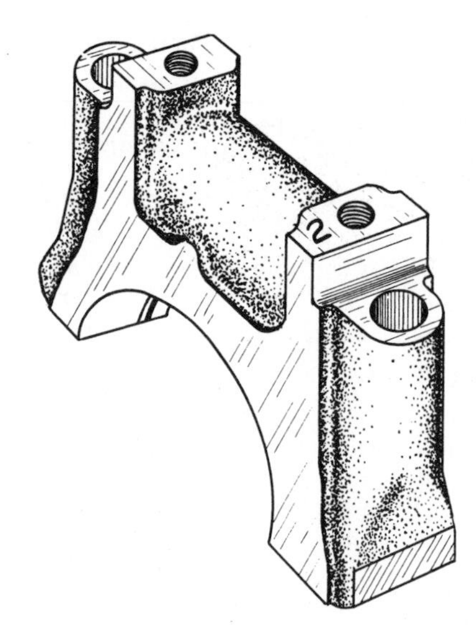

Fig. 12-64. Main bearing cap used in GMC series 71 trucks. Numeral aids when replacing the cap in its original position.

Steel back silver lined bearings with a babbitt overlay have had successful application on hardened shafts. Advantages claimed for aluminum steel backed bearings include high fatigue resistance (taking loads up to 10,000 psi), corrosion resistance, good embedability and conformability.

Main bearing shells are held in position in the cylinder block by means of main bearing caps, Fig. 12-64, and cap bolts.

As bearing caps support the crankshaft and consequently the loads of compression and combustion, as well as inertia of the connecting rods and pistons, the caps must be carefully made. Material used is mostly malleable alloy cast iron.

The main bearing bores and caps of the International DV-462/550 engine, in addition to the usual bearing cap bolts, employs transverse tie bolts, Fig. 12-65. These tie bolts are designed to load the area around the main bearing bores in compression, and eliminate the tensile stress at the oil hole and corner areas of the bearing caps.

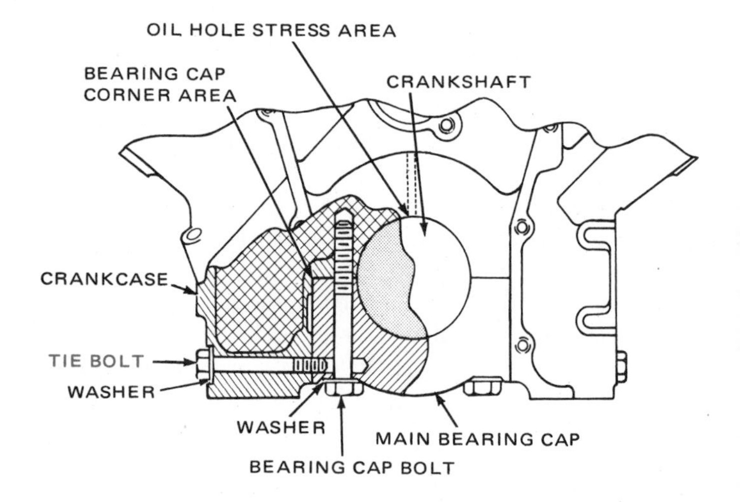

Fig. 12-65. Transverse tie bolt designed to relieve stresses. (International Harvester Co.)

REVIEW QUESTIONS – CHAPTER 12

1. In the automotive field, the cylinder block and crankcase form a single ____________.
2. Most diesel engine cylinder blocks are provided with ____________ liners or sleeves.
3. Vertical movement of the cylinder sleeve is prevented when:
   a. It is bolted in position.
   b. It is provided with clamps.
   c. It has a shoulder at the top.
4. Most cylinder sleeves are of the ____________ for easy replacement.
5. The temperature at the top of a cylinder sleeve may be as high as:
   a. 500 deg. F.
   b. 750 deg. F.
   c. 1000 deg. F.
   d. 1100 deg. F.
   e. 2500 deg. F.
6. Large diesel engine pistons are sometimes assembled with a ____________ made of a steel forging.
7. Two advantages of aluminum as piston material are:
   a. ____________________.
   b. ____________________.
8. The strength of aluminum decreases faster than ____________ as the temperature is raised.
9. Aluminum melts at a temperature of__________deg. F. while cast iron melts at __________deg. F.
10. The purpose of the alloy inserts in the upper ring grooves of pistons provides:
    a. Better heat conduction.
    b. Reduced wear and provides strength.
    c. Larger ring usage.
11. The purpose of the oil passages in the crown of a piston is to:
    a. Provide lubrication.
    b. Reduce temperature.
    c. Carry off carbon accumulations.
12. An unusual type piston is the__________ type.
13. The major advantage of a composite piston is:
    a. Improved lubrication.
    b. Reduced weight.
    c. Stronger construction.
14. The skirt of the Detroit diesel crosshead piston is free from the ____________.
15. The most usual method of cooling a piston is by:
    a. Spray.
    b. Shaker.
    c. Circulation.
16. The purpose of giving the piston a coating such as tin plating is to:
    a. Reduce possibility of scoring.
    b. Better heat conduction.
    c. Reduce weight.
17. Longer piston skirts reduce cylinder____________.
18. In a crosshead engine, the ____________takes the thrust of the connecting rod.
19. The two main classifications of piston rings are:
    a. ____________________.
    b. ____________________.
20. The conventional material used for piston rings is:
    a. Cast iron.
    b. Aluminum.
    c. Phosphur bronze.
    d. Babbitt.
21. A compression ring with inside bevel utilizes gas pressure for which of the following purposes:
    a. Increase pressure against the cylinder wall.
    b. Improve lubrication.
    c. Reduce wear.
22. Engine cylinders_________form a true circle.
23. The purpose of hollow stem valves is to:
    a. Reduce weight.
    b. Reduce temperature.
    c. Increase strength.
24. The purpose of valve seat inserts is to:
    a. Extend valve life.
    b. Improve heat dissipation.
    c. Make reconditioning easier.
25. Three advantages are listed for the use of valve rotators, they are:
    a. ____________________.
    b. ____________________.
    c. ____________________.
26. The crankshaft throws of a six cylinder two-cycle engine are ____________ deg. apart.
27. Piston pins fasten the piston to the ______________.
28. The piston rod in a crosshead engine has a ____________ reciprocating motion.
29. The most usual metal for backing a precision type engine bearing is:
    a. Silver.
    b. Steel.
    c. Bronze.
    d. Babbitt.

Chapter 13
EXHAUST SYSTEMS

The diesel exhaust system carries the products of combustion from the engine to an area where they will not be an annoyance. As diesel engine use has grown, several new features have been added. The catalytic converter, however, currently is not being used because of the excessive amount of oxygen in the exhaust.

Modern exhaust systems are designed to muffle the noise of the explosions in the engine without introducing any significant back pressure, and quench any spark that might occur. Diesel exhaust systems are designed to furnish energy to drive turbochargers, and in the case of industrial and marine installations, they may supply heat to evaporators for making fresh water, or for heating purposes. That considerable heat is developed can be appreciated by realizing that temperatures in the exhaust manifold close to the valve ranges between 700 to 850 deg. F. Temperatures vary with engine design, fuel and load on the engine, Fig. 13-1.

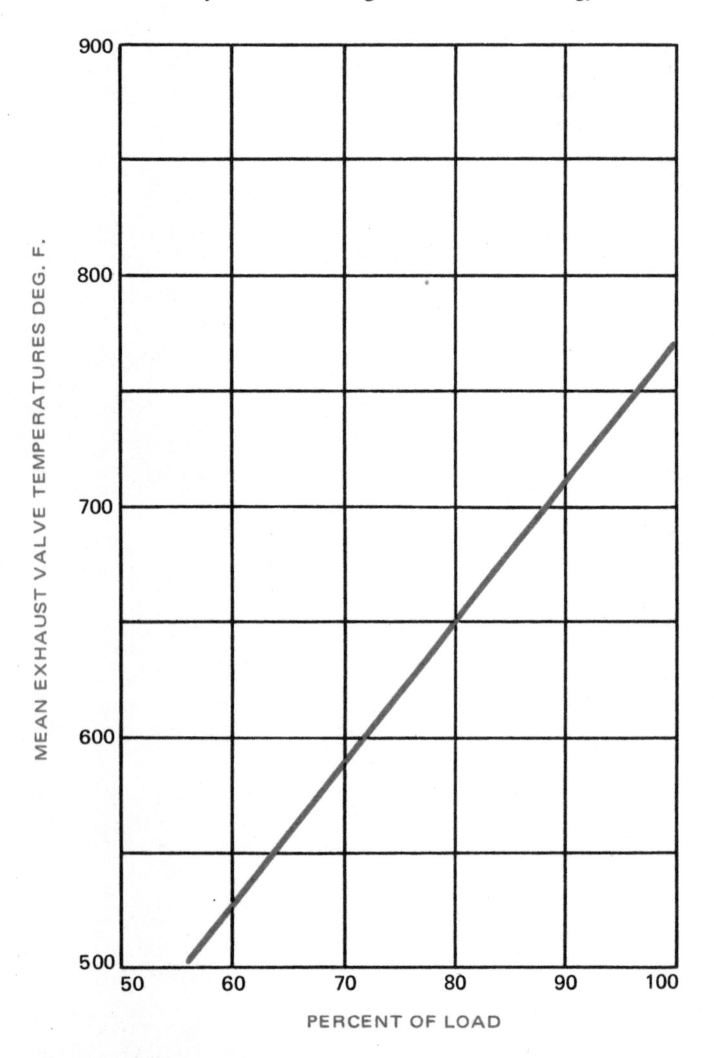

Fig. 13-1. How exhaust valve temperature varies with the load.

In the automotive field, the exhaust system is relatively simple and consists of the exhaust manifold, exhaust pipe, muffler and tail pipe, Figs. 13-2, 13-3 and 13-4. Note in Fig. 13-4, there is a flexible connection between the muffler inlet pipe and the muffler. This reduces the transmission of vibration and also takes care of expansion due to heat. Some automotive exhaust systems provide a resonator in the system, Fig. 13-4. This is designed to reduce noise by tuning the system.

Fig. 13-2. Exhaust stack on Dodge diesel truck. (Dodge Div., Chrysler Motor Corp.)

An exhaust system for a large industrial plant is shown in Fig. 13-5. This shows a Waukesha installation, and illustrates some of the important points to be considered. Back pressure is built up by bends in the system. Welded tubing is desirable because of reduced resistance to the flow of exhaust gases.

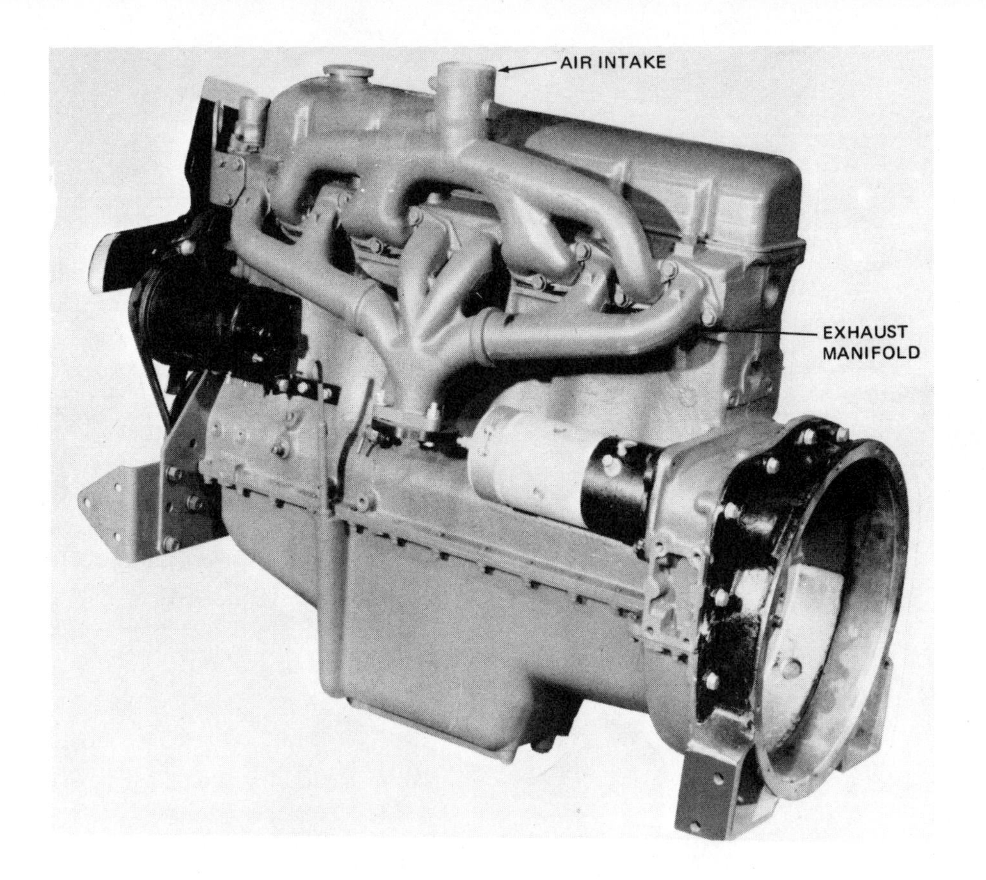

Fig. 13-3. Above. Exhaust manifold on Ford diesel engine. (Ford Motor Co.)
Fig. 13-4. Below. Vertical exhaust system on Cummins V-6, F and T series diesel truck.

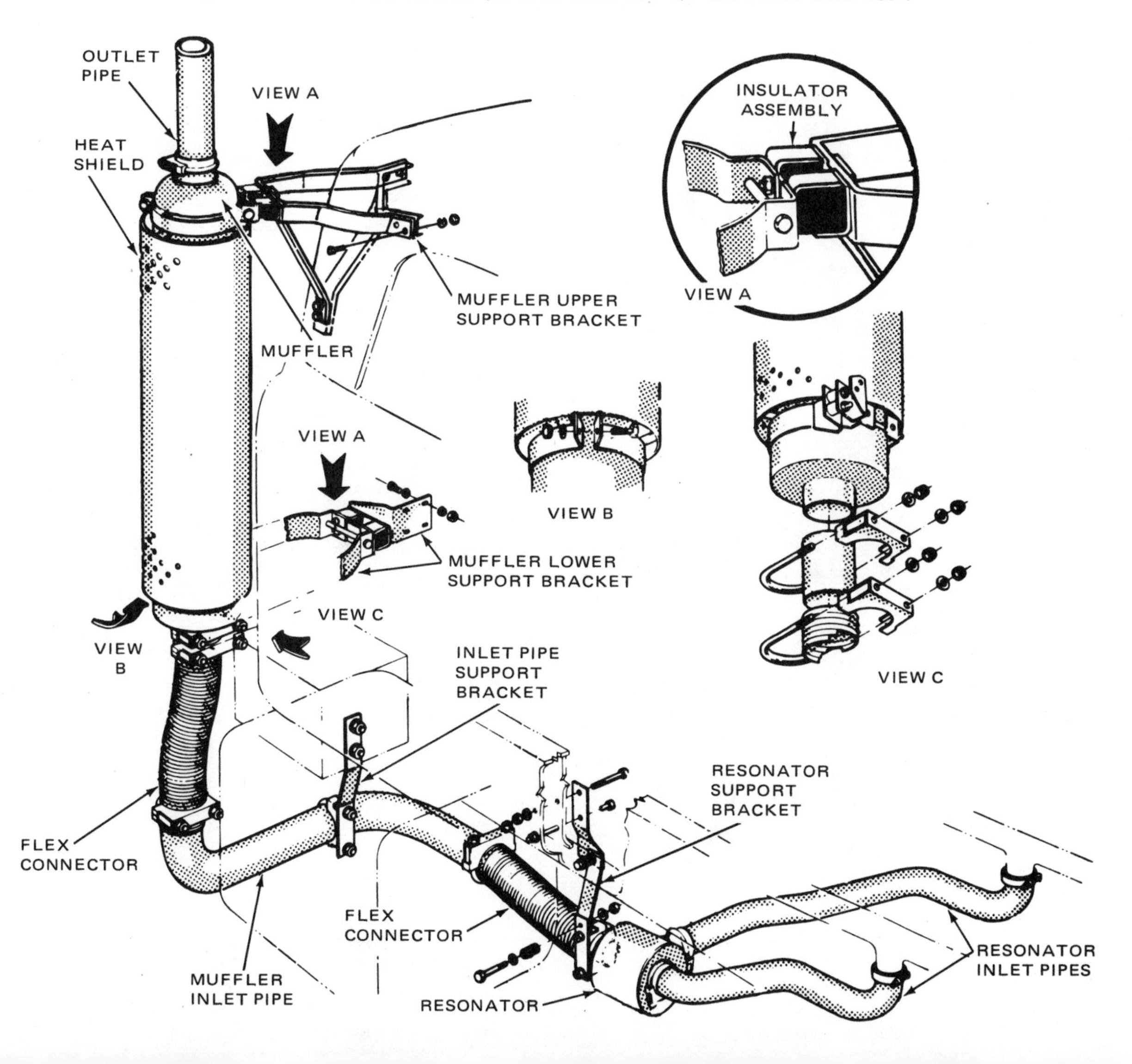

Fig. 13-5. Exhaust system on Waukesha industrial installation. (Waukesha Motor Co.)

When bends are necessary, the radius should be at least four or five times the diameter of the tubing. Exhaust pipes get hot and should be provided with slip joints or flexible metallic connections.

To dispose of heat radiated by exhaust pipes, sheet metal ducts should be provided for the exhaust pipe with an inch or two spaced between the duct and pipe. This will provide a chimney effect and help carry away a portion of the heat. Water-cooled exhaust manifolds and pipes are used particularly in the case of marine installations. Such features are important as they reduce the possibility of fire.

In V-type engines, particularly those in industrial and marine fields, separate exhaust systems are usually provided for each bank of cylinders. Two manifold branches should not be brought together in a single pipe with a T connection, as this leads to pulses of exhaust from one bank interfering with exhaust from the other. Extreme high back pressure is developed.

By placing the muffler as close to the engine as possible, interference resulting from pulsing effect is reduced.

An example of a water-cooled exhaust system is shown in Fig. 13-6. This is a Fairbanks Morse opposed piston engine installation. The coolant in the closed system is recirculated.

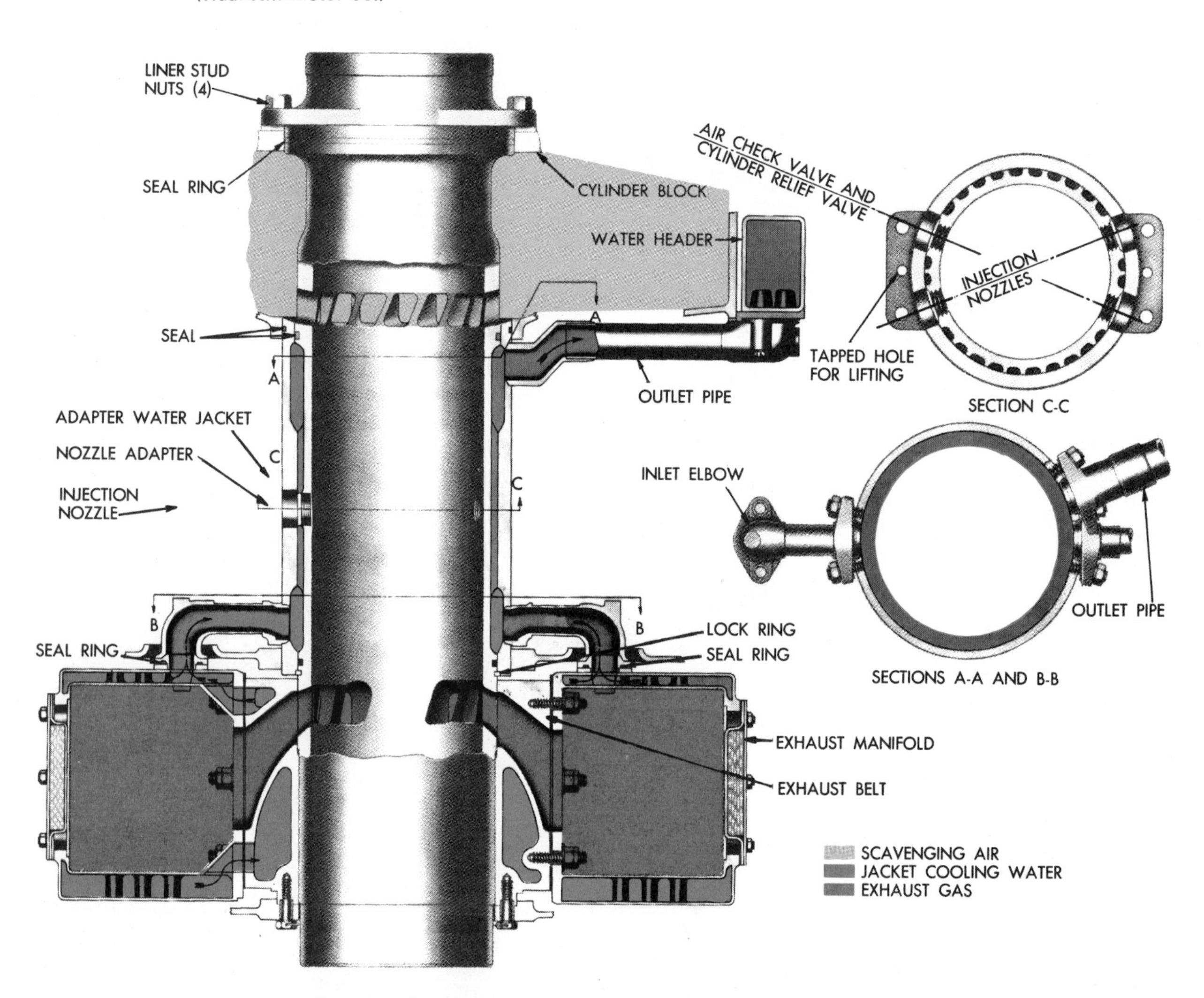

Fig. 13-6. Exhaust system on Fairbanks Morse opposed piston engine. (Fairbanks Morse Power Systems Div., Colt Industries, Inc.)

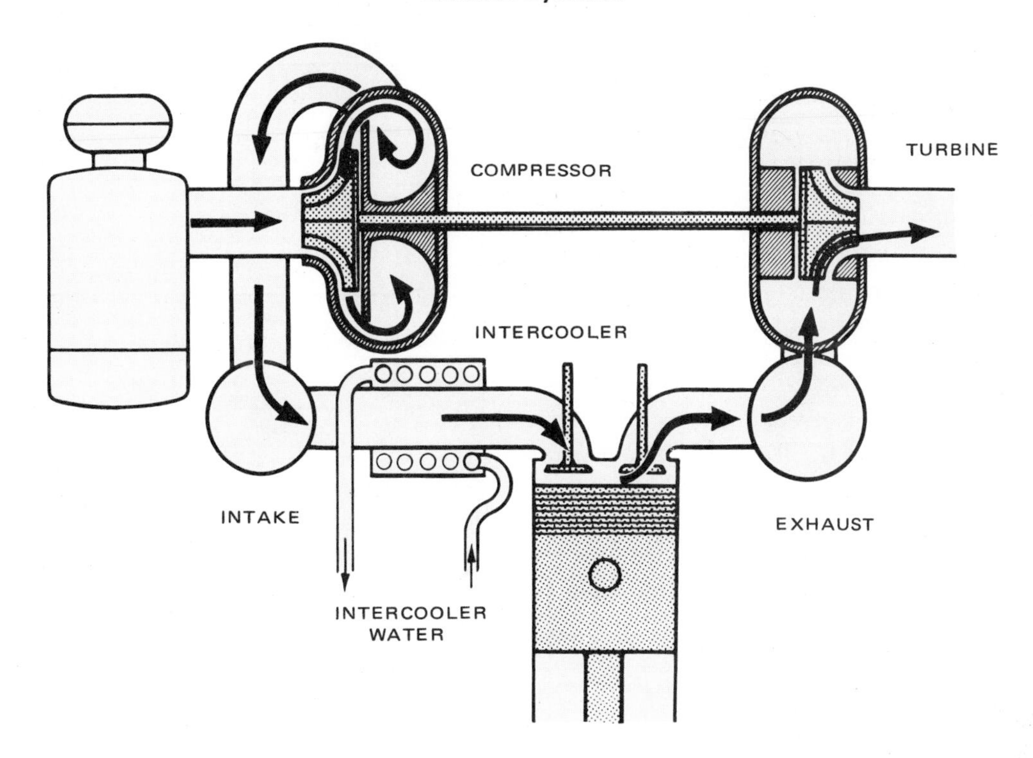

Fig. 13-7. Turbocharger with intercooler. (Waukesha Motor Co.)

A schematic drawing of turbocharged engines with the exhaust gases driving a turbine is shown in Fig. 13-7. In this Waukesha installation, it will be noted that exhaust gases pass directly from the exhaust manifold to the turbine and are then exhausted into the atmosphere.

Every effort must be made to reduce exhaust back pressure. As back pressure increases, power drops, fuel economy decreases, valve temperatures soar and cooling system temperatures rise.

Special care must be taken in selecting the exhaust pipes to be sure they are of sufficient diameter, there are no sharp bends, and they are adequately supported. Pipes of small diameter will cause back pressure as will sharp bends. Unless the exhaust system is well supported, excessive strain will be placed on the engine manifolds and turbochargers, which are not designed to carry such loads.

Moisture will condense in the exhaust system. In installations with vertical exhaust pipes it is important that a condensate trap be provided. Such traps should be drained at regular intervals to eliminate the possibility of condensate draining back into the engine. Should that occur, valves would stick. In severe cases there is a possibility of hydraulic lock and consequent breakage of pistons and cylinders.

Industrial exhaust pipes are usually made of either cast iron, wrought iron or steel. Cast iron has the advantage of greater resistance to corrosion. Because of lighter weight, wrought iron and steel are frequently used. Flexible hose has the advantage of eliminating sharp bends in the line and also reducing the transmission of heat and vibration. In addition, it corrects minor errors in misalignment of connecting pipes. Installation time is also greatly reduced by the use of flexible tubing.

Heat transfer from the exhaust system to the engine room is kept to a minimum by covering the pipe lines with insulating material such as magnesia compounds. In addition, metal shields of aluminum are also used to protect specific areas, Fig. 13-5. Special hose of an insulated type, consisting of an inner and outer hose with insulating material such as asbestos and a dead air space between is also available.

MUFFLERS

In general mufflers are expansion chambers into which the exhaust gases pass from the engine, they expand slowly and become cooler before they are discharged into the atmosphere. Muffler designs vary considerably. Mufflers, to efficiently reduce the sound of the explosions, must decrease the velocity of the exhaust gases and either absorb the sound waves or cancel them by interference with other waves. The usual design practice is to have the volume of the muffler equal to six to eight times the displacement of the engine.

One basic type of muffler is known as the straight through type and consists of through tubes to which closed cavities are connected through small holes along the tube, Fig. 13-8.

Loss in engine power resulting from back pressure is indicated in Fig. 13-9. As the speed of the vehicle increases back pressure also increases. Note that for a given car speed

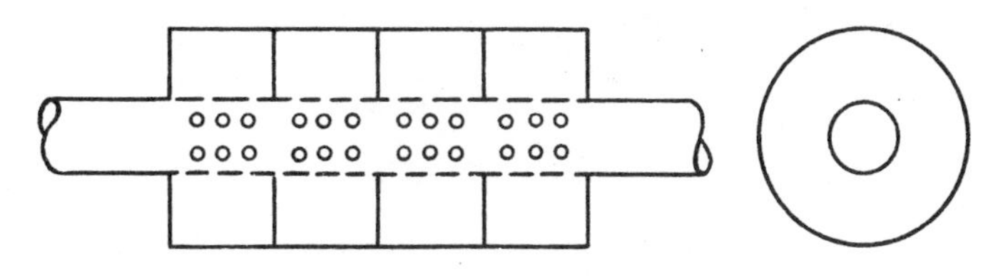

Fig. 13-8. Sectional view of straight through muffler.

the loss in power increases rapidly with an increase in back pressure. Similarly, fuel consumption increases rapidly as exhaust back pressure increases, Fig. 13-10.

A certain amount of expansion and cooling of the products of combustion takes place in the exhaust manifold and exhaust pipe. Usually these are designed to provide from two to four times the volume of a single cylinder of the engine. Additional expansion is then provided for in the muffler.

In some industrial type mufflers special provision is made to eliminate sparks. This is particularly important in areas where sparks would be a fire hazard.

Wet-type mufflers are used in constructions which are designed for marine use. Water is mixed with exhaust gases

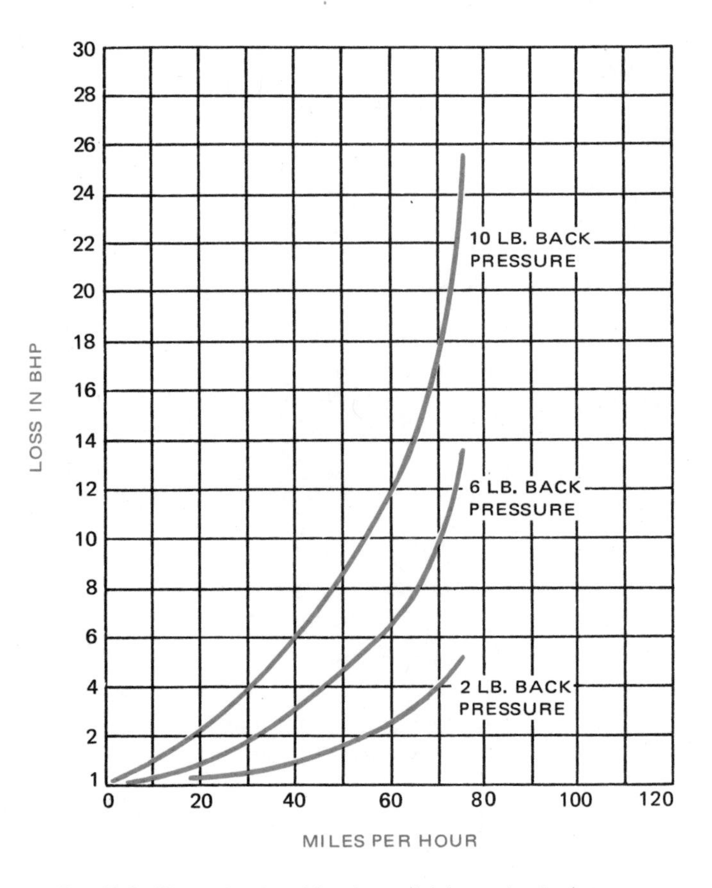

Fig. 13-9. Illustrating how bhp drops with increasing back pressure.

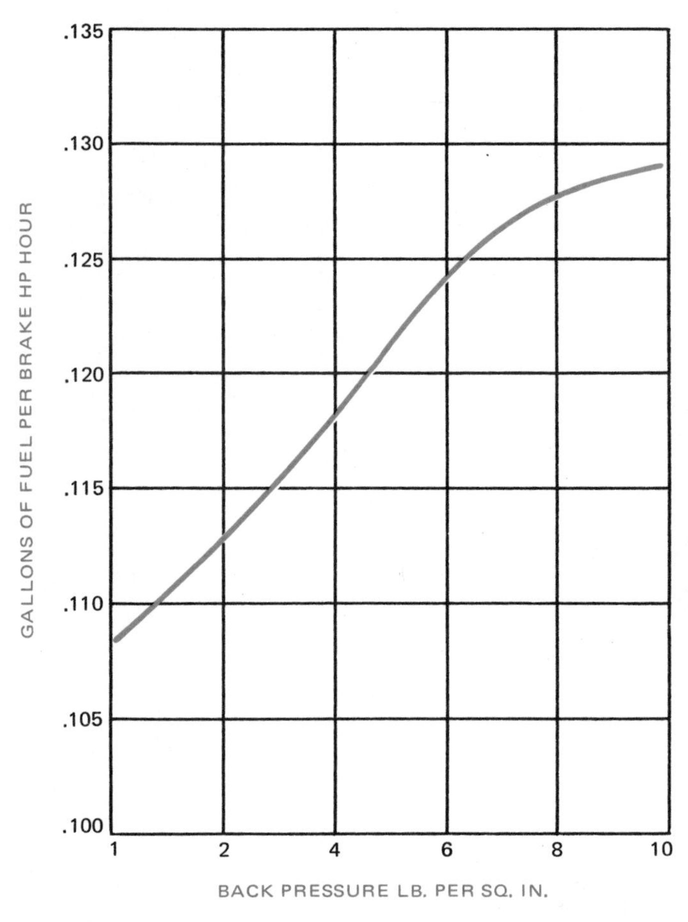

Fig. 13-10. Fuel consumption increases with back pressure increase.

which quenches sparks, cools the exhaust, contributes to silence, and eliminates the need for insulation.

Industrial installations of large diesel engines are usually provided with pyrometers to measure the temperature of the exhaust gases of the individual cylinders. When an engine is in good condition, there will be little variation of the temperature of the cylinders. Any temperature variation from the norm is a sure indication that trouble exists.

Manometers are used to check back pressure. These instruments, which measure the back pressure in inches of mercury, are connected to the exhaust manifold.

REVIEW QUESTIONS – CHAPTER 13

1. Exhaust back pressure in a diesel ____________ the power of the engine.
2. The temperature range in the exhaust manifold close to the exhaust valve is:
   a. 700 to 850 deg. F.
   b. 500 to 750 deg. F.
   c. 850 to 1250 deg. F.
3. The advantage of welded tubing in an exhaust system is:
   a. Low cost.
   b. Easy installation.
   c. Reduced resistance to flow of exhaust gases.
4. Two manifold branches should ____________ be brought together in a single pipe to form a T connection.
5. Condensate traps should always be installed in exhaust systems with ____________ pipes.
6. The purpose of covering exhaust pipes with insulating material is to:
   a. Keep the exhaust pipes warm.
   b. Reduce heat transfer to the engine room.
   c. Reduce formation of condensate.
7. Expansion of gases in the muffler should be as ____________ as possible.
8. As the speed of the vehicle increases, back pressure ____________.
9. Fuel consumption increases very rapidly as exhaust back pressure ____________.

Chapter 14
COOLING SYSTEMS

Two major methods are used to cool compression ignition engines:

1. Water cooling.
2. Air cooling.

Oil circulating through the engine also contributes to engine cooling. Both major methods of cooling are effective. Water cooling was used exclusively during the early development of the diesel engine. However, during the second world war, air cooling received considerable attention and today many diesel engines are being air cooled. Air cooling is used primarily in cooling small engines but it has been used successfully in engines up to 300 hp.

Cooling systems are needed to dissipate the waste heat of the total heat supply to the cylinder of an engine by the burning fuel. Approximately one-third of the fuel's heat energy is transformed into useful work. An equal amount is lost in the exhaust gases. The remainder must be absorbed by the cooling system, Fig. 14-1. Without the cooling system, engine parts would be damaged as the engine might reach temperatures in the range of 2000 deg. F. Since the diesel engine operates at such high temperatures, the cylinders must be cooled in order to retain proper lubrication.

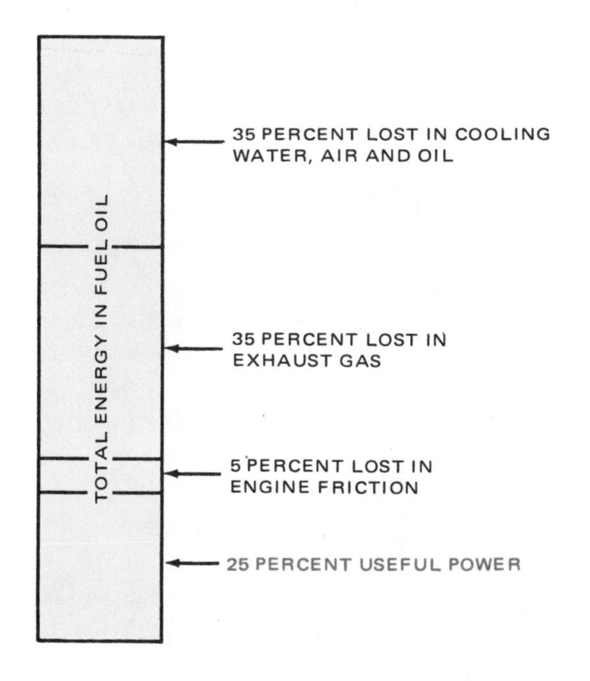

Fig. 14-1. Of all the energy in diesel fuel, only 30 percent is available for work and a large portion of that is used in overcoming friction in the engine.

The lubricating system also contributes to the cooling of the engine. In many installations, oil coolers are provided to reduce oil temperature. Such equipment serves to maintain better lubricating conditions and also keeps the engine temperature within prescribed limits.

HEAT BALANCE

As shown in Fig. 14-1, approximately 25 percent of the available heat of the fuel is turned into useful work and 35 percent is disposed of by the cooling system. Such division will vary with different engines, the design of the combustion chamber, the ignition system, degree of supercharging or the condition of the various parts. According to the Diesel Engine Manufacturers Association, the amount of heat lost to the cooling system will range from 2100 Btu per bhp hour to 2700 Btu at full-load for the turbocharged four-cycle engine. In the case of a two-cycle loop scavenged engine, the heat loss to the cooling system would range from 2100 Btu to 2850 Btu for valve-in-head engines of the two-cycle type.

It is a simple matter to calculate the amount of heat which would be lost to the cooling system after first calculating the heat input by means of the following formula:

Heat Input = BHP x F x H
Where BHP = Brake horsepower
F = Fuel consumption per BHP hour
H = Heat value of the fuel

Assume a 400 bhp engine running at full-load, with fuel consumption of 0.4 lb. per bhp hour and using fuel having a heat value of 19,600 Btu per lb., the following calculation is made.

Heat input = 400 x 0.4 x 19,600 = 3,136,000
Percent lost in cooling system = 35%
35% x 3,136,000 = 1,090,000
Expressed in BTU per BHP hour, this becomes

$$\frac{1{,}090{,}000}{400} = 2725 \text{ BTU per BHP}$$

An adequate cooling system must be designed to keep the engine at a temperature that lubrication is not destroyed. In this way, engine parts will not have been heated to such a degree that expansion has not caused clearances to be materially reduced. It is necessary that the water jacket be of sufficient size with an adequate supply of coolant. In the case of an air-cooled engine, the cooling fins should be adequate and supplied with an ample amount of air.

WATER COOLING

There are several factors which must be considered in the design of a water-cooled system. The water jackets must be of adequate size to permit the flow of sufficient water to maintain the engine at the correct operating temperature. Water jackets must be correctly located so no area of the engine will become overheated. There must be sufficient water to perform the cooling operation, and there must be a means of cooling the water after it has removed the heat from the engine.

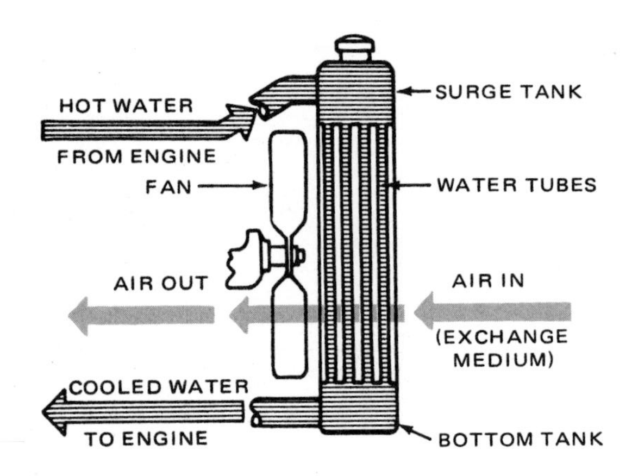

Fig. 14-2. Details of a typical radiator.

TYPES OF WATER COOLING SYSTEMS

There are two major types of water cooling systems:

1. Open system.
2. Closed system.

In the open system, water passes through the water jackets and is returned to city mains, a river, lake or ocean. It may also be returned to a cooling tower or spray pond, where, after it has been cooled, it is again pumped through the engine. In the latter case, water must be added to replace that which was lost as the result of evaporation. In both instances, calcium and other minerals will be deposited in the water jackets and it is necessary that they be removed periodically.

In the closed system of water cooling a diesel engine, the same coolant is recirculated, Fig. 14-2. In automotive and construction services, the water, after absorbing the heat of the engine, is passed through a radiator. Air passing through

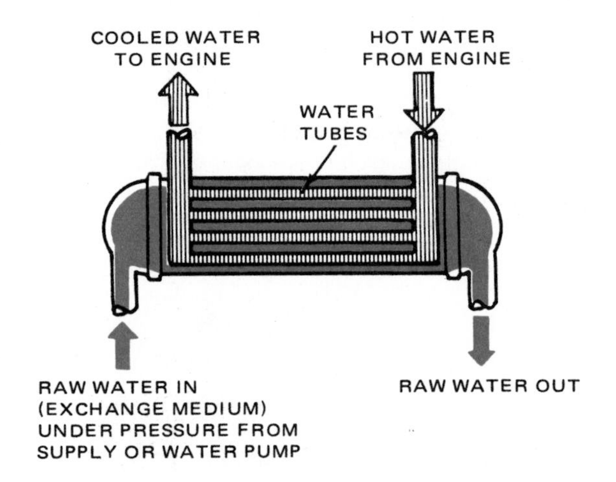

Fig. 14-3. Schematic drawing of heat exchanger.

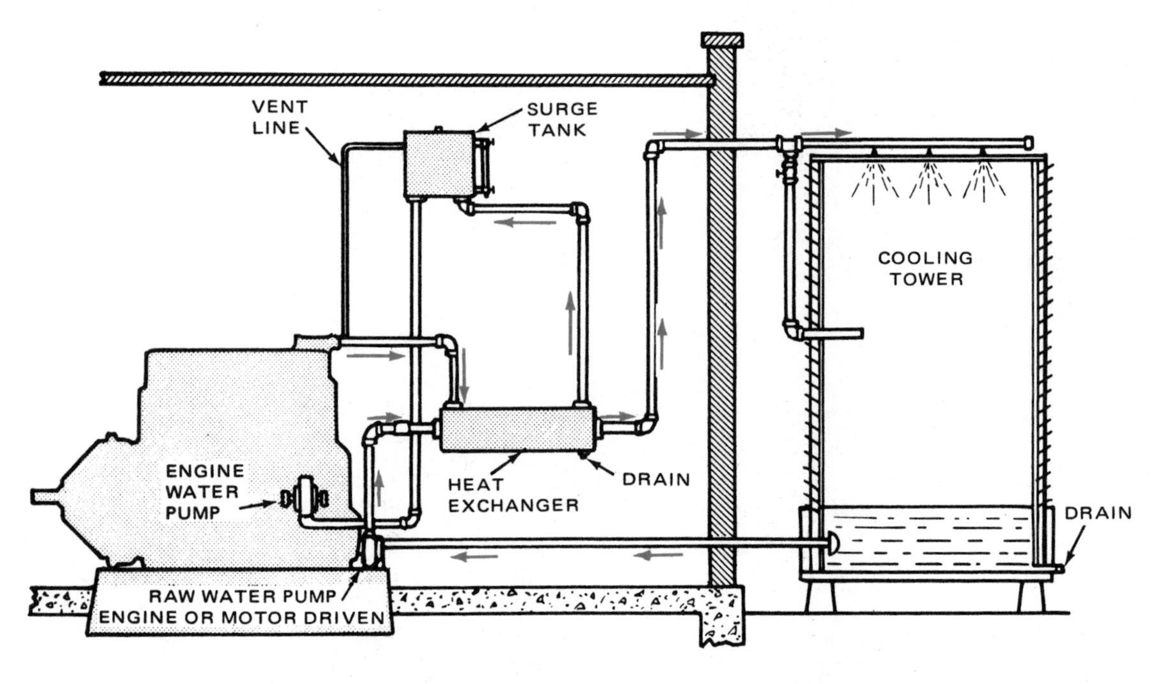

Fig. 14-4. Details of an open cooling system. (Waukesha Motor Co.)

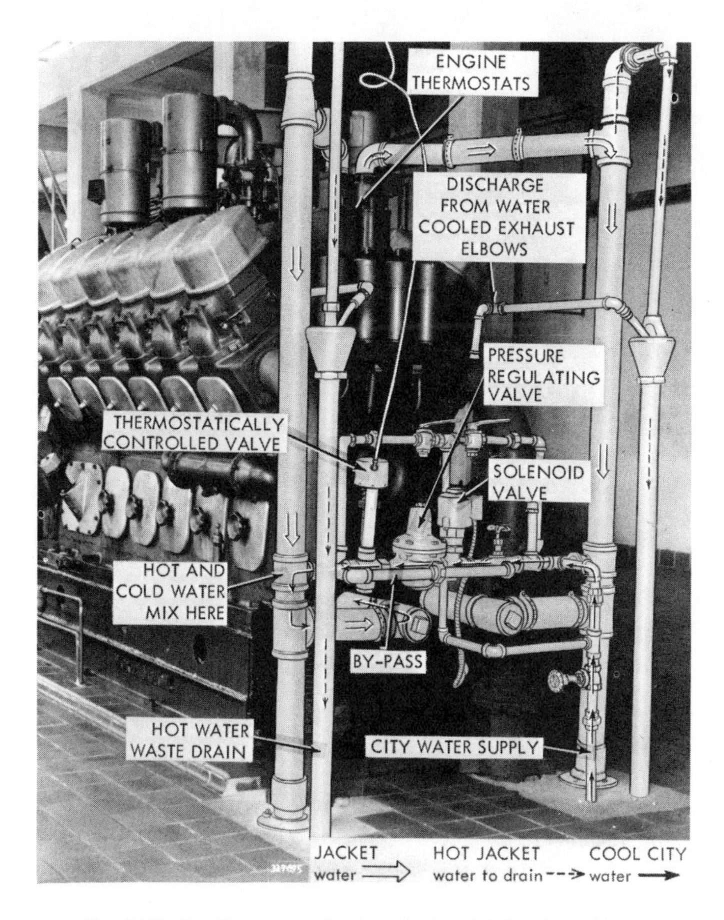

Fig. 14-5. Cooling system for large industrial diesel installation. (Waukesha Motor Co.)

the radiator cools the water. In marine service, sea, lake or river water is passed around a heat exchanger, Fig. 14-3, thereby reducing the temperature of the coolant. The same method is used in industrial diesels where the water is obtained from city mains or artesian wells.

A major advantage of the closed system is that purified water or water that has been chemically treated can be used and recirculated, and there will be little or no scale formed in the water jackets.

OPEN COOLING SYSTEM

In some open cooling systems, particularly those which draw the water from a river, lake or ocean, the water is drawn from the source by a pump, then discharged through the lubricating oil cooler. The oil cooler acts as a heat exchanger. Heat from the oil circulating through the cooler is transferred to the water passing through the cooler. The temperature of the oil is lowered and the temperature of the intake water is raised before the water passes into the water jackets of the engine. After cooling the engine, the water passes through the exhaust silencer water jackets and is discharged overboard.

An example of the open cooling system is shown in Fig. 14-4. The arrows show the direction of flow. Note that the coolant, after passing through the engine water jacket, passes through the heat exchanger and then to the cooling tower.

The basic components for a city water cooling system are shown in Fig. 14-5. In this system, laid out by the Waukesha Motor Co., the pressure regulating valve is designed to regulate

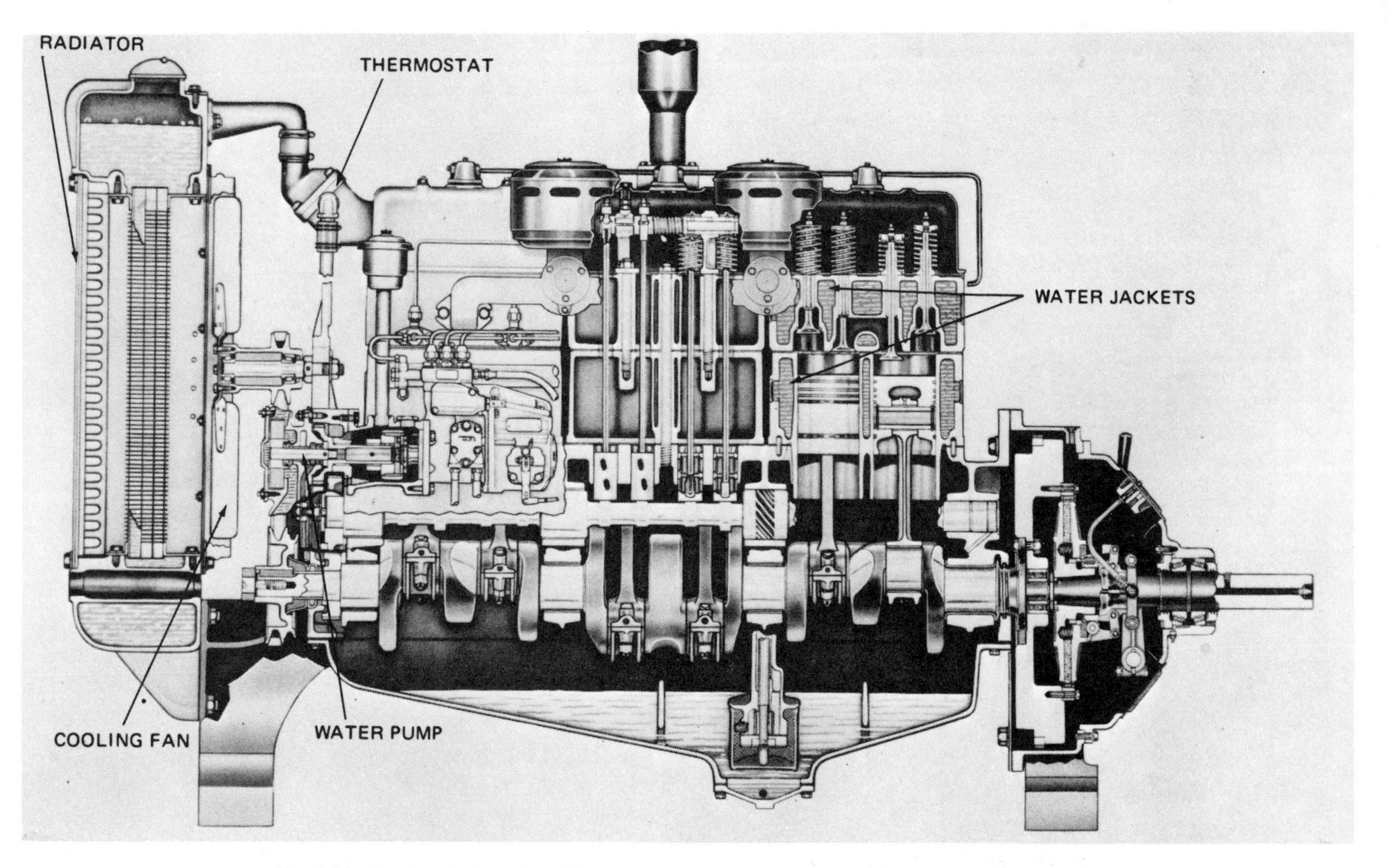

Fig. 14-6. Sectional view of cooling system and tractor engine. (Minneapolis Moline Inc.)

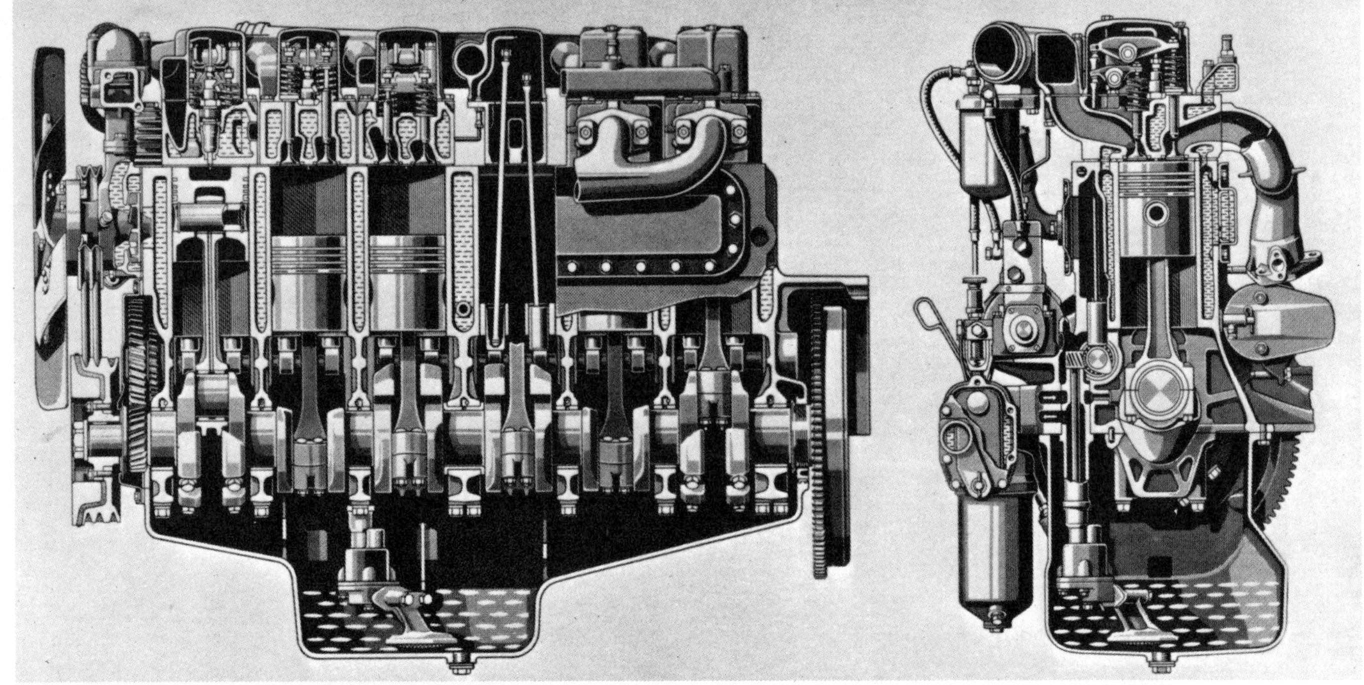
Fig. 14-7. Sectional view of Mercedes-Benz OM 355 direct injection engine showing large capacity water jackets. This 707 cu. in. engine develops 265 hp at 2200 rpm.

the pressure from the city mains. Excess pressure can be troublesome, and will impair the operation of some control valves and cause leakage of cylinder sleeve seals. An open system employing a heat exchanger and a cooling tower is shown diagrammatically in Fig. 14-4. When an adequate supply of water is not available, various forms of cooling towers, or cooling ponds may be used. In these systems the cooling water is recirculated through the engine after having been cooled in the tower or the pond.

CLOSED COOLING SYSTEM

The simplest and most familiar closed cooling system is that used in the automobile field. Portable, semiportable and farm equipment also make use of this system. One such system is shown in Fig. 14-6. In such systems, the coolant is forced through the system by means of a water pump. The coolant circulates through the engine until it reaches the operating temperature, at which point the thermostat opens and the coolant passes through to the radiator. After passing through the radiator, the cycle is repeated. It uses little water and presents a minimum of service problems.

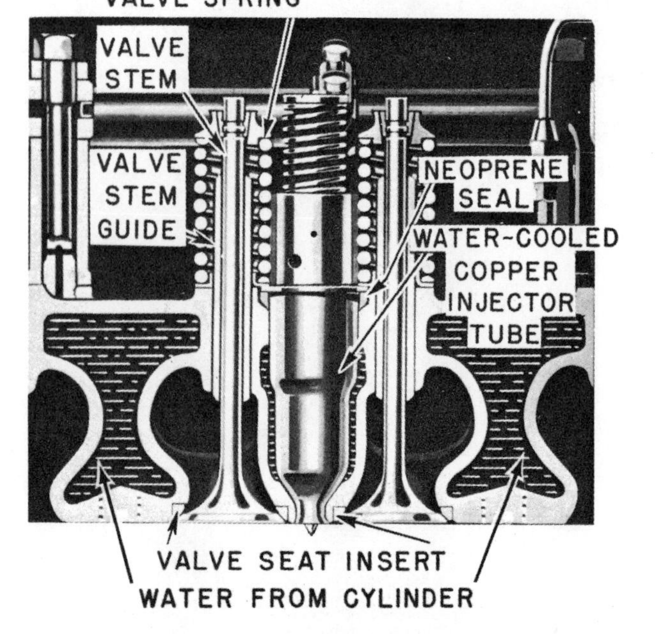

Fig. 14-8. Water passages in cylinder head of Gray Marine engine.

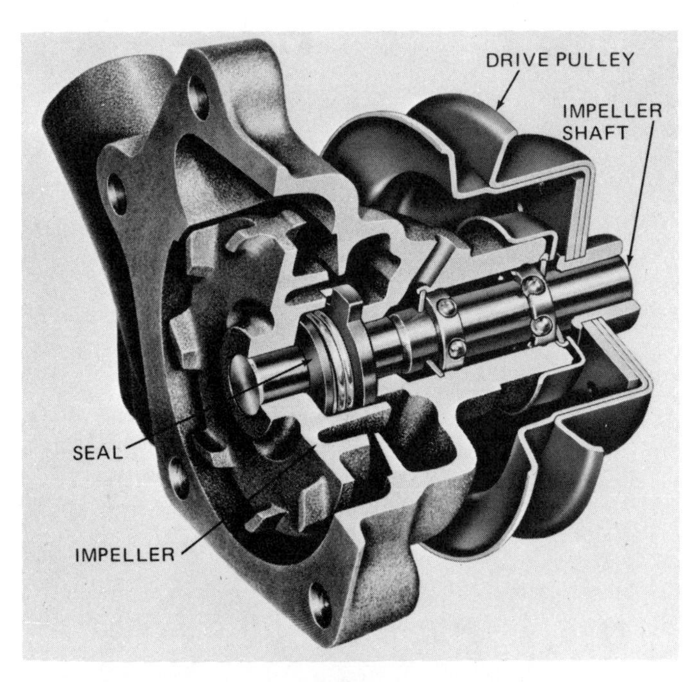

Fig. 14-9. Automotive type water pump. (Sealed Power Corp.)

In Fig. 14-6, note the ample water jacket surrounding the cylinders and valves in the cylinder head. This is also brought out in Fig. 14-7. It should be noted that the water jackets extend the full length of the cylinders. This helps insure against distortion which results from uneven cooling. Also note that the water jackets provide cooling for the injector.

Cooling for the Gray Marine engine valves, cylinder head and injector is shown in Fig. 14-8. An automotive type water pump is shown in Fig. 14-9. Note that the impeller shaft is carried on ball bearings. Such pumps are belt driven from the crankshaft. When the vehicle is being driven at low speeds, a belt driven fan is mounted directly behind the radiator to assist in cooling.

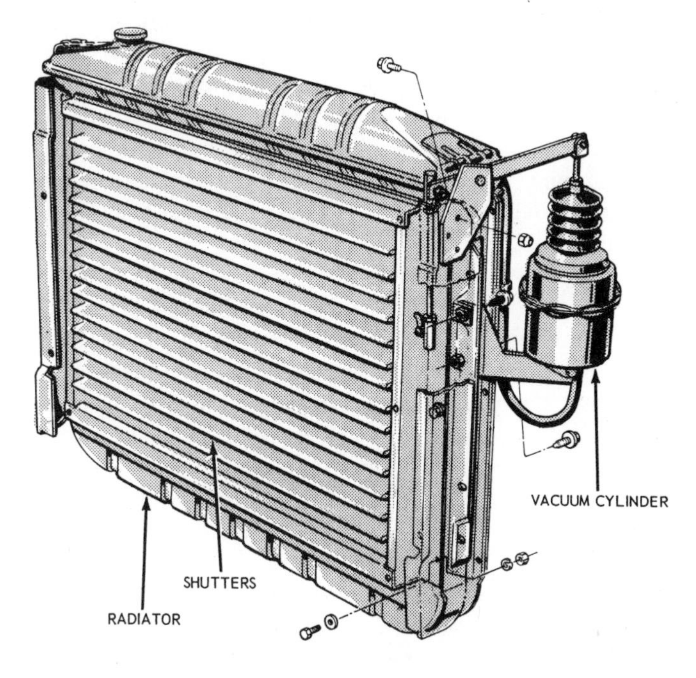

Fig. 14-10. Vacuum operated radiator shutter assembly. (Ford Motor Co.)

In cold weather operation it is not unusual to control coolant temperatures by the use of shutters placed in front of the radiator. Shutters are used in addition to the thermostat in the engine and are usually controlled by a vacuum cylinder, Fig. 14-10. A typical automotive type thermostat is shown in Fig. 14-11. Some radiators are extremely large. Fig. 14-12 shows a radiator for cooling a 700 hp diesel of an 85 ton payload ore hauler.

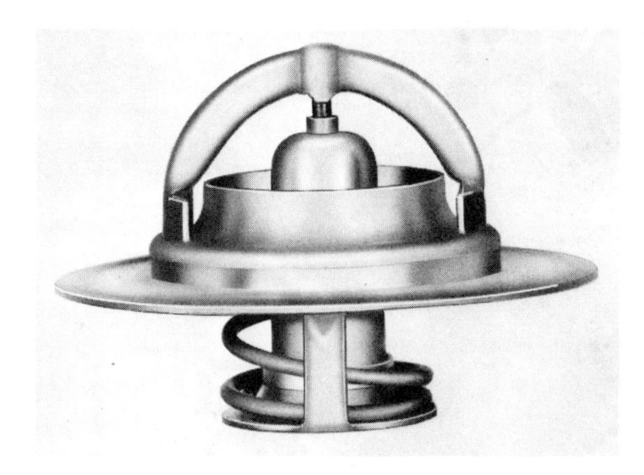

Fig. 14-11. Typical cooling system thermostat. (Standard Thomson Corp.)

Fig. 14-12. Large radiator for use on large ore carrier. (Young Radiator Co.)

Large radiator installations are also used in connection with stationary industrial engines. In one design, the radiators are placed to form a large square with the outer walls made of louvers. A power driven fan at the top draws air through the louvers and radiators. Additional sections are provided for cooling oil when desired.

HEAT EXCHANGERS

When used in large marine and industrial installations, certain changes in the cooling system are needed since a much greater amount of heat must be dissipated. In these cases, a

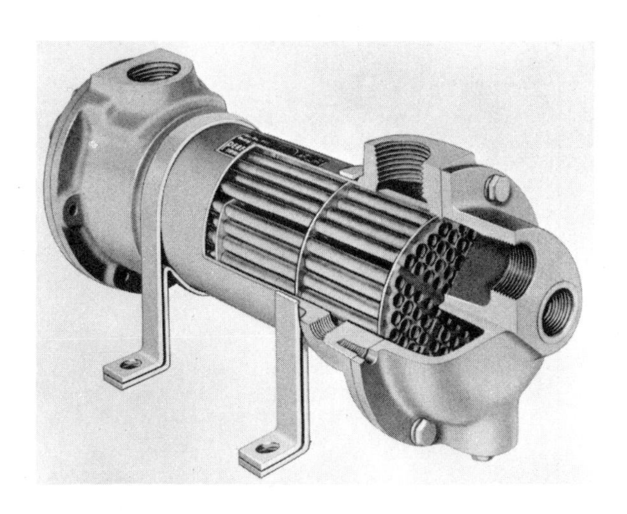

Fig. 14-13. Shell and tube heat exchanger. (Perfex Corp.)

Fig. 14-14. Heat exchanger installation on 1500 KW generator set. Arrows indicate oil and water coolers. (Perfex Corp.)

heat exchanger, Fig. 14-3, is often used instead of a radiator. The hot coolant is passed through the core or pipes of the heat exchanger and water from the city water supply or other source is passed around the core and pipes to reduce the temperatures of the coolant. A typical heat exchanger is shown in Fig. 14-13, and an installation of heat exchangers is shown in Fig. 14-14.

Heat exchangers may be used for heating or cooling. However, in diesel engine installations they are generally used to cool the engine coolant and lubricating oil. In some instances, heat exchangers are used to reduce the temperature of engine intake air and generator cooling air. Coolers or heat exchangers may be classified in several ways: (1) By the relative direction of flow of the two fluids (parallel flow, counterflow and cross-flow types). (2) By the number of times either fluid passes the other fluid (single pass and multipass types). (3) By the path of heat (indirect or surface type) and direct contact type. (4) By general construction features of the unit (shell and tube type and jet or mixer type). The coolers of diesel engines are usually identified on the basis of construction features.

In jet type coolers, the hot and cold fluids enter the unit where they are mixed and discharged as a single fluid. Since this feature is not desirable in engine cooling, the exchangers used with diesel engines are of the shell and tube type, Fig. 14-13. In the shell and tube type of cooler, the hot and cold fluids are prevented from mixing by the thin walls of the element.

The shell and tube cooler consists of a bundle of tubes encased in a shell, Fig. 14-13. The liquid to be cooled flows through the tubes while the cooling water enters the shell and flows around the tubes, being directed by baffles. In some designs the cooling water flows through the tubes and the fluid to be cooled flows around the tube.

Many heat exchangers used in marine engines are of the strut tube type. The strut tube type, Fig. 14-15, has the advantage of providing considerable heat transfer in a compact space. However, the shell and tube type heat exchanger will withstand a higher degree of scaling without clogging the cooling system.

Each of the tubes in the strut tube unit, is composed of two sections or strips. Both sections of each tube contain either a series of formed dimples or cross tubes braised into the tubes. These "struts" (sometimes referred to as baffles) increase the inside and outside contact surfaces of each tube. This creates turbulence in the liquid flowing through the tube. Hence, the heat transfer from the cooled liquid to the cooling liquid is increased.

Tubes in a strut type tube cooler are fastened in place with a header plate at each end. The entire assembly is then

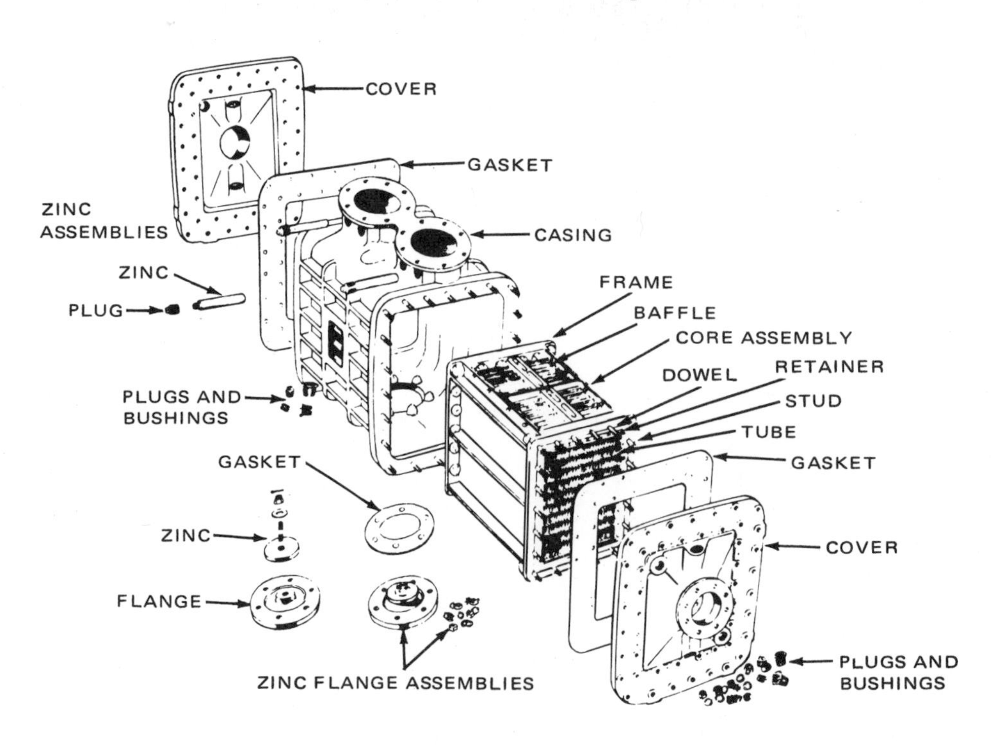

Fig. 14-15. Parts of strut type cooler. (Harrison Radiator Div. General Motors Corp.)

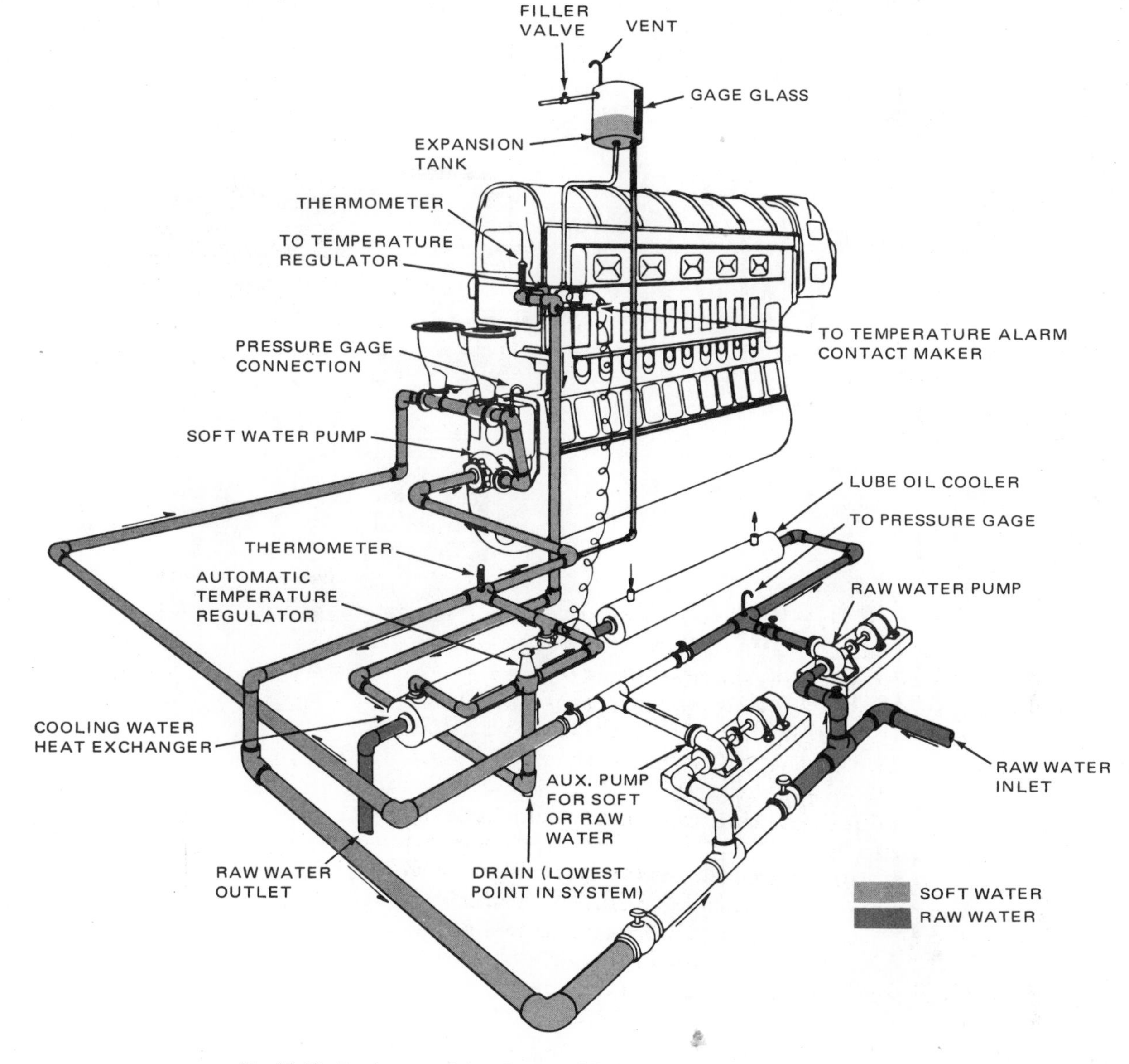

Fig. 14-16. Fresh water piping diagram of Fairbanks Morse opposed piston engine.

mounted in a bronze frame and placed in a cast metal housing. Header plates at the ends of the tube separate the cooling liquid space in the casing from the cooled liquid parts in the end covers. The cooled liquid flows through the tubes in a straight path from cover to cover. The intermediate tube plate acts as a baffle to create a U-shaped path for the cooling liquid flowing around the outside of the tubes.

Circulation of the coolant (fresh water) and the raw water of a closed system is shown in Fig. 14-16. The equipment varies with different installations. Fig. 14-13 shows a typical heat exchanger used in stationary installations. Marine installations are similar and use raw water available from river, lake or ocean.

STATIONARY ENGINE SYSTEM

After passing through the engine, Fig. 14-16, the coolant or soft water (as it is also termed) passes to the heat exchanger where the heat is absorbed by the raw water. A bypass, controlled by an automatic temperature regulator, is placed around the heat exchanger so the temperature of the coolant can be regulated. The regulator may be the diverting or mixing type. After being cooled in the heat exchanger, the soft water flows back to the engine pump suction and is again forced through the cooling passages of the engine.

A vent pipe, Fig. 14-16, leads from the water outlet to an expansion tank. Another pipe leads from the pump suction line to the expansion tank. This arrangement enables the closed cooling system to accommodate variations in water volume resulting from expansion and contraction of heating and cooling. The static head, the connection between the pump section line to the expansion tank, increases pump efficiency.

The raw water pump, Fig. 14-16, draws water from the city mains, river, lake or ocean. It is discharged through the lubricating oil cooler, the heat exchangers and then out of the system. Full flow of raw water is carried through the lubricating oil cooler and the heat exchanger. The auxiliary pump is connected in the system so it can take the place of either the raw water or soft water pump. The auxiliary pump may also be used as an after cooling pump. Operating the pump for a few minutes after the engine is shut down will

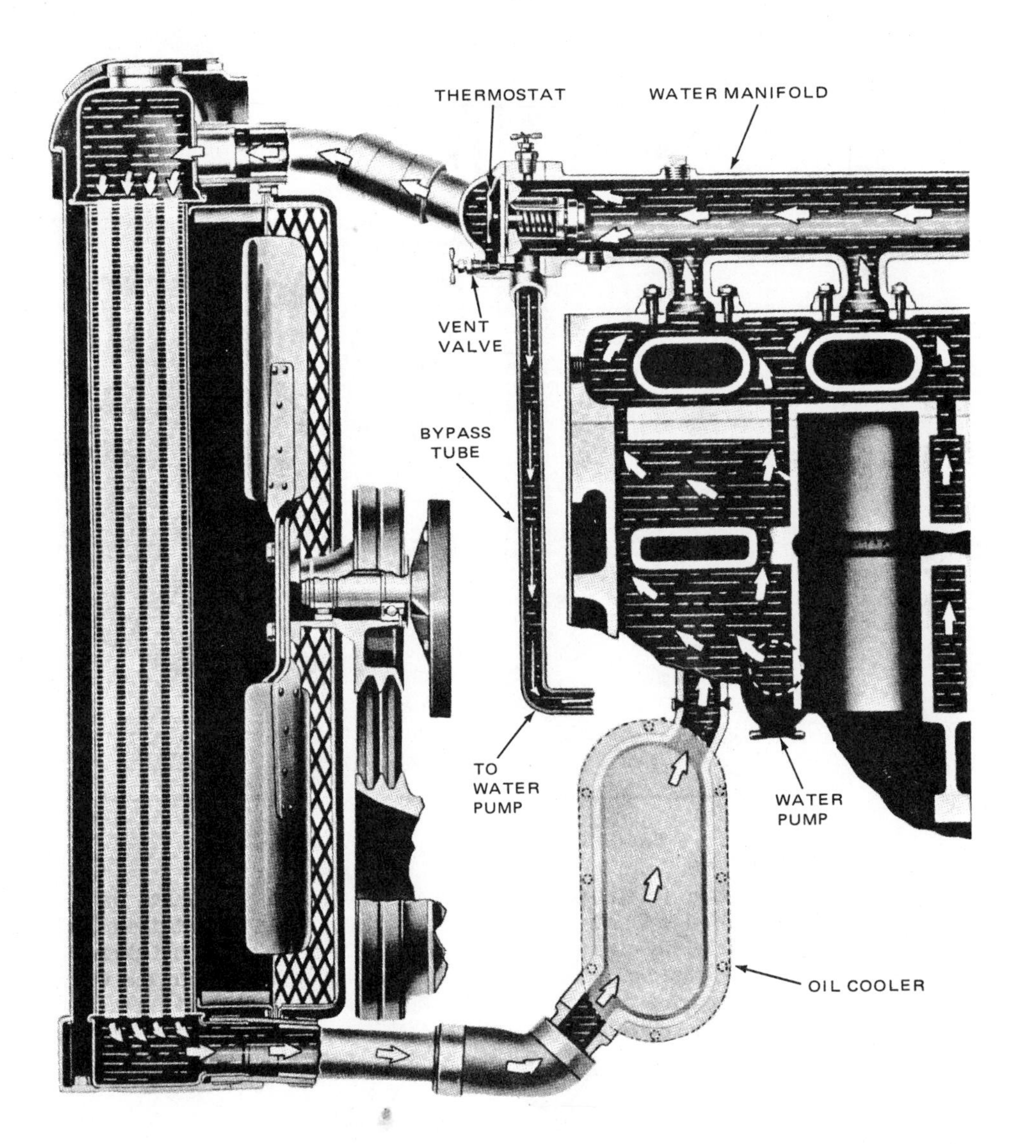

Fig. 14-17. Cooling system with radiator and fan of GMC series 71 engine.

remove residual heat.

The types of heat exchangers which have been discussed include those using water to cool the coolant or the coolants were sprayed into a pond or cooling tower. Another method resembles the familiar automotive radiator, since tubes carrying the coolant are assembled with upper and lower tanks. A large power driven fan is used to draw cooling air through the core.

KEEL SYSTEM

A system for cooling water from the water jacket which is used to some extent in the marine field is known as the keel cooling system. Heat transfer takes place in a unit attached to the keel of the vessel and flow to the cooler is usually thermostatically controlled. Coolant is drawn to the water pump from the bottom of an expansion tank, through the oil cooler, engine cylinder block, cylinder head and into the water manifold. A portion of the coolant is bypassed from the rear of the water manifold into the rear of the jacket surrounding the exhaust manifold, then into the expansion tank. When the thermostat is open, the major portion of the coolant in the water manifold passes through the thermostat housing and flows directly to the keel cooling coils. When the thermostat is closed, that portion of the coolant entering the thermostat housing is bypassed directly to the engine water pump inlet. Here it remixes with the coolant from the exhaust manifold jacket. The coolant is then circulated through the cylinder block and cylinder head. With the thermostat closed a quick warm-up is assured, since the coolant does not pass through the keel cooling coils.

TEMPERATURE REGULATION

The temperature in the cooling system must be regulated to meet existing conditions. The principal factor affecting the proper cooling of an engine is the rate of flow through the cooling system. The more rapid the flow, the less danger there is of scale deposits and hot spots. As the velocity of the circulating water is reduced, the discharge temperatures of the cooling water increases and more heat is carried away by each gallon of cooling water circulated.

The temperatures of the engine coolant can be controlled by two methods. First, the temperature is controlled by regulating the amount of water discharged by the pump into the engine. The second method of temperature control is by regulating the amount of water which passes through the fresh water cooler.

The first method may be accomplished by means of a manually operated throttling valve while the latter method usually employs thermostats. When the manually operated throttle valve is located in the pump discharge, the valves may cause the coolant to pass through the engine slowly or rapidly. If the pump is driven by an electric motor, the same effect can be obtained by altering the speed of the motor.

In modern diesel engine installations, automatic temperature control by means of thermostatically operated bypass valves is standard.

A typical thermostatically controlled system is shown in Fig. 14-17. This illustration shows the cooling system as installed on the GMC series 71 diesel engine. Water is circulated by means of the water pump. When the coolant is cold, the thermostat is closed so that the coolant bypasses the radiator and circulates only through the engine. As the temperature rises, the thermostat opens and the coolant passes from the engine cylinder head or manifold to the top of the radiator. Air blowing through the radiator cools the water which is drawn from the bottom of the radiator by the pump action. The coolant passes through the oil cooler (if installed) and enters the engine water jacket. The action of the engine driven fan assists in cooling the water. In many installations, the action of the fan is thermostatically controlled. If so, the fan does not function at low temperatures. Operating temperatures vary with different engines. However, the usual range is from 165 to 195 deg. F. (74 to 90.6 C).

Two different thermostats (both operating on the same principle) are shown in Figs. 14-18 and 14-19. These thermostats do not use a diaphragm. They depend on the expansion of confined wax under heat, Fig. 14-20, uniformly squeezing an elastic boot with pressure up to 2000 psi, causing it to

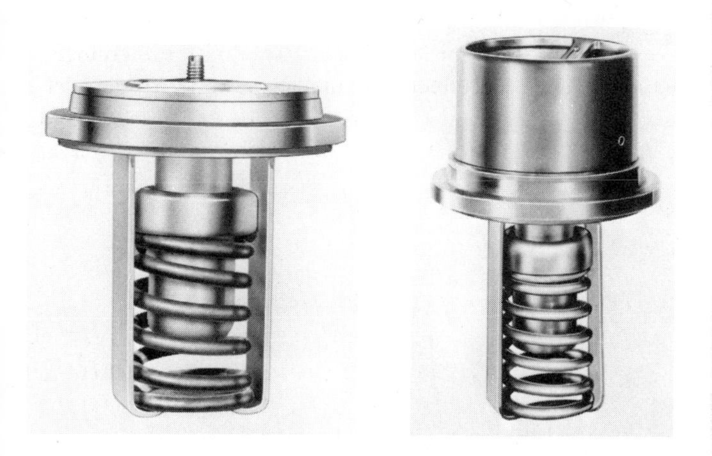

Fig. 14-18. Left. Cooling system thermostat as installed in some Cummins engines. Fig. 14-19. Right. Cooling system thermostat as installed in some GMC diesel engines. (Standard Thomson Corp.)

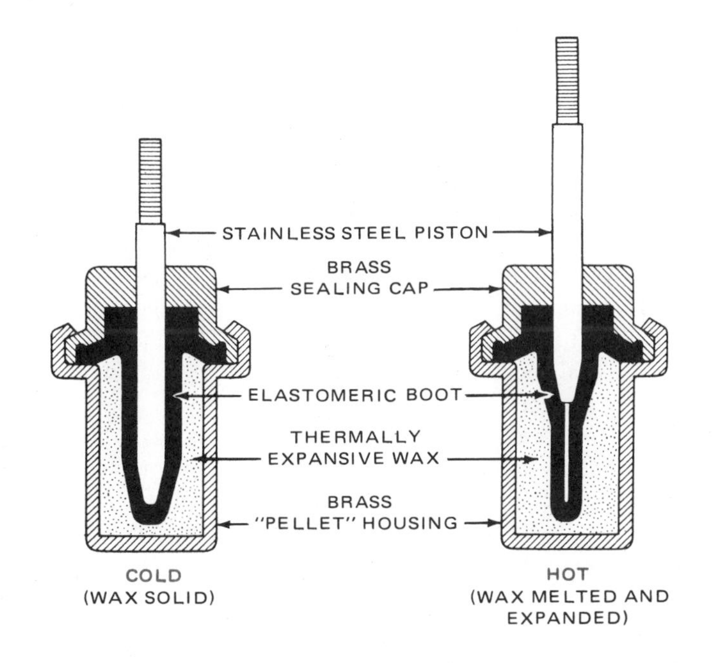

Fig. 14-20. Operation of one type of thermostat element. (Standard Thomson Corp.)

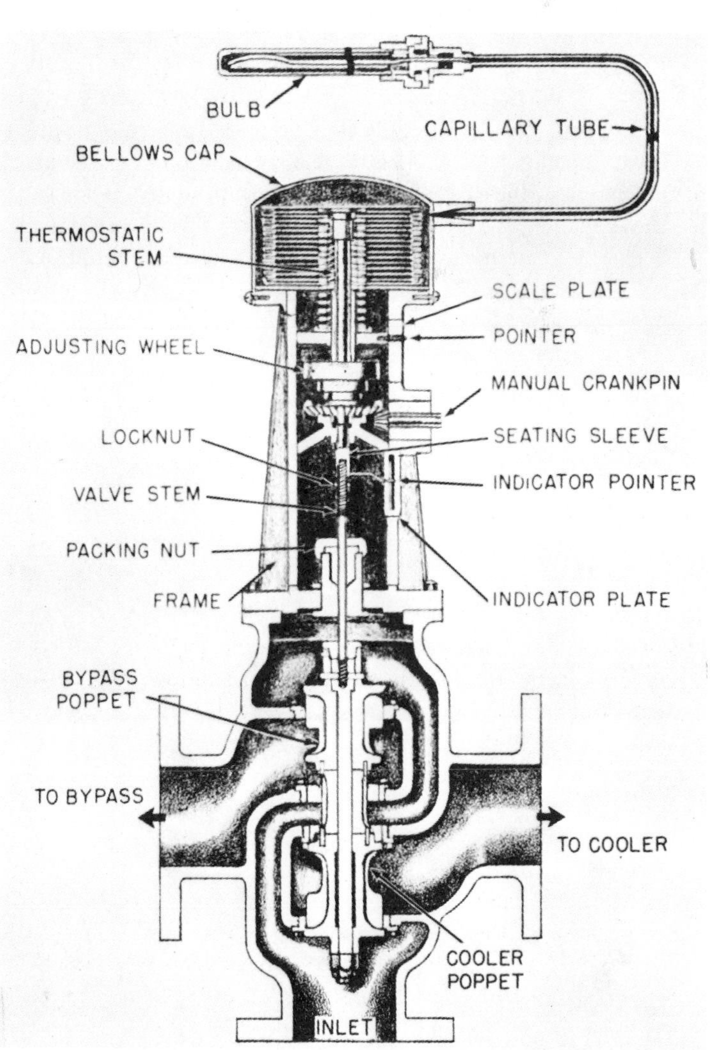

Fig. 14-21. Details of automatic temperature regulator as used in some U. S. Navy installations.

extrude a stainless steel piston which controls the movement of the valve.

Another type of coolant control is the automatic regulating valve which maintains the fresh water temperature at the desired level by passing a portion of the water around the water cooler. The regulator, shown in Fig. 14-21, is commonly used in marine engines. A bellows type thermostat is used in this device, but provision is also made for manual operation.

The temperature regulator consists of a valve upon which a thermostatic control unit is mounted. The thermostatic control unit consists of two parts, the temperature control element and the control assembly.

The temperature control element consists of a bellows connected by a flexible armored tube to a bulb mounted in the engine coolant discharge line. The control element is essentially two sealed chambers. One chamber is formed by the bellows and caps which are sealed together at the bottom. The other chamber is the bulb. The entire system (except for a small area at the top of the bulb) is filled by a mixture of ether and alcohol, which vaporizes at a low temperature. When the bulb is heated, the liquid vaporizes. Pressure within the bulb increases, forcing the liquid out of the bulb and through the tube. This pushes the bellows down to operate the valve.

COOLANT LEVEL

Coolant temperature indicators are of extreme importance in the operation of a diesel engine. Equally important is the indication of the coolant level in the radiator upper tank. Tests show that in the event of a sudden loss of coolant, actual engine temperature would rise so quickly that it could cause complete failure of the engine.

Special equipment is available which monitors the coolant level in the radiator. According to the manufacturer, sufficient warning is provided so that corrective measures can be taken. A similar device indicates the level of the lubricating oil in the engine oil pan. When the oil level drops, the condition is indicated by the device.

WATER PUMPS AND FANS

Water pumps for circulating the coolant are usually of the centrifugal type. Figs. 14-9, 14-22 and 14-23 are examples of this type of pump. The General Motors' Unit used in the 71 in-line diesel engines, shown in Fig. 14-23, has the impeller (3) pressed into the one end of the shaft (23), and a pump drive coupling (24) with an oil thrower (25) pressed on the opposite end. The oil thrower shrouds the inner end of the pump body flange to prevent oil from creeping along the shaft and through the shaft bearing. The shaft is supported at the drive end by a sealed double row combination radial and thrust ball bearing (22). An oil slinger (21) is fitted between the seal and the ball bearing. The pump shaft (23) and bearing (22) form a single assembly.

The fans used on automotive and mobile diesels are usually mounted on an extension of the water pump shaft, Fig. 14-22. They are run at approximately crankshaft speed, generally have four blades, and are belt driven.

Some installations are equipped with special fans which operate only when the engine is at full operating temperature. The fan does not operate when the engine is cold, saving fuel required to drive the fan. To control the operation of the fan,

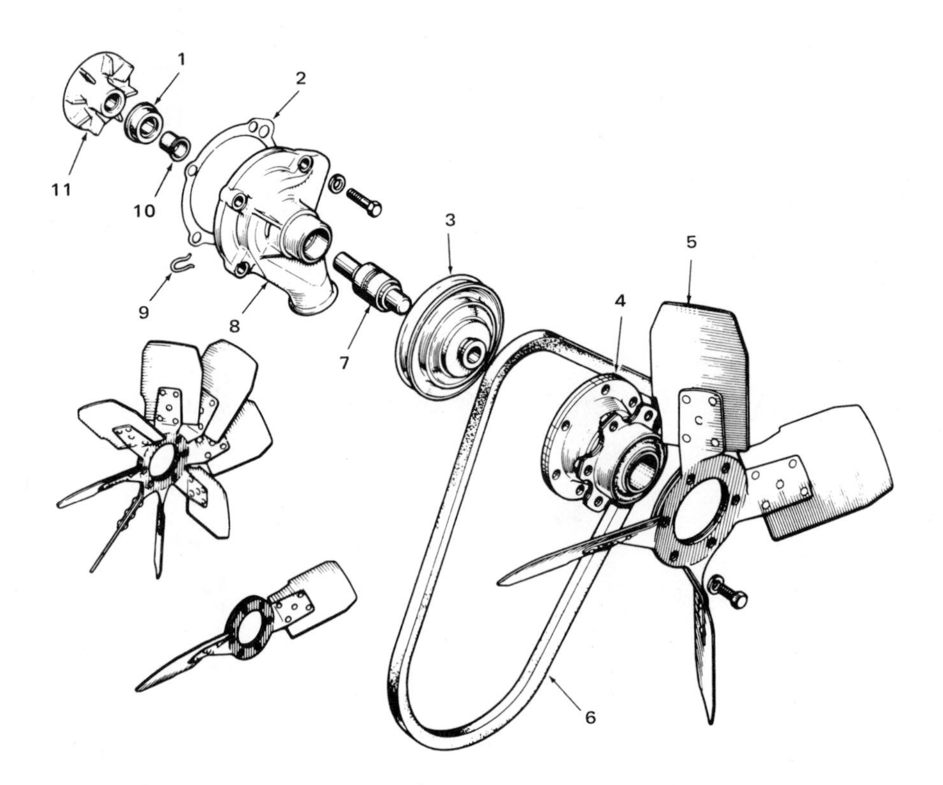

Fig. 14-22. Exploded view of water pump and fan. 1—Seal. 2—Gasket. 3—Pulley. 4—Fan hub. 5—Fan blade assembly. 6—Fan belt. 7—Bearing and shaft assembly. 8—Water pump housing. 9—Bearing retainer. 10—Shaft slinger. 11—Impeller. (Ford Motor Co.)

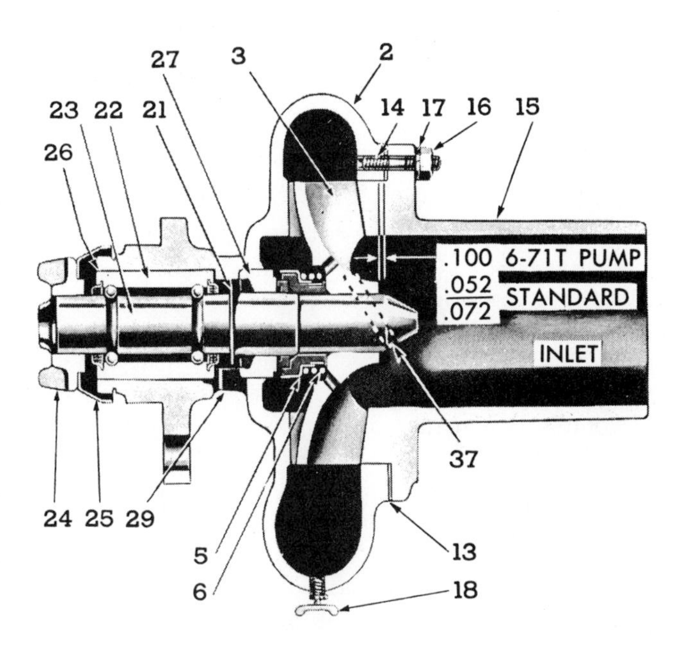

Fig. 14-23. Water pump assembly used on GM model 71 diesel engine: 2-Pump body. 3-Impeller. 4-Fan hub. 5-Seal assembly. 6-Spring seal. 13-Gasket. 14-Stud. 15-Pump cover. 16-Nut. 17-Lock washer. 18-Drain valve. 21-Shaft slinger. 22-Ball bearing. 23-Shaft. 24-Drive coupling. 25-Thrower. 26-Stake points. 27-Insert. 29-Drain cavity. 37-Pin. (Detroit Diesel Allison Div., General Motors Corp.)

a bimetallic strip or helical coil spring is used, Fig. 14-24. Another type of fan provides varying amounts of air depending on the speed of the engine. This flex-blade fan has riveted flexible blades. As the engine speed increases, the blades flex. Less power is required to drive the fan and less cooling is provided.

Exhaustive tests of highway trucks indicate that full operation of the cooling fan is required probably less than 10 percent of the time. Reports by Eaton Corp. engineers showed that, under test conditions with the fan operating, fuel consumption averaged 5.226 mpg. With the fan off, the fuel consumption totaled 5.6 mpg or a saving of 7.16 percent.

It must be remembered that fuel consumption is seriously affected by the weight of the load being carried, velocity of the existing winds, character of the terrain, engine efficiency, altitude, barometric pressure, ambient temperature, engine temperature and wind resistance created by the frontal design of the vehicle.

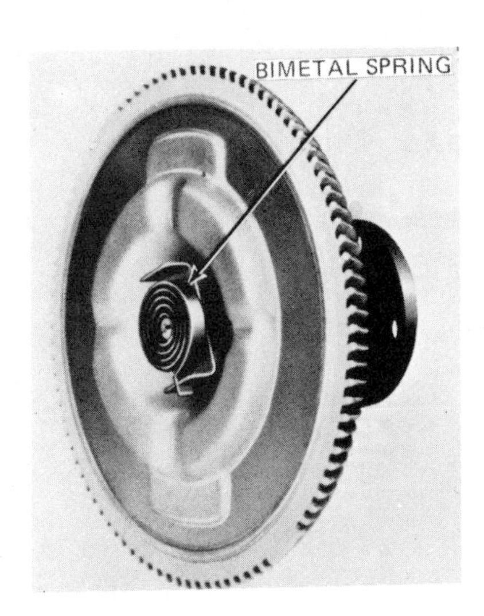

Fig. 14-24. Details of fan drive clutch. (Ford Motor Co.)

COOLING RAILWAY DIESELS

A method of cooling railway diesel engines, which was developed in Germany, varies the speed of the fan to control the temperature of the coolant. This is a flexible system. It can be mounted at the front, on the roof or on the side of the locomotive. The side mounted system is shown in Fig. 14-25. The fan speed is thermostatically controlled and can be varied from zero to maximum at any time. The speed is determined solely by the prevailing temperature of the coolant. This is important since locomotives are frequently motionless for long periods with the engine operating.

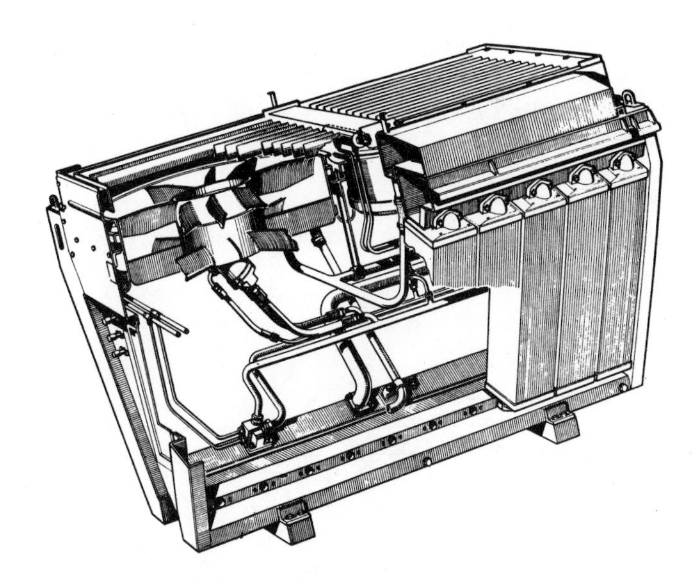

Fig. 14-25. Temperature of coolant is controlled by varying the speed of the fan in this system designed for diesel locomotives. (Julius Behr, Germany)

Speed of the cooling fan is independent of variations in engine load and ambient temperature. The desired operating temperature can be maintained ± 3 deg. F.

Power consumption of the fan is determined by the coolant requirements. The thermostat is of the impact and pressure resistant type.

AIR COOLING

The original diesel engines were water-cooled and it was not until World War II that any progress was made in air cooling diesel engines.

When studying air-cooled engines, it is important to remember that heat is a form of energy in transit. Whenever there is a difference in temperature between two adjacent bodies or two areas of the same body, a flow of heat is established from the higher to the lower temperature. Such

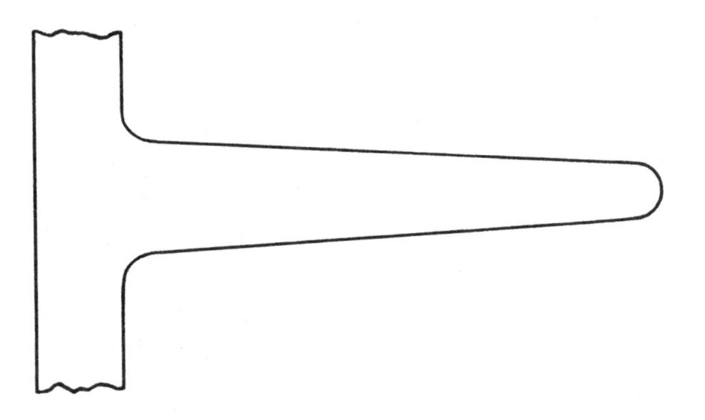

Fig. 14-26. Typical shape used for cooling fins.

heat can be transferred by three different methods; conduction, convection, or radiation. At times, two or all three of these methods combine in varying degrees to transfer heat from one area to another.

As pointed out previously, approximately one-third of the heat of combustion is dissipated by the cooling system. Some of this is received by the walls of the combustion chamber and piston by direct radiation. Part is transmitted through the metal walls by conduction of heat from the hot gases in the vicinity of the walls. The turbulence that occurs during compression and combustion increases the rate of heat transfer to the metal surfaces. It brings fresh particles of hot gases into each part of the surface replacing those which have already given up some of their heat.

AIR-COOLED SURFACES

When a blast of cool air passes over a hot metallic surface, the air close to the surface is set into eddies. Because of surface friction, the velocity of the moving air is reduced. For this reason, considerable research has gone into the shape of cooling fins. Fig. 14-26 shows a typical fin that gives good cooling, is easy to manufacture, and has sufficient mechanical strength to withstand rough handling.

Figs. 14-27 and 14-28 show the cooling fins on air-cooled cylinders. Note that the diameter of the fins increases in the area of greatest heat, around the valves.

Size and spacing of cooling fins are dependent on the amount of heat to be disbursed, the material of which the fins are made, the spacing or pitch of the fins, the diameter of the cylinder, and the speed and temperature of the cooling air.

A large number of short fins is theoretically better than a smaller number of larger fins. However, consideration must be given to providing sufficient rigidity to the cylinder barrel to prevent fin distortion. Also, if the fins have too short a pitch, the effectiveness of air flow is reduced.

With increased pitch or spacing of the fins, a larger air flow

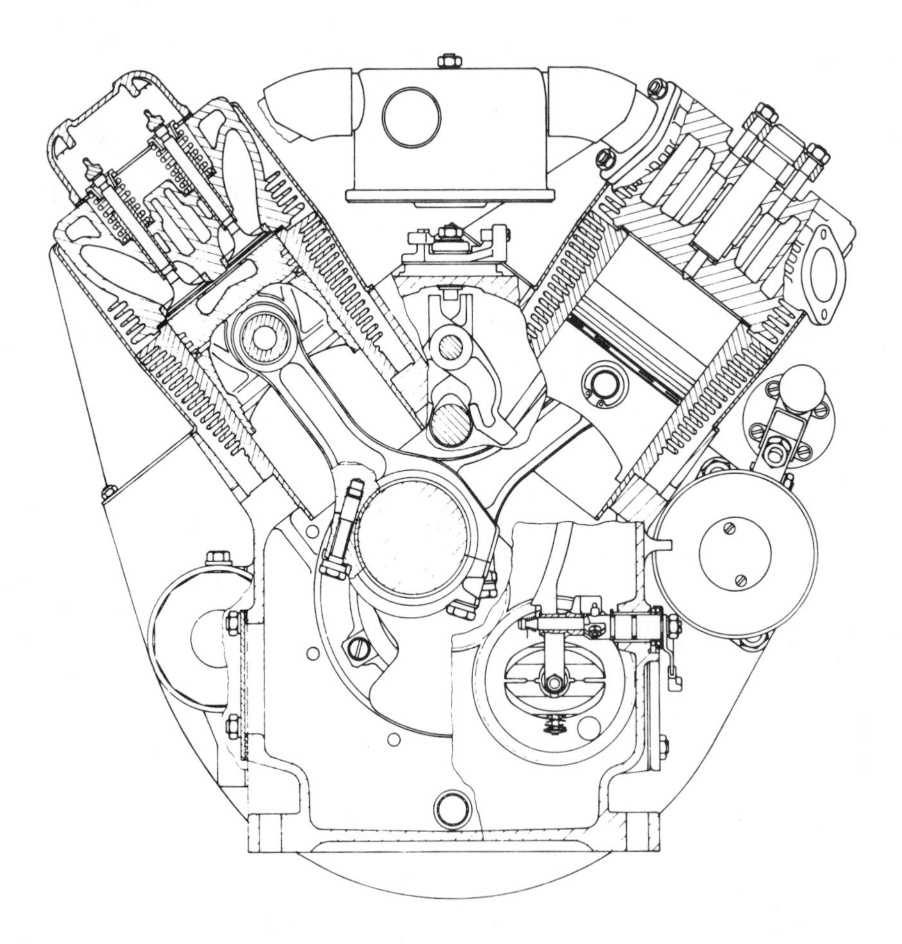

Fig. 14-27. Sectional view of air-cooled diesel V-type engine. (Farymann Diesel)

area is provided and a greater quantity of air is needed. However, the air does not become as hot. With smaller fins and reduced pitch, the cooling air is heated to a greater extent but less is required.

Higher air pressure is required as space between fins is decreased. Minimum spacing between fins is generally set at 0.10 in. for fins of .04 to .06 in. For more effective cooling, two separate air streams are often provided.

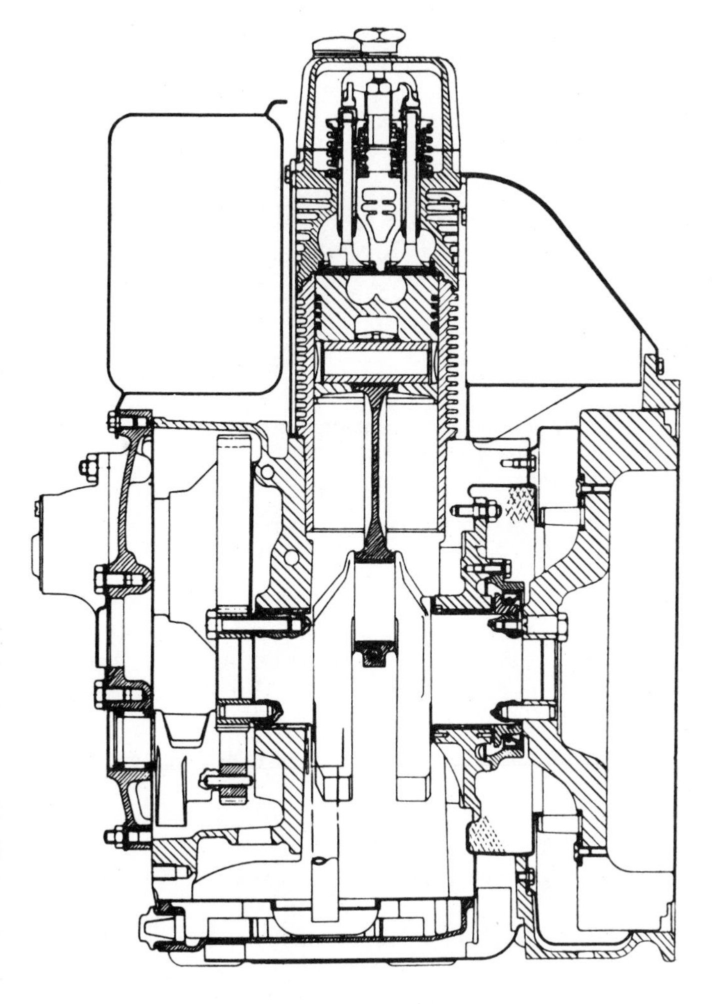

Fig. 14-28. Note cooling fins on this air-cooled diesel. (Petters, Ltd.)

Engine driven fans, Fig. 14-29, for both automotive and stationary installations are provided together with thermostatic control to supply the needed volume of air.

Desired air temperature is maintained by supplying a large quantity of air. In practice, about 1700 to 2000 cu. ft. of air at normal temperature and pressure are required per brake horsepower hour.

ADVANTAGES OF AIR-COOLED ENGINES

An air-cooled diesel power plant weighs less than a water-cooled unit because it requires only an engine driven fan and cooling fins, together with sheet metal ducts and thermostats. A water-cooled engine requires a radiator, coolant, circulating pump, coolant passages, piping fan and thermostat. Because of the weight of the coolant, the air-cooled engine weighs about 10 percent less than a comparable water-cooled diesel engine.

Fig. 14-29. Engine driven fan on Ford diesel. (Ford Motor Co.)

An air-cooled engine is also much more compact, with no possibility of coolant leakage as in a water-cooled diesel engine. It can also operate under a wide range of atmospheric conditions.

Wear rate of the air-cooled engine compares favorably with that of water-cooled diesel engines, yet maintenance costs are no higher. The air-cooled engine may be somewhat longer because of the cooling fins. In general, they are also noiser.

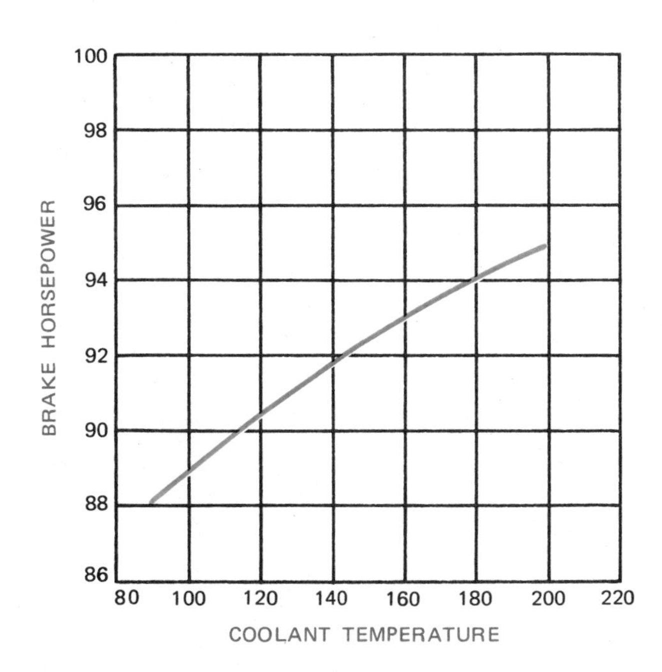

Fig. 14-30. Up to approximately 200 deg. F., bhp increases.

EFFECT OF COOLING TEMPERATURE

Various researchers have shown that the operating temperature of internal combustion engines has a marked effect on the power produced and economy. Tests made by Armstrong-Sauer Ltd. Fig. 14-30, shows that when operating temperatures of the coolant were increased from 100 to 200 deg. F., brake hp jumped from 89 to 95 bhp or 6.7 percent. There was also a corresponding improvement in fuel economy. Such figures clearly demonstrate the importance of maintaining accurate control of cooling temperatures.

Naturally there is a limit to engine operating temperature. Since excessive temperatures result in excessive expansion of reciprocating and rotating parts, there would be considerable loss due to frictional hp. Excessive temperatures cause rapid deterioration of lubricating oil and also increased wear and deterioration of parts.

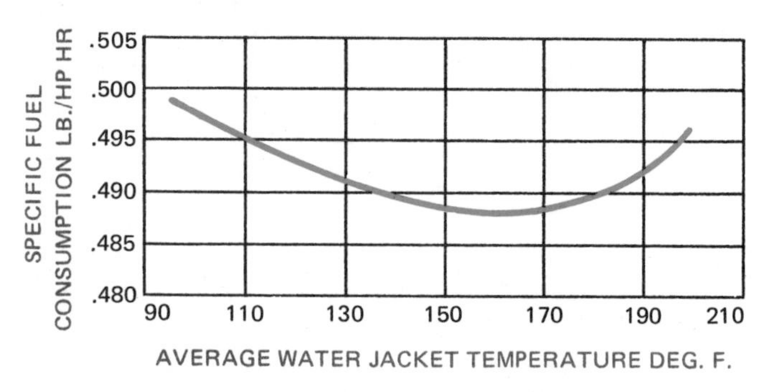

Fig. 14-31. On this particular engine, specific fuel consumption dropped and then started to rise after water temperature reached 160 deg. F. On other engines, the fuel consumption will continue to decrease until higher temperatures are reached.

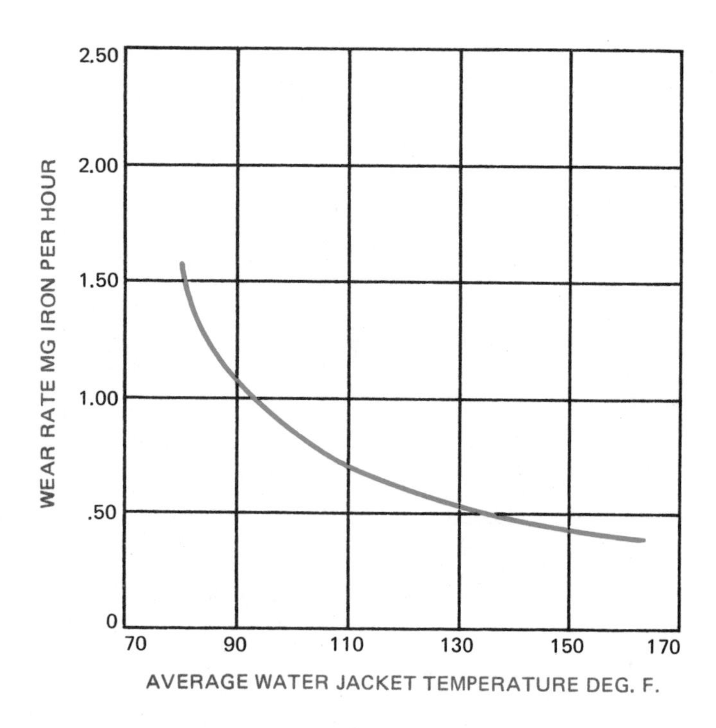

Fig. 14-32. Note that wear rate keeps decreasing within the range of the operating temperatures indicated.

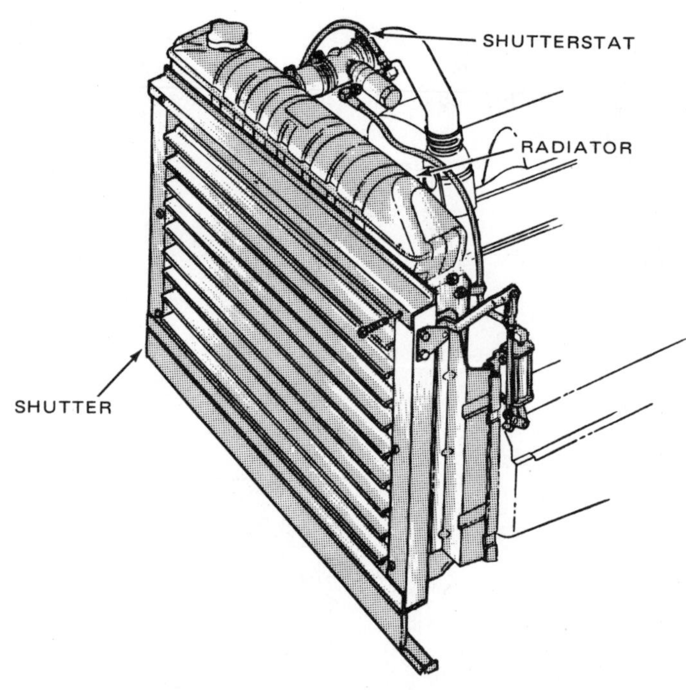

Fig. 14-33. Typical air operated shutter assembly as installed on N-series Ford Trucks. (Ford Motor Co.)

RADIATOR SHUTTERS

It is important that temperature variations of the coolant be maintained within close limits. Then wear may be kept to a minimum and power at a maximum together with low fuel consumption. In Fig. 14-30, it was shown that the bhp varied considerably with the temperature of the coolant. The specific fuel consumption also varies. Fig. 14-31 gives the data for a

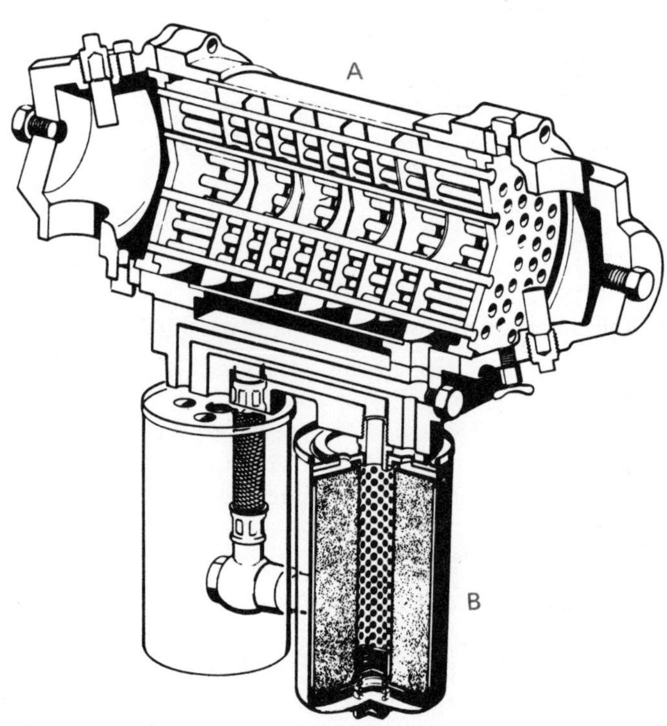

Fig. 14-34. A Volvo Penta single unit oil cooler with oil filters.

typical V-8 engine. Similarly the rate of wear increases as the operating temperature decreases, Fig. 14-32. Such data would of course vary with different engines and operating conditions.

In general there are three methods of operating the shutters in front of the radiator which control the flow of air.

1. Thermostat operation.
2. Air operation for vehicles with air brakes.
3. Vacuum operation for vehicles equipped with vacuum powered brakes.

The thermostat operated system consists of a shutter control assembly, a shutter control rod and a shutter assembly. The shutter control assembly has a power element which works like a thermostat. It controls the movement of the shutter in accordance with the temperature.

The air operated shutter system, Fig. 14-33, consists of the shutter assembly, an air cylinder which operates the shutter, and a thermostatically controlled air valve called a shutterstat. With no air in the system, springs in the shutter assembly will hold the shutter blades in the open position. With normal air presure and the engine below operating temperature, the shutter blades close. As the coolant reaches the operating temperature setting of the shutterstat, an air valve in the shutterstat closes cutting off the air to the air cylinder. The air in the air cylinder is then exhausted through the shutterstat and the shutter springs open the shutter blades.

The vacuum operated shutter system, Fig. 14-10, consists of the shutter assembly, a vacuum power cylinder which operates the shutter and a thermostatically controlled vacuum valve called a shutterstat. Operation of this system is the same as the air system except that vacuum power is the operating medium.

OIL COOLERS

In diesel engines, a great deal of the temperature is absorbed by the lubricating oil. Oil coolers are often provided to maintain the oil at a satisfactory temperature. Fig. 14-34 shows an oil cooler with oil filters forming a single unit. This Volvo Penta design is built primarily for large industrial installations. A, is the oil cooler, while the full flow oil filters are shown at B, in Fig. 14-34.

REVIEW QUESTIONS – CHAPTER 14

1. The approximate amount of the total heat of the fuel used in diesel engines that is transferred into useful work is:
   a. 15 percent.
   b. 20 percent.
   c. 25 percent.
   d. 35 percent.
2. The ____________ system also contributes to cooling the diesel engine.
3. The type of engine that will lose the greatest amount of heat to the cooling system is the:
   a. Two-cycle.
   b. Four-cycle.
4. In the ____________ system of water cooling an engine, the same coolant is recirculated.
5. The main disadvantage of using "raw" water in the cooling system is:
   a. That the temperature varies.
   b. Because it contains impurities.
   c. Because of incorrect pressure.
6. The advantage of using full length water jackets in an engine is:
   a. To keep the engine cooler.
   b. To reduce cylinder distortion.
   c. To reduce temperature of crankcase oil.
7. Heat exchangers may be used either as heaters or ____________.
8. The type of heat exchanger that will withstand the greater amount of scale formation is:
   a. Strut type.
   b. Shell and tube type.
   c. Jet type.
9. The more rapidly the coolant flows through the radiator, the ____________ danger there is of scale formation.
10. Typical thermostats depend on a ____________ system for their operation.
11. Water pumps used for circulating coolant are usually of the ____________ type.
12. Cooling fans are usually driven by an extension of the ____________ shaft.
13. Three methods of transferring heat are by:
    a. ____________________________.
    b. ____________________________.
    c. ____________________________.
14. Heat is transmitted from one end of a metal bar to the other by:
    a. Convection.
    b. Conduction.
    c. Radiation.
15. An air-cooled diesel engine weighs ____________ than a water-cooled engine of the same horsepower.
16. If the operating temperature of an engine is increased from 100 to 200 deg. F., the power is ____________.
17. The advantage of radiator shutters in addition to water flow control with a thermostat is:
    a. Closer control of temperature.
    b. More heat converted into useful work.
    c. Specific fuel consumption increased.
18. A cooling fan is needed approximately what percent of the time?
    a. 5 percent.
    b. 7.5 percent.
    c. 10 percent.
    d. 12 percent.

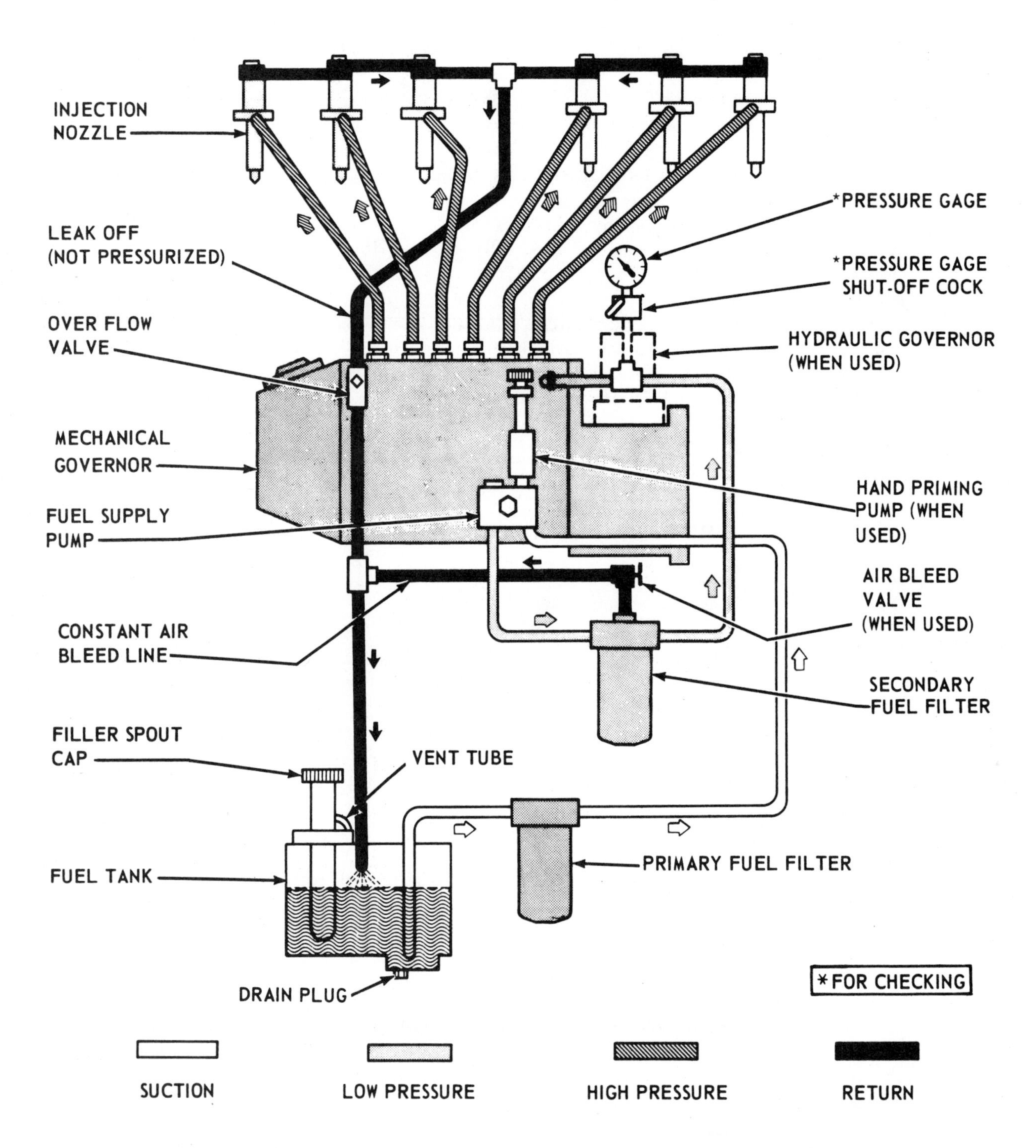

Schematic diagram showing typical diesel fuel system which filters the fuel, meters the amount of fuel injected into the combustion chamber, and injects it at the proper time to obtain peak performance. (Mack Trucks, Inc.)

Chapter 15
FILTERS, AIR AND FUEL

Keeping air, oil and fuel clean is one of the most important phases of diesel engine operation. Unless air is cleaned before it enters the engine, wear of engine cylinders and other parts will be seriously accelerated. Maintenance costs will also be increased. Similarly, engine lubricating oil must be kept clean to keep repair costs to a minimum. All foreign matter should be removed from the fuel oil before it enters the injectors. If not, operation will be erratic and loss of power will result. This chapter will deal with air and fuel filters, while oil filters will be discussed in the following chapter on Lubricating Systems.

AIR CLEANERS

While the unit which removes the dirt from the intake air is generally known as an air cleaner, in many cases it also acts as a silencer and a flame arrester.

In stationary installations of large industrial engines, the air is often passed through a thick curtain of water which removes much of the dust from the air and cools it at the same time. Such cooling results from the fact that the water is cooler than the air and because the water evaporates and carries off some of the heat from the air. After passing through the curtain of water, the air is then filtered at the engine air intake.

There are two main types of air cleaners:

1. Dry type.
2. Oil type.

Many air filters reverse the direction of air flow one or more times as the air enters the unit. Centrifugal action throws out a large percentage of foreign matter. This action is used in both dry and oil bath types of air cleaners, as the first cleaning stage. In the case of the oil bath cleaner, the incoming air is usually directed downward striking the surface of the oil where much of the dirt and dust collects. Such centrifuging removes 90 percent or better of the foreign matter which the air contains.

There are a number of dry filtering elements. They include a screen mesh, crimped paper, metal, spun glass and cloth. Paper and most other elements are descarded after their period of service.

Fig. 15-1 shows a United air cleaner known as the Tri-Phase. It is claimed that up to 97 percent of dust is centrifuged out. As outside air is drawn through the filter

Fig. 15-1. Installation of Tri-Phase air cleaner on deep well turbine pump. (United Air Cleaner Div., United Filtration Corp.)

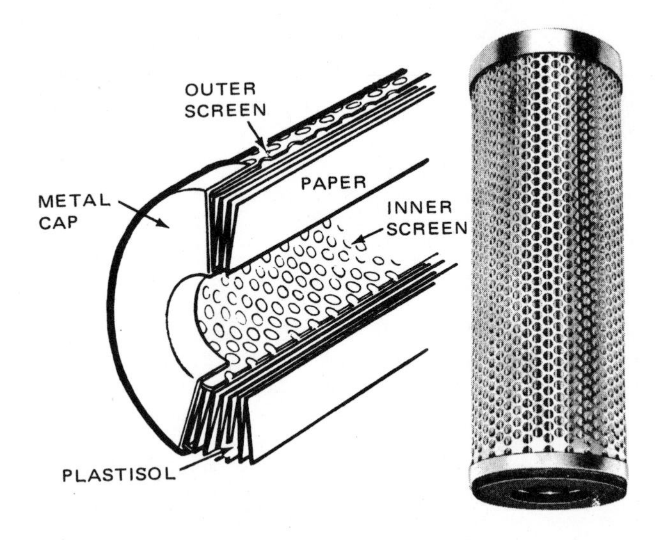

Fig. 15-2. Details of tubular type air cleaner showing construction of folded paper element.
(United Air Cleaner Div., United Filtration Corp.)

intake pipe, it is set into circular motion. The spinning action of the dirty air throws out the heavier particles where they are collected in the dust port and then expelled through the automatic unloader. The air then passes through the filter element, Fig. 15-2. The element may be cleaned by tapping, reverse flow cleaning with compressed air, or washing in water with a non-sudsing detergent.

The Donaldson FHG Cyclopac air cleaner, Fig. 15-3, is a dry type air filter. Air first enters the pre-cleaner fin assembly (1), where the angled veins give a twist to the entering air, spinning out the larger contaminants. The centrifuged contaminants are carried along the wall of the cleaner (2), and are ejected at the baffled dust cup. Dust remaining in the pre-cleaned air is removed by the primary filter (3). This chemically treated filter element is pleated and embossed for maximum surface area. A perforated metal shell protects the element, giving it rigidity and strength. Air flows through both the primary and safety elements. In the case of accidental perforation of the primary filter, safety element (4) protects the engine. It also guards the engine during the cleaning of the primary element. The clean air outlet duct (5) has fittings for mounting the restriction indicator. The vacuator valve (6) ejects dust and water continuously, eliminating regular cup servicing.

Another dry type air filter is shown in Fig. 15-4. This is a paper element type filter and is highly efficient. The filter element consists of special paper which has been formed into an accordion treated ring and sealed top and bottom with a plastic ring. This element can be cleaned by removing it from the housing and tapping against a hard surface to jar off accumulated dirt.

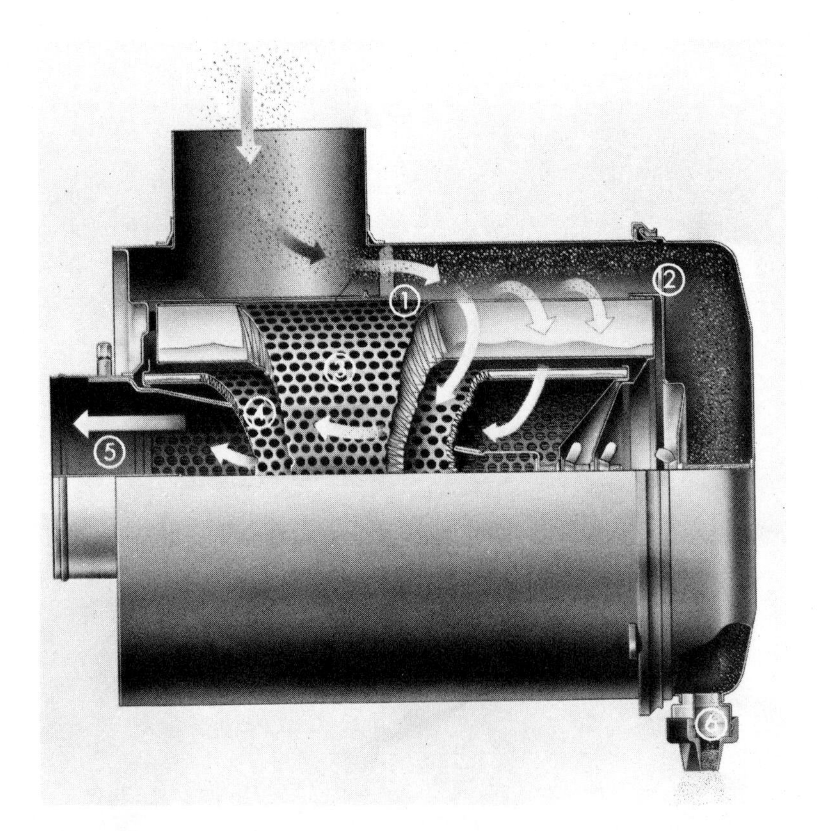

Fig. 15-3. Showing direction of air in Cylopac air cleaner. (Donaldson Co., Inc.)

1. PRE-CLEANER FIN ASSEMBLY.
2. END WALL.
3. PRIMARY FILTER.
4. SAFETY ELEMENT.
5. CLEAN AIR OUTLET DUCT.
6. VACUATOR VALVE.

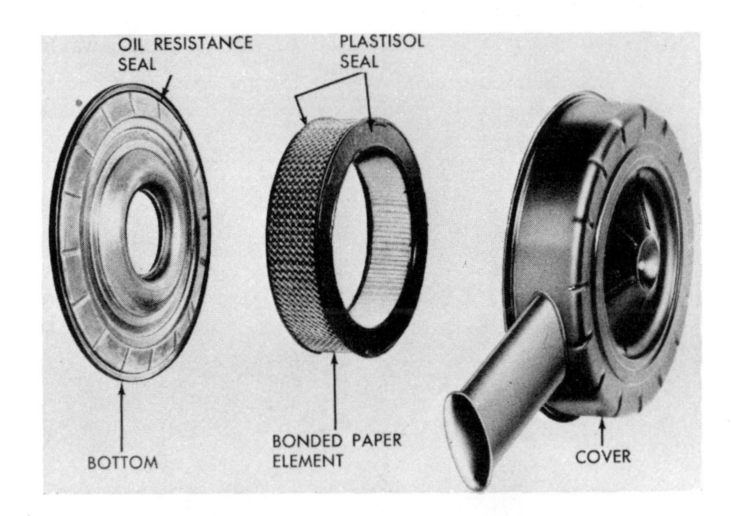

Fig. 15-4. Dry type air cleaner with bonded paper element.

An oil bath type cleaner is shown in Fig. 15-5. In this filter, the direction of the incoming air is reversed and directed over the surface of the oil bath. As a result, a large portion of the oil is retained in the oil bath. The air, in some designs, is then passed through an oil wetted copper mesh screen. In other designs, a treated paper filter is used.

Foamed polyurethane is a sponge-like material used as a filtering element in some air cleaners. The material is moistened with light mineral oil and is classed as an impingment type filter. The filtering element is supported on a perforated metal ring and can be cleaned by rinsing in mineral spirits.

Air filters used in large industrial sized engines are basically similar to those which have just been described. They are larger in size to provide the amount of air required. For example, Fig. 15-6 shows 16 dry type filters assembled to form a bank having large air capacity.

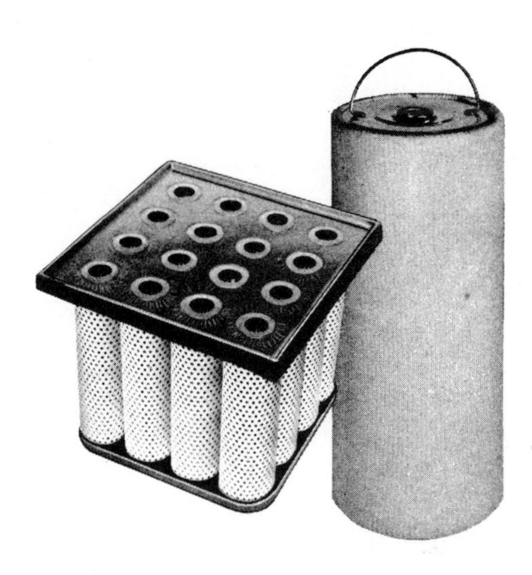

Fig. 15-6. Bank of dry type air filters. (United Air Cleaner Div., United Filtration Corp.)

A roll type air filter is shown in Fig. 15-7. This is an industrial type filter which uses a moving filtering element of fiber glass as the filtering element. The clean filtering element

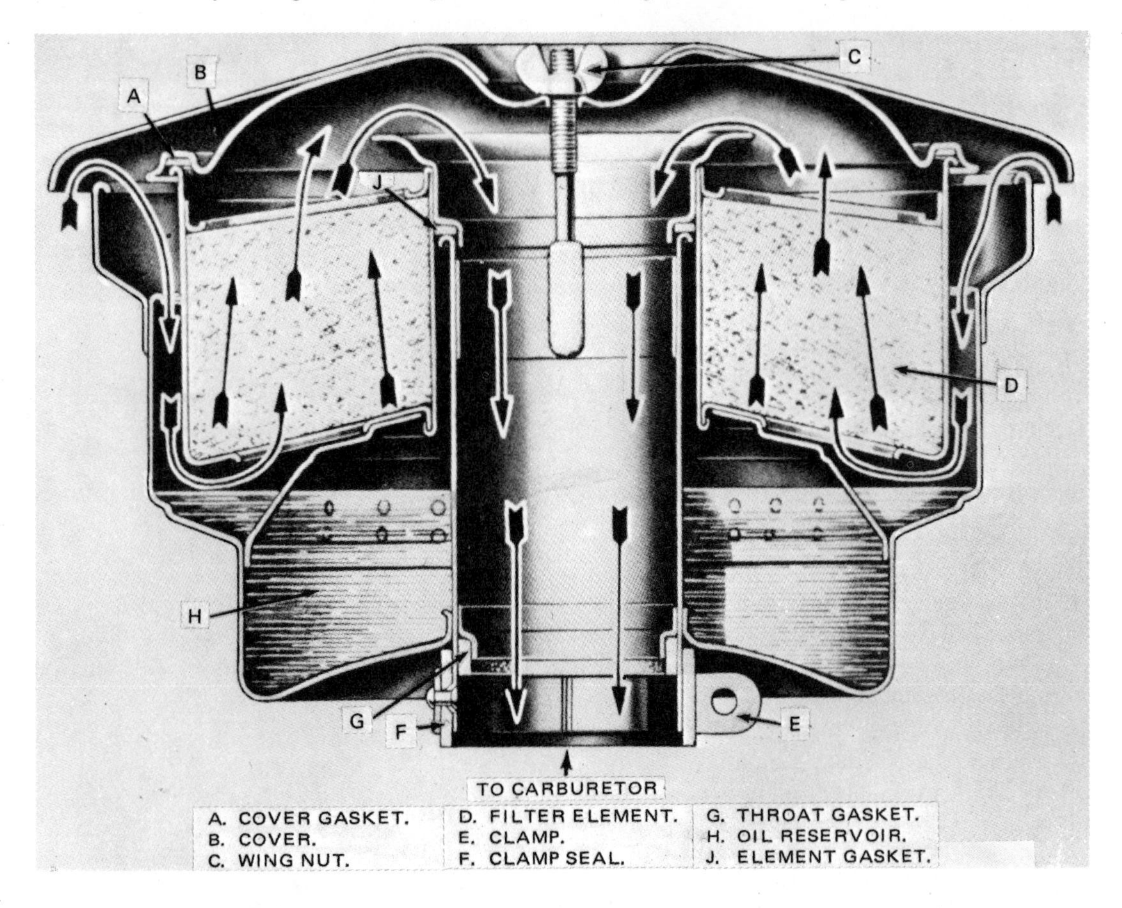

Fig. 15-5. Wet type air cleaner showing oil bath and reversal of air.

is on a roll at the top of the unit and accumulates the dust load as it moves along the face of the filter. It is rolled onto the rewind spool at the bottom of the unit. Methods of operation include pressure, timer, manual, and electric.

Fig. 15-7. Roll type filter equipped with warning light and pressure differential control. (Filter Products Div., North American Rockwell)

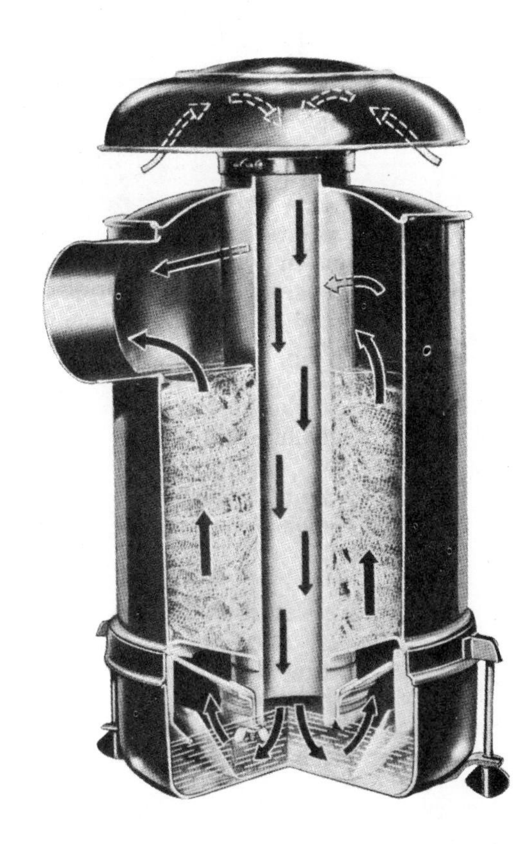

Fig. 15-8. Combined air cleaner and silencer. (Detroit Diesel Engine Div., General Motors Corp.)

SILENCERS

In smaller installations, silencers are usually incorporated in the design of the air cleaner, Fig. 15-8. In larger diesel engine installations, it is often a separate unit, Fig. 15-9.

Silencers are particularly important in large industrial engines which have a greater air intake. Pulsations in the system may set up strong vibrations and noise. Factors involved are the size and length of air intake ducts, material from which the ducts are made, and sharpness of bends. Altering such factors will help minimize noise. Covering metal ducts with sound deadening material may also solve the noise problem. A separate silencer, Fig. 15-9, is used in many installations.

When turbochargers are used, an air cleaner is installed at the air intake. A diagram of an installation made on an International Harvester engine is shown in Fig. 15-10. Note that the incoming air is spun, giving it a centrifugal effect to cast out the heavier particles.

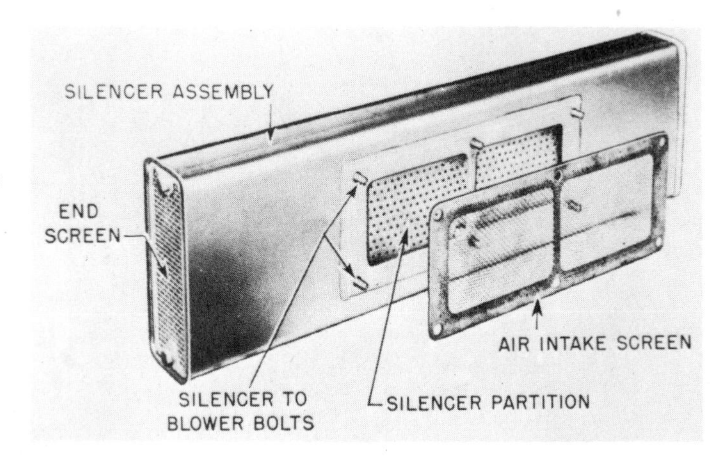

Fig. 15-9. Air intake silencer assembly as used on some large industrial type diesel engines.

FUEL LINE FILTERS

It is imperative that all abrasives and foreign matter be removed from the fuel because of the extremely close tolerances used in the manufacture of diesel engine injectors.

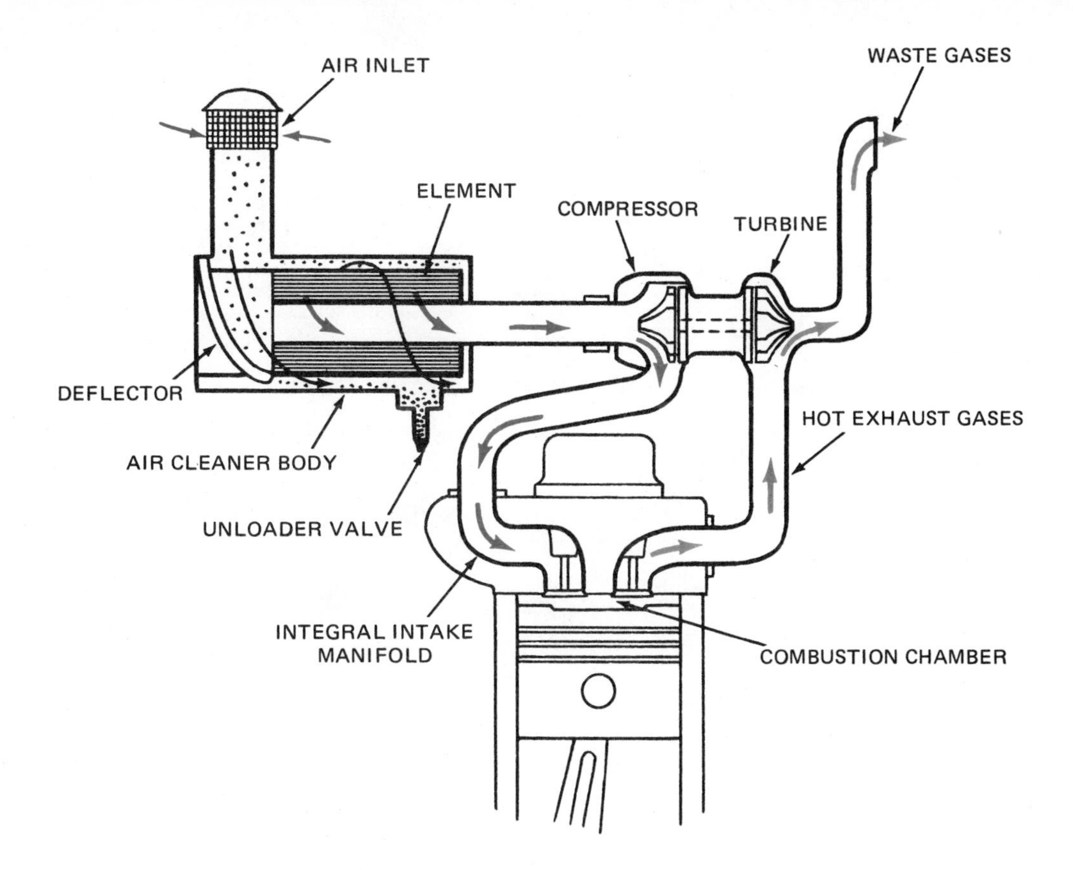

Fig. 15-10. Air cleaner installed on turbocharger air intake.
(International Harvester Co.)

If this is not done, wear on the injector plungers will be accelerated and precision calibration will be destroyed. The fuel system usually contains two or more filters to provide clean fuel. Fig. 15-11 shows the primary and secondary filters as installed on an Allis-Chalmers turbocharged intercooled diesel. This engine develops 435 hp with a displacement of 844 cu. in. Fig. 15-12 shows a complete fuel system of a Detroit Diesel Allison engine used on a Ford truck. In each of these installations, two filters are provided between the fuel supply tank and the injectors. In addition, filters are often placed in the main storage tank fill line and immediately before each injector to clean the fuel.

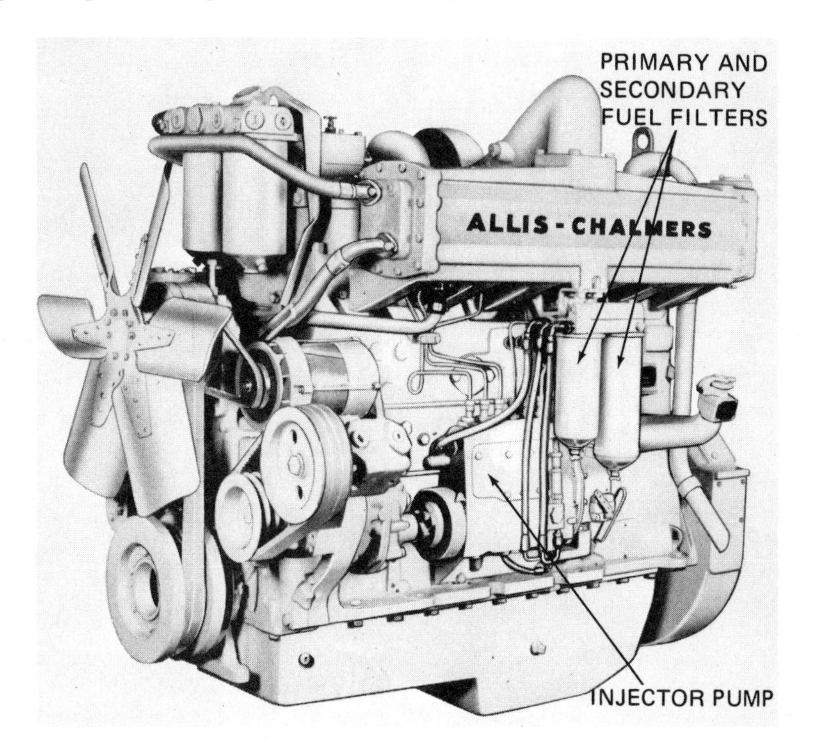

Fig. 15-11. Installation of primary and secondary fuel filters.
(Allis-Chalmers Mfg. Co.)

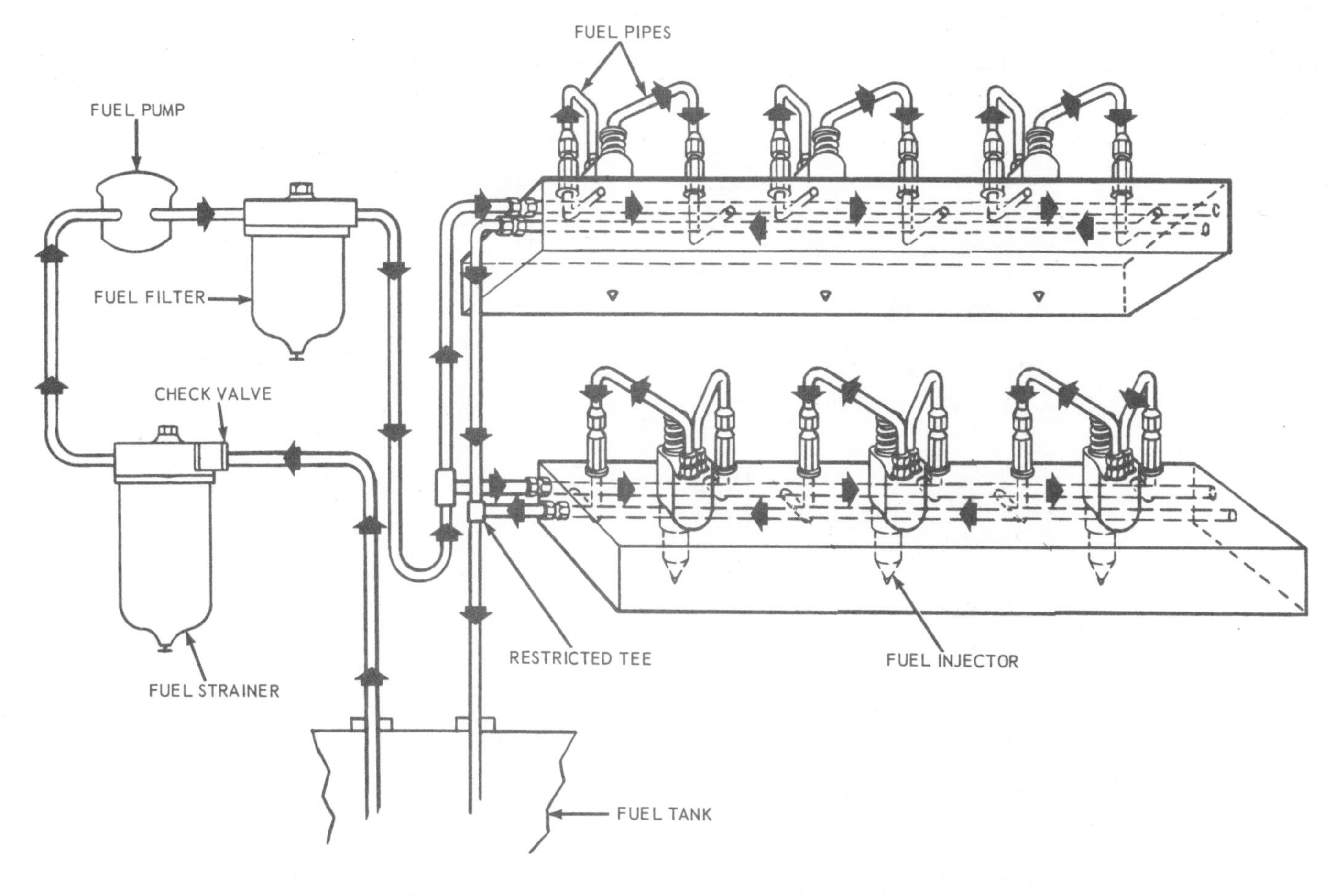

Fig. 15-12. Detroit Diesel Allison fuel system. Note fuel strainer and fuel filter together with fuel return to tank.

Fuel filter elements are made of materials which include:
1. Fiber discs.
2. Pleated and plastic impregnated paper.
3. Cotton fiber or waste.
4. Fuller's earth.
5. Bonded wool fiber.
6. Wire mesh.
7. Metal edge.

In the fiber disc type filter, as produced by the William W. Nugent Engine Co., Inc., the filtering element consists of a stack of laminated fiber discs, each rotated 45 deg. from the position of the adjacent disc, Fig. 15-13. Each disc exposes a maximum surface for filtering. The fuel oil to be filtered flows from the exterior through adjacent discs to the interior of the filter stack. In this way, greater area is exposed to the flow. The nap on the element traps the very fine particles of dust and builds up a thick porous filtering mass on the corrugated cellulose.

Gaining in popularity is the accordion pleated plastic impregnated filter, Fig. 15-14. Such filters come in all sizes and the small one shown in Fig. 15-14 is being removed from the fuel pump inlet.

Wire mesh screens are also used. In Fig. 15-15, a filter screen is shown being installed in the inlet to the injector on a Cummins diesel. The fineness of filtration is controlled by the spacing between the metal discs in metal edge filters. In one type, fineness of filtration is controlled by the difference in

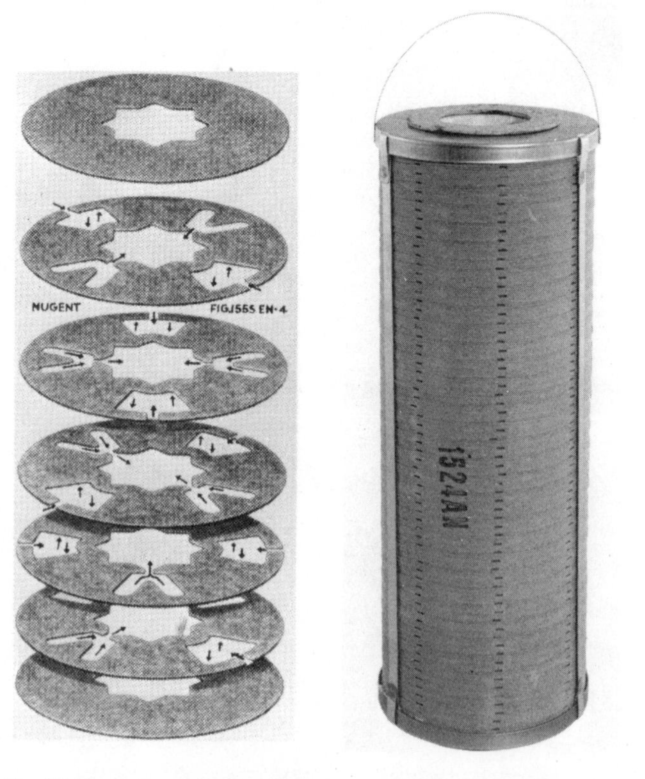

Fig. 15-13. Laminated disc filter element. The complete assembly is shown on the right and individual discs at the left. (Wm. W. Nugent and Co.)

the outside diameter of the minor discs and inside diameter of the major discs. Another type of metallic filter element is made by winding a metal ribbon edgewise on a corrugated cylinder. Filtering passageways of .0005 in. are provided. In metal edge filters, the fuel oil flows radially from the outer edge to the inner, impurities being collected on the surface.

Fig. 15-14. Filter screen element. The illustration shows a Cummins engine as installed in a Ford truck.

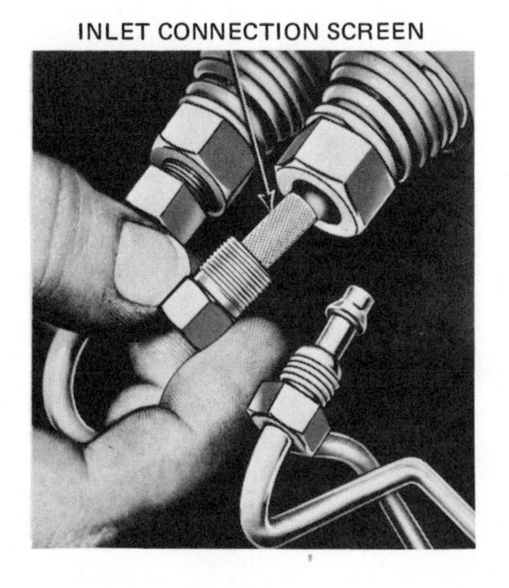

Fig. 15-15. Inlet injector screen in Cummins installation in Ford Truck.

The accordion pleated plastic paper filter will remove foreign particles as small as one micron (one-thousandths of a millimeter). A large filtering area is provided in a small space, Fig. 15-16, and at the same time there is less resistance to the flow of fuel.

Fig. 15-16 shows the details of the CAV paper element filter. In this filter, the paper element is contained in a metal canister and the paper strips forming the element are wound around a cylindrical core in the form of a spiral. The strips are cemented together at the top and bottom so as to form a series of continuous V shaped coils. This provides a large filtering area. The filter unit consists of three main parts, the filter head, the element and base. The filter is the cross-flow type. Inlet and outlet connections are arranged on the filter head which also incorporates the mounting bracket.

Fuller's Earth filters will remove acid from the fuel by adsorption and solid matter by absorption. It is claimed that

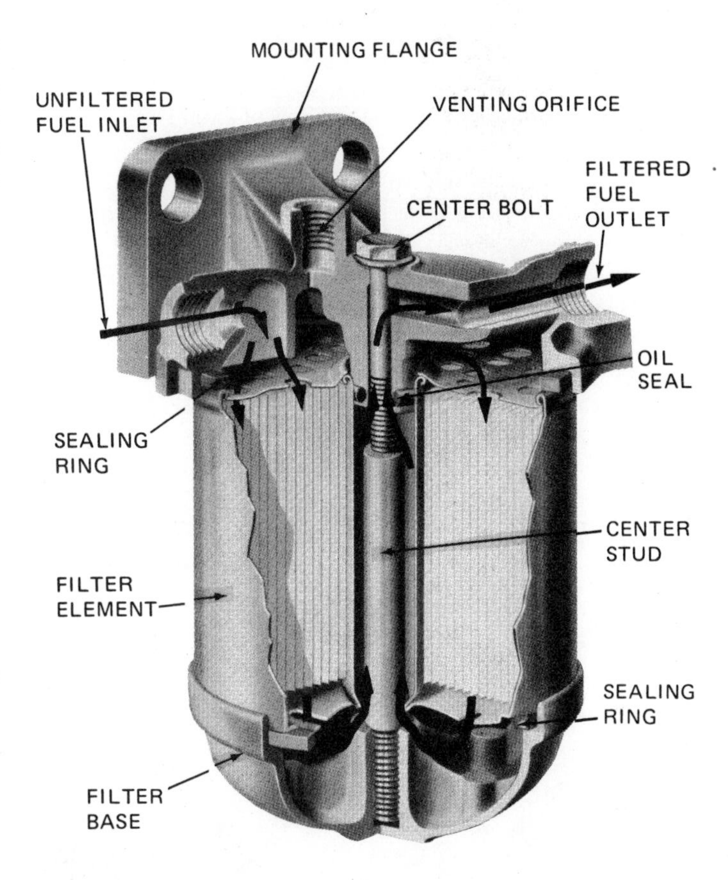

Fig. 15-16. Sectional view of CAV paper filament filter showing fuel flow through the element. (C.A.V. Div. Lucas Electrical Service Inc.)

solids as fine as 1/2 micron are retained. In general, filters with Fuller's Earth should not be used for fuels which contain additives.

Absorption is the taking up or sucking up of a substance by another. Adsorption is the adhesion of the molecules of a fluid in a very thin layer to the surface of a solid body.

Cotton waste filters consist of cotton waste which has been tightly packed into a cylindrical container. In many cases the filter has a few layers of impregnated paper inside the cotton waste and the fuel flow passes radially through the waste.

Another filter which employs cotton is the honeycomb filter. Cotton yarn is wound in a honeycomb fashion around a cylindrical wire screen. Fuel flow is toward the center and large particles are entrapped to the outer layers of yarn, and the smaller particles closer to the center of the unit.

FILTER REPLACEMENT

Some filters are designed so that is is only necessary to replace the element which is installed in a housing permanently attached to the diesel engine. Other types of filters are designed so that the entire unit is discarded. Metal edge filters do not require replacement since the filter is cleaned by rotating the blade.

Frequence of replacing the filter element is largely determined by the character of service in which the engine is being operated and the type of fuel and oil being used.

REVIEW QUESTIONS – CHAPTER 15

1. The advantage of reversing the flow of air in an air cleaner is to:
   a. Increase the velocity of the air.
   b. Throw out a large percentage of the foreign matter.
   c. Reduce the velocity of the air.
2. Air filters used in large industrial engines are basically the same as those used on ____________.
3. The size and length of air intake ducts are factors in the amount of ____________ made by the air intake.
4. Two filters are usually installed between the fuel supply tank and the ____________.
5. List seven different materials used for filter elements.
   a. ____________.
   b. ____________.
   c. ____________.
   d. ____________.
   e. ____________.
   f. ____________.
   g. ____________.
6. Adsorption is the adhesion of the molecules of a fluid in a very thin layer to the ____________ of a solid body.
7. Metal edge filters do not require ____________ since the filter is cleaned by rotating the blade.
8. When turbochargers are used, an air cleaner is installed at the ____________.
9. A plastic sponge-like material used as a filtering element in some air cleaners is ____________.
10. The two main types of air cleaners are:
    a. ____________.
    b. ____________.

A diesel-powered tractor discing a field. Frequent air cleaner maintenance is a "must" in farming applications. (Deere & Co.)

Chapter 16
LUBRICATING SYSTEMS

The purpose of the lubricating system of a diesel engine is to keep friction between bearing surfaces to a minimum and assist in dissipating some of the heat that is developed by fuel combustion.

The engine system which supplies the oil is the pressure type in most modern diesel engines. Variations may exist in the details of engine lubricating systems, but the parts and their operation are basically the same.

The lubricating system of a diesel engine, Fig. 16-1, includes the following items: oil pumps, passages for conducting the oil, oil reservoirs, filters, strainers, oil coolers, pressure gauges and pressure regulators.

While these parts are found in virtually all systems, they may be connected in many ways so that there are different types of systems. These include the full flow, shunt, sump filtering and bypass filtering systems.

The full flow system is found in most new designs. In this system, the filter elements are designed with high flow rate permitting the entire pump delivery to pass through the filters. A bypass valve is provided in case the flow of oil through the filters is obstructed. If so, the filter will be bypassed.

In the shunt type filtering system, the pressure pump

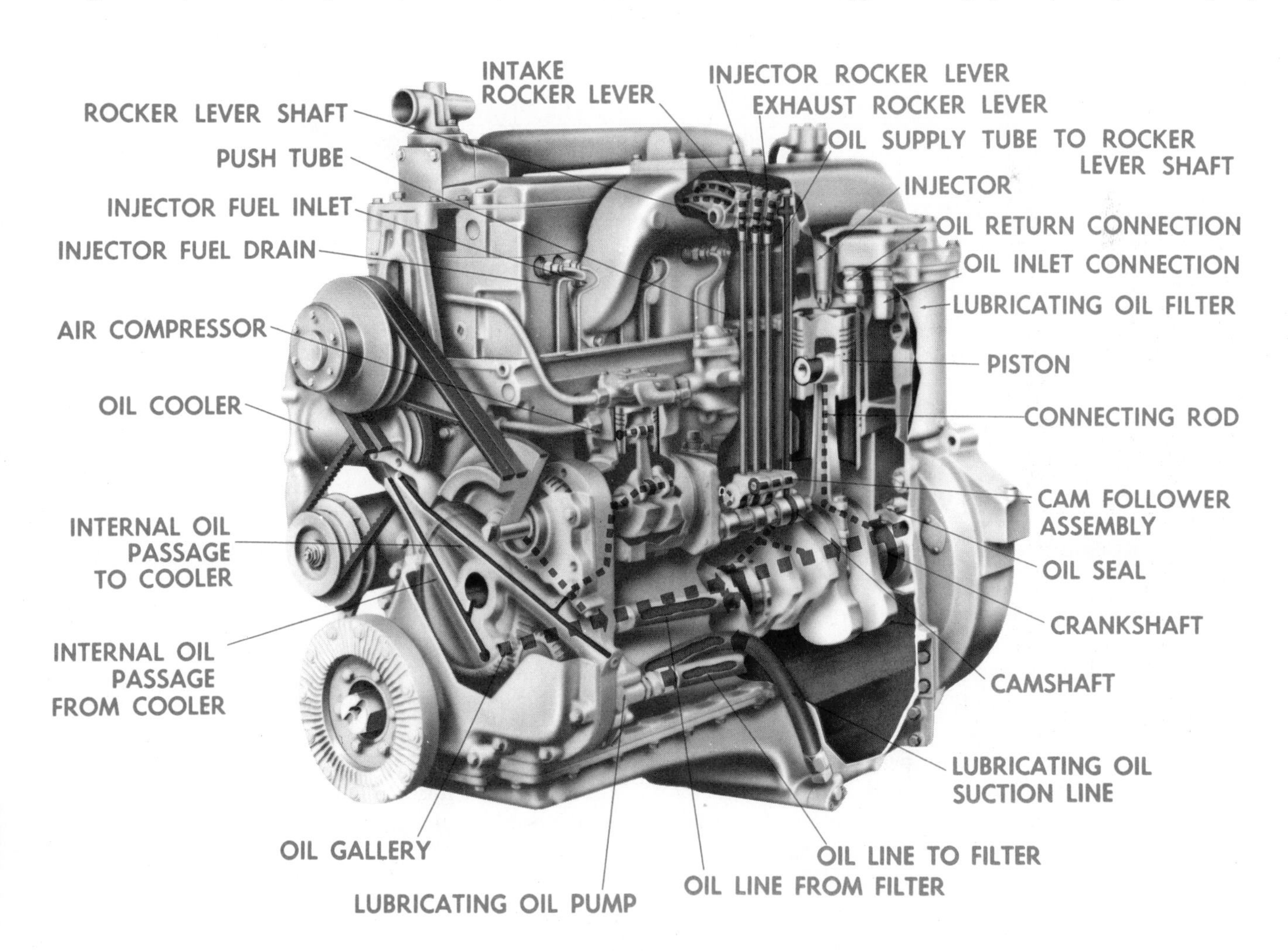

Fig. 16-1. Lubricating oil flow in Cummins NRTO diesel engine.

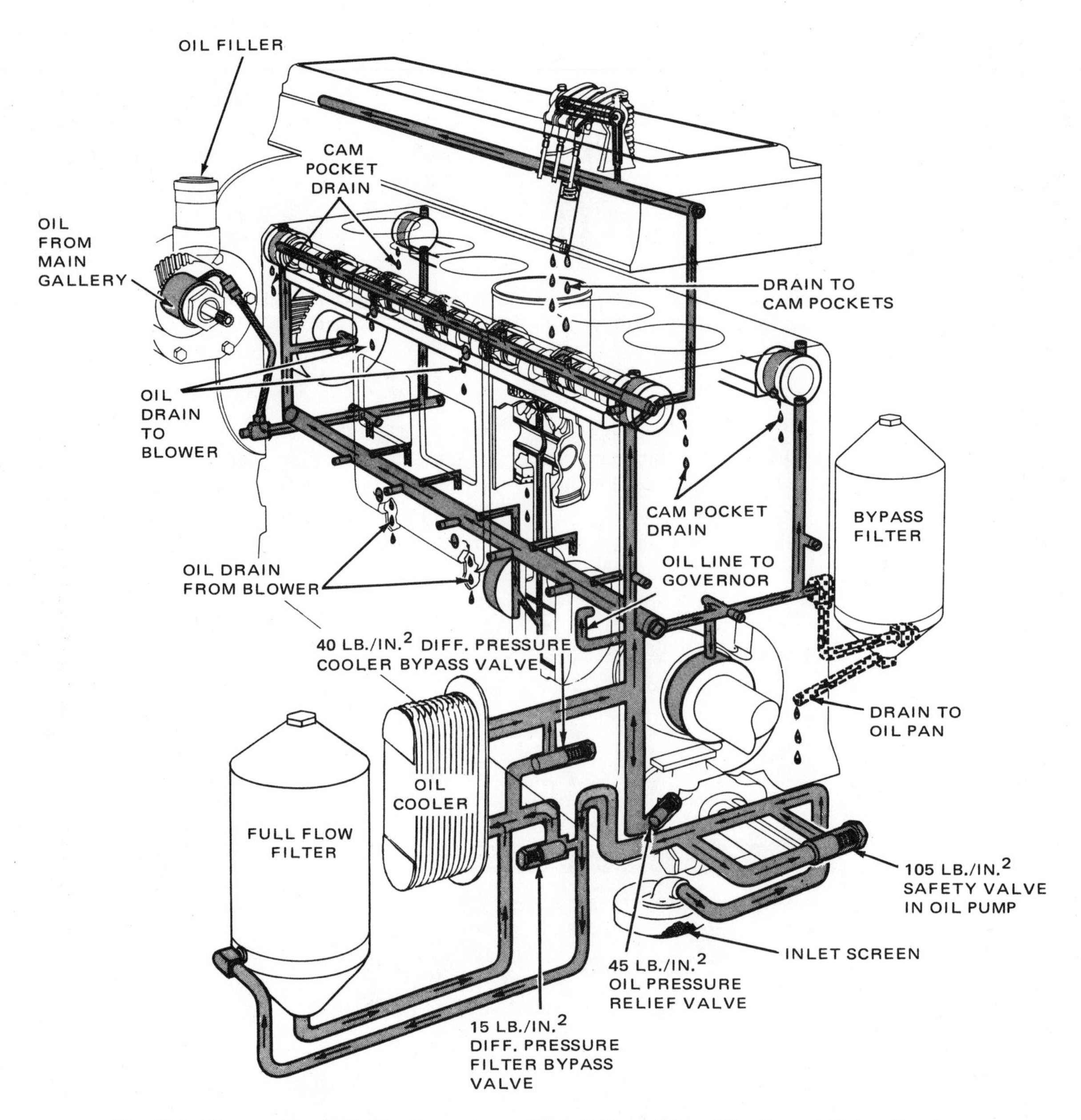

Fig. 16-2. Diagram of typical lubrication system. (Detroit Diesel Allison Div., General Motors Corp.)

discharges the oil through the filters, then through a cooler to the engine. In order that an adequate flow of oil will be delivered under all conditions, the pump is designed for excess capacity and the filters are provided with a bypass through which a portion of the oil flows. If the filter becomes clogged, or if the oil is cold, a relatively large portion of the lubricating oil is shunted through the bypass.

The sump filtering system is similar to the shunt system, except that the filter is placed in a separate circulating system including a separate motor driven pump. The sump system permits the lubricating oil to circulate through the filter, even when the engine is not operating. Oil to be filtered is taken from the sump by the motor driven pump, forced through the filter and then discharged back to the sump. Oil to the engine is taken from the sump by the engine driven pump and forced through the cooler and filter to the engine.

The bypass lubricating system is similar to the shunt system, except that a portion of the oil is continuously bypassed back to the sump through the filters. In order that sufficient oil will be supplied to the engine, the amount of oil permitted to flow through the filter is limited by the size of the piping or an orifice. A schematic diagram of the lubricating system used on the General Motors two-cycle Series 71 engine is shown in Fig. 16-2. The Series 71 engine, depending on the number of cylinders, is designed for use in transportation, marine and stationary service. Fig. 16-2 illustrates the arrangement of the various units and control valves of the full flow lubricating system used on this engine. The complete system consists of an oil pump, oil cooler, a full flow oil filter with a bypass valve, together with a suitable relief valve to the pump, a bypass valve to the oil cooler and a pressure regulator valve in the cylinder block oil gallery. A bypass oil filter is also included in some engines. Oil for lubrication of the connecting rod bearings, piston pins, and for cooling the piston head is provided through the drilled crankshaft from the adjacent forward main bearing. The gear train is lubricated by the overflow oil from the camshaft pocket through a communicating passage into the flywheel housing. A certain amount of oil spills into the flywheel housing from the camshaft, balancer shaft and idler gear bearings. The blower drive gear bearing is lubricated through an external pipe from the rear horizontal oil passage of the cylinder block.

Oil overflows into two holes, one at each end of the blower housing, providing lubrication for the blower drive gears at the rear end of the governor mechanism at the front of the engine.

Fig. 16-3 shows the engine lubricating system of the GM 5.7 litre diesel engine produced by the Oldsmobile Division. In this system, a gear type oil pump is bolted to the rear main bearing cap. It is driven by the camshaft gear through a hexagonal drive shaft.

During operation of the 5.7 litre diesel, the oil passes through an oil filter on the right side of the engine block. (A by-pass will open at 5 1/2 to 6 1/2 psi if the filter is restricted.) Next, the engine oil flows to the oil cooler located in the radiator. From there, it is directed back to the engine oil galleries. A by-pass valve is also built into the filter base to allow a continuous flow of oil to the engine in case of a restriction in the cooler or cooler lines. This filter by-pass opens at about 12 psi.

In the lubricating system of an International series DV engine, oil is supplied by a 20 gpm (gallons per minute) capacity pump which features recirculation of bypassed oil in the pump's body. Full length oil galleries in the crankcase are eliminated by incorporating the main gallery, and all passages from the oil pump, into the filter cooler base die casting. Lubrication of the main, rod and cam bearings is by drilled holes from the main gallery. Cross drilling in the camshaft front and rear journals provides an intermittent supply of oil to the rocker arms, timing gears, fuel pump cam and tachometer drive bushing. Guided run-off from the tachometer drive bushing provides lubrication for the pump drive. Lubrication for the injection pump and the air compressor is supplied by an external line from the main gallery. Oil flowing from the injection pump lubricates the automatic advance mechanism, pump drive and pump drive gear.

In the Nordberg Series 13 four-cycle turbocharged in-line engine, which is available as a diesel, dual fuel or spark ignition gas engine, Fig. 16-4, oil is supplied by an oil manifold outside the engine crankcase. From the manifold, it passes to a channel surrounding a tie bolt where the oil is distributed to the crankshaft connecting rod bearings. Note that the connecting rods have drilled oil passages which conduct the oil to the piston pin. A jet of oil from the piston pin boss sprays the underside of the piston for improved cooling. Some Nordberg engines lubricate the cylinder by multifeed lubricators which supply oil to the cylinder box at a number of points near the top of the stroke. Each cylinder has a lubricator with a separate pump for each feed. It delivers a metered amount of

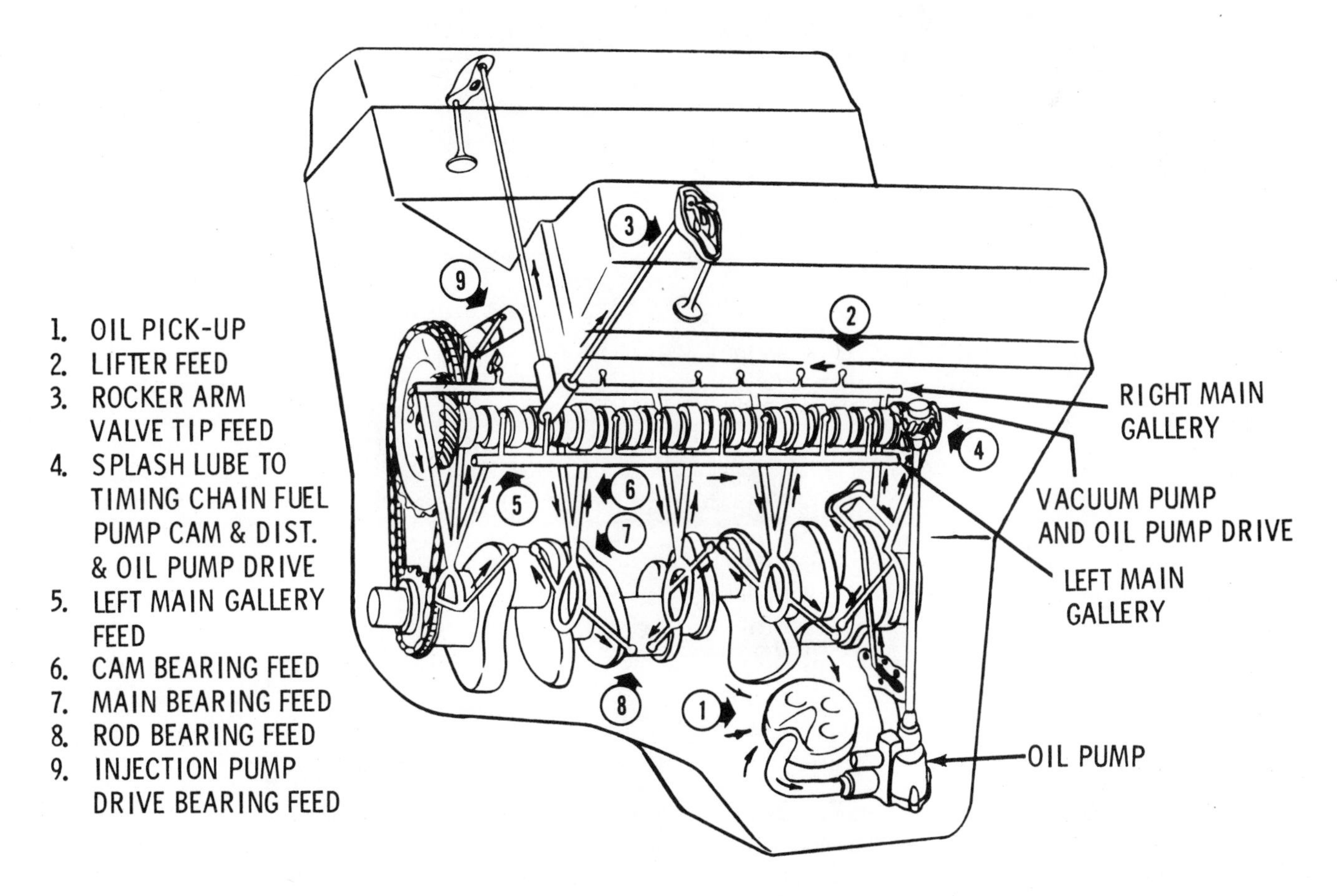

Fig. 16-3. Engine lubrication system of the GM 5.7 L diesel engine. (Oldsmobile Division, General Motors Corp.)

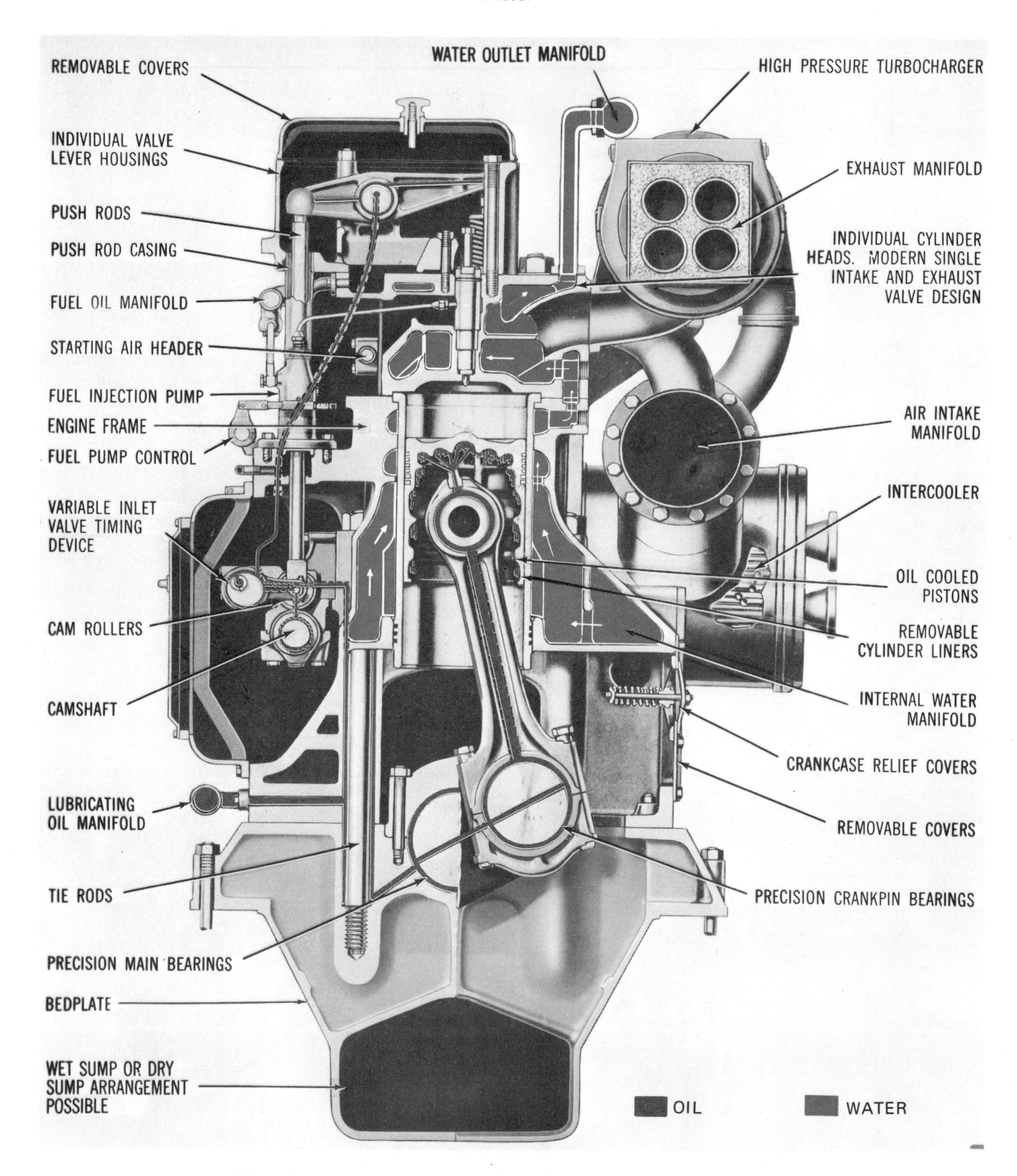

Fig. 16-4. Cross section of Nordberg series 13 in-line engine showing lubricating system.

lubricating oil to the liner wall. When the piston rings pass the oil feed holes in the inner circumference of the liner wall, the lubricators deliver a metered amount of lubricating oil.

Caterpillar's new 5.4 bore V-12 engine lubricating system, Fig. 16-5, includes four filters and a scavenging pump. This system includes a positive displacement gear type pump which is mounted to the bottom of the cylinder block within the oil sump. The pump for the vehicular engine is the double section type. The main pump section supplies oil to the engine for lubrication. The scavenging pump section transfers this oil

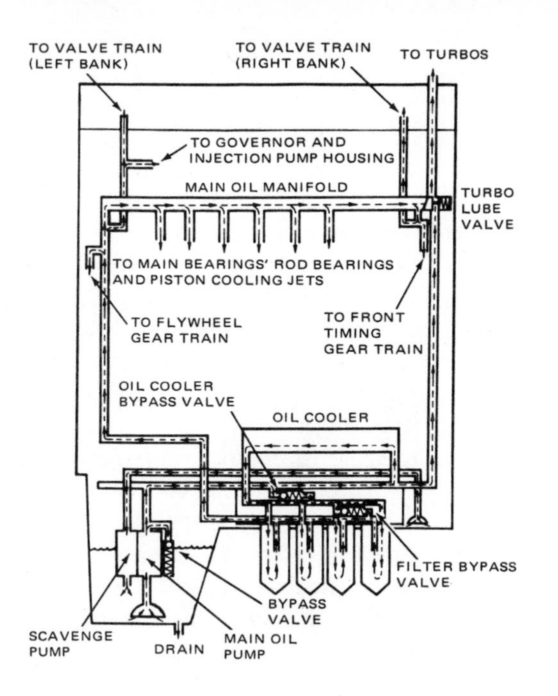

Fig. 16-5. Note the bank of four filters. In addition to main oil pump a scavenge pump is also used. (Caterpillar Tractor Co.)

from the front shallow pan to the rear sump. This arrangement allows engine operation when tilted in any direction of 235 deg. without a significant loss in oil pressure. At rated engine speed, the capacities of the main oil pump and scavenge pump are 112 gpm and 88 gpm, respectively.

Oil from the pump enters the block at the right rear of the engine. A cored passage carries the oil to the oil filter base, which is mounted at the right side of the engine. A shell and tube bundle type oil cooler is mounted below the filter base. A valve in the oil filter base bypasses oil around the oil cooler until the oil has reached operating temperature.

The four oil filters are 5 in. paper elements in individual housings. A valve in the filter base permits oil to bypass the filter in the event the element becomes clogged. An additional cored passage takes the oil from the cooler and filter to a drilled passage in the flywheel housing. This passage takes oil to the rear timing gear train and three manifolds, one located in the center vee and the other on each side of the vee. The center manifold supplies oil to the main and connecting rod

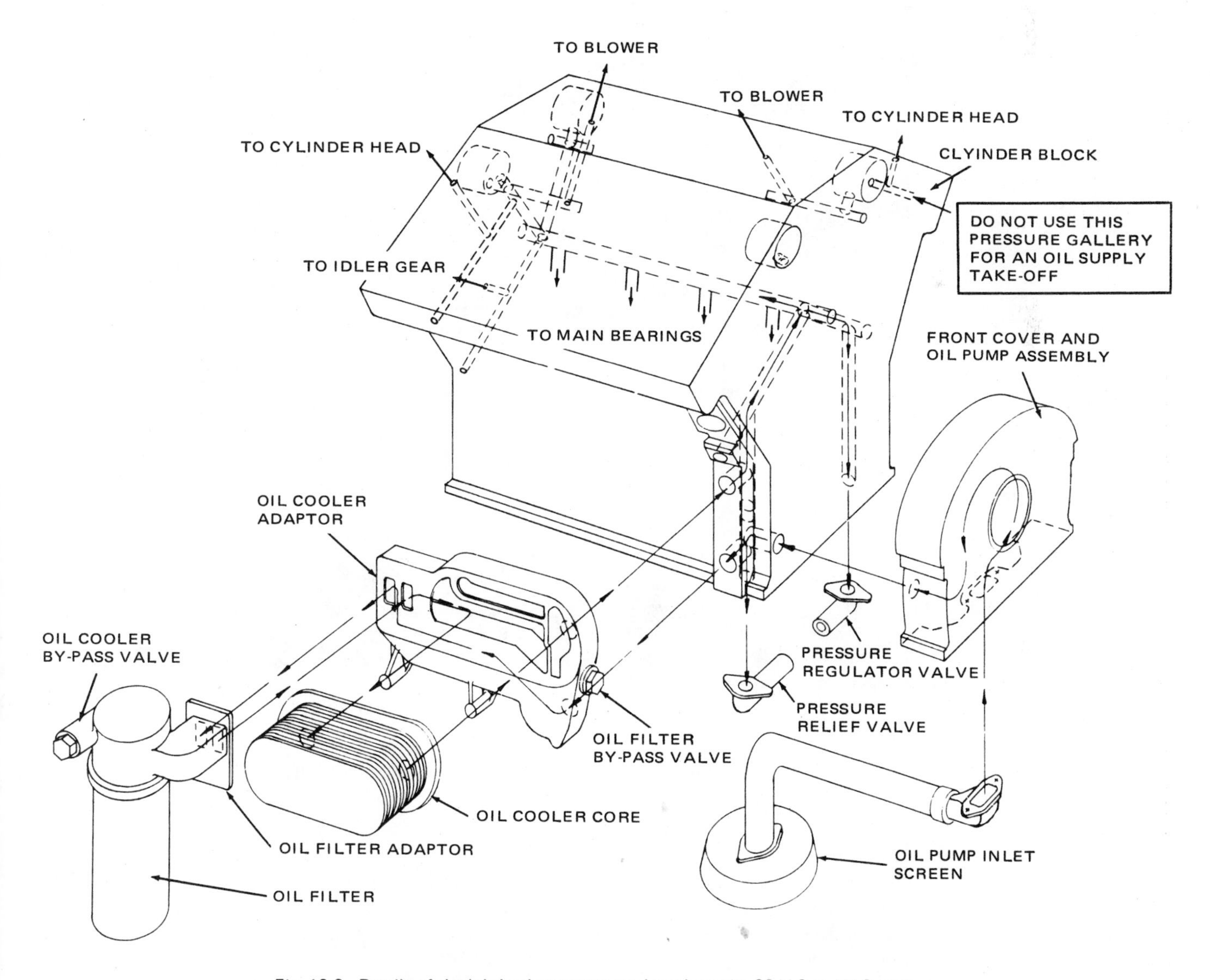

Fig. 16-6. Details of the lubricating system used on the series 92 V-6 and V-8 engines. (Detroit Diesel Allison Div., General Motors Corp.)

bearings, governor, fuel pumps and connects to the front timing gear housing and turbochargers. The other two manifolds provide oil for the jets which spray the pistons.

An oil passage leads directly to the front timing gear housing from the oil filter base, through the turbo valve to the turbocharger. This insures lubrication of the turbocharger while the engine is being started. When oil reaches the turbocharger valve, Fig. 16-5, through the center manifold, the valve shifts and allows oil to flow normally to the supercharger. Surplus oil drains from the turbocharger to the oil pan by way of the front timing gear housing.

The oil system for Caterpillar industrial, marine and electric set engines is similar to the vehicular engine just described, except the oil pump does not include a scavenging pump. The oil pump is mounted horizontally across the engine front.

The lubricating system installed on Detroit Diesel Allison filter, through oil cooler, then back to cylinder block.

In the block, a short vertical gallery and a short diagonal gallery carry oil to main longitudinal oil gallery through the middle of the block. Valves are also provided to bypass the oil filter and oil cooler should either one become plugged.

An auxiliary lubricating oil pump is provided in case the built-in pump fails. The auxiliary pump may also be used as a prelubricating pump.

A sump type lubricating oil system is shown in Fig. 16-7. In this Cooper Bessemer engine, oil is drawn from the sump by a pump operated by an electric motor. After passing through the oil cover, one line passes the oil to the metal edged strainer and onto the engine. Another line from the cooler passes oil through the oil filter and returns the filtered oil to the engine sump. The oil can be circulated while the engine is stopped since the pump is independent of engine operation.

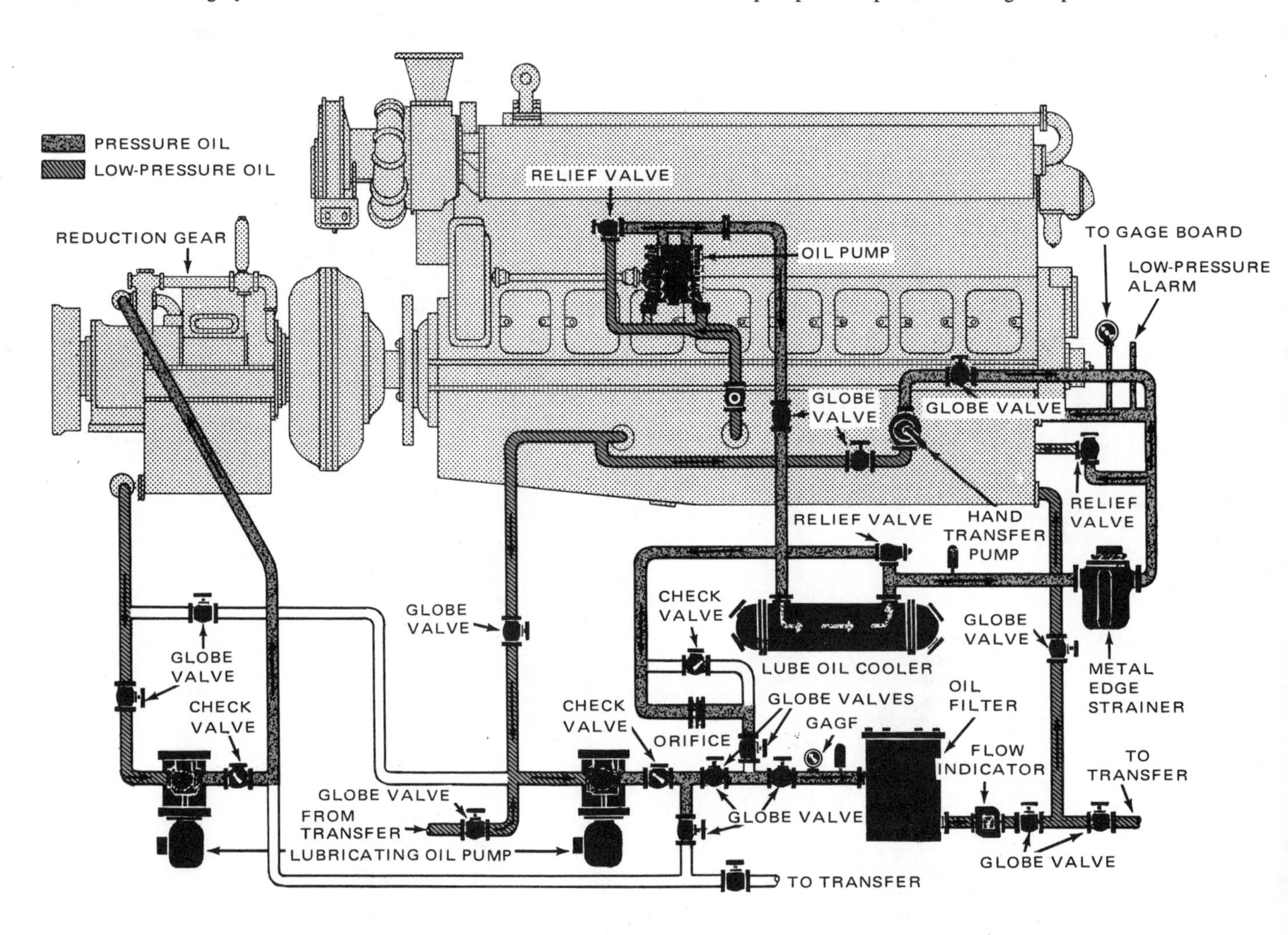

Fig. 16-7. Sump type lubricating system. (Cooper Bessemer Div. of Cooper Industries)

V-6 and V-8 engines is shown in Fig. 16-6. Components include the oil pump, oil filter, oil cooler, by-pass valves, pressure regulator and pressure relief valve.

The gear type oil pump is placed in crankshaft front cover. From the pump, oil passes through a short gallery in cylinder block to oil cooler adapter plate and, at the same time, to pressure relief valve. From adapter plate, oil passes to full flow

The lubricating system of the two-cycle Nordberg radial engine is of particular interest. Details are shown in Fig. 16-8. Two systems of lubrication are used on this engine, mechanical force feed and circulating pressure feed. Motor driven multi-feed lubricators deliver oil to two points of each cylinder for piston lubrication. The pressure system serves the dual purpose of providing oil to all parts requiring lubrication and piston

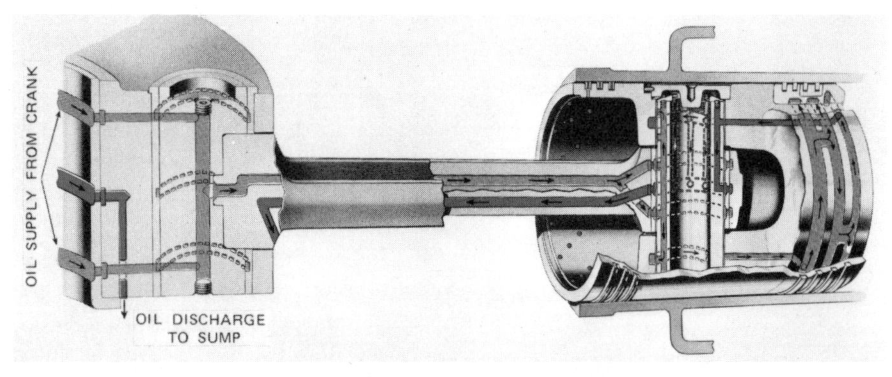

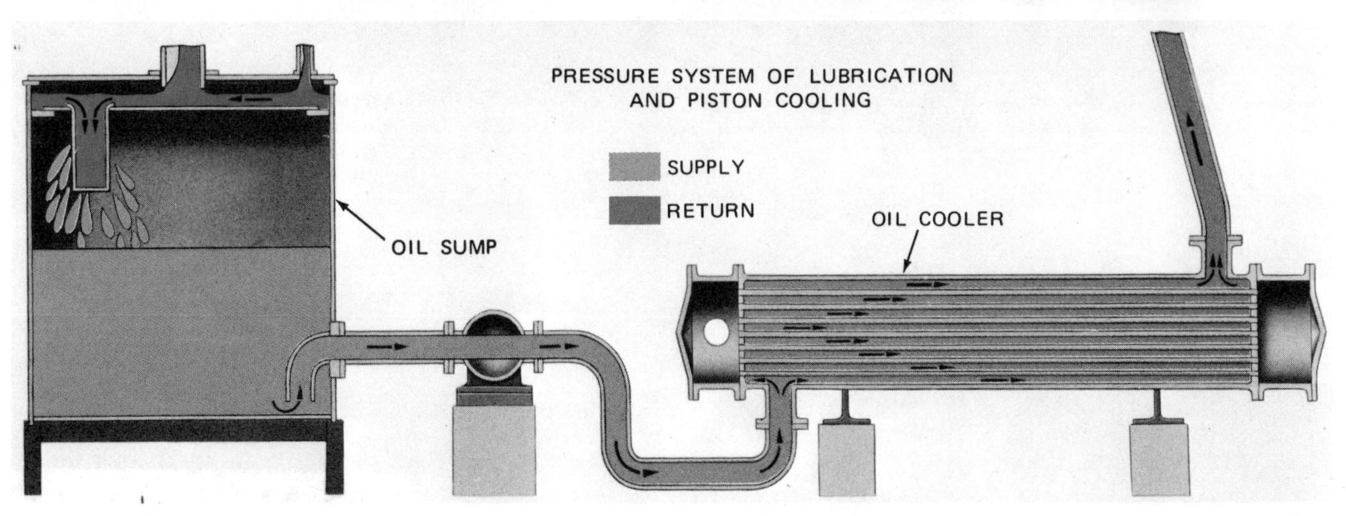

Fig. 16-8. Lubricating system on radial engine. (Nordberg Mfg. Co.)

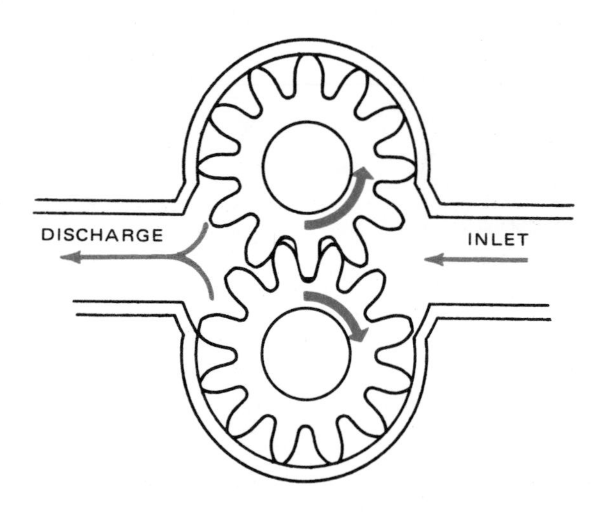

Fig. 16-9. Typical spur gear oil pump design. (Thompson Products)

cooling. After circulating through the pistons, the oil returns to the sump for recirculation.

OIL PUMPS

The pumps used in the lubricating system of the diesel engine are of the positive displacement type and are classed as follows:

1. Spur gear pump.
2. Herringbone gear pump.
3. Vane pump.

Positive displacement pumps utilize the force of a mechanical member confined within a chamber to force fluid from the inlet port to the outlet port, Fig. 16-9. The parts are so arranged that their movement creates an expanding fluid chamber. Suction of fluid into the chamber and a contracting chamber causes fluid to be expelled to the exit port.

The rate of discharge is directly proportional to the speed of rotation. Positive displacement pumps continue pumping against a higher pressure head until the parts fail or until driving torque fails. For this reason, a pressure relief valve is usually included in the design.

The overall efficiency of a positive displacement pump is high. Major loss is due to friction. If pump clearances are high, or if the parts are not sufficiently lubricated, high friction losses result. If internal clearances are low, there will be considerably leakage from the high pressure side to the low pressure side and efficiency will drop.

In Fig. 16-9, oil enters on the right side of the pump, is carried around the periphery of the gears in the spaces between the teeth, and discharged on the opposite side. As the teeth mesh on the discharge side, oil is forced out of the spaces between the teeth. When the teeth come out of mesh on the inlet or suction side, oil flows into the spaces between the teeth. Since the meshing teeth form a line contact seal, the oil is prevented from returning from the pressure side to the suction side of the pump.

A spur gear type of oil pump, Fig. 16-10, is used on the Waukesha F 7D six cylinder engine, with a bore and stroke of

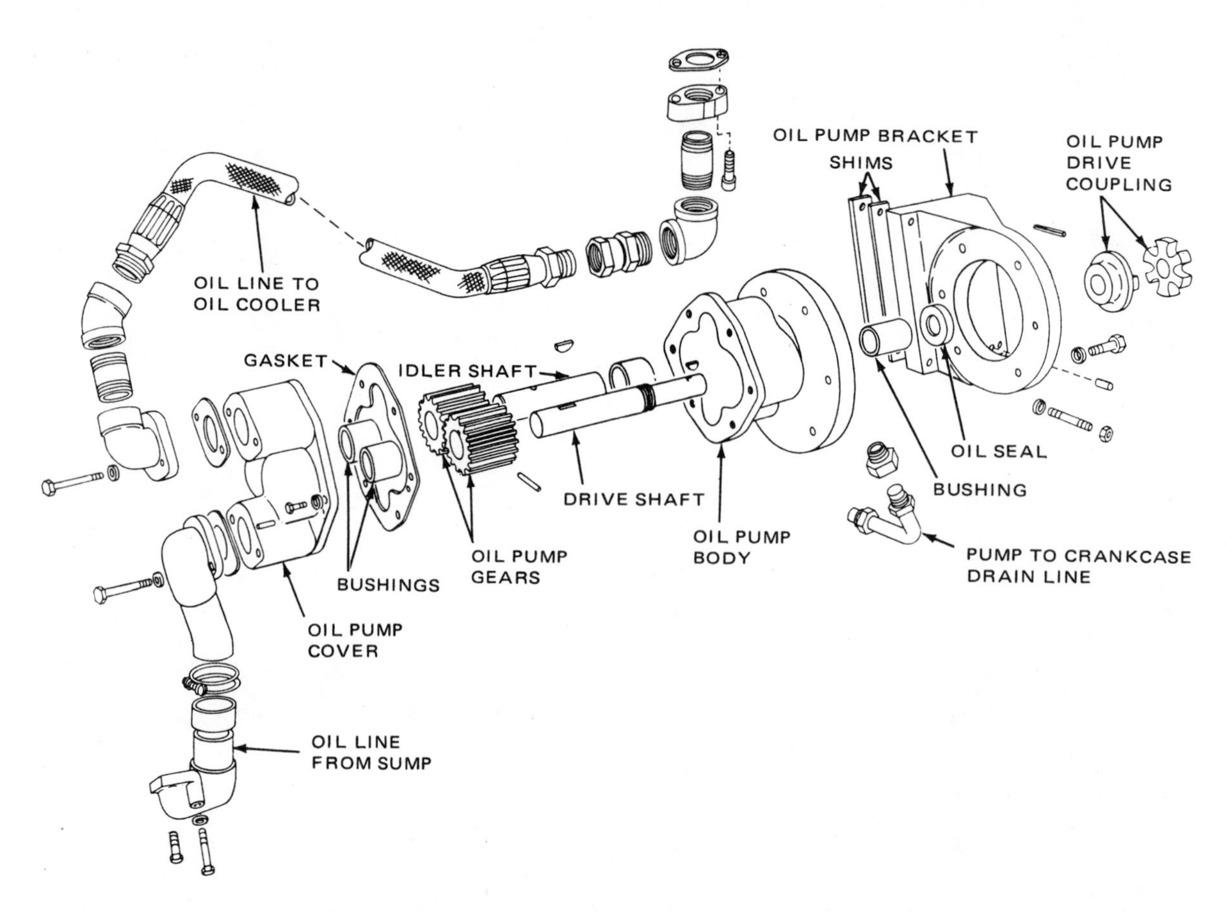

Fig. 16-10. Exploded view of gear type oil pump. (Waukesha Motor Co.)

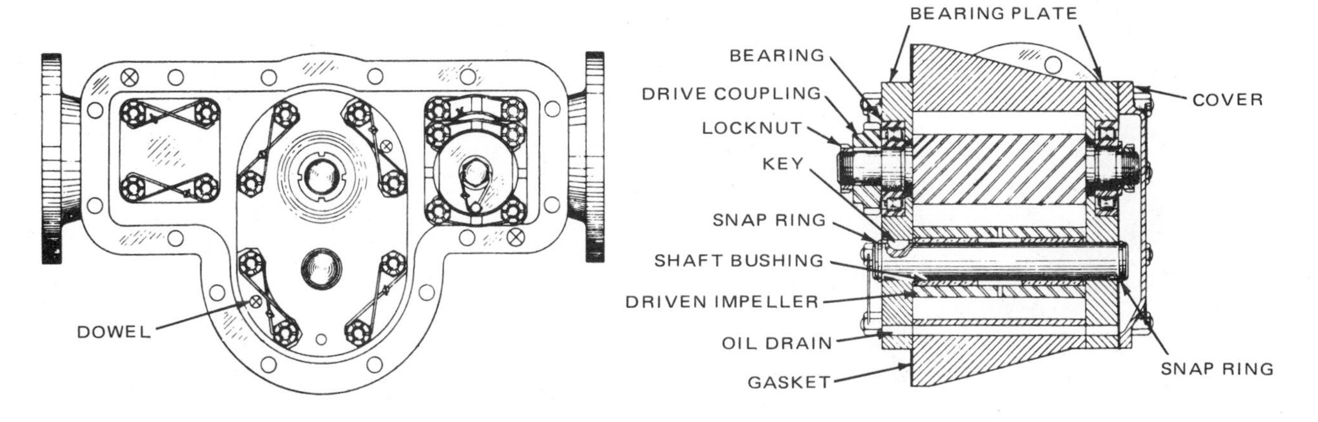

Fig. 16-11. Herringbone gear type oil pump. (Fairbanks Morse Power Systems Div., Colt Industries Inc.)

6 1/4 and 6 1/2 in., giving a displacement of 1197 cu. in. This is an externally mounted oil pump with a cast body unit and a removable cover. The oil pump driven gear is keyed and pinned to the heat-treated shaft which runs in replaceable bushings in the housing cover. The drive shaft is sealed at the drive end by a spring-loaded seal pressed into the oil pump body. The idler gear is pinned to the idler shaft.

A herringbone gear type of oil pump is shown in Fig. 16-11. This is used on a Fairbanks Morse Model 38D18 diesel engine, and is driven through gears and a flexible drive on the lower drive shaft.

A sliding vane type of oil pump, used on the Ford 2700 range of four and six cylinder diesel engines, is shown in Fig. 16-12. It consists of a rotor housed eccentrically in a bored hole machined in the pump body. Four sliding vanes are located in the grooves, machined in the periphery of this rotor, and positioned on either side of the rotor. The vanes are held against the pump body bore by centrifugal force while the pump is operating. The inlet port is connected to the sump and the outlet port is connected to the oil galleries in the engine. As the rotor revolves, the vanes pass over the inlet port and oil is drawn into the space between the rotor and the pump housing. Oil is carried between the vanes to the outlet port where it is forced out into the engine oil galleries, since the space between the rotor and pump bore decreases.

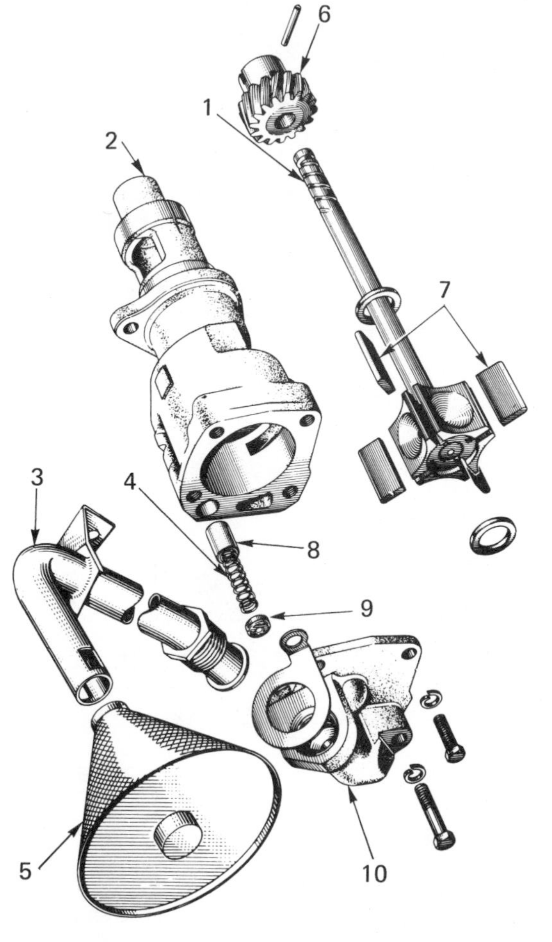

Fig. 16-12. Vane type oil pump. (Ford Motor Co.)

1. ROTOR AND SHAFT ASSEMBLY.
2. PUMP BODY.
3. INLET PIPE.
4. SPRING.
5. GAUZE SCREEN.
6. SKEW GEAR.
7. VANES.
8. OIL PRESSURE RELIEF VALVE PLUNGER.
9. RETAINER.
10. COVER.

OIL COOLERS

The lubricating system of most diesel engines, especially the large ones, must include some method of cooling the lubricating oil so that it is maintained at the most efficient operating temperature.

Unless some provision is made to remove the heat, the oil temperatures will rise to excessive values. At high temperatures, oil tends to oxidize rapidly and form carbon deposits. In addition, additives will be dissipated and there will be a great reduction in oil viscosity. Furthermore, oil consumption will be increased. By cooling the oil, its viscosity and other lubricating qualities are maintained and the temperature of the engine is kept within limits.

To keep lubricating oil temperatures down, some small car engines have cooling fins on the oil pan (Peugeot). Larger engines pass the oil through oil coolers. See Figs. 16-1, 16-2, 16-5 and 16-7. Oil coolers (heat exchangers) usually are shell and tube construction. Oil passes through the tube and

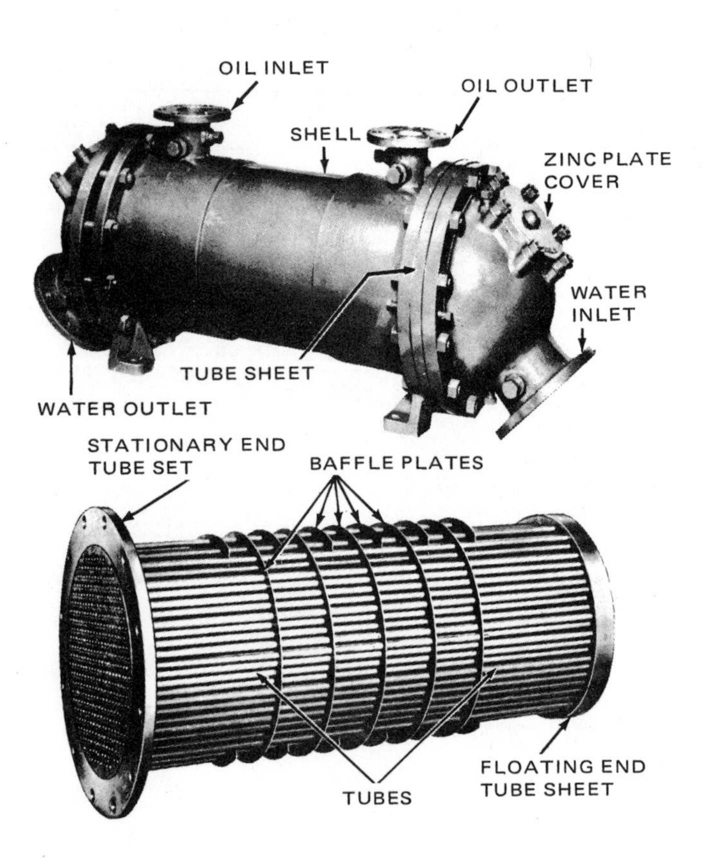

Fig. 16-13. Tube type heat exchanger. (Cooper-Bessemer Div. Cooper Industries)

the coolant circulates outside the tubes. The size of the unit depends upon the amount and temperature of the oil to be cooled. Important factors are the location of the heat exchanger, surrounding temperature, velocity of the oil and the temperature of the cooling water. Coolers usually are placed beyond the oil filter and, in the case of passenger cars and trucks, it is often located in the radiator.

OIL FILTERS

To do an efficient job of filtration, the temperature of the oil should be in excess of 165 deg. F. (74 C) for SAE 20, 175

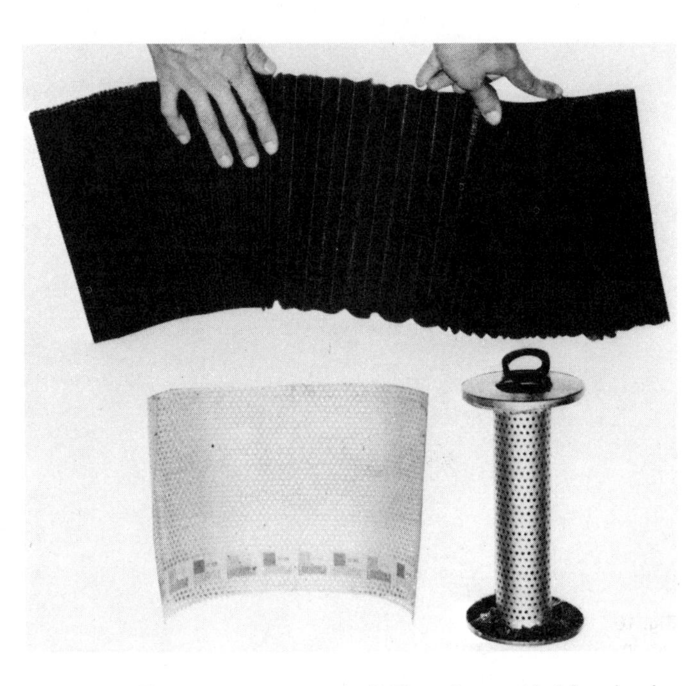

Fig. 16-15. Pleated paper type of oil filter disassembled for cleaning. (Cummins Engine Co.)

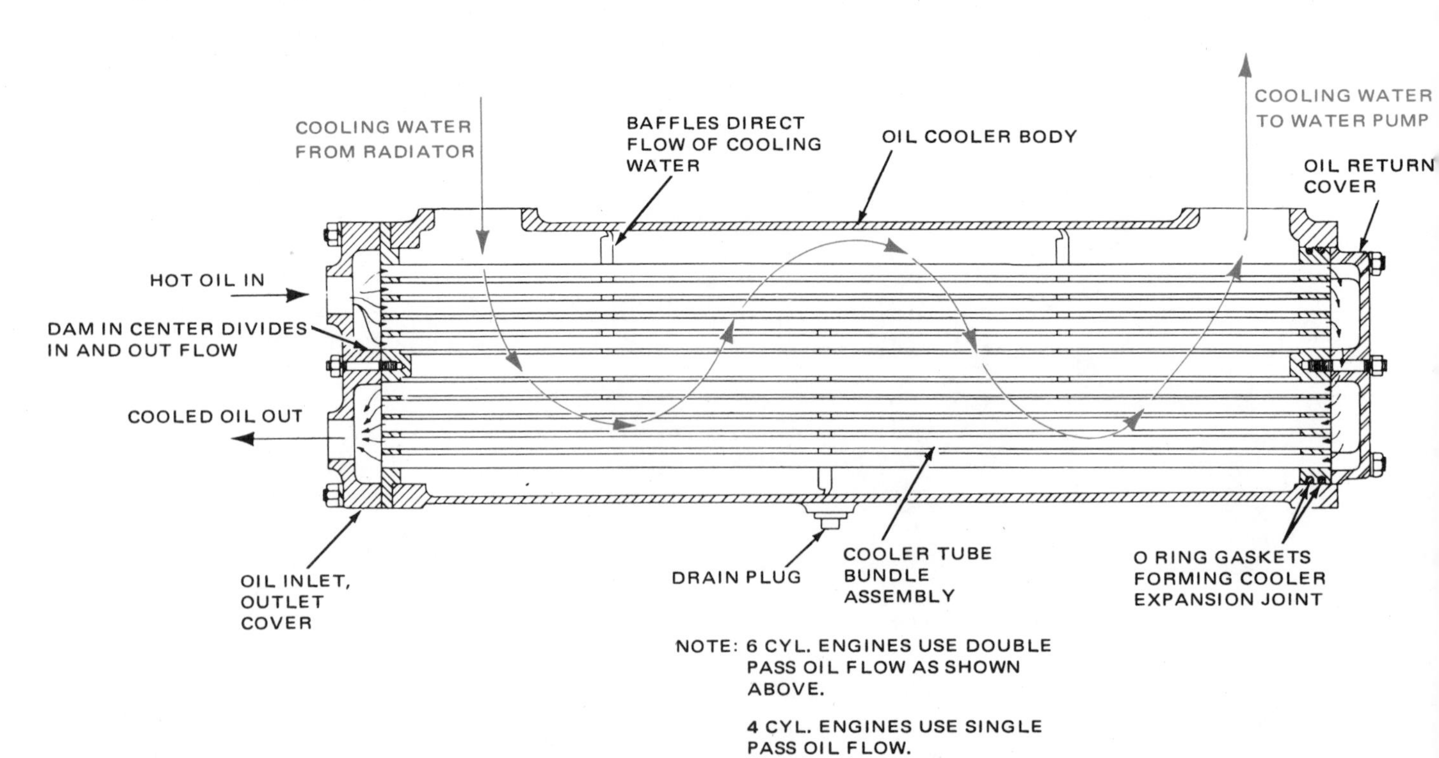

Fig. 16-14. Oil cooler showing direction of cooling water flow. (Murphy Diesel Co.)

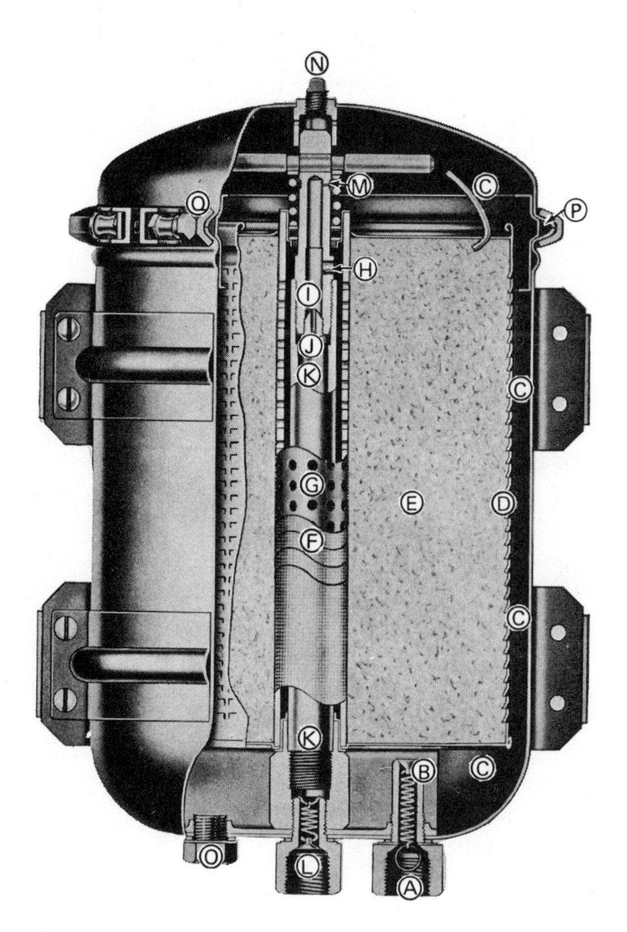

Fig. 16-16. The filter media in this lubricating oil filter is designed for use in filtering detergent oil. A—Inlet. B—Inlet check valve. C—Filter chamber. D—Perforated side of pack. E—Filter media. E—Fibrous wrappings. G—Perforated tube. H—Center pack tube opening. I—Basic hold-down assembly. J—Orifice. K—Outlet tube. L—Outlet check valve. M—Hold-down spring. N—Top plug. O—Drain plug. P—Basket. Q—Cover band. (Luber-Finer Inc.)

deg. F. (79.4 C) for SAE 30, 185 deg. F. (85 C) for SAE 40 oil. However, crankcase oil temperatures usually are in excess of these values.

As noted in the chapter on fuels, filtration processes include absorption, adsorption and mechanical straining. Several different types of materials are used for filtering lubricating oil. These materials include treated paper, Fuller's earth, and synthetic materials. Such materials are used singly and in combination.

In some designs, the complete filter is discarded after a sufficient service interval. In other cases, only the element is replaced. Depending on the size of the engine and the size of the filter, one or more may be used for each engine. Fig. 16-5 shows four filter units installed on a Caterpillar tractor engine.

In most cases, all the oil passes through the filter. This is known as a full flow design. That is, all oil passes through the filter before it reaches the bearings. However, in the event the filtering element becomes clogged or obstructed, a by-pass valve is provided so that oil will continue to reach the bearings.

Another system, known as the shunt or by-pass system, filters a portion of the oil and bypasses the remainder. Eventually, all the oil passes through the filter.

A pleated paper type oil filter is shown in Fig. 16-15. In this particular design, the filtering element can be removed for inspection and cleaning.

The filter shown in Fig. 16-16 was designed especially for use with detergent oils. Oil enters the unit through inlet (A), passes through a check valve (B) and fills the chamber (C). Engine oil pressure forces the oil through the perforated side of the pack (D) and through filter media (E). The cleaned oil then passes through several layers of fibrous wrappings (F) around the perforated tube (G), which prevents migration of any filter media. The clean oil then rises in the center pack tube and flows through the opening (H) into the hollow body of the pack hold down assembly (I), then through the orifice (J) which controls the rate of flow. It then passes downward through the outlet tube (K) and returns through the outlet check valve (L).

Another replaceable type of lubricating oil filter element is shown in Fig. 16-17. The key design feature of this element is that two different density filtering materials are used. Also worth noting, these materials are placed in series rather than in parallel. See Fig. 16-17 for construction details.

This dual oil filter element offers certain other advantages. For example, the flow rate through the filter media varies constantly since the inner and outer depths change in thickness. Also, early surface plugging is generally eliminated because of the gradual dual density design. A perforated center tube prevents media migration.

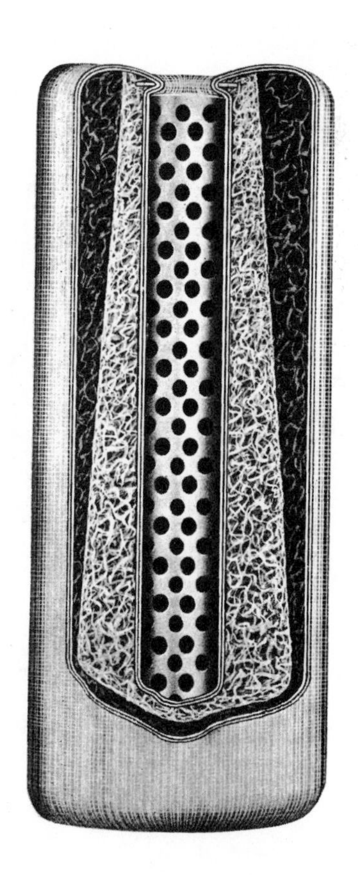

Fig. 16-17. Two different density elements are a feature of this filter. (Nefco Filter Corp.)

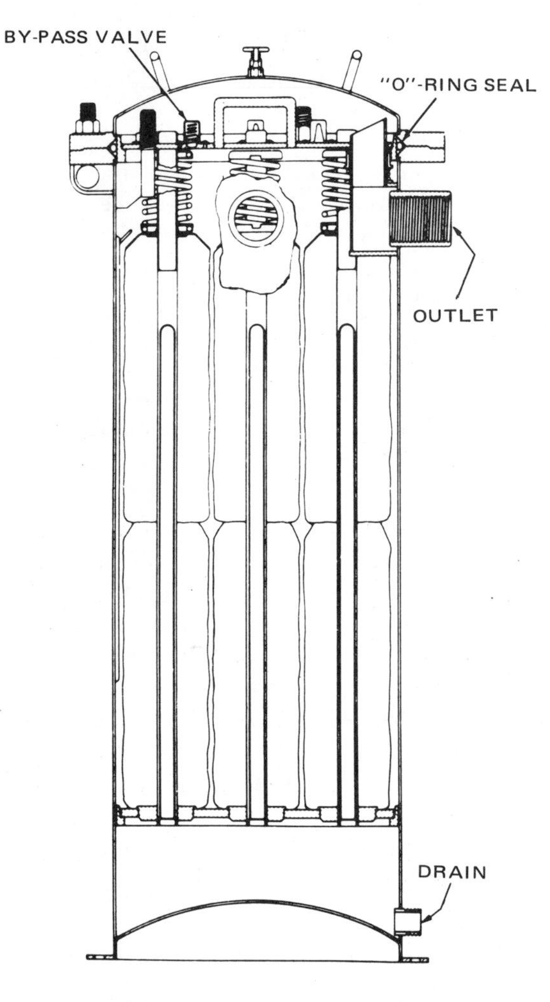

Fig. 16-18. A 14 element full flow oil filter.

In an unusual design, a 14 element full flow oil filter is installed on some Waukesha industrial engines. A cross-sectional view is shown in Fig. 16-18. Note the by-pass valve in the top of the filter.

REVIEW QUESTIONS – CHAPTER 16

1. The purpose of the lubricating system is to:
   a. ______________________.
   b. ______________________.
2. The full flow system of diesel engine lubrication is designed with a __________ in case the flow of oil through the filters is obstructed.
3. Piston pins are usually lubricated by oil passing through __________ holes in the connecting rods.
4. Some diesel engines are fitted with a complete oil filter which is __________ after a sufficient service interval.
5. Oil pumps used in the lubricating system of diesel engines are of the __________ type.
6. The rate of discharge of a positive displacement pump is __________ proportional to the speed of rotation.
7. The preferred location of an oil cooler is:
   a. Before the filters.
   b. After the filters.
   c. Between the filters.
   d. Before and after the filters.
8. If an oil cooler were not used on an engine and oil temperature became extremely high, it would cause:
   a. ______________________.
   b. ______________________.
   c. ______________________.

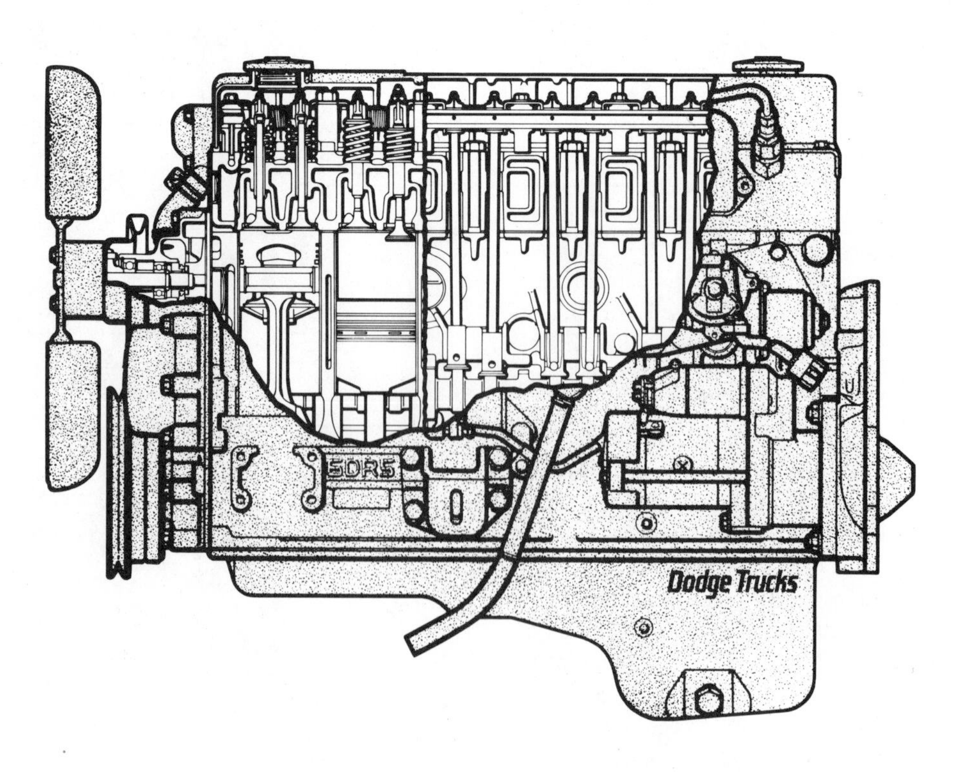

Mitsubishi six cylinder diesel engine used in Dodge power wagons.

Chapter 17
STARTING AND CONTROL SYSTEMS

Starting a diesel engine is not as easy as starting a gasoline spark ignition engine. Normally, faster cranking speeds are required for the diesel engine. At low cranking speeds, there is too much time for the required heat of compression to dissipate. While the cranking speeds needed for starting a diesel engine vary with different types, a cranking speed of about 200 rpm is required to obtain a compression pressure of 400 psi, Fig. 17-1.

As the cranking speed is increased, less heat is lost to the cylinder walls and compression temperatures rise. Cold weather intensifies starting problems and is discussed later in this chapter.

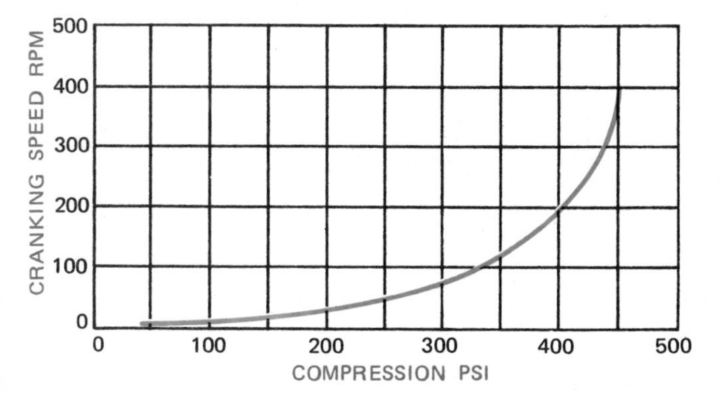

Fig. 17-1. Showing relation of compression pressure to cranking speed.

TYPES OF STARTING SYSTEMS

There are several different methods used for starting diesel engines. The four major types are: electric motors, gasoline engines, hydraulic cranking motors and compressed air cranking systems. Regardless of the type of motor or engine used to crank the diesel, engagement is made with a ring gear on the flywheel of the engine.

ELECTRIC MOTORS

Electric cranking motors are used extensively in small and large diesel engines and almost exclusively in the automotive field. These are direct current series or compound wound motors, and operate on 12, 24 or 32 volts, depending on the size of the diesel engine. For larger diesel engines, the starting motors are provided with reduction gears to provide the greater torque required.

Current for the starting motor is supplied by lead-acid type storage batteries. They are kept in a charged condition by a

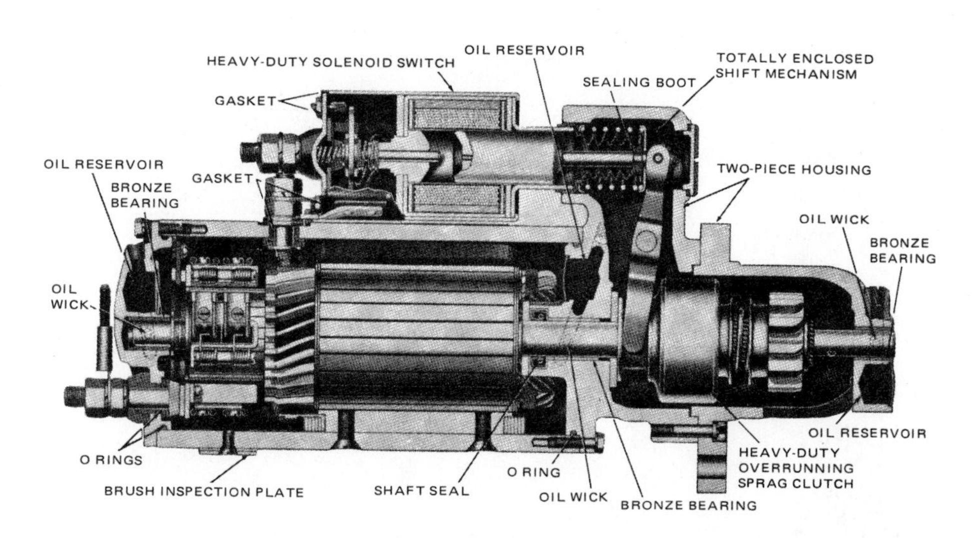

Fig. 17-2. Typical electric cranking motor. (Delco Remy Div. General Motors Corp.)

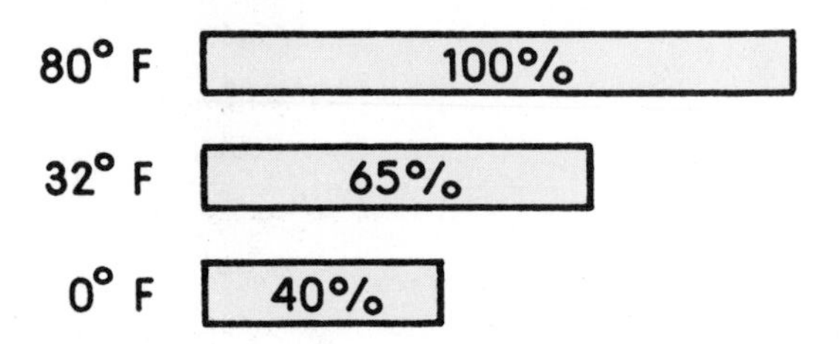

Fig. 17-3. Comparing cranking power of fully charged battery at different temperatures.

direct current generator or an alternator with suitable rectifying equipment. An accessory shaft from the diesel engine provides the necessary drive.

In the automotive field, the generator or alternator is the 12 volt type which conforms to the requirements of the 12 volt lighting system and other automotive accessories. When a 24 volt cranking motor is used with a 12 volt generator, two 12 volt batteries are used. They are connected in series for starting and lighting and connected in parallel for charging. This is accomplished by means of a series parallel switch which is either magnetically or manually operated. This 12 volt – 24 volt system is also used extensively for installations designed for automatic starting and lighting. A typical electric starting motor is shown in Fig. 17-2.

When diesel engines are used to drive direct current generators, like those used on some diesel electric locomotives, the main generator can be used as a starting motor. Power is derived from large 110 volt storage batteries.

A disadvantage of the electric motor for starting is that the efficiency of the starting battery drops rapidly as the temperature drops. For example, the efficiency of a starting battery at 0 deg. F. is only 40 percent of what it is at 80 deg. F., Fig. 17-3. Not only is the efficiency of the battery lower but, because of the increased viscosity of the oil, the internal friction of the engine is greater. Consequently, greater torque is required for cranking.

Another difficulty with electric starters is the fire hazard resulting from sparks. This is a particular problem in oil fields, refineries, gas fields and mines. Special provisions can be made to overcome many of these problems and will be discussed later.

GASOLINE ENGINES

A small gasoline engine can crank a diesel at much higher speeds than is possible with an electric motor and battery. In many cases, magnetos are provided for ignition and the need for batteries and other servicing is eliminated. Gasoline starting engines are usually two or four cylinders and the horsepower is dependent on the size of the diesel to be started. Gasoline engines are primarily used on small diesel engines, such as tractors and graders.

Gasoline engines used for starting require slightly more space than the equivalent electric motor and battery. Engines used are the water-cooled type. Provision is usually made so that the gasoline engine coolant, after reaching operating temperature, can be passed through the cooling system of the diesel. This improves starting conditions.

The gasoline engine can be started either electrically or by hand cranking. It is usually mounted directly on the side of the diesel and drives the engine through a clutch, gear box and drive pinion. Dyer and Bendix drives are also used.

In some engines, the starting engine exhaust passes through a tube in the diesel air intake pipe. This warms the air charge and improves starting conditions during cold weather.

Some tractor engines (International Harvester) are designed to start as gasoline spark ignition engines and then after full operating temperature is reached, are switched over to diesel operation. While operating with gasoline, compression is reduced and carburetion is provided by suitable equipment.

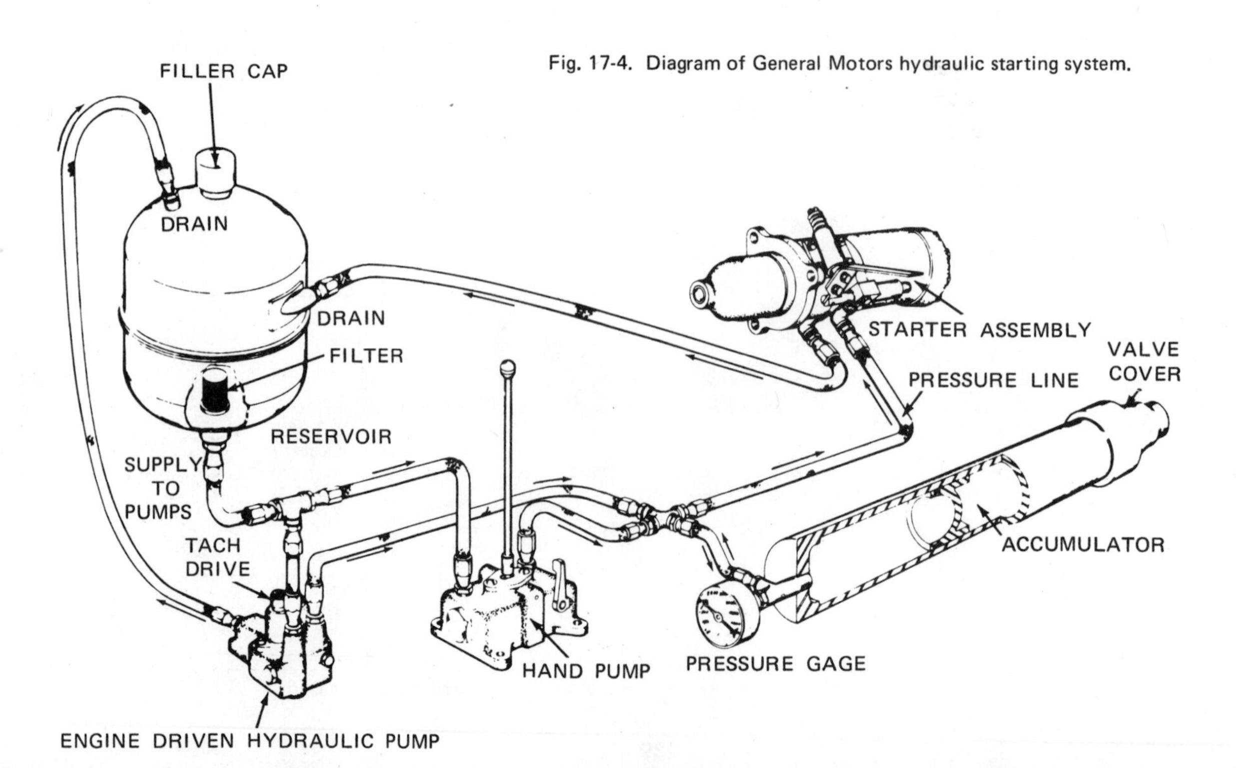

Fig. 17-4. Diagram of General Motors hydraulic starting system.

HYDRAULIC CRANKING MOTORS

There are several different types of hydraulic starting systems available. In most installations, the system consists of a hydraulic starting motor, a piston type accumulator, a manually operated hydraulic pump and a reservoir for the hydraulic fluid. One type used extensively on General Motors' diesel engines is shown in Fig. 17-4.

In this system, hydraulic pressure is obtained in the accumulator by manually operated hand pump or from the engine driven pump while engine is operating. After pressing starting lever, control valve allows hydraulic fluids under pressure from the accumulator to pass through hydraulic starting motor, thereby cranking engine. When starting lever is released, spring action disengages starting pinion and closes control valve. This stops the flow of hydraulic fluid from accumulator. Starter is protected from high speeds of the diesel engine by action of an overrunning clutch.

Operation of the hand pump will charge the accumulator to a maximum of 2900 to 3300 psi. The recommended pressure is maintained by the engine driven pump. When the ambient temperature is above 40 deg. F., an accumulated pressure of 1500 psi will provide adequate pressure for starting.

Between 40 deg. F. and 0 deg. F., a pressure of 2500 psi is usually sufficient. Below that temperature the accumulator should be charged to the maximum recommended pressure.

The accumulator is comprised of a heavy-duty cylinder and piston designed to hold a charged pressure for an extended period of time. The accumulator is charged with nitrogen through a small valve and sealed by the manufacturer. Oil enters the accumulator under pressure from either the engine driven pump or from the hand pump. This forces the piston back compressing the nitrogen gas and storing the energy to operate the system. Accumulators are either 1 1/2 gal. or 2 1/4 gal. capacity. If larger cranking periods are required, two or more accumulators can be connected in parallel.

Some points of similarity will be noted between the General Motors system just described and the American Bosch Hydrotor cranking system. This system, Fig. 17-5, can be installed as a completely independent means of cranking an engine without the need of any auxiliary power, such as electricity or compressed air. It consists of a cranking motor and accumulator, a pump or pumps, an oil reservoir and a starter valve. These various components are connected with flexible hose and steel fitting assemblies.

Hydraulic oil stored in the reservoir is pumped into the accumulator against a pre-charge of nitrogen gas and held under a pressure of approximately 3000 psi until needed. When the engine is to be cranked, the starter control valve is open to permit the oil under high pressure to flow to the hydraulic cranking motor. System pressure can be restored manually in an emergency.

The accumulator is a thick-walled cylinder closed at both ends. It contains a free floating piston equipped with O ring seals. Initially, the accumulator is given a pre-charge of nitrogen to a pressure of 1500 psi which forces the piston to one end of the cylinder bore. When hydraulic fluid is pumped into the accumulator, it forces the piston back compressing the nitrogen gas to approximately 3000 psi. This provides the stored hydraulic energy to operate the cranking motor.

The cranking motor is a piston type positive displacement motor approximately the size of other cranking motors. It consists of a rotating cylinder carrying pistons and is splined to the drive shaft. These pistons, when subjected to the high-pressure flow of oil, move against a tilted thrust bearing which allows them to slide around the races causing the cylinders and shaft to rotate. As the cylinder rotates, the pistons on the other side are being forced backwards exhausting oil at low-pressure back to the reservoir.

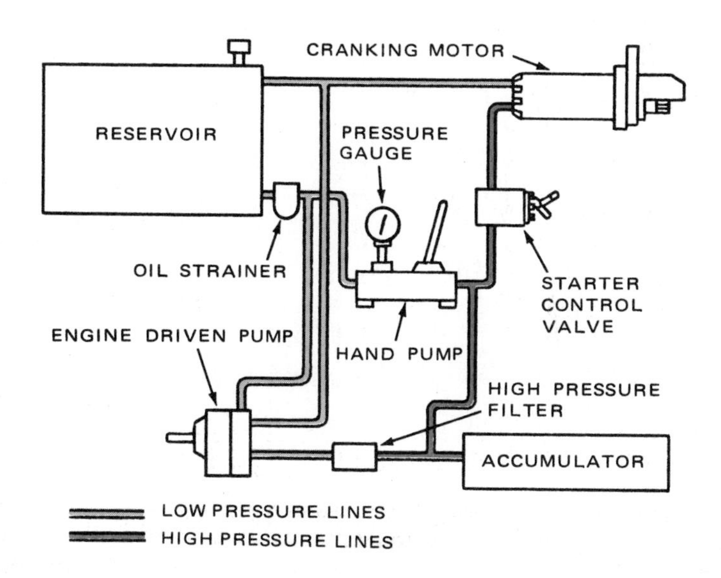

Fig. 17-5. Diagram of American Bosch Hydrotor cranking system.

The hand pump is a double acting piston type and can build up pressures to 3000 psi. It is used initially to fill the accumulator after installation. It may also be used in emergency when starting trouble exhausts the accumulator.

The engine driven pump is also a piston type and operates to maintain constant accumulator pressure while the engine is running. However, it works under load only while filling the accumulator. When the desired operating pressure is reached, an interval unloading valve diverts the oil back to the reservoir at reduced pressure.

COMPRESSED AIR CRANKING

Large diesel engines are often provided with air starting systems. One method is to admit compressed air into the cylinders at a pressure capable of cranking the engine. The process is continued until the pistons have built up sufficient compression to cause combustion. The pressure used in such starting systems ranges from 250 to 600 psi.

The Fairbanks Morse engine starting mechanism uses this system which includes the air control valve, air start distributor, the air header, the pilot air tubing and the air start check valves at the individual cylinders.

Another method is to use motors driven by compressed air to crank the diesel engine. There are two major types of air

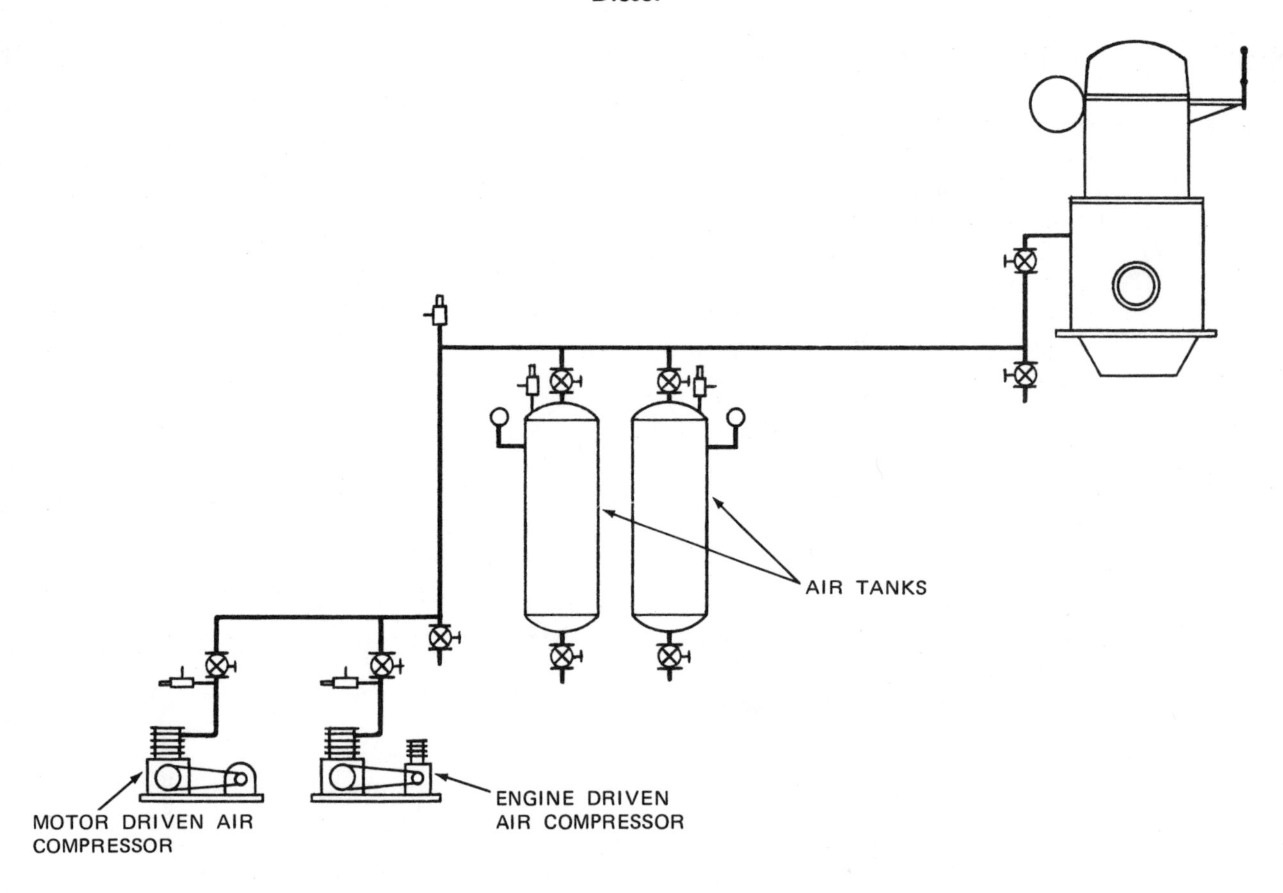

Fig. 17-6. Typical air starting system. (Diesel Engine Manufacturers' Assn.)

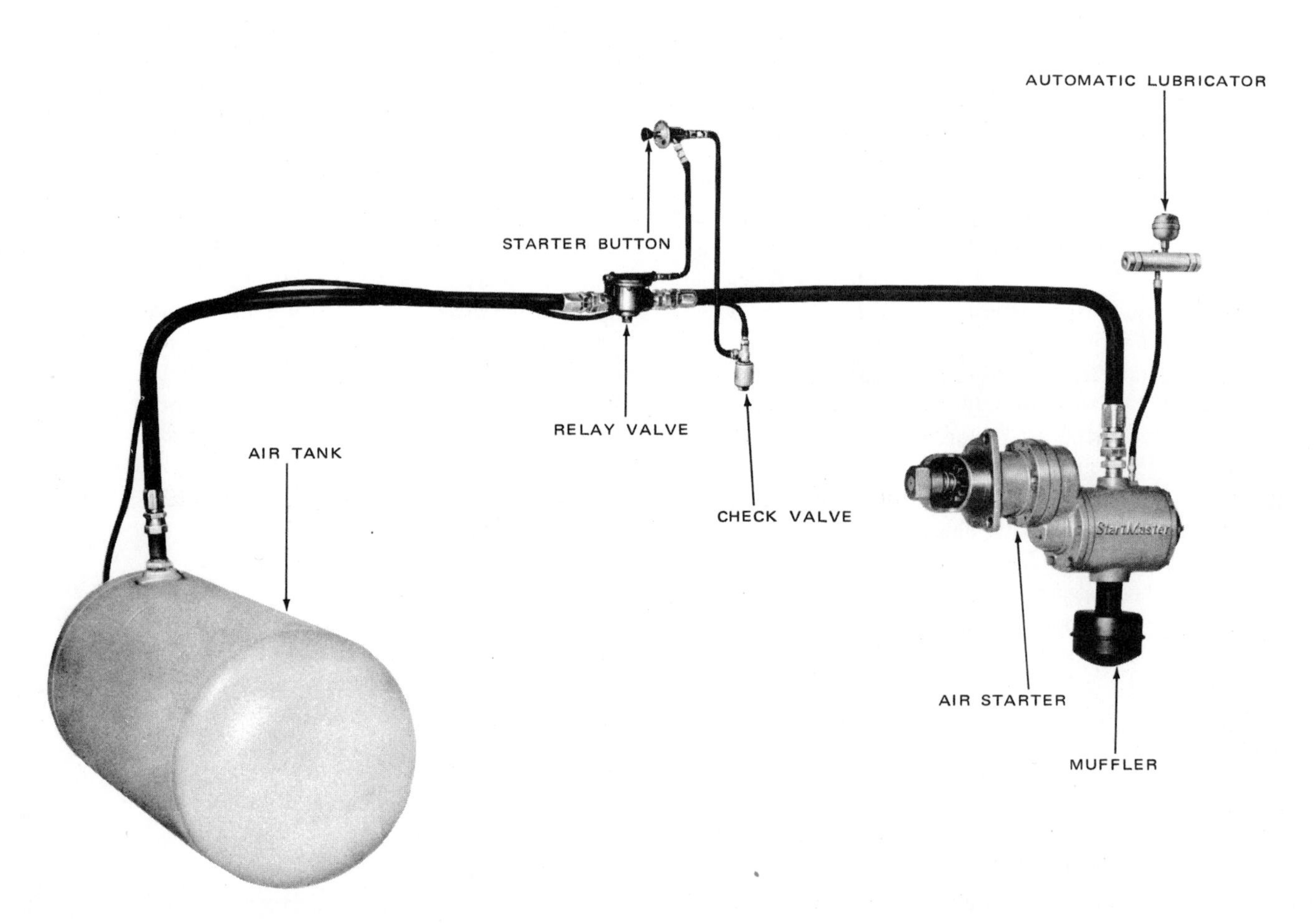

Fig. 17-7. Details of Startmaster air starting system. (Hartford Div., Stanadyne Inc.)

motors designed to crank diesels. One is a piston type while the other is the vane type. Air motors are available in sizes to replace electric starting motors and are used on diesel engines of relatively small displacement. These air motors are powered by compressed air from a storage tank.

In the Ingersoll-Rand vane type motor, phenolic vanes are carried by a steel rotor mounted in an eccentric housing. Centrifugal force keeps the vanes in contact with the eccentric bore of the housing. In the Gardner-Denver type motor, air is distributed to the cylinders by a rotary valve which is directly connected to the motor crankshaft.

Fig. 17-6 shows a typical air starting system. In many installations two or more air tanks are used, the number being dependent on the size of the engine and the length of time required for starting. In Fig. 17-6, two compressors are shown, one motor driven and the other engine driven. This method makes air available for emergency operation.

In the Startmaster air starting system, Fig. 17-7, compressed air is supplied to the air starter by the vehicle's air brake system or from a compressor and tank driven by a stationary engine. The engine is started by operating a quick opening valve between the air tank and the starter. Pressure is maintained by the compressor as soon as the engine is operating. If the air supply is depleted before the engine is started, an emergency start cartridge is supplied which can be inserted in the air tank.

The starting motor, Fig. 17-8, is designed to crank diesel engines up to 1800 cu. in. and will turn at 40 psi.

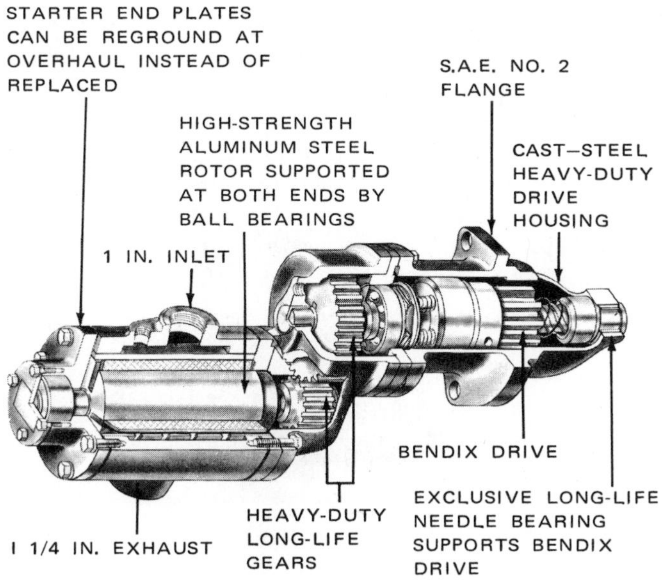

Fig. 17-8. Sectional view of air starting motor. (Hartford Div., Stanadyne Inc.)

COLD WEATHER STARTING

The problem of starting a diesel engine in cold weather is extremely difficult and usually impossible if adequate accessories have not been provided to assist in the starting process. However, with auxiliary equipment, starting time of an automotive type diesel engine ranges from 3.7 seconds to 7.5 seconds at temperatures ranging from 25 deg. F. to -25 deg. F. The type of accessory equipment required to start a diesel engine may vary with the temperature. As the temperature drops, battery efficiency is reduced and cranking load becomes high. The increased cranking load results from greater oil viscosity. The cold cylinder walls also chill the incoming air so that it does not reach the temperature required for combustion.

The methods used to aid starting under cold weather conditions include: (1) use of special fuels, (2) heating the coolant, (3) heating the lubricating oil, (4) air heating the engine, (5) heating the intake air, (6) auxiliary heat from glow plug, and (7) heating the starting battery.

These methods are used singly or in combination, depending on the surrounding temperature. There is no problem if the vehicle is stored in a warm garage or the engine is in a warm power house.

The method used most frequently to start diesel engines under cold weather conditions is by injection of ethyl ether. Methods of injecting ethyl ether during recent years have been highly developed. With such systems it is possible to start a diesel at sub-arctic temperatures in three to twelve seconds.

Ether is not a petroleum product and is not affected by temperature. By using it for starting purposes ignition temperatures are materially reduced. Cranking requirements to reach a temperature conductive to starting are also lowered.

The use of auxiliary heat in the intake manifold or glow plugs in the combustion chamber is not necessary or permitted. With the higher volatility of ether, such additional heat could cause ignition in the intake manifold or preignition in the combustion chamber. Ether can be used as an aid to starting by spraying it into the engine air cleaner from an

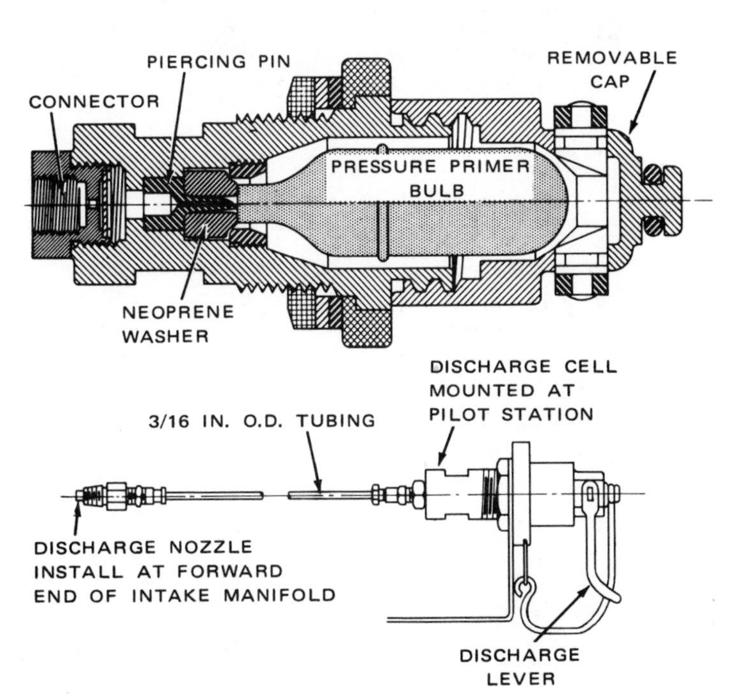

Fig. 17-9. Details of an ether capsule primer.

aerosol can. With more sophisticated equipment, it can be injected into the intake manifold by manipulating a control on the instrument panel. One piece of equipment using an ether capsule is shown in Fig. 17-9.

Care must be taken so that excessive amounts of ether are not used in starting, since serious damage to the engine may result.

COOLANT HEATERS

One of the more popular methods to improve starting conditions is the coolant heater. When cold fuel enters a cold combustion chamber and the air from the intake manifold is also cold, the fuel fails to evaporate. Instead, it collects in the combustion chamber, washes the lubricant from the cylinder walls, and dilutes the oil in the crankcase. These problems are overcome by heating the coolant. Some of the coolant heaters used are: (1) external tank circulating type, (2) freeze plug type heater, (3) head bolt type heater, (4) drain plug type heater, (5) water jacket type heaters, and (6) boiler system heaters.

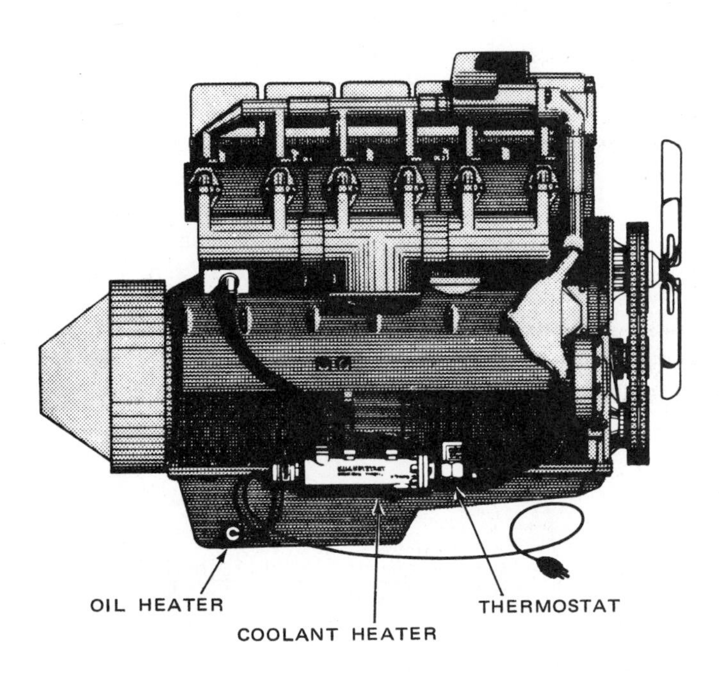

Fig. 17-10. Crankcase oil and coolant electrically heated. (Kim Hotstart Mfg. Co.)

The external tank circulating system is basically a thermosiphon (inducing warm water circulation) method of heating. The unit is mounted low on the side of the engine and is connected to the lower radiator hose or to the drain cock in the cylinder water jacket. The outlet is connected to the water jacket of the cylinder head. While a circulating pump is provided for extremely cold weather, most of the systems depend on the thermosiphon action of the warm water to provide circulation. Heat for the coolant is provided by propane gas or fuel oil. Such heaters can be left in operation overnight, or as long as the diesel engine is inoperative.

Plug type heaters are also known as block type heaters. Some heaters have electric heating elements that are screwed into the freeze (core) plug opening. Others replace a cylinder head bolt or water jacket drain plug. These heaters are connected to the electric power circuit. Some diesel engines have a cover plate on the side of the water jacket and can be replaced with an electric heating element. Freeze plug type heaters have the advantage of low cost and ease of installation.

Boiler system heaters are generally used where fleets of vehicles are stored in a parking area. In this system, a boiler heats a large quantity of coolant. This heated coolant is conducted to the parked vehicles through pipes, and after passing through the cooling system of each vehicle, is returned to the boiler where it is reheated. Fuel oil is generally used to heat the coolant in the boiler system.

Air heating is another method used to aid in starting diesel engines in cold climates. In this system, warm air is blown into the engine crankcase and circulates through the crankcase and under the pistons. This causes the oil and complete engines to maintain a temperature sufficient to insure easy starting at temperatures as low as -55 deg. F. Warm air is often produced from the exhaust of a small gasoline engine.

Lubricating oil heaters are also a popular method to improve starting conditions. In small installations, bayonet type electric heaters are often used. Large stationary installations employ steam or hot water in the jackets of oil tanks where circulation is provided by positive displacement pumps.

BATTERY HEATERS

The efficiency of a battery drops rapidly with the temperature, Fig. 17-3. Therefore, special equipment is available to warm batteries, maintain their efficiency at the highest levels, and provide desired cranking speed. One design includes an electric heating element which may be placed under the battery, or fastened to the side of the battery. Insulated battery boxes with self contained electric heating elements are also used.

One type of battery box heater contains two heating elements. One is connected to the vehicle 12 volt electric system which keeps the battery warm while on the road. The other element operates on 110 volts and is connected to commercial circuits when the vehicle is parked at night.

The Kim Hotstart equipment is designed to keep the engine and lubricating oil at normal operating temperature under conditions as low as -40 deg. F. The equipment, Fig. 17-10, is attached to the side of the engine. One heating element supplies heat to the oil in the crankcase while the other supplies heat to the coolant in the water jacket.

Fig. 17-11. Typical glow plug. (Champion Spark Plug Co.)

GLOW PLUGS

The glow plug, Fig. 17-11, is another device designed to improve starting conditions for diesel engines. It is a low voltage heating element inserted in the combustion chamber, Fig. 17-12, or in the intake manifold. It is controlled by a switch on the instrument panel. The glow plug is used only briefly before the cold engine is cranked. The length of time the glow plug is used is dependent largely on the surrounding temperature. In general, the time limit for the use of a glow plug is about two minutes. Operating temperature of a glow plug is approximately 1652 to 1832 deg. F.

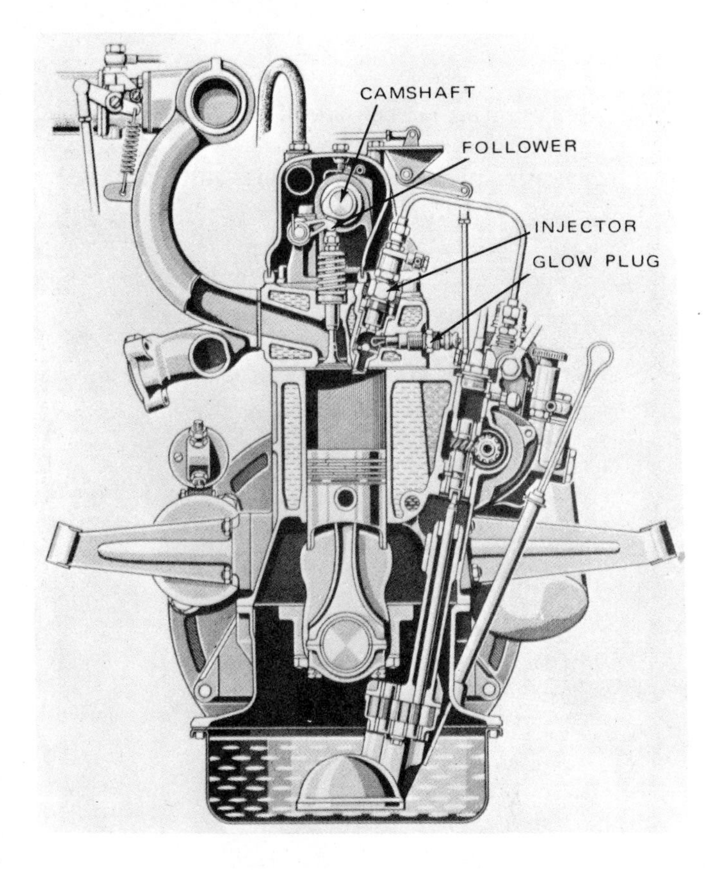

Fig. 17-12. Showing location of glow plug in precombustion chamber of Mercedes-Benz OM 621 diesel engine. (Daimler-Benz Aktiengesellschaft)

OPERATION CONTROL EQUIPMENT

Special operation control equipment is available to protect diesel engines from abnormal operating conditions. These monitor oil and water temperature and pressure; speed and load; coolant and lubricating oil level; frequency and voltage of generated power.

Such equipment is designed to sound an alarm or to stop the engine before any damage is caused. Some systems also provide automatic start and stop systems. This control equipment is usually mounted in a separate panel. Of particular interest are control panels designed to automatically bring into operation electric generating units during periods of peak power.

GARRETT STARTING SYSTEM

The Garrett air starting system, Fig. 17-13, is designed for use on industrial diesel engines. It consists of an air inlet assembly, a turbine section, a reduction gear system, an overrunning clutch and an output shaft.

The turbine section includes a single-stage, axial flow or radial flow turbine wheel with a full admission stator and a containment ring. To reduce the high-speed, low-torque of the turbine to the low-speed, high-torque required for cranking, a low-speed, high-torque reduction gear is provided. This consists of a two-mesh, compound gear system.

The Garrett air turbine starter has been designed for use with compressed air or natural gas in models ranging from 25 to 600 hp.

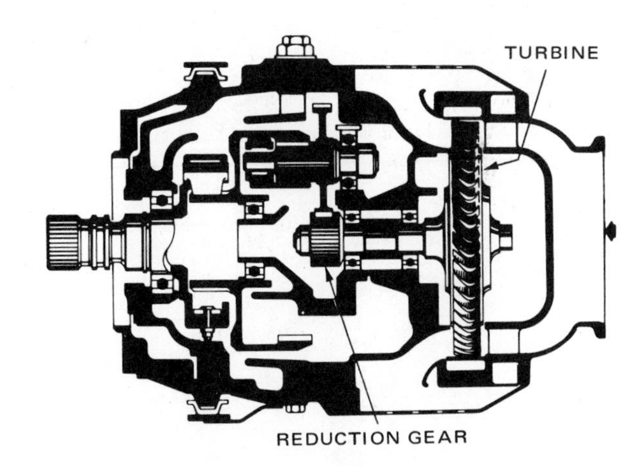

Fig. 17-13. Sectional view of turbine air starting system for diesel engines. (Garrett Corp.)

REVIEW QUESTIONS – CHAPTER 17

1. The approximate compression pressure that results when a diesel engine is cranked at 200 rpm is:
   a. 200 psi.
   b. 300 psi.
   c. 400 psi.
   d. 500 psi.
2. Electric starting motors are direct current series or ____________ motors.
3. Diesel engines in the automotive field are equipped with which of the following voltage alternators.
   a. 12 volt.
   b. 24 volt.
   c. 32 volt.
4. Starting battery efficiency increases as the temperature ____________.
5. Gasoline engines used for starting require slightly more ____________ than the equivalent electric motor and battery.

6. The gasoline engine can be started either____________ or by hand cranking.
7. In hydraulic starting systems, most installations consist of:
   a.______________________________.
   b.______________________________.
   c.______________________________.
   d.______________________________.
8. At a temperature of 40 deg. F., an adequate pressure in the accumulator of a hydraulic system for starting is:
   a. 1,000 psi.
   b. 1,500 psi.
   c. 2,000 psi.
   d. 2,500 psi.
9. In air pressure starting systems where air is admitted into the cylinders, the usual range of pressure is from _________ to _________ psi.
10. Two major types of air motors designed to crank diesel engines are:
   a.______________________________.
   b.______________________________.
11. Methods used to aid starting diesel engines under cold weather conditions include:
   a.______________________________.
   b.______________________________.
   c.______________________________.
   d.______________________________.
   e.______________________________.
   f.______________________________.
   g.______________________________.
12. Ether is not a petroleum product and is not affected by ____________.
13. Four different types of coolant heaters are:
   a.______________________________.
   b.______________________________.
   c.______________________________.
   d.______________________________.
14. The boiler system for heating diesel engines is primarily used for:
   a. Large fleets in a parking area.
   b. Individual vehicles.
   c. Industrial engines in power houses.
15. Two reasons for using battery heaters on diesel engines are:
   a.______________________________.
   b.______________________________.
16. A glow plug is a low voltage heating element inserted in the____________of a diesel engine.

Volvo series TD 40 is a four-stroke cycle, in-line, six cylinder, turbocharged diesel engine with precombustion chamber design. It develops 120 hp at 3600 rpm and is used in small trucks, fork lifts, construction and forestry machines, tractors and combine harvesters.

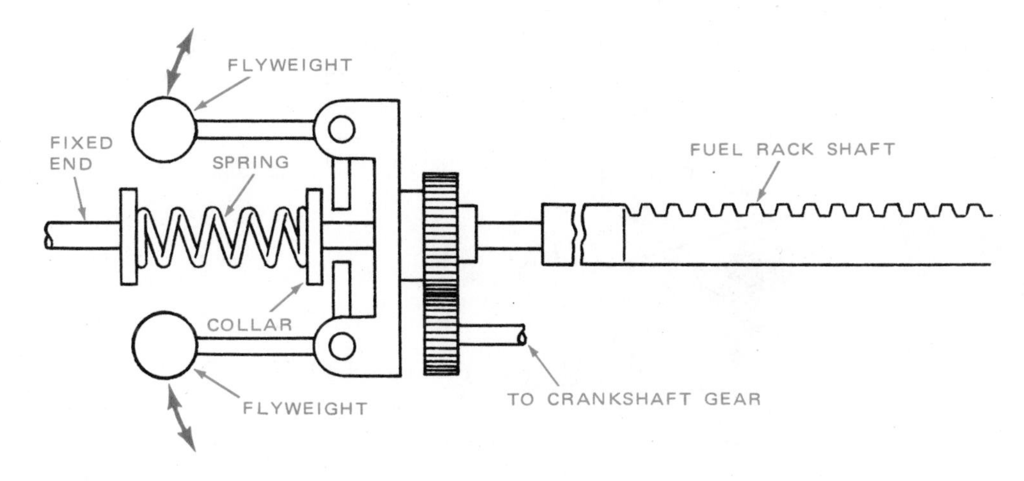

Fig. 18-1. This diesel engine governor (speed-sensing and controlling device) consists of rotating weights called flyweights that spin and fly outward as they spin, a spring, and a sliding shaft called a fuel rack. (Caterpillar Tractor Co.)

Chapter 18
GOVERNORS

A governor is a device designed to control the speed of an engine. It does this by varying the flow of fuel in accordance with the requirements of the load, speed, electrical frequency, and other conditions.

Governors are often included in the design of the fuel injection pump. Several were described in Chapter 8.

Three basic types of governors are:

1. Mechanical.
2. Pneumatic.
3. Hydraulic.

Governors used in the automotive or trucking field are designed primarily to provide sufficient fuel for idling and to cut off fuel above the maximum rated rpm.

There are also governors or speed control devices different in design which are used in special applications. For example, it is necessary when pumping certain fluids, that the engine speed be controlled by the pumping pressure.

When generating electric power, it is essential that the voltage and frequency be maintained within close limits. Fuel quantity for the diesel engine is controlled by the electrical load on the alternator, or the frequency of the current being generated.

Engine overloading is controlled by a load limiting governor which prevents overloading at a predetermined speed. This is done by limiting the maximum flow of fuel.

GOVERNOR CHARACTERISTICS

It is desirable that diesel engine governors have certain qualities or characteristics. One of the qualities is for certain operations, engine speed should be maintained at a constant

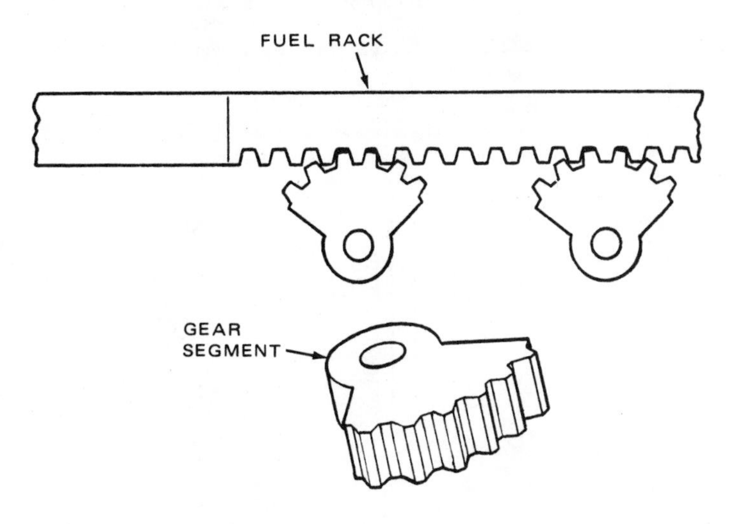

Fig. 18-2. The governor fuel rack has gear teeth on one side. The teeth mesh with partial gears are called gear segments.

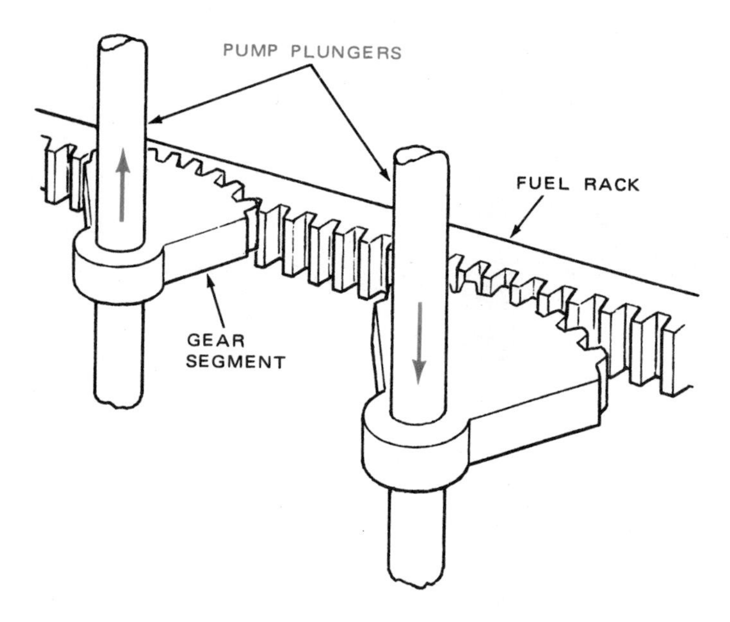

Fig. 18-3. The gear segment which is fastened to the base of the pump plunger, moves up and down with the plunger. The gear remains engaged with the rack.

value regardless of the load. This is known as "isochronous."

Governors must also be able to maintain the desired speed without variation. This quality is known as stability. Without stability, a condition known as "hunting" occurs where the speed will swing back and forth around the desired value. A governor with a high degree of precision or stability is known as a "dead beat" governor.

The speed variation from no load to full load is known as "steady state regulation" or "speed droop." It is expressed as a percent of rated speed.

$$\text{Speed Droop} = \frac{S_1 - S_2}{S_2} \times 100$$

Where S_1 = No load speed

S_2 = Rated full load speed

For example, if the no load speed is 3200 rpm and the rated speed is 3000 rpm, the speed droop is

$$\frac{3200 - 3000}{3000} \times 100 = 6.67\%$$

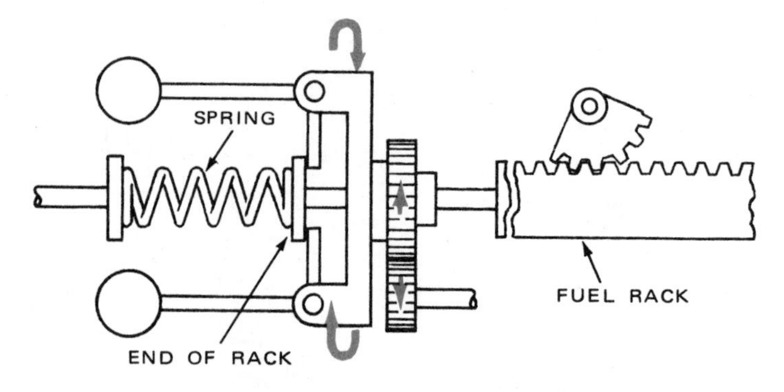

Fig. 18-4. At low speed (idle) the governor spring pushes against the end of the rack. The centrifugal force developed by the revolving flyweights is not sufficient to compress the spring.

A somewhat similar calculation is made for the "degree of unbalance" (difference in speed under variation of load) at maximum speed. This is calculated as follows:

$$\text{Percent of Unbalance} = \frac{S_1 - S_2}{S_3} \times 100$$

Where S_1 = No load speed

S_2 = Rated full load speed

$$S_3 = \frac{S_1 + S_2}{2}$$

For example, let's assume an engine has a full load speed (S_2) of 3000 rpm and a no load speed (S_1) of 3200 rpm.

$$\text{Percent of Unbalance} = \frac{3200 - 3000}{\frac{3200 + 3000}{2}} \times 100 = 6.45\%$$

The percent of speed change required to make a corrective movement of the fuel control mechanism is known as "sensitivity."

When there is a gradual variation of the governed speed above and below the desired governed speed, the condition is known as "speed drift." This is similar to "hunting."

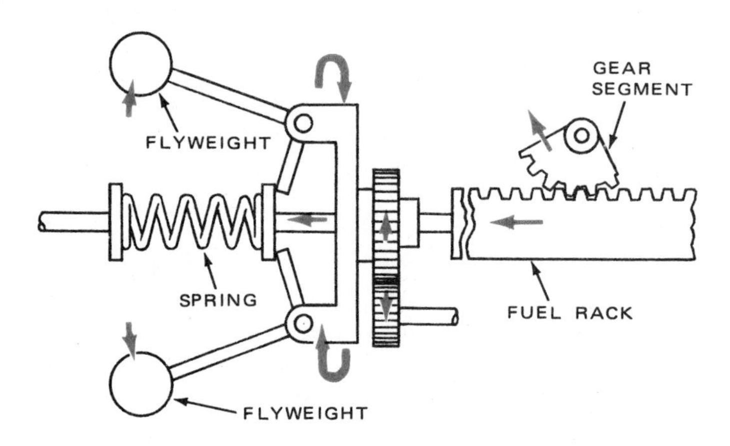

Fig. 18-5. As speed increases the flyweights "fly" outward. Centrifugal force develops pressure which compresses the spring. This causes the fuel rack shaft to move and the gear segments to turn.

BASIC DESIGN OF GOVERNORS

The spring loaded flyweight, also known as a centrifugal ball head, is one of the simplest and most sensitive of the speed sensing devices. Because of these qualities and its rugged construction this is used extensively in the design of diesel engine governors. The ball head depends for its operation on the fact that a centrifugal force causes the ball head mass to move outward and follow a circular path. See Figs. 18-1 to 18-6 inclusive.

DIRECT MECHANICAL GOVERNOR

In a direct mechanical governor the collar is connected to the throttle and by careful calibration of weights and limiting the travel of the collar, the speed of the engine is controlled.

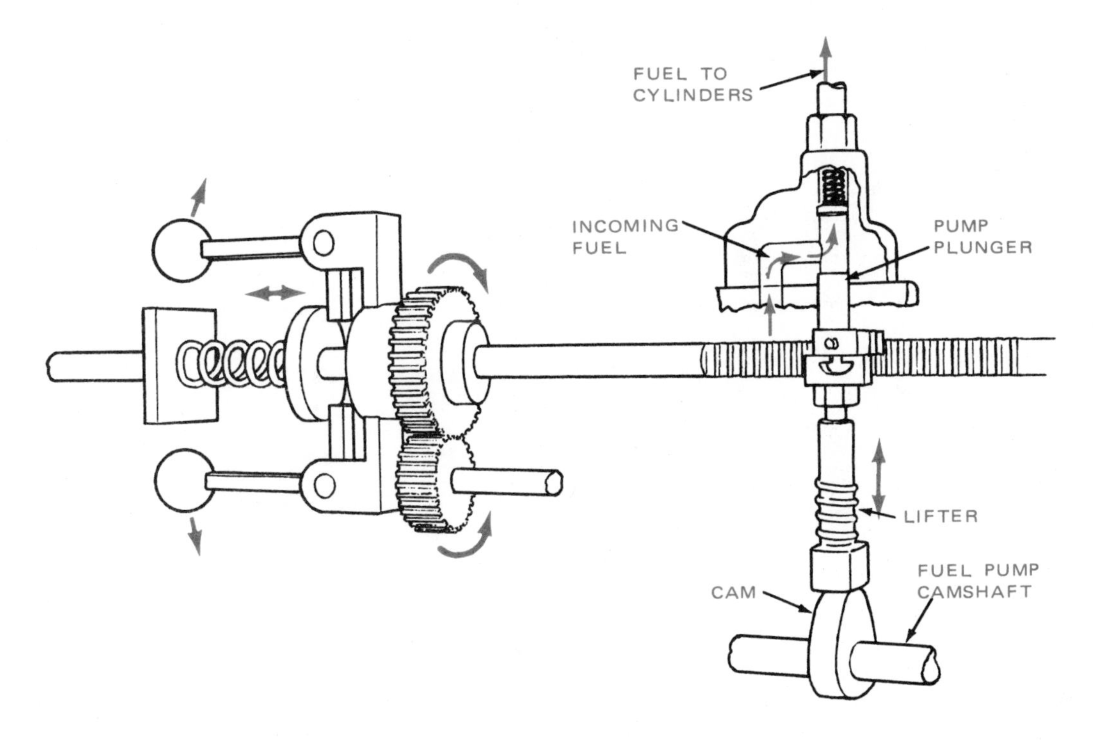

Fig. 18-6. The quantity of fuel pumped to the cylinders changes as the rack moves in and out and the gear segments turn. The pump plunger is moved up and down by a cam and lifter. The segment gear at the base of the plunger stays in mesh with the rack.

There are many variations in the design of the ball head. In current designs of Woodward governors, the design usually consists of two weights with their centers of mass approximately the same distance from the axis of rotation as the pivots about which they swing. The so-called toes are arranged at right angles to the body of the flyweight. When the weight moves outward, and away from the axis of rotation, the toes convert this motion to an axial movement of a pilot valve or other control. The movement of the speeder rod, Fig. 18-7, is opposed and balanced by the speeder spring.

WOODWARD PG GOVERNORS

Woodward PG governors are used to control the speed of diesel, gas and dual fuel engines in a variety of fields. PG governors all have similar basic elements regardless of how simple or complex the complete control might be. The following elements found in each PG governor are sufficient to maintain a constant engine speed as long as the load does not exceed the capacity of the engine:

1. An oil pump, storage area for oil under pressure, a relief valve to limit maximum oil pressure.
2. A centrifugal flyweight head-pilot valve assembly which controls the flow of oil to and from the governor power cylinder assembly.
3. A power cylinder assembly (also called a servomotor) which repositions the fuel racks, fuel valve of the engine.
4. A compensating system which gives stability to the governed system.
5. A means of adjusting the governor.

A PG type Woodward governor is shown in Fig. 18-7. The governor drive shaft, driven at a speed proportional to engine speed by a mechanical connection to the engine, rotates the pump drive gear and the governor pilot valve bushing. As the

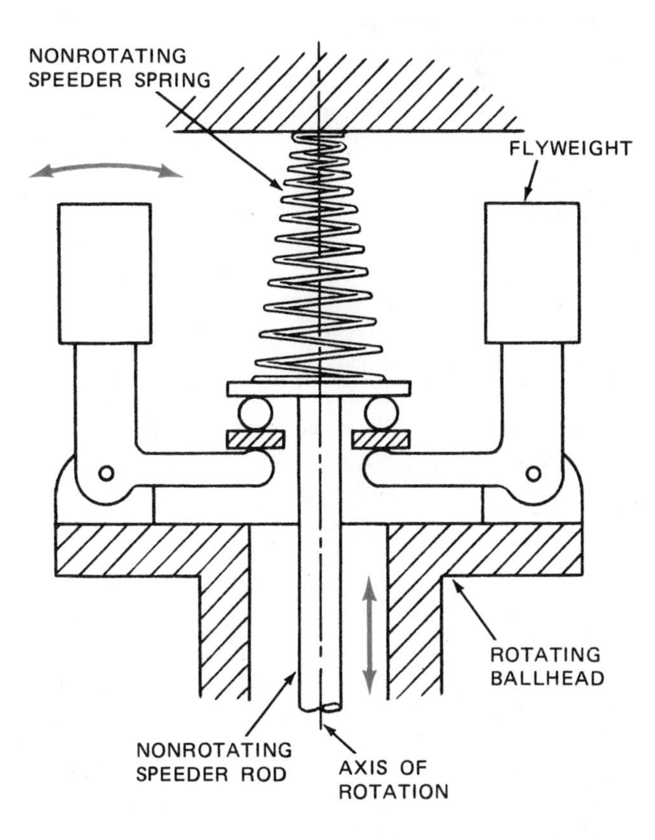

Fig. 18-7. Variation of the typical flyweight governor design. (Woodward Governor Co.)

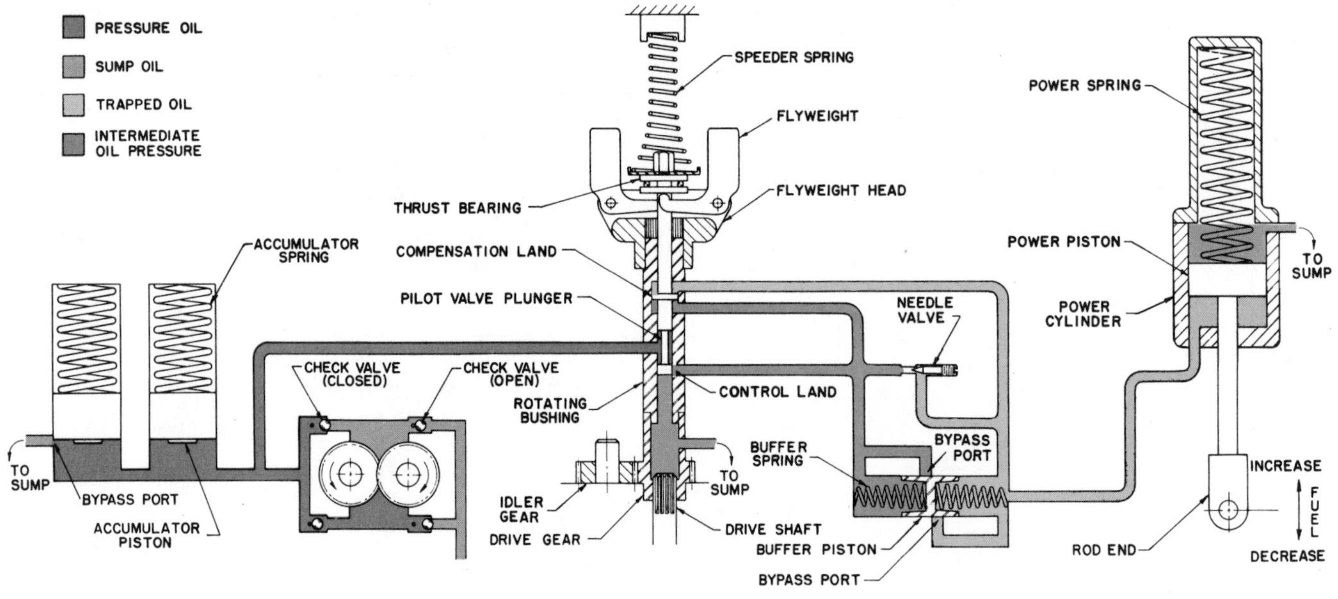

Fig. 18-8. Basic elements of Woodward type PG governor.
(Woodward Governor Co.)

rotating drive gear turns the idler gear, oil is drawn from the oil pump. It is carried to the space between the gear teeth and walls of the gear pocket to the discharge side of the pump. The oil is forced from the space between the gear teeth as the drive and idler gears mesh.

Assume that all control valves are closed. Oil from the oil pump first fills the various oil passages. It then forces the accumulator pistons up against the downward force of the accumulator springs. When the piston uncovers the bypass hole, excess oil from the governor pump returns to the sump. The accumulators not only provide a reservoir for pressure oil but also act as a relief valve to limit maximum pressure in the hydraulic circuit.

The arrangement of the four check valves on the suction and discharge sides of the oil pump permits the governor drive shaft to be rotated in either direction without any changes being made in or to the governor. The direction of pump rotation does not affect the oil pressure system or governor operation. Were the pump gears rotated in the directions opposite those shown in Fig. 18-8, the open check valves would be closed and the closed check valves opened.

The pilot valve plunger, Fig. 18-8, moves up and down in the rotating pilot valve bushing to control the flow of oil to or from the power cylinder assembly. When the pilot valve plunger is centered (i.e. the control land of the plunger covers the control port of the bushing), no oil flows to or from the power cylinder assembly.

The greater of two forces moves the pilot valve plunger up or down. The centrifugal force developed by the rotating flyweights is translated into an upward force which tends to lift the plunger. The centrifugal force is opposed by the downward force of the speeder spring. When the opposing forces are equal, the pilot valve plunger is stationary.

With the pilot valve plunger centered and the engine running on-speed, a change in either of the two forces will move the plunger from its centered position. The plunger will be lowered if the governor speed setting is unchanged, but an additional load slows the engine and governor (decreasing the centrifugal force developed by the rotating flyweights), or if the engine speed is unchanged but the speeder spring force is increased to raise the governor speed setting. Similarly, the pilot valve plunger will be raised if the governor speed setting is unchanged, but load is removed from the engine causing an increase in engine and governor speed. It will also be raised if the engine speed is unchanged, but the speeder spring force is reduced to lower the governor speed setting.

The governor pilot valve plunger controls the movement of the power piston in the power cylinder assembly. The power piston, acting through the connecting linkage, controls the flow of oil to the engine. The power spring continually urges the power piston down in the "decrease fuel" direction. However, the power piston will not move down unless the pilot valve plunger is raised above its centered position. Only when the pilot valve plunger is above center can the oil trapped in the circuit between the plunger and power piston escape to the sump. If the pilot valve plunger is lowered, pressure oil from the governor pump will be directed to the power piston. It will push the piston up against the force of the power spring in the direction to increase fuel.

Note that the power piston will move only when the pilot valve plunger is not centered to permit the oil flow required. With the plunger centered, the power piston is in effect locked.

Stability of the model PG Woodward governor system is achieved by the use of a temporary negative feedback signal which biases the speed signal to the pilot valve plunger. This temporary signal is in the form of a pressure differential applied across the compensating land of the pilot valve plunger. The pressure differential is derived from the buffer compensating system and is dissipated as engine speed returns to the set speed.

The buffer piston, buffer springs and needle valve in the hydraulic system between the control land of the pilot valve plunger and the power piston comprise the "buffer compensating system" of the governor.

Lowering the pilot valve plunger permits a flow of oil pressure into the buffer cylinder and power cylinder to move the power system up to increase fuel. Raising the pilot plunger permits oil to flow from the buffer cylinder and power cylinder to the governor sump. The power piston spring moves the power piston down to decrease fuel.

This flow of oil in the buffer system – in either direction – carries the buffer piston in the direction of flow, compressing one of the buffer springs and releasing the other. The buffer piston movement creates a slight difference in oil pressures on the two sides of the buffer piston. The higher oil pressure is on the side of the piston opposite the spring being compressed. The difference in oil pressure is proportional to buffer piston displacement.

Oil pressure on one side of the buffer piston is transmitted to the lower side of the compensation land on the pilot valve plunger. Pressure on the other side of the piston is transmitted to the upper side of the compensation land. The difference in oil pressure (often called a compensating force) – upward or downward as the case may be – which assists the flyweights or speeder spring in recentering the pilot valve plunger whenever a fuel correction is made.

Operation of the buffer system can be seen by following the sequence of operations when the engine slows down because of the addition of a relatively light load.

The decrease in centrifugal force developed by the rotating flyweights permits the speeder spring to push the flyweights in, lowering the pilot valve plunger and opening the control port.

As the buffer piston moves in the direction of the oil flow – from the pilot valve to the power cylinder – the right-hand buffer spring is compressed and the left-hand spring is relieved. Oil displaced by the buffer piston as it moves to the right forces the power piston up, thereby increasing fuel to the engine, and the engine begins to accelerate. The buffer piston moves to the right, moving the power piston up, until the upward force created by the pressure differential across the buffer piston and compensating land is sufficient, when added to the centrifugal force from the rotating flyweights, to recenter the pilot valve plunger. As soon as the pilot valve plunger is recentered, the power piston movement stops. When the governor is properly adjusted, this new piston position corresponds to the fuel increase needed to operate at a set speed with the new load, even though the engine has not yet returned to the set speed.

OPERATION OF WOODWARD PSG GOVERNOR

Oil supply for the Woodward PSG governor, Fig. 18-9, is supplied through the relief valve from the engine lubricating system. This oil is supplied to the governor oil pump where its pressure is boosted to 175 psi above the inlet pressure. Four check valves, two of which are shown, are used to permit rotation of the governor in either direction. Relief valve discharge is back to supply, so unused oil is recirculated within the governor.

Oil under pressure is carried through ducts to the pilot valve. This is a three-way spool valve arranged to connect the area below the governor power piston to the pressure oil supply upon an underspeed signal, or to discharge upon overspeed. The governor flyweights are carried on pivot pins in the rotating bushing which forms both the outer member of the valve and the drive shaft. The flyweights act upon a thrust bearing attached to the pilot valve plunger and the centrifugal force is translated to axial force at the flyweight toes and opposed by the speeder spring. Speeder spring compression, and the speed at which the governor must run so the flyweight force will balance that of the spring, is changed by the position of the speed adjusting lever.

The isochronous feature of the Woodward PSG governor is provided through the use of a compensating system which establishes temporary speed droop stability and then dissipates this droop so that engine speed is constant under steady state conditions regardless of the load. This compensating system consists of a buffer piston floating between two springs to establish a pressure differential as oil flows to or from a section of the power cylinder. It is combined with a compensating land on the pilot valve plunger across which this differential pressure is applied, and a needle valve through which the pressure difference is dissipated.

Upon a reduction in engine speed from its set value, the speeder spring force overcomes the reduced centrifugal force of the flyweights and the pilot valve plunger moves downward in its bore. This movement uncovers the port at the lower end of the plunger, permitting oil under pressure to enter the passage leading to the power cylinder. The power cylinder has two concentric areas, both of which are exposed to the control oil metered by the pilot valve. The lower, being smaller in diameter, is acted upon directly. The upper annulus is connected through the bore in the power piston in which the buffer piston is carried. Flow of oil into the power cylinder forces the power piston up against the return spring (not shown in the diagram) and some of the oil displaces the buffer piston to force oil into the upper annulus. Flow into the upper annulus establishes a pressure differential across the buffer piston, which is transmitted to the spaces above and below the compensating land on the pilot valve plunger. Higher pressure on the lower side of the land acts in the direction to supplement the flyweight force, causing the closure of the pilot valve before the original speed has been regained. As oil leaks across the needle valve, this false speed signal is dissipated. The buffer piston recenters in its bore with the engine speed returning to normal.

Action under the influence of an overspeed is similar but in the reverse direction. The increased centrifugal force of the flyweights, due to the increased speed, overcomes the speeder spring force and lifts the pilot valve plunger. Upward movement of the pilot valve plunger opens the regulating port to drain. This permits the power piston to be forced in the reduced fuel direction by the return spring. At the same time, flow of oil out of the annular space between the two diameters

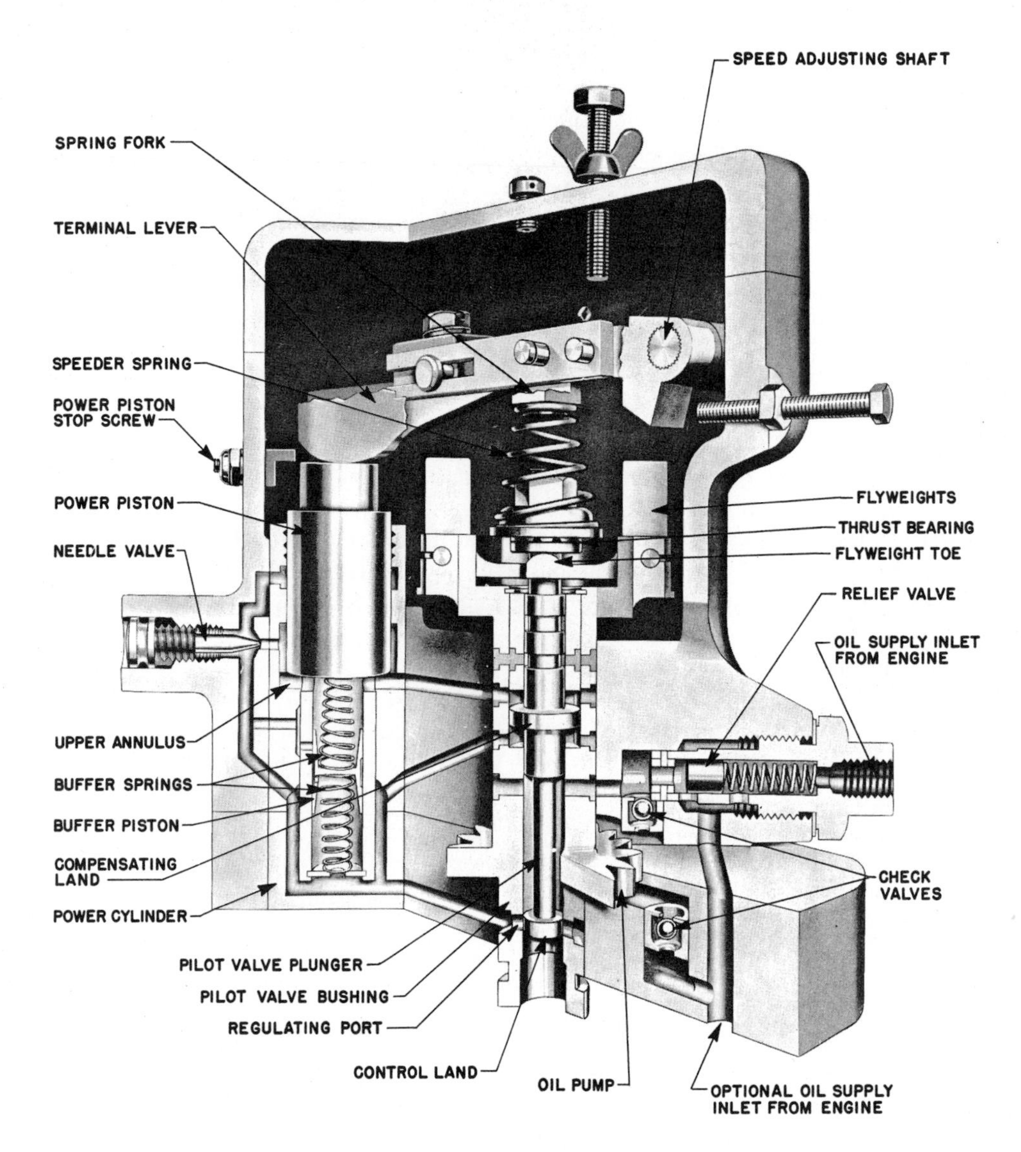

Fig. 18-9. Sectional view of Woodward PSG governor.

of the power piston, uncenters the buffer piston in the downward direction. The pressure difference thus created across the buffer piston, acting on the compensating land, recenters the pilot valve plunger. As oil leaks across the needle valve, this pressure difference is dissipated and the return of speed to normal brings the flyweight force back to normal.

CUMMINS IDLING AND HIGH-SPEED MECHANICAL GOVERNOR

The Cummins idling and high-speed mechanical governor, Fig. 18-10, sometimes called an "automotive governor" is identical on Cummins PT (type R) and PT (type G) fuel pumps. This governor is actuated by a system of springs and weights, Fig. 18-10, and has two functions. First, the governor maintains sufficient fuel for idling with the throttle in the idling position. Second, it cuts off fuel to the injectors above maximum rated rpm. Idle springs in the governor spring pack positions the governor plunger so the idle fuel port is opened enough to permit passage of fuel to maintain idle speed.

During operation between idle and maximum speeds, fuel flows through the governor to the injectors in accord with the engine requirements. This is controlled by the throttle and limited by the pressure regulator of PT (type R) fuel pumps and by the size of the idle spring plunger counterbore on PT (type G) fuel pumps. When the engine reaches governed speed, the governor weights move the governor plunger, and fuel passages to the injectors are shut off. At the same time another passage opens and dumps the fuel back into the main pump body. Engine speed is controlled and limited by the governor regardless of throttle position. Fuel leaving the pump flows through the shut down valve, inlet supply lines and on into the injectors.

CUMMINS TORQUE CONVERTER GOVERNOR

A Cummins PT (type R) fuel pump is usually supplied when a torque converter is used to connect the engine with its driven unit. The auxiliary governor section is driven off the torque converter output shaft to exercise control over the engine governor and to limit converter output shaft speed. The engine governor and the converter (auxiliary) governor must be adjusted to work together.

The PT torque-converter governor consists of two mechanical variable-speed governors in series. One is driven by the engine and the other by the converter, Fig. 18-11.

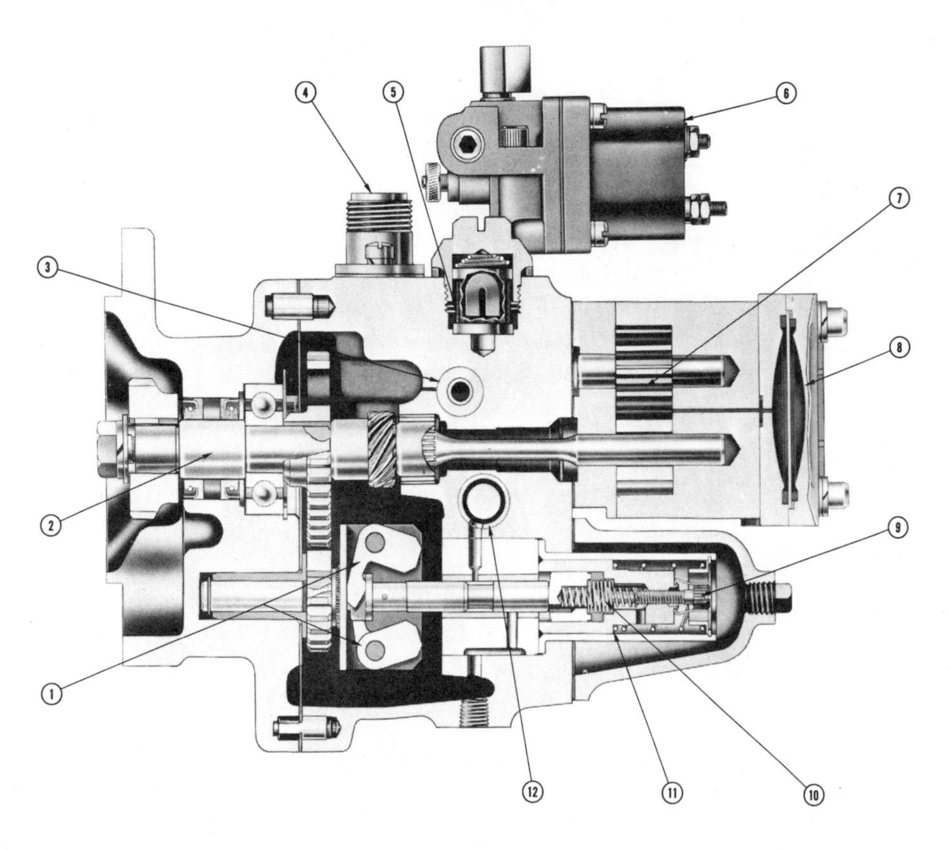

① GOVERNOR WEIGHTS
② MAIN SHAFT
③ PRESSURE REGULATOR
④ TACHOMETER CONNECTION
⑤ FILTER SCREEN
⑥ SHUT-DOWN VALVE
⑦ GEAR PUMP
⑧ PULSATION DAMPER
⑨ IDLE SPEED SCREW
⑩ IDLE SPRINGS
⑪ MAXIMUM SPEED SPRING
⑫ THROTTLE SHAFT

Fig. 18-10. Above. Cummins idling and high speed mechanical governor as used on type R injector.
Fig. 18-11. Below. Cummins torque converter governor with type R injector.

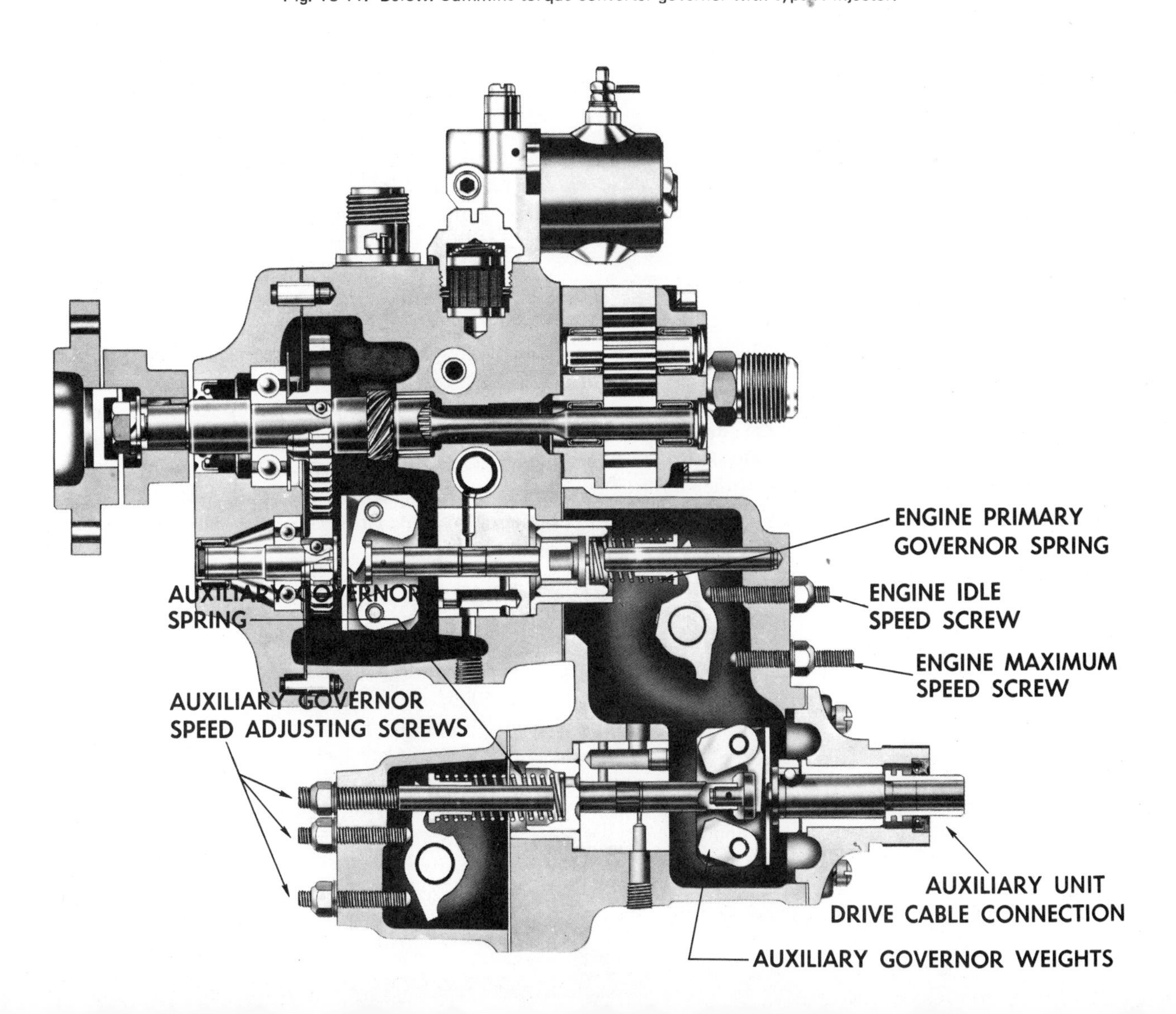

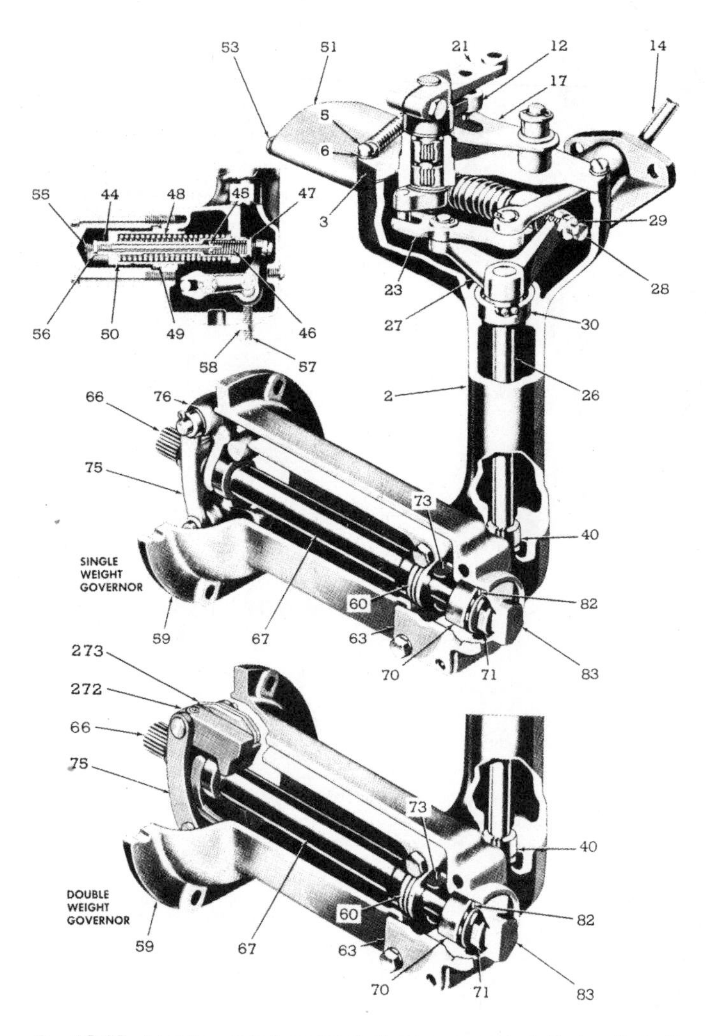

Fig. 18-12. Limiting speed mechanical governor as installed on Detroit Diesel. (Detroit Diesel Engine Div., General Motors Corp.)

The engine governor, in addition to providing a variable engine speed, acts as an over-speed and idle-speed governor while the converter driven governor is controlling the engine. Each governor has its own control lever and speed adjusting screws.

The converter driven governor works on the same principle as the standard engine governor except it cannot cut off fuel to the idle jet in the engine driven governor. If the converter tail shaft overspeeds it will not stop the engine.

LIMITING SPEED MECHANICAL GOVERNORS

The limiting speed mechanical governor, Fig. 18-12, as installed on some Detroit Diesel engines performs the following functions:

1. Controls engine idling speed.
2. Limits the maximum operating speed of the engine.
3. Regulates fuel input at part throttle.

The limiting speed governor used on some series 71 Detroit Diesel engines may be either single weight or double weight, Fig. 18-12.

The governor consists of three subassemblies.

1. Control cover housing.
2. Control housing.
3. Weights and housing.

The governor controls fuel for starting when the control lever is in the idle position. Immediately after starting, the governor moves the injector racks to the position required for idling. In general, the series 71 Detroit diesel is used in marine propulsion and for the generation of electricity.

Operation of the single weight governor is as follows: The centrifugal force of the revolving governor weights is converted into linear motion which is transmitted through the lever (67), Fig. 18-12, and operating shaft (26) to the operating lever (27). One end of the lever (27) operates against the high and low speed springs (48 and 46) through the spring cap (47). The other end provides a moving fulcrum on which the differential lever (23) pivots.

When the centrifugal force of the revolving governor weights balances out the tension on the high or low speed spring (depending on the speed range), the governor stabilizes the speed for a given setting of the governor control lever.

In the low speed range, the centrifugal force transmitted operates against the low speed spring. As the engine speed is increased, the centrifugal force compresses the low spring (46) until spring cap (47) is tight against the high speed plunger (44). This removes the low speed spring from operation and the governor is then in intermediate speed range. In this range, the centrifugal force is operating against the high speed spring and the engine speed is manually controlled.

As the engine speed is increased to a point where the centrifugal force overcomes the preload of the high speed spring, the governor will move the injector racks out to that position required for maximum no-load speed.

A fuel rod (14), connected to the differential lever and injector control tube lever, provides a means for the governor to change the fuel settings of the injector control racks.

The engine idle speed is determined by the centrifugal force required to balance out tension on the low speed spring. Adjustment of the engine idle speed is accomplished by changing the tension on the low speed spring by means of the idle adjusting screw (55). The maximum no-load speed is determined by the centrifugal force required to balance out the tension of the high speed spring. Adjustment of the maximum no-load speed is accomplished by the high speed spring retainer (50). Movement of the high speed spring retainer nut will increase or decrease the tension on the high speed spring.

OPERATION OF DOUBLE WEIGHT GOVERNOR

The centrifugal force of the revolving governor double weights (272 and 273), Fig. 18-12, is converted into linear motion. This motion is transmitted through the riser (67) and operating shaft (26) to the operating shaft lever (27). One end of the lever (27) operates against the high and low speed springs (48 and 46) through the spring cap (47), while the other end provides a moving fulcrum on which the differential lever (23) pivots.

When the centrifugal force of the revolving governor weights balances out the tension on the high or low speed spring (depending on the speed range), the governor stabilizes the engine speed for a given setting of the governor control lever.

In the low speed range, centrifugal force of the low speed weights operates against the low speed spring. As the engine speed increases, the centrifugal force of the low speed weights compresses the low speed spring until the weights are against their stops, thus limiting their travel. The low speed spring is then fully compressed and the low speed cap is against the high speed plunger.

Throughout the intermediate speed range the operator has complete control of the engine speed because both the low speed spring and the low speed weights are not exerting enough force to overcome the high speed spring.

As the speed continues to increase, the centrifugal force of the high speed weights increases until this force can overcome the high speed spring and the governor again takes control of the engine. This limits the maximum engine speed. The fuel rod (14) connected to the differential lever and injector control tube lever, provides a means for the governor to change the fuel settings of the injector control racks.

The engine idle speed is determined by the centrifugal force of the low speed weights (272) required to balance out tension on the low speed spring.

The maximum no-load speed is determined by the centrifugal force of the high speed weights (273) required to balance out the tension on the high speed spring.

FUEL MODULATOR

One type of fuel modulator is shown in Fig. 18-13. This is installed on some Detroit Diesel series 71 engines equipped with turbocharged units to improve fuel economy and reduce exhaust smoke. The modulator maintains proper fuel to air ratio in the lower speed ranges where the governor would normally act to provide maximum injector output. It operates in such a manner that, although the engine throttle may be moved into the full speed position, the injector racks cannot advance to the "full fuel" position until the turbine speed is sufficient to provide proper combustion.

The modulator connected to the engine air box is sensitive to the air box pressure. It has a counteracting spring which is overcome by air box pressure when the engine nears its maximum speed. As the air box pressure drops with a decrease in engine speed, the fuel modulator counteracts piston spring (281), Fig. 18-13, and gradually moves the injector racks into a decreased fuel position by means of the cam (284) action against the lever and roller assembly (285). The modulator torsion spring (266) on the end of the control tube permits the control tube to be rotated away from the "full fuel" position while the governor remains at its "full fuel" position.

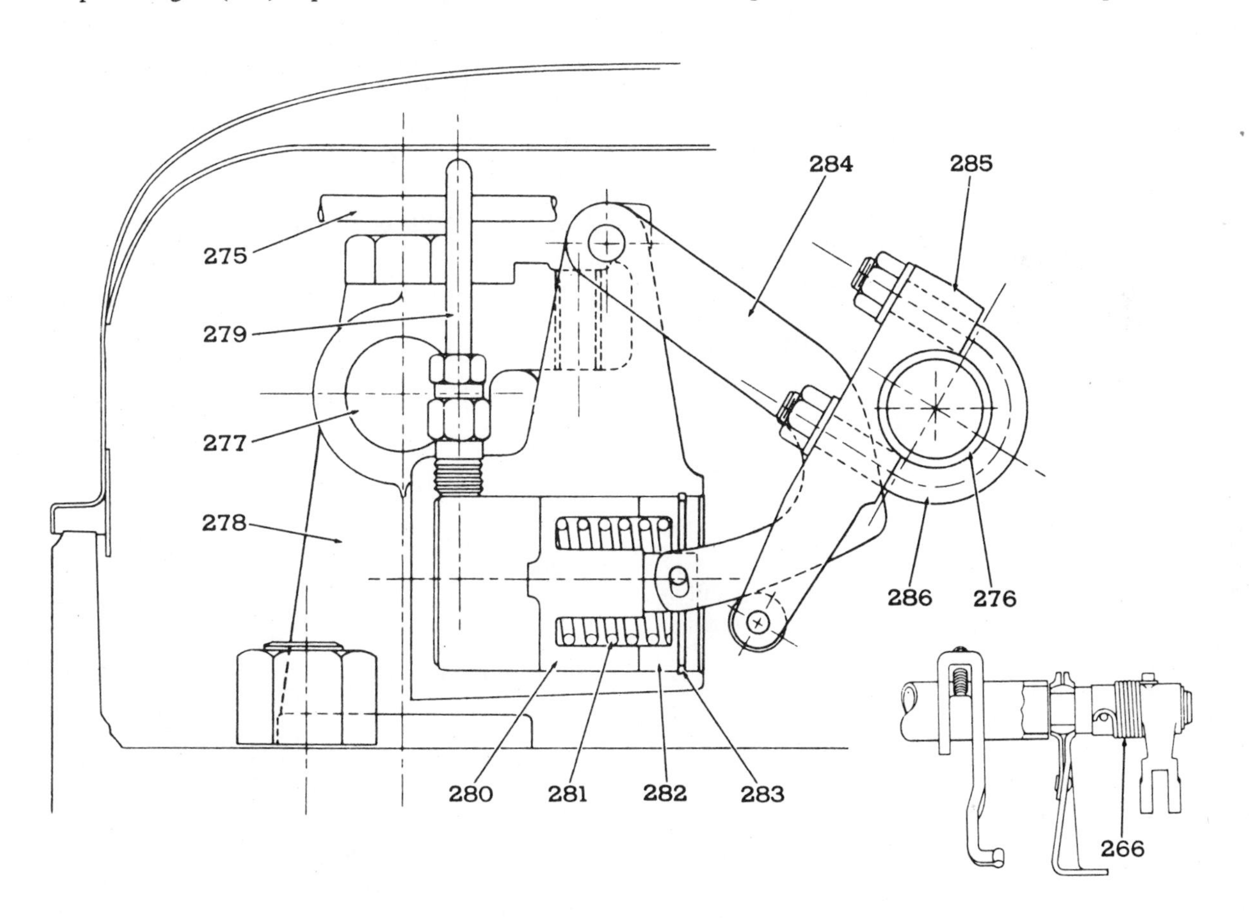

Fig. 18-13. Fuel modulator. (Detroit Diesel Allison Div., General Motors Corp.)

ENGINE LOAD LIMITING DEVICE

Engines may be equipped with a load limiting device to reduce the maximum horsepower of an engine by reducing the fuel output. Such a device, Fig. 18-14, consists of a load limit screw threaded into a plate mounted between two adjacent rocker shaft brackets and a load limit lever clamped to the injector control tube. When properly adjusted for the maximum horsepower desired, this device limits the travel of the injector racks and fuel output of the injectors.

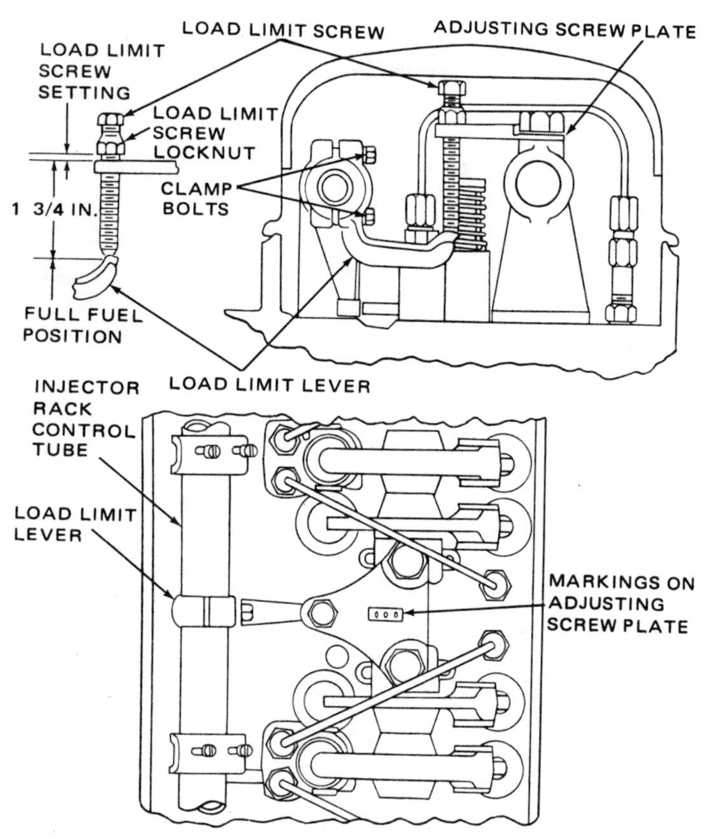

Fig. 18-14. Load limiting device. (Detroit Diesel Engine Div., General Motors Corp.)

PNEUMATIC TYPE GOVERNOR

One type of pneumatic governor is shown in Fig. 18-15. This governor consists of a throttle unit in the inlet manifold connected by suitable tubing to a diaphragm unit mounted on the fuel injection pump.

When the engine is stopped, the governor spring pushes the diaphragm and the control rod to the left, which is the maximum fuel delivery position. As soon as the engine starts, the high air speed past the nearly closed throttle valve and suction pipe orifice creates a high vacuum in the diaphragm chamber. This causes the diaphragm and fuel pump control rod to be drawn toward the right, reducing fuel delivery.

When the engine is operating under load, with the throttle valve fully open, vacuum in the diaphragm chamber is low, due to the low air speed past the throttle valve, and the diaphragm and control valve rod are held in the maximum fuel delivery position by the governor spring.

Any variation in the setting of the throttle valve or engine load causes a variation in the air speed past the throttle valve.

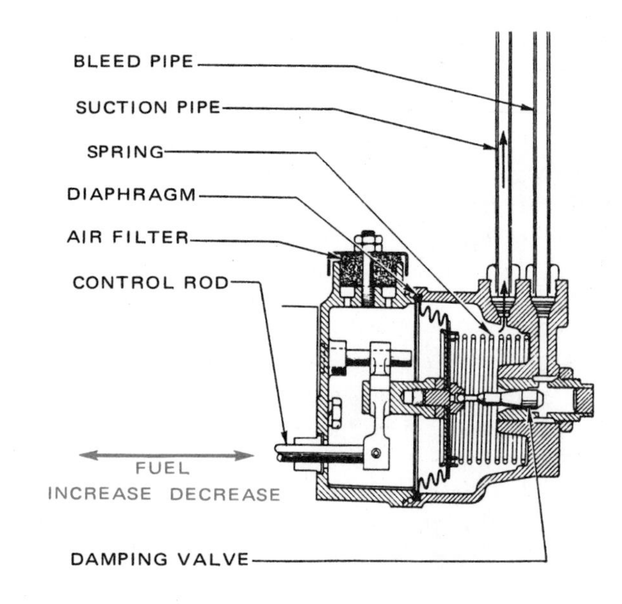

Fig. 18-15. Details of pneumatic type governor as installed on some Ford engines.

This changes the vacuum in the diaphragm chamber and varies the fuel delivery.

To obtain instant response to throttle variations when a pneumatic governor is used, it is essential that the governor spring be as light as possible. This sometimes results in speed fluctuations when the engine is idling. A dampening valve is generally used to prevent such fluctuations.

VACUUM TYPE GOVERNOR

A governor which depends on the vacuum in the intake manifold for its operation is used on some Mercedes-Benz engines (OM 636 for example).

This governor consists of two major parts; the throttle duct,

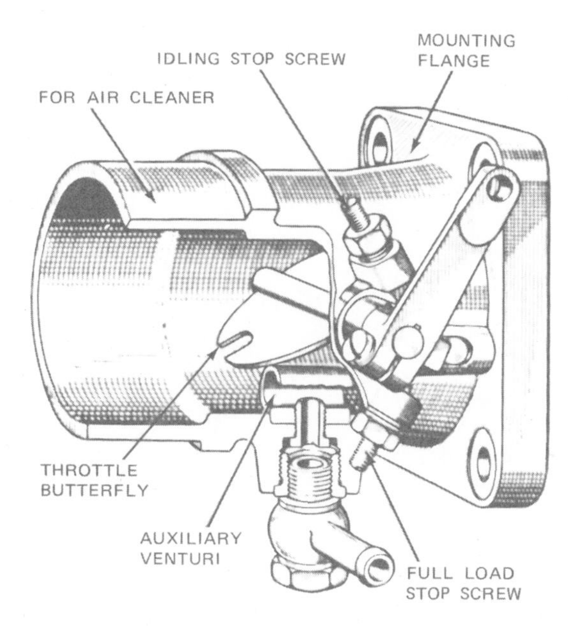

Fig. 18-16. Sectional view of throttle duct, used in pneumatic type governor. (Daimler-Benz Aktiengsellschaft)

Fig. 18-16, and the diaphragm unit, Fig. 18-17. The vacuum or pneumatic governor, as it is also called, is designed to prevent the engine from exceeding the permissible maximum speed and, during coasting, with throttle closed stops fuel injection. At the same time it provides satisfactory idle speed.

The throttle duct is designed to transfer manifold vacuum to the diaphragm unit and control the engine speed by the position of the throttle butterfly. The throttle duct is a venturi. At the narrowest section of the venturi is an auxiliary venturi connected to the vacuum source. The throttle butterfly is connected to the accelerator and to the adjusting lever.

A calibrated air jet for the vacuum connection is installed in the auxiliary venturi. Air velocity in the throttle duct increases or decreases according to the position of the throttle butterfly and the speed of the engine. The same is true for the vacuum behind the throttle duct. When the vacuum is strong enough to overcome the pressure of the control spring, governing sets in.

When the engine is not in operation, the diaphragm (3), Fig. 18-17, is pressed against the full stop (11) by the control spring (6). In order to keep the control rod (1) in the full load position during the starting of the engine, the throttle butterfly can be opened completely by means of the accelerator pedal. Due to the large cross section of the throttle duct, only a weak vacuum is produced. This will not provide power to adjust the diaphragm and control rod (2) to a low injection rate immediately, and the engine will start easier.

During starting, by operating the starting switch the full load stop bolt (9) is pressed out of its normal position for an additional increase of fuel discharge by way of the adjusting lever and the double lever (10). As a result the highest possible injection rate is provided.

When releasing starting switch, spring presses starting lever and double lever (10) immediately back to normal full load position by way of stop bolt (9) and adjusting lever.

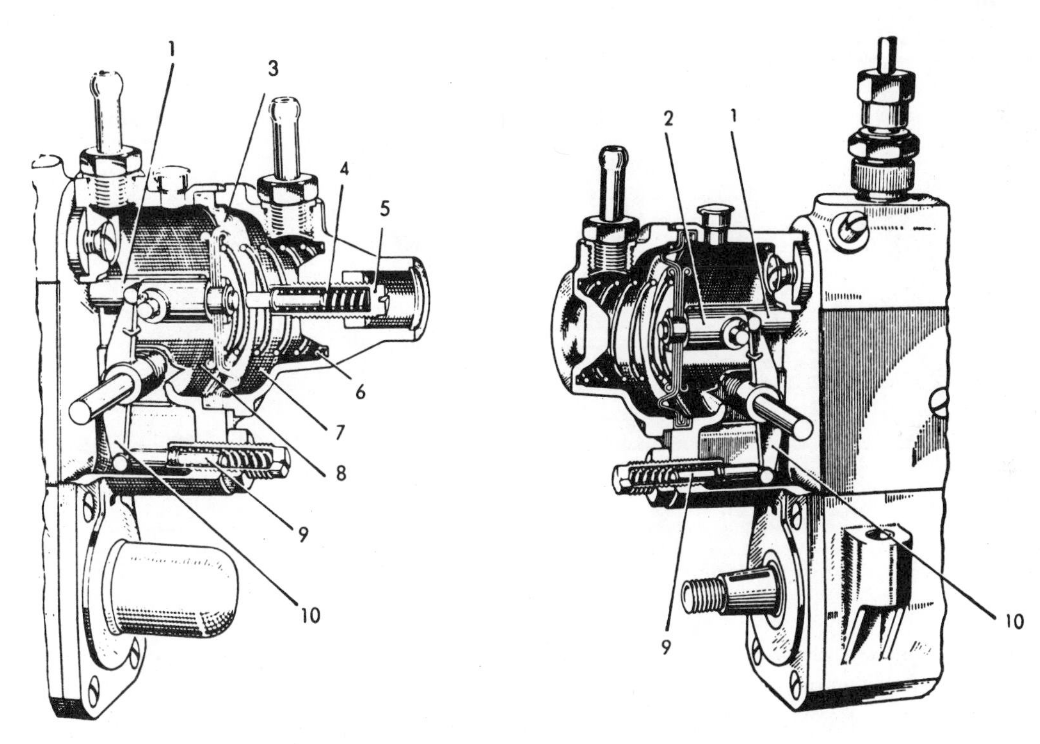

Fig. 18-17. Diaphragm unit of pneumatic type governor. (Daimler-Benz Atiengesellschaft)

1. CONTROL ROD.
2. DIAPHRAGM ROD.
3. DIAPHRAGM.
4. HELPER SPRING.
5. ADJUSTING SCREW.
6. CONTROL SPRING.
7. VACUUM CHAMBER.
8. ATMOSPHERIC CHAMBER.
9. STOP BOLT.
10. DOUBLE LEVER.
11. FULL LOAD STOP.
12. SPRING.
13. DIAPHRAGM BOLT.
14. CONTROL SPRING.

The degree of vacuum controls the position of the diaphragm and the control rod. The auxiliary venturi serves to boost the vacuum and protects the engine from racing during reverse operation.

The diaphragm unit mounted on the front side of the injection pump, is subdivided by the diaphragm (3), Fig. 18-17, into two chambers, the vacuum chamber (7) and atmospheric chamber (8).

Under full load and low speed, the throttle butterfly is fully opened, the speed is low and the vacuum is weak. The control spring (6) presses the diaphragm (3) with the diaphragm bolt by way of the double lever (10) against the stop bolt (9) of the sprung full load stop, Fig. 18-17. The adapting spring (4) in the diaphragm is still compressed by the force of the control spring (6) and the engine receives the full injection rate. Minor adjustments can be made by turning the adjusting screw (5).

HYDRAULIC GOVERNOR

In the hydraulic governor, the pressure of the engine lubricating oil system acting on the governor piston is utilized to open or close the injectors as required by variations in engine load.

When the engine is operating, the flyweights move outward due to centrifugal force. This force is transmitted through the flyweight linkage and the connected plunger to the end of the governor pilot valve, Fig. 18-18.

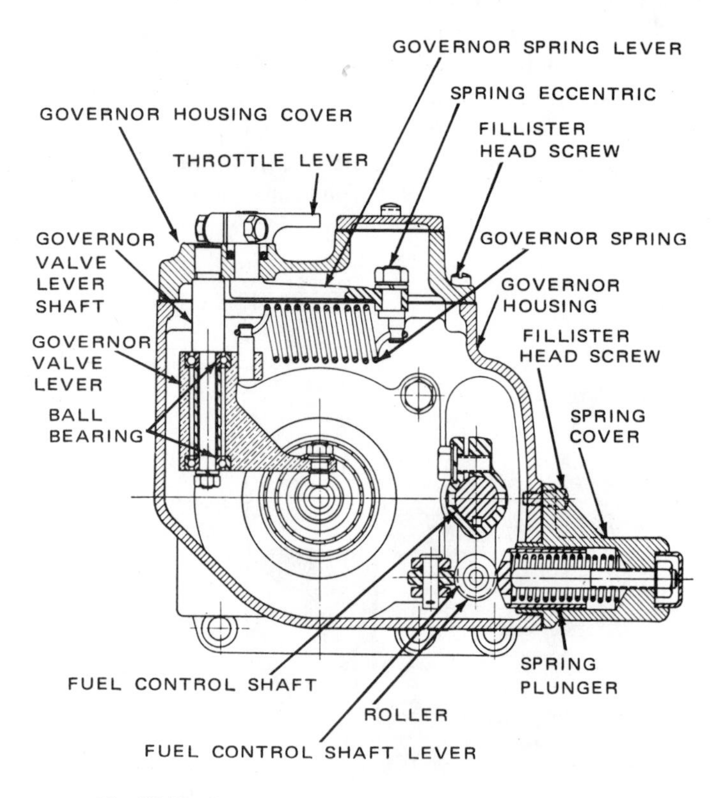

Fig. 18-18. Hydraulic type governor. (Murphy Diesel Co.)

Centrifugal force of the flyweights is opposed by the force of the governor spring acting on the pilot valve in the opposite direction. The pilot valve is free to move in either direction depending on which of the forces imposed on it is greater. In so doing the pilot valve controls the flow of oil under pressure into, or out of, the space behind the governor piston, thus opening the injectors or permitting them to close.

If the engine load increases, the engine speed will decrease, reducing the centrifugal force of the flyweights on the governor pilot valve. The force exerted by the governor spring will then be greater than the two forces. This will move the pilot valve and permit oil under pressure to flow to the space behind the governor piston. As the pressure behind the piston increases, it will move, transmitting movement through the linkage to the injectors causing them to open and supply more fuel to the engine.

With the increase in power, the engine speed will increase until the centrifugal force of the flyweights and the governor spring are again in balance.

REVIEW QUESTIONS – CHAPTER 18

1. The governor on an engine controls the engine speed by ___________the flow of fuel.
2. The governors used on diesel powered trucks are designed to control the_____________ throughout the entire speed range.
3. For certain operations, diesel engine speed should remain at a constant value regardless of the load. This is known as being ____________ .
4. A dead beat governor has a high degree of___________ .
5. Speed droop is the same as:
   a. Speed reduction.
   b. A drop in engine load.
   c. Steady state regulation.
6. The provision made in some Woodward governors to permit operation in both directions is through:
   a. Idler gears.
   b. Additional check valves.
   c. The governor pilot valve.
7. The method used on some Woodward governors to attain stability is by a:
   a. Buffer compensating system.
   b. Special gearing system.
   c. Low pressure pumping system.
8. The Cummins PT torque converter governor consists of two mechanical variable speed governors in _________.
9. The limiting speed mechanical governor used on some Detroit Diesel Allison engines functions as follows:
   a. ______________________________________.
   b. ______________________________________.
   c. ______________________________________.
10. The fuel modulator installed on some Detroit Diesel Allison engines with turbochargers is used to:
    a. Improve fuel economy.
    b. Improve speed droop.
    c. Increase stability.

Chapter 19
FUEL HANDLING

The importance of clean fuel in the operation of a diesel engine cannot be overemphasized. Maximum power and operating economy is difficult to attain unless the fuel is free from water and other foreign material. Contamination will quickly damage the precision made parts of the injectors.

Fuel oil as received from the refinery is usually free from water and sediment. However, if special care and precautions are not taken in delivery and handling of the fuel, wear producing contaminants may find their way into the containers.

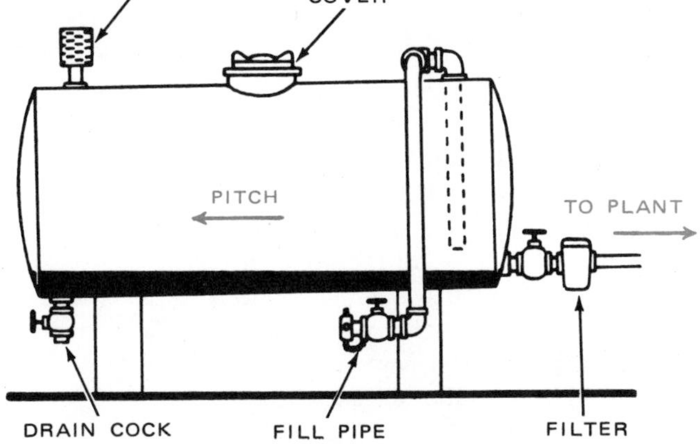

VENT PIPE AND AIR FILTER
MANHOLE COVER
PITCH
TO PLANT
DRAIN COCK
FILL PIPE
FILTER

Fig. 19-1. Outdoor storage tank showing the vent pipe with air filter and drain cock for removing condensate. (American Bosch Div., Ambac Ind.)

FUEL SUPPLY TANKS

Industrial installations and diesel powered fleets use large quantities of fuel. Storage facilities of large capacity are required. Depending on the time of delivery, many operations find it advantageous to store only a weekly supply. Storing larger quantities is generally not desirable. Stored fuel has a tendency to form gum, corrosive compounds and deteriorate in other ways. However, refineries are using an increasing number of additives which help materially in stabilizing diesel fuel and reducing any tendency toward gum formation. When fuel is stored in large tanks, there is also a greater tendency for moisture condensation within the containers. This tendency can be reduced by using a series of small tanks and injecting them into the line as needed.

Fuel oils do not all age at the same rate, since oils from different areas have different stabilizing characteristics. Additives have been developed to help this condition.

Fuel stored in above ground tanks will deteriorate more rapidly than fuel stored in underground tanks. Above ground tanks, Fig. 19-1, are subject to wider variations in temperature and humidity and tend to collect a greater degree of condensation within the tank.

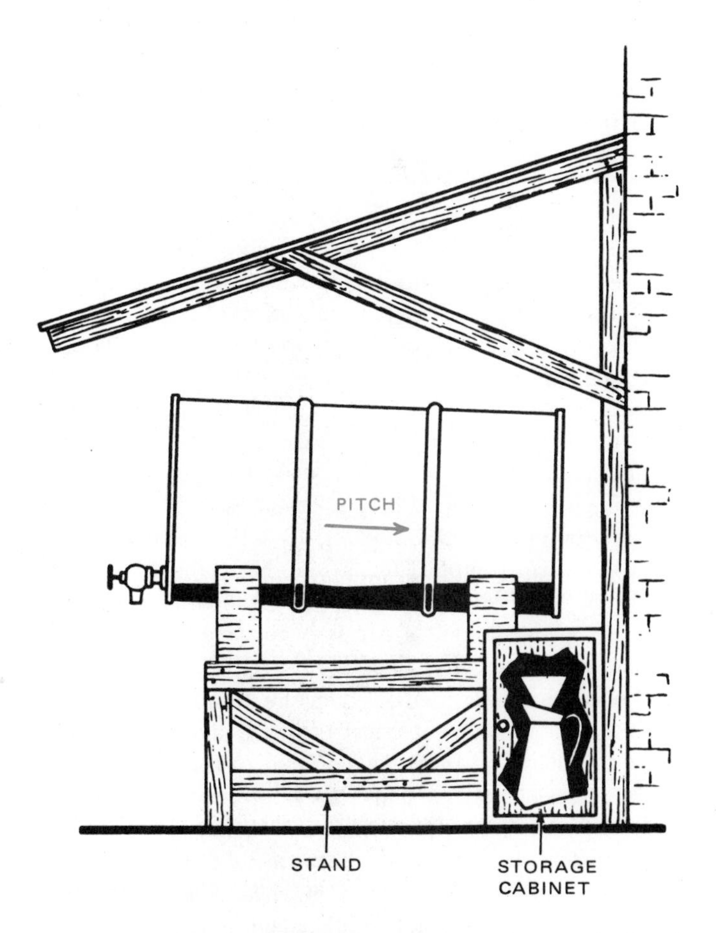

Fig. 19-2. When barrels are used for storage, the front of the barrel should be higher than the rear so contaminants would collect at the back and only clean fuel would be taken from the tank. (American Bosch Div., Ambac Ind.)

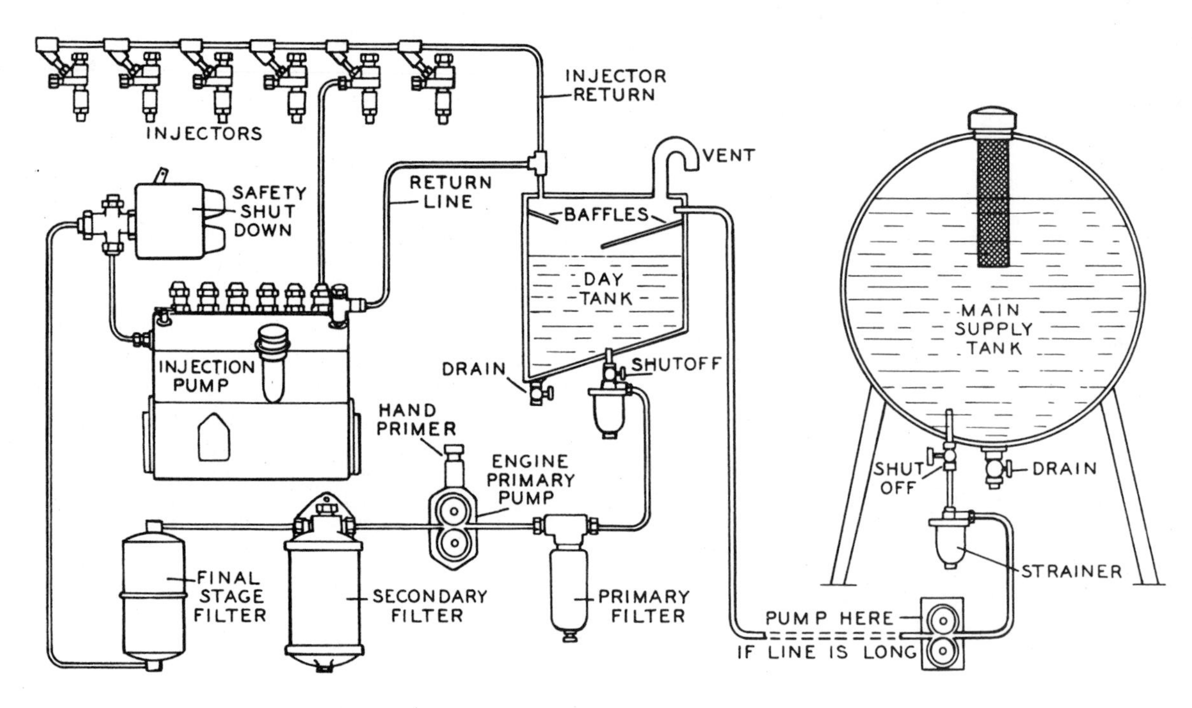

Fig. 19-3. Fuel system layout with storage tank and day tank. (Waukesha Motor Co.)

If long storage is contemplated, special additives are available which will reduce aging tendencies. Adding fresh fuel to a tank partly filled with old fuel is a poor practice since the old fuel will produce more rapid aging of the new fuel.

Before building any storage facilities, the National Bureau of Fire Underwriters as well as local authorities must be consulted to insure that the installation conforms to their requirements.

Special precautions are needed for the construction of storage tanks located within buildings. Such precautions are set forth in the underwriters specifications. Regulations cover the size of tanks, location, ventilation of enclosures, foundation, and piping.

Fuel is often stored in drums, particularly on construction jobs in isolated areas. When this is necessary, every precaution should be taken to keep out dirt and water when handling the fuel. A preferred method is to place the tanks in a horizontal position on racks, as shown in Fig. 19-2. Drums should also be kept in a shed to protect them from direct sunlight and rain.

A new drum should not be tapped until the preceding drum has been emptied. If it is necessary to stand tanks on end, a suction pump is required to draw out the fuel.

The suction pipe for the pump should not come close to the bottom of the tank. Normally there should be three inches from the open end of the suction pipe to the bottom of the tank. This will provide ample space for condensed moisture and solid foreign matter to collect. If the oil is dispensed by gravity from a horizontal tank, the rear end of the tank should be lower than the front, thereby providing a sump for foreign matter, Fig. 19-2. The filler pipe should have an 80 to 100 mesh screen and the vent pipe should have a cap with an air filter to prevent dust from entering.

In large industrial installations of diesel engines, a day tank is often included in the fuel system, Fig. 19-3. The day tank is provided to measure the fuel consumption of each engine. In some installations, the day tank has sufficient capacity to hold a day's supply of fuel. However, the size is restricted by the recommendations of the Bureau of Fire Underwriters which limits the quantity of fuel which may be stored in one engine room.

The day tank in some installations is placed below the level of the fuel injectors so that excess fuel can drain back to the tank. Other designs place the day tank higher than the engine so that fuel will flow by gravity to the injectors.

Fig. 19-3 shows the location of the main supply tank above ground. In many cases the supply tank is below ground and the transfer pump must have sufficient capacity to raise the fuel to the desired level of the engine day tank.

FUEL METERS

Upon delivery, fuel is usually measured by gallons and a meter is placed between the tank car and the storage tank. If delivered from a tank truck, the fuel is measured directly by a meter on the truck. Such meters are usually certified for accuracy by local authorities. In large installations of diesel engines, it is also customary to meter the fuel supplied to the individual engines.

TRANSFER PUMPS

Pumps used to transfer fuel from the main storage tanks to the day tanks are usually motor driven. They may also be driven by a power takeoff from the diesel engine, or an extension of the fuel injection pump.

In many installations these are positive displacement pumps

of the rotary gear type. Vane type pumps or helical screw type pumps are also used.

The positive displacement spur gear pump was discussed and illustrated in the chapter on lubrication systems. A similar type pump is used for fuel oil transfer.

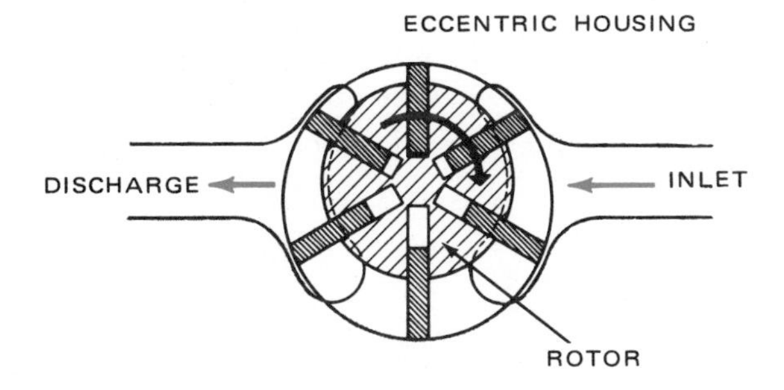

Fig. 19-4. Simple vane type pump. Vanes are free to slide radially. (TRW Thompson Products, Inc.)

A simple vane type pump is illustrated in Fig. 19-4. The rotor contains the sliding vanes or blades that are free to move outwardly against an eccentric housing. Spaces between adjacent vanes expand on the suction side, filling with oil, and contract on the discharge side, forcing the oil through the discharge port. Suction and discharge ports may be kidney-shaped openings at the end face, as shown in Fig. 19-4, or they may be radial openings. Vane pumps are available in constant or variable displacement output. Vane pumps generally have high efficiency and a high output for the relatively small space occupied.

A rotary, positive displacement helical screw type pump is illustrated in Fig. 19-5. This design has three meshing screws shaped so they form a tight seal relative to each other and to the surrounding housing. As the screws rotate, they form sealed enclosures filled with fuel oil, which moves axially from inlet to discharge. This rotational motion is transformed into continuous and uniform axial motion. Rotor threads are shaped so that the volume enclosed by the threads remains constant. There is no tooth nor end clearance volume to pocket the fluid being pumped. Consequently, there are no hydraulic compressive forces so no liquid shock occurs between the threads. Fluid pressure is used to drive the rollers which serve as sealing elements.

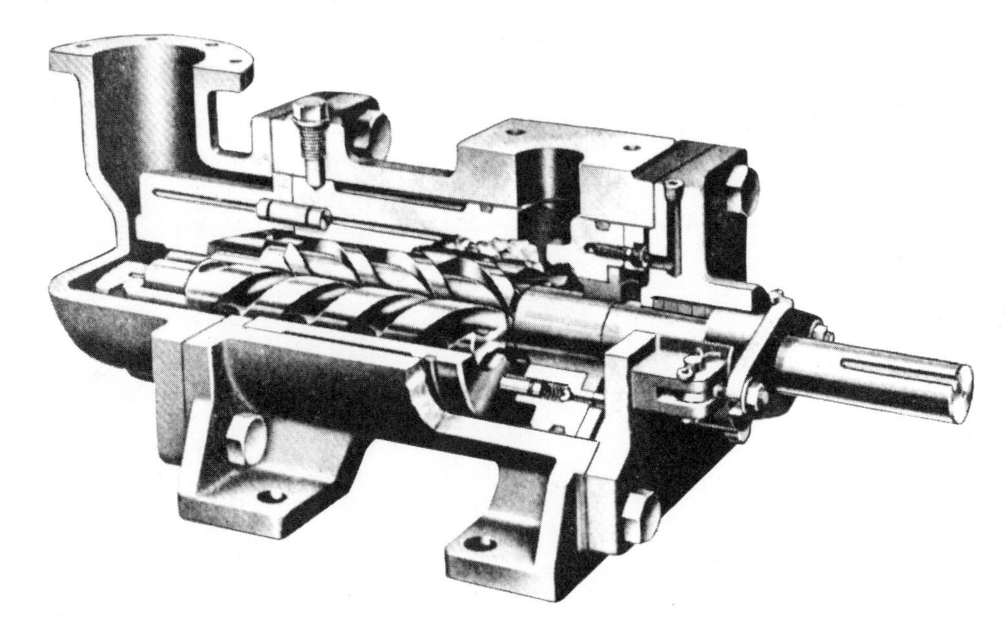

Fig. 19-5. Helical screw type pump. (De Laval Turbine Co.)

REVIEW QUESTIONS – CHAPTER 19

1. Fuel is received from the refinery is usually free from ___________ and ___________ .
2. Fuel oil stored over long periods of time tends to:
   a.___________________________.
   b.___________________________.
   c.___________________________.
3. Variations in temperature and humidity of stored fuel oil tends to cause ____________ in the storage tanks.
4. When removing oil from a tank with a suction pipe, the pipe should come no closer to the bottom of the tank than:
   a. 1 in.
   b. 3 in.
   c. 4 in.
5. A day tank is often placed below the level of the fuel injectors so that excess fuel can ____________ back to the tank.
6. Diesel fuel oil is usually measured in ____________.
7. Three methods of driving transfer pumps that deliver oil from main storage tanks to day tanks are:
   a.___________________________.
   b.___________________________.
   c.___________________________.

Chapter 20
RECONDITIONING DIESEL ENGINES

INSPECTING AND CLEANING THE ENGINE

The importance of cleaning an engine before starting any reconditioning operation cannot be overemphasized. Even before cleaning it pays to make a thorough examination of the engine, since the external appearance of the cylinder head and engine block will often reveal the basic cause of trouble. For example, if the cylinder head and rocker arms do not show any signs of oil and have a dry, heated appearance, it is usually safe to assume that oil has not been reaching the head and it has been running hot. As a result, valve springs will have lost their temper and valves will have been burned. Examine the outside of the head and engine block for evidence of leaking core plugs or cracks which would be indicated by rust streaks. After preliminary examination, the engine may then be cleaned externally. It can then be disassembled and the individual parts cleaned. There are several different methods of cleaning that are used by diesel shops. These methods include the following:

1. Steam cleaners.
2. Cold tank.
3. Hot tank.
4. Mechanical washers.
5. Jet cleaning.
6. Untrasonic cleaning.
7. Glass bead cleaning.

Fig. 20-1. Cleaning a cylinder block with steam and a detergent. (Oakite Products Inc.)

The steam cleaner, Fig. 20-1, was one of the first methods developed for cleaning parts and complete engines. This equipment consists of a steam generator and the necessary hose and gun so that the jet of steam can be directed against the parts to be cleaned. One difficulty of this method is the relatively large amount of space required. When used with a detergent, the steam method is used extensively for removing road dirt and oil sludge from engines.

The cold tank method, Fig. 20-2, as the name implies, provides a large tank containing a detergent together with a pump and hose which directs the detergent against the parts to be cleaned. This method is used extensively for cleaning individual parts.

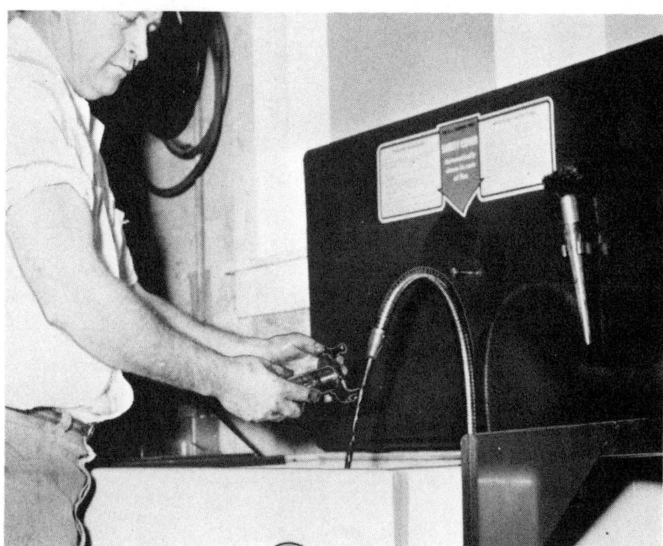

Fig. 20-2. Cleaning parts by the cold tank method.

The hot tank method, Fig. 20-3, is a popular and effective method of cleaning diesel parts. The tank is of sufficient size to accommodate a complete automotive engine so that it is completely immersed. A hot detergent is sprayed directly on the desired areas. Using proper detergents, this method does not require brushing. After cleaning, water is used to wash off the detergent.

Fig. 20-3. Inspection of a diesel part after removal from a hot cleaning tank.

In the mechanical type washer, the work is agitated in a detergent solution. When washing is completed, the parts are rinsed in water.

When using the jet or high velocity cleaner, the detergent is sprayed from a large number of jets which direct a high-pressure spray on parts from all sides, Fig. 20-4. In some types, the parts are also rotated, Fig. 20-5.

Fig. 20-4. Illustrating the high velocity spray method of cleaning parts. (Peterson Machine Tool Co.)

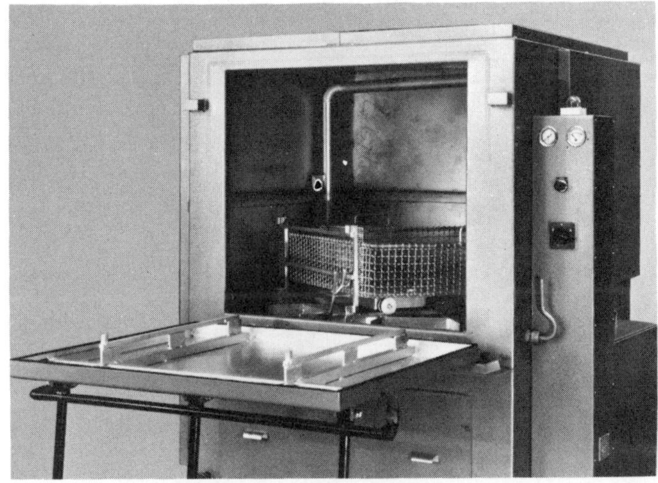

Fig. 20-5. High velocity cleaning machine. (Van Norman Machine Co.)

The ultrasonic method of cleaning parts is relatively new. The term ultrasonic pertains to acoustic frequency above the range audible to the human ear, which is approximately 20,000 cycles per second. An ultrasonic cleaner, Fig. 20-6, consists basically of three parts:

1. A generator which converts line electric current into high frequency electric energy.
2. A transducer which converts the electric energy into mechanical energy.
3. A reservoir containing a liquid through which the ultrasonic energy is passed. This creates cavitation, (a swiftly moving, agitated liquid).

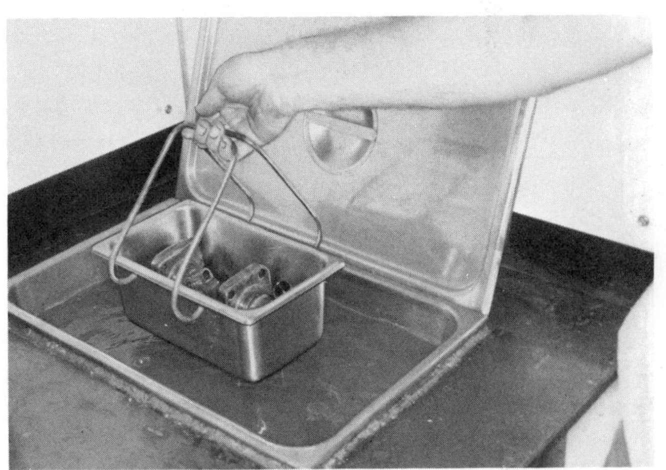

Fig. 20-6. The ultrasonic method of cleaning parts is one of the newer methods.

When a dirty part is placed in the tank, the action of cavitation blasts away the particles. This is one of the newer methods of cleaning parts and is very effective.

A blast of compressed air carrying round glass beads is another effective method of cleaning parts, Fig. 20-7. Both wet and dry blast machines are available and clean parts down to the bare metal. Since the glass beads are round, they do not embed themselves in the surface being cleaned. They leave a uniform indented surface which imparts strength by compressing the metal. Critical dimensions are not altered. Parts to be cleaned must be dry and free from grease, since grease will contaminate the beads and reduce their effectiveness.

Fig. 20-7. A popular method of cleaning parts is by the use of glass beads. (Van Norman Machine Co.)

TYPES OF DIRT

Dirt that accumulates on diesel engines generally consists of oil, mud, road tar and salt from deicing roads. It may also include antifreeze that has leaked from cracks in the engine cooling system.

Such dirt is not easy to remove. It should be remembered that a detergent that is satisfactory for steel and cast iron will not be safe to use on aluminum or die cast metal. Some effective cleaning detergents may also have annoying or harmful fumes.

When aluminum parts and engine bearings are to be cleaned, special detergents should be used. A thorough rinsing in water should follow the cleaning.

CRANKSHAFT RECONDITIONING

The first step in reconditioning a crankshaft, even before cleaning, is a visual inspection. Such an inspection should be made to see if there is any evidence of burning or overheating around the bearing journals. Also check to see if there are any cracks which sometimes are more visible before cleaning. Damage resulting from a broken connecting rod may also be found. An examination will often disclose whether the shaft can be simply reconditioned by regrinding or whether it must be built up by welding or some similar method before grinding the journals to size.

The equipment needed for reconditioning crankshafts includes:

1. Crankshaft grinder.
2. Steady rests.
3. Balancing equipment.
4. Extra hubs and wheels.
5. Magnetic particle inspection.
6. Cleaning equipment.
7. Straightening equipment.
8. Engine lathe.
9. Radius dresser.

The crankshaft can then be cleaned and any plugs in the shaft removed so that the oil passages can be thoroughly cleaned. These plugs are often difficult to remove. It may be necessary to apply heat from a torch then quench in cold water in order to loosen the plugs.

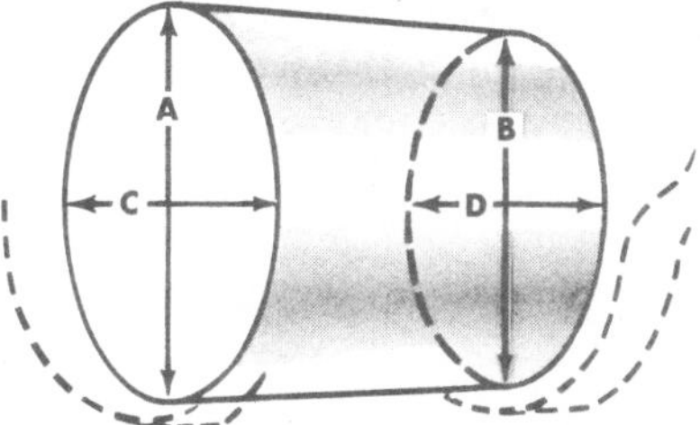

Fig. 20-8. Measuring a bearing journal for taper and out-of-round.

Fig. 20-9. Checking a crankshaft for straightness between V-blocks.

After cleaning, measure the bearing journals for out-of-round and taper, Fig. 20-8. Obtain specifications for tolerances from the engine manufacturer. Typically, connecting rod journal and main bearing journal tolerance in the 400 cu. in. displacement area is limited to .0005 in. out-of-round.

Then place the crankshaft between centers or on V-blocks and check for alignment, Fig. 20-9. If the centers have been damaged, it will be necessary to recenter the shaft. Also check the flywheel flange and nose of the shaft to be sure they are running true.

Having made these checks, determine whether the shaft can be restored to standard undersize dimensions by regrinding. If regrinding is not possible because of damage, build up the journals by welding. Then grind the journals to size.

Another check to be made is a close examination to determine if there are any cracks in the crankshaft. Many shops use the magnetic particle process, other shops use the Magnaflux method, Fig. 20-10. Some cracks are more serious

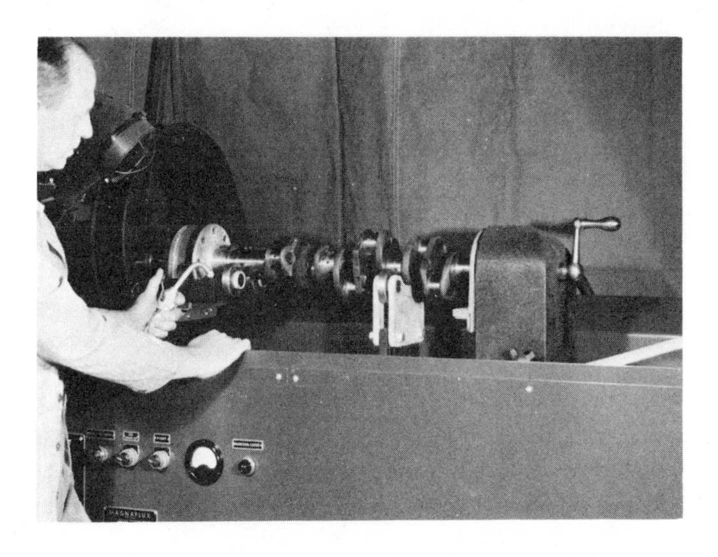

Fig. 20-10. Locating cracks by the Magnaflux method. (Magnaflux Corp.)

than others. For example, cracks which run around the fillet area are usually cause for rejection of the shaft. Similarly, if cracks are deep into the oil holes, the shaft should be discarded. Some cracks are fine and small and can be ground out. Cracks running parallel with the journal may be ground out if not too deep.

Another method of locating cracks is to use dye penetrants. The parts must first be cleaned with kerosene or other grease removing solvent. Then supply the dye penetrant. Allow time for the dye to penetrate the flaws which usually takes about 15 minutes. Remove any excess penetrant. Then apply developer which will make the flaw more prominent.

Straightening the shaft is not difficult. With the shaft mounted between centers, the extent of misalignment can be measured with a dial gage, Fig. 20-9. Readings are taken at the end bearing journals and also at the center bearing. Any difference in the readings is the amount of misalignment. However, measurements should also be taken at the other main bearings. Having checked those journals, the crank throws should also be checked, since a single crank throw can be out of alingment as the result of a broken connecting rod.

The actual straightening can be performed in a lathe between centers, Fig. 20-11, in a press, or in a press while peening. The amount of pressure to be applied will depend on the size and type of crankshaft and the skill of the operator. Pressure is usually applied to spring the shaft slightly beyond the straight position, allowing it to spring back to its original straight axis. Skilled operators indicate that the shaft should be trued four or five times until it is straight.

Fig. 20-11. Straightening a crankshaft in a lathe.

When using the peening method to straighten shafts, the shaft should be peened on the low side. As much as 1/16 in. can be straightened by using a round nosed chisel and a hammer. A similar method is to apply pressure to the shaft and peen it at the same time.

When a shaft is bent at one of the crank throws, due to a broken connecting rod, the straightening force is applied to the misaligned area. The shaft should be supported at the adjacent main bearings.

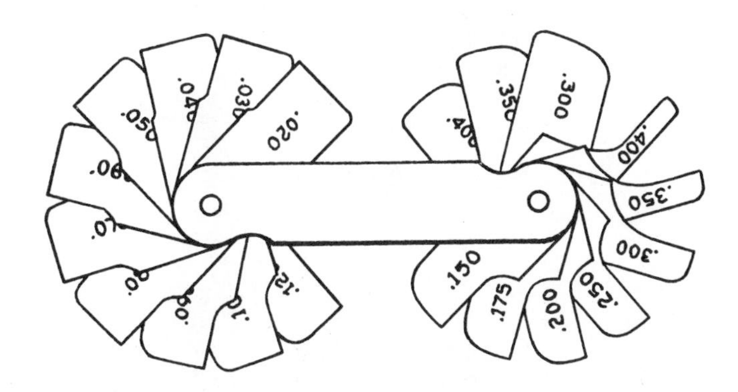

Fig. 20-12. Typical fillet or radius gage.

Fig. 20-13. Dressing a grinding wheel preparatory to grinding crankshaft. (Van Norman Machine Co.)

Not all shafts can be straightened by applying pressure. The "Elotherm" crankshaft used in some International Harvester engines is an example.

After having passed these inspections, the next step is to measure the radii of the fillets, using a radius gage, Fig. 20-12. The radii should be recorded. In most cases the radius of the connecting rod journals will differ from that of the main bearing journals and it is essential that these radii are not altered in the regrinding or polishing processes. If they are altered, the bearing shells which are to be installed will not fit properly and early bearing failure can be expected.

DRESSING THE GRINDING WHEEL

Having these radii, the next step is to dress the grinding wheel to conform to the radius of the journal to be ground. A wheel must be selected that is suitable for the material from which the shaft is made. Grinding wheel manufacturers will help in the proper selection of the grinding wheel.

The sides of the grinding wheel are dressed first and then the face. Following that, the radius is ground, Fig. 20-13. After dressing the wheel, it must be balanced.

If a new grinding wheel is being installed, the hubs should be true and free from any roughness. The wheels should slide down freely over the center portion of the hub without any binding. The blotter washers should be placed on each side of the wheel between the flange and the wheel. The flange should be at least one-third of the diameter of the wheel. Excessive tightening may cause breakage.

There are a number of very important safety precautions which must be followed when using grinding wheels. Gloves should always be worn when handling the wheels. Guards should always be in place around the wheel, Fig. 20-14, and the operator should always use goggles, Fig. 20-15. The operator should always stand to one side of the wheel, particularly when the wheel is first started. The manufacturer's instructions as to the speed of rotation must always be followed. The wheels should always be balanced with a high degree of accuracy. It must be remembered that as rotational speed increases, centrifugal forces tend to pull the wheel apart. Should that occur, pieces of the wheel will be hurled at terrific speed, sufficient to maim or kill any individual they might strike.

SELECTING THE GRINDING WHEEL

The type of grinding wheel used for regrinding is generally dependent on whether the crankshaft has been cast or forged. Surface hardening is also a factor to be considered.

Fig. 20-14. Large size crankshaft grinder. (Van Norman Machine Co.)

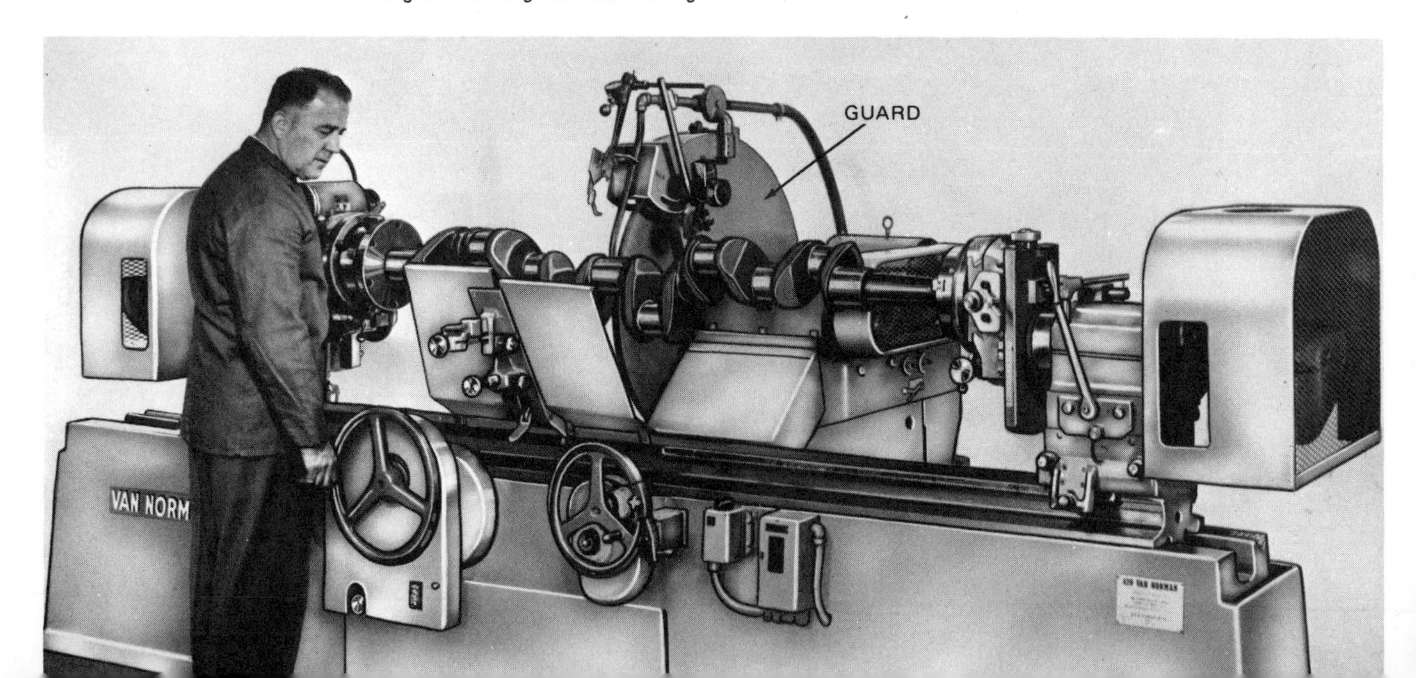

Fig. 20-15. Face guards or goggles should always be worn when doing any grinding operation.

Basically, there are two types of surface hardening. One is known as the Toco hardening method in which the main journal and crankpins are hardened by either induction or flame hardening. In this process, the depth of hardening is not less than 0.050 in. and, therefore, crankshafts of this type can be reground.

The other method of hardening crankshafts includes Tufftriding or nitriding. By these methods the crankshaft is hardened over its entire surface, but only to a depth of 0.0003 to 0.0005 in. Crankshafts hardened by these methods must be rehardened after being reground.

One method of determining what type of hardening has been done on a crankshaft is to file a small area on a crank cheek (not a journal). If metal can be removed with a slight pressure, the shaft has been hardened by the Toco method. If no metal is removed by filing, the crankshaft has been hardened by either the Tufftriding or nitriding method. Currently, such crankshafts are being retreated only by specialists. After Tufftriding, it is usually necessary to straighten the crankshaft again.

An exception to the above is found in the crankshafts used on the V-8 Detroit diesel engine. Recent models of these engines are now equipped with crankshafts which are fully induction hardened.

Shafts treated with the Elotherm method have the bearing surfaces and fillets hardened to a depth of 0.100 in. Special precautions are necessary when grinding these shafts. First, the shafts should not be straightened. When grinding, care should be taken to avoid burning by reducing the rate of wheel infeed. An automatic wheel dresser is mandatory to prevent burning, chattering or poor surface finish. Grinding wheel hardness should be in the "M" hardness range with 50 grit size. Ample coolant should be used while grinding.

A Peterson–Barco crankshaft grinder set up with a Cummins crankshaft is shown in Fig. 20-16.

While opinions differ as to the direction of rotation in the grinding operation, many authorities recommend that the crankshaft be ground in a counterclockwise rotation and polished in a clockwise rotation. This polishing operation will remove all the grinding fuzz, Fig. 20-17.

REBUILDING THE CRANKSHAFT

Not every shaft is in good enough condition that it only requires regrinding of the journals so that it can be placed back in service again. In many cases, bearing journals have been so damaged or worn that if reground, the diameter of the journal would be smaller than next available undersize bearing.

Fig. 20-16. Regrinding a model NH Cummins diesel crankshaft on a crankshaft grinder. (Peterson Machine Tool Co.)

Fig. 20-17. After grinding, the crankshaft journals are polished.

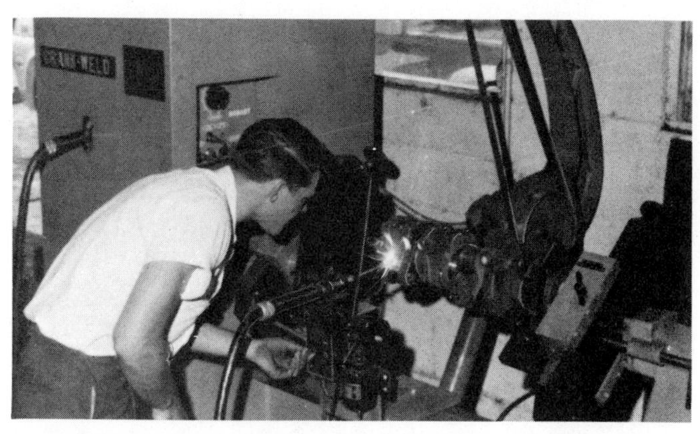

Fig. 20-18. Building up a worn crankshaft journal by arc welding.

When that occurs, the journals must be built up by:

1. Arc welding.
2. Gas welding.
3. Electric stick electrode welding.
4. Tungsten inert gas (TIG) welding.
5. Submerged arc welding.
6. Metallic inert gas (MIG) welding.
7. Chrome plating.

The Hobart electric arc welder is shown in Fig. 20-18, and the Gleason submerged arc welder in Fig. 20-19.

The method used to reclaim a crankshaft will depend largely on the material from which the crankshaft is made. Materials range from low carbon steel to cast iron and may be be either cast or forged.

Crankshaft welding machines are automatic. The crankshaft is rotated at a specified speed and the welding torch is moved

Fig. 20-19. Illustrating the submerged arc method of reconditioning a crankshaft journal. (W. R. Gleason Engineering)

Fig. 20-20. Close up of crankshaft journal in process of being reconditioned by the submerged arc method. (W. R. Gleason Engineering)

over the bearing journal at a speed designed to insure a coating of constant thickness and penetration.

While the procedure varies with different methods, the procedure, in general, is as follows: The shaft is first cleaned, usually in an alkaline solution. It is then placed in the crankshaft grinder and the worn journal accurately aligned. All scores must be removed by grinding or by an abrasive belt. The welding torch is then positioned and the oil holes in the shaft are plugged with ceramic plugs to prevent the welding material from entering the opening. If there is any oil in the hole, it may run out when heated and spoil the weld. The welding process is then started, as shown in Figs. 20-19 and 20-20.

When the welding process is completed, the shaft must be stress relieved. This can be done with a butane burner. A temperature indicating crayon is used to mark the shaft. As heat is applied, the color of the mark changes and will indicate when the desired temperature is reached.

Shafts that have been welded usually require straightening, Fig. 20-11. One procedure for the modular type shaft is to straighten it while the shaft is still warm. If it has cooled it must be reheated to 450 deg. and then straightened.

Crankshafts of low alloy steel as used in many industrial diesel engines, can be stress relieved by rough grinding, before straightening, after which they are finish ground.

BALANCING DIESEL ENGINES

Precision balancing of the diesel engine, including the crankshaft, pistons and connecting rods, can increase bearing life as much as 25 to 100 percent and at the same time step up horsepower 10 percent.

In any rotating object there may be two types of unbalance. One is known as kinetic unbalance and the other dynamic unbalance.

Static unbalance exists when only weight or gravity is involved. This is the situation when a shaft is placed on knife edges. It will then rotate and come to rest when the heaviest part of the shaft is at the bottom, Fig. 20-21.

Kinetic unbalance occurs when a shaft is rotated. The unbalance weight sets up a centrifugal force which will tend to throw the mass outward and cause a bending or flexing of the shaft.

Dynamic unbalance in a rotor or shaft will result when there are two weights in separate planes, Fig. 20-21. These weights will set up separate forces which will cause the rotor or shaft to rock and twist on an axis perpendicular to the axis of rotation.

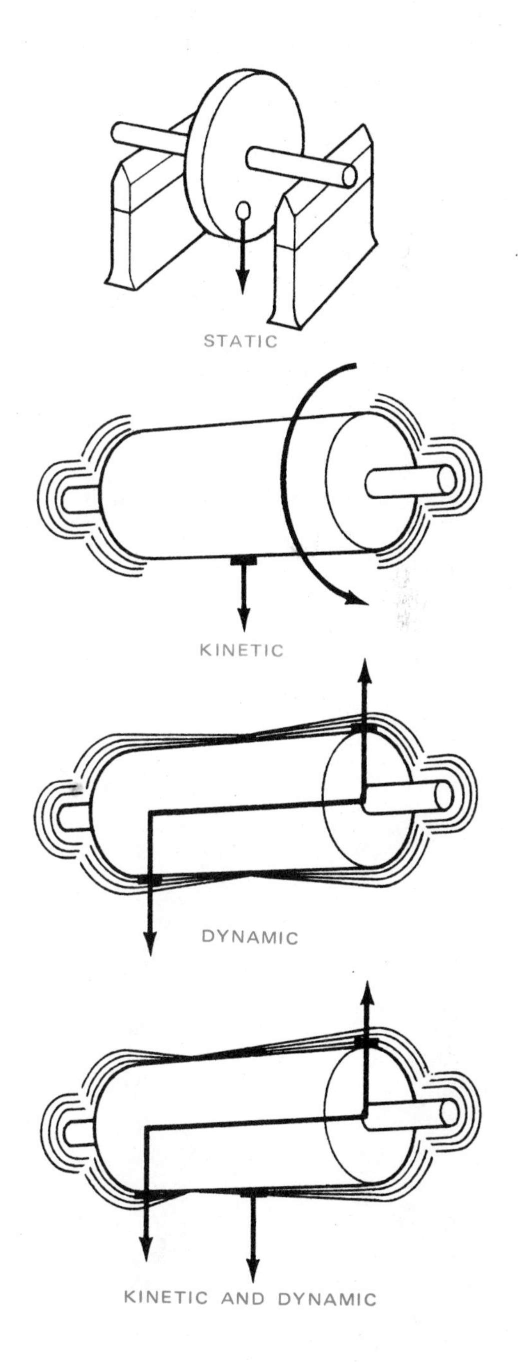

Fig. 20-21. Illustrating the different types of balance. (Stewart-Warner Corp.)

Fig. 20-22. Special type of scale used to measure each end of a connecting rod separately.

In an engine crankshaft and piston assembly, both kinetic and dynamic forces develop as soon as the engine starts. These forces increase rapidly as the speed of rotation increases. To have a smooth running engine, the various weights must be made as equal as possible in order to reduce vibration and load on the bearings. Keeping these weights in balance will reduce such forces and consequently engine life will be materially increased.

Precision balancing involves making the weight of each piston assembly the same as the weight of the lightest piston assembly with a tolerance of 0.5 grams. Diesel engine piston, pin and ring assemblies usually have a tolerance of 5 to 10 grams. The pistons are brought to the same weight by removing metal from the inside of the piston skirt. Special precision scales are available for weighing the pistons.

Connecting rods must be balanced as individual units. They must also be balanced from end to end. The rotating end (big end) must be brought to within 0.5 grams of each other and the piston pin ends must be brought to the same tolerance. Special scales are available for measuring the weight of each end of the rod separately, Fig. 20-22. Each end must be balanced separately because the reciprocating end has a different affect on the crankshaft than the large rotating end.

The variation in weight of new commercial connecting rods ranges from 7 to 15 grams. To bring them into balance, metal is removed from the balance pads found at each end of the rod.

In the case of in-line engines such as a six cylinder engine, the crankshaft is in inherent balance since the crank throws are equally spaced at 120 deg. Therefore, in these engines it is only necessary to be sure that the weights of the piston assemblies and rods are the same to bring the engine into balance. Any unbalance in the crankshaft due to variations in the forging or casting surfaces which are not machined are eliminated in the balancing procedure.

In V-8 engines, the crank throws are in two planes usually 90 deg. apart. The forces due to the reciprocating parts also act in these planes. Therefore, counterweighting must be properly designed for these forces which do not nullify each other. When balancing the forces in the V-8 engine, the crankshaft must be balanced with "bob" weights attached to each crank throw, Fig. 20-23. These bob weights must be accurate to weight and simulate the effective weight of each piston and rod assembly.

The bob weight used on a V-8 is equal to 100 percent of the rotational weight about each crank throw, plus 50 percent of the reciprocating weight.

Fig. 20-23. Equipment used to balance crankshafts. (Bear, The Wheel Service Div., Applied Power, Inc.)

The 100 percent rotational weight consists of the rotating weight (lower end) of two rods, their bearing inserts and the weight of the oil in the crank throw. The 50 percent reciprocating weight is the weight of the upper end of one connecting rod plus the weight of one piston with its rings and pin locks. The sum of these weights is known as the bob weight, and is clamped to each throw of the V-8 crankshaft, Fig. 20-23.

In the case of the V-6, such as the GMC V-6, the bob weight formula is 50 percent of the reciprocating weight and 100 percent of the rotating weight of each piston and rod assembly.

SERVICING DIESEL CYLINDER HEADS

The first step in reconditioning a cylinder head is to remove it from the engine. However, this should not be done while it is still warm since it may cause warpage and distortion. After the engine has cooled, all fuel lines should be disconnected and sealed with cups or plugs and all linkage disconnected. Manifolds and accessory equipment must also be removed.

The reconditioning starts with a careful examination of the cylinder head. This is done before any actual cleaning, since it is possible in some cases to observe the effects of compression leakage due to a blown gasket, a warped cylinder head or cylinder block. The evidence of leaking core plugs or cracks in the water jacket may also be noted. Burned exhaust valves and their location (particularly in the case of four-cycle engines) will also be apparent.

The cylinder head should then be cleaned. In many shops, the head is first cleaned with steam and a detergent, Fig. 20-1, or in a hot tank, Fig. 20-3. Some will use a hot tank followed by a hot rinse.

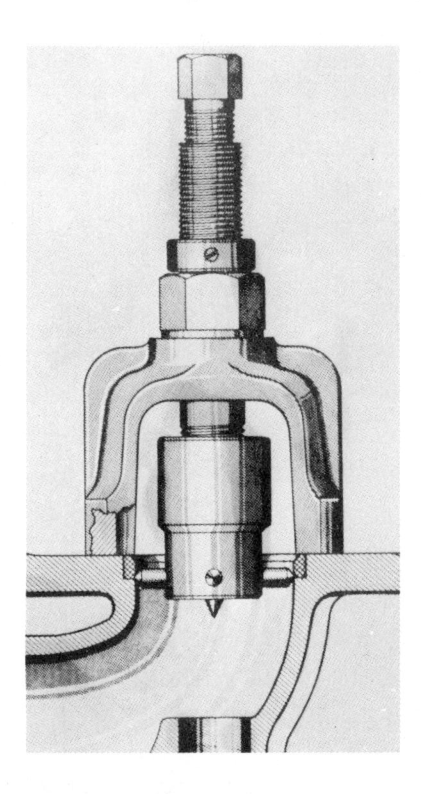

Fig. 20-24. One method of removing a valve seat insert.

Following cleaning, the head is disassembled. When removing the various parts, they should all be identified in relation to their location on the engine. This will aid in tracing troubles and in replacing each part to its original position. Injectors are removed as well as valves and springs, rocker arms, crossheads, rocker shafts and push rods. If valve seats are in poor condition or loose, they should also be removed, Fig. 20-24. Some cylinder heads, Cummins for example, are equipped with fuse plugs which melt if the engine is overheated. These should be removed and inspected for evidence of overheating. Core plugs should also be removed.

If injector nozzles are to be removed from the engine before the cylinder head is removed, the area around the nozzle should be carefully cleaned to prevent the entrance of dirt into the engine. All fuel lines must be carefully plugged or capped.

Removal of the injectors varies with the type and make. In the case of the GMC, a lever type tool is used, Fig. 20-25. The injectors are mounted in the cylinder heads with their spray tips projecting slightly below the lower face of the cylinder head, Fig. 20-26. A clamp bolted to the cylinder head and fitting a machined recess on each side of the injector body, holds the injector in place in a water-cooled copper tube which passes through the cylinder head. A dowel pin in the injector body registers with a hole in the cylinder head for accurately locating the injector assembly. The copper tube is installed in the cylinder with a sealed ring at the flanged upper end. The lower end is peened into a recess of the cylinder head. The tapered lower end of the injector seats in the copper tube, forming a tight seal to withstand the high pressure within the combustion chamber. Removal of the hold-down clamp, Fig. 20-25, permits the removal of the GMC unit fuel injector with the lever type tool.

Fig. 20-25. Removing an injector from a GMC series 71 cylinder head. (Detroit Diesel Allison Div., General Motors Corp.)

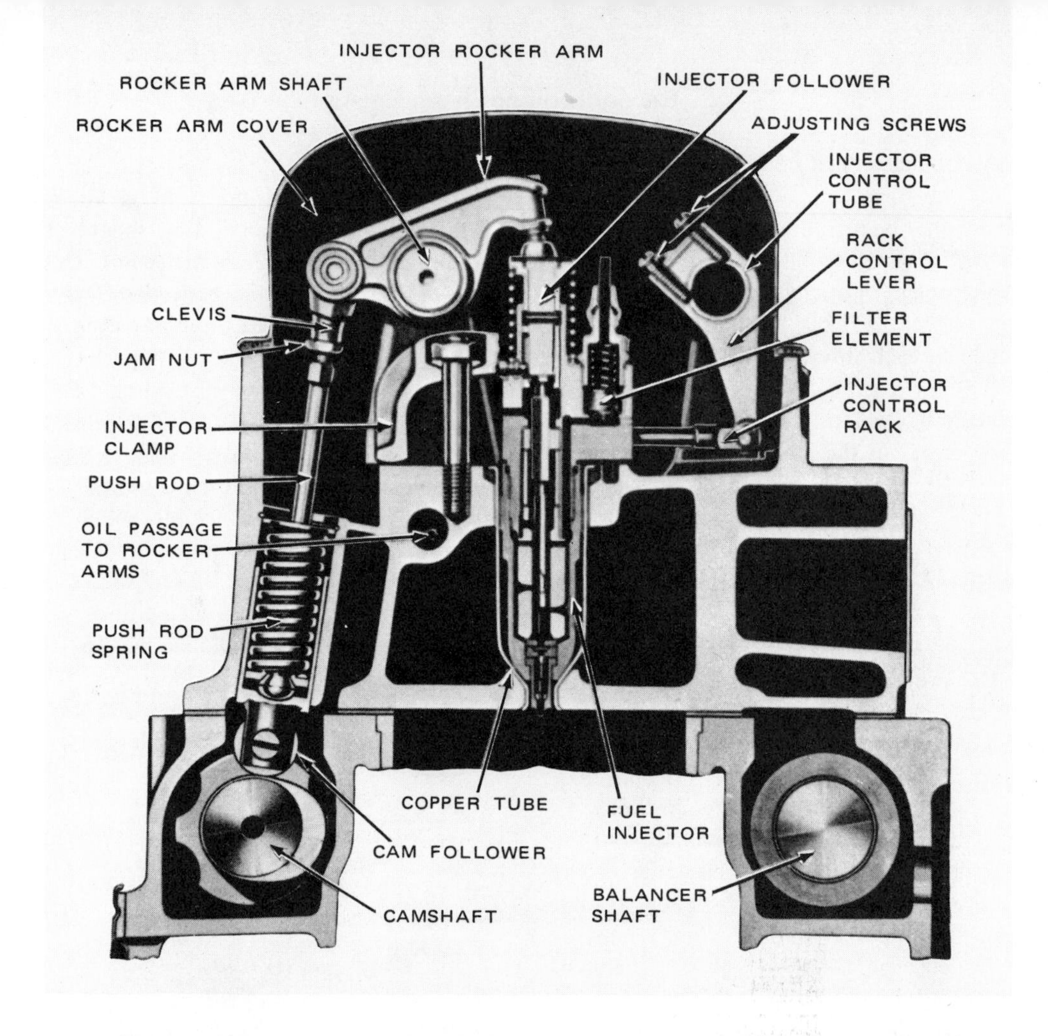

Fig. 20-26. Sectional view of GMC two-cycle engine. (Detroit Diesel Allison Div., General Motors Corp.)

Removal of the copper tube which seals the injector from the water jacket requires a special drift. With the drift in place, the tube can be driven out with a hammer. A neoprene sealing ring should be used when replacing the tube. At the bottom, the tube is sealed by upsetting into a flare on the lower side of the cylinder head. A special tool is required for that operation.

ROBERT BOSCH INJECTOR

The Robert Bosch injector, Fig. 20-27, installed in some Mercedes-Benz passenger cars can be removed as follows. First, disconnect the fuel lines at the nozzle and at the injector pump. Seal the lines and pump with suitable plugs or caps. Loosen the oil leak line cap screws a turn or two. Remove the

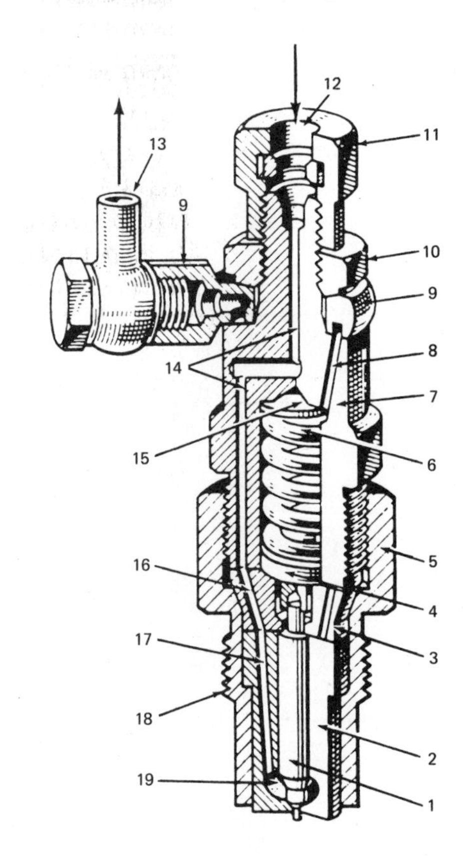

1. NOZZLE NEEDLE.
2. NOZZLE HEAD.
3. NOZZLE HOLDER INSERT.
4. PRESSURE BOLT.
5. CAP NUT TO SECURE INJECTION NOZZLE.
6. TENSION SPRING.
7. NOZZLE HOLDER.
8. DRIP OIL PASSAGE FOR INJECTION NOZZLE.
9. ADAPTER WITH ANNULAR GROOVE FOR DRIP OIL CONNECTOR.
10. HEX NUT TO FIX ADAPTER.
11. CAP NUT TO FIX INJECTION LINE.
12. FUEL INLET.
13. DRIP OIL OUTLET BACK TO FUEL TANK.
14. PRESSURE PASSAGE TO NOZZLE HOLDER.
15. WASHER OF TENSION SPRING.
16. ANNULAR GROOVE AND INLET HOLES IN NOZZLE HOLDER INSERT.
17. ANNULAR GROOVE AND PRESSURE PASSAGES IN NOZZLE HEAD.
18. MOUNTING THREAD.
19. PRESSURE CHAMBER IN NOZZLE HEAD.

Fig. 20-27. Details of Bosch fuel injector. (Robert Bosch Allgemeine Geselshaft)

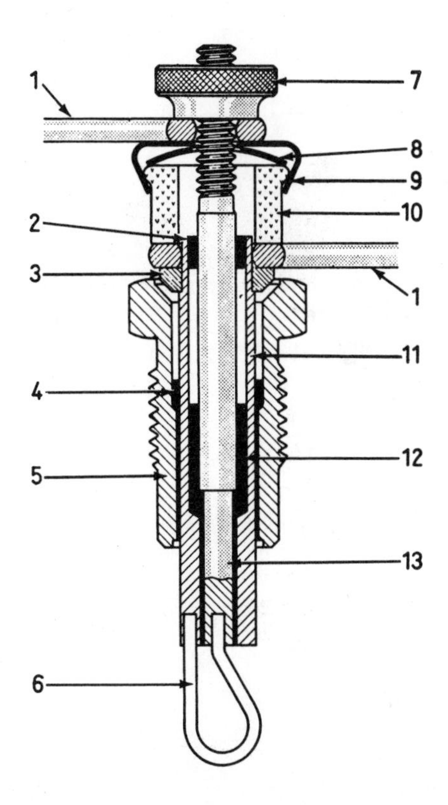

Fig. 20-28. Type of glow plug used in Mercedes-Benz.

1. POWER LEAD.
2. PLASTICS INSULATOR.
3. RING SHOULDER (ON THE OUTER).
4. INSULATING COMPOUND.
5. PLUG BODY.
6. FILAMENT.
7. KNURLED NUT.
8. SPRING WASHER.
9. SHEET METAL CAP ON THE CC INSULATOR.
10. CONNECTOR INSULATOR.
11. OUTER ELECTRODE.
12. INSULATING COMPOUND.
13. CENTER ELECTRODE.

nuts holding the oil leak line at the cross fitting on the fuel filter, and then remove the oil leak lines and fittings. Then the nozzles can be removed from the precombustion chamber. A typical glow plug installed on the Mercedes-Benz is shown in Fig. 20-28, and is easily removed after being disconnected by using a wrench on the plug body.

CUMMINS INJECTORS

Cummins injectors in current use are of different types, Fig. 20-29. The removal procedure as applied to the Cummins NH series engines (except NH 250) is as follows: After removing the rocker housing cover, disconnect the fuel manifold and fuel return line from the injector inlet and drain connections. Remove the fuel inlet screens, Fig. 20-30. Remove the fuel and drain connections from the injector, Fig. 20-31. Loosen the injector rocker lever locknut and unscrew the adjusting screw until the push rod can be disengaged. Disengage the push rod and tip the rocker lever away from the injector. Remove the injector retaining screws and lift the flanged injector from the cylinder head. Take care not to bruise the injector tip. Do not turn the injector bottom side up since the plunger may fall out. Leave the plunger in the injector since the plungers are not interchangeable.

On Cummins V-6, V-8 and NH250 engines, the procedure for removing the cylindrical injector is as follows: Remove the valve cover. Remove the steering pump, if so equipped. Loosen the locknut and back off the injector rocker lever adjusting nut until the push rod can be disengaged. Hold the push rod to one side and tip the rocker lever away from the injector. Remove the injector clamp retaining screws and clamps. Pry

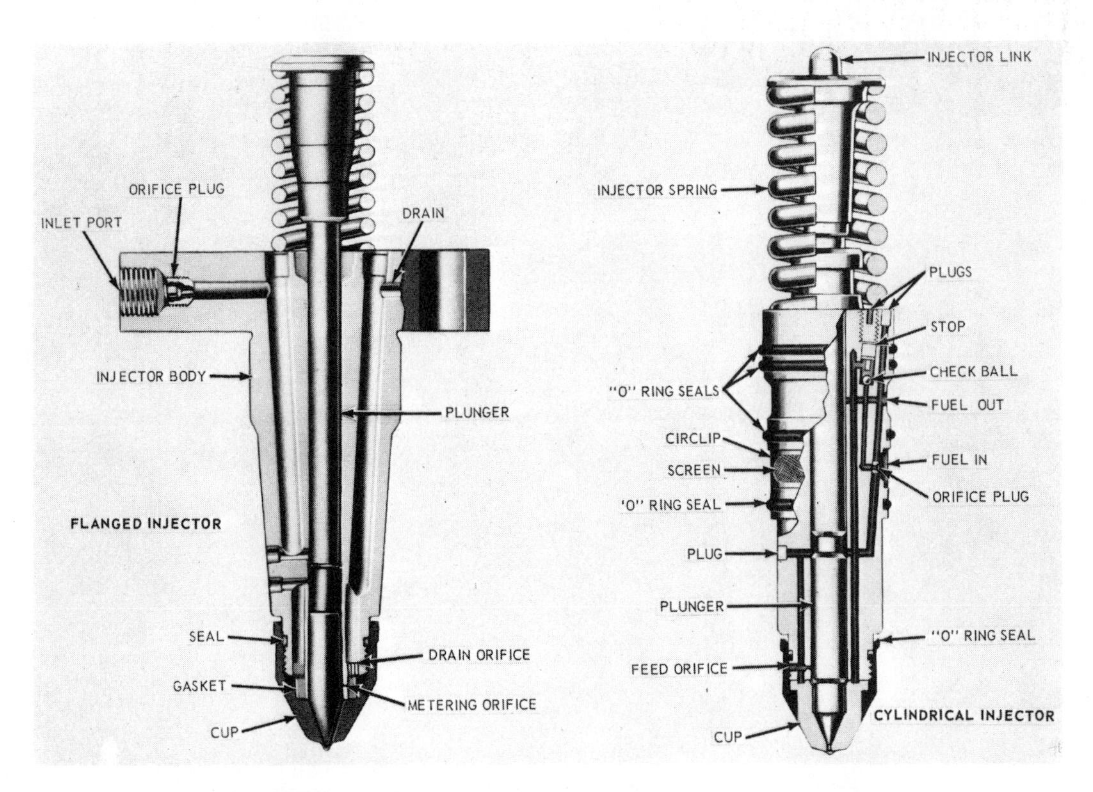

Fig. 20-29. Left. Cummins flanged injector. Right. Cylindrical injector. (Cummins Engine Co.)

Fig. 20-30. Removing Cummins injector screen. (Ford Motor Co.)

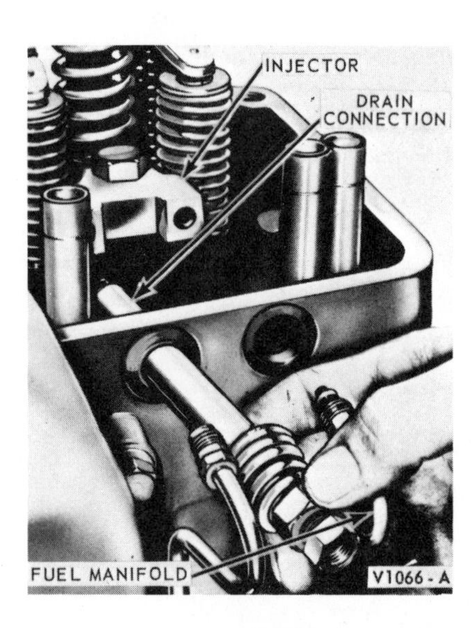

Fig. 20-31. Fuel inlet and drain connections on Cummins engine. (Ford Motor Co.)

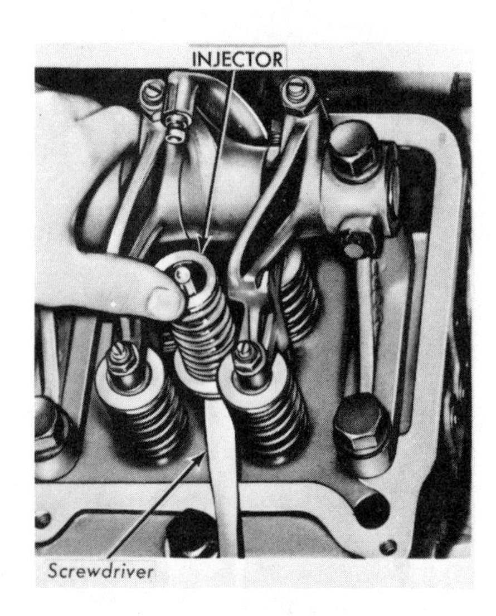

Fig. 20-32. Removing a cylindrical injector from Cummins engine. (Ford Motor Co.)

the injector from the cylinder head, Fig. 20-32. Take care not to bruise injector tips and keep unit upright, since plunger might drop out. Plungers are not interchangeable.

CATERPILLAR TRACTOR

No regulation is needed with Caterpillar injection valves or pumps. Each is precision calibrated during manufacture and is interchangeable with any other Caterpillar injector or pump. Injectors are changed as easily as spark plugs after disconnecting the fuel lines.

AMERICAN BOSCH

The American Bosch injector, Fig. 20-33, is available in several different types. Some designs are provided with a

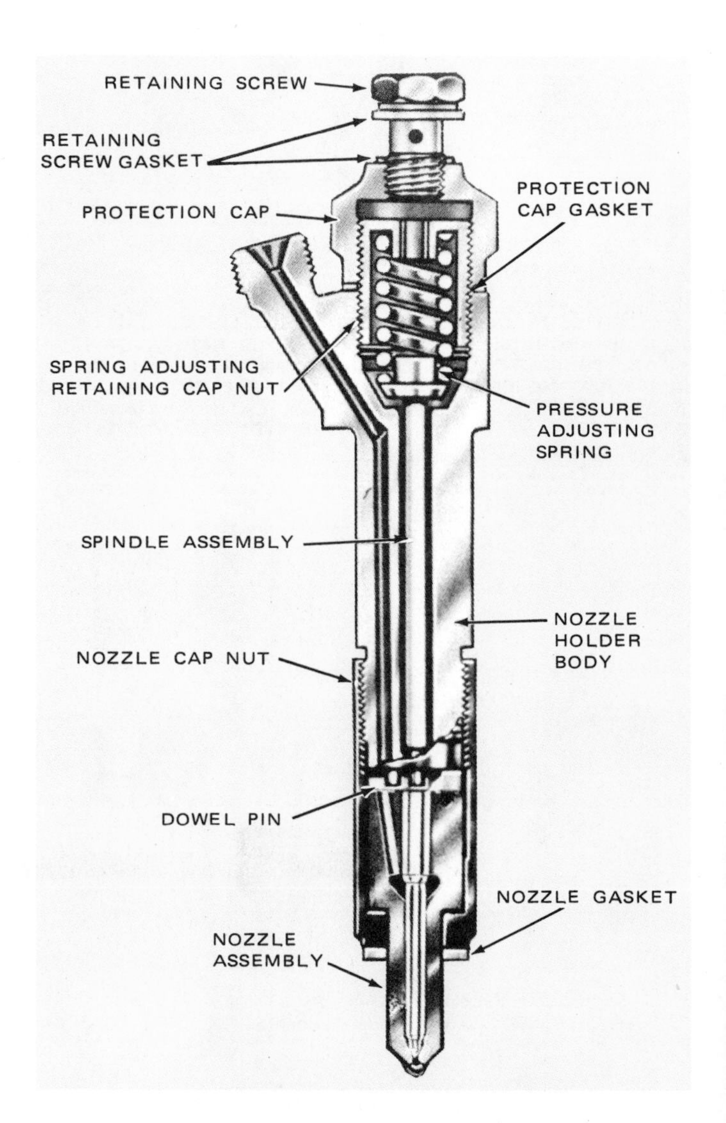

Fig. 20-33. Sectional view of injector. (A M B A C Industries, Inc., American Bosch Div.)

hold-down yoke which is bolted to the cylinder head. Another type is provided with a clamp which is bolted to the cylinder head.

Fig. 20-34. Device for locating cracks. (George Olcott Co.)

C.A.V. INJECTORS

The C.A.V. injectors are secured to the cylinder head by two round headed, slotted screws which pass through a flange forming an integral part of the nozzle holder.

CLEANING WATER JACKETS

After removing the injectors, valves and other parts from the cylinder head, the heads should be placed in a bath to dissolve all water jacket deposits. This is done after all core plugs have been removed. In the case of some diesel engines, International Harvester for example, it is also necessary to remove the precombustion chambers.

After cleaning, the heads are checked for cracks, Figs. 20-34 and 20-35. This is one of the most important parts of rebuilding a cylinder head. When making such a test, the pole pieces of the magnetic type tester are placed in position over the suspected area. A special magnetic powder is sprinkled over the area between the pole pieces. The excess power is blown away and if a crack is present, the powder will follow the lines of the crack, Fig. 20-34.

PRESSURE TESTING

Equally important to the test for cracks by the magnetic method is to test the heads under pressure, Fig. 20-35. When making this test, all of the water jacket openings are sealed or plugged and water is applied under pressure. Water will seep out of any cracks. A variation of this method is to apply air pressure and submerge the head in water. Air leakage will quickly be noticed since air bubbles will rise to the surface. Heating the water will tend to open any cracks that may be present and make them more noticeable.

Some cracks around valve seats can be repaired by installing new seats. If cracks are severe, threaded type seal is generally installed. In less severe cases a press-fit type seat can be used.

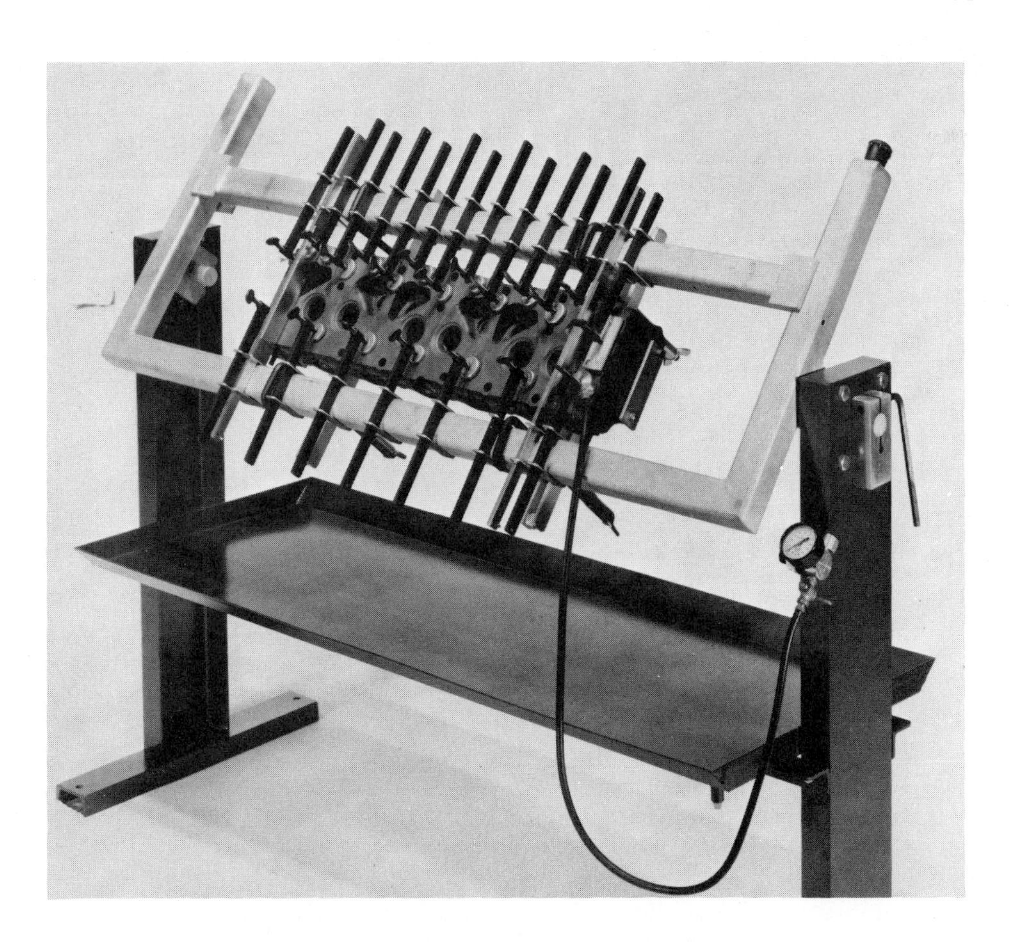

Fig. 20-35. Testing a cylinder head for leaks by applying pressure. (Van Norman Machine Co.)

WARPED CYLINDER HEADS

When the cylinder head has passed all these tests, the next step is to inspect the head for warpage. A straightedge and thickness gage are all that is required. The straightedge is laid progressively in the positions indicated in Fig. 20-36. The

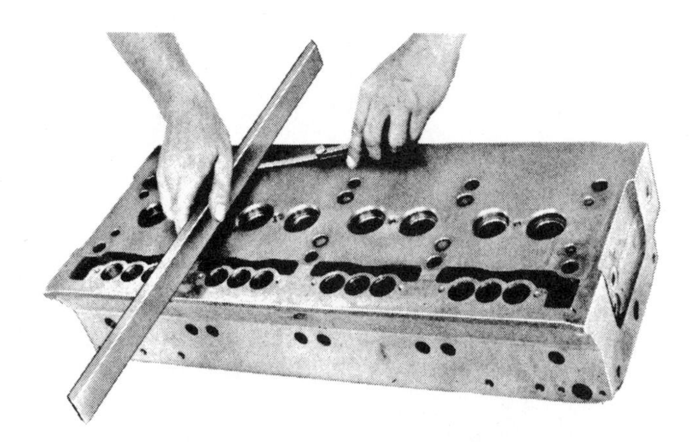

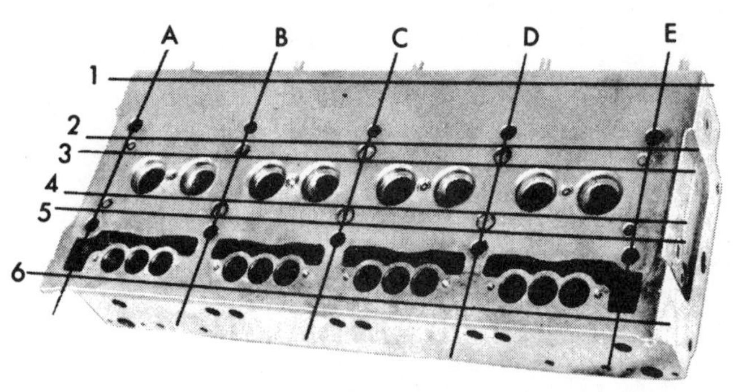

Fig. 20-36. Checking a cylinder head for flatness with a straightedge and feeler gage. (Detroit Diesel Allison Div., General Motors Corp.)

thickness gage is then inserted between the straightedge and the cylinder head. The allowable longitudinal variation will depend on the size of the cylinder head and the manufacturer's specifications. As an example, the GMC V-6 and V-8 71 specifications for warpage or flatness are: longitudinal warpage .0055 in. for the V-6 and .008 in. for the V-8. The transverse warpage limit is .004 in. for both models.

When inspecting the head, particular attention should be paid to the gasket sealing areas and for any scratched or worn areas around the water holes. If such areas are eroded, the holes can be reamed and bushing installed.

RESURFACING CYLINDER HEADS

Cylinder heads can be refinished on a broach, Fig. 20-37, on a grinder or they can be milled. Limits are placed on the various cylinder heads as to the amount of metal that can be removed. When a cylinder head is resurfaced, an equal amount of metal should be removed from the valve seat counterbores and new valve seats installed. The reason for limiting the amount of metal to be removed from the cylinder heads is that

Fig. 20-37. Reconditioning a cylinder head on a Van Norman broach.

there is usually very little clearance between the piston at the top of its stroke and the cylinder head. Removing too much metal would increase the compression ratio. There is also the possibility of the valves striking the pistons when they are open.

In addition to checking the thickness of the cylinder head after resurfacing, it is necessary to correct the protrusion of the valve inserts. A check should also be made of the projection of the valves, injector hole tubes and injector spray tips, Fig. 20-38, from the gasket surface of the cylinder head. Push rod length must also be considered in order to prevent the valves from striking the top of the pistons when the cylinder head is installed.

Some heads are grooved for the top of the cylinder sleeve. Grooves are also provided in some designs for special gaskets. Any grooves must be remachined to the recommended depth before replacing the head in service. Repair shops make a practice of marking the thickness of the cylinder head after reconditioning so that the next shop to work on the engine

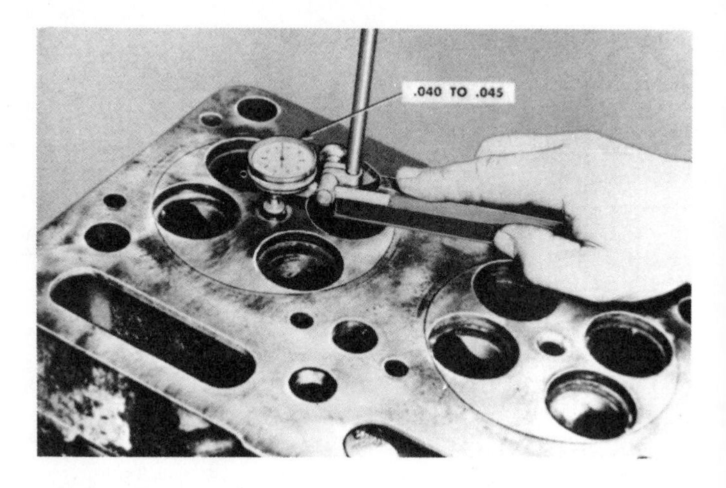

Fig. 20-38. Measuring the protrusion of an injector tip from the cylinder head. The dimension indicated applies only to specified engine.

will be warned that the head thickness has been reduced from the original dimension.

Some cylinder heads, Cummins and GMC for example, have hard Stellite valve seat inserts. As a result, many shops prefer to resurface such heads by grinding rather than using a broach since the hard metal would tend to nick the cutters of the broach. Some Cummins heads are resurfaced with a circular pattern around the valve chamber.

VALVES, GUIDES AND SEATS

After resurfacing of the head is completed, the valves, valve seats and guides are reconditioned.

When inspecting the valves, check for worn stems, Fig. 20-39, worn keyways, stretched stems, bent stems, Fig. 20-40, cracks and evidence of burning. Critical valve dimensions are illustrated in Fig. 20-41. The valve seat angle is obtained from manufacturer's specifications and the valve is faced to the specified angle. If the edge of the valve after refacing is less than the specified amount for that particular engine (usually 1/32 in.), the valve should be discarded. When refacing a valve, it should always be chucked as close to the valve head as possible to eliminate flexing of the stem. Aluminum coated valves should not be refaced.

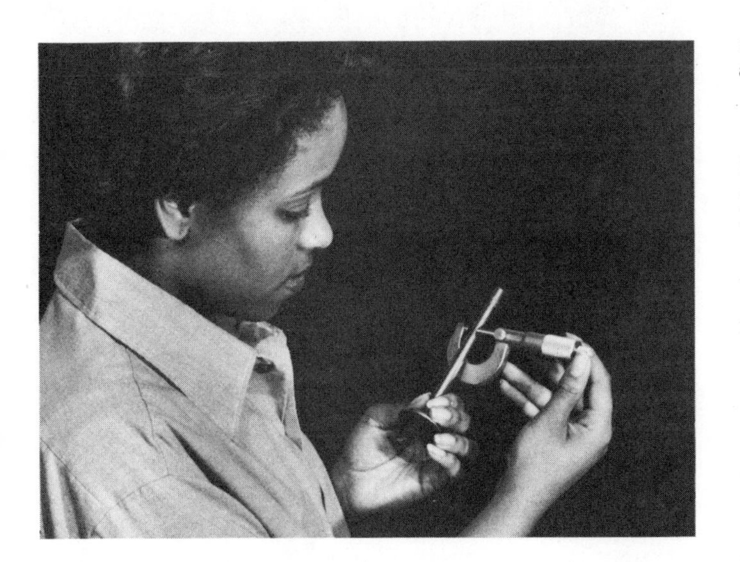

Fig. 20-39. Checking a valve stem for wear with micrometers.

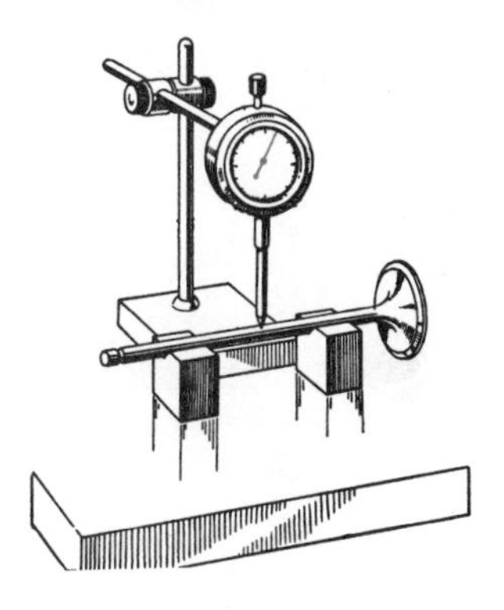

Fig. 20-40. Method of mounting a dial gage to check a valve stem for straightness.

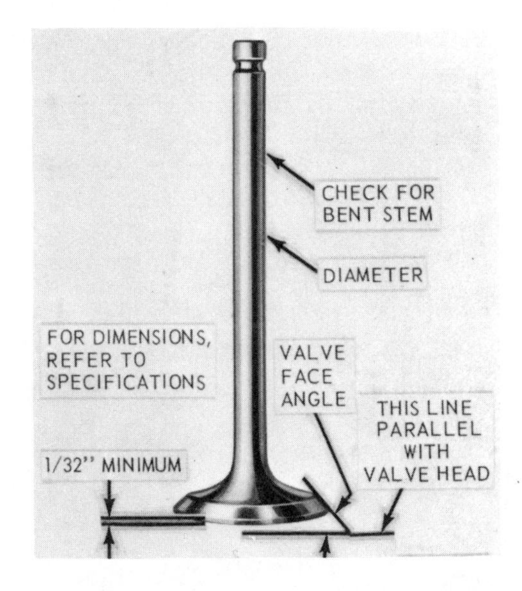

Fig. 20-41. Points on a valve that should be checked for wear. Some authorities recommend valve head thickness greater than 1/32 in. (0.794 mm) as an aid in dissipating heat.

After refacing, check the valve face for runout, Fig. 20-42. Valve face runout is usually limited to .0015 in. for automotive size valves. This is a check on the accuracy of the valve refacer. If the valve stem is not bent or worn, the valve face after refacing should be well within the specified limits.

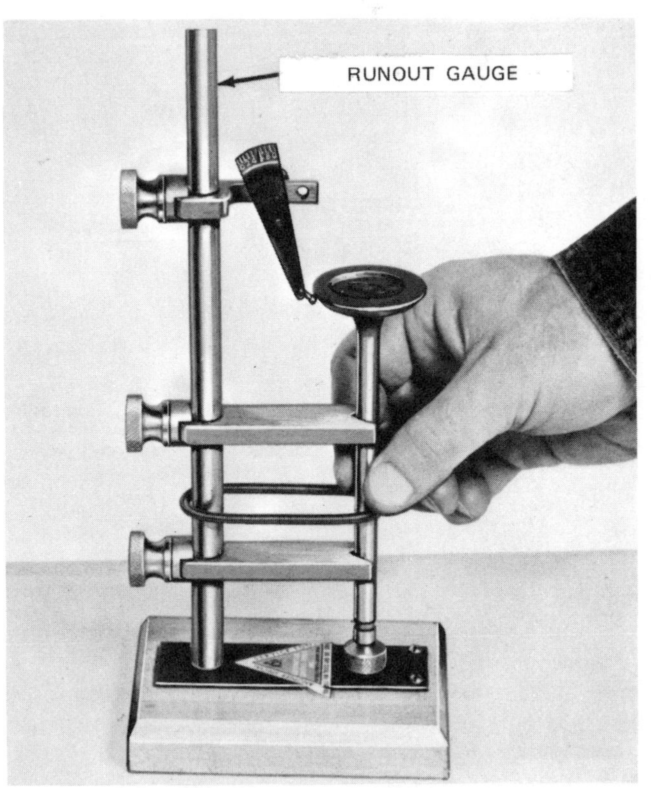

Fig. 20-42. Checking a valve face for runout. (Ford Motor Co.)

The next step is to check the condition of the valve guides. One method of checking the amount of side motion is to install a new valve in the guide and then measure the side motion with a dial indicator, Fig. 20-43. A better and more

Fig. 20-43. Using a dial indicator to measure wear in a valve stem and guide. (Ford Motor Co.)

accurate method is to measure the wear of the valve guide by means of split ball type gage and a pair of micrometers, Fig. 20-44.

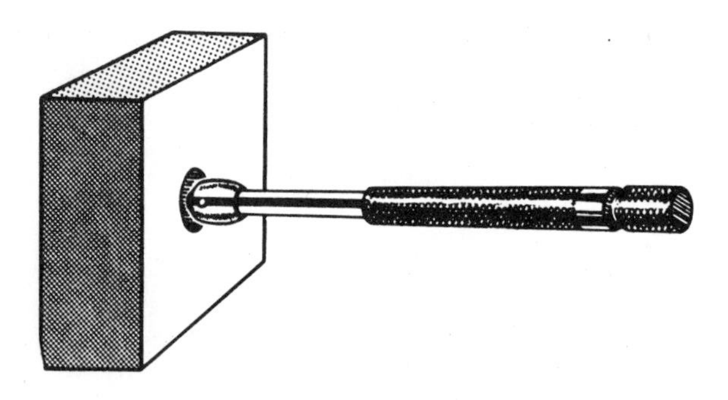

Fig. 20-44. Illustrating the method of using a split ball gage to measure the diameter of a small hole.

In a major overhaul, it is the usual practice to install new valve guides if they are of the replaceable type. If the guides are integral with the cylinder head, the guides are reamed, Fig. 20-45, and valves with oversize stems are installed.

Another method of reconditioning the valve guides is known as knurling. This method can recondition guides worn up to .006 in. However, hardened or chilled guides like those used in some International Harvester engines cannot be reconditioned by this method. The first step in the knurling procedure is to clean the worn guides. Then after applying a lubricant, a knurling tool is used, Fig. 20-46. This operation is

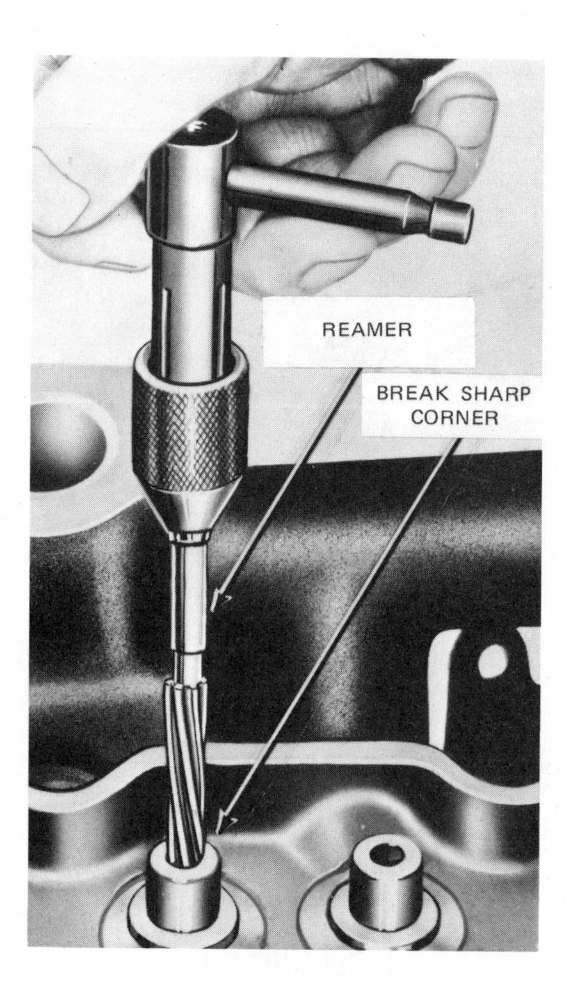

Fig. 20-45. Reaming a valve guide.

Fig. 20-46. Reconditioning a valve guide by means of knurling.

performed at a relatively slow speed. The knurled surface is then reamed to the desired size. The knurling process forms a series of grooves which serve as pockets for the lubricant. In that way, the life of the valve and guide is prolonged.

Guides can also be repaired by the installation of repair sleeves. Another method is to install a bronze spiral throughout the length of the guide, in much the same manner as a thread repair.

If new replacement guides are installed, the old ones are first driven out and the new ones pressed in. Guides must be removed from the rocker arm side of the cylinder head and installed from the combustion chamber side. Some mechanics make a practice of chilling the replacement guides in dry ice. They are then more easily pressed into position and, on returning to normal temperature, will fit tight. The amount of guide protrusion is important. This varies with different engines and care must be taken to follow the manufacturer's specifications.

VALVE SEATS

Many diesel engines are provided with valve seat inserts to insure longer seat and valve life. These can be checked for runout after the guides have been reconditioned, Fig. 20-47.

Fig. 20-47. Checking a valve seat for runout.

These seats are made of a special steel to better withstand the stresses of hammering and temperature. Seats can be removed by driving through or using a special puller which is available for this purpose, Fig. 20-24. One type of valve seat insert is screwed into position and then pinned. When installing valve seat inserts, it is customary to first shrink them in dry ice and then press them in position. Manufacturers give specific instructions on the procedure to be followed when installing their valve seat inserts.

Before grinding the valve seat inserts, the grinding stone should be dressed. The stone must be of the correct angle for that particular engine, the correct size and correct grit. The stone selection will depend on the material of which the insert is made. In general, a soft stone is used for hard seats and a hard stone for soft seats such as cast iron.

After grinding the seat to the correct angle, Fig. 20-48, the width and valve contact area are adjusted. The seat width is measured as shown in Fig. 20-49. To determine the seat location, apply a light coating of Prussian blue to the valve face and rotate the valve one-quarter turn on its seat. The blue will indicate accurately the location of the seat which should be centered on the valve face. If the seat is too low or too narrow, further grinding is needed. A 15 deg. stone will reduce its width and at the same time lower it in the head. A 70 deg. stone will also narrow it, but raise it in the head.

Fig. 20-48. Regrinding a valve seat.

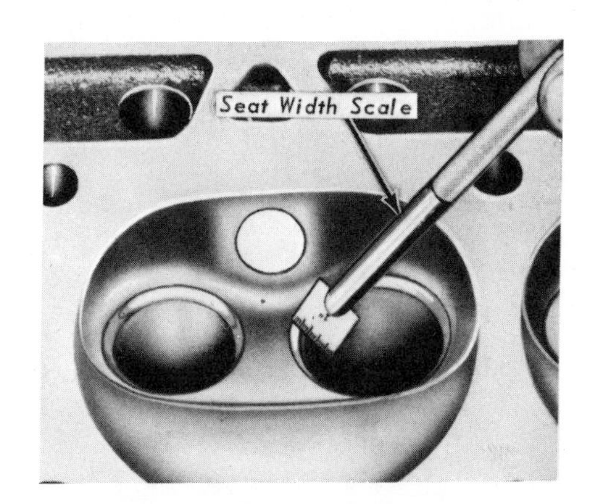

Fig. 20-49. Measuring the width of a valve seat.

Valve seat inserts should be checked for looseness by using a small pry bar. Seats should be replaced if any looseness is noted. If a puller is not available for removing worn valve seats, they can be removed by arc welding a small bead on the inner diameter of the seat. This will cause the seat to shrink and make it easier to be pried out. Another method of removing valve seat inserts is to weld a core plug to the insert. The insert can then be driven out by inserting a drift through the valve guide. Be sure to clean all counterbores before installing new seats.

VALVE SPRINGS

Valve springs should be carefully examined for evidence of overheating. If the ends are worn or polished, it indicates that there has been spring surge and rotation. The latter would result from weak springs plus excessive engine speed.

Springs should always be tested for pressure, Fig. 20-50, and also tested for squareness, Fig. 20-51.

Fig. 20-50. Valve springs must be checked for pressure.

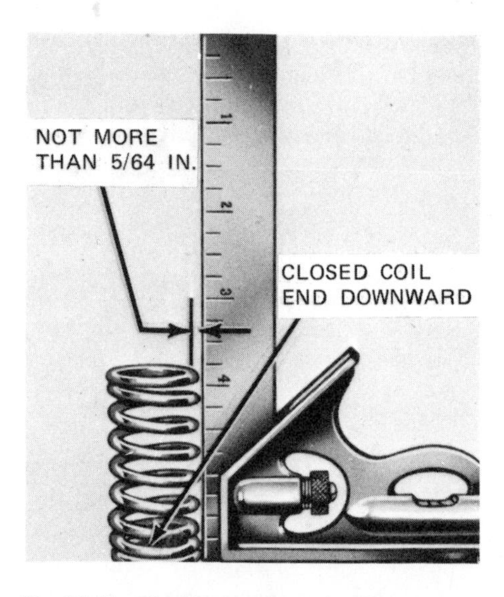

Fig. 20-51. Checking a valve spring for squareness.

VALVE OPERATION

Rocker arms, crossheads, shafts and bushings must be checked dimensionally for wear and cracks. Crossheads are used on engines with dual exhaust and intake valves to insure that they open and close simultaneously. Wear will destroy timing and affect performance. These parts should also be checked magnetically for cracks. A rocker arm is being rebored in Fig. 20-52.

Push rods must be inspected for straightness, Fig. 20-53, and for evidence of wear on the ends.

Fig. 20-52. Reconditioning a rocker arm. (Cedar Rapids Engineering Co.)

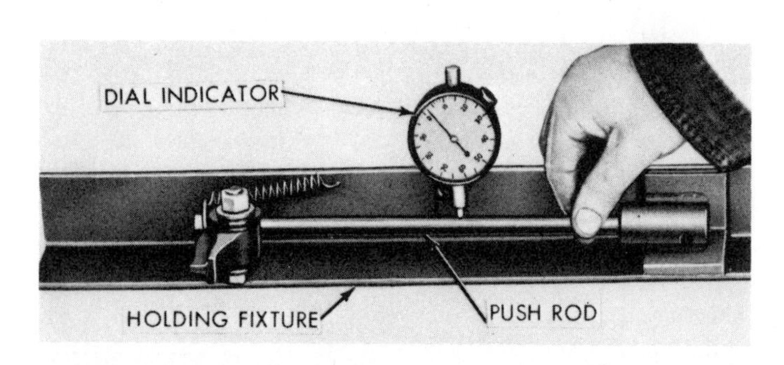

Fig. 20-53. Checking a push rod for straightness with a dial indicator.

After the valves and springs are installed, it is important to check the height of the spring, Fig. 20-54. If the height is greater than the factory specifications, the effective pressure of the spring is reduced and valve action will be sluggish. If necessary, spacers can be installed between the spring and the cylinder head to restore normal pressure. Spring condition results from the spring seat having been lowered as the result of excessive grinding and excessive refacing of the valves. If necessary, new parts should be installed.

Fig. 20-54. Measuring the height of the installed valve spring.

INSTALLATION TIPS

When installing the cylinder head, care must be taken to be sure the correct gasket is used and is installed with the correct side to the cylinder block. If an adhesive is advised, care must be taken that it is applied evenly so none gets in the cylinder bores or in the valve ports. Many cylinder heads are secured with bolts of different sizes. These must be installed in their correct positions.

Tightening the cylinder head bolts must be done progressively in the order indicated in the shop manual for that particular engine. If that is not available, start at a center bolt, then tighten the one directly opposite. Proceed in this manner until all are tightened finger tight. Then proceed with torque wrench tightening to approximately one-third of the specified torque. Tighten to the specified torque on the final tightening, Fig. 20-55.

Fig. 20-55. Using a torque wrench to tighten the cylinder head bolts on a Caterpillar diesel engine.

RECONDITIONING THE CYLINDER BLOCK

Reconditioning the cylinder block of a diesel engine is a job requiring extreme precision. There are many different types of cylinder blocks used in diesel engines and reconditioning presents a variety of problems not encountered in the reconditioning of gasoline engines. There are in-line engines, V-type engines, radial engines and blocks that have dry sleeves, dry flanged sleeves and wet sleeves. They have one thing in common; extreme accuracy is required when reconditioning.

The first step in reconditioning any type of diesel cylinder block is to determine what work is needed and what work may have been previously done on the engine.

The condition of the cylinder walls is the first part to be examined. Dark spots on the cylinder walls indicate low areas. A typical wear pattern is shown in Fig. 20-56. When measuring the bore, the difference between the diameter at the top of the ring travel and the diameter at the bottom is the taper. These diameters are taken at right angles to the crankshaft. Allowable taper varies with different makes and models of diesel engines. However, in automotive sizes some manufacturers give a limit of 0.007 in. for taper. If the taper exceeds the specified amount, the cylinder should be reconditioned. The bores are reconditioned after the top of the cylinder block has been reconditioned.

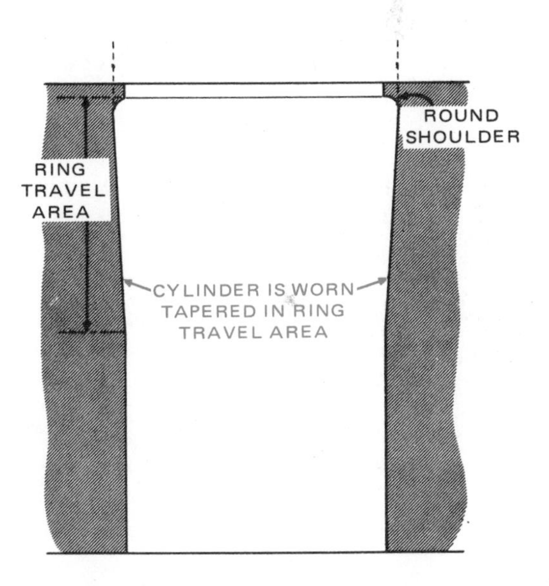

Fig. 20-56. Note that cylinders wear in a taper form.

Cylinder bores must also be checked for out-of-round condition. The out-of-round is the difference in diameter as measured transversely and longitudinally across the cylinder block. Limits of out-of-round are in the area of 0.005 in. for most engines.

The top of the cylinder block must be checked for flatness

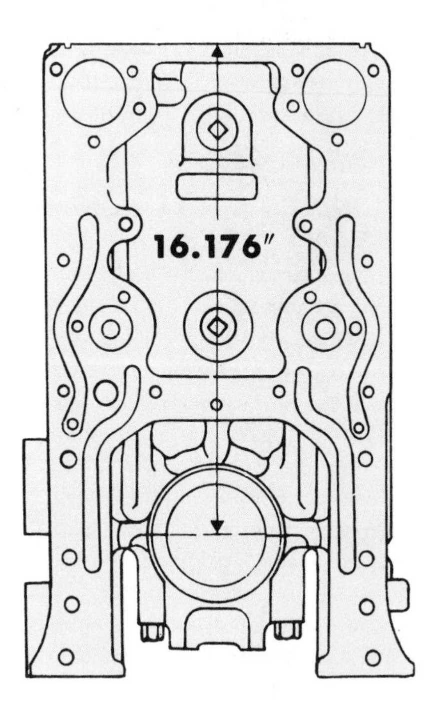

Fig. 20-57. The measurement from the top of the cylinder block is critical. Care must be taken not to remove too much metal. The dimension shown applies only to one GMC engine. (GMC Truck and Coach Div., General Motors Corp.)

with a straightedge and feeler gage as described for the cylinder head check. If the error exceeds 0.003 in. transversely and 0.009 in. longitudinally on small engines, the top of the cylinder block must be reconditioned. A greater tolerance is usually permitted on larger engines. The condition of the top of the cylinder block is critical, since any unevenness or roughness will result in compression loss or water damage.

The height of the diesel engine block is measured from the top of the block to the center line of the crankshaft, Fig. 20-57. If the block has been previously reconditioned it may not be possible to recondition it again. Most maintenance shops will stamp the top surface of the cylinder block with the amount of metal that has been removed or with the distance from the top of the block to the center line of the crankshaft. The dimensions given in Fig. 20-57 apply only to the in-line 71 GMC diesel.

Making this measurement often presents a problem. One method is to bolt an extension plate in place of the main bearing cap and another plate to the top of the cylinder block. The distance between the two plates can then be measured with a micrometer.

The top surface of the cylinder block can be resurfaced on a broach, grinder or milling machine. A Caterpillar four cylinder block is shown on a Peterson grinder in Fig. 20-58. A GMC six cylinder block is being resurfaced on a Storm-Vulcan milling machine in Fig. 20-59.

The cylinder block should be located on the surfacer in reference to the bearing pads and not on the oil pan ledge. Only light cuts of 0.001 to 0.0015 in. should be taken since heavier cuts may cause warpage or a rough surface. When the resurfacing is completed, the amount of metal removed should be stamped on the top of the block.

When it is necessary to resurface the top of the cylinder

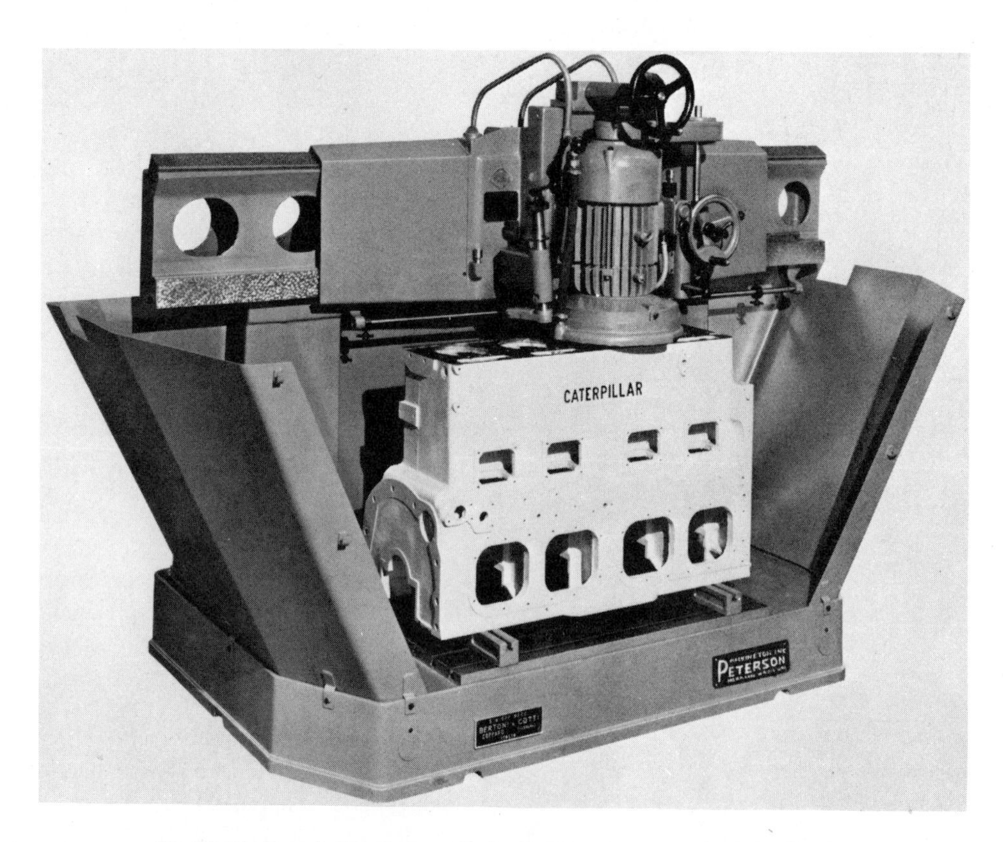

Fig. 20-58. Resurfacing the top of a Caterpillar cylinder block with a grinder. (Peterson Machine Tool Inc.)

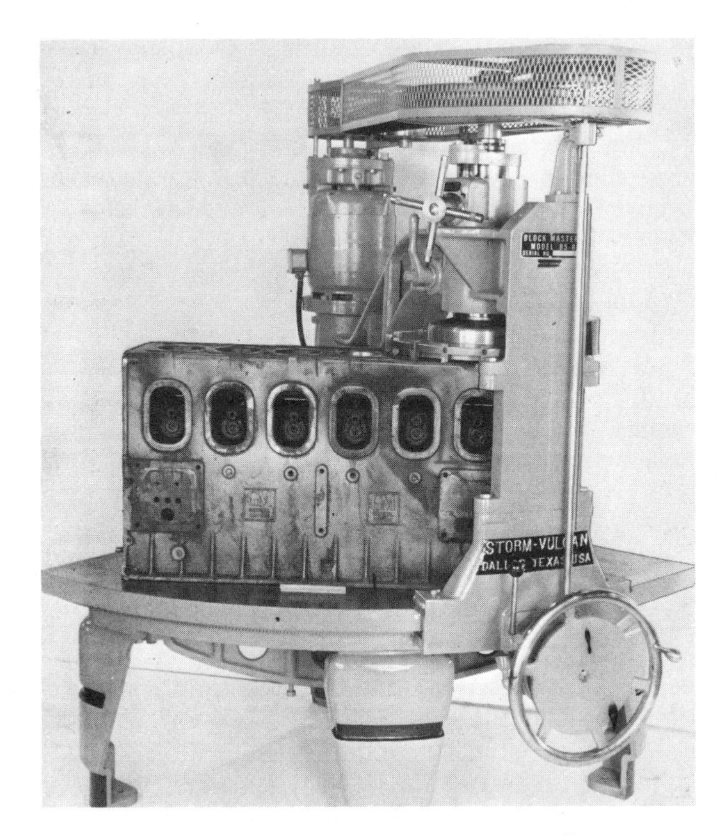

Fig. 20-59. Milling the top of a cylinder block. (Storm-Vulcan Inc.)

block, it is also necessary to machine any grooves in the top of the block the same amount. This refers to the cylinder sleeve counterbore depth and any other groove in the top surface of the cylinder block. In many cases, special tools are required to machine these grooves. The depths of the sleeve counterbore and other grooves can easily be measured with a dial indicator, as shown in Fig. 20-60.

The diameter of the counterbore must be accurately measured. The diameter and depth are both held to limits of 0.002 in. in most automotive size engines. This is important, since any variation would result in a poor fit for the cylinder sleeve when installed. If the counterbore is too deep, the sleeve will have vertical movement. If it is too high, the sleeve will strike the cylinder head and compression leakage will result. To insure a good seat for the sleeve, the counterbore must be smooth and perpendicular to the liner bore.

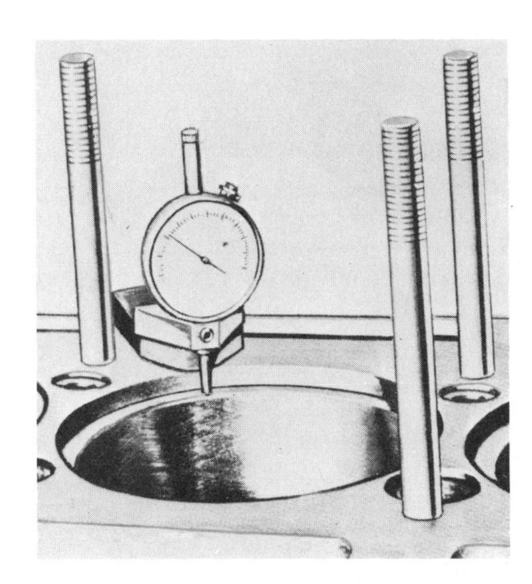

Fig. 20-60. Checking the height of counterbore for cylinder liner.

Some diesel engines are designed to have the cylinder sleeves protrude a specified amount above the cylinder bore. If sleeves show little signs of wear, they can be reconditioned by honing. If sleeves are badly worn (in excess of 0.005 in. out-of-round and 0.007 in. taper) they can be removed and new ones installed.

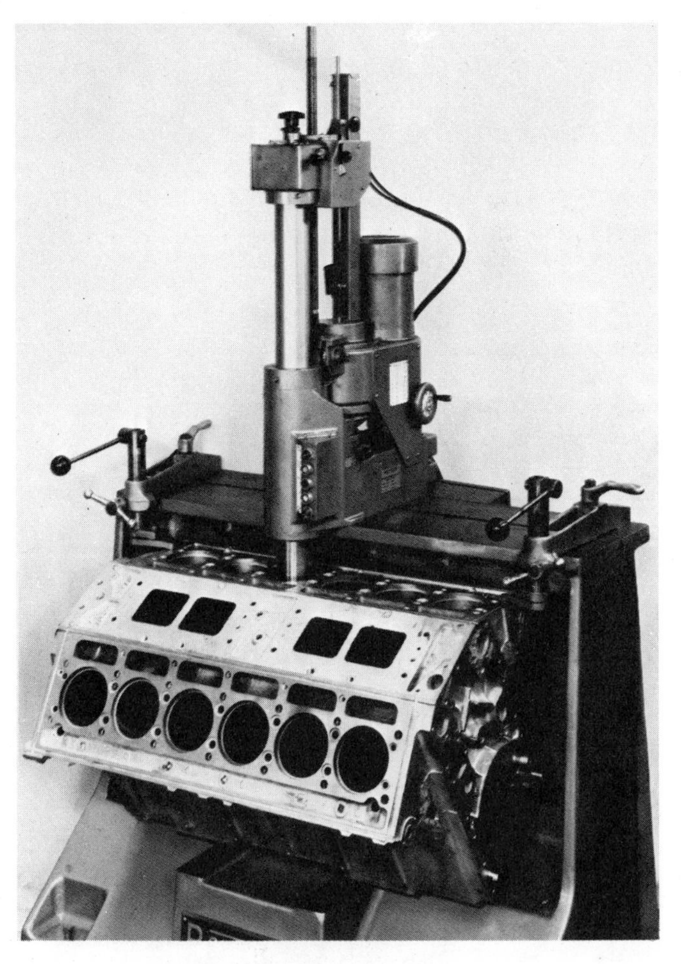

Fig. 20-61. Reboring a V-12 diesel cylinder block. (Rottler Boring Bar Co.)

Special pullers are available for removing sleeves. In some installations they can be driven out from below.

The liner to block clearance in many installations is excessive. In such cases, the block will have to be rebored for oversize sleeves. A reboring operation is shown in Fig. 20-61, and honing operatings in Figs. 20-62 and 20-63.

In the case of the wet type cylinder sleeve, sealing O rings are provided for the outside of the sleeve to prevent leakage of the coolant. If the grooves for these O rings have been

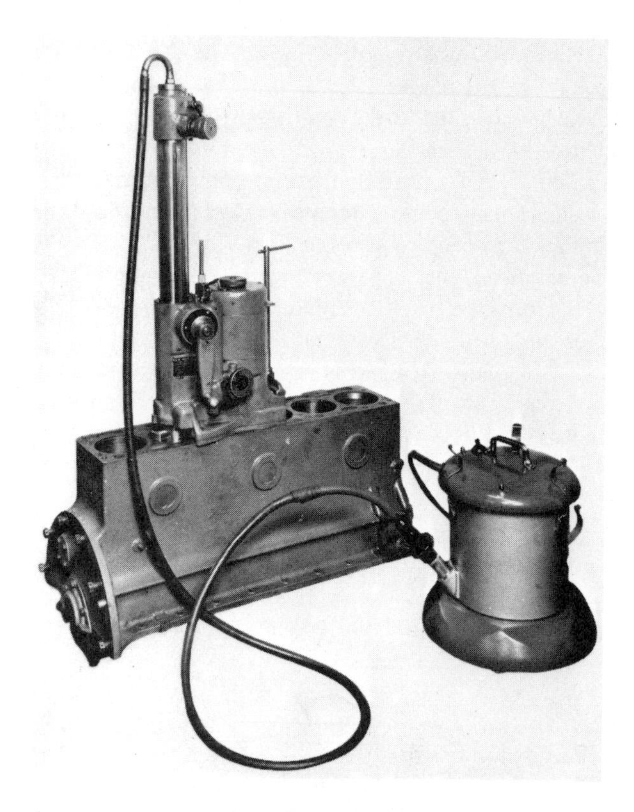

Fig. 20-62. Portable type boring bar with vacuum attachment to collect metal particles and abrasives. (Van Norman Machine Co.)

Fig. 20-63. Automatic cylinder resizing machine.

damaged, or if the cylinder block has been rebored for an oversize sleeve, the grooves for the sleeves will have to be restored to the specified size.

When installing sleeves with O rings it is often difficult to make installation without twisting the O rings. Using hydraulic brake fluid as a lubricant on the rings will usually overcome the difficulty.

CYLINDER FINISH

To obtain the desired cylinder finish special precautions are necessary. If the cylinder or cylinder sleeve is to be bored, it is essential that the tool be correctly sharpened. The manufacturers of the equipment supply sharpening fixtures which should be used in accordance with their instructions. The depth of the cut and speed of feed are critical.

When honing the cylinders, a cross hatch pattern as shown in Fig. 20-64 is desired. By using a honing speed of about 50 strokes per minute, a cross hatch pattern of 35 to 45 deg. is obtained. This pattern enables the piston rings to shear the sharp projections during break-in. The pattern also holds the lubricating oil. If the pattern is vertical, there will be little or no shearing and a path is provided for blow-by.

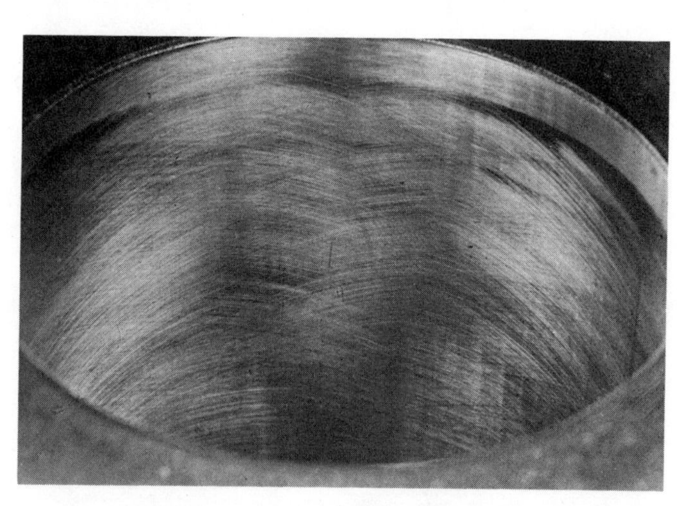

Fig. 20-64. Desirable type of cylinder wall finish.

After all honing and boring operations are completed, the cylinder blocks and sleeves should be carefully cleaned. Many shops scrub the surface with soap and water in order to remove all traces of abrasives and metal cuttings.

When doing any boring operation, it is important that the cylinder block surface be clean and smooth before mounting the boring equipment on the face of the block or on the table of a boring machine. In production type machines, either boring or honing, the manufacturer's instructions for mounting must be carefully followed.

MAIN BEARING BORES

Main bearing bores should also be checked. In some cases they may have become damaged, the caps stretched or the

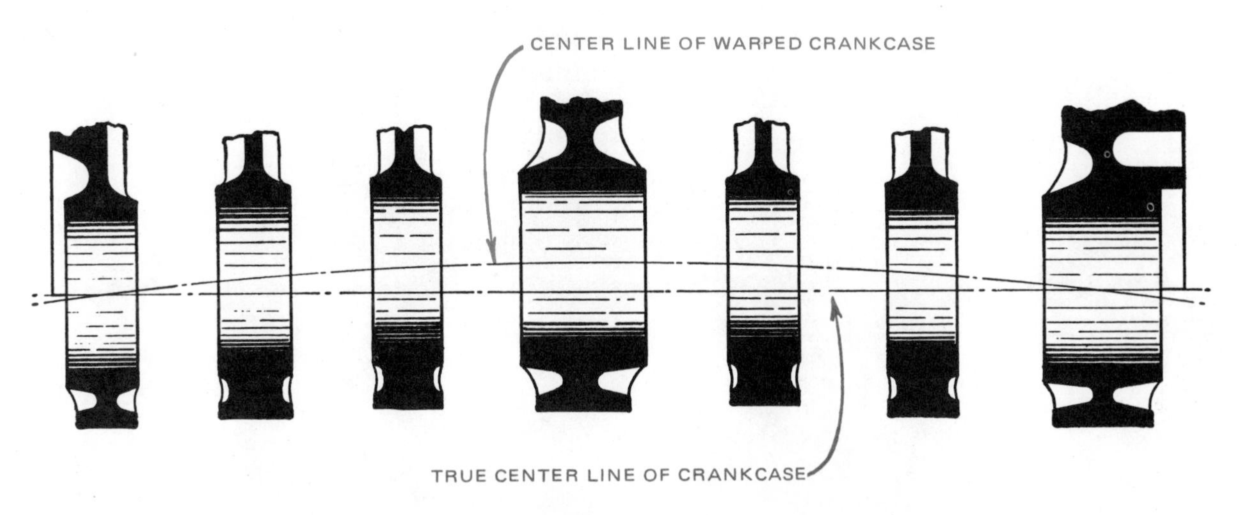

Fig. 20-65. Illustrating a bowed condition of a crankcase. (Federal Mogul Corp.)

block has become warped or bowed, Fig. 20-65, so that the center line of the main bearing bores is no longer at right angles to the center line of the cylinder bores.

One method of checking for both roundness and alignment of the main bearing bores is to use an arbor. It should be the length of the engine and .002 in. less in diameter than the main bearing bores. The arbor is inserted in the bearing bores where it should turn freely. If not, there is misalignment present. When a 0.002 in. feeler gage is inserted between the arbor and the bearing bore, it should lock up. The feeler gage should pass around the circumference of the arbor at each bearing bore if there is no misalignment.

If a reground or new shaft is installed in a cylinder block which has warped or stretched bearing caps, the shaft will bind and maximum bearing contact will not be obtained. Early failure of shaft and bearings will result.

To eliminate such trouble, the main bearing bores are rebored or honed to restore their original circularity and their alignment with the cylinder bores. An align-boring machine setup to recondition the main bearing bores of a large Caterpillar block is shown in Fig. 20-66. An align-honing setup is shown in Fig. 20-67. In the latter case, the bearing bore caps are first ground on a cap and rod grinder after which the bores are honed back to size.

CAMSHAFTS

If the main shaft bores are out of alignment, the camshaft bores will also need reconditioning. The same general equipment is used to recondition the camshaft bores. The procedure

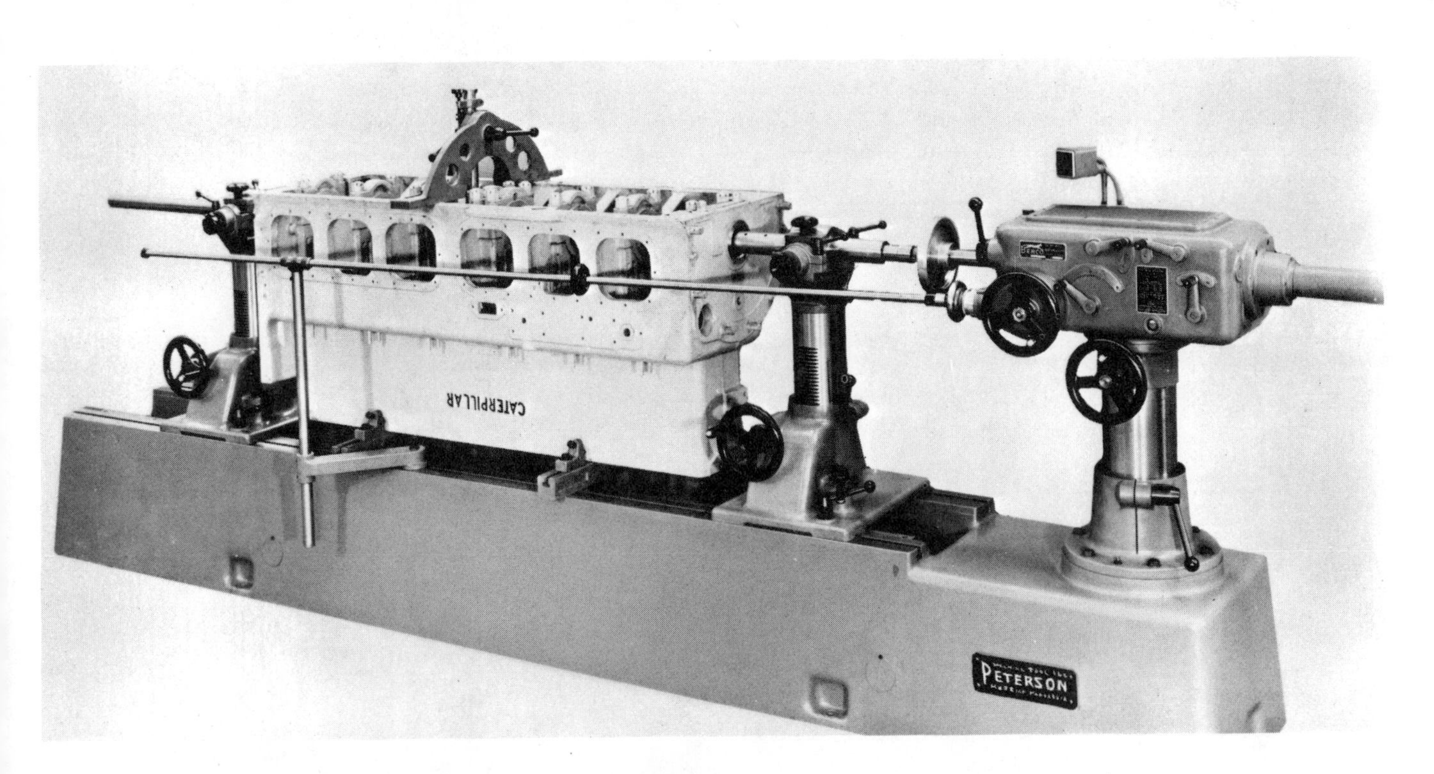

Fig. 20-66. Align-boring a Caterpillar cylinder block. (Peterson Machine Tool Inc.)

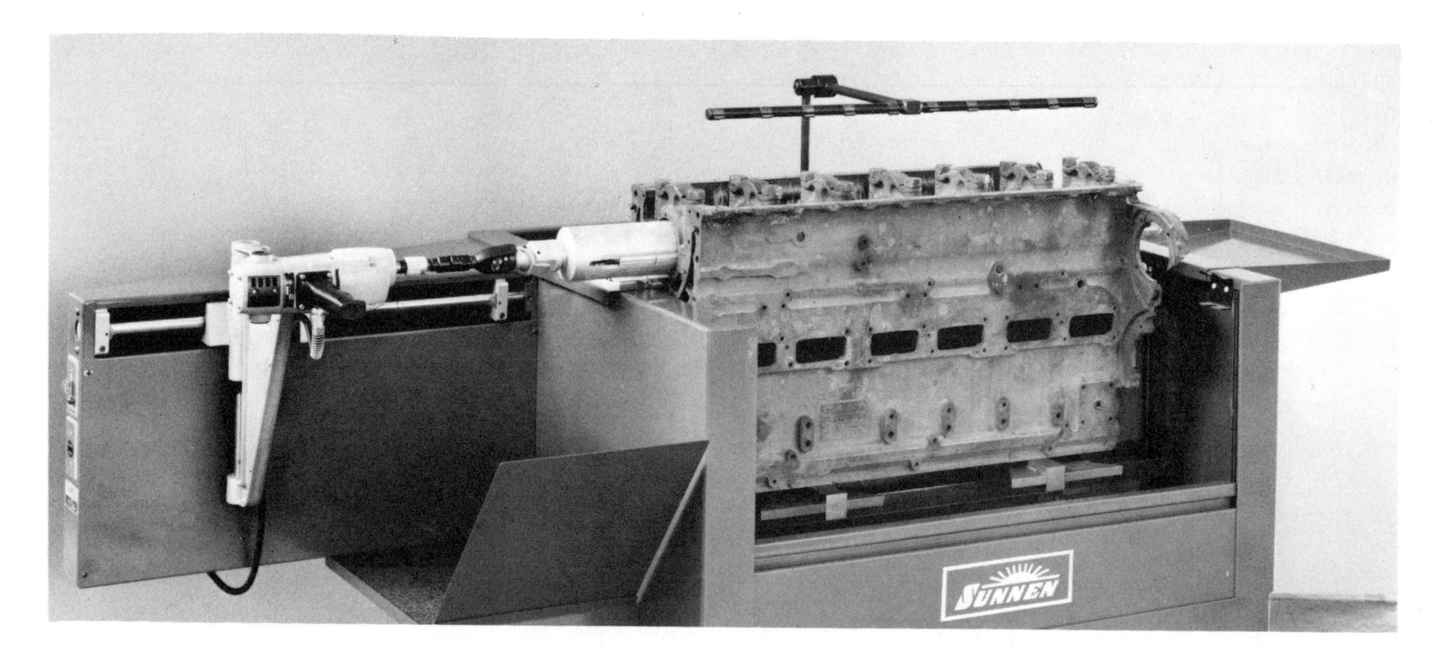

Fig. 20-67. Align-honing the main bearing bores in a cylinder block.

is to install new camshaft bearings which are align reamed or bored. Camshaft bearings are usually the sleeve type. They may be removed by a special puller called a slide hammer. New bearings are installed in a similar manner and are bored or honed to the desired size.

Camshafts should be checked between centers or on V-blocks to be sure they are straight, Fig. 20-68, and the cams should be measured to be sure they are not worn and have the specified lift, Fig. 20-69.

GEAR TRAINS

An important part of reconditioning the cylinder block of diesel engines is the inspection of the timing and accessory gears and their bearings. In the case of the two-cycle GMC engine, the balance shaft and drive gear should be inspected.

Clearance between the gear teeth should be measured. This

Fig. 20-68. Checking a camshaft between centers on a lathe.

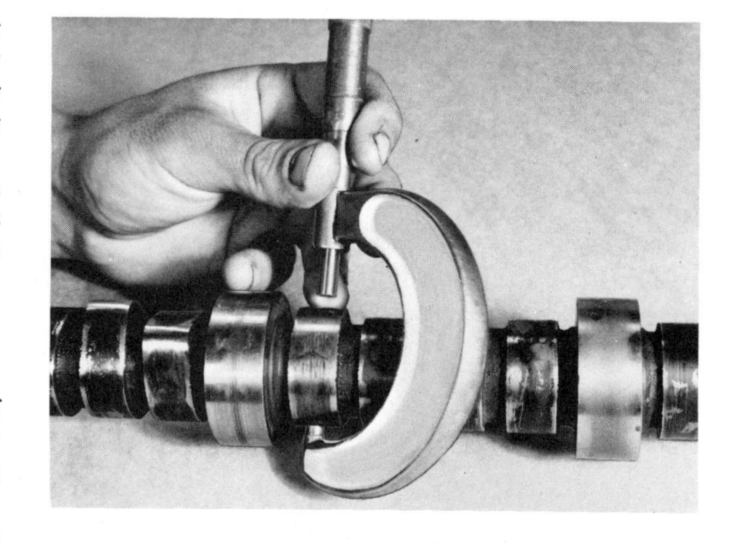

Fig. 20-69. Checking a cam for wear.

Fig. 20-70. Installing timing gears on a GMC engine.

can be done with a thickness gage or a dial indicator. The amount of clearance or backlash varies with different installations, but usually is in the range of 0.005 in. Bearing clearance also varies, but 0.0025 to 0.004 in. is normal. In some installations, taper roller bearings are used for idler gear bearings. Typical front end timing gear trains are shown in Figs. 20-70 and 20-71.

Fig. 20-71. Timing gear layout on a GMC series 53 engine. 1—Balance shaft gear, left hand helix. 2—Governor drive gear. 3—Crankshaft gear. 4—Idler gear. 5—Blower rotor lower gear. 6—Blower rotor upper gear. 7—Camshaft gear. (GMC Truck and Coach Div., General Motors Corp.)

MAIN AND ROD BEARINGS

Main and rod bearings are the replaceable type and nonadjustable. They are prevented from endwise and radial movement by a tang at the parting line on one side of each shell.

Bearing caps are stamped with the number of the corresponding cylinder and a matching number is stamped on the rod or bottom of the crankcase for the main bearings.

When dismantling an engine, it is important to place each bearing on the workbench in numerical order. This will help in locating any trouble that may be present and also aid in installing them in their original position.

If the engine is in the chassis, the main bearing inserts can be removed by first removing a bearing cap. The upper half of the bearing is then removed by inserting a special pin, Fig. 20-72, in the oil hole in the crankpin. This pin protrudes slightly above the surface of the crankpin and as the crankshaft is rotated, the pin will push the bearing shell ahead of it. A new bearing shell is installed by reversing the procedure.

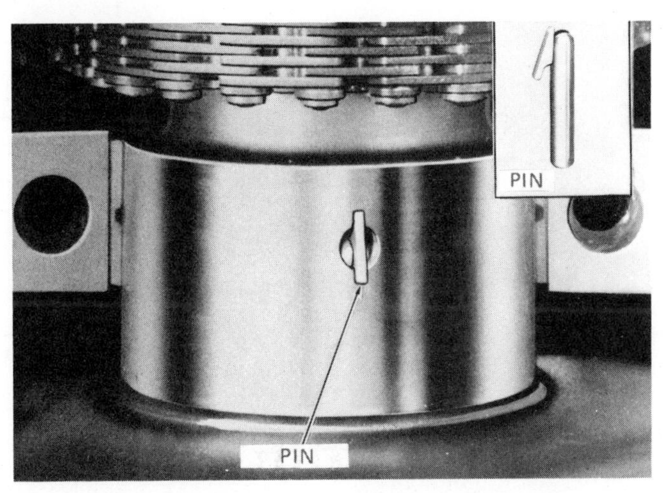

Fig. 20-72. Special pin inserted in oil hole which will push out upper bearing shell as crankshaft is rotated. (Federal Mogul Corp.)

Both front and back surfaces of the bearing shell should be carefully examined, since the condition of the surface will reveal the cause of the failure, Figs. 20-73 and 20-74. According to Federal Mogul, about 43 percent of bearing failure is caused by dirt and other foreign material.

Fig. 20-73. Note wear on these bearing shells caused by abrasive dirt.

MAJOR CAUSES OF BEARING FAILURE	
Dirt	42.9 percent
Lack of lubricant	15.3 percent
Misassembly	13.4 percent
Misalignment	9.8 percent
Overloading	8.7 percent
Corrosion	4.5 percent
Other	5.4 percent

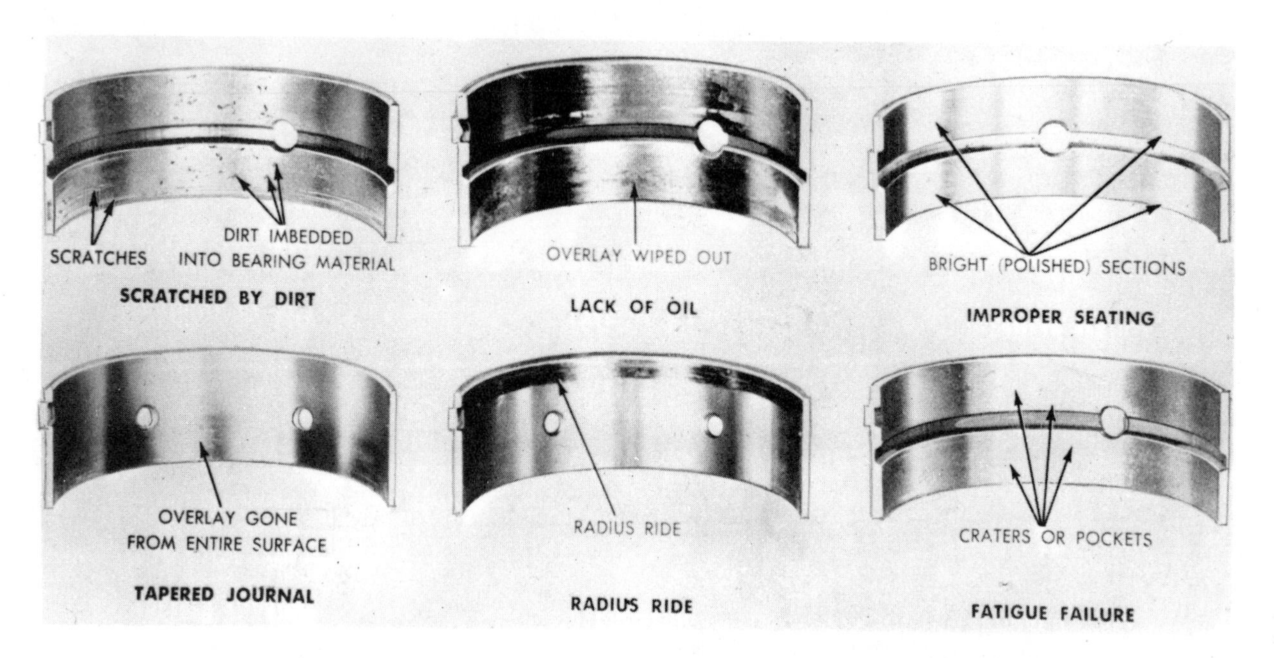

Fig. 20-74. Illustrating different types of bearing failure. (Ford Motor Co.)

A bent connecting rod or a warped cylinder block will result in unusual pressures being placed on the bearings so that wear areas will be distorted. Fig. 20-75 shows a bent connecting rod where the arrows indicate the points of high pressure and maximum wear. The affect of a tapered journal on a connecting rod bearing is shown in Fig. 20-76. A similar situation would result if the cylinder block was warped or bowed, Fig. 20-65. Aligning the connecting rod will overcome the first condition and align-boring or honing will overcome the condition of a warped cylinder block.

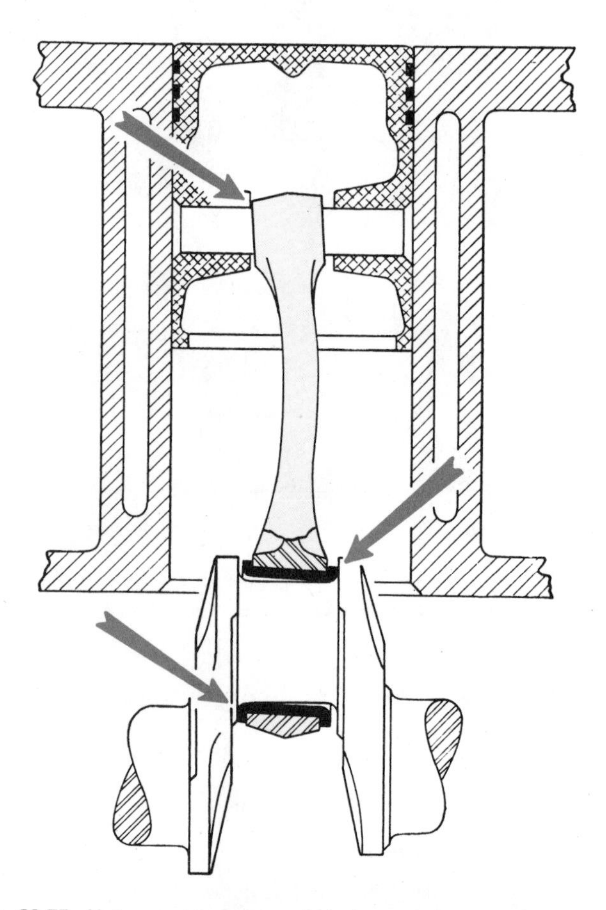

Fig. 20-75. Note pressure areas resulting from a bent connecting rod.

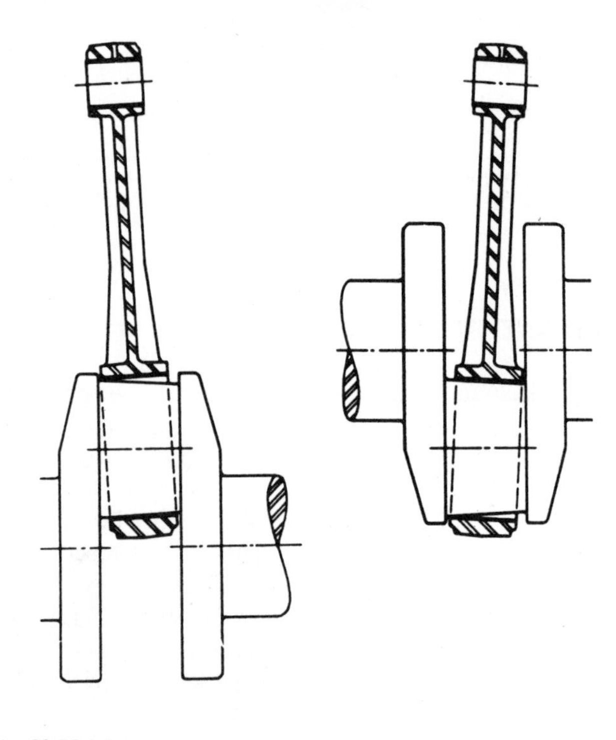

Fig. 20-76. Tapered crankpins throw connecting rods out of alignment.

INSTALLATION OF BEARINGS

The first step in installing the bearing shells is to clean the connecting rod or main bearing bore and measure the bore for size and circularity, Figs. 20-77 and 20-78. The bearing journal

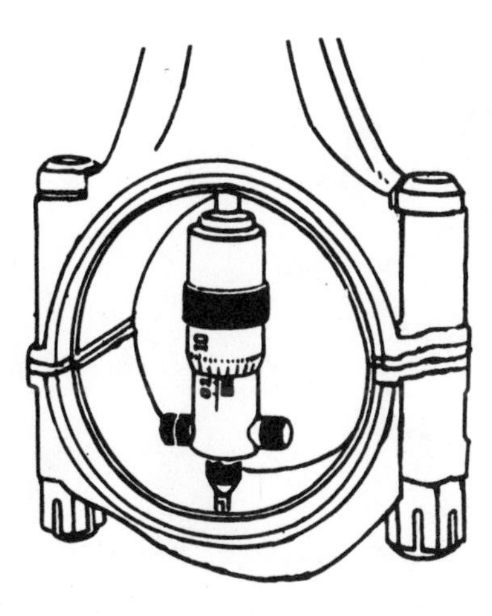

Fig. 20-77. Measuring the bore of a connecting rod bearing.

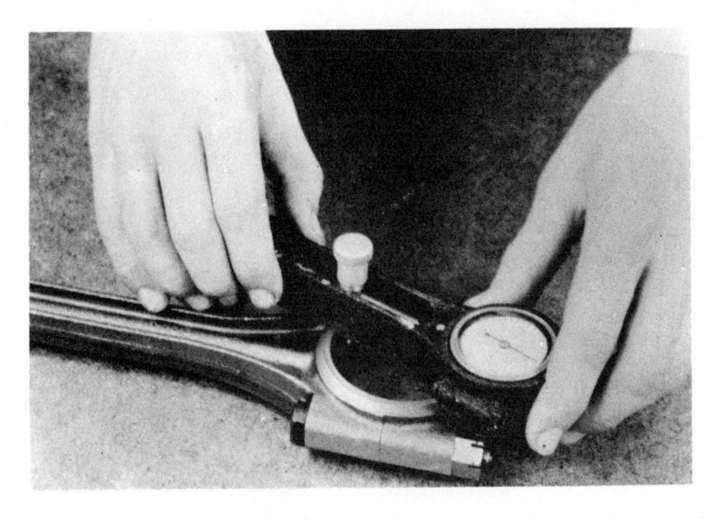

Fig. 20-78. Special gage used to measure circularity of connecting rod bore. (Federal Mogul Corp.)

should also be measured for diameter and taper. When ordering bearings, it is necessary to give the diameter of the journal and the diameter of the bearing bore.

Before installing the bearing shells the bearing bores and shells must be wiped clean. The bearing shells are then pressed into position, taking care that the oil holes in the shells match up with the oil supply holes.

There must be intimate contact between the inside of the bearing seat bore and the outside of the bearing shell. If close contact is not made, there will be poor heat transfer and excessive temperature will shorten the life of the bearing.

To make sure that the bearing shell will have a close fit in the bearing bore, the bearing is manufactured with a slight spread so it is necessary to force it into the bearing bore. The bearing shell is so designed that the edge of the shell will protrude slightly beyond the bearing seat, Fig. 20-79. This is known as "crush." In effect, each bearing half is slightly larger than a full half circle. When the bearing cap is installed and the bolts tightened, the bearing shells are forced tightly into the bearing bore. Not only is there better heat transfer, but there is also less possibility of the shell turning in the bore.

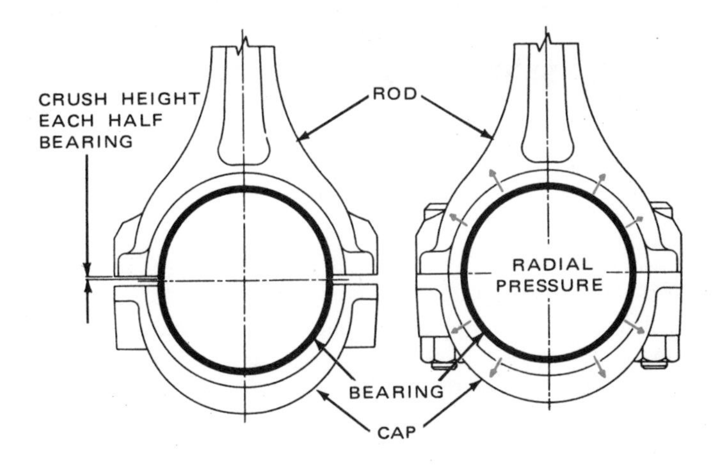

Fig. 20-79. Edge of bearing shells should extend slightly above the bearing seat.

BEARING CLEARANCE

The bearing clearance is the difference between the diameter of the bearing journal and the diameter of the bearing shell. This can easily be determined by making the necessary measurements with micrometers. Another method of measuring bearing clearance, used extensively in automotive shops, is the Plastigage method, Fig. 20-80. In this method, a piece of special small diameter plastic rod is placed on the surface of the bearing. The bearing and cap are installed and torqued to specifications. This crushes the plastic rod. The degree to which the plastic rod is flattened is proportional to the clearance of the bearing. A gage is provided to measure the width and translates the width into the amount of clearance in thousandths of an inch.

RECONDITIONING CONNECTING RODS

The reconditioning of the connecting rod is a very important operation and contributes a great deal to long engine life. If not done with maximum precision, compression ratio will be affected. There is also the possibility of the piston striking the valves, early failure of bearings and scoring of the crankshaft journals. As a result of misalignment, scoring of pistons and cylinders may also occur.

The objective to be attained when reconditioning a diesel connecting rod is to restore it as nearly as possible to its original conditions. That includes journal bore and finish, alignment, center to center length, piston pin fit, weight and width of the large end. When measuring the diameter of the large end, the cap should be in place and torqued to specifications.

The need for reconditioning the connecting rods results from distortion caused by the forces of combustion, compression and centrifugal force which act on the rod. Such forces tend to change the shape of the large end from circular to oval, to twist and bend the rod and deform the shape of the piston pin bore.

Checking the connecting rod to see if its dimensions and condition will permit its being restored to the manufacturer's

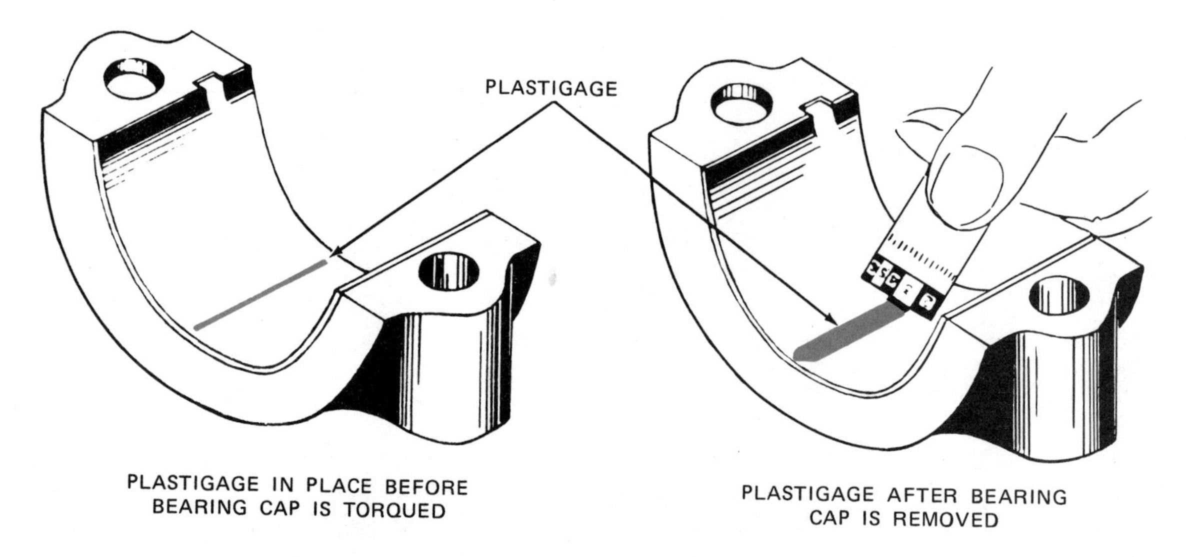

Fig. 20-80. Measuring bearing clearance by the Plastigage method.

original specifications is the first step in the reconditioning operation. Be sure the parting surfaces of the large end are true and smooth and have not been filed. The notches in the rod which are designed to receive the bearing lips should not be damaged. Check the bolt holes for abrasion and distortion. If the holes are of the blind type, make sure they are clean and of sufficient depth to receive the bolt. If the rod is of the rifle drilled type to carry oil to the piston pin, make sure it is unobstructed. The connecting rod should also be checked magnetically for cracks.

It will generally be found that the bearing bore of the connecting rods, which have seen extended service, will be oval rather than a true circle, Fig. 20-81. If the rod is bent, the bearing bore may also be tapered.

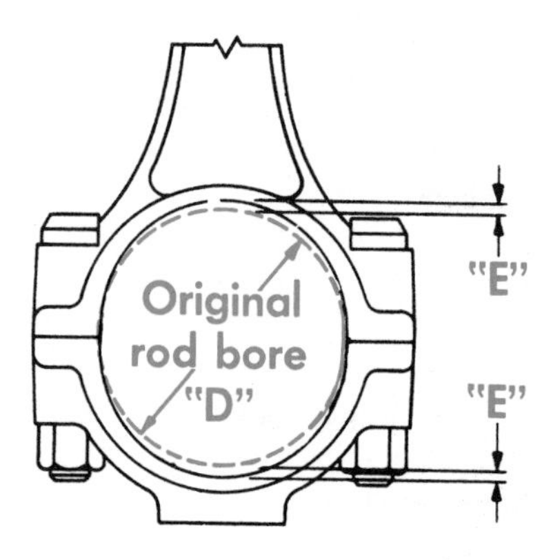

Fig. 20-81. As the result of pressures, the large end of connecting rods become oval. "D" is the original bore and "E" is the amount of stretch at top and bottom.

To overcome these conditions, the parting faces of the rod and the cap are both accurately machined, Fig. 20-82. This will reduce the vertical dimension of the bore diameter. The bore of the journal end of the rod is then honed or bored to the desired dimension, Fig. 20-83. By changing the hone, the same equipment is used to restore the piston pin bore. Fig.

Fig. 20-82. Machine designed to grind the parting areas of connecting rods. (Sunnen Products)

20-84 shows the piston pin bore being gaged for size after having been honed. A complete pin fitting machine is shown in Fig. 20-85. It also checks the rod for center to center length and twist. A simple fixture for checking the alignment of a connecting rod is shown in Fig. 20-86.

All connecting rods for an individual engine must be of the same length, otherwise each cylinder would develop a different amount of power causing rough operation.

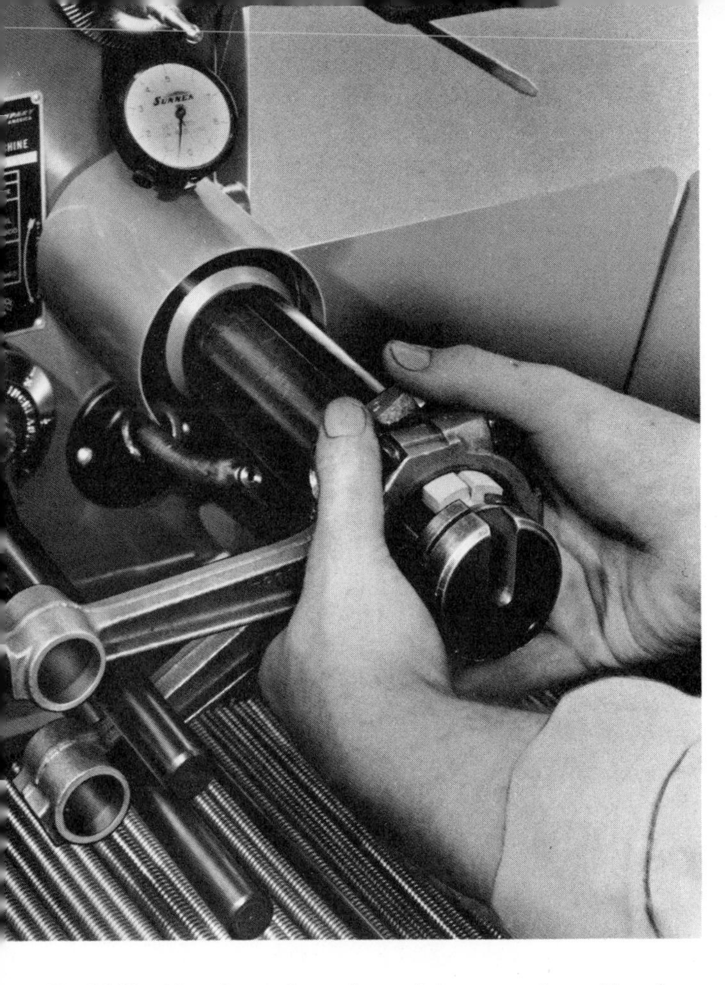

Fig. 20-83. After the parting surfaces of the connecting rod have been reconditioned the journal end is honed to specified diameter. (Sunnen Products)

Some connecting rods have serrated caps. While it is possible to machine such surfaces, the time required would be excessive and it would be more economical to install new connecting rods.

Fig. 20-84. Reconditioning the piston pin bore of a connecting rod.

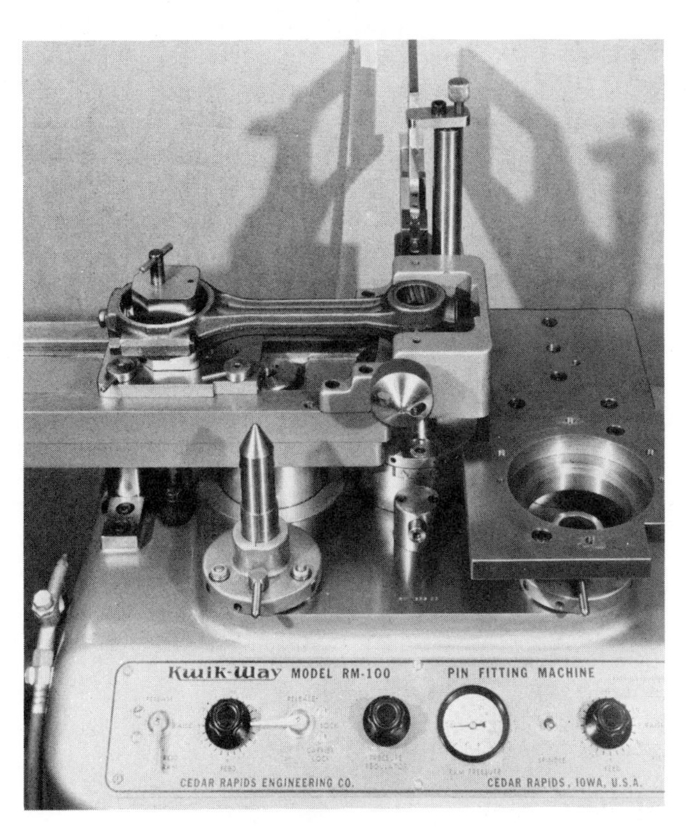

Fig. 20-85. Center to center distance of connecting rod bores must be matched. (Cedar Rapids Engineering Co.)

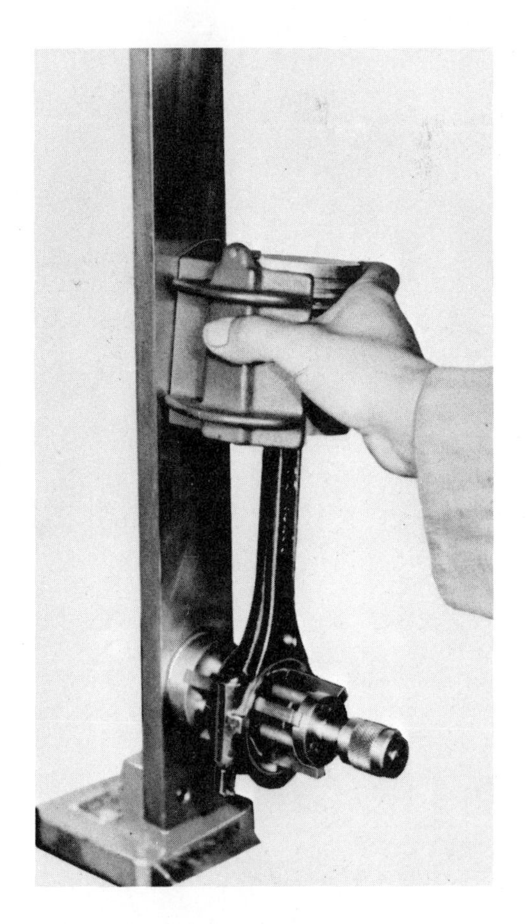

Fig. 20-86. Checking a connecting rod and piston assembly for alignment.

PISTON CHECKING AND FITTING

If the old pistons are to be used again, it is necessary to inspect them carefully. It should be made certain that they will provide satisfactory clearance in the cylinders bores, that there are no cracks present, that the ring grooves are in good condition and that the piston pin bores can be reconditioned if necessary.

If the piston is made of cast iron, check it magnetically for cracks, if made of aluminum, use one of the other crack detecting methods.

When cleaning some pistons (Cummins N and NH models for example) the temperature of the cleaning solution should not exceed 212 deg. since the piston skirts are coated with a plating that may blister if overheated. In the case of aluminum pistons, special detergents that will not attack the aluminum should be used.

After cleaning the piston, measure the diameter of the skirt with a micrometer at right angles to the piston pin, Fig. 20-87. Some pistons are tapered and should be measured at a point directly below the ring lands and also at the bottom of the skirt. Barrel ground pistons should be measured at the center of the skirt.

Fig. 20-87. Measuring diameter of a piston.

Piston clearance in the sleeve or cylinder is determined by measuring the diameter of the bore and then the diameter of the piston with micrometers. The difference between these two measurements is the piston clearance. Another method is to insert a feeler gage between the piston and the cylinder wall. Some mechanics will judge the clearance by the "feel" required to pull the specified feeler gage from between the piston and the cylinder wall. In other cases, the pull is measured by a spring scale, Fig. 20-88.

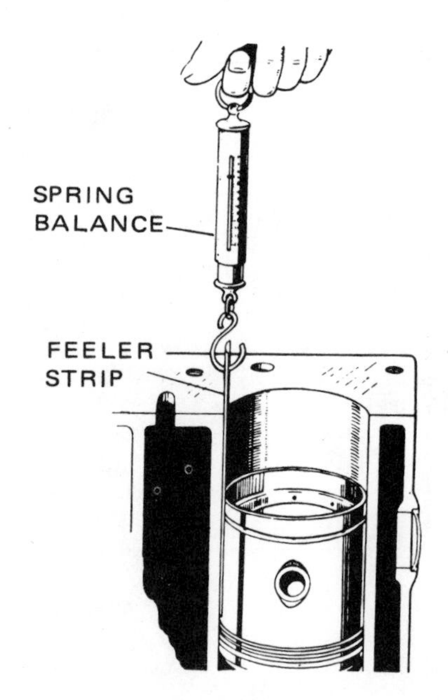

Fig. 20-88. Checking piston clearance with a feeler gage and balance scale.

RINGS AND GROOVES

It is important to inspect the condition of the ring grooves. Before measuring the width of the grooves, they should be cleaned with a scraper type tool, Fig. 20-89. Special gages are available for checking the width of the ring grooves. This is particularly important since worn ring grooves result in excessive oil consumption and blow-by. If such gages are not immediately available, a new ring of the desired size can be rolled around the groove. The ring should roll freely around the groove without looseness or bind. Actual clearance between the ring and the groove is about 0.003 in., Fig. 20-90, but will vary with the size and type of piston.

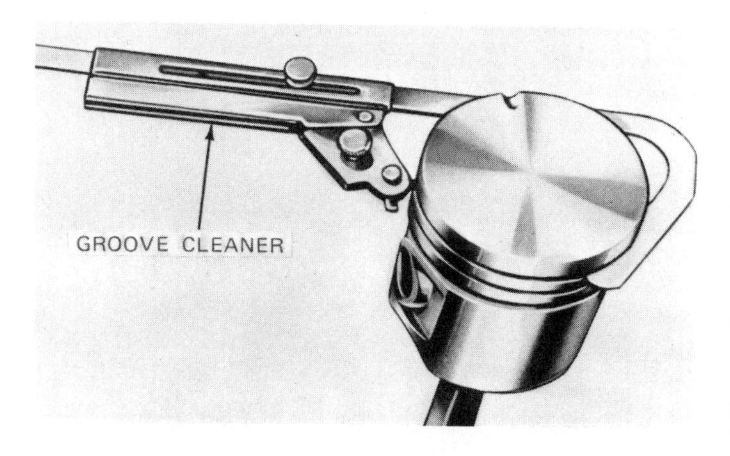

Fig. 20-89. Cleaning ring grooves on a piston with a special tool.

Piston rings should be installed on the pistons so that ring gaps are not in alignment but are staggered 180 deg. apart on alternate rings.

Piston rings must be given sufficient end gap that they will

Fig. 20-90. Measuring side clearance of piston ring with feeler gage.

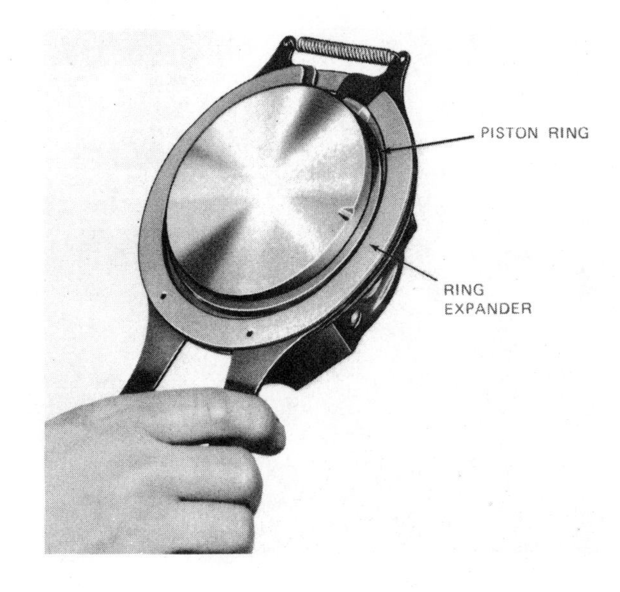

Fig. 20-92. Using a piston ring expander to install rings on a piston.

not abut when reaching operating temperature. Clearance will vary with different engines, but should be approximately 0.003 in. per inch of bore diameter. To make this measurement, place a piston ring in the cylinder and position it by forcing it down with a piston. Measure the gap with a thickness gage, Fig. 20-91.

Fig. 20-91. Measuring end gap of piston ring.

When installing rings on a piston, a ring expander should be used since hand expansion will tend to distort the rings, Fig. 20-92. To install the piston assembly in the cylinder, the rings are compressed in the grooves by means of a ring compressor, Fig. 20-93, so the assembly can be easily tapped into the cylinder bore with the aid of a hammer handle.

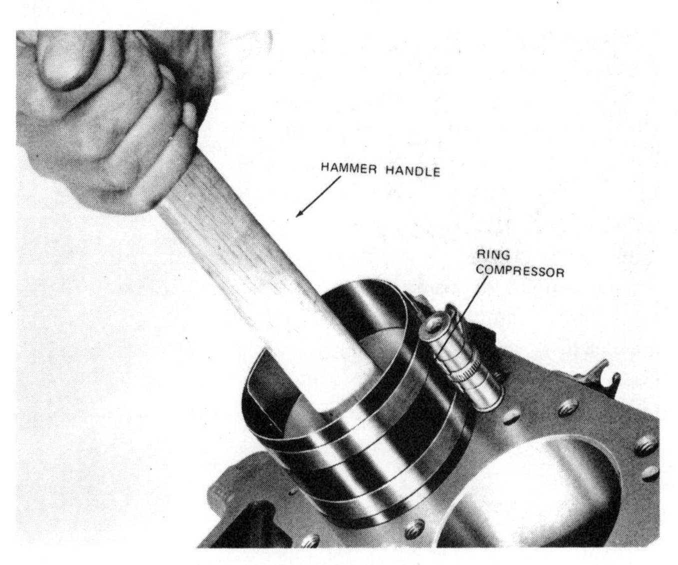

Fig. 20-93. Installing piston, ring and rod assembly in cylinder.

FITTING PISTON PINS

Fitting the piston pins requires the utmost precision. Compression pressure on a four inch diesel piston is in excess of two tons. On the power stroke it is much higher and all pressure must be taken by a small diameter piston pin and its bushing in the piston. Consequently great care must be taken when fitting the piston pins. In that way, full advantage may be taken of all the bearing area that is available.

There are many different methods of preventing the piston pin from moving sideways in the piston. Some piston pins are clamped in the rod. Other designs provide snap rings at the

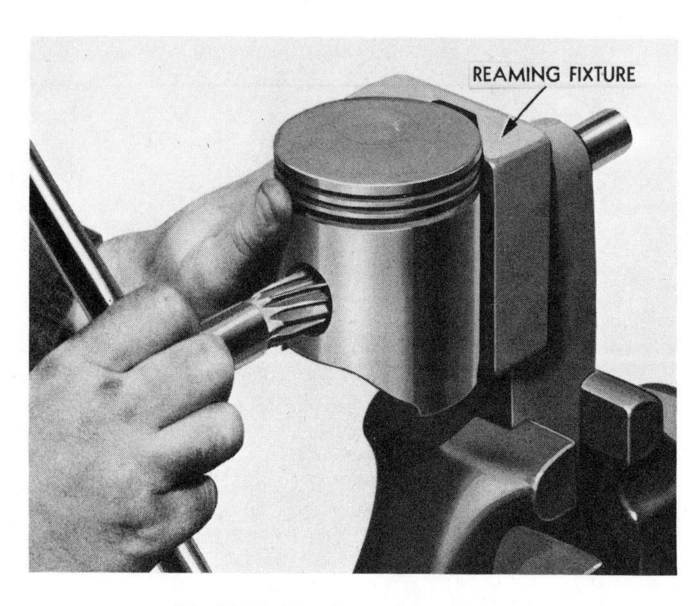

Fig. 20-94. Reaming a piston pin boss.

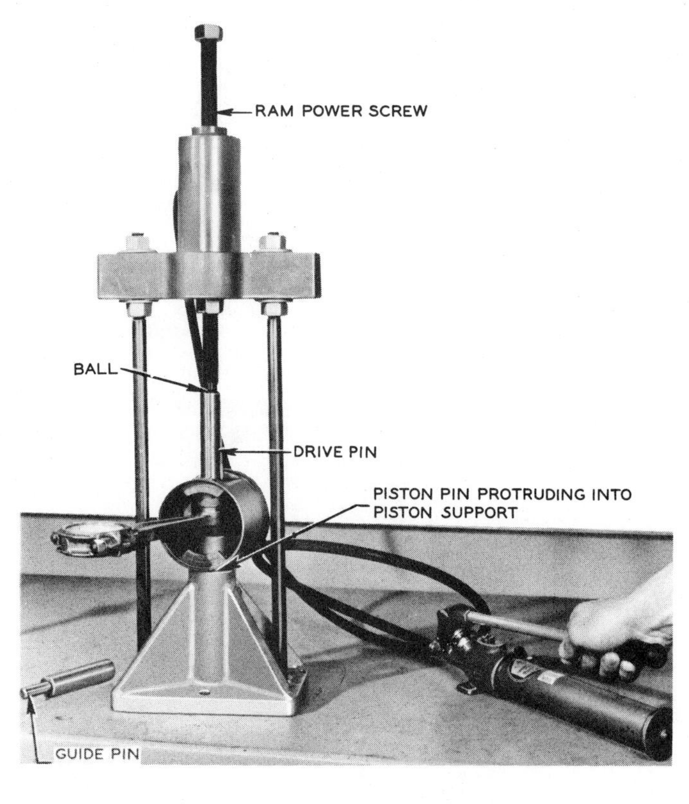

Fig. 20-96. Pistons must be well supported when pressing in the piston pins.

outer end of each piston pin boss and is known as the full floating pin. Another method is to lock the pin in one piston boss. The pins may also oscillate in the piston and be press fit in the rod, or locked in the piston with a setscrew.

When disassembling a piston from a connecting rod, the piston should be supported on a suitable arbor. After removing the snap rings (if installed), the pin is pressed from the piston. This procedure will avoid distorting the piston.

Piston pins are fitted to the piston by reaming, broaching or honing the bosses in the piston. Before resizing the hole, make sure that oversize pins are available. Bushings will also be needed in some cases.

When reaming a piston pin boss, misalignment of the bosses can be avoided by using a fixture designed for the purpose, Fig. 20-94. Clamp the fixture in a vise, then insert the guide bushing in the fixture and secure it with a setscrew. Place the piston in the fixture and insert the pilot end of the reamer through the clamping bar, piston pin bushing and into the guide bushing. Turn the reamer in a clockwise direction. Remove chips with compressed air.

The honing method of fitting pins is very popular, Fig. 20-95. The hone is power driven at the desired honing speed and a gage is located on the machine so the diameter of the hole being reamed can be checked as frequently as desired.

After the pin has been fitted and the pin, piston and rod assembled, the assembly should be checked for alignment. When assembling the piston pin to the piston, the piston bosses must be well supported to avoid distortion of the piston, Fig. 20-96.

Lubrication of the piston pin is often provided by passing the oil through the rifle drilled connecting rod, Fig. 20-97.

Fig. 20-95. Fitting a piston pin on a honing machine. (Sunnen Products)

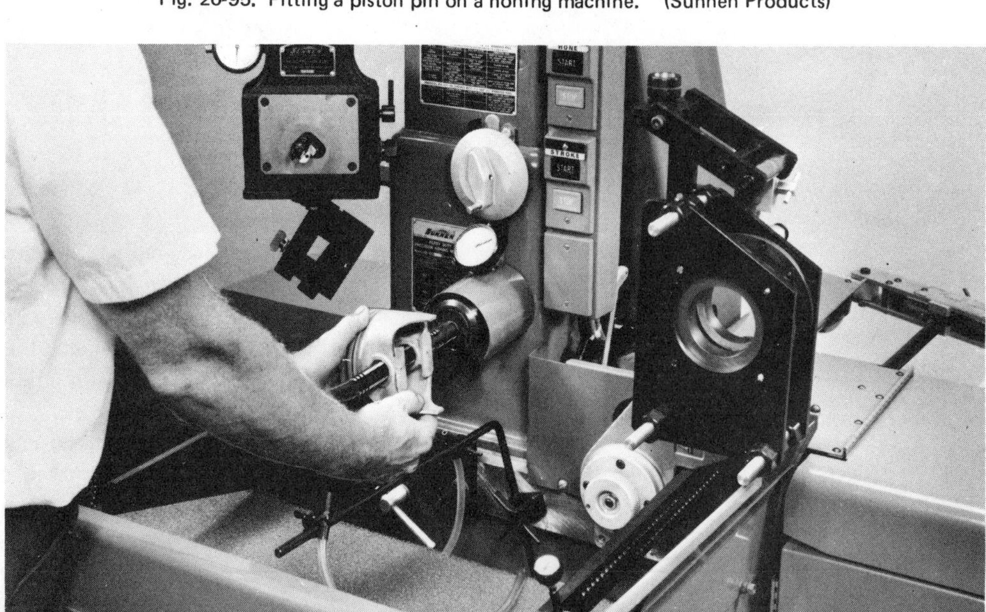

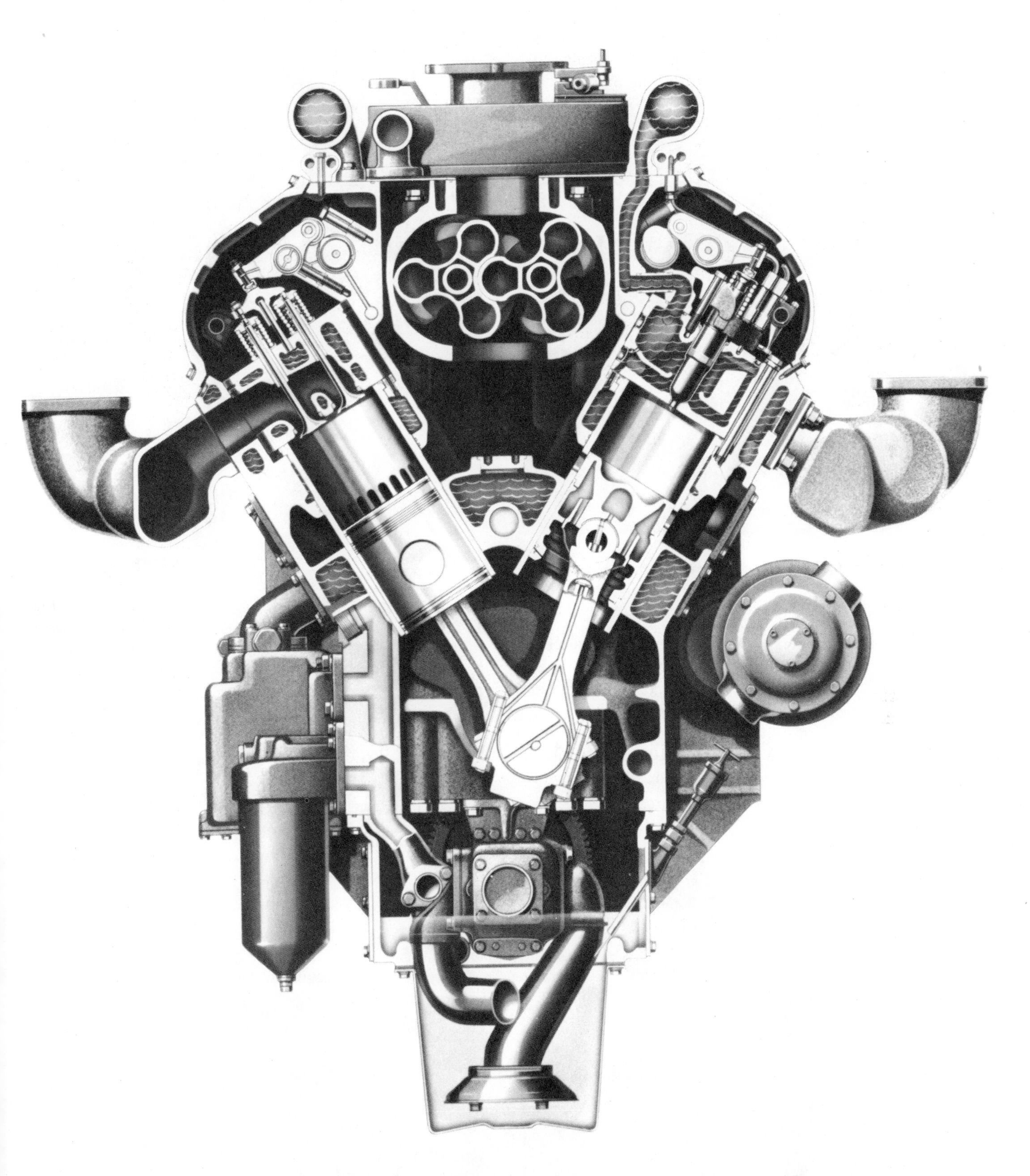

Fig. 20-97. Pictured is a Detroit Diesel Allison two cycle, series 149 diesel engine. Rifle drilled connecting rods provide lubrication for the piston pin, and also direct a spray of oil to the underside of the pistons to aid in cooling. Note the Roots type blower in the Vee of the engine. (Detroit Diesel Allison Div., General Motors Corp.)

SERVICING PUMPS AND INJECTORS

As in all maintenance work, the first step in servicing a diesel injection pump or injector is a thorough cleaning, Fig. 20-98. The illustration shows an American Bosch injector pump in the first cleaning tank where the "field" dirt is being removed.

Fig. 20-98. Cleaning exterior of American Bosch injector pump with detergent.

Following disassembly, Fig. 20-99, the parts are placed in an ultrasonic cleaning solution, Fig. 20-100. Sonic energy passes through the detergent and cavitation (extreme agitation) removes all foreign matter. Next, the parts go to a neutralizing tank in 180 deg. hot water, Fig. 20-101. Then the parts are dried with compressed air, Fig. 20-102, which must be perfectly clean so there is no chance of any oil or dust in the drying air.

Fig. 20-99. After exterior dirt is removed the injector pump or injector is disassembled.

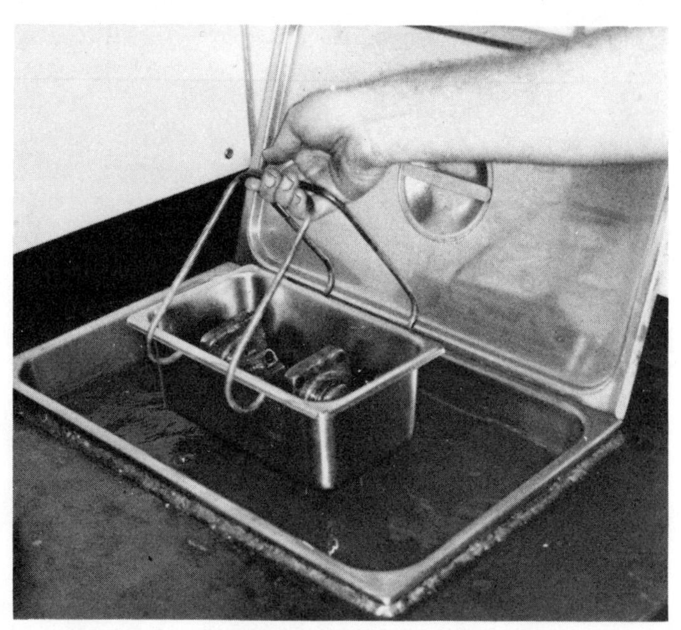

Fig. 20-100. Removing parts from ultrasonic cleaner.

Fig. 20-101. After the ultrasonic cleaning, parts are placed in a neutralizing bath of water and ammonia at 180 deg. F.

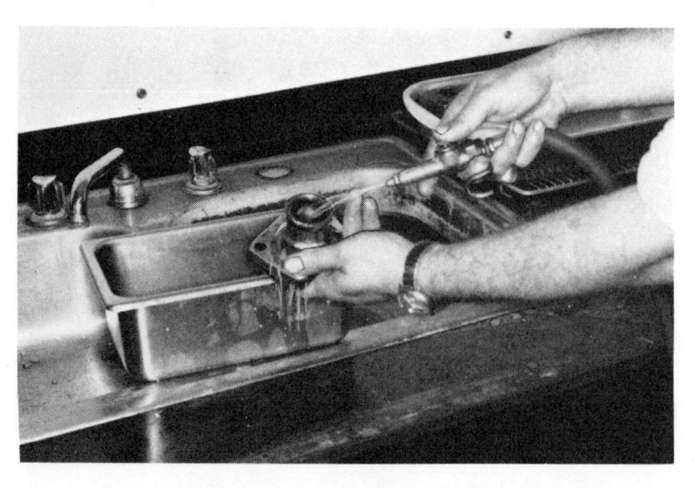

Fig. 20-102. Drying the parts with clean compressed air.

Having cleaned and dried the parts, whether injector or injector pump, they are then taken to the assembly bench, Fig. 20-103.

Fig. 20-103. Parts are now ready for assembly.

After assembly, the pumps or injectors are checked on various types of testing equipment to make sure they meet factory specifications for quantity of fuel being sprayed, pressure, spray pattern, timing, duration of spray and cutoff.

Fig. 20-104. Technician is taking test readings before calibrating a diesel fuel injection pump. (Hartridge Equipment Corp.)

Two types of equipment used primarily for testing the operation of diesel injection pumps are shown in Figs. 20-104 and 20-105. Such equipment will test the complete operation of the injection pumps including such features as quantity of fuel sprayed, timing, etc.

Fig. 20-105. Another type of tester used on some injector pumps.

The unit type injector used on the GMC 71 is often tested on equipment shown in Fig. 20-106, while the spray pattern is checked on equipment of the type shown in Fig. 20-107. Good and poor types of spray patterns are shown in Fig. 20-108.

Fig. 20-106. Special tester for checking the operation of unit injectors.

Fig. 20-107. Analyzer and tester used for testing spray pattern of multiple and pintle injector nozzles.

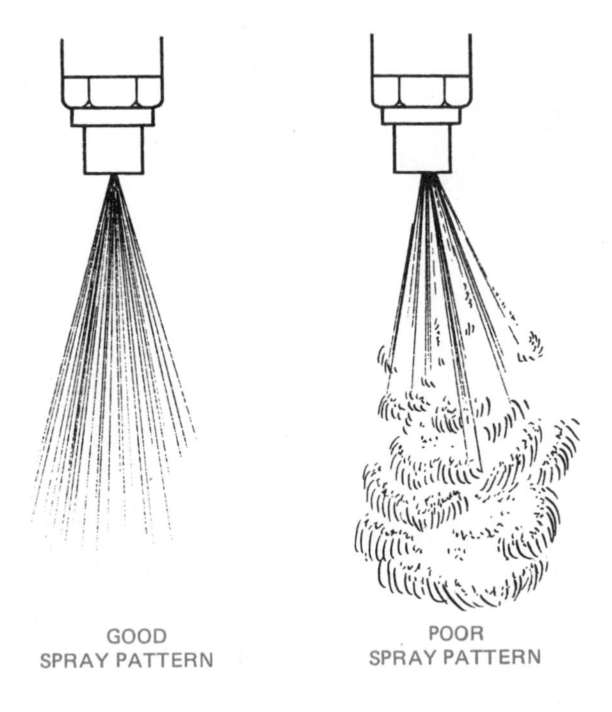

Fig. 20-108. Comparison of good and poor spray patterns.

Details of injector and injection pump testing procedures will vary with the type of equipment being used to make the tests. In each case the manufacturer's instructions should be carefully followed.

To aid in meeting the more stringent standards covering the maximum allowable exhaust emissions, more specialized test equipment is being marketed for checking fuel injection pumps and injectors. Fig. 20-109, for example, shows an injector comparator scientifically designed for checking Cummins injectors. Fig. 20-110 shows a calibrator used for checking Detroit Diesel Allison injectors.

Specialized equipment is also available for testing the amount of smoke in the exhaust.

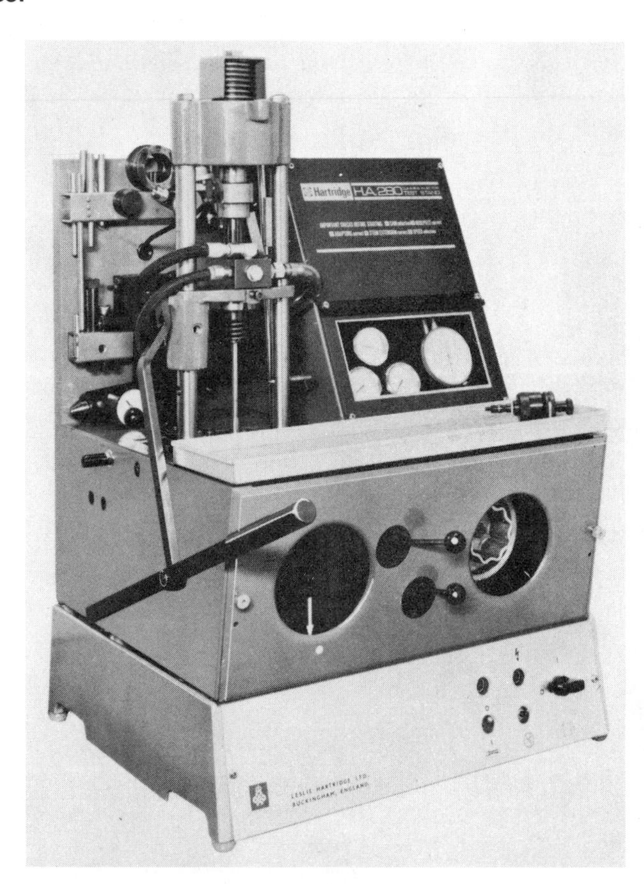

Fig. 20-109. New type of injector comparator designed to test the complete range of Cummins PY injectors. (Hartridge Equipment Corp.)

Fig. 20-110. Calibrator for checking Detroit Diesel Allison injectors includes an electronic trip counter to insure accuracy in counting strokes. (Hartridge Equipment Corp.)

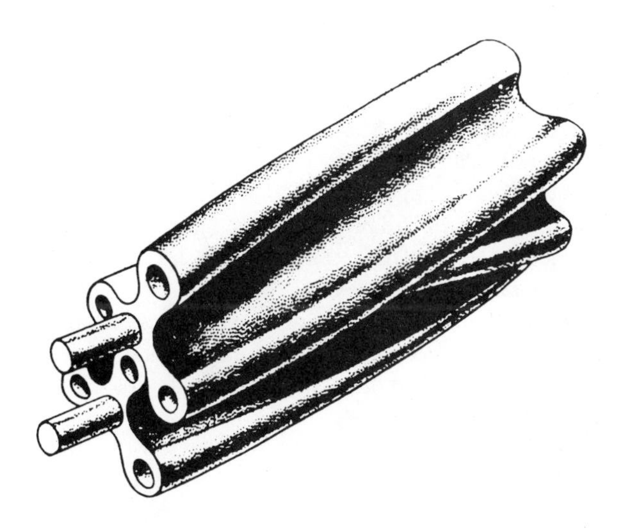

Fig. 20-111. Type of rotors used in GM blowers.

SERVICING BLOWERS, SUPERCHARGERS AND TURBOCHARGERS

While the terms blower and supercharger are often used interchangeably, the unit used on the two-cycle engine for scavenging is generally called a blower, while superchargers and turbochargers are designed to supply more air to the cylinders than would be obtained by natural aspiration. The supercharger and blower are mechanically driven devices, while the turbocharger is driven by the engines exhaust system pressure. Special care must be taken to maintain accuracy and cleanliness when servicing the blowers, superchargers and turbochargers.

The design of these units is not particularly complicated but parts are fitted to close tolerances, not only in linear dimensions but also as to weight. Since the rotors of these units revolve at speeds of 5000 rpm or more, the importance

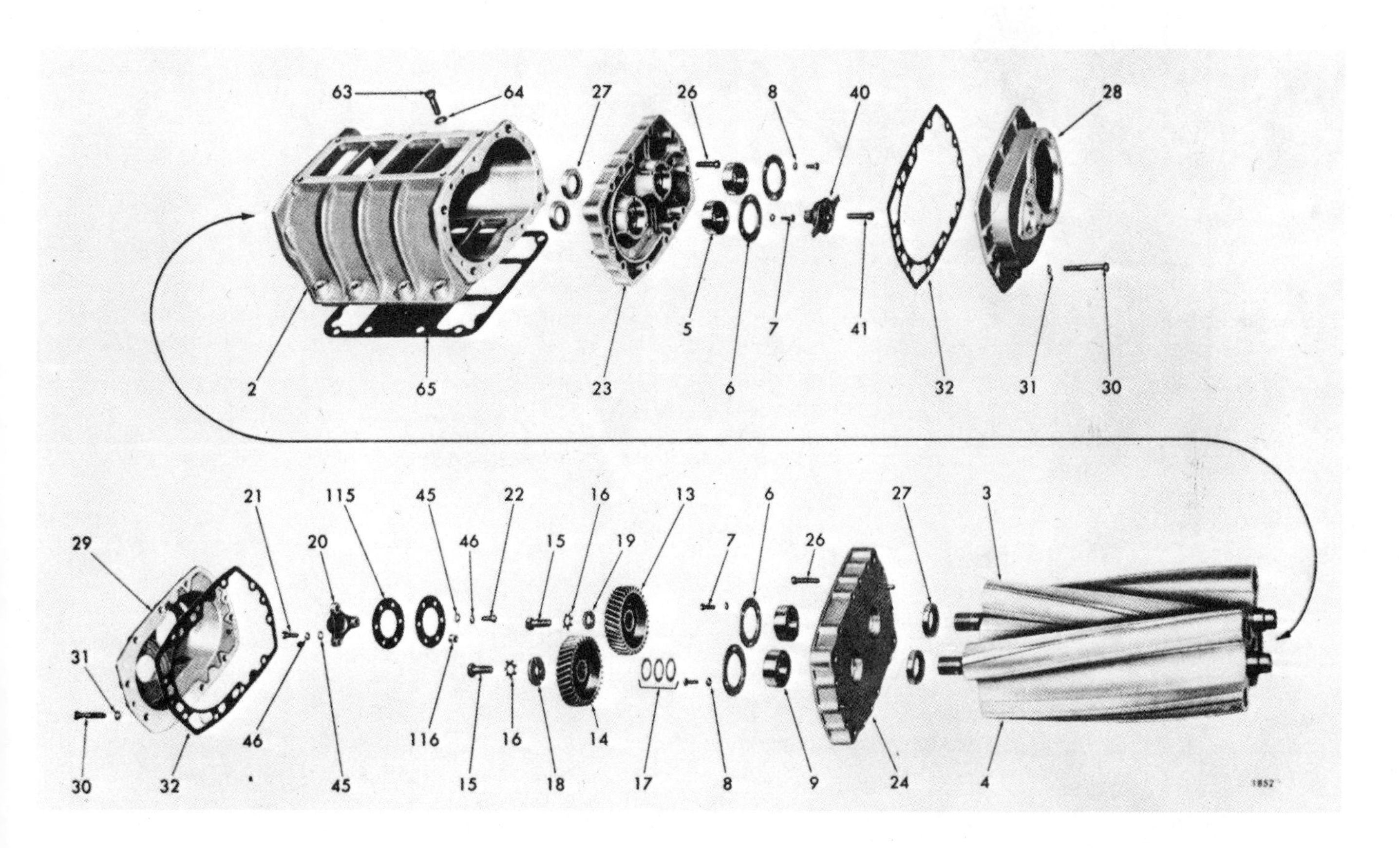

2. BLOWER HOUSING.
3. BLOWER ROTOR, UPPER, RIGHT HAND HELIX.
4. ROTOR BLOWER, LOWER, LEFT HAND HELIX.
5. FRONT ROLLER BEARING.
6. BEARING RETAINER.
7. BEARING RETAINER BOLT.
8. LOCK WASHER.
9. REAR DOUBLE ROW BALL BEARING.
13. UPPER ROTOR GEAR, RIGHT HAND HELIX.
14. LOWER ROTOR GEAR, LEFT HAND HELIX.
15. ROTOR GEAR BOLT.
16. LOCK WASHER.
17. GEAR TO BEARING SHIM, FOR TIMING ROTORS.
18. FUEL PUMP DISC COUPLING.
19. ROTOR GEAR RETAINING WASHER.
20. ROTOR DRIVE HUB.
21. PLATE TO GEAR BOLT.
22. PLATE TO HUB BOLT.
23. FRONT END PLATE.
24. REAR END PLATE.
25. HOUSING TO END PLATE DOWEL PIN.
26. END PLATE BOLT.
27. END PLATE OIL SEAL.
28. END PLATE FRONT COVER.
29. END PLATE REAR COVER.
30. END PLATE COVER BOLT.
31. LOCK WASHER.
32. END PLATE COVER GASKET.
40. WATER PUMP DRIVE COUPLING ASSEMBLY.
41. ALLEN HEAD BOLT.
45. PLAIN WASHER.
46. LOCK WASHER.
63. BLOWER MOUNTING BOLT.
64. BLOWER HOUSING GASKET.
65. BLOWER HOUSING GASKET.
115. BLOWER ROTOR DRIVE HUB PLATE.
116. PLATE TO GEAR SPACER.

Fig. 20-112. Typical details and location of parts of blower used on Detroit Diesel Allison six cylinder series 53 engine.

of not altering the weight of any part cannot be overemphasized.

GMC BLOWER

The blower used on many GMC two-cycle engines supplies fresh air needed for combustion and scavenging. Its operation is not unlike that of a gear type oil pump. To provide continuous and uniform displacement of air, the rotors are made in a helical form, Fig. 20-111.

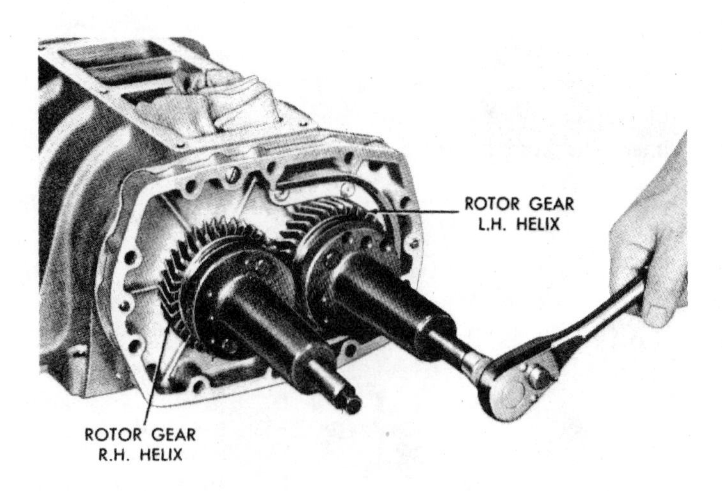

Fig. 20-113. Pulling rotor gears. Both rotor gears must be pulled at the same time. (Detroit Diesel Allison Div., General Motors Corp.)

Two timing gears located at the drive end of the rotor shaft, Fig. 20-112, revolve at about twice the camshaft speed. The complete gear train is shown in Fig. 20-71. The rotor shafts are carried on antifriction bearings, usually ball bearings at the drive end and roller bearings at the other end.

Should the bearing bores ever get damaged and require machining, it is essential that the original center lines be maintained, otherwise the clearance between the rotors will be altered and damage may result.

Oil seals located in the blower end plates prevent oil leakage and also keep the lubricating oil from entering the rotor compartment. Lip type oil seals or metal ring type oil seals are installed in the blower end plates.

The blower rotors are timed by the two rotor timing gears, Fig. 20-71, at the rear end of the blower shafts. This timing must be correct, otherwise the required clearance between the rotor lobes will not be maintained.

As the teeth of the timing gears wear, clearance between the lobes of the rotors will be affected. While the rotor lobe clearance can be adjusted by means of shims behind the gears, the gear backlash cannot be adjusted. The GMC specification for backlash limit is 0.004 in. Excessive backlash between the blower timing gears usually results in the rotor lobes rubbing throughout their entire length.

BLOWER INSPECTION

The blowers may be inspected for any of the following

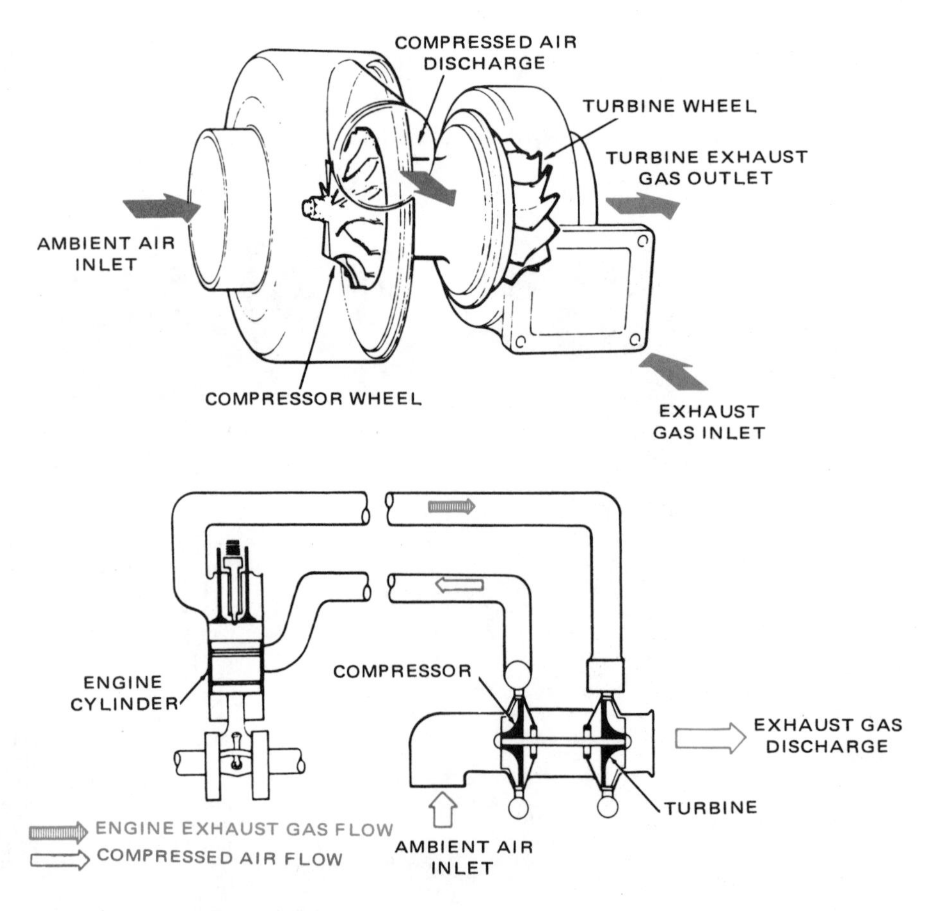

Fig. 20-114. Schematic drawing of Airesearch turbocharger.

conditions without removal from the engine. However, to make the inspection, the air silencer or air shut-down housing must be removed.

As a safety measure when inspecting the blower, be careful to operate the engine only at slow speeds. Keep fingers and clothing away from the moving parts of the blower.

Foreign matter, such as dirt or metal chips, that may be drawn through the blower will make deep scratches in the rotors and rotor housing. Burrs may be formed around the scratches. If such burrs cause interference between the rotors or between the rotors and the housing, the blower should be removed from the engine and the parts dressed down to eliminate interference. In severe cases it will be necessary to install new rotors and possibly a new housing.

Leaking oil seals will be indicated by the presence of oil on the blower end plates and on the inside of the housing. This condition can be checked by running the engine at a slow speed and directing a light into the rotor housing. A thin film of oil radiating away from the seals will be noted if the seals are leaking.

A rattling noise on the inside of the blower will be caused by a worn blower drive. The condition may be verified by firmly grasping the top rotor and attempting to rotate it. Rotors may move 3/8 to 5/8 in. as measured at the rotor crown. When released, the rotor should move back at least 1/4 in. If the rotor cannot be moved in this manner, or if the rotor moves too freely, inspect the flexible rotor drive coupling and replace if necessary.

Loose rotor shafts or damaged bearings will cause rubbing and scoring between the crowns of the rotor lobes and the mating rotor roots and the end plates, or between the rotors and the housing.

The blower inlet screen should be inspected periodically for accumulation of dirt. The frequence of such inspection is dependent on the conditions under which the engine is being operated.

BLOWER SERVICE TIPS

Should it be necessary to disassemble the blower as installed on some GM engines, there are several precautions that should be followed.

Both gears must be pulled from the rotor shafts at the same time, Fig. 20-113.

When performing a major overhaul, all oil seals should be replaced. Rotor lobes should be carefully inspected for burrs and scoring, especially the sealing ribs. Rotor surfaces must be smooth for efficient operation. Slight scores or roughness may be removed with emery cloth.

Check rotor shaft serrations for wear, burrs or peening. Inspect bearing and oil seal contact surfaces of the shaft for wear and scoring. The inside of the rotor housing must also be checked for roughness.

A special tool should be used when installing the oversize seals in the blower end plates. Both gears must be installed on the rotor shafts at the same time when reassembling. A center punch mark placed in the end of each rotor shaft will assist in aligning the gears on the shafts.

AIRESEARCH TURBOCHARGER

The Airesearch turbocharger used on some GM in-line 71 diesel engines is designed to increase engine efficiency and power output. It consists of a radial inward flow turbine, centrifugal compressor wheel, center housing, bearings seals, turbine housing and compressor housing, Fig. 20-114.

After the engine is started, the flow of the exhaust gases from the engine pass through the turbine housing and cause the shaft assembly to rotate, Fig. 20-114. The compressor wheel mounted on the opposite end of the shaft draws ambient air into the compressor housing, compresses the air and delivers it to the engine.

During operation, the turbocharger responds to the engine load demands by reacting to the flow of the exhaust gases. As the power output of the engine increases, the flow of the exhaust gases also increases and the speed and output of the rotating assembly increases proportionately.

The turbocharger is cooled by water passing through a water jacket within the turbocharger, the water being obtained from the engine water jacket.

Lubricating oil for the turbocharger is supplied through an external oil line extending from the engine cylinder block to the top of the center housing.

Drilled passages carry the oil to the shaft bearings, thrust ring, thrust bearing and thrust plate. Minimum oil flow to the turbocharger, with the engine at idle, is 10 psi at an oil temperature of 200 deg. F.

TURBOCHARGER OPERATION

New units that have not been operated for a long period of time or units that have overhauled require lubrication of the shaft bearings, thrust ring, thrust bearing and thrust plate before starting the engine.

To be sure the parts are lubricated and to prevent any subsequent bearing damage, the following procedure is recommended by General Motors.

Thoroughly clean the area around the turbocharger oil inlet line. Remove the oil inlet line from the top of the center housing. Pour approximately four ounces of lubricating oil in the inlet opening of the center housing. Turn the rotating assembly by hand to coat the bearings, thrust ring and thrust plate bearing with oil. Fill the oil inlet line with lubricating oil and connect the oil inlet line to the top of the center housing.

After the turbocharger has been running long enough to permit the unit and the oil to reach operating temperature, the rotating assembly, Fig. 20-114, should coast freely to a stop after the engine is turned off. If the rotating assembly jerks to a sudden stop, the cause should be determined and corrected.

TURBOCHARGER INSPECTION

Periodic inspection of the Airesearch turbocharger, as installed on some GMC 71 in-line engines, should include inspection of the oil inlet and return lines to make sure there is no leakage. Make sure there is no restriction of the flow of oil to and from the center housing.

Air ducts should be inspected for possible leakage while the engine is shut down. Similar inspection should be made at the turbine inlet and exhaust manifold gasket.

If any leakage exists in the air inlet or outlet ducts, or if the air cleaner is dirty and not filtering efficiently, the turbocharger should not be operated. Leakage or poor operation of the filter may allow dust to enter the turbocharger and engine, causing damage to both units.

A check should be made of the compressor inlet. This is done after removing the compressor air inlet duct. If any foreign material has accumulated in that area, it should be removed and the cause determined and eliminated.

A caustic solution should never be used when cleaning parts of the Airesearch turbocharger. A wire bristle brush or metal scraper should never be used to clean these parts.

Fig. 20-115. Showing crack in water jacket. (United States Casting Repair Corp.)

REPAIRING CRACKS

The repairing of cracks in diesel engine cylinder blocks and cylinder heads plays an important part of diesel engine maintenance. By making such repairs, an expensive part may be salvaged and the engine placed back in service at a relatively small cost.

Cracks in cylinder heads and cylinder blocks result from excessive temperatures and rapid cooling. They also result from the coolant freezing, which may cause enough expansion to crack the water jacket. Excessive heat around the valves and valve seats causes cracks to form. Cracks in the valve seat area are not unusual in diesel engines because of extended engine operation. Their occurrence can be greatly reduced by idling the engine for about 15 minutes after a long run before shutting it off. In that way, the engine temperature is normalized and the tendency toward forming cracks is greatly reduced.

Fig. 20-116. Drilling holes and recesses for bridges to prevent crack from spreading.

PEENING AND PINNING

There are several methods used to repair cracks. Peening and pinning of some sort form the basis of most methods.

One method is to drill and tap a series of overlapping holes along the length of the crack. Threaded rods are then screwed into the holes. When the entire crack has been treated in that manner, the exposed ends of the threaded rods are peened over.

A variation of this method is to first bridge the crack, Fig. 20-115, with specially formed bridges or locks which prevent the crack from spreading wider, Fig. 10-116. Special jigs are used to drill the holes which form the recesses for the bridges. The first bridge is inserted about 1/4 in. from the end of the crack and the rest are evenly spaced along the crack.

Each bridge is hammered in place, as shown in Fig. 20-117. With the bridges or locks in place, overlapping holes are drilled between the bridges, Fig. 20-118. These holes are tapped and threaded pins of special alloy are screwed in place. The ends of these pins are then peened, Fig. 20-119. This is followed by grinding with a wheel or disc and another light peening, Fig. 20-120.

Fig. 20-117. Driving in Seal Locks.

Fig. 20-118. Overlapping holes are drilled and tapped between the locks.

Fig. 20-119. Peening the heads of the threaded pin inserts.

Fig. 20-120. Surface peening of the total area around the crack.

REVIEW QUESTIONS – CHAPTER 20

1. The eight different methods listed for cleaning diesel engines and diesel engine parts are:
   a. ______________________.
   b. ______________________.
   c. ______________________.
   d. ______________________.
   e. ______________________.
   f. ______________________.
   g. ______________________.
   h. ______________________.
2. In the glass bead method of cleaning parts, the beads produce a:
   a. Mirror finish.
   b. Indented surface.
   c. Corrugated surface.
3. Should the same detergent used for cleaning steel and cast iron parts be used for cleaning aluminum parts?
   a. Yes.
   b. No.
4. The approximate tolerance for a main bearing journal is:
   a. .0005 in.
   b. .0050 in.
   c. .0010 in.
5. Which is the preferred method of checking crankshaft for alignment?
   a. In V-blocks.
   b. Between centers.
6. Gloves should always be worn when handling grinding wheels.
   a. True.
   b. False.
7. Should the engine be warm or cold when removing a cylinder head?
   a. Warm.
   b. Cold.
8. Fuse plugs are used in some cylinder heads as protection against ____________.
9. When a cylinder head is resurfaced should an equal amount of metal be removed from the valve seat counterbore?
   a. Yes.
   b. No.
10. In the case of cylinder heads with hard Stellite seats, the preferred method of resurfacing is:
    a. Grinding.
    b. Broaching.
    c. Milling.
11. Is the knurling method used for reconditioning valve guides?
    a. Yes.
    b. No.
12. When a 70 deg. stone is used to narrow a valve seat, will the seat be raised or lowered?
    a. Lowered.
    b. Raised.
13. Dark spots on a cylinder wall indicate a ________ area.

14. In an automotive size diesel engine, a normal allowable taper for cylinder bore is:
    a. .0007 in.
    b. .0070 in.
    c. .0004 in.
    d. .0040 in.
15. Unevenness or roughness of the top surface of a cylinder block may result in ____________ loss.
16. The distance from the top of the cylinder block to the center line of the ____________ is critical.
17. When reconditioning the top of the cylinder block, it is advisable to take __________ cuts.
18. The diameter and depth of the cylinder sleeve counterbore must be held within ____________ limits.
19. O rings are provided on the outside of cylinder sleeves to prevent:
    a. Leakage of compression.
    b. Leakage of the coolant.
    c. Vertical movement of the sleeves.
20. When honing a cylinder, the hone should be moved rapidly up and down at a rate of about:
    a. 25 strokes per minute.
    b. 50 strokes per minute.
    c. 75 strokes per minute.
21. The major cause of engine bearing failure is:
    a. Overloading.
    b. Corrosion.
    c. Dirt.
    d. Lack of lubrication.
22. When measuring the diameter of a piston, the measurement should be taken ____________ to the piston pin.
23. The clearance between a piston ring and the ring groove is normally about:
    a. .001 in.
    b. .003 in.
    c. .005 in.
    d. .010 in.
24. In the ultrasonic method of cleaning, sonic energy is passed through the detergent and causes ____________.
25. The most important characteristic of the compressed air used when overhauling diesel pumps and injectors is:
    a. Pressure.
    b. Volume.
    c. Cleanliness.
26. The blower used on many GMC two-cycle engines supplies fresh air needed for combustion and ____________.
27. A rattling noise on the inside of the blower will be caused by a ____________ blower drive.

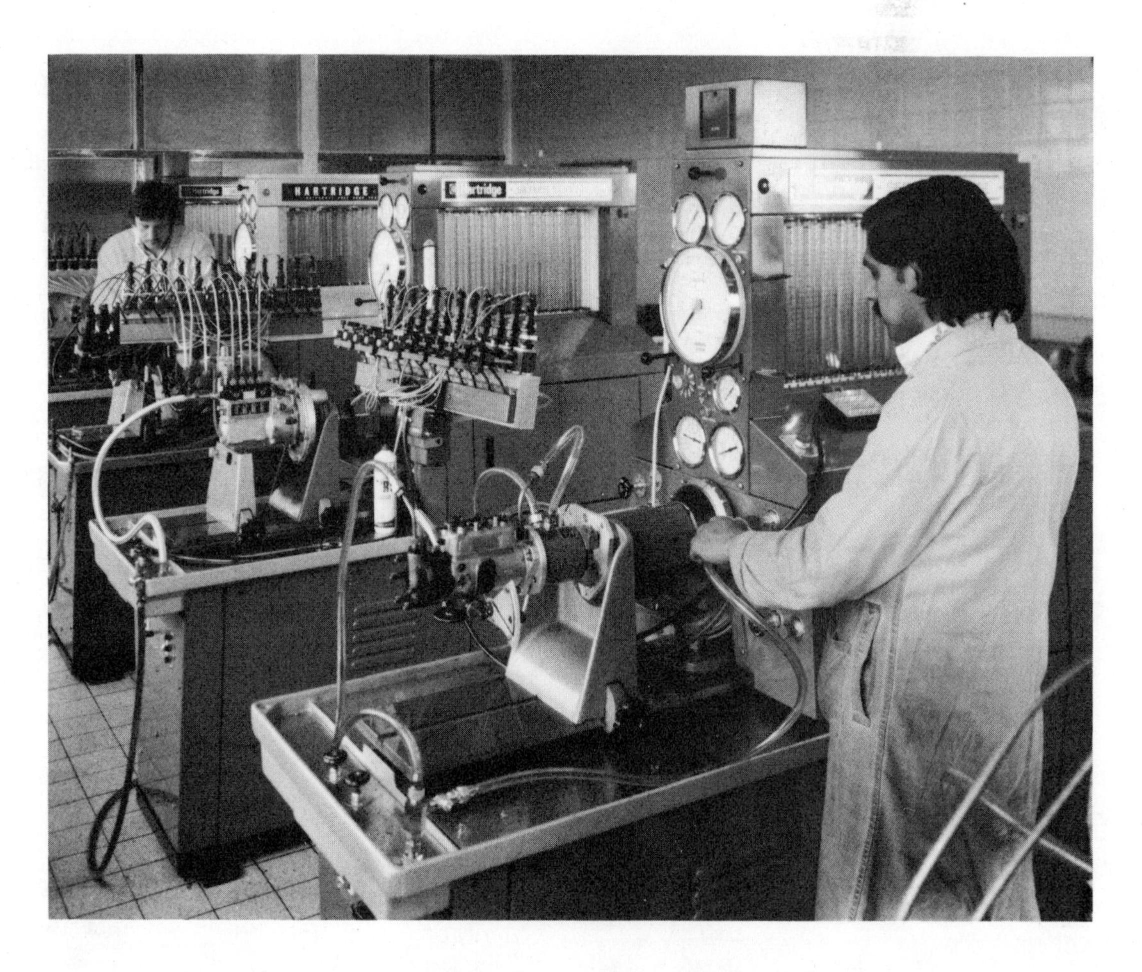

Technician in foreground is preparing to check a fuel injection pump for fuel pressure, fuel delivery, governor and advance operation. (Hartridge Equipment Corp.)

Chapter 21
TUNE-UP AND TROUBLE SHOOTING

The general procedure for tuning a diesel engine should include the following items:

1. Check compression.
2. Service the air filter.
3. Check for air leaks.
4. Inspection of turbocharger or blower.
5. Inspect oil lines to turbocharger or blower.
6. Inspect Aneroid (Cummins).
7. Inspect external fuel lines.
8. Inspect fuel supply pump.
9. Adjust valve tappets and bridges.
10. Adjust unit injector tappets.
11. Adjust fan belt.
12. Inspect cooling system.
13. Service oil filters.
14. Inspect exhaust system.
15. Inspect fuel injection pump or injectors.
16. Check timing.
17. Check adjustment and operation of governor.

Tune-up procedure will vary with different makes and models of diesel engines and on the type of the service for which the engine is being operated. For example, the tune-up procedure for a two-cycle engine would differ from that of a four-cycle engine. A different procedure would also be used for an engine used in truck delivery service from that in an over-the-road vehicle. An engine used in road construction work would require still a different procedure than would be followed for diesel engines used in generating plants or in a railway engine. The fuel being used would also affect tune-up procedure. However, some general instructions can be given.

General Motors, in their tune-up procedure covering their in-line 71 engine, indicates there is no scheduled interval for performing an engine tune-up on that engine. No tune-up should be needed as long as engine performance is satisfactory. Adjustment of valves and unit injector may be required periodically to compensate for normal wear of parts. Each of the four different types of governors used on the 71 in-line engines have different characteristics and, therefore, tune-up procedure would vary considerably. The governors used include:

1. Limiting speed.
2. Variable speed.
3. Constant speed mechanical.
4. Hydraulic.

However, to make a complete tune-up on a 71 in-line engine equipped with a mechanical governor it would be necessary to include the following:

1. Adjust the exhaust valve clearance.
2. Time the fuel injectors.
3. Adjust the governor gap.
4. Position the injector rack control lever.
5. Adjust the maximum no-load speed.
6. Make idle speed adjustment.
7. Make buffer speed adjustment.
8. On variable speed governor, adjust the throttle booster spring.
9. If fuel modulator is installed, adjust.

If a hydraulic governor is installed, the following tune-up adjustments should be made:

1. Adjust exhaust valve clearance.
2. Time fuel injectors.
3. Adjust fuel rod.
4. Correctly position the injector rack control lever.
5. Adjust the load limit screw.
6. On PSG governor, make compensation adjustment.
7. Adjust the speed droop.
8. Adjust maximum no-load speed.

Fig. 21-1. Adjusting valve clearance on Detroit diesel. (Detroit Diesel Allison Div., General Motors Corp.)

When adjusting valve clearance on the 71 in-line GMC engine, first clean the valve rocker cover and remove it. Then place the governor stop lever in the "no fuel" position. Crank the engine until the injector follower is fully depressed on the cylinder to be adjusted and loosen the push rod locknut, Fig. 21-1. On engines with two valves per cylinder, the clearance is measured between the valve stem and the rocker arm. In the case of an engine with four valve cylinder heads, the clearance is measured between the end of the valve stem and the valve bridge adjustment screw, Fig. 21-2, (spring-loaded bridge only), or between the valve bridge and the valve rocker arm pallet (unloaded bridge only).

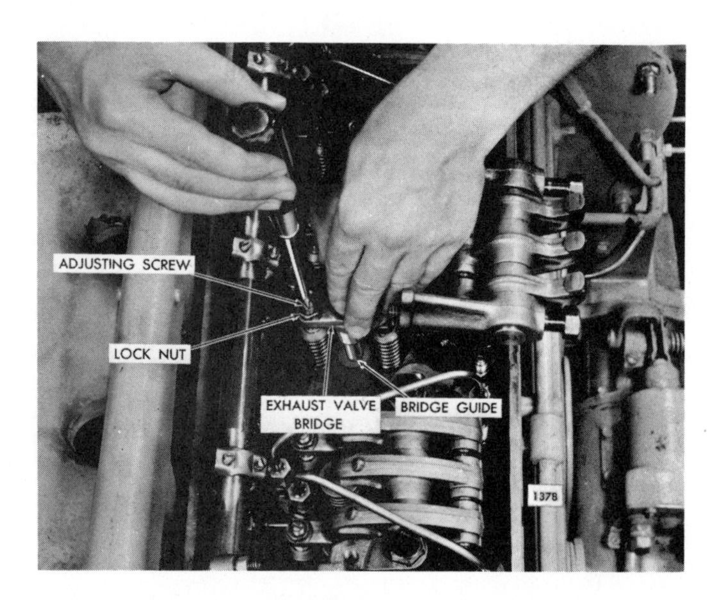

Fig. 21-2. On a Detroit diesel with a four valve cylinder head, the exhaust valve bridge assembly is adjusted and the adjustment screw locked securely after the cylinder head is installed. (Detroit Diesel Allison Div., General Motors Corp.)

Adjust the valves to the specified clearance. The foregoing instructions apply to a cold engine. The adjustment should be rechecked with the engine at operating temperature. Valve clearance of a hot engine differs from that of a cold engine. The specified clearance should be used in each instance.

TIMING FUEL INJECTORS

Timing the unit fuel injector on the GMC 71 in-line engine is an extremely important operation and must be done with precision, particularly if the engine is to meet government regulations concerning exhaust emissions. On the GMC 71 in-line engine, use the following procedure. The injector must be adjusted to the specified height in relation to the injector body. All the injectors can be timed in firing order sequence during one full revolution of the crankshaft on a two-cycle engine. Special gages are available and the procedure is as follows:

Clean and remove the rocker cover. Place the governor stop lever in the "no fuel" position. Rotate the crankshaft until the exhaust valves are fully depressed on the cylinder being timed.

The small end of the special injector timing gage is placed in the hole provided in the top of the injector body with the flat of the gage toward the injector follower, Fig. 21-3. Loosen the push rod locknut and turn the push rod. Adjust the injector rocker arm until the extended part of the gage will just pass over the top of the injector follower. Hold the push rod and tighten the locknut. Recheck adjustment.

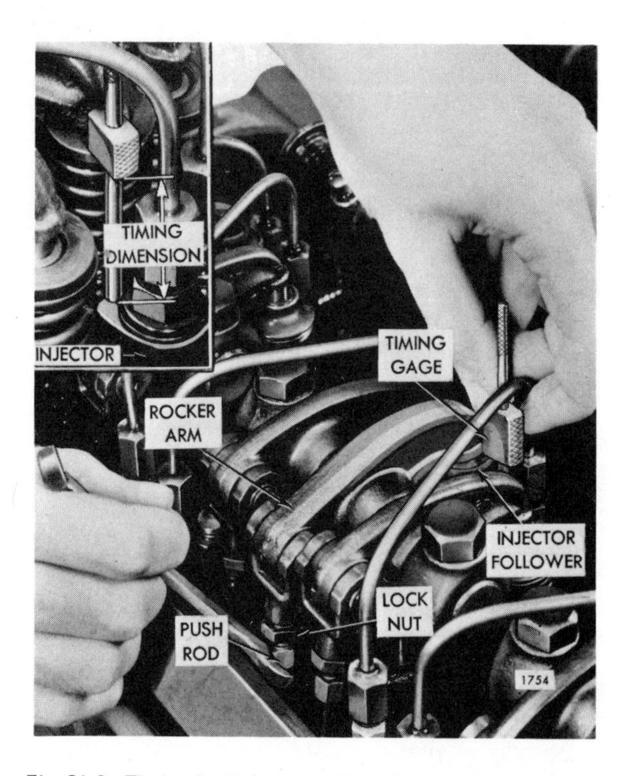

Fig. 21-3. Timing fuel injector on Detroit diesel 71 series engine. (Detroit Diesel Allison Div., General Motors Corp.)

CUMMINS TUNE-UP

The following instructions cover the in-vehicle adjustments of the Cummins PT fuel injection system as installed on their model N and NH engine. While applying particularly to the installation on Ford trucks, they will also serve as a guide to other installations.

FUEL PUMP

In-vehicle adjustments of the fuel pump are based on fuel manifold pressure and fuel supply vacuum. Determine these pressures by snap readings which provide a method of finding maximum fuel manifold pressure and fuel supply vacuum without a dynamometer. Obtain manifold pressure and fuel supply vacuum readings when the engine is at full-load and rated speed. When the engine is accelerated to full throttle, full engine load is momentarily attained when the engine reaches rated speed. Fuel pressure and vacuum gage readings increase with engine speed until rated speed is attained. At this point, the gage hand hesitates and falls off due to governor action at speeds above rated speed. It is necessary to read the maximum swing of the pressure gage hand quickly. The snap reading

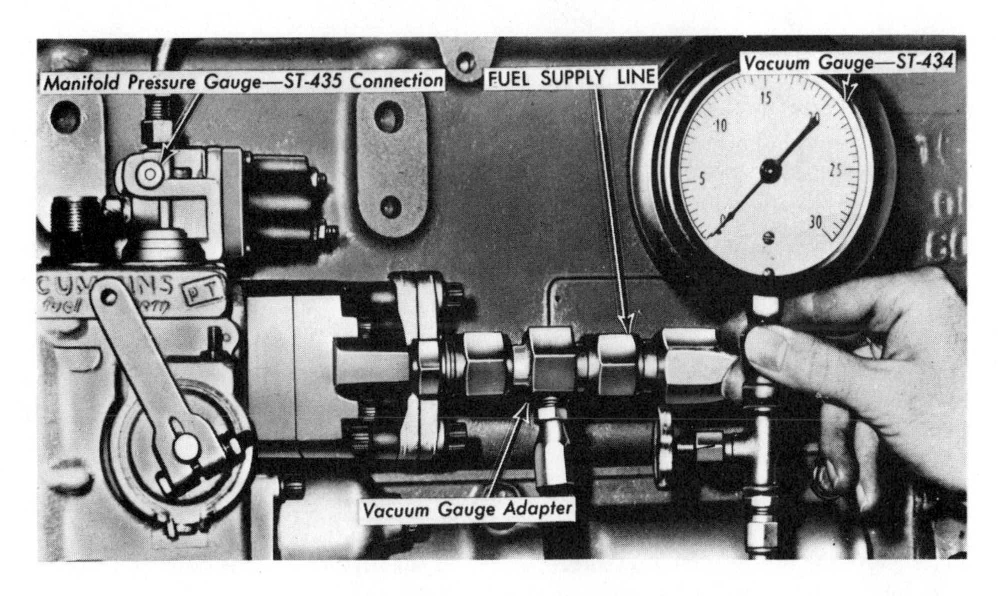

Fig. 21-4. Checking fuel manifold pressure and fuel supply vacuum on Cummins installation on Ford truck.

technique is not as accurate as measurements attained with a dynamometer.

Snap readings give an indication of fuel manifold pressure and fuel supply vacuum. For greater accuracy use the average of five snap readings.

Install a vacuum gage in the fuel supply line, Fig. 21-4. Check the suction restriction of the fuel supply line. If the vacuum exceeds 8 in. of mercury, determine the cause and correct. Connect a manifold pressure gage at the manifold pressure gage connection, Fig. 21-4. Check the manifold pressure at rated speed and full-load. If the manifold pressure is not within specifications, correct as necessary to obtain the correct specification.

The following adjustment procedures should be performed before adjusting the forward throttle screw and rear throttle screw. The forward throttle screw is closer to the drive end of the fuel pump, Fig. 21-5.

Operate the engine until the engine reaches 140 to 160 deg. F. Make all fuel pump or injector settings on a hot engine. Install a vacuum gage in the fuel supply system, Fig. 21-4. Check the suction restriction of the fuel supply line. If vacuum reading exceeds 8 in. of mercury determine the reason and correct. Adjust valve and injector lash to specifications. Check high idle or maximum governed speed.

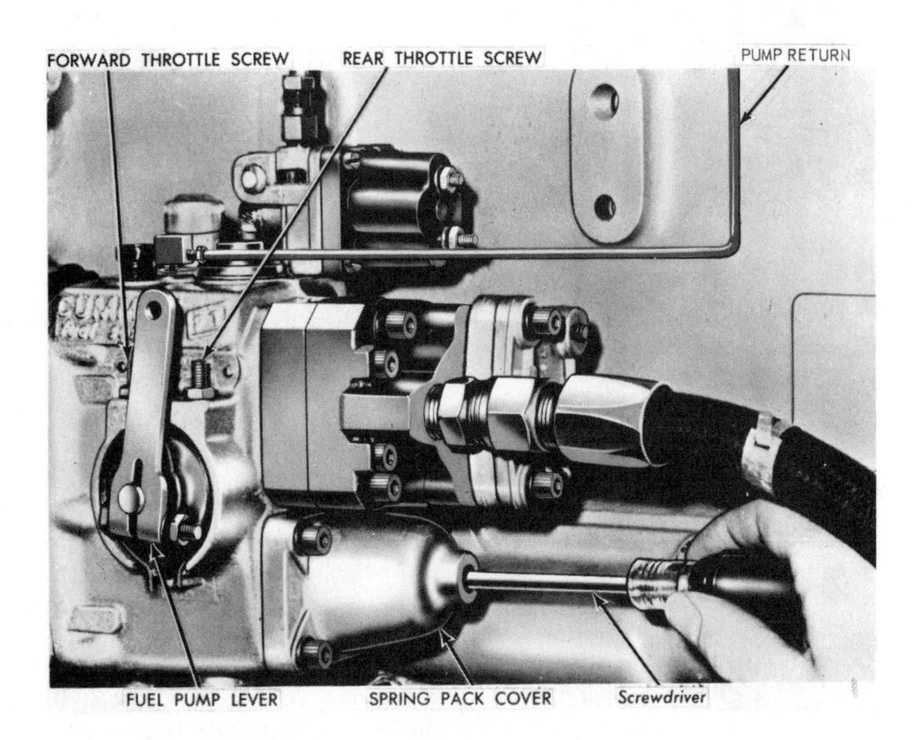

Fig. 21-5. Adjusting the governor screw on Cummins installation on Ford truck.

REAR THROTTLE SCREW ADJUSTMENT

Connect a manifold pressure gage at the manifold pressure gage connection, Fig. 21-4. Check the fuel manifold pressure at rated speed and full-load. Turn the rear throttle screw out until the highest fuel manifold pressure reading is obtained. Turn the rear throttle screw in to reduce the fuel manifold pressure 5 psi.

Do not increase the rear throttle adjustment more than 5 psi or the engine speed at which peak torque occurs will change. Throttle restriction over 5 psi allows unauthorized adjustment to raise fuel manifold pressure, thereby overfueling the engine.

FORWARD THROTTLE SCREW ADJUSTMENT

The adjustment of the forward throttle screw or throttle leakage adjustment is as follows: with the engine operating at idle speed, turn in the forward throttle screw until the engine gains speed. Back the screw out two full turns. Stop the engine and remove the spring pack cover plug and adjust the idle governor screw, Fig. 21-5, to idle the engine at specified speed. Operate the engine until it is purged of the air let in. Check the idle speed. Check the deceleration. If it is too slow, the throttle leakage is excessive. If the throttle faulters or dies when the idle speed is reached, the throttle leakage is too low. Adjust the screw to obtain permissible deceleration time. Any change in the desired idle speed is made with the governor idle screw. Follow the sequence under "Idle Speed Adjustments."

Fig. 21-6. Removing fuel pump screen.

IDLE SPEED ADJUSTMENTS

With the engine stopped, remove the pipe plug from the spring pack cover, Fig. 21-5. The idle adjustment screw is held in position with a spring clip. Turn the screw in to increase, or out to decrease the speed, Fig. 21-5. Replace the pipe plug. With the engine running, air which collected in spring pack housing will cause a rough idle temporarily. This will stabilize when the housing fills with oil.

FUEL FILTER ELEMENT

The fuel filter has disposable type element. Replace it at specified intervals, even though it may appear serviceable.

FUEL PUMP SCREEN

Remove the retainer cap from the fuel pump. Remove the spring and lift out the filter screen assembly, Fig. 21-6. Wash the screen and magnet in cleaning solvent and dry with compressed air. Install the top screen retainer. Install the filter screen, with the hole down in the fuel pump. Install spring and retainer cap. Torque retainer cap to the specified amount.

C.A.V. MODEL DPA INJECTION PUMP TUNE-UP

Adjust the idle speed stop screw on the C.A.V. model DPA fuel injection pump, Fig. 21-7, to obtain an engine speed of 450-550 rpm. Momentarily increase engine speed and recheck idle speed to ensure a consistent return to the desired idle speed. A cold engine with correct idling adjustment may stall, but it will run satisfactorily after about 30 seconds warm up. Do not increase the idling speed to compensate for this temporary condition.

MAXIMUM SPEED

With the engine at normal operating temperature, check and adjust the engine governed speed at no-load with a tachometer as follows:

With the engine operating and the transmission in neutral, depress the accelerator fully to hold the governor control lever against the maximum speed stop screw. Loosen the locknut and adjust the stop screw to obtain the desired speed. Tighten the locknut and seal the stop screw. With the governor control lever in the maximum speed position, the accelerator pedal should be against the pedal stop. Adjust the accelerator pedal as needed.

BLEEDING THE SYSTEM

If any part of the fuel system is disconnected, or air has entered the system, it will be necessary to remove all air from the fuel and to prime the injection pump by bleeding. Make sure that all fuel line connections are tight and there is sufficient fuel in the supply tank. Operate the lift pump priming lever. The filter will be bled automatically by the permanent bleed in the filter head. After installing a new element, loosen the plugged connection on the filter head and operate the lift pump priming lever until fuel flows from this connection to indicate the filter is primed. Tighten the plug. If the lift pump cam on the engine camshaft is on maximum lift, it will render the fuel lift pump priming lever inoperative. If this occurs, rotate the engine until the priming lever can be operated. Repeat this procedure at the injection pump inlet connection. After running out of fuel, additional priming of the injection pump may be required.

Loosen the bleed valve on the governor housing and operate

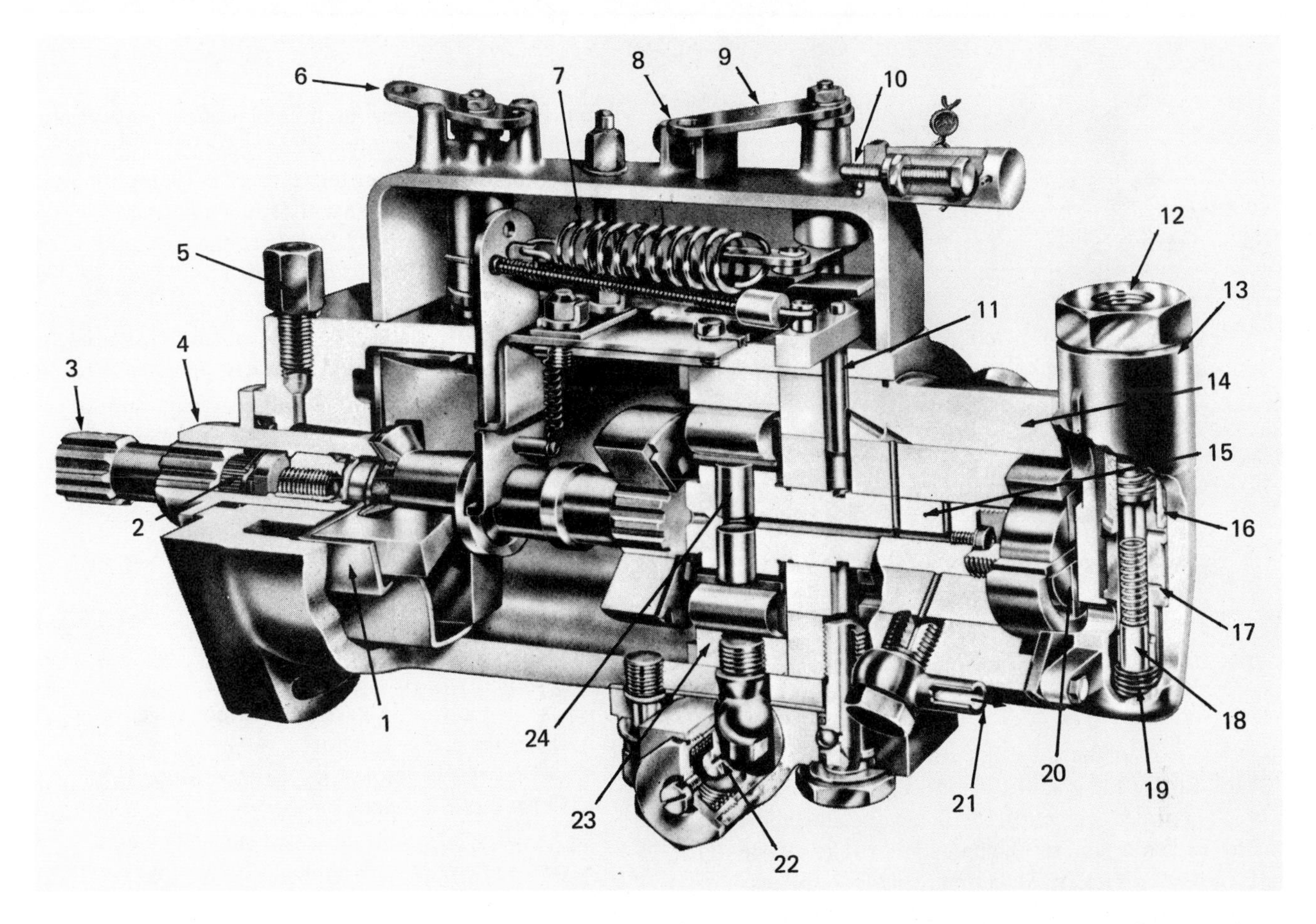

Fig. 21-7. Sectional view of C.A.V. distributor type fuel injection pump with automatic advance device. 1—Governor weights. 2—Drive hub securing screw. 3—Quill shaft. 4—Drive hub. 5—Back leak connection. 6—Shutoff lever. 7—Governor spring. 8—Idling stop. 9—Control lever. 10—Maximum speed stop. 11—Metering valve. 12—Fuel inlet. 13—End plate assembly. 14—Hydraulic head. 15—Rotor. 16—Nylon filter. 17—Regulating valve sleeve. 18—Regulating piston. 19—Priming spring. 20—Transfer pump. 21—To injector. 22—Advance device. 23—Cam ring. 24—Plungers.

the lift pump priming lever until the fuel flow is free of air. Tighten the governor housing bleed valve.

Unscrew any two injector inlet adapter seal nuts and loosen the corresponding inlet adpaters. Make sure the stop control is pushed in and, with the throttle lever in the full open position, operate the starter motor until the engine fires on the remaining cylinders. Tighten the injector inlet adapters and the seal nuts.

Wipe all surplus fuel from the injection pump and filter body. Be sure to expel all air from the system, or misfiring and erratic operation will result.

C.A.V. INJECTION PUMP TIMING

Timing marks are machined on the pump drive plate and cam ring. Since these are fixed, no timing adjustment is required. However, the injection pump should be set in the timed position to facilitate assembly on the engine.

Remove the inspection cover from the injection pump body. Turn the injection pump until the timing marks on the drive plate and cam ring are in alignment, Fig. 21-8. Remove the sealing bolt from the pump flange. Substitute the timing locking bolt, Fig. 21-9, for the sealing bolt and tighten to 25 in. lbs. Lock the pump in the timed position. The bracket attached to the locking bolt covers the two upper holes in the pump mounting flange to make sure of removal of the locking bolt when installing the pump on the engine. Install the inspection cover. Tighten the bolts and seal with a wire and lead seal. The timing locking bolt and bracket should not be removed until the injection pump is on the engine.

To install the injection pump, first locate it on the mounting plate and retain with the lower bolt, but do not tighten. Install the drive gear on the drive hub. Locate the

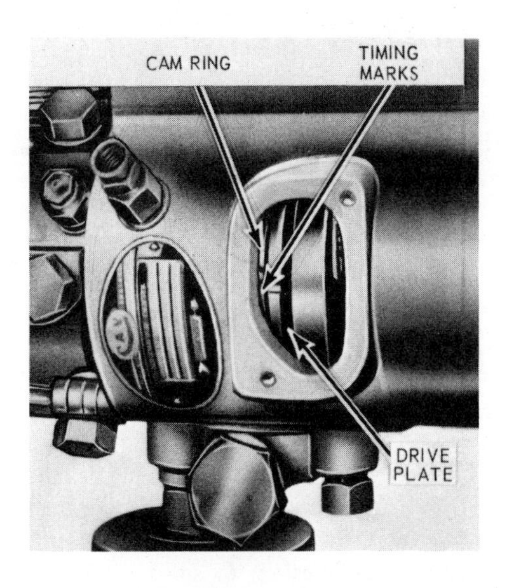

Fig. 21-8. Injection pump timing marks. C.A.V. injector pump.

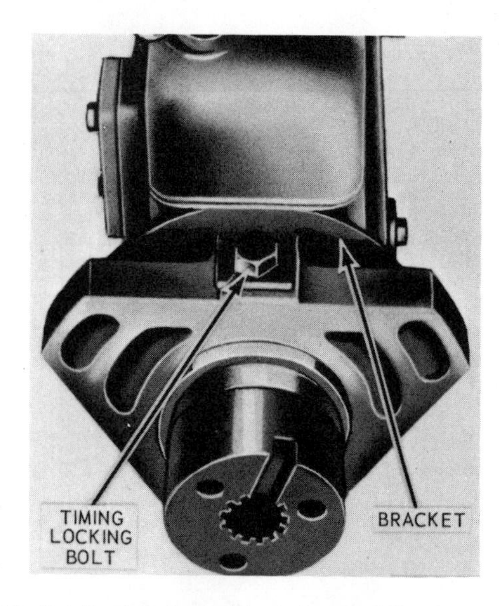

Fig. 21-9. Details of timing locking bolt on C.A.V. injector pump.

dowel in the slot in the drive hub and retain the gear with three bolts. Rotate the crankshaft until the timing mark is in line with the correct degree mark with No. 1 piston approaching top dead center on the compression stroke. When the engine is on the correct stroke, a timing mark on the rear face of the camshaft gear will be visible in the injection pump mounting opening. Install the injection pump mounting plate and gear assembly on the engine. Turn the injection pump clockwise on the mounting plate so that the single retaining bolt is at the end of the slotted hole in the flange to allow for the gear teeth helix. Locate a new "O" ring on the mounting plate and install the assembly on the engine. Secure the mounting plate with four bolts, one nut and lock washer. If the injection pump is not locked in the timed position, it will be necessary to remove the inspection cover on the pump body and assemble the pump to the engine so that the timing marks on the drive plate and cam ring are in alignment.

Install the injection pump on the mounting plate. Tighten one bolt finger tight to hold the pump in the timed position and remove the timing locking bolt and bracket. Install the two remaining bolts, spring and plain washer and tighten. Install a sealing bolt with a copper washer in the pump body flange. Connect the accelerator linkage and solenoid and connect the tubing which had been disconnected.

While the foregoing applies particularly to the C.A.V. installation on a Ford truck, it is typical of the installation of that unit on other equipment.

TROUBLE SHOOTING

Trouble Shooting a diesel engine has much in common with the same jobs performed in the familiar gasoline engine. It starts with information received through the complaints of the driver of the vehicle. This is followed by listening to the operation of the engine, driving the vehicle and then applying knowledge acquired through experience and study.

First of all is the question of smoke. Is there excessive smoking and under what condition does it occur? Are there any unusual noises? Does the trouble occur when accelerating, decelerating, under a steady pull or under idling conditions? Is the clutch engaged or disengaged? Is oil consumption normal? Is the specified fuel being used? Is fuel consumption normal, excessive or below normal? What was the operating temperature of the engine? Was the engine idled for 15 minutes before shutdown?

The following questions provide a guide for a beginning analysis of trouble shooting a diesel.

1. Is acceleration normal?
2. How does it start when cold?
3. How does it start when hot?
4. Is there any misfiring?
5. Under what condition does it misfire?
6. Is exhaust not normal?
7. Does the engine surge at any speed?
8. Is there any oil leakage?
9. Is there any coolant leakage?
10. Is there any fuel leakage?
11. Is there any air leakage from turbocharger or blower?
12. Does the engine run hot?
13. When was the last service work performed?
14. What work was done on the engine?
15. Under what conditions is the engine operated?
16. Is the operator new, skilled, trustworthy?
17. Are any knocks apparent?
18. Under what conditions are the knocks apparent?
19. What is operating temperature of oil, of bearings, of exhaust?
20. Is exhaust gas analysis normal?
21. Is there any restriction in air duct? Any leakage?
22. What is the compression pressure of various cylinders?
23. Has the brand of fuel been changed recently?

After the engine is disassembled, a lot can be learned from a detailed examination of the various parts. Much of this information has been detailed in the chapters on engine bearings, pistons, connecting rods, valves and piston rings.

Fig. 21-10. Checking compression pressure on a diesel engine. (Bacharach Instrument Co.)

The use of a pressure gage is basic, Fig. 21-10. Pressure gages used for checking compression pressures of diesel engines are calibrated up to 1000 psi and are provided with suitable fittings so they can be screwed into fuel injector openings in the cylinder head.

The manometer is another instrument that is used extensively in diesel shops. It has many uses in checking the operation of diesel engines since it can be used for measuring pressures in the engine crankcase and also the pressure in the exhaust system. It can be used for measuring both positive and negative pressures (vacuum).

Excessive crankcase pressure can result in severe lubricating oil leakage around the crankshaft seals. A negative pressure (less than atmospheric) will result in dust being drawn into the crankcase causing accelerated wear of engine parts.

A simple method of measuring such pressures can be devised by drilling and tapping a small hole in the crankcase to receive a brass fitting to which a plastic hose can be attached. This can be formed into a manometer. Bend the plastic tube into a "U" shape and clip to a wooden board as shown in Fig. 21-11. Water added to the tube should be at equal height when the engine is not operating. This point is marked as zero and then the balance is calibrated in inches and fractions thereof. When the engine is operated, pressure should be within limits of plus 1/8 to minus 1/4 in. This same manometer can be used to measure the pressure in the exhaust system.

The pyrometer is another instrument used to trouble shoot a diesel engine. This instrument is used to measure the temperature of the exhaust of the individual cylinders. Any variation in temperature from normal will indicate trouble.

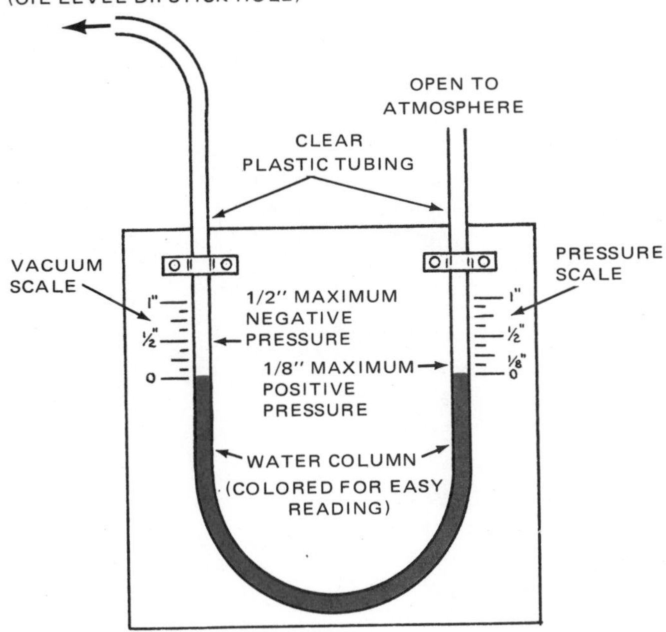

Fig. 21-11. Details of a manometer used to measure pressure in crankcase and exhaust system.

TROUBLE SHOOTING PERKINS DIESEL EQUIPPED WITH C.A.V. INJECTION SYSTEM

ENGINE WILL NOT CRANK

a. Starting motor at fault.
b. Weak battery.
c. High resistance in circuit.
d. Defective starting switch.
e. Internal seizure.
f. Engine oil too heavy.
g. Low ambient temperature.

ENGINE CRANKS BUT WILL NOT START

a. Injection pump faulty.
b. Air intake restricted.
c. Battery low in charge.

ENGINE DOES NOT DEVELOP FULL POWER

a. Moisture in fuel.
b. Fuel filter clogged.
c. Clogged air filter.
d. Poor fuel.
e. Injector pump not timed correctly.
f. Worn injector pump.
g. Air leaks around intake manifold.
h. Restricted exhaust system.
i. Low compression.
j. Faulty fuel supply pump.
k. Fuel systems needs bleeding.

EXHAUST SMOKE

a. Engine overloaded.
b. Wrong type lubricating oil.
c. Wrong type fuel.
d. Worn valve stems and guides.
e. Worn piston, rings and cylinder walls.
f. Defective fuel nozzle.
g. Excessive oil in air filter.
h. Injection pump not timed correctly.
i. Clogged air cleaner or air system.

ENGINE OVERHEATS

a. Insufficient coolant.
b. Cooling system clogged.
c. Fan belt slipping.
d. Poor air circulation through radiator.
e. Insufficient lubrication oil.
f. Engine overloaded.
g. Water pump defective.

HARD STARTING

a. Throttle not in starting position.
b. Energizing shut-off valve at blower partly closed.
c. Fuel tank empty.
d. Fuel pump needs priming.
e. Incorrect injector timing.
f. Incorrect rack setting.
g. Wrong type fuel.
h. Contaminated fuel.
i. Air in fuel system.
j. Compression low.
k. Defective injector.
l. Defective blower.

UNEVEN RUNNING

a. Incorrect injector timing.
b. Incorrect rack setting.

INSUFFICIENT FUEL

a. Governor adjustment incorrect.
b. Coolant temperature too low.
c. Leaking engine valves.

ONE OR MORE CYLINDERS CUTS OUT

a. Incorrect valve lash.
b. Faulty injector.
c. No fuel at injector.
d. Low compression.

AIR IN FUEL SYSTEM

a. Leaking fuel line connections.
b. Defective gaskets.

ENGINE STALLS

a. Low idling speed.
b. Operating temperature too low.
c. Cylinders misfiring.
d. Governor surging.
e. Clogged fuel filters
f. Defective injectors.
g. Governor maladjusted.
h. Incorrect linkage adjustment.
i. Air in fuel system.

LOSS OF POWER

a. Injector racks incorrectly set.
b. One or more cylinders cutting out.
c. Choked air cleaner.
d. Blowers not delivering air.
e. Fuel supply insufficient.
f. Choked fuel filter.
g. Injectors faulty.
h. Governor adjustment incorrect.
i. Low compression.
j. Fuel tank vent plugged.
k. Governor low speed spring gap incorrect.

BLACK SMOKE IN EXHAUST

a. Incorrect fuel for engine.
b. Injector timing incorrect.
c. Wrong type injector.
d. Insufficient air.
e. Blower not delivering enough air.

BLUE SMOKE IN EXHAUST

a. Injector rack set incorrectly.
b. Cylinder cutting out.
c. Engine lube oil reaching combustion chamber.
d. Oil leakage into blower.
e. Oil level too high in air filter.

DETONATION

a. Fuel oil in air intake.
b. Leaking injector.
c. Leaking fuel connection in cylinder head.
d. Oil pull over from air filter.
e. Leaking blower gasket or seals.
f. Plugged air box drains.

UNEVEN ENGINE OPERATION

a. Water in fuel.
b. Low fuel pressure.
c. Clogged air filter.

LOW STARTING RPM

a. Engine oil wrong viscosity.
b. Starting battery low in charge.
c. Low ambient temperature.
d. Defective starter or solenoid.

LOW ENGINE COMPRESSION

a. Leaking engine valves.
b. Incorrect valve lash.
c. Worn piston rings.
d. Defective head gasket.

LACK OF FUEL

a. Engine stop mechanism not released.
b. Accelerator linkage not operating correctly.
c. Low fuel supply.
d. Clogged vent in fuel tank cap.
e. Break in fuel supply line.
f. Water or ice in fuel supply line.
g. Clogged fuel filters.
h. Overflow valve defective.
i. Fuel line shut-off valve closed.

DETONATION

a. Incorrect injector timing.
b. Defective fuel nozzles.

UNEVEN RUNNING

a. Low idle speed.
b. Engine misfiring.
c. Sticking engine valves.
d. Engine temperature too high or too low.

LACK OF POWER

a. Incorrect valve lash.
b. Insufficient fuel.
c. Insufficient air.
d. Excessive exhaust back pressure.
e. Incorrect engine operating temperature.
f. Incorrect timing.
g. Incorrect calibration.

EXCESSIVE SMOKE – BLACK OR GREY

a. Insufficient air.
b. Exhaust back pressure.
c. Injection pump incorrectly timed.
d. Injection nozzles defective.
e. Wrong type of fuel.
f. Engine overheating.
g. Low engine compression.
h. Engine over fueled.
i. Worn cylinders, rings, pistons.

EXCESSIVE SMOKE – BLUE OR WHITE

a. Oil leak.
b. Oil in air cleaner too high.
c. Engine oil in crankcase too high.
d. Engine operating temperature too low.
e. Low compression.
f. Wrong type fuel.
g. Worn cylinders, pistons, rings.

TROUBLE SHOOTING CUMMINS ENGINE WITH PT FUEL PUMP

ENGINE WILL NOT START
a. Initial start-line and filter dry.
b. Tanks out of fuel.
c. Fuel shut-off valve closed.
d. Suction leaks.
e. Broken gear pump shaft.
f. Preheater not operating.
g. Protective black gasket under intake or exhaust connections.
h. Governor plunger in high idle position.
i. Broken pump plunger.
j. Excessive suction lift or lines drained back.
k. Batteries low.
l. Malfunction in safety shut-down equipment.
m. Pressure regulator cap "O" ring defective.
n. Pressure regulator plunger stuck open.
o. Governor weight carrier broken.
p. Wrong type pump installed.
q. Water in fuel.

HARD STARTING
a. Oil too heavy for weather conditions.
b. Substandard fuel.
c. High fuel suction lift and drain back.
d. Governor plunger stuck at main passage position.
e. Restricted air intake.
f. Worn gear pump.
g. Malfunction in automatic shutdown.
h. Loose regulator cap.
i. Shutdown valve leaking or open.
j. Poor ground on shutdown valve.

ENGINE STOPS
a. Broken wire in shut-off valve circuit.
b. Foreign objects in fuel tank.
c. Suction hose clogged.
d. Suction leaks.
e. Saddle tanks not equalizing.
f. Governor plunger stuck in overrun position.
g. Malfunction of automatic shutdown equipment.

ERRATIC LOSS OF POWER
a. Suction leaks in fuel lines.
b. Throttle opening not uniform.
c. Sticking injector.
d. Clogged fuel line.
e. Cracked pulsation damper diaphragm.
f. Floating obstruction in fuel tank.
g. Brakes slow to release.
h. Loose baffles in muffler.
i. Rag left in air cleaner basin.
j. Road gradient unrecognized by driver.
k. Seals leaking on type G PT pump.
l. Plugged air cleaner.
m. Collapsing air intake hose
n. Sticking pressure regulator.

SUDDEN LOSS OF POWER
a. Lining collapsed in fuel suction line.
b. Cracked flange or abraded hose in suction line.
c. Throttle lever slipped on shaft.
d. Sticking injectors.
e. Lack of air accompanied by heavy smoke.
f. Ice in fuel lines or filter.
g. Leaking pressure regulator cap.
h. Low rail pressure.
i. High rail pressure.
j. Throttle not opening.
k. Low weight assist setting.
l. Restriction in fuel passage in heads.
m. Too much oil in crankcase.
n. Suction leaks.
o. Aeration of fuel from leaking injector seal rings.

SLOW DECELERATION
a. Linkage sticking.
b. Throttle linkage set too high.
c. Throttle shaft and sleeve worn.
d. Suction leaks.
e. Restriction in injector drain line.
f. Injector drain orifice obstructed.
g. Fuel tank vent plugged.
h. Check balls leaking in cylindrical injectors.

ENGINE DIES DURING DECELERATION
a. Throttle linkage set too low.
b. Leaks in fuel suction lines.

ENGINE SURGES AT IDLE SPEED
a. PT (type R) fuel pump large idle spring out of seat.
b. PT (type G) fuel pump weight assist plunger stuck.
c. Air in fuel system.
d. Weight assist spring not functioning.
e. Idling on throttle linkage.
f. Cracked pulsation damper diaphragm.
g. Leaking check valve in fuel bypass line.

EXCESSIVE EXHAUST SMOKE
a. Injectors dirty or eroded.
b. Air cleaner dirty.
c. Turbocharger impeller dirty or bearing worn.
d. High temperature or altitude.
e. Poor fuel.
f. Torque rise set too high.

(CONTINUED)

g. Over fueling.
h. Over calibration.
i. Worn engine valves.
j. Improper injector cups.
k. Loose injectors.

HIGH FUEL CONSUMPTION

a. Rough driving technique.
b. Poor quality fuel.
c. Injectors worn or dirty.
d. Exhaust back pressure.
e. Brakes dragging.
f. Wind resistance or road gradient.
g. Fuel rate too high.
h. Excessive idling.
i. High load factor.
j. Improper gearing.

ENGINE MISFIRING

a. Dirty or eroded injectors.
b. Leaking or sticking valves.
c. Suction leaks.
d. Obstructed fuel manifold.
e. Foreign objects in fuel passages in cylinder head.
f. Broken crosshead.
g. Leaking injector ball check.
h. Cracked injector cup.
i. Sticking injector.
j. Plugged spray holes.
k. Plugged injector metering device.
l. Defective cup gasket.
m. Cold engine operation.
n. Low compression.
o. Damaged injector O rings.
p. Injector cup over torqued.

ENGINE WILL NOT SHUT DOWN

a. Manual shut-off valve in open position.
b. Restriction in injector drain system.
c. Injector drain discharge higher than engine.
d. Small suction leaks.
e. Poor plunger-to-cup seating.
f. Shut-down valve rubber seal deteriorated.
g. Check balls in cylindrical injectors leaking.

ERRATIC GOVERNOR ACTION

a. Low governed speed.
b. Varying governed speed.

HYDRAULIC LOCK-UP OF ENGINE

a. High fuel tank.
b. Fuel entering cylinder from injector return while parked.
c. Injector return line connected to fill neck on fuel tank.
d. Vehicle roll over in accident.

TROUBLE SHOOTING FUEL INJECTION SYSTEM IN VW RABBIT DIESEL ENGINE

Trouble-shooting guide for diesel fuel injection system with Robert Bosch distributor injection pump VE in VW Rabbit diesel

It is assumed that the engine is in good working order and properly tuned, and that the electrical system has been checked and repaired if necessary.

SYMPTOM

Starting Problem	Engine surges at idle	Rough idle when engine is warm	Engine misses under load	Low Power	Excessive Fuel Consumption	Engine cannot be shut off	Poor performance **or** black smoke **or** low power	Fog-like exhaust in full-load range (white or blue)	Incorrect idle or maximum speed	Engine does not rev up	Injection pump runs hot	CAUSE	REMEDY
●		●	●	●			●	●		●		Improper fuel (gasoline) in tank	Drain tank, flush system, fill with proper fuel
●	●		●	●				●		●		Tank empty **or** tank vent blocked	Fill tank/bleed system, check tank vent
●	●	●	●				●	●		●		Air in the fuel system	Bleed fuel system, eliminate air leaks
	●	●										Pump rear support bracket loose	Replace as necessary
●						●						Low voltage, no voltage **or** stop solenoid defective	Correct electrical faults/replace stop solenoid
●			●	●				●		●		Fuel filter blocked	Replace fuel filter
●			●	●			●	●				Injection lines blocked/restricted	Drill to nominal I.D. or replace
●			●	●				●		●		Fuel-supply lines blocked/restricted	Test all fuel supply lines – flush or replace
●			●	●								Loose connections, injection lines leak or broken	Tighten the connection, eliminate the leak
●			●					●		●		Paraffin deposit in fuel filter	Replace filter, use Diesel Fuel no. 1
●			●	●	●		●	●		●		Pump-to-engine timing incorrect	Readjust timing
		●	●	●	●		●					Injection nozzle defective	Repair or replace
				●	●		●					Engine air filter blocked	Replace air filter element
●												Pre-heating system defective	Test the glow plugs, replace as necessary
●		●		●	●		●					Injection sequence does not correspond to firing order	Install fuel injection lines in the correct order
					●				●			Low idle misadjusted	Readjust idle stop screw
				●					●			Maximum speed misadjusted	Readjust maximum speed screw
	●		●	●				●	●	●		Overflow fitting interchanged with inlet fitting	Install fittings in their proper positions
											●	Overflow blocked	Clean the orifice or replace fitting
●										●		Cold-start device not operating	Check bowden cable and lever movement
●		●		●			●			●		Low or uneven engine compression	Repair as necessary
●	●	●	●	●	●		●	●	●	●		Fuel injection pump defective or cannot be adjusted	Replace

TROUBLE SHOOTING
MACK DIESEL

Mack engines are equipped with American Bosch fuel injection systems. Before trouble shooting, make sure the following items have been checked and found to be in good operating condition.

1. All electrical connections are tight and in good condition and the batteries in fully charged condition.
2. Check the starter and starter switch making sure they are in good operating condition.
3. Make sure there is fuel in the supply tank and the system is primed.
4. Make sure the correct type and viscosity of engine oil is in the crankcase and is at the proper level.
5. Make sure there is no restriction in the air intake system.
6. Make sure the engine cooling system is in good condition.
7. Make sure that fuel line connections are tight and fuel is being delivered to the engine.
8. Make sure all gages and meters read correctly.

Check the following trouble shooting procedure as conditions exist.

ENGINE CRANKS – WILL NOT START

a. Fuel pressure normal.
b. Emergency shut-off cock closed or partly closed.
c. Air intake manifold valve closed or partly closed.
d. Low ambient temperature, use starting aid.
e. Governor throttle shaft linkage binding or sticking governor parts.
f. Injection pump rack stuck.
g. Hydraulic governor faulty.
h. Internal adjustment of injection pump incorrect.
i. Injection pump shaft bearings worn.
j. Loose injection pump drive assembly, timing incorrect.

FUEL PRESSURE LOW

a. Air intake manifold closed or partly closed.
b. Fuel tank vent restricted.
c. Fuel supply pump inoperative.
d. Air leaks or restriction on suction side of fuel system.
e. Fuel filters clogged.
f. Fuel tank or filters cracked, drain plugs loose.
g. Filter air bleed too large or missing.
h. Filter air bleed plugged.
i. Injection pump overflow valve leaking, stuck open or closed.
j. Broken governor throttle spring or linkage disengaged.

ENGINE STARTS, ERRATIC FIRING, ALL CYLINDERS

a. Low fuel pressure.
b. Emergency shut-off cock closed or partly closed.
c. Fuel tank vent restricted.
d. Fuel supply pump inoperative.
e. Air leaks or restrictions on suction side of fuel system.
f. Fuel filters clogged.
g. Fuel filters cracked, drain plugs loose.
h. Filter air bleed too large or missing, air bleed plugged.
i. Injection pump overflow valve leaking, stuck open or closed.

ERRATIC OPERATION AT IDLE AND HIGH SPEEDS
FUEL PRESSURE ADEQUATE – SMOKE

a. Intake manifold valve closed or partly closed.
b. Restricted air intake, clogged air cleaner on turbocharger.
c. Injection pump internal adjustment incorrect.
d. Worn injection pump shaft bearings.
e. Injection pump drive assembly loose, timing incorrect.
f. Low engine compression.
g. Filter air bleed clogged.
h. Faulty nozzle spray.
i. Nozzle cap nut incorrectly torqued.
j. Nozzle pintle stuck.
k. Bent or broken spindle or spring.
l. Nozzle improperly installed in cylinder head.
m. Blow-by at crankcase breather.
n. Incorrect valve lash, camshaft or lifters worn.
o. Defective valve springs.
p. Defective valves or valve guides.

ERRATIC OPERATION AT IDLE AND HIGH SPEEDS,
FUEL PRESSURE ADEQUATE – NO SMOKE

a. Fuel tank vent restricted.
b. Fuel supply pump defective.
c. Air leaks or restriction on suction side of fuel system.
d. Fuel filters clogged.
e. Fuel tank or filters cracked or drain plug loose.
f. Filter air bleed inoperative.
g. Governor throttle shaft linkage binding, sticking governor parts.
h. Improper setting of governor throttle linkage.
i. Governor drive assembly binding.
j. Broken or weak speeder springs.
k. Solenoid stuck.
l. Loose governor ball-head drive assembly, excess ball-head tang clearance on hydraulic governor.
m. Improper rack, gap, cam angle or cam plate setting on mechanical governor.
n. Incorrect speed droop setting, or length of speed follower spring on hydraulic governor.
o. Injection pump overflow valve leaking, stuck open or closed.
p. Injection pump plunger stuck, rack stuck, pump plungers worn.

(CONTINUED)

q. Injection pump internal adjustment incorrect.
r. Worn injection shaft bearings.
s. Loose injection pump drive assembly, timing incorrect.
t. Leaking cylinder head gasket.

ENGINE MISSING ON ONE OR MORE CYLINDERS – AT IDLE SPEED ONLY

a. Improper setting of governor idle speed.
b. Low compression.
c. Delivery valve stuck or leaking.
d. Faulty nozzle spray.
e. Nozzle cap nut improperly torqued.
f. Nozzle pintle stuck.
g. Bent or broken spindle or spring.
h. Nozzle improperly installed in head.
i. Injection pump faulty.
j. Injection pump rack stuck.

MISSING AT HIGH AND IDLE SPEED – SMOKE

a. Injection pump internal adjustment incorrect.
b. Worn injection pump shaft bearings.
c. Loose injection pump drive assembly, timing incorrect.
d. Faulty spray nozzle, opening incorrect.
e. Nozzle cap incorrectly torqued.
f. Nozzle pintle stuck.
g. Bent or broken pintle or spring.
h. Nozzle incorrectly installed in cylinder head.
i. Low compression.
j. Leaking cylinder head gasket.
k. Incorrect tappet clearance.
l. Worn camshaft or lifters.
m. Defective valve springs.
n. Valves sticking or leaking.

MISSING AT HIGH AND IDLE SPEED – NO SMOKE

a. Emergency shut-off cock closed or partly closed.
b. Fuel tank vent restricted.
c. Fuel supply pump defective.
d. Air leaks or restriction on suction side of fuel system.
e. Fuel filters restricted.
f. Injection pump overflow valve closed or defective.
g. Nozzle cap nut incorrectly torqued.
h. Nozzle incorrectly installed in cylinder head.
i. Injection pump defective.
j. Injection pump incorrectly adjusted.
k. Incorrect valve lash.
l. Camshaft worn, lifters worn.
m. Valves sticking.
n. Leaking cylinder head gasket.

ENGINE STALLS – LOW FUEL PRESSURE

a. Supply tank vent restricted.
b. Supply pump faulty.
c. Air leaks on suction side of fuel system.
d. Fuel filters clogged.
e. Filter air bleed defective.
f. Injection pump overflow valve defective.
g. Air in governor oil feed.

ENGINE STALLS – GOOD FUEL PRESSURE – NO SMOKE

a. Incorrect setting of governor idle speed.
b. Governor throttle shaft linkage binding or incorrectly set.
c. Governor drive assembly binding.
d. Defective speeder springs.
e. Solenoid stuck.
f. Mechanical governor setting faulty.
g. Air in governor oil feed.
h. Restriction in lubricating oil system.
i. Injection pump internal setting incorrect.
j. Injection pump plunger stuck, pump rack stuck, plungers worn.

ENGINE STALLS – BLACK SMOKE

a. Governor throttle shaft linkage not functioning correctly.
b. Turbocharger passing oil.
c. Restricted air filter, nozzle ring restricted, binding impeller.
d. Injector pump internal adjustments incorrect.
e. Incorrect valve lash, worn engine cams, worn lifters, sticky valves, weak valve springs.
f. Faulty nozzle spray, opening pressure incorrect.
g. Nozzle pintle stuck, bent or broken spindle.

ENGINE SURGES

a. Incorrect setting of governor idle speed.
b. Air leaks or restriction on suction side of fuel system.
c. Air bleed faulty.
d. Air in governor oil feed.
e. Governor throttle shaft linkage binding, broken or weak speeder springs.
f. Loose governor ball-head drive.
g. Excess ball-head tang clearance (hydraulic governor).
h. Worn sliding sleeve, flyweight fingers, linkage or excessive end play, (mechanical governor).
i. Injection pump internal adjustment faulty, worn pump shaft bearing.
j. Incorrect timing.
k. Injection pump plungers stuck or worn.

ENGINE WILL NOT ATTAIN NO-LOAD GOVERNED RPM

a. Emergency shut-off cock closed or partly closed.
b. Air intake manifold valve partly closed.
c. Fuel tank vent restricted.

d. Low ambient air temperature.
e. Incorrect setting of governor or high speed stop.
f. Governor throttle shaft linkage faulty.
g. Incorrect rack, gap, cam angle or cam plate setting (mechanical governor).
h. Worn mechanical governor.
i. Pressure relief valve spring broken or relief valve stuck open (hydraulic governor).
j. Air leaks or restrictions in suction side of fuel system.
k. Fuel filters clogged, filter air bleed faulty.
l. Fuel supply pump faulty.
m. Injection pump internal adjustment incorrect.
n. Injection pump worn.

ENGINE OVERSPEEDS

a. Incorrect setting of governor or high speed stop.
b. Faulty governor.
c. Injection pump rack stuck.
d. Turbocharger passing oil.

ENGINE LACKS POWER – NO SMOKE

a. Emergency shut-off cock partly closed.
b. Air intake manifold valve partly closed.
c. Low ambient air temperature.
d. Air leaks or restriction on suction side of fuel system.
e. Fuel not reaching engine sufficiently.
f. Leaking cylinder head gaskets.
g. Injection pump plunger stuck.
h. Injection pump internal adjustment incorrect, worn pump shaft bearings, timing incorrect.
i. Incorrect setting of governor.
j. Governor shaft linkage binding.
k. Incorrect speed droop or length of speed follower spring.
l. Improper rack, gap, cam angle or cam plate setting (mechanical governor).
m. Air in hydraulic governor oil feed.
n. Restricted turbocharger air intake, clogged air cleaner, nozzle ring restricted, faulty turbocharger impeller.

ENGINE LACKS POWER – BLACK OR BROWN SMOKE

a. Emergency shut-off cock closed or partly open.
b. Air intake manifold valve closed or partly open.
c. Fuel tank vent partly closed.
d. Low ambient air temperature.
e. Defective spray nozzle, leaking, opening pressure incorrect, nozzle nut incorrectly torqued, nozzle pintle stuck, bent or broken spindle, nozzle incorrectly installed in cylinder head.
f. Injection pump internal adjustment faulty, timing incorrect.
g. Incorrect rack, gap, cam angle or cam plate setting (mechanical governor).
h. Low engine compression.
i. Incorrect valve lash, engine cams or lifters worn, defective valve springs, engine valves leaking.
j. Leaking cylinder head gasket.
k. Excess oil passing pistons.
l. Restriction in oil lube system.
m. Turbocharger passing oil.
n. Restricted air intake to turbocharger.

ENGINE LACKS POWER – BLUE OR WHITE SMOKE

a. Leaking cylinder head gasket.
b. Excessive oil passing pistons.
c. Turbocharger passing oil.
d. Restricted supply of air to turbocharger.
e. Slow charger rpm.
f. Water leak in combustion chamber.

ENGINE WILL NOT SHUT DOWN

a. Governor throttle shaft linkage binding, solenoid stuck.
b. Incorrect speed droop or length of speed follower spring (hydraulic governor).
c. Faulty mechanical governor.
d. Injection pump plunger stuck, pump rack stuck.

ENGINE OVERHEATS

a. Leaking cylinder head gasket.
b. Restricted air intake to turbocharger, clogged air cleaner.
c. Injection pump internal adjustment incorrect.
d. Nozzle pintle stuck.
e. Turbocharger passing oil.
f. Excessive fuel consumption.
g. Injection pump internal adjustment incorrect.
h. Air to turbocharger restricted.
i. Low compression.
j. Incorrect setting of governor or high speed stop.
k. Clogged air cleaner.
l. Overloaded vehicle.
m. Excessive speeds of vehicle.

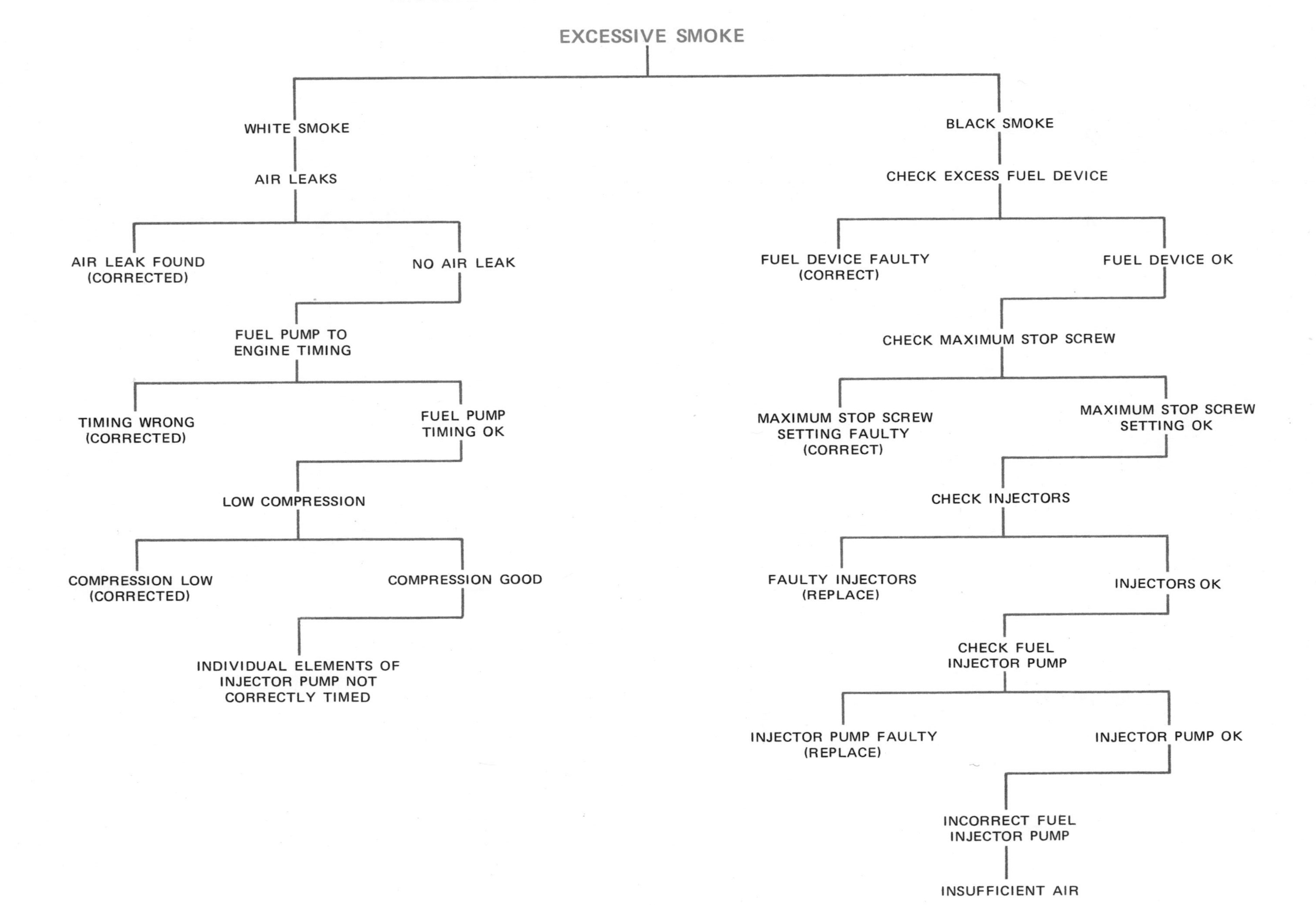

TROUBLE SHOOTING CHARTS FOR FUEL INJECTORS
EXCESSIVE SMOKE
WHITE SMOKE
AIR LEAKS
AIR LEAK FOUND (CORRECTED)
NO AIR LEAK
FUEL PUMP TO ENGINE TIMING
TIMING WRONG (CORRECTED)
FUEL PUMP TIMING OK
LOW COMPRESSION
COMPRESSION LOW (CORRECTED)
COMPRESSION GOOD
INDIVIDUAL ELEMENTS OF INJECTOR PUMP NOT CORRECTLY TIMED
BLACK SMOKE
CHECK EXCESS FUEL DEVICE
FUEL DEVICE FAULTY (CORRECT)
FUEL DEVICE OK
CHECK MAXIMUM STOP SCREW
MAXIMUM STOP SCREW SETTING FAULTY (CORRECT)
MAXIMUM STOP SCREW SETTING OK
CHECK INJECTORS
FAULTY INJECTORS (REPLACE)
INJECTORS OK
CHECK FUEL INJECTOR PUMP
INJECTOR PUMP FAULTY (REPLACE)
INJECTOR PUMP OK
INCORRECT FUEL INJECTOR PUMP
INSUFFICIENT AIR

TROUBLE SHOOTING CHART

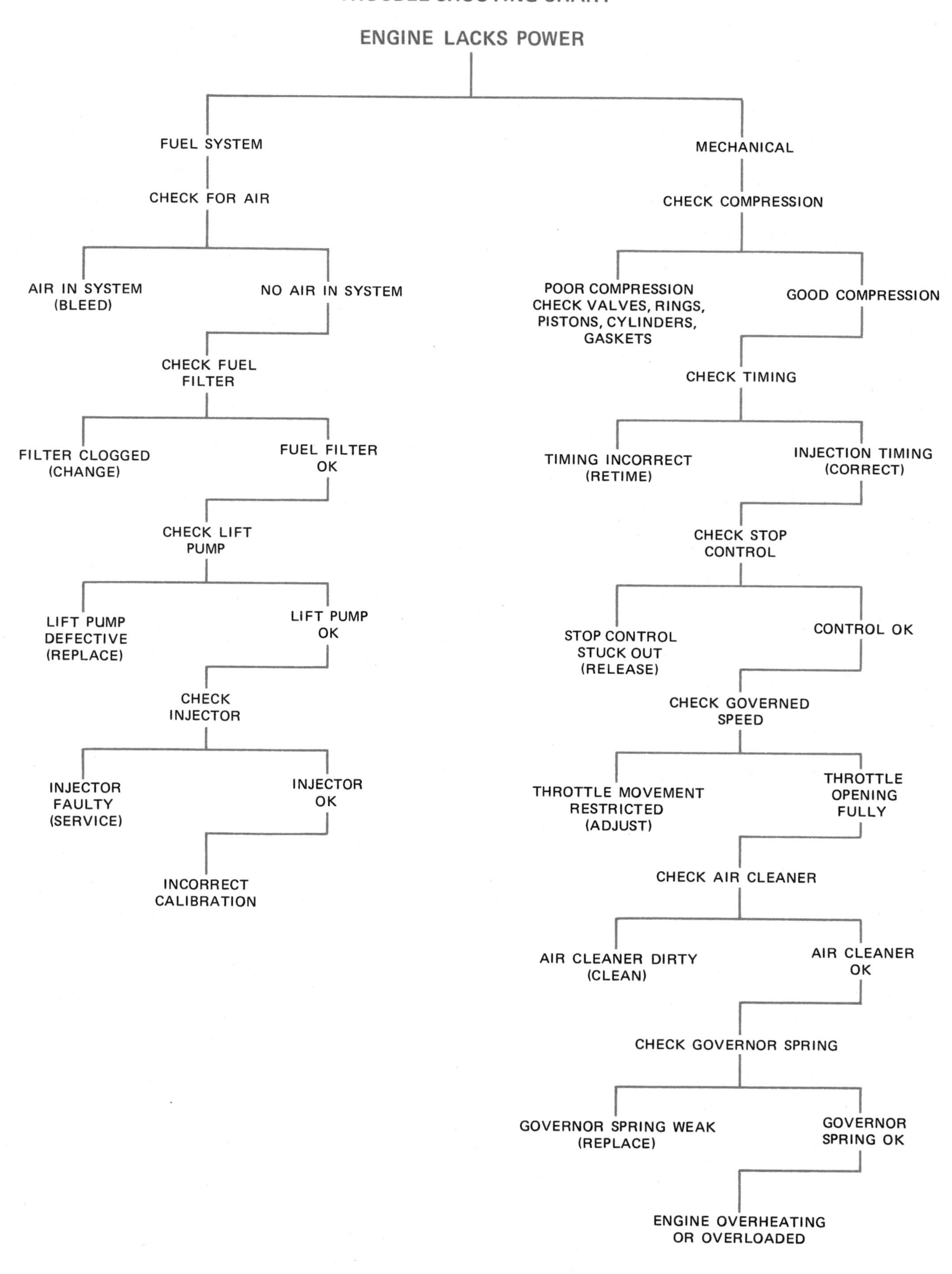

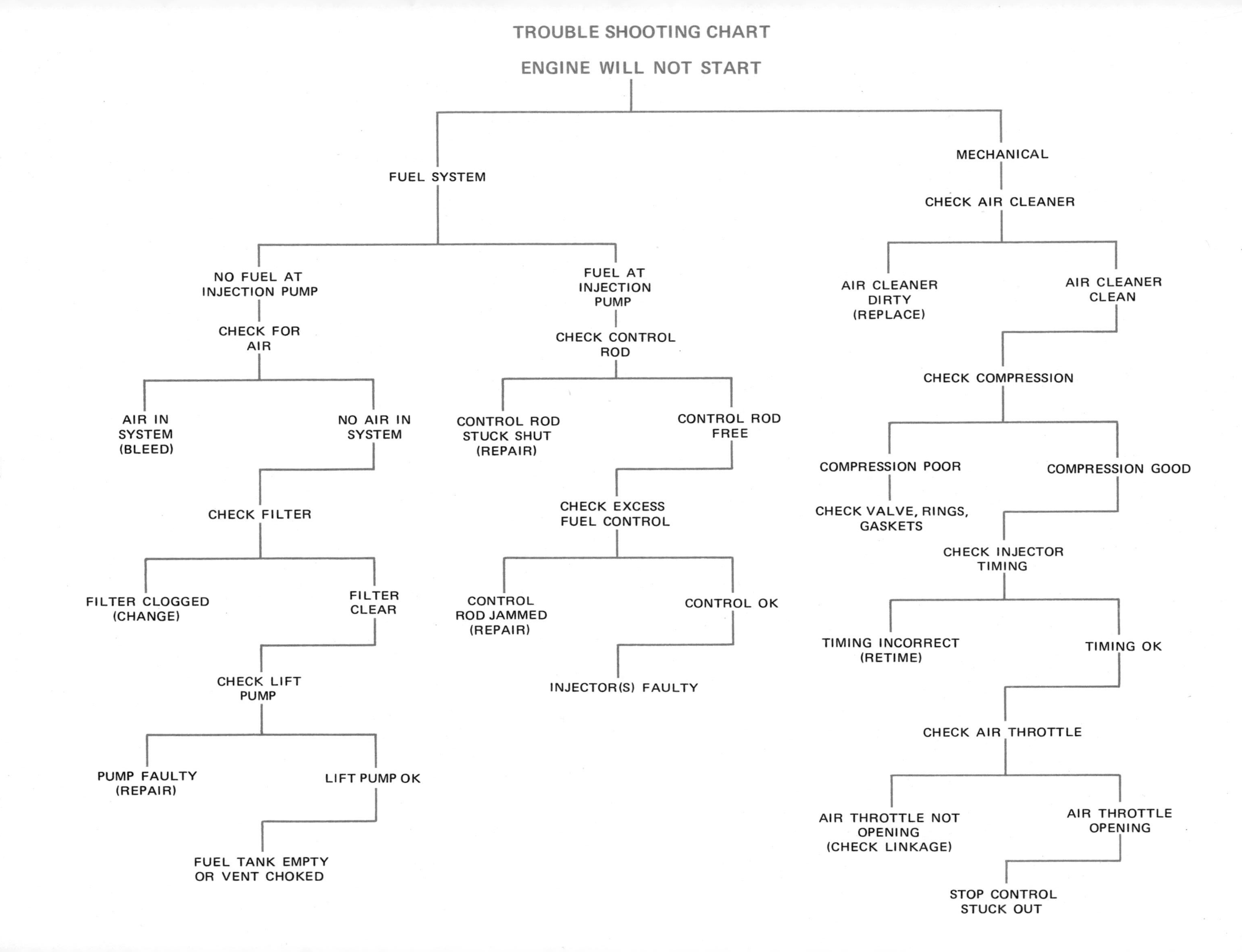
TROUBLE SHOOTING CHART
ENGINE WILL NOT START
FUEL SYSTEM
NO FUEL AT INJECTION PUMP
CHECK FOR AIR
AIR IN SYSTEM (BLEED)
NO AIR IN SYSTEM
CHECK FILTER
FILTER CLOGGED (CHANGE)
FILTER CLEAR
CHECK LIFT PUMP
PUMP FAULTY (REPAIR)
LIFT PUMP OK
FUEL TANK EMPTY OR VENT CHOKED
FUEL AT INJECTION PUMP
CHECK CONTROL ROD
CONTROL ROD STUCK SHUT (REPAIR)
CONTROL ROD FREE
CHECK EXCESS FUEL CONTROL
CONTROL ROD JAMMED (REPAIR)
CONTROL OK
INJECTOR(S) FAULTY
MECHANICAL
CHECK AIR CLEANER
AIR CLEANER DIRTY (REPLACE)
AIR CLEANER CLEAN
CHECK COMPRESSION
COMPRESSION POOR
CHECK VALVE, RINGS, GASKETS
COMPRESSION GOOD
CHECK INJECTOR TIMING
TIMING INCORRECT (RETIME)
TIMING OK
CHECK AIR THROTTLE
AIR THROTTLE NOT OPENING (CHECK LINKAGE)
AIR THROTTLE OPENING
STOP CONTROL STUCK OUT

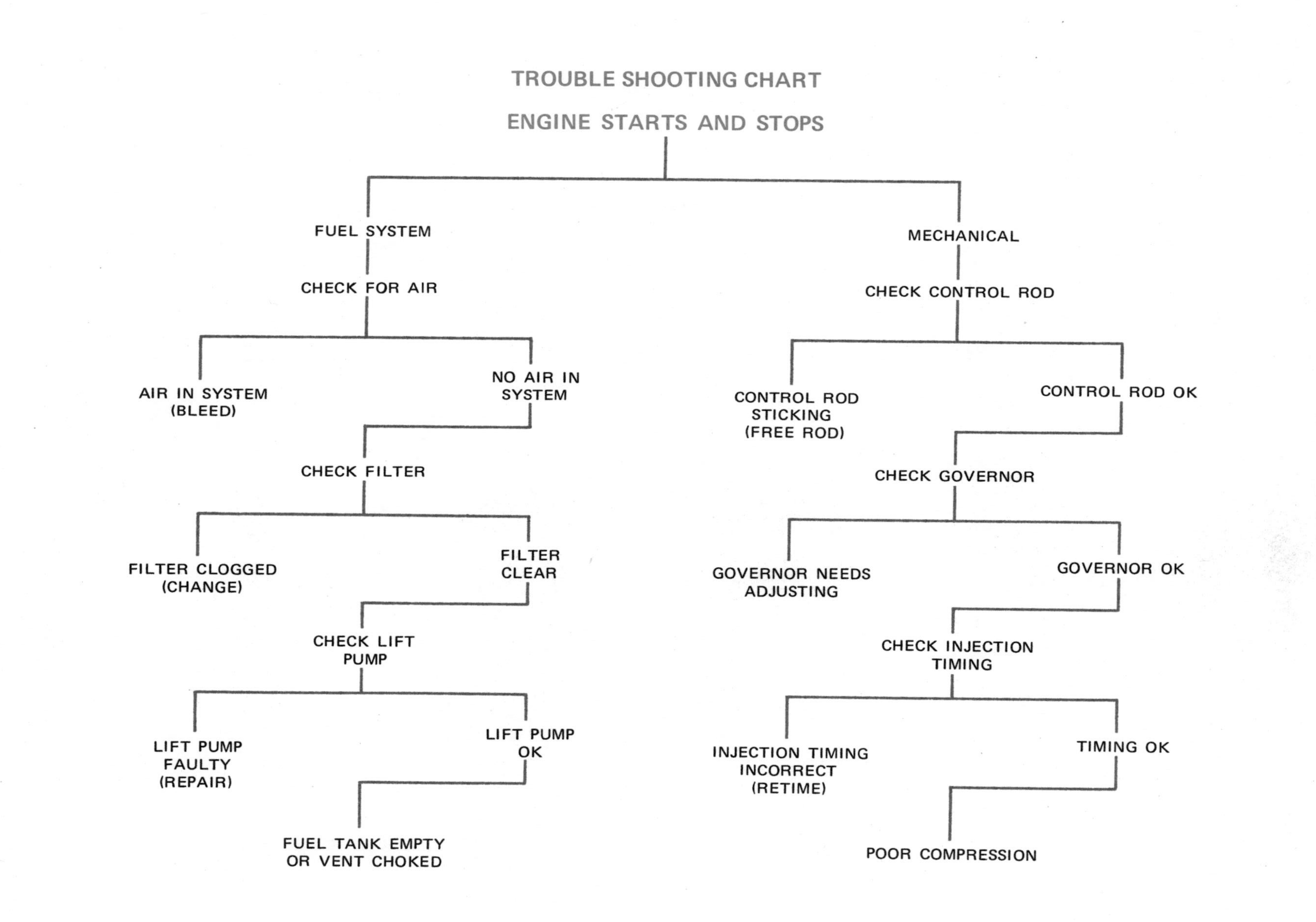
TROUBLE SHOOTING CHART
ENGINE STARTS AND STOPS
FUEL SYSTEM
CHECK FOR AIR
AIR IN SYSTEM (BLEED)
NO AIR IN SYSTEM
CHECK FILTER
FILTER CLOGGED (CHANGE)
FILTER CLEAR
CHECK LIFT PUMP
LIFT PUMP FAULTY (REPAIR)
LIFT PUMP OK
FUEL TANK EMPTY OR VENT CHOKED
MECHANICAL
CHECK CONTROL ROD
CONTROL ROD STICKING (FREE ROD)
CONTROL ROD OK
CHECK GOVERNOR
GOVERNOR NEEDS ADJUSTING
GOVERNOR OK
CHECK INJECTION TIMING
INJECTION TIMING INCORRECT (RETIME)
TIMING OK
POOR COMPRESSION

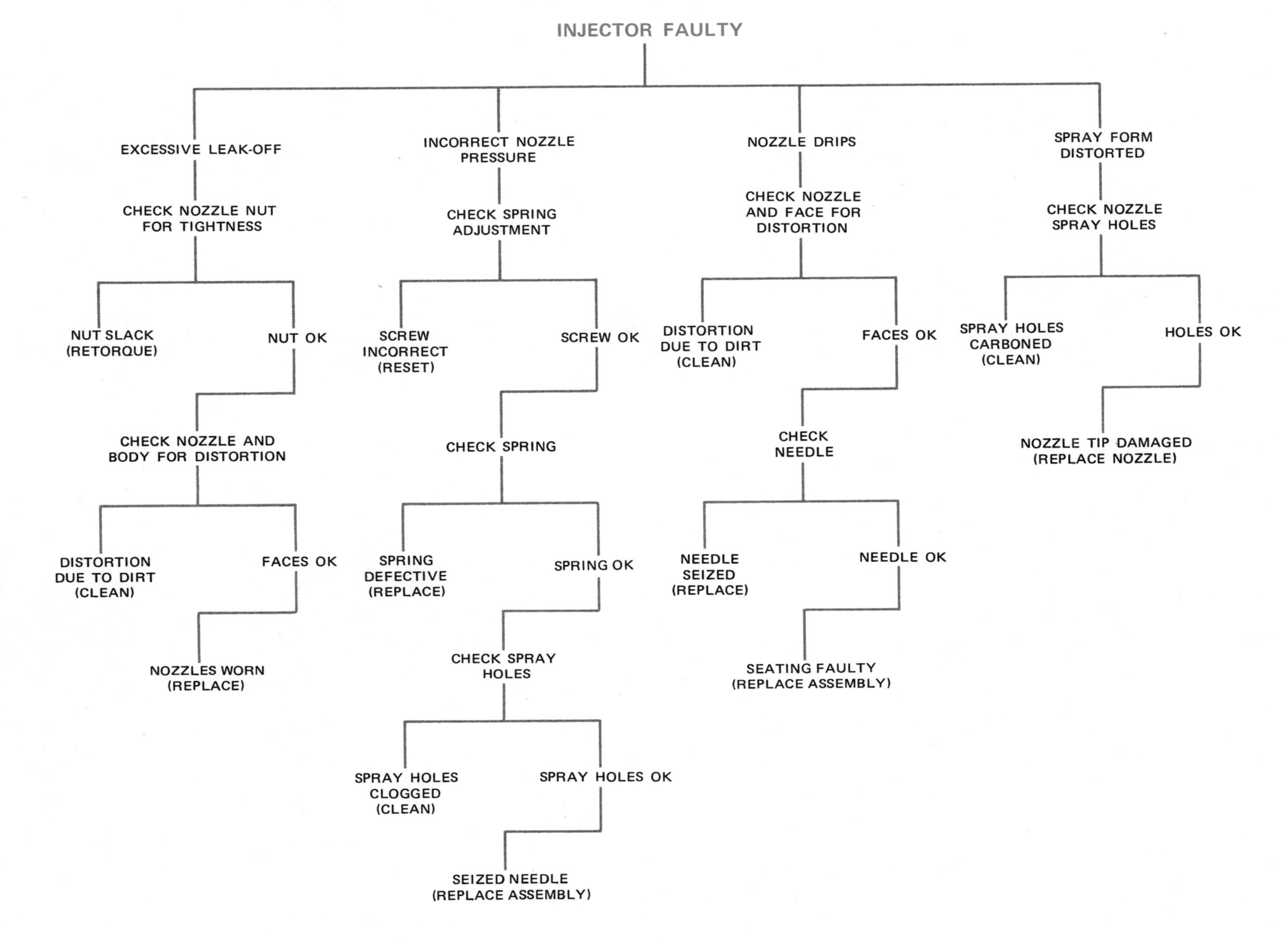
TROUBLE SHOOTING CHART
INJECTOR FAULTY
EXCESSIVE LEAK-OFF
CHECK NOZZLE NUT FOR TIGHTNESS
NUT SLACK (RETORQUE)
NUT OK
CHECK NOZZLE AND BODY FOR DISTORTION
DISTORTION DUE TO DIRT (CLEAN)
FACES OK
NOZZLES WORN (REPLACE)
INCORRECT NOZZLE PRESSURE
CHECK SPRING ADJUSTMENT
SCREW INCORRECT (RESET)
SCREW OK
CHECK SPRING
SPRING DEFECTIVE (REPLACE)
SPRING OK
CHECK SPRAY HOLES
SPRAY HOLES CLOGGED (CLEAN)
SPRAY HOLES OK
SEIZED NEEDLE (REPLACE ASSEMBLY)
NOZZLE DRIPS
CHECK NOZZLE AND FACE FOR DISTORTION
DISTORTION DUE TO DIRT (CLEAN)
FACES OK
CHECK NEEDLE
NEEDLE SEIZED (REPLACE)
NEEDLE OK
SEATING FAULTY (REPLACE ASSEMBLY)
SPRAY FORM DISTORTED
CHECK NOZZLE SPRAY HOLES
SPRAY HOLES CARBONED (CLEAN)
HOLES OK
NOZZLE TIP DAMAGED (REPLACE NOZZLE)

TROUBLE SHOOTING CHART

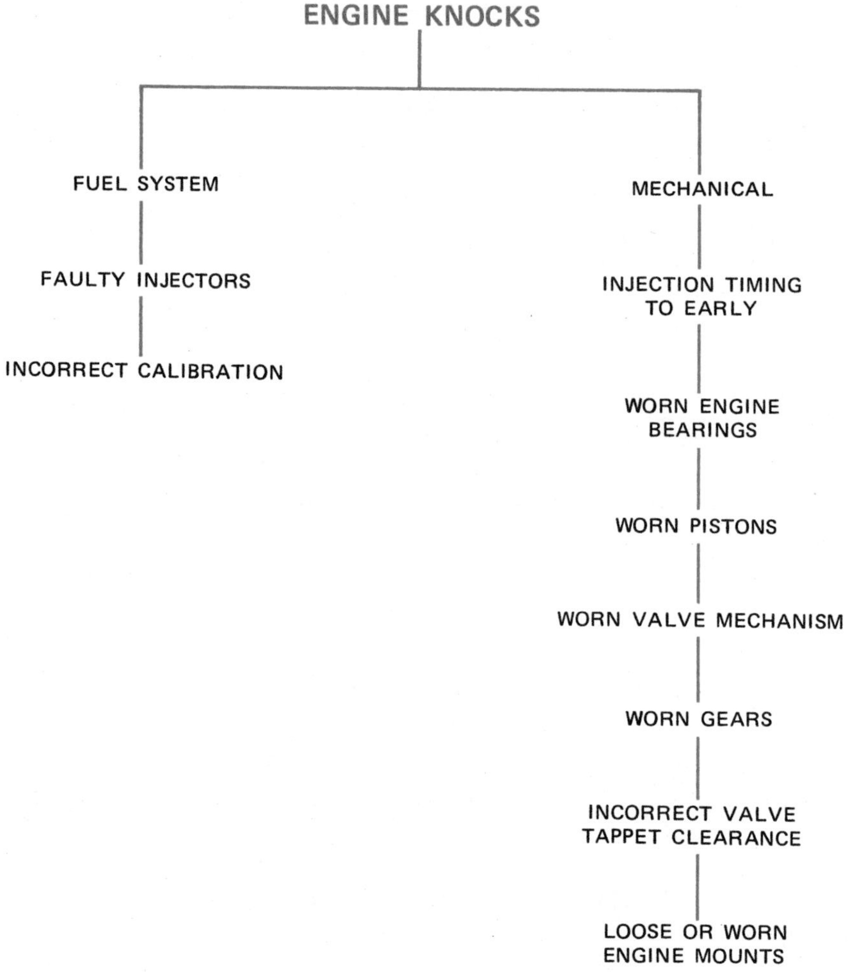

ENGINE KNOCKS

- FUEL SYSTEM
  - FAULTY INJECTORS
  - INCORRECT CALIBRATION
- MECHANICAL
  - INJECTION TIMING TO EARLY
  - WORN ENGINE BEARINGS
  - WORN PISTONS
  - WORN VALVE MECHANISM
  - WORN GEARS
  - INCORRECT VALVE TAPPET CLEARANCE
  - LOOSE OR WORN ENGINE MOUNTS

REVIEW QUESTIONS – CHAPTER 21

1. Tune-up procedure on two-cycle and four-cycle engines is ________ (the same/different).
2. The type of governor used on a diesel injection system (does/does not) affect the tune-up procedure.
3. On a two-cycle engine with a unit injector, the injector can be timed in firing order sequence in how many revolutions of the engine?
   a. One.
   b. Two.
   c. Three.
   d. Four.
4. On a Cummins engine should the engine be hot or cold when setting the fuel pump and injector?
   a. Hot.
   b. Cold.
5. A manometer is used to measure the ____________ in an engine crankcase.
6. Governors used on the 71 in-line engines include:
   a.________________________________.
   b.________________________________.
   c.________________________________.
   d.________________________________.
7. A pyrometer is used on a diesel engine to:
   a. Measure engine speed.
   b. Measure temperature of exhaust.
   c. Measure pressure developed in turbocharger.
8. Excessive crankcase pressure can result in lubricating oil leakage around the _____________ seals.
9. Timing marks are machined on the C.A.V. injection pump drive plate and_____________.
10. Fuel pressure and vacuum gage readings____________ with engine speed until rated speed is attained.

Chapter 22
DIESEL EXHAUST EMISSIONS

As a result of steadily increasing gasoline prices, more attention is being paid to diesel engines, not only in commercial vehicles but also in passenger cars. In addition to using lower priced fuels, diesels provide better fuel economy than spark ignition engines.

In regard to exhaust emissions, diesel engines are said to be superior to spark ignition engines in terms of fewer hydrocarbons and less carbon monoxide. They are approximately equal in the emission of oxides of nitrogen. However, when smoke, odor and particulate emissions are considered, diesel engines pollute more than spark ignition engines.

There are many variables that affect exhaust emissions from diesel engines. In general, tests show that the direct injection engine is superior to the precombustion chamber type of engine from the standpoint of toxic emission levels.

The General Products Division of Teledyne Continental Motors conducted some experimental work on an Opel engine equipped with variable compression ratio pistons and variable injection timing. The engineers found that emissions were reduced to .41 HC/63.4 CO and .40 NO_x. Also, fuel economy tests showed that this experimental engine gave 55 percent more miles per gallon than a comparable gasoline engine. In

Fig. 22-1. Injector fuel pump test stand for checking Cummins injectors. (Hartridge Equipment Corp.)

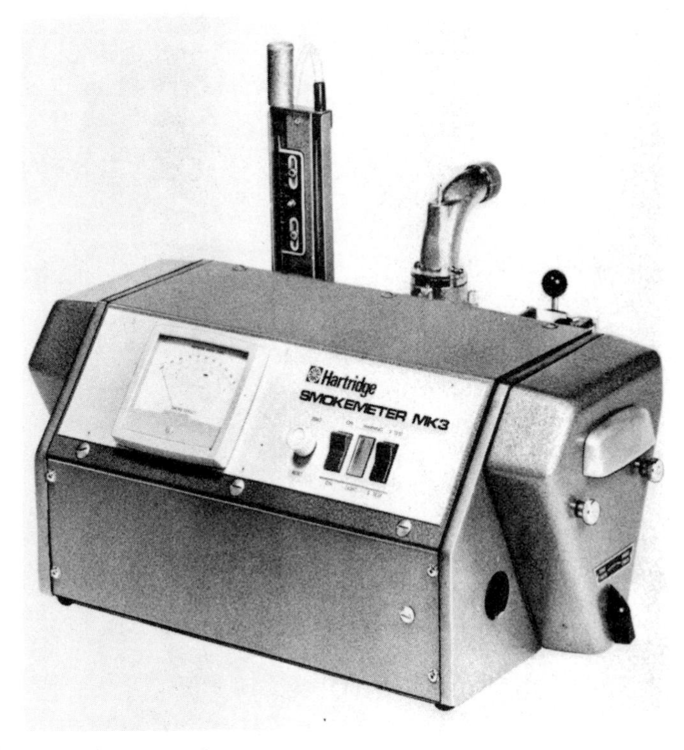

Fig. 22-2. Diesel engine smoke meter and analyzer.

addition, the experimental engine had a broad fuel tolerance and was smaller in size and comparable in weight to the gasoline engine.

Another factor influencing exhaust emissions is the extent of the area of the combustion chamber. The greater the area, the greater the amount of hydrocarbons and oxides of nitrogen in the exhaust.

As pointed out by the engineers of CAV Limited, the performance of a diesel engine combustion system is determined basically by the pattern of air and fuel mixing, and the timing of the injection of the fuel. Also noted was the fact that increased mixing of fuel and air results in higher efficiency, more noise and greater nitrogen oxides. Reduced mixing of fuel and air results in increased hydrocarbon and carbon monoxide formation.

Unburned hydrocarbons result in lubricating oil, quenched combustion at the cylinder walls and the combustion process. The hydrocarbons arising from the combustion process increase as the firing point is approached. Therefore, a retarded ignition timing to top dead center or beyond (depending on the speed of the engine) will result in a huge increase in hydrocarbons.

Fig. 22-3. Diesel engine injector nozzle analyzer.

CAV Limited engineers also reported:

1. The introduction of turbocharging leads to the reduction of smoke and hydrocarbons arising from the combustion process.
2. Nitrogen dioxide decreases or increases, depending on whether or not there is any air charge cooling.
3. An increase in compression ratio decreases the ignition delay period due to the increase in compression temperatures, which increases smoke and nitrogen dioxide.

Tests made by engineers of the Southwest Research Institute and the Environmental Protection Agency on four diesel-powered passenger cars (Datsun, Mercedes-Benz 220D, Peugeot 504D and the Opel Rekord) showed that 1977 standards for hydrocarbons were met by all cars except Peugeot.

Carbon monoxide emissions from all four cars were below the desired limit. Similarly, oxides of nitrogen emissions from all four cars were below the 1977 specified limit. It is interesting to note that the four cars showed the following fuel economy records:

1. Datsun 11.9 km/l (28.0 mpg).
2. Mercedes-Benz 11.9 km/l (28.0 mpg).
3. Peugeot 9.8 km/l (23.05 mpg).
4. Opel Rekord 10.1 km/l (23.7 mpg).

From the standpoint of service, remember that precision maintenance is of the utmost importance if the diesel engine is to conform to the Federal standards for exhaust emission levels. Injection pressure, timing, spray form, compression pressure, operating temperature and the mechanical condition of the engine are all critical if the Federal standards are to be met.

Not only must the fuel injection system be properly maintained, but also the engine cylinders, pistons and rings, together with the cylinder head and its valves and the engine cooling system.

Assisting in the maintenance procedures are: injector test stands, Fig. 22-1; smoke meters, Fig. 22-2; injector nozzle analyzers, Fig. 22-3.

REVIEW QUESTIONS – CHAPTER 22

1. Diesel engines give ____________ fuel economy than spark ignition engines.
2. With regard to exhaust emissions, diesel engines are superior to spark ignition engines in terms of fewer ____________ and less ____________.
3. However, diesel engines emit more ____________, ____________ and ____________ than spark ignition engines.
4. A direct injection diesel engine pollutes more than a precombustion chamber type engine from the standpoint of toxic emission levels. True or False?
5. Give five critical factors in maintaining a diesel engine so that it conforms to Federal standards for exhaust emission levels.

Chapter 23
SAFETY IN THE SHOP

Safety is everybody's business. No one can afford to be careless or overlook any of the basic requirements of safety.

Carelessness can lead to an injury and lost time. Carelessness can result in anything from a small scratch to the loss of sight, hearing, a finger, an arm, a leg or even life itself. It pays to be careful and follow a few basic rules of safety in the shop.

CAUTIONS AND PRECAUTIONS

While diesel fuel does not ignite as readily as gasoline, it does burn. Therefore, care must be taken when handling or storing fuel and other combustibles. These include cleaning fluids and paints. Obey "no smoking" signs. Make sure the work area is well ventilated, preferably by means of an exhaust ventilation system.

Fig. 23-1. Wear protective glasses or goggles when doing lathe work, particularly when grinding. Note shield on grinding wheel.

Always wear clothing suitable for shop work. Do not wear loose clothing, a necktie or a hat with a brim. Remember, too, that long hair is dangerous. Like the necktie, it can catch in a rotating shaft or gear. Do not wear an identification tag around your neck. Remove your wrist watch and other jewelry.

Wear protective goggles or a face shield when doing any grinding, Fig. 23-1, or when using a hack saw, since all hard blades will shatter and produce flying chips. Also wear goggles when using a chisel or when welding. In fact, it pays to wear goggles or a face shield when doing any kind of shop work.

Never run in the shop. Keep shop floors clean, dry and free of litter. Put all oily and dirty rags in a covered container. Keep combustible liquids in specified containers and limited to permissible quantities.

TOOLS AND EQUIPMENT

Carefully follow all operating instructions and cautions covering the use of tools and equipment. Make sure that all protective guards are in position, particularly on grinders and other rotating equipment.

Be sure that all electric wiring conforms to Underwriters' requirements. Check service outlets for proper grounding. See that extension cords are not frayed or damaged. Protect light bulbs with wire grilles. Make sure that there are ample fire extinguishers handy and in good operating condition.

REVIEW QUESTIONS – CHAPTER 23

1. Carelessness in the shop can lead to ____________ and ____________.
2. Put all oily rags in a ____________.
3. When operating rotating equipment, make sure all ____________ are in position.
4. Wear loose clothing on the job. True or False?

Chapter 24
CAREER OPPORTUNITIES

Having studied the many aspects of the diesel engine, from basic components through operation, trouble shooting, and repair, have you considered a career in the diesel field? The large diesel industry offers many career opportunities in the general catagories of marine, railway, industrial (including farm and construction), and transportation, Fig. 24-1. Many careers are available in the diesel industry for people with diversified skills. These jobs provide good pay and rewarding experience.

Fig. 24-1. Typical diesel powered bus. Continuous operation requires constant service and repair by the diesel mechanic. (GMC Truck and Coach Div., General Motors Corp.)

THE DIESEL INDUSTRY

Over the past several years the diesel industry has grown tremendously. The prime reason for this growth has been the increasing demand for equipment primarily in the trucking and construction industries. Equipment in the diesel industry is usually large and expensive. If the operation is to be profitable, the equipment must be properly maintained for years of extensive use. This not only involves many jobs in production of new diesel engines, but increases the demand for highly skilled service and repair technicians.

Those companies manufacturing diesel engines employ thousands of workers. A limited knowledge of the diesel engine is necessary for most of these positions. The diesel industry uses a production line system similar to the automotive industry. They mass produce parts and assemble engines. Assembly line jobs provide one answer for entrance into the industry. However, further study and experience are necessary for advancement.

DIESEL MECHANIC

Service and repair of diesel engines is the field that requires the largest number of employees. Persons working in this area are generally called diesel mechanics or technicians.

Diesel mechanics do repair and maintenance work on diesel engines that power construction equipment such as earthmovers, Fig. 24-2, bulldozers, graders and cranes. They also service transportation vehicles including ships, locomotives, busses and heavy trucks, Fig. 24-3. Stationary equipment, such as pumps and generators, require the skills of the diesel mechanics for periodic servicing and repair.

Fig. 24-2. This diesel powered earth and gravel loader is driven by a 192 hp engine.

Diesel mechanics have many trade responsibilities. They do engine inspections and trouble shooting. They do preventive maintenance, repair and replace defective parts and adjust engine components. These people usually are referred to as

Fig. 24-3. A 175 hp truck diesel in a concrete transit mixer. (Allis-Chalmers)

general mechanics, since they do all types of diesel repair, Fig. 24-4.

Specialization has brought about a number of specific job responsibilities. Some diesel mechanics specialize in the repair of fuel injection systems or other technical jobs such as engine rebuilding, reconditioning cylinder heads, repairing turbochargers or starting engines. Experience and further training often lead to more specific work and job titles.

HEAVY EQUIPMENT MECHANICS do all of the work previously described except they usually specialize in bulldozers, earthmovers and other construction equipment.

TRUCK MECHANICS are assigned to maintenance and repair of truck fleets or service departments of trucking companies.

Fig. 24-4. A diesel mechanic and service manager often discuss the necessary maintenance or repair before work is begun. Here they discuss service on a 340 hp diesel tractor.

SHOP SUPERVISOR

Shop supervisors are responsible for the assignment of jobs to diesel mechanics. They must know the skills and abilities of the mechanics they supervise. When special problems arise, the shop supervisor should be able to assist the mechanic in locating the difficulty and make suggestions for testing and evaluating. Ordering supplies, tools and equipment may also be a responsibility of the shop supervisor, who must work closely with the service manager in running the shop.

SERVICE MANAGER

Many large companies sell and service their own equipment. In such cases, a service manager usually is employed to coordinate the complete service department. Responsibilities include taking customer service orders, seeing that time schedules are met, working with the shop supervisor and sales manager for the distributor. Often, the service manager is responsible for the inventory of parts and equipment.

EDUCATIONAL REQUIREMENTS

If you have an interest in working as a diesel mechanic as a career, there are many avenues open to you to secure the necessary skills and knowledge to enter this field.

Many young people start as helpers to experienced automotive or diesel mechanics. Usually in 3 to 4 years they have gained the necessary background to do all types of general diesel service and repair. If they are employed by companies that use or service diesel powered equipment, they are often given a year or more additional training on the equipment of the individual company.

Formal apprenticeship programs are often operated by companies that manufacture diesel operated equipment. These programs usually consist of four years of practical experience in diesel repair combined with classroom study.

Another method of entry into the diesel field is through full-time study. Many vocational-technical schools and community colleges offer comprehensive programs in diesel engine service and repair. Most of these provide one or two years in specialized areas of study. Upon graduation the individual usually needs considerable on-the-job training before becoming a skilled diesel mechanic.

Most companies in the diesel industry prefer to employ apprentices and trainees who have a high school education. High school courses in auto mechanics, diesel mechanics (if available), electricity, drawing and machine shop, as well as courses in science and mathematics, are highly desirable.

EMPLOYMENT AVAILABILITY

Opportunities for employment of diesel mechanics and technicians is expected to increase rapidly. This is primarily due to the expansion of industries that use diesel engines in large numbers. The diesel engine is expected to replace gasoline engines on a wide variety of equipment in the future.

Some of the equipment that is being replaced by the diesel engine includes farm machinery, small delivery trucks, heavy industrial machinery and transportation vehicles.

The outlook for employment appears attractive. New industries and service repair facilities are on the increase. Therefore, job opportunities for the skilled diesel mechanic will be readily available.

Wages for diesel mechanics, technicians, supervisors and service managers are equal or better than those of other comparable service industries. Most mechanics earn a good living. With dedication to their job, they may advance to positions of responsibility and respect in their given career.

TEACHING

A career of teaching in the field of diesel engine technology is a highly rewarding experience, both in salary and working conditions. For those who are qualified, teaching positions are available in high schools, trade and vocational schools, factory and apprentice programs, community colleges and universities. See Fig. 24-5.

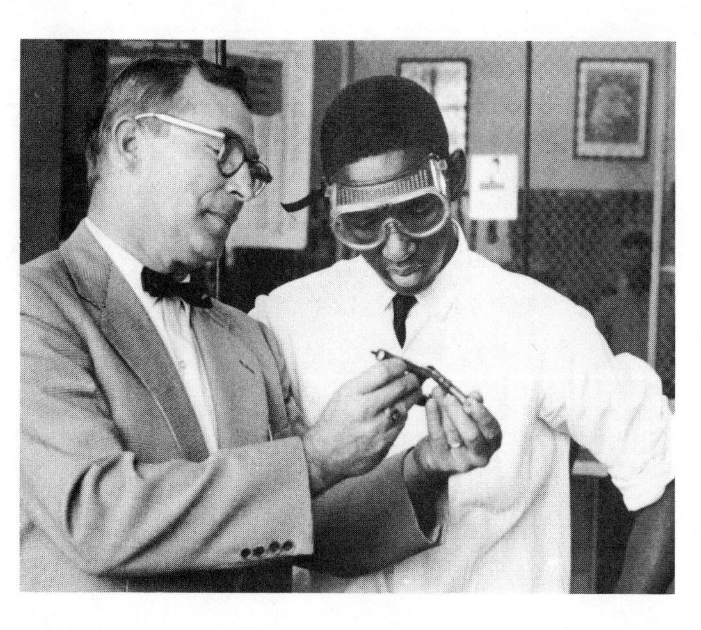

Fig. 24-5. There will be a great demand for teachers in diesel mechanics for years to come. Here the instructor assists the student in precision measurement of a diesel engine component.

Educational requirements generally include from two to five years of occupational experience plus a bachelor's or master's degree in a technical program at a college or university. Information on teaching may be obtained by writing for college catalogs.

ADDITIONAL INFORMATION

If you seriously consider a career in the field of diesel engine technology, you may want further information. Your school guidance office will probably have a great deal of resource material available for you to review. You may also want to write to the following sources for materials describing positions available and training requirements:

International Association of Machinists
and Aerospace Workers
1300 Connecticut Ave. N.W.
Washington, D. C. 20036

International Union, United Automobile,
Aerospace and Agricultural Implement Workers
of America
8000 East Jefferson Ave.
Detroit, Michigan 48214

U. S. Department of Labor
Bureau of Apprenticeship and Training
Washington, D. C. 20006

SUGGESTED ACTIVITIES

1. Prepare a report on specific job responsibilities of the diesel mechanic. Write to trade and vocational schools and industrial associations for literature. Also, use your technical library for resource material.
2. Look through local newspapers and trade magazines for job opportunities related to the diesel industry. Prepare a listing of jobs available and place each one under the general job classifications discussed. Note which job classification appears to be in need of the most employees.
3. Write to companies, vocational schools, and technical training programs for information on educational requirements necessary to become employable in the diesel industry. Prepare a written report on the specific training necessary for each job classification.
4. If possible, make a visit to a diesel engine manufacturer or diesel repair shop. Discuss job opportunities with their personnel director or service manager. Also talk with diesel mechanics to find out how they obtained their position and what additional training was required for advancement.
5. Make arrangements for an experienced diesel mechanic to speak to your class. Ask your instructor to plan for a question and answer session on job opportunities, educational requirements, working conditions, and how to seek employment.
6. Write a letter of application for employment as a diesel mechanic. Ask your instructor to offer guide lines for good letter writing and have your application evaluated when completed. Be sure to include all the details of your educational and mechanical experience that an employer would want to know about.

GLOSSARY

AC: Alternating current.

ACCELERATION: When the velocity of a body is increased it is said to have accelerated.

ADDITIVE: A material added to fuels and lubricants which is designed to improve their properties such as viscosity, cetane, pour point, film strength, etc.

ADIABATIC: Pertains to a reversible thermodynamic process occurring without the gain or loss of heat.

AIR CLEANER: A device for filtering and removing dust and other foreign matter from the air before it reaches the engine cylinders.

ALTERNATOR: A generator which produces alternating current.

ALTERNATING CURRENT: An electric current that reverses its direction of flow at regular intervals.

AMBIENT: Surrounding on all sides.

ANODE: The positive pole of an electric current.

ANTIFREEZE: A chemical placed in the liquid coolant of an engine which reduces its freezing point.

ASPIRATED: To draw by suction, such as the removal of contaminated fuel from an injector.

ATMOSPHERIC PRESSURE: The weight of air, usually measured at sea level; about 14.7 lbs. per sq. in.

ATOM: The smallest unit of an element and composed of electrons, protons and neutrons.

BACKFIRE: Ignition of a mixture in the intake manifold by flame from the cylinder.

BACK PRESSURE: A resistance to free flow, such as restriction in the exhaust system.

BALL BEARING: An antifriction bearing consisting of a series of balls placed between circular inner and outer races.

BLOW-BY: A leakage or loss of pressure past the piston into the crankcase.

BMEP: Abbreviation for brake mean effective pressure.

BOILING POINT: That temperature of a liquid which causes vapor to be formed.

BORE: The diameter of an engine cylinder.

BRAKE HORSEPOWER: (BHP) A measurement of the developed power of an engine in actual operation.

BRAKE SPECIFIC FUEL CONSUMPTION: For liquid fuels it is (a) pounds of fuel consumed per brake horsepower for the period of one hour, or (b) heating value of fuel (Btu per lb.) times the pounds of fuel consumed per brake horsepower hour. For gaseous fuels the BSFC is the heating value of the gas (Btu per cu. ft.) times the cubic feet of gas consumed per brake horsepower for a period of one hour. For electric generating plants the BSFC may be expressed as kilowatt hours instead of BHP hr.

BRINELL HARDNESS: A scale for designating the degree of hardness of a substance.

BTU: Abbreviation of British thermal unit.

BUTANE: A petroleum hydrocarbon compound which has a boiling point of approximately 32 deg. F. Sometimes referred to as liquid petroleum gas. Often combined with propane.

CALIBRATE: To check and adjust the graduations of a measuring instrument.

CALIPER: An adjustable gage for determining the thickness by contact and retaining such measurement for comparison with a scale or other part.

CALORIFIC VALUE: A measurement of heating value of fuel.

CALORIMETER: An instrument designed to measure the amount of heat given off by a substance when it is burned.

CALORIE: Metric measurement of amount of heat required to raise one gram of water from zero degree to one degree centigrade. 251.996 calories = 1 Btu.

CAMSHAFT: Shaft containing lobes or cams which operate engine valves.

CARBON DIOXIDE: One of the products of combustion. Compressed into solid form this material is known as dry ice, and remains at that temperature of -109 deg. F. It goes directly into a vapor state CO_2.

CARBON MONOXIDE: A gas formed by incomplete combustion. It is colorless, odorless and very poisonous. Chemical symbol is CO.

CAVITATION: Pitting in a cooling system surface caused by the sudden formation and collapse of bubbles formed in the coolant.

CENTER OF GRAVITY: Point of a body from which it could be suspended or on which it could be supported and be in balance. The center of gravity of a body is also defined as that point at which the weight of the body may be represented as a single force, no matter what angle the body is turned.

CENTRIFUGAL FORCE: A force which tends to move a body away from its center of rotation.

CENTIGRADE: A measurement of temperature at which

water freezes at 0 deg. and boils at 100 deg.

CETANE: A colorless liquid, $C_{16}H_{34}$.

CETANE NUMBER: A measure of the ignition quality of diesel fuel. It influences both ease of starting and combustion roughness of the fuel.

CLOCKWISE ROTATION: Rotation in the same direction as the movement of the hands of a clock.

CLUTCH: A device for connecting and disconnecting an engine from the device being driven.

COEFFICIENT OF FRICTION: The ratio of the force required to slide a body, to the force pressing the surfaces together.

COMBUSTION: A burning. Chemistry – A chemical change, especially oxidation, accompanied by the production of light and heat.

COMBUSTION CHAMBER: The volume of space above the piston when the piston is at top dead center.

COMMUTATOR: A ring of adjacent copper bars, insulated from each other and to which the wires of the armature of a direct current generator are attached.

COMPRESSION RATIO: The ratio of the maximum volume in an engine cylinder with the piston at bottom dead center to the minimum volume in the cylinder with the piston at top dead center.

CONNECTING ROD: The connecting link between the piston and the crankshaft.

COOLANT: The liquid used to cool an engine and is contained in the water jacket.

CORE PLUG: A metal cap or plug placed in the outer surface of the engine water jacket and covers the hole from which the core was removed after the engine cylinder block was cast.

CRANKCASE: The housing within which the crankshaft is supported and rotates.

CRANKCASE DILUTION: The unburned fuel which finds its way past the piston rings into the crankcase oil causing dilution.

CRANKSHAFT: The main shaft of an engine which, together with the connecting rods, changes the reciprocating motion of the pistons to rotary motion.

CRANKSHAFT COUNTERBALANCE: Series of weights attached to the crankshaft throws designed to eliminate any forces tending to produce vibration.

CRUDE OIL: Liquid petroleum oil that has not been refined.

CU. IN.: Abbreviation for cubic inch.

CYCLE: A series of events which are repeated.

CYLINDER: A round hole in which the piston reciprocates. Also known as the engine bore.

CYLINDER BLOCK: The mass of metal which includes the cylinder and water jackets or cooling fins in the case of an air-cooled engine. That part of the engine which includes the cylinder bores.

CYLINDER HEAD: A detachable portion of an engine which covers the upper end of the cylinder bores and includes the combustion chamber and, in the case of overhead valve engines, also includes the valves.

CYLINDER SLEEVE: A liner or sleeve interposed between the piston and the cylinder wall or water jacket to provide an easily replaceable surface for the cylinders.

DAMPER: A device for nullifying or reducing vibrations in a crankshaft.

DC: Direct current.

DEAD BEAT GOVERNOR: A governor with a high degree of precision or stability.

DEAD CENTER: Also top or bottom dead center – extreme upper or lower position of the piston within the cylinder.

DENSITY: The density of any subject is defined as the mass per unit volume of the substance.

DEPRESSANT: Substance which lessens undesirable properties.

DETERGENT: An additive used in engine oil to remove and hold in suspension any foreign matter that finds its way into the oil.

DETONATION: When detonation occurs, unburned gases are compressed ahead of the flame front in the combustion chamber, raising the pressure of such gases and their temperature until self ignition point is reached. This causes a knock or "pinging" sound.

DIESEL: An engine designed to convert chemical energy of fuel into mechanical power, the fuel being ignited by the heat of the air which was compressed by the motion of the piston within the cylinder.

DIODE: An electronic device that permits current to flow through it in one direction only.

DISPLACEMENT: The volume displaced by the movement of a piston as it moves from one end of its stroke to the other.

DISSIPATED: Scattered in various directions.

DROOP: Sometimes referred to as steady state speed regulation. It is the speed variation from no load to full load.

ECCENTRIC: The degree of being off center. The distance between the center of an eccentric and its axis; the throw. One circle within another circle not having the same center.

ELECTRODE: An electric conductor through which an electric current enters or leaves a medium such as an electrolyte, a nonmetallic solid, a gas or a vacuum.

ELECTROMAGNET: A magnet which derives its magnetic power from a current of electricity which passes through a coil of wire wound around a central core of iron.

EMISSIONS: Gases produced from exhaust, crankcase and fuel tanks and their contribution to smog.

EN BLOC: Refers to a cylinder block of an engine cast as a single piece.

ENERGY: Any agency which can do work in overcoming a resisting force possesses energy. The energy of a body is a measure of its ability to do work.

ENGINE DERATING: Reducing maximum flow of fuel.

ENGINE DISPLACEMENT: Sum of the displacement of the individual cylinders which compose the engine. (See displacement.)

EVAPORATION: Process of changing from a liquid to a vapor.

EXHAUST EMISSIONS: The products of combustion that are discharged from the engine exhaust.

EXHAUST GAS ANALYZER: An instrument designed to measure the quantities of the different gases which form the exhaust and thereby determine the engine's efficiency.

EXHAUST SYSTEM: Consists of the piping and muffling devices which carry the exhaust gases from the engine.

EXTREME PRESSURE LUBRICANTS: (EP) A lubricant with additives designed to increase its capacity to withstand high pressures.

FAHRENHEIT: A scale of temperature measurements in which the freezing point of water is 32 degrees and its boiling point is 212 degrees. System used mostly in English speaking countries.

FERROUS METALS: Metals composed primarily of iron.

FIELD: That region of an electric motor or generator which produces an electromagnetic force.

FILTER: A device designed to remove suspended impurities and other foreign matter from intake air, fuel or lubricating oil.

FLASH POINT: The flash point of an oil or fuel is the temperature to which the oil or fuel has to be heated until sufficient flammable vapor is given off to flash when brought into contact with a flame.

FLOW METER: An instrument designed to measure the quantity of fuel used in an engine.

FLUIDS: Fluids are distinguished from solids by their capacity to flow readily and have the tendency to assume the shape of the container.

FOOT POUND: This is the unit of mechanical energy of work, being equivalent to the energy of a force of one pound acting through a distance of one foot.

FORCE: Force may be defined as a push or pull acting on an object.

FRICTION HORSEPOWER: The power needed to overcome friction within the engine. Friction results from pressure of the piston and rings against the cylinder walls, rotation of the crankshaft and camshaft in their bearings, and friction developed by other moving parts.

FUEL: Any substance consumed to produce energy. In the case of a diesel engine, it is usually fuel oil obtained from petroleum.

FUSE: A piece of wire which will melt when a specified amount of electric current is passed through it, thereby protecting the circuit from damage of excessive current.

GAS: A fluid substance that can be changed in volume and shape according to temperature and pressure applied to it.

GENERATOR: A device consisting of an armature, field coils and other parts which generates electricity when the armature coils are rotated in the magnetic field produced by the field windings.

GOVERNOR: A device designed to control the speed of an engine within specified limits.

GRAM: Metric unit of weight which equals 0.03527 oz.

HP: Abbreviation for horsepower.

HEAT: A form of energy associated with the motion of atoms or molecules and capable of being transmitted by conduction, convection, and radiation.

HEAT EXCHANGER: A device which utilizes normally wasted heat for useful purposes.

HEAT LOSS: Heat from burning fuel that is lost in the cylinder without doing useful work.

HELICAL GEAR: A gear design in which the gear teeth are cut on a helical curve.

HERRINGBONE GEAR: A pair of helical gears designed to operate together, the angle of the teeth forming a "V."

HUNTING: Engine speed swings back and forth around a desired value.

HYDROCARBON: A compound composed entirely of hydrogen and carbon. Petroleum is a hydrocarbon.

HYDROMETER: An instrument for determining the specific gravity of a solution. Used extensively in measuring the state of charge of electric storage batteries and also the protection afforded by the antifreeze in the coolant.

IGNITION: The combustion of the fuel mixture in the combustion chamber.

IGNITION DELAY: The period which covers the time from initial injection to actual ignition of fuel and air.

INDICATED HORSEPOWER: (HP) Indicated horsepower is the power developed within the cylinders and is equal to the brake horsepower plus the horsepower lost to friction.

INERTIA: Inertia is the characteristic of a body to continue in a state of rest or uniform motion unless acted on by some other force.

INHIBITOR: A material used to restrict or hinder some unwanted action. e.g. rust inhibitor added to the coolant in the cooling system to stop the formation of rust.

INJECTOR: A device for injecting fuel oil into the combustion chamber of an engine against the pressure of air within the chamber.

INLET PORT: The opening in the cylinder of a two-cycle engine through which the air enters the cylinder.

INLET VALVE: The valve which permits air to enter the cylinder of a diesel engine.

INSOLUBLE: Cannot be dissolved.

INSULATION: Materials which do not readily conduct electricity or heat.

INTERNAL COMBUSTION ENGINE: An engine which derives its power from fuel which is burned within the engine, as contrasted to the steam engine in which the fuel is burned externally.

INTERNAL ENERGY: Energy which a body possesses due to its condition, such as pressure and temperature.

ISOCHRONOUS: Maintaining constant engine speed independent of the load being carried.

JOURNAL: That part of a shaft or axle in actual contact with the bearing.

KILOMETER: A metric measurement of distance. One kilometer equals 0.6214 mile.

KILOWATT: A measure of electric power equal to 1000 watts.

KILOWATT HOUR: The work accomplished by an agent operating at a power level of a kilowatt for one hour. 1 kw. hr. = 1000 watts x 3600 seconds or 3,600,000 watt seconds.

KINETIC ENERGY: Energy developed by a moving body as a result of its motion.

LAPPING: A process of fitting one surface to another by rubbing them together with an abrasive material between the two surfaces.

LB.: Abbreviation for pound.

LINKAGE: Any series of rods, yokes, pedals and levers used to transmit motion from one unit to another.

LIQUID: The state of matter in which a substance exhibits a characteristic readiness to flow, but with little or no tendency to disperse, and relatively high compressibility.

LITER: Metric unit of volume equal to 61.027 cu. in. or 1.057 liquid quart.

LOBES: Ear-shaped projections on a rotating camshaft.

LPG: Abbreviation for liquid petroleum gas.

LUBRICANT: A material designed to reduce friction between surfaces.

MALLEABLE CASTING: A casting which has been toughened by annealing.

MANIFOLD: Pipes connecting a series of openings or outlets to a common opening, (e.g. the exhaust manifold).

MANOMETER: An instrument used for measuring the pressure of liquids and gases.

MASS: A measure of inertia; quantity of matter. (Weight is a force.)

MEP: Abbreviation for mean effective pressure.

MECHANICAL EFFICIENCY: The ratio of brake horsepower output of an engine to the indicated horsepower in the cylinders.

MELTING POINT: The temperature at which solid materials become liquid.

METER: Metric unit of length equal to 39.37 in.

MICROMETER: A precision instrument for measuring either internal or external dimensions.

MILL: To cut or machine metals with rotating tooth cutters.

MILLIMETER: Metric unit of length equal to 0.03937 in.

MOLE: The term mole is an abbreviation of one pound molecular weight or pound mole and is applied to a substance which has a weight in pounds equal to its molecular weight.

MOLECULAR WEIGHT: The molecular weight of an element is its atomic weight times the number of atoms in a molecule of the element.

MPG: Abbreviation for miles per gallon.

MPH: Abbreviation for miles per hour.

MULTIFUEL ENGINE: An engine designed to start and operate on any one of a variety of fuels without engine modification.

NATURAL GAS: A colorless and odorless gas produced from crude oil or obtained directly from wells in the earth.

NEEDLE BEARING: An antifriction bearing of the roller bearing type, but the diameter of the rollers is very small.

NEGATIVE POLE: Point from which an electric current flows as it passes through the circuit.

NEUTRON: That portion of an atom which carries no electric charge and with the protons, form the central core of the atom about which electrons rotate.

NICKEL STEEL: Nickel alloyed with steel to form a heat and corrosion resistant alloy.

NONFERROUS METAL: Metals such as copper and brass which contain no iron.

NORMALLY ASPIRATED: An engine without supercharger which operates with cylinder air charge at start of compression at pressure very near or slightly below atmospheric.

NORTH POLE: That pole of a magnet which tends to point toward the north and from which magnetic lines of force start.

OCTANE NUMBER: A numerical measure of the antiknock properties of a motor fuel based on the percentage by volume of the iso-octane in a standard reference fuel.

OD: Abbreviation of outside diameter.

OHM: A unit of resistance to the flow of electric current. It is equal to that of a conductor in which a current of one ampere is produced by a potential of one volt across the terminals.

OHMMETER: An electrical instrument designed to measure the resistance of a circuit in ohms.

OHM'S LAW: The resistance in ohms is equal to the voltage divided by the current in amperes flowing in the circuit.

OILINESS: The property of an oil which causes a difference in friction when lubricants of the same viscosity, at the same temperature and pressure on the film, are used with the same bearing surfaces.

OSCILLOSCOPE: An electrical testing device which shows a pattern wave form of spark ignition action on a viewing screen.

OTTO CYCLE: When the complete cycle of events in the cylinder requires four strokes of the piston, it is known as the Otto or four-cycle engine.

OVERHEAD VALVE: An engine designed with the valves in the cylinder head. Also known as a valve in head engine.

PANCAKE ENGINE: A design in which the engine cylinders are placed in line and horizontally in order to obtain minimum height.

PEEN: To stretch or clinch over by pounding with the peen end of a hammer.

PETROLEUM: A natural, yellow to black, thick flammable liquid hydrocarbon mixture found principally beneath the earth's surface and processed for fractions including natural gas, gasoline, diesel fuel, kerosene, lubricating oil, etc.

PHOSPHOR BRONZE: An alloy consisting of copper tin and lead. Sometimes used as material for heavy-duty engine bearings.

PINION: A small gear driving a larger gear.

PISTON: A cylindrical part closed at one end which has a close sliding fit in the engine cylinder. It is connected to the connecting rod by means of the piston pin. The force of the exploding fuel against the closed end of the piston forces the piston down in the cylinder causing the connecting rod to rotate the crankshaft.

PISTON BOSS: That portion of the piston which supports the piston pin.

PISTON DISPLACEMENT: The volume of air moved or displaced by movement of the piston from bottom dead center to top dead center.

PISTON HEAD: The closed end of the piston above the piston rings.

PISTON LANDS: That area of the piston between the piston rings.

PISTON PIN: A cylindrical piece of alloy steel which passes through the piston bosses and upper end of the connecting rod so that movement of the piston is transmitted to the connecting rod.

PISTON RING: A split ring of the expanding type placed in the grooves of the piston. Designed to prevent passage of air from the combustion chamber to the crankcase.

PISTON RING EXPANDER: A ring placed behind the piston ring. Designed to increase the pressure of the ring against the cylinder wall.

PISTON RING GAP: The clearance between the ends of the piston ring.

PISTON RING GROOVE: Grooves formed in the piston into which the piston rings are fitted.

PISTON SKIRT: That portion of the piston below the piston pin which is designed to take the side thrust of the piston.

PNEUMATIC: Pertaining to air pressure.

POLARITY: Refers to positive and negative terminals of an electric battery or circuit; also to the north and south poles of a magnet.

POPPET VALVE: A disc type valve used to open and close the valve ports, which is opened by the action of the cam and closed by the action of the valve springs.

PORT: As applied to a diesel engine it is the opening in the cylinder head or cylinder through which air or exhaust gas pass.

POSITIVE TERMINAL: That terminal in a circuit to which current flows.

POTENTIAL DIFFERENCE: A difference of electrical pressure which sets up a flow of electrical current.

POTENTIAL DROP: A loss of electrical pressure due to resistance.

POTENTIAL ENERGY: Energy possessed by a body as a measure of its ability to do work.

POUR POINT: The pour point of an oil is that temperature at which oil will just flow under prescribed conditions.

POWER: The rate at which work is being done.

PREIGNITION: Ignition of the fuel in the combustion chamber prior to the timed injection of the fuel.

PRIME MOVER: A source of mechanical power, (e.g. an internal combustion chamber).

PRODUCER GAS: A gas made by burning coal in insufficient air for complete combustion.

PRONY BRAKE: A machine for measuring the brake horsepower of an engine.

PROPANE: A petroleum hydrocarbon gas, C_3H_8 which has a boiling point of about -44 deg. F. Often mixed with butane.

PROTON: That portion of an atom which carries a positive charge.

PSI: Abbreviation of pounds per square inch.

PUSH ROD: A connecting link in an operating mechanism, (e.g. a cam operated rod that operates the rocker arm which opens an engine valve).

RACE: The inner and outer portions of an antifriction bearing (either ball or roller type) which provide a surface for the balls or rollers.

RADIAL ENGINE: An engine with its cylinders arranged in a circle around the crankcase.

RADIATION: Thermal radiation is energy in the form of wave motion. It is not heat and may pass through glass or any transparent substance. If it falls on some material object where it is absorbed, it produces an increase in internal energy and results in a rise in temperature.

RATIO: A fixed relationship between things in number, quantity or degree. Also, the relative size of two quantities expressed as a quotient of one divided by the other. The ratio of 9 to 5 is written 9:5 or 9/5.

ROCKER ARM: An arm used to change upward movement of the cam operated push rod to downward motion to open an engine valve on an overhead valve engine.

ROLLER BEARING: An antifriction bearing utilizing a series of straight or barrel shaped rollers placed between annular inner and outer races.

ROLLER TAPPETS: Valve lifters that have a roller placed on the end contacting the cam.

RPM: Abbreviation for revolutions per minute.

SAE: Society of Automotive Engineers.

SAFETY VALVE: A valve designed to open and relieve the pressure within a container when the pressure exceeds a predetermined value.

SAYBOLT VISCOSIMETER: An instrument used to determine the viscosity of a liquid.

SCAVENGING: Process of removing exhaust gases from the cylinders of an engine.

SCAVENGING EFFICIENCY: In a two-cycle engine scavenging efficiency is the ratio of a new air charge trapped in a cylinder to the total volume of air and exhaust gases in the cylinder at port closing position.

SCORE: A scratch, ridge or groove marring a finished surface.

SEAT: A specially machined surface upon which another part rests (e.g. surface upon which a valve face rests).

SEIZE: Grab and fail to move due to expansion.

SEMI-DIESEL: A diesel which utilizes injection of fuel, but also uses electric spark ignition.

SENSITIVITY: The percent of speed change required to produce a corrective movement of the speed control mechanism of an engine.

SEWAGE GAS: A gas produced from sewage sludge as it decomposes in sewage disposal plants.

SHIM: Thin sheets of metal used as spacers between two parts.

SHRINK FIT: An exceptionally tight fit, (e.g. a shaft or rod slightly larger than the hole in which it is to be fitted). The shaft is chilled which will reduce its diameter. It can then be inserted in the hole. On returning to normal temperature the shaft cannot be withdrawn from the hole.

SILENCER: A muffler.

SILICONE: A special lubricant having a wide range of thermal stability and great lubricating qualities.

SINTERED BRONZE: Tiny particles of bronze pressed together to form a solid piece. It is highly porous and is often used as a filtering element.

SKIRT: That portion of a piston below the piston pin which takes the side thrust of the piston.

SLIP-IN BEARING: These bearings are used for crankshaft and connecting rods and are made so accurately that they can be slipped into position without additional fitting. Also known as precision bearings.

SOHC: Abbreviation for single overhead camshaft.

SOLENOID: An iron core about which is wound a coil of wire

designed to move linkage when a current is passed through the coil.

SOLVENT: A solution which dissolves another material.

SOUTH POLE: That pole of a magnet to which magnetic lines of force flow. Opposite to the north pole.

SPECIFIC GRAVITY: The specific gravity of a substance is defined as the ratio of the density of the substance to the density of an equal quantity of water at the same temperature.

SPECIFIC HEAT: The quantity of heat (Btu) required to raise the temperature of one pound of water 1 deg. F.

SPEED CHANGER: A device for adjusting the speed governing system to change the engine speed.

SPEED DRIFT: A very gradual deviation of the mean governed speed from the desired governed speed.

SPEED DROOP: Engine speed variation from no load to full load.

SPUR GEAR: A gear on which the teeth are cut parallel to the axis of the shaft.

SQ. FT. Abbreviation for square foot.

SQ. IN.: Abbreviation for square inch.

SQUISH: An increase in turbulence of air in the cylinder induced by the shape of the piston head, usually consisting of a wide rim around the circumference of the piston head.

STABILITY: The ability of a governor to maintain desired engine speed without speed fluctuations.

STAMPING: A piece of metal cut and formed by means of a die.

STATOR: Stationary portion of an alternator which provides the magnetic field.

STELLITE: Nonferrous metal containing cobalt, tungsten, chromium, carbon.

SUPERCHARGER: A blower or pump designed to force air into the cylinders.

SWIRL: Rotation of mass of air as it enters the cylinder is known as "swirl." This is one form of turbulence.

SYNCHRONIZE: To cause two events to occur at the same time.

TACHOMETER: An instrument designed to measure the rotational speed of an engine in revolutions per minute.

TANDEM: One back of another, or acting in conjunction.

TAPPET: An adjusting screw for varying the distance or clearance between the valve stem and cam. May be built into the valve lifter or rocker arm.

TDC: Abbreviation for top dead center and refers to the position of the piston in the cylinder.

TEMPER: To change the metallurgical characteristics of metal (usually steel) by applying heat and quenching in water or oil.

TEMPERATURE: The measurement of the hotness or coldness of an object or substance. When two objects are placed in close contact, the one which is hotter will begin to heat the other, and it is said to have the higher temperature.

TENSION: Effort that is devoted to elongation or stretching of material.

THERMOCOUPLE: A thermoelectric device used to measure temperature accurately, especially one consisting of two dissimilar metals joined together so that a potential difference is generated between the points of contact. The amount of potential difference is a measure of the heat at the point of contact of the two dissimilar metals.

THERMOSTAT: A heat control valve used in a cooling system which controls the flow of coolant in relation to the temperature.

THERMOSIPHON: A method of controlling the flow of liquid coolant which utilizes the difference of specific gravity in hot water and cold water, causing the heated water to rise.

TOLERANCE: A permissible variation between two extremes of specifications or dimensions.

TORQUE: An effort devoted toward twisting or turning.

TORSIONAL VIBRATION: Vibration resulting from the variation in torque set up in a crankshaft.

TURBINE: A wheel upon which a series of angled vanes are affixed so that a moving column of liquid will impart rotational movement to the wheel.

TURBOCHARGER: Violent irratic movement of fluid or gas. Used particularly in reference to movement of air in the cylinder and combustion chamber.

TURBULENCE: Violent swirling motion. Fuel injection provides some turbulence. Additional turbulence is provided by the design features of the combustion space.

TWO-STROKE CYLE ENGINE: An engine requiring only one complete revolution of the crankshaft to complete the cycle of events.

UNBALANCE: Difference in engine speed under variation of load.

VACUUM: An enclosed area in which the pressure is less than atmospheric pressure.

VACUUM GAGE: A device designed to measure the degree of vacuum existing in a chamber.

VALVE: A device designed to open and close an opening.

VALVE CLEARANCE: Air gap between end of valve stem and valve lifter or rocker arm.

VALVE TRAIN: The various parts making up the valve and valve operating mechanism.

VAPORIZATION: The breaking up of fuel into fine particles as it is sprayed into the combustion chamber.

VELOCITY: Distance traveled in a given amount of time.

VENTURI: Two tapering streamlined tubes joined at their small ends.

VIBRATION DAMPER: A device attached to the crankshaft designed to reduce vibrations.

VISCOSITY INDEX: (VI) is an emperical system for expressing the rate of change of viscosity of an oil with change in temperature.

VISCOSIMETER: An instrument designed to measure the viscosity of a liquid.

VOLATILE: Easily evaporated.

VOLT: A unit of electrical pressure. The international system unit of electrical potential and electromotive force.

VOLTMETER: An instrument designed to measure voltage.

VOLUMETRIC EFFICIENCY: A comparison between the actual volume of fuel mixture drawn into the cylinder and that which would be drawn in if the cylinder were completelv filled.

WATER JACKET: The space around the cylinder and cylinder head that is filled with coolant for the purpose of maintaining an even temperature.

WEIGHT: Weight is a force. Mass is the measure of inertia.

WET SLEEVE: A cylinder sleeve so installed that the coolant contacts a major portion of the sleeve.

WORK: Mechanical work is done when some object is moved against a resisting force. It is the product of force which acts to produce displacement of body and the distance through which body is displaced in the direction of force.

WORK CAPACITY: A device for adjusting the speed governing system to change the engine speed or power output relationship with the generator units when on parallel operation.

ACKNOWLEDGMENTS

Airesearch Industrial Div., Garrett Corp.
Alco Engines, White Industrial Power Inc.
Allis-Chalmers Corp.
Aluminum Company of America
American Bosch Div., AMBAC Industries, Inc.
American Honda Motor Co.
American Society of Testing Materials
Avco Lycoming Industrial Products Operations
Bacharach Instrument Co., AMBAC Industries, Inc.
Bear, The Wheel Service Div., Applied Power, Inc.
Behr-Thomson Dehnstoffregler GmbH
Bendix Corp., Electrical Components Div.
Bohn Aluminum and Brass Div., Gulf & Western Mfg. Co.
Brown Boveri, Sulzer Turbomachinery Ltd.
Carborundum Co., Filters Div.
Caterpillar Tractor Co.
CAV/Joseph Lucas, Inc.
Champion Spark Plug Co.
Chicago Pneumatic Tool Co.
Chrysler Corp., Industrial Products Div.
Colt Industries Inc., Fairbanks Morse Engine Div.
Cooper Industries Inc.
Cummins Engine Co., Inc.
Daimler-Benz AG
Deere & Co.
Delaval Turbine Inc.
Delco-Remy Div., General Motors Corp.
Detroit Diesel Allison Div., General Motors Corp.
Deutz Corp.
Diesel Engine Manufacturers Association
Dodge Div., Chrysler Corp.
Donaldson Company, Inc.
Electro-Motive Div., General Motors Corp.
Elliott Company
EX-CELL-O Corp.
Farymann Diesel Corp.
Federal-Mogul Corp.
Fiat Motors of North America, Inc.
Ford Motor Co.
GEA Airexchangers Inc.
General Electric Co., Transportation Systems Business Div.
George Olcott Co.
Gleason Engineering International
GMC Truck and Coach Div., General Motors Corp.
Gray Marine Co.
Harrison Radiator Div., General Motors Corp.
Hartridge Equipment Corp.
Hispano-Suiza (THM) Div. of Snecma
Hobart Bros. Co.
Ingersoll-Rand
International Harvester Co.
Isotta Fraschini Motori Breda
Julius Behr
Kim Hotstart Mfg. Co.
Koppers Co., Inc.
Kwik-Way Mfg. Co.
Leece Neville Div., Sheller Globe Corp.
Lombardini of Italy
Mack Trucks, Inc.
Magnaflux Corp.
MAN Maschinenfabrick
McQuay-Perfex Inc.
Mercedes-Benz of North America, Inc.
Minneapolis Moline, Inc., Div. White Motor Corp.
Murphy Diesel Co.
N.V. Motorenfabriek Thomassen
Nefco, Div. of Neiman Industries
Nordberg Mfg. Co.
Oakite Products, Inc.
Oldsmobile Div., General Motors Corp.
Onan Corp.
Paxman Engine Div., English Electric Diesels Ltd.
Perkins Engines, Inc.
Peterson Machine Tool, Inc.
Petters Ltd.
Peugeot, Inc.
Purolator Filter Div., Purolator, Inc.
Renault U.S.A., Inc.
Robert Bosch GmbH
Rockwell International
Rolls-Royce Motors Inc.
Roosa Master Div., Stanadyne
Roots Blower Operations, Dresser Industries, Inc.
Rottler Boring Bar Co.
Schwitzer Engineered Components, Wallace-Murray Corp.
Sealed Power Corp.
SIGMA Ste Industrielle Generale de Mecanique Appliquee
Simms Motor Units Ltd.
Society of Automotive Engineers
Stanadyne, Hartford Div.
Standard-Thomson Corp.
Stewart-Warner Corp.
Stork-Werkspoor
Storm-Vulcan, Inc.
Sulzer Bros., Inc.
Sun Electric Corp.
Sunnen Products Co.
TRW, Inc.
Teledyne Continental Motors, Inc.
United Air Cleaner Div., Halle Industries Inc.
United States Casting Repair Corp.
Volkswagen of America, Inc.
Volvo of America Corp.
Waukesha Engine Div., Dresser Industries, Inc.
Westinghouse Air Brake Div., WABCO
White Engines, White Motor Corp.
Winona Van Norman Machine Co.
Wm. W. Nugent & Co.
Woodward Governor Co.
Worthington Compressors, Inc.
Young Radiator Co.

TABLES

CONVERSION FACTORS

TO CONVERT	INTO	MULTIPLY BY
Btu	Horsepower-hours	3.931×10^{-4}
Btu	Kilowatt-hours	2.928×10^{-4}
Btu/hr	Horsepower-hours	3.929×10^{-4}
Btu/hr	Watts	0.2931
Celsius	Fahrenheit	(C x 9/5) + 32
Centimetres	Inches	0.3937
Centimetre-grams	Pound feet	7.233×10^{-5}
Centimetres per second	Feet per minute	1.1969
Cubic centimetres	Cubic feet	3.531×10^{-5}
Cubic centimetres	Cubic inches	0.06102
Cubic centimetres	Gallons (U. S. Liquid)	2.642×10^{-4}
Cubic feet	Cubic inches	1728.0
Cubic feet	Cubic metres	0.02832
Cubic feet per minute	Gallons per second	0.1247
Cubic inches	Cubic centimetres	16.39
Cubic inches	Cubic feet	5.787×10^{-4}
Cubic inches	Cubic metres	1.639×10^{-5}
Cubic inches	Litres	0.0164
Cubic metres	Cubic feet	35.31
Feet	Centimetres	30.48
Feet	Metres	0.3048
Feet	Millimetres	304.8
Feet per minute	Centimetres per second	0.5080
Feet per second	Metres per minute	18.29
Foot pounds	Btu	1.286×10^{-3}
Foot pounds	Horsepower-hours	5.050×10^{-7}
Foot pounds	Kilograms-metres	0.1383
Foot-pounds per minute	Btu per minute	1.286×10^{-3}
Foot-pounds per minute	Horsepower	3.030×10^{-5}
Gallons	Cubic centimetres	3785.0
Gallons	Cubic inches	231.0
Gallons	Litres	3.785
Gallons (Liquid Br. Imp.)	Gallons (U. S. Liquid)	1.20095
Gallons (U.S.)	Gallons (Imp.)	0.83267
Grams per litre	Pounds per cubic foot	0.062427
Grams per square centimetre	Pounds per square foot	2.0481
Horsepower	Btu per minute	42.44
Horsepower	Foot pounds per minute	33,000
Horsepower	Foot pounds per second	550.0
Horsepower	Kilowatts	0.7457
Horsepower hours	Btu	2547.0
Horsepower hours	Kilowatt hours	0.7457
Inches	Centimetres	2.540
Inches	Metres	2.54×10^{-2}
Kilograms	Pounds	2.205
Kilograms	Tons (long)	9.842×10^{-4}
Kilograms	Tons (short)	1.102×10^{-3}
Kilograms per cubic metre	Pounds per cubic foot	0.06243
Kilograms per metre	Pounds per foot	0.6720
Kilopascals	Pounds per square inch	6.895
Kilometres	Feet	3,281
Kilometres	Miles	0.6214
Kilometres per hour	Feet per minute	54.68
Kilowatts	Btu per minute	56.92
Kilowatts	Horsepower	1.341
Kilowatt hours	Btu	3,413
Litres	Cubic inches	61.02

CONVERSION FACTORS (CONT'D.)

TO CONVERT	INTO	MULTIPLY BY
Litres	Gallons (U. S. Liquid)	0.2642
Litres per minute	Cubic feet per second	5.886×10^{-4}
Metres	Feet	3.281
Metres	Inches	39.37
Metres per minute	Miles per hour	0.03728
Miles per hour	Centimetres per second	44.70
Miles per hour	Kilometres per hour	1.609
Millimetres	Inches	0.03937
Ounces	Grams	28.349527
Ounces (fluid)	Litres	0.02957
Pounds	Kilograms	0.4536
Pound feet	Metres kilograms	0.1383
Pounds per cubic feet	Grams per cubic centimetre	0.01602
Pounds per cubic inch	Grams per cubic centimetre	27.68
Square centimetres	Square feet	1.076×10^{-3}
Square centimetres	Square inches	0.1550
Square inches	Square millimetres	645.2
Square metres	Square feet	10.76
Watts	Btu per hour	3.4192
Watts	Btu per minute	0.05688
Watt hours	Btu	3.413
Watt hours	Horsepower hours	1.341×10^{-3}

METRIC CONVERSION TABLE
MILLIMETRES TO INCHES

MILLIMETRES	INCHES
.01	.0003937
.02	.0007874
.03	.00118
.04	.00157
.05	.00197
.06	.00236
.07	.00276
.08	.00315
.09	.00354
.10	.00394
.11	.00433
.12	.00472
.13	.00512
.14	.00551
.15	.00591
.16	.00630
.17	.00669
.18	.00708
.19	.00748
.20	.00787
.21	.00827
.22	.00866
.23	.00905
.24	.00944
.25	.00984
.26	.01024
.27	.01063
.28	.01102
.29	.01141
.30	.01181
.31	.01221
.32	.01259
.33	.01299
.34	.01338
.35	.01378
.40	.01575
.45	.01772
.50	.01968
1.00	.03937

METRIC EQUIVALENT OF ENGLISH CLEARANCE DIMENSIONS

INCHES	MILLIMETRES
.001	.0254
.002	.0508
.003	.0762
.004	.1016
.005	.1270
.006	.1524
.007	.1778
.008	.2032
.009	.2286
.010	.2540
.011	.2794
.012	.3048
.013	.3302
.014	.3556
.015	.3810
.016	.4064
.017	.4318
.018	.4572
.019	.4826
.020	.5080
.021	.5334
.022	.5588
.023	.5842
.024	.6096
.025	.6350
.030	.7620
.035	.8890
.040	1.0160

CONVERSION TABLE
METRIC TO ENGLISH

WHEN YOU KNOW	MULTIPLY BY: * = Exact		TO FIND
	VERY ACCURATE	APPROXIMATE	
LENGTH			
millimetres	0.0393701	0.04	inches
centimetres	0.3937008	0.4	inches
metres	3.280840	3.3	feet
metres	1.093613	1.1	yards
kilometres	0.621371	0.6	miles
WEIGHT			
grains	0.00228571	0.0023	ounces
grams	0.03527396	0.035	ounces
kilograms	2.204623	2.2	pounds
tonnes	1.1023113	1.1	short tons
VOLUME			
millilitres		0.2	teaspoons
millilitres	0.06667	0.067	tablespoons
millilitres	0.03381402	0.03	fluid ounces
litres	61.02374	61.024	cubic inches
litres	2.113376	2.1	pints
litres	1.056688	1.06	quarts
litres	0.26417205	0.26	gallons
litres	0.03531467	0.35	cubic feet
cubic metres	61023.74	61023.7	cubic inches
cubic metres	35.31467	35.0	cubic feet
cubic metres	1.3079506	1.3	cubic yards
cubic metres	264.17205	264.0	gallons
AREA			
square centimetres	0.1550003	0.16	square inches
square centimetres	0.00107639	0.001	square feet
square metres	10.76391	10.8	square feet
square metres	1.195990	1.2	square yards
square kilometres		0.4	square miles
hectares	2.471054	2.5	acres
TEMPERATURE			
Celsius	*9/5 (then add 32)		Fahrenheit

CONVERSION TABLE
ENGLISH TO METRIC

WHEN YOU KNOW	MULTIPLY BY: * = Exact		TO FIND
	VERY ACCURATE	APPROXIMATE	
LENGTH			
inches	* 25.4		millimetres
inches	* 2.54		centimetres
feet	* 0.3048		metres
feet	* 30.48		centimetres
yards	* 0.9144	0.9	metres
miles	* 1.609344	1.6	kilometres
WEIGHT			
grains	15.43236	15.4	grams
ounces	* 28.349523125	28.0	grams
ounces	* 0.028349523125	.028	kilograms
pounds	* 0.45359237	0.45	kilograms
short ton	* 0.90718474	0.9	tonnes
VOLUME			
teaspoons		5.0	millilitres
tablespoons		15.0	millilitres
fluid ounces	29.57353	30.0	millilitres
cups		0.24	litres
pints	* 0.473176473	0.47	litres
quarts	* 0.946352946	0.95	litres
gallons	* 3.785411784	3.8	litres
cubic inches	* 0.016387064	0.02	litres
cubic feet	* 0.028316846592	0.03	cubic metres
cubic yards	* 0.764554857984	0.76	cubic metres
AREA			
square inches	* 6.4516	6.5	square centimetres
square feet	* 0.09290304	0.09	square metres
square yards	* 0.83612736	0.8	square metres
square miles		2.6	square kilometres
acres	* 0.40468564224	0.4	hectares
TEMPERATURE			
Fahrenheit	* 5/9 (after subtracting 32)		Celsius

MEASURING SYSTEMS

ENGLISH	METRIC
LENGTH	
12 inches = 1 foot	1 kilometre = 1000 metres
36 inches = 1 yard	1 hectometre = 100 metres
3 feet = 1 yard	1 dekametre = 10 metres
5,280 feet = 1 mile	1 metre = 1 metre
16.5 feet = 1 rod	1 decimetre = 0.1 metre
320 rods = 1 mile	1 centimetre = 0.01 metre
6 feet = 1 fathom	1 millimetre = 0.001 metre
WEIGHT	
27.34 grains = 1 dram	1 tonne = 1,000,000 grams
438 grains = 1 ounce	1 kilogram = 1000 grams
16 drams = 1 ounce	1 hectogram = 100 grams
16 ounces = 1 pound	1 dekagram = 10 grams
2000 pounds = 1 short ton	1 gram = 1 gram
2240 pounds = 1 long ton	1 decigram = 0.1 gram
25 pounds = 1 quarter	1 centigram = 0.01 gram
4 quarters = 1 cwt	1 milligram = 0.001 gram
VOLUME	
8 ounces = 1 cup	1 hectolitre = 100 litres
16 ounces = 1 pint	1 dekalitre = 10 litres
32 ounces = 1 quart	1 litre = 1 litre
2 cups = 1 pint	1 decilitre = 0.1 litre
2 pints = 1 quart	1 centilitre = 0.01 litre
4 quarts = 1 gallon	1 millilitre = 0.001 litre
8 pints = 1 gallon	1000 millilitre = 1 litre
AREA	
144 sq. inches = 1 sq. foot	100 sq. millimetres = 1 sq. centimetre
9 sq. feet = 1 sq. yard	100 sq. centimetres = 1 sq. decimetre
43,560 sq. ft. = 160 sq. rods	100 sq. decimetres = 1 sq. metre
160 sq. rods = 1 acre	10,000 sq. metres = 1 hectare
640 acres = 1 sq. mile	

TEMPERATURE

FAHRENHEIT		CELSIUS
32 degrees F	Water freezes	0 degree C
68 degrees F	Reasonable room temperature	20 degrees C
98.6 degrees F	Normal body temperature	37 degrees C
173 degrees F	Alcohol boils	78.34 degrees C
212 degrees F	Water boils	100 degrees C

USEFUL CONVERSIONS

WHEN YOU KNOW:	MULTIPLY BY:	TO FIND:

	TORQUE	
Pound - inch	0.11298	newton-metres (N-m)
Pound - foot	1.3558	newton-metres
	LIGHT	
Foot candles	1.0764	lumens/metres^2 (lm/m^2)
	FUEL PERFORMANCE	
Miles/gallon	0.4251	kilometres/litre (km/L)
	SPEED	
Miles/hour	1.6093	kilometres/hr (km/h)
	FORCE	
kilogram	9.807	newtons (n)
ounce	0.278	newtons
pound	4.448	newtons
	POWER	
Horsepower	0.746	kilowatts (kw)
	PRESSURE OR STRESS	
Inches of water	0.2491	kilopascals (kPa)
Pounds/sq. in.	6.895	kilopascals
	ENERGY OR WORK	
BTU	1055.0	joules (J)
Foot - pound	1.3558	joules
Kilowatt-hour	3600000.0	joules (J = one W/s)

INDEX

D

E

F

G

H

I

K

L

M

N

O

P

R

U

V

W